HEMOPARASITES OF THE REPTILIA

COLOR ATLAS AND TEXT

HEMOPARASITES OF THE REPTILIA

COLOR ATLAS AND TEXT

SAM ROUNTREE TELFORD, JR.

The Florida Museum of Natural History
University of Florida
Gainesville, Florida

CRC Press
Taylor & Francis Group
Boca Raton London New York

CRC Press is an imprint of the
Taylor & Francis Group, an **informa** business

CRC Press
Taylor & Francis Group
6000 Broken Sound Parkway NW, Suite 300
Boca Raton, FL 33487-2742

Printed in the United States of America on acid-free paper
10 9 8 7 6 5 4 3 2 1

International Standard Book Number-13: 978-1-4200-8040-7 (Hardcover)

Library of Congress Cataloging-in-Publication Data

Telford, Sam.
Hemoparasites of the Reptilia : color atlas and text / Sam R. Telford, Jr.
p. cm.
Includes bibliographical references.
ISBN 978-1-4200-8040-7 (alk. paper)
1. Reptiles--Parasites. 2. Reptiles--Parasites--Atlases. I. Title.

SF997.5.R4T45 2008
639.3'9--dc22 2008019817

Visit the Taylor & Francis Web site at
http://www.taylorandfrancis.com

and the CRC Press Web site at
http://www.crcpress.com

CONTENTS

PREFACE

Morphological variability is a basic characteristic of all species from Protista through the vertebrates and is the critical factor allowing their evolution as their environment changes. Probably, most of the known species of hemoparasites have been poorly described in terms of their variable morphological characters visible under light microscopy. This is unlikely to change as variation among genomes replaces visible morphological characters as the basis for taxonomic distinction. Yet, the need for the ability to distinguish one taxon from another by visible characters will not completely disappear.

The blood parasites of reptiles are both diverse and morphologically variable to an extent not present, or at least not as well reported, among mammalian and avian hosts. Perhaps this is related to the larger size of reptilian erythrocytes and the presence of prominent nuclei within them that can strongly influence the appearance of the parasite within the cell. The student of the mammalian malarial parasites has a far simpler job of identifying what he or she is looking at than does one who samples populations of reptiles host to *Plasmodium* and its relatives. The diversity of reptilian hemoparasites is greater than that of mammals and birds in the numbers of genera and species, although all three of the tetrapod classes are host to the same important groups of unicellular parasites (i.e., plasmodiids, hemogregarines, and trypanosomatid flagellates). The lower vagility of terrestrial reptiles and their more restricted or isolated habitats are major factors in the increased taxonomic diversity of their parasites, probably influenced considerably by the greater phyletic age of reptiles.

The intent of this work is to compile in a single location all of the published data on the morphology of the unicellular parasites of reptilian blood. It is supplemented by the data acquired but unpublished during my 45 years of research, collected primarily from the field while resident in North, Middle, and South America; eastern, southeastern, and southern Asia; and East Africa, as well as from material sent to me for identification by numerous students, veterinarians, and colleagues. The species accounts also contain host and geographic distribution, with precise localities when possible; prevalence, life cycles, and vectors where known; effects on the host; and ecology of the host–parasite relationship. Not all of the published reports have been read because of inability to obtain the original papers (or in a few cases, to read them), but most of the literature on reptilian unicellular hemoparasites is cited. No attempt has been made to survey the veterinary literature, as this consists largely of reports of individual "disease" cases from captive reptiles, with little demonstrated significance to the natural populations, or reviews based on previous reviews, many of which are now outdated.

The scope of the species considered varies according to the taxonomic group. All reptilian species of the Plasmodiidae are described. Only those hemogregarines, the most speciose group of reptilian blood parasites (over 300 spp.), for which at least partial development in a vector is known are included, which reduces the number of species accounts to less than 50. Trypanosome species for which the descriptions are sufficient in terms of dimensions and locations of structures to permit identification, and a few of the leishmanial species known from reptiles, for which morphology is available and useful, are described. Only a general account has been possible of the several species of uncertain classification, except if ultrastructural characters indicate their bacterial or viral natures. Several new species are described, mostly from slides collected decades ago for which additional material has not become available. The most recent classification published by the Society of Protozoologists (2000) has been followed except for my recognition of the genus *Haemocystidium* within the Plasmodiidae, containing those species parasitic in lizards that were previously considered to be *Haemoproteus*. Tissue meront morphology demonstrates generic affinity with *Plasmodium* rather than *Haemoproteus*, and this is supported by recent molecular phylogeny, as cited here. Subgenera of the reptilian *Plasmodium* species are those defined by Telford (1988a). The host taxonomic names used are those most familiar to herpetologists and have not been updated to reflect the genomic analyses of recent years, which have synonymized many names based on characters derived from visible morphological or ecological characters. Usually, the host name stated in original descriptions of hemoparasites has been used, except when the designation is long outdated, with little or no use in the last half of the past century. Many of the recent changes proposed from DNA analysis are not yet generally accepted, and some already have joined the taxonomic synonymy of the taxa studied.

Materials and Methods

Morphometric data from the slides of haemosporidiids, hemogregarines, hemococcidia, and trypanosomatids were originally obtained by measurement of adequate series of parasites from slides using a calibrated ocular micrometer with a Nikon compound microscope. Perhaps ten of the several hundred samples measured over the years were obtained with a Zeiss microscope; the remainder were made using the same Nikon microscope that survived my many international moves from 1965 to 1985. A minimum of 25 parasites of each stage needed for description (i.e., meronts, gametocytes, gamonts, sporozoites, trypomastigotes, and amastigotes) were measured with data recorded on a standard sheet, which also contained observations on immature forms, locations of important individual parasites on the particular slide, and, of considerable importance, the grid locations of the vertical paths searched on the individual slide. When these individual parasites were photographed, a notation to that effect was almost always made. If a single slide contained too few parasites to meet the sample desired, a second or more slides made on the same date, if available, was searched. At times, it was necessary to use slides from subsequent dates to meet the standard desired. Rarely, it was necessary to base a description on smaller samples, then usually because of age of infection, which often affects parasitemia and the stages present. When available, infections of the same parasite from at least three different individuals of the same host species were studied. When a parasite species infected additional host species, the measurements obtained were never combined into a single sample from all hosts but were analyzed separately. Total sample sizes comprised up to several hundred in some plasmodiid species.

Initially, because more than one sample often was used for description of a parasite, the mean values were reported as mean plus or minus standard error of the mean, but in later years the more usual mean plus or minus standard deviation was stated. The laborious calculations necessary before the arrival of desktop computers involved rather primitive (by today's standards), glorified adding machines and handheld scientific calculators. With the advent of appropriate computer programs, all data from the large number of data sheets were recorded in Lotus 1-2-3 spreadsheets, then exported to a statistical program called Microstat© (1984, Ecosoft, Indianapolis, IN) for analysis. When samples did not require more complex techniques, they were simply summarized using the Lotus procedures. With the availability of Excel spreadsheets, all of the original Lotus sheets were copied into that system. Individual infections for each host species were combined by host species for the descriptions presented in this book, and the statistics given, recalculated, represent the entire sample for that parasite, again with separations by host species and, if logical, geographic origin. Significant differences between samples were based on one-way analysis of variance (ANOVA) comparisons, with significance taken at $P \leq .05$. All of the original data sheets as well as the computerized formats are deposited in the herpetology collection of the Florida Museum of Natural History, where most of the host specimens obtained by the author are preserved, and field notes, when recorded, are on file. Wherever in the text the author's name appears as (Telford) with no date citation, especially within Other Localities or Prevalence sections, this indicates unpublished data of the author. Prevalence of a parasite in a sample size of less than ten is not expressed as a percentage.

Except on rare occasions when material was prepared by others, all of the blood slides I collected were fixed in absolute methanol and stained by the Giemsa technique for at least 55 minutes or more at pH 6.8 during residence in Japan (1965–67) or 7.0 thereafter. Many slides stained over 40 years ago have retained the original results, but sadly, many more have partially or largely destained. A great error was committed when I mounted much of the type material in a supposedly neutral euparal mounting medium, which usually resulted in rapid, near total destaining. Results were mixed when Permount® (Fisher Scientific) was the mounting medium, but were generally much better than with euparal. Slides of some value with nearly or completely vanished stain were sometimes destained in slightly acidic ethanol and neutralized with basic ethanol, then restained using the Kimsey (1992) Giemsa staining technique. Again, results were mixed, with some slides successfully restained almost to the original colors, while often the results did not proceed beyond shades of basophilia. A slide of *Plasmodium minasense carini* gametocytes digitally photographed for **Plate 17B, e–l**, prepared by C. M. Wenyon in 1915 on Trinidad, has retained the original staining remarkably well, while slides I made within the last 10 years are already destaining. It is impossible to generalize what the duration and quality of stain on a given slide will be. Hapantotype slides of all species I described in the past will be deposited as time permits in the U.S. National Parasite Collection in Beltsville, Maryland. Although some type material of plasmodiids was deposited in the Garnham collection in London or in the Muséum National d'Histoire Naturelle, Paris, many were retained for "eventual deposition with the Telford collection" because of my opinion of the postal system in developing countries where I lived. The entire Telford collection is well under way in cataloguing and will be offered to an appropriate depository when completed. The preparation of tissue samples other than blood was reported in the various articles in which they were used, but standard techniques were always employed.

More than 10,500 digital images were obtained with a Nikon Coolpix© digital camera during about a year and one-half following the completion of most of the text. Most were unsuitable for publication, but those that best showed the details of the species involved were in sufficient numbers to provide some idea of the considerable variation normal to reptilian hemoparasite species.

Many of the infections from which hemoparasite species were described were followed for considerable intervals after capture, even exceeding 4 years. The host animals were maintained by appropriate measures on similar diets, in most cases, to those that were natural to the species. I almost always made host species identifications, utilizing my knowledge and experience in herpetology, which now exceeds 60 years. Voucher specimens of most host species, often in very large series (~2000 *Takydromus tachydromoides*, for example), are on deposit in the herpetology collection of the Florida Museum of Natural History.

ACKNOWLEDGMENTS

In an earlier work (Telford, 1988a), I acknowledged the help of numerous colleagues who contributed specimens, information, and advice toward the several studies that formed the basis of this work, and their assistance continues to be appreciated. M. B. Markus and E. McClain must be added to that group. Two colleagues in particular more recently sent me slides that facilitated the inclusion and preparation of species accounts that would otherwise have not been included. Juergen Stein, during his doctoral studies in Australia, referred well over 100 slides from the host he studied, *Egernia stokesii*, to me for identification of the several hemoparasite species present. David Modry provided slides of *Haemogregarina stepanowi* from the type host, *Emys orbicularis*, and of *Hemolivia mauritanica* in *Testudo marginata*. Their contributions are greatly appreciated.

My sons have contributed continuously to my studies since they dispersed. Sam collected material from Venezuelan Amazonas, Sabah, and Russia during projects in those areas. Randy's lifelong enthusiasm has obtained dozens of snakes in south Florida, from which at least 10 undescribed *Hepatozoon* species became available for sporogonic studies. Robert assisted in the acquisition of computer components and in the necessarily repetitive explanations of how to use them. And, all three sons contributed digital equipment and other logistical support. Jim Schram struggled with my incompetence in matters digital and eventually succeeded in enabling me to overcome some of my computer ignorance.

Finally, throughout the course of a 50-year marriage, Michiko Miyazawa Telford put up with a lot of strange activities by her husband, with the presence of hundreds of reptiles over time in and around her home, wherever we lived in seven countries abroad and in the United States, some of which required frequent admonitions on her part to avoid widowhood. A good example is the occasion in December 1981 on which she watched her eldest son and his father manhandling an 8-foot black mamba on the highway south of Tanga in Tanzania. No one was bitten, the blood slide (*Hepatozoon* positive) is in the Telford collection, and the snake is deposited in the Florida Museum of Natural History. Her competence in the handling of household affairs and financial matters provided the needed stability of the home environment for my research. None of my hemoparasite studies ever received institutional or grant support, apart from my various professional assignments, which happily placed me into situations that permitted a fascinating utilization of my spare time. It has been a long and fun journey for a boy who once roamed barefoot the scrubs, flatwoods, hammocks, and swamps of central Florida, developing his herpetological expertise, and exercising his unlimited curiosity for the natural world. Finally, I must thank my deceased parents, Sam Rountree Telford and Ann Marion Frances Schiller, who never really understood, but did not forbid, when once they could.

ABOUT THE AUTHOR

Sam R. Telford, Jr., a fourth-generation Floridian, was born and raised in Winter Haven, Florida, and attended local schools. He was active in scouting for 7 years, camping throughout much of Florida before it became the urban/suburban mess that it is today, and achieved the rank of Eagle Scout. Fortunate to grow up on a large tract of land comprised of pine woods, swamp, and citrus that fronted on one of the area's many lakes, he began learning herpetology at the age of 11, and during high school maintained a large collection of snakes at his home on Lake Shipp. Sam's first article, reporting a litter of 101 green water snakes, was published in *Herpetologica* around the time of his 16th birthday.

His first 2 years of college were at the University of Florida, where he spent much of the time, when he should have been studying, in the field with zoology graduate students, collecting reptiles and amphibians. Realizing the need for a more intensive academic environment, he transferred to the University of Virginia, graduating with a B.A. in biology in 1955. After a semester in law school at the University of Florida, he entered the U.S. Army, was trained at the Army Intelligence Center, and served for 3 years with two military intelligence units near Tokyo, Japan. Here, he met and married his lifelong companion, Michiko Miyazawa, who also worked for one of his intelligence units. Her first realization of what kind of biologist she had married was when she discovered a couple of preserved Japanese rat snakes in bottles that had previously contained bourbon.

When they returned to the States in 1959, Sam entered graduate school at the University of Florida, intending to pursue herpetology. After taking a stimulating first course in parasitology from George W. Hunter III, he realized that virtually nothing was known of reptilian parasitology apart from scattered reports. Under the influence of Eugene C. Bovee, his master's thesis, "Studies on the Incidence of Intestinal Protozoan Inquilines in Snakes and Lizards of the Southeastern United States," was completed in 1961. Simultaneously with his graduate research, he reviewed the taxonomy of the southeastern crowned snakes, genus *Tantilla*. Sam accepted a National Institutes of Health (NIH) Predoctoral Fellowship at the University of California at Los Angeles (UCLA) and conducted his doctoral research under Gordon H. Ball, completing "A Comparative Study of Endoparasitism Among Some Southern California Lizard Populations" in 1964.

While at UCLA, he necropsied countless snakes and lizards in return for identifying them for a dealer in exotic reptiles. After finding high prevalence of filarial worms in Mexican boa constrictors, he obtained a grant from Sigma Xi and took his wife and two small boys to Manzanillo in Colima, Mexico, for 6 weeks in 1963, obtained blood from 100 snakes, and began his career with reptilian hemoparasites. After obtaining his Ph.D. in 1964, he received a 3-year NIH postdoctoral fellowship in the Institute for Infectious Diseases, University of Tokyo, Japan, in the laboratory of Professor Manabu Sasa. Sam studied the population biology of a Japanese lizard and its relationship to its 20 species of symbiotes, published 30 years later as *The Ecology of a Symbiotic Community* in two volumes, and began his studies of reptilian malarial parasites and other blood parasites that continue today.

After the birth of their third son in Tokyo, Sam and Michiko moved to Panama, where he worked at Gorgas Memorial Laboratory from 1967 to 1970 as a vertebrate ecologist on studies of leishmaniasis in forest mammals. In 1970, they returned to Florida, where Sam taught biology to undergraduates for 3 years until he was recruited by the World Health Organization (WHO) for a project on Chagas' disease in Venezuela. After nearly 2 years there, WHO sent him to Karachi, Pakistan, to study rodent-borne diseases (1975–77), then to WHO Headquarters in Geneva, Switzerland, for administrative work on leishmaniasis and Chagas (1977–78). Boring quickly of the desk job, he transferred to Rangoon, Burma, to pursue rodents responsible for plague. Sam resigned his WHO appointment in 1980 and returned briefly to Florida and then was selected by the Danish International Development Agency as project leader for their research and training program in rodent biology and control on the campus of Sokoine University in Morogoro, Tanzania.

After 4 years in Morogoro, Sam and Michiko tired of living abroad and returned to Gainesville, Florida, in 1985, where they presently live. The nearly 20 years abroad provided the material for authoring or coauthoring well over 200 research papers and several book chapters on reptilian, avian, and mammalian blood and intestinal parasites; Florida's endemic reptile fauna; cutaneous leishmaniasis; Chagas' disease; and, strangely, mesostigmatic mites. Sixty years after publication of his first paper, Sam continues his research and writing with no intention of quitting and taking up fishing or golf.

1
THE PLASMODIID PARASITES

In the first decade of the 20th century the first reptilian malarial parasites were recognized, joining those reported from humans and birds within the previous 20 years. Wenyon (1909a), during his tenure as traveling protozoologist for the Wellcome Research Laboratories, found *Plasmodium agamae* and *P. mabuiae* in agamid and scincid lizards of the Sudan. In the same year, and with priority to Wenyon's discovery, two species, *Plasmodium diploglossi* and *P. tropiduri*, were described from Brazil in anguid and tropidurid lizards (Aragão and Neiva, 1909). The pace of species discovery and description rose slowly until the 1960s, with only 29 species and subspecies recognized by Garnham from reptiles in his classic *Malarial Parasites and Other Haemosporidia*, which appeared in 1966. At the end of that decade, the recognition of *Plasmodium* species and species of related genera began to rise in seemingly geometric progression, with 87 taxa known by 1989 (Telford, 1994), then slowed in the 1990s, with 101 species and subspecies of *Plasmodium sensu stricto* described or under description by 2007, as well as 37 other related species of plasmodiids: *Garnia* (10), *Fallisia* (10), *Haemocystidium* (14), *Saurocytozoon* (2), and *Progarnia* (1). To a considerable extent, this proliferation resulted from long-term residence in endemic areas by parasitologists interested in these organisms, in contrast to brief visits with the limited collections possible by traveling scientists or physicians, often with other primary interests.

Most articles dealing with reptilian plasmodiids have been taxonomic until recently, when interest arose in using plasmodiids as examples supporting ecological theory. Of late, genomic analysis has begun, but until a broad spectrum of known species has been studied, the results will remain tantalizing but far from definitive in tracing systematic/phylogenetic relationships, which must remain based on morphological and life history traits.

Morphology and Life Cycles

Vectors

Until 1970 the complete sporogony of a saurian *Plasmodium* species remained unknown. There were occasional reports of oocysts present on the midguts of culicine mosquitoes from experimental feedings on infected lizards. Huff (1941) found an oocyst on the midgut of an *Aedes aegypti* that had ingested a saurian malaria parasite later described as *Plasmodium floridense* (Thompson and Huff, 1944b), but subsequent attempts to infect *A. aegypti* and *Culex pipiens* by *Plasmodium mexicanum* and *Plasmodium rhadinurum* were unsuccessful. In the case of *P. rhadinurum*, all fed mosquitoes died within a day of feeding on the infected lizards, but those fed on an uninfected lizard survived, leading the investigators to suggest that the parasite was lethal to the mosquitoes. In Liberia, Baker (1961) fed four *Aedes* species on *Agama agama* infected by *Plasmodium giganteum* or *Plasmodium agamae*. Male and female gametes and ookinetes of *P. giganteum* were seen in several *A. aegypti*, but no other sporogonic stages were found.

Greater success was obtained by Jordan (1964) when she fed eight mosquito species on *Anolis carolinensis* and *Sceloporus undulatus* infected by *P. floridense*. Four species could not be infected, although they took blood meals (*Aedes atlanticus-tormentor, A. triseriatus, Mansonia perturbans, Psorophora* sp.), and another ten mosquito species common in the area refused to feed on the lizards. All *Aedes aegypti* and *Culex quinquefasciatus*, nearly 500, that fed on *A. carolinensis* had a negative result. One of 80 *A. aegypti* fed on *S. undulatus* had an oocyst on its midgut when dissected, and on two occasions single *C. quinquefasciatus* from 150 dissected showed one and three oocysts present. Four of 75 *Culex territans* had 1, 2,

16, and 23 oocysts present on midguts. Three individuals of a small, dark, unidentified *Culex* species were fed on an infected lizard, and one had 70 oocysts present when dissected. Salivary gland dissections all were negative for sporozoites.

Speculation concerning the natural vectors of saurian malarial parasites took a different direction in 1970 when Stephen C. Ayala described complete sporogony of *Plasmodium mexicanum* in two species of phlebotomine sand flies, *Lutzomyia vexator* and *L. stewarti*, in California. Some parasitologists even assumed that all saurian malarias utilized sand flies for their invertebrate hosts, relying on the dictum that the first host found to support sexual reproduction of the parasite inevitably would prove to be the characteristic vector group for all related species. The oocysts on midguts, sporozoites in salivary glands of *L. vexator*, and transmission of *P. mexicanum* by inoculation of sporozoites into laboratory hatched or uninfected, wild-caught juvenile lizards were convincing evidence that the sand fly could indeed transmit the parasite (Ayala and Lee, 1970). Sand flies collected from ground squirrel burrows were fed on *Sceloporus occidentalis* infected with *P. mexicanum*, and within the next 7–10 days, oocysts in profusion formed on midguts, and sporozoites matured within oocysts, which then ruptured, releasing hundreds of sporozoites into the sand fly hemocoel. Some of them entered the salivary glands of the sand fly. Experimental infections in the young lizards became patent within 22 days postinoculation with sporozoites. Vector competence was confirmed by Klein (1985), who fed colony-raised *L. vexator* of North Florida origin on *S. occidentalis* from California, infected with *P. mexicanum*, and after sporogony was completed, infected North Florida *Sceloporus undulatus* with *P. mexicanum* by sand fly bite (Klein et al., 1987b). More recently, Fialho and Schall (1995) explored the effect of temperature on development of *P. mexicanum* in *L. vexator* and the relationship of temperature to the ecology of the vector, both within the ground squirrel burrows in which it shelters and in the laboratory. They concluded that *P. mexicanum* enhanced transmission success through its comparatively rapid rate of development at an optimum temperature range of 22.9–24.9°C, and suggested that the parasite adaptively manipulates the thermoregulatory behavior of the sand fly. Further confirmation of the vector role of *L. vexator* for *P. mexicanum* resulted from inoculation of a crushed midgut with mature oocysts into an uninfected lizard, producing an infection within 2 weeks.

Involvement of another dipteran family in possible transmission of saurian *Plasmodium* species was suggested when Petit et al. (1983) obtained mature oocysts containing sporozoites of the African *Plasmodium agamae* from the gut of the European ceratopogonid fly *Culicoides nubeculosus*. Perhaps because of the unnatural host, oocysts did not rupture and release sporozoites capable of entering the salivary glands. When Klein (1985) fed *Lutzomyia vexator* on lizards infected with *Plasmodium floridense*, sporozoites were produced by the oocysts on the sand fly midgut but were retained within the oocysts and did not enter the salivary glands within an observation period of 14 days.

As described, Jordan (1964) found 70 oocysts on the midgut of a small, dark *Culex* species that she fed on lizards infected with *P. floridense* in Georgia. Klein (1985) demonstrated that *P. floridense* in both species of lizard hosts in northern Florida, *Anolis carolinensis* and *Sceloporus undulatus*, readily underwent complete sporogony in *Culex erraticus*, an abundant local mosquito that could be described as small and dark, similar to the unidentified species in which Jordan found the largest number of oocysts of *P. floridense*. The sporogonic pattern of *P. floridense* was typical of the other, nonreptilian *Plasmodium* species that develop in mosquito hosts. Mosquitoes infected in the laboratory that took second blood meals 2 weeks or so following their infective meal transmitted *P. floridense* to uninfected local lizards, with patency of infection evident in 18–40 days following the second feeding (Klein et al., 1987a). *Culex erraticus* readily entered mosquito traps baited with *A. carolinensis* and fed on them. Additional implication as a vector of *P. floridense* came from the light trap collections that demonstrated that the peak of abundance of *C. erraticus* in north Florida occurs in late summer, just preceding the appearance of new infections in both young and mature anoles. *Culex territans*, however, in which Jordan found up to 23 oocysts, reaches its peak of abundance 2 months earlier than *C. erraticus*, and Klein was unable to infect the species in the laboratory. Occasional transmission of *P. floridense* by *C. territans* possibly occurs in the spring and early summer when *C. erraticus* is not present in abundance.

Another plasmodiid, *Saurocytozoon tupinambi* in Brazil, produced mature oocysts containing sporocysts in *Culex pipiens* (Landau et al., 1973), but sporozoites were retained within oocysts and did not infect salivary glands, in a similar manner to the sporogony of *P. floridense* in *Lutzomyia vexator*, and *P. agamae* in *Culicoides nubeculosus*. Retention of the sporozoites within the oocysts is probably due to the dipteran being an unnatural and not completely capable invertebrate host. There is another dipteran family proven to transmit a haemosporidian parasite of reptiles, *Chrysops callidus* (Tabanidae), the natural vector of *Haemoproteus "metchnikovi"* in the northern United States (DeGiusti et al., 1973). Transmission by acarinids of reptilian plasmodiids has not been clearly demonstrated, but Peláez and Perez-Reyes (1952) thought that *P. mexicanum* infections that appeared in newborn *Sceloporus torquatus* caged with infected adult lizards infested by mites, a *Hirstiella* sp., might have been acquired from the mites. Similar transmissions occurred in two additional trials (D. Peláez, personal communication, 1986). Transmission of avian *Plasmodium* by ingestion of infected vectors is possible (Et. Sergent and Sergent, 1912; Et. Sergent, 1937; Young, 1941), and this may explain the observations of Peláez on *P. mexicanum*. Another possibility is congenital transmission from infected females of ovoviparous species, such as *S. torquatus*, to the developing young, as can happen in *Hepatozoon* species of snakes, but neither of these alternative methods of transmission, presumably mechanically at least in the latter method, has been critically studied.

Course of Infection in the Vertebrate Host

Preerythrocytic Phase No one has studied the development of sporozoites in the lizard host once experimental infection has occurred. Restained tissue sections from natural infections of *Plasmodium sasai* in four lizards were studied by Telford (1989), which provided information for an outline of this portion of the life history that might be pertinent to other saurian *Plasmodium* species. Uninucleate parasites were present in hepatic parenchymal cells of a lizard captured immediately after emergence from hibernation, which may have been infected late in the fall, immediately preceding hibernation. The lizard showed an acute parasitemia at capture. The uninucleate parasites were contained within parasitophorous vacuoles, as were binucleate and multinucleate meronts. It was suggested that the uninucleate stages were comparable to the hypnozoites of a primate malarial parasite, *Plasmodium cynomolgi*, reported by Krotoski et al. (1982a). These would enable sporozoites introduced just prior to cessation of lizard activity before winter to begin preerythrocytic merogony immediately following hibernation before the first generation of vectors could appear. The merogonic stages present in hepatic parenchymal cells probably represent the cryptozoic generation. Multiply infected macrophages present in hepatic sinuses apparently were host to the metacryptozoite generation, which then gave rise to phanerozoites in the endothelium and connective tissues of heart, lungs, femoral muscles, testes, and brain. Telford (1994) summarized the preerythrocytic phase of infection thus:

> Sporozoites inoculated by the vector appear to enter parenchymal cells of the liver. Some of them may remain inactive, as hypnozoites, whereas others may undergo at least one merogony as cryptozoites before becoming inactive. Some of the progeny may enter macrophages and form metacryptozoic meronts. Merozoites from the merogony or subsequent merogonies then are likely to parasitize the capillary endothelium and connective tissues of various organs to begin the phanerozoic merogonic cycles.

In some species of *Plasmodium* from lizards, macrophages containing developing meronts have been found when erythrocytic parasitemia has begun, suggesting some continuation of the metacryptozoic cycle after phanerozoites have appeared and erythrocytes were invaded. Alternatively, these could represent a later invasion of macrophages by the progeny of phanerozoites.

Phanerozoic meronts of *P. mexicanum* were termed "*gallinaceum* type" by Thompson and Huff (1944a). They have been observed in sections of endothelium and connective tissue in most organs during the course of active and chronic *Plasmodium* infections, and have been found in the following species, in addition to *P. mexicanum*: *P. sasai* (Telford, 1989, 1996b, 1998b); *P. agamae* (Telford, 1994); *P. michikoa-P. gologoloense* mixed infection (Telford, 1988b); *P. pitmani* (Garnham, 1950; Telford, 1994); *P. mackerrasae* (Telford and Stein, 2000); *P. floridense, P. aurulentum, P. loveridgei, P. holaspi, P. cordyli, P. lionatum,* and *P. fischeri-P. acuminatum* mixed infection (Telford, unpublished). In the host of *P. sasai* from mainland Japan, *Takydromus tachydromoides*, a lizard that must hibernate during winter, some phanerozoites occur with a cyst wall around them (Telford, 1989, 1996b), which may be an adaptation for survival of the infection through winters exceptionally prolonged or perhaps, although doubtful, as a defense against immune response. The encysted phanerozoites were termed "chronozoites" by Telford (1989) and are similar to the encysted "relapse schizonts" reported for *Leucocytozoon simondi* (Desser et al., 1968).

Erythrocytic Phase The course of infections has been studied in experimental infections of only four *Plasmodium* species of lizards: *P. mexicanum* (Thompson, 1944; Thompson and Huff, 1944a; Jordan, 1970a; Klein, 1985); *P. floridense* (Thompson, 1944; Goodwin, 1951; Goodwin and Stapleton, 1952; Jordan, 1975; Klein, 1985); *P. tropiduri* (Scorza, 1970a); and *P. sasai* (Telford, 1972a). The prepatent period, time required to reach peak of parasitemia, maximum parasitemia, and duration of infection are similar in each species. Although the prepatent or incubation periods have been reported to range from 2 to 45 days, infection in less than 5 to 7 days resulting from inoculation of infected blood probably represents the presence in circulating blood of infected erythrocytes from the inoculum. The appearance of tiny trophozoites, clearly resulting from merogony in the new host, is a better indicator of the onset of parasitemia than simply finding parasites of various stages in the blood. The use of intracardial or intravenous inoculation, sometimes utilized by Thompson and

Huff (1944a) and by Goodwin and Stapleton (1952), would be expected to result in circulating parasites more quickly than using intraperitoneal infection as they did in some cases and as used by Scorza (1970a) and Telford (1972a). Klein (1985) found that prepatent periods for sporozoite-induced infections by *P. mexicanum* and *P. floridense* were similar to those resulting from inoculation of infected blood, 23–40 days and 13–25 days, respectively. The length of time is determined by both the time required for preerythrocytic development and the need for invasion of an adequate number of infected erythrocytes for detection of infection by examination of slides.

It is evident now that the rather long periods of prepatency in infections induced by inoculation of infected blood may be due, at least in part, to the development of phanerozoites, possibly several generations of them, within the endothelial and connective tissue cells of the recipient lizard. Following intraperitoneal inoculation of blood from lizards infected with *P. sasai*, phanerozoites were found in capillary endothelium and connective tissue of most organs examined of juvenile *Takydromus tachydromoides* killed at 48, 72, and 96 hours postinoculation (Telford, 1998b). Erythrocytic infections in the series of lizards, at 0.04% parasitemia, were not detected until the 6th day postinoculation. The inocula were massive, 450,000 parasites per 10^4 erythrocytes. Other *T. tachydromoides*, both juveniles and adults, infected by inocula at levels of less than 15,000 to 108,000 parasites per 10^4 erythrocytes, that died 21–296 days postinoculation were examined histologically, and all had phanerozoites present in their tissues. The length of the prepatent period is also affected by host species and the level of inoculum (Thompson, 1944; Thompson and Huff, 1944a; Telford, 1972). In the *P. sasai* study (Telford, 1972a), the mean prepatent periods were 34.7, 27.2, and 11.3 days when inocula were less than 4,000, 40,000–46,000, and 57,000–68,000 parasites per 10^4 erythrocytes, respectively. The prepatent period was no shorter when the inoculum varied from 108,000 to 167,000 than when 57,000–68,000 parasites per 10^4 erythrocytes were used. Given the prepatent period of 6 days in the series of juveniles given 450,000 parasites per 10^4 erythrocytes, this is perhaps the minimum time for one generation of phanerozoic merogony in *Plasmodium sasai*, in its type host, that is adequate to provide erythrocytic infection detectable by microscopy.

Erythrocytic Merogony Meronts of the reptilian plasmodiid species undergo from two to seven nuclear divisions, resulting in 4–130 merozoites. In two subgenera, *Sauramoeba* and *Garnia*, the largest meronts occur in mature erythrocytes, but in many species of other subgenera, meronts are larger and contain more merozoites in immature erythroid cells, in particular proerythrocytes, than in mature cells. There may be no difference in size of the meronts in mature and immature cells, but they contain fewer merozoites when mature. During the course of infection, as parasitemia increases, there is a tendency by some species, such as *P. floridense* (Jordan, 1975) and *P. tropiduri* (Scorza, 1970a), to produce more merozoites during the period of acute rise in the infection. This may result from greater utilization of immature host cells as erythropoiesis increases. When peak parasitemia is attained, crisis forms may appear in these two species, and as the infection enters chronic phase, meronts again largely parasitize mature erythrocytes. In natural, active-phase infections of *P. colombiense*, larger meronts with more merozoites occurred in immature cells than in chronic-phase meronts (Ayala and Spain, 1976). Telford (1994) suggested that mature erythrocytes perhaps possess less of some resource essential for nuclear division than do immature cells, despite the presence of less hemoglobin in the latter group.

In some but not all saurian malaria species, there is a distinct peak of infection following acute rise in parasitemia, and this can vary by host species. Jordan (1975) inoculated a pathogenic strain of *P. floridense* from *Sceloporus undulatus* into *Anolis carolinensis*. In *S. undulatus,* the period of acute rise in parasitemia required about 65 days; peak occurred at 13,400 parasites per 10^4 erythrocytes; mean merozoite number per meront during acute rise was 10.7; but following peak, as parasitemia declined, meronts averaged 8.8 merozoites. The hosts died 4 days after peak. In *A. carolinensis*, however, the period of acute rise took 41 days, peak parasitemia reached only 2,400 parasites per 10^4 erythrocytes, and patent parasitemia lasted for 81 days, with all hosts surviving the infection. The mean numbers of merozoites before and after peak were higher than in *S. undulatus*, 13.7 and 11.2, respectively.

Geographically different strains of *P. sasai* were found to produce different patterns of infection in their natural hosts (Telford, 1972a). In the host of *P. sasai* in central Honshu, *Takydromus tachydromoides*, patency required 7–45 days, with peak parasitemia occurring 25–96 days postinoculation, usually between 2,500 and 6,000 parasites per 10^4 erythrocytes, and varying from 730 to 12,700 parasites per 10^4 erythrocytes. The acute phase of infection was usually about 80 days but ranged from 15 to 183 days. Infections became chronic following the peak parasitemia, with some lizards surviving up to 291 days postinoculation, during which one or two relapses occurred in which the level of parasitemia was far lower than the previous peaks of those infections. The course of infection by the strain of *P. sasai* in *Takydromus smaragdinus* from Amami Oshima in the Ryukyu Islands had a more benign effect on the hosts. Infections were patent in 11–50 days at low parasitemia, about 200 parasites per 10^4 erythrocytes, and about half of the infections showed no peak parasitemia. Acute infections averaged about 16 days, ranging from 5 to 24 days, and developed between days 30 and 85 postinoculation. Peak parasitemias occurred between 35 and 90 days and averaged 879 parasites per 10^4 erythrocytes, varying from 332 to 2,278 parasites per 10^4 erythrocytes. Infections persisted up to 150 days, with no host mortality. The Ryukyu strain of *P. sasai* retained its characteristics when inoculated into *T. tachydromoides* from Honshu, producing a low-level parasitemia without clear peaks of infection. One infection only of the Honshu strain resulted from inoculation into *T. smaragdinus*, lasted only 1 month, and produced a peak of only 680 parasites per 10^4 erythrocytes. The mild infections produced by the Ryukyuan strain of *P. sasai* in *T. tachydromoides* suggests that the intrinsic characteristics of the strain limited the infection, producing mild parasitemias and short duration of infection, regardless of development in a host where parasitemias by the Honshu strain achieved high levels and lasted for many months.

In nature, the course of infection by saurian *Plasmodium* species may not show the characteristics of experimental infections in the laboratory. Bromwich and Schall (1986) did not observe an acute rise and a distinct peak followed by a sharp decline in a field study of *P. mexicanum* in California lizards. The course of infection as determined from recaptures of infected lizards remained relatively stable during the summer. Although these observations could have been a sampling artifact (samples were taken only during the summer), other *Plasmodium* species characteristically show stable infections over time. Certainly in laboratory studies of *P. tropiduri* (Scorza, 1971b), *P. floridense* (Jordan, 1975; Thompson, 1944; Goodwin and Stapleton, 1952; Klein et al., 1987a), *P. mexicanum* (Thompson, 1944; Jordan, 1970; Klein et al., 1987b), and *P. sasai* (Telford, 1972a), the classical characteristics of infection (i.e., acute rise, distinct peak, and decline [often sharp]) into chronicity were observed. The inoculation of sporozoites that begins natural infections may well produce different patterns of infection than that resulting from the large numbers of parasites inoculated to establish infections in the laboratory.

Synchronicity of Merogony Synchronous division of meronts was suggested by Jordan (1975) for *P. floridense*, and Scorza (1970a) found a merogonic cycle of 48 hours in a single *Tropidurus hispidus* infected with *P. tropiduri* and followed for 9 days. Telford and Ball (1969) found clear evidence of synchronous division in three *Takydromus tachydromoides* infected with *P. sasai*, followed for a period of 114 hours. Merozoites were released usually between midnight and 0600 hours, with peaks occurring at 24-hour intervals under normal day-night cycle, and at ambient temperatures of 25–28°C. Thompson and Huff (1944a, 1944b), however, found no evidence of synchronicity in the division of *P. mexicanum* in an unnatural host species, *Crotaphytus collaris*, and in one *Sceloporus undulatus*, a natural host of *P. floridense*.

Paraerythrocytic Merogony Meronts utilize both fixed cells of various organs and circulating leukocytes and thrombocytes in addition to erythrocytes. Nonerythrocytic cells may provide some protection against immune responses directed at parasites of erythroid cells, but this has not been established. The most commonly infected cells are lymphocytes, monocytes, and thrombocytes, but a variety of granulocytes can also be infected. *Plasmodium mexicanum* produces exoerythrocytic (EE) meronts in both circulating nonerythroid cells and fixed cells of organs, a pattern described by Thompson and Huff (1944a) as "both *elongatum* and *gallinaceum* types," respectively, of exoerythrocytic merogony from the patterns described originally in two species of avian *Plasmodium*. This combination was called the "*mexicanum*" type by Garnham (1966), who described it as distinct from both avian and mammalian patterns. Merogony in endothelium suggested a close relationship between avian *Plasmodium* and *P. mexicanum*. The evidence available from 12 of the saurian *Plasmodium* species listed above as host to "*gallinaceum*" type EE merogony suggests that the "*mexicanum*" type may be characteristic of most reptilian *Plasmodium* species inasmuch as all but the mixed infection of *P. acuminatum/fischeri* also had *elongatum*-type meronts in circulating cells. Exoerythrocytic meronts were produced by *P. mexicanum* in all species of hosts infected by blood inoculation (Thompson and Huff, 1944a). *Plasmodium sasai* infections induced by inoculation of infected blood produced phanerozoic meronts that persisted throughout the course of infection and for the remaining lifetime of the lizard hosts (Telford, 1998b). It is probable that both the initial erythrocytic infection and subsequent relapses are derived from phanerozoic meronts in the endothelium and connective tissues of the hosts.

In addition to the evidence that phanerozoic and erythrocytic infection by *P. sasai* persisted for the remaining life of the experimental lizards, there are several examples cited by Telford (1994) of infections remaining for some years following capture of naturally infected lizards. Ayala (1977) reported asexual stages of a *Plasmodium* species persisting for 3 years. In the absence of asexual stages of *P. chiricahuae*, gametocytes remained present continuously for 495 and 369 days in *Sceloporus jarrovii* (Telford, 1970b), and it was suggested that exoerythrocytic meronts may have produced them directly, as reported for a rodent parasite, *Plasmodium berghei yoeli* (Killick-Kendrick and Warren, 1968). Two *Cordylus cordylus tropidosternum* were positive at capture for asexual stages of *P. cordyli* (Telford, 1989), and the lizards remained positive for 49 months in one case and for 36 months in the other. In each lizard, there was a rise in parasitemia during May and June each year, and gametocytes appeared briefly during this period. In the longest surviving lizard, phanerozoites were abundant in the heart and other organs at death. In the very early spring, *Sceloporus occidentalis* with barely detectable gametocytemias and no asexual parasites present had many phanerozoic meronts of *P. mexicanum* present in tissues. The phanerozoites are probably the direct source for the "spring relapse" of gametocytes in *P. mexicanum* (Ayala, 1970b; Bromwich and Schall, 1986).

Gametocytes In many, if not most, saurian *Plasmodium* species, gametocytes appear early in the course of infection, simultaneously with or soon after the appearance of asexual stages, and usually all of the stages are present throughout the course of active infection and in some species during the chronic phase as well. There may be changes in the morphology of gametocytes correlated with the infection phase. Chronic-phase gametocytes may be either smaller or larger than those in active phase and are often rounder (see individual species accounts). In some species, gametocytes are the dominant stage present in established infections, and in others their appearance seems transitory. Gametocytes are seldom seen in infections of *P. rhadinurum* and *P. minasense carinii* of *Iguana iguana* (Thompson and Huff, 1944b; Ayala, 1977) and in *P. cordyli*, as described above. At least in the latter species gametocyte appearance appears to be seasonal, and it is reasonable to suggest that their appearance is related to the seasonal abundance of whatever vector is responsible for the transmission of *P. cordyli*. In two species with marked dominance of gametocytes following the initial acute infection, *P. mexicanum* and *P. chiricahuae* (Jordan, 1970a; Ayala, 1970b; Telford, 1970b), chronic-phase gametocytes are much larger than those in active infection. In its natural host in California, *Sceloporus occidentalis*, *P. mexicanum* gametocytes appeared late in the active phase, 40 and 58 days postcapture, but in experimental sporozoite-induced infections of the unnatural host *Sceloporus undulatus* (Klein, 1985), asexual stages were found on average on day 31, ranging from 26 to 40 days postinfection, while the appearance of gametocytes averaged 39 days, ranging from 32 to 44 days. *Plasmodium mexicanum* infections induced in *S. undulatus* of north Florida produce very high parasitemia and inevitably kill the host. In the natural Mexican host of *P. mexicanum*, *Sceloporus torquatus*, young gametocytes appeared 8 days following asexual stages, but they grew slowly, attaining two-thirds the size of mature gametocytes in 23 days (Peláez et al., 1948). In *P. chiricahuae* of *S. jarrovii*, gametocytes appeared at the same time as asexual stages and quickly attained the length of mature gametocytes but remained more slender for some time (Telford, 1970b). Klein (1985) found asexual stages of *P. floridense* in sporozoite-induced infections of its natural host, *Anolis carolinensis*, present on day 24 postinfection, with gametocytes appearing on days 32 and 36. In blood-induced infections of *P. sasai* in *Takydromus tachydromoides*, asexual parasites appeared on average on day 25 postinoculation, with a range of 7–45 days, while gametocytes were present on average on day 33, ranging from 7 to 51 days and often (19%) present on the day asexual stages appeared (Telford, unpublished).

Few studies have given data on sex ratio of gametocytes. Ayala and Spain (1976) found a ratio of 1.57 in favor of macrogametocytes in a sample of over 900 *P. colombiense* gametocytes. Schall (1989) also found macrogametocytes to be more common in three species of *Plasmodium*: The proportions of microgametocytes in 54 *P. mexicanum* infections of *S. occidentalis* averaged 0.474 (0.25–0.73), in 30 infections of *P. agamae* in *Agama agama* at a proportion of 0.401 (0.29–0.63), and in 30 infections of *P. giganteum* in *A. agama* at a proportion of 0.371 (0.22–0.51), with no evidence of a characteristic proportion of microgametocytes for any species. The sex ratio, however, could remain constant within individual infected lizards over time regardless of changes in gametocytemia.

Ultrastructure of Reptilian Plasmodiids

Sporogonic Stages

Plasmodium In *Plasmodium agamae* (Boulard et al., 1983) developing in the laboratory vector *Culicoides nubeculosus* (Ceratopogonidae), oocysts form between intestinal epithelial cells into which they are invaginated and the basal membrane. A trilaminate wall is present: The outer layer has a fibrous structure, and the innermost layer, the plasma membrane, is thin, and surrounds the cytoplasm of the oocyst. The round or oval nuclei are situated near the periphery of the oocyst. Numerous mitochondria and about ten vesicles containing a granular, electron-dense material surround a clear central zone. Several well-defined pigment grains are visible in oocysts. With maturation, the inner plasma membrane moves away from the outer fibrous wall, and the cytoplasm contained within condenses. Cryptomitosis is centered on a centriolar plate, while sporozoites differentiate in several sites within the oocyst. They have a trilaminar pellicle and two large, elongate rhoptries and lack a conoid in the apical complex. There are 26 subpellicular microtubules distributed along the two halves of the circumference in two groups, one-half containing 18 microtubules, and the other half 8. The large number of subpellicular microtubules resembles the numbers described for *Leucocytozoon* species, 29–35, and *Haemoproteus* species, 22–23, in contrast to 11–18 found in the mammalian and avian *Plasmodium* species that have been studied (Boulard et al., 1983).

In *Plasmodium mexicanum* (Klein et al., 1988b), oocysts formed extracellularly between epithelial midgut cells and in contact with their basal membrane. As in *P. agamae*, large nuclei are scattered within the cytoplasm but are not especially peripheral in position. Vacuolation of the oocyst indicates the beginning of its differentiation, and sporoblastoids form early. As vacuoles form in the sporoblastoid body, its plasma membrane contracts away from the oocyst capsule. Large vacuoles within the sporoblastoid form and coalesce, producing clefts as they extend to the surface of the sporoblastoid. The clefts divide the cytoplasm of the oocyst into sporoblasts. Sporozoites then bud off the sporoblastoid and become elongate, stout, and crescent-shaped bodies enclosed within a pellicle composed of outer and inner plasma membranes and bilaminate, in contrast to the trilaminate pellicle described by Boulard et al. (1983) for sporozoites of *P. agamae*. Structures present in sporozoites of *P. mexicanum* include polar rings, a cytostome, 2 rhoptries, micronemes, and 14 subpellicular microtubules. As in *P. agamae*, the distribution of the microtubules is asymmetrical, with 9 of 14 contained in one-half of the circumference of the sporozoite, and the remaining 5 in the other half. Although this unequal distribution is similar in both species, there are nearly twice as many (26) subpellicular microtubules present in the sporozoites of *P. agamae* as in *P. mexicanum* (14), with the latter species more consistent with the lower number of microtubules recorded from avian and mammalian *Plasmodium* species.

Oocysts of *P. floridense*, as described by Klein et al. (1988c), have a thick, trilaminate wall, similar to that of *P. agamae*. Oocyst position with regard to the midgut is variable; most oocysts protrude slightly into the hemocoel, but some are "tightly packed between midgut epithelial cells or protruded into the midgut lumen" (Klein et al., 1988c). The structure of undifferentiated oocysts is similar to that described for other *Plasmodium* species. Differentiation of the sporoblastoid and formation of sporozoites resembles that of *P. mexicanum* (Klein et al., 1988b), but daughter sporoblastoids do not form from the sporoblastoid until somewhat later than in *P. mexicanum*. Some sporoblastoids remain after sporozoites complete their differentiation, and the mature oocysts rupture. Sporozoites, elongate and thin as in other *Plasmodium* species transmitted by mosquitoes, have comparable structure to other species but differ in number and arrangement of subpellicular microtubules from those present in the short, stout sporozoites of *P. mexicanum* and *P. agamae*. While within oocysts, there are up to 10 or 11 microtubules arranged asymmetrically in the anterior third of the sporozoite. Two microtubules are located along one surface of the sporozoite, and the remainder are along the opposite surface. Subpellicular microtubules of sporozoites within the salivary glands are infrequently visible in the anterior third of the sporozoite and are clumped together in small groups. Sporozoites of *P. floridense* are intracellular in the salivary gland, similar to those of other mosquito-transmitted *Plasmodium* species, in contrast to the extracellular position of *P. mexicanum* sporozoites, which occupy the lumen of the gland. The extracellular position of *P. mexicanum* sporozoites "may be related to the morphological structure of the salivary gland because the gland of *L. vexator* is only a fluid-filled sac" (Klein et al., 1988b).

Haemoproteus The only ultrastructural studies of a reptilian *Haemoproteus* species are those on *H. "metchnikovi"* by Sterling (1972) and Sterling and DeGiusti (1972, 1974), in which the fine structures of sporogonic stages, merogony, and gametocytes were described.

Sterling and DeGiusti (1974) obtained oocysts of *H. "metchnikovi"* from the midgut wall, and sporozoites from the salivary glands of the natural vector *Chrysops callidus* (Tabanidae). Oocysts form beneath the basement membrane of the midgut epithelium, "deep within the tortuous folds of the fly midgut." The oocyst capsule appears to form from the basement membrane. Young oocysts, 7–9 µm in diameter, have a deeply infolded marginal cell membrane, and their cytoplasm contains abundant ribosomes and some pigment granules along with aggregates of crystalloid material. As oocysts mature, they reach a maximum diameter of 18–20 µm, contain many nuclei, and bulge outward from the midgut epithelium. When differentiation into sporozoites begins, cytoplasm contracts away from the capsule, and regions of thickened membrane appear beneath the cytoplasmic-limiting membrane. Cellular components and nuclei become arranged beneath the thickenings. Sporozoites form from a single sporoblastoid body. The membrane thickenings form part of the double inner pellicular membrane of the sporozoite. A nucleus, mitochondrion, and other cytoplasmic elements move into the sporozoite buds as they form, after which the buds pinch off from the sporoblastoid cytoplasm. This becomes the residual body that contains the persistent remains of the anterior-end complex of the ookinete.

The sporozoite pellicle is comprised of the outer limiting membrane and a double inner membrane complex. A cytostome is present midway between the anterior end of the sporozoite and the subcentrally located nucleus. There are three dense concentric polar rings at the anterior end of the sporozoite, followed posteriorly by a pellicular cavity. The subpellicular microtubules, 23 in number and apparently originating near the cavity, are arranged at regular intervals around the sporozoite circumference and extend its full length. Two and sometimes three or four rhoptries are present in the anterior portion of the sporozoite, along with small micronemes, near which "crystalloid" material is present in the sporozoite cytoplasm. A mitochondrion and a spherical body 0.3 μm in diameter are present in the cytoplasm near the cytostome. Tubular cristae are contained in the mitochondrion. A double-membrane envelope surrounds the sporozoite nucleus, the outer membrane of which is continuous with Golgi and endoplasmic reticulum located anterior to the nucleus. No nucleolus was observed.

The oocysts and sporozoites of *H. "metchnikovi"* are similar in their morphology to both *Plasmodium* and *Leucocytozoon*, but many of the shared features differ between the last two genera. As in *Plasmodium*, the mature oocysts of *H. "metchnikovi"* lie external to the basement membrane of the midterm epithelium, although their early site lies beneath the membrane, as in *Leucocytozoon* and probably *Plasmodium* as well (Sterling and DeGiusti, 1974). In comparison to the much larger, expanding oocysts of *Plasmodium* species, both *Leucocytozoon* and *Haemoproteus* species of birds form mature oocysts of relatively small size, and *H. "metchnikovi"* is similar to these last genera in this character. The formation of sporozoites around a single sporoblastoid body, as in *Leucocytozoon*, contrasts with the multiple sporoblastoids present in *Plasmodium* oocysts. Although avian *Haemoproteus* species vary in the number of sporozoites formed within oocysts, the 100–200 sporozoites per oocyst of *H. "metchnikovi"* far exceed the numbers produced by oocysts of *Leucocytozoon* and most *Haemoproteus* species. *Plasmodium* sporozoites arise from an initial subcapsular vacuolation, followed by cytoplasmic cleft formation, processes not described for *Leucocytozoon* (Sterling and DeGiusti, 1974), but present in *H. "metchnikovi."* Crystalloid material, although present initially, is not found in differentiated oocysts and sporozoites of *Plasmodium* but is present in these stages of *Leucocytozoon* and *H. "metchnikovi."* The subpellicular microtubules are distributed regularly around the entire circumference of *H. "metchnikovi"* sporozoites and those of *H. columbae* and *L. simondi* (Sterling and DeGiusti, 1974), in contrast to the arrangement found in sporozoites of reptilian, avian, and mammalian *Plasmodium* species, in which most of the microtubules occur in one-half to two-thirds of the circumference, with fewer than half present in the remaining circumference of the sporozoite. In most other structural features, the sporozoites of *Plasmodium, Haemoproteus,* and *Leucocytozoon* species are similar. Although a cytostome is present in sporozoites of *H. "metchnikovi,"* its presence is inconsistent in avian *Haemoproteus* species and in *Plasmodium*, and it is absent in *L. simondi* (Sterling and DeGiusti, 1974). One character apparently unique to the sporozoites of *H. "metchnikovi"* is the presence of a spherical body associated with the mitochondrion.

Intrinsic Stages

The fine structure of reptilian plasmodiid stages within blood cells has been described for four pigmented *Plasmodium* species, two unpigmented *Plasmodium* (*Garnia*) species, one *Fallisia*, *Haemoproteus "metchnikovi,"* and five species of *Haemocystidium*. All later studies support the conclusion indicated by Aikawa and Jordan (1968) that reptilian plasmodiid stages do not differ essentially in either structure or sequence of development from the avian and mammalian *Plasmodium* species. There are some differences, as would be expected among the various species studied, but the relevance of these differences to the systematics of the plasmodiids cannot be given much importance until many more species have been examined, a situation comparable to the use of genome analysis, by which fewer than 10% of the known reptilian *Plasmodium* species have been studied.

Plasmodium In *P. floridense*, the dedifferentiation of merozoites following entry into erythrocytes, ingestion of host cell cytoplasm within food vacuoles, nuclear division, and migration of newly formed organelles into the budding areas of meronts to form merozoites occur as in avian and mammalian *Plasmodium* species; gametocytes have trilaminate membranes as do avian parasites; and in macrogametocytes the greater abundance of ribosomes and rounded osmiophilic bodies distinguish them from microgametocytes (Aikawa and Jordan, 1968). Subsequently, Aikawa et al. (1969) distinguished avian and reptilian *Plasmodium* species by the trilaminate pellicle of gametocytes from those of mammals, in which it is apparently bilaminate; by the distinct mitochondria with cristae of tubular form and rounded osmiophilic bodies in contrast to the mammalian condition of elongated osmiophilic bodies and for which a double membrane-bound structure possibly represents the mitochondria; and by the presence of nucleoli in trophozoites and gametocytes of the avian and reptilian species, which do not occur in mammalian *Plasmodium* species.

The ultrastructure of *Plasmodium tropiduri* was described in detail by Scorza (1971a). Only two important differences were found between *P. tropiduri* and *P. floridense*, as described by Aikawa and Jordan (1968). Merozoites in the latter species formed a stellate structure, with their anterior ends projecting toward the cytoplasm of the host erythrocyte, which was not typical of *P. tropiduri*, and the rounded structures present in merozoites of *P. floridense* were not present in *P. tropiduri*. In gametocytes of *P. tropiduri*, the clear vacuole, commonly seen in other saurian *Plasmodium* species, was associated with the endoplasmic reticulum, thereby possibly connecting to the cell surface, and Scorza suggested that its function might be osmoregulatory or as a reservoir of water. Merozoites containing similar vacuoles might become gametocytes following completion of merogony. Gametocytes of *P. tropiduri* contained an apparent exflagellatory apparatus, similar to coccidian and avian *Plasmodium* species. The *P. floridense* material examined by Aikawa and Jordan (1968) contained few gametocytes, which might account for their failure to report the exflagellatory apparatus. Scorza considered *P. tropiduri* to be closely related to *P. floridense* and to avian *Plasmodium* species. Both asexual and sexual stages, apparently of *P. tropiduri*, occur in thrombocytes of *Tropidurus hispidus*, in addition to the far more common erythrocytic infections (Scorza, 1970b). Except for the absence of pigment, the thrombocytic parasites were similar in ultrastructure to those occupying erythrocytes, and the structure of the large, uninucleate thrombocytic forms clearly indicated their identity as gametocytes.

Moore and Sinden (1973) studied the ultrastructure of *P. mexicanum* and found close similarity to both *P. floridense* and *P. tropiduri*. The rounded, moderately electron transparent bodies in merozoites of *P. floridense* were not present in *P. mexicanum* and *P. tropiduri*, and the merozoites of the last two species did not project beyond the original area occupied by the meront into the host cell cytoplasm, which was reported for *P. floridense*. There was greater similarity of the micropores in the trophozoites of *P. mexicanum* to those of *P. floridense* than to those present in *P. tropiduri*, but their size was about one-half the internal diameter in *P. mexicanum* than in *P. floridense*. *Plasmodium mexicanum*, and apparently *P. floridense* as well, lacked the umbrella-shaped structure associated with gametocyte micropores in *P. tropiduri*. In all three species, there was greater density of ribosomes in macrogametocytes than in microgametocytes.

In their study of a *P. tropiduri*-like species from *Kentropyx calcarata* in Brazil, Paperna and Lainson (2002) clarified the origin of the outermost membrane surrounding all stages of the parasite as the boundary of the parasitophorous vacuole "which becomes detached in processing" (Aikawa, 1971). In gametocytes, the two inner membranes were of equal size, in comparison to other species in which the middle membrane may be very thin. Although the "general ultrastructure" is in conformity with the reptilian and avian species of *Plasmodium* so far studied, Paperna and Lainson found that their transmission electron microscopic (TEM) images did not "clearly resemble" the ultrastructural description of *P. tropiduri* as described by Scorza (1971a), suggesting that the *P. tropiduri*-like parasite differs taxonomically. Cytoplasm of the trophozoites contained a mitochrondrion and cytoplasmic clefts. Of possible importance to the studies cited below, a large food vacuole remained in the meront residuum after merozoites were formed, as in *P. mexicanum* (Moore and Sinden, 1974), and the emerging merozoites contained a "large, conspicuous cytostome as described for *P. tropiduri* (Scorza, 1971) and *P. floridense*" (Aikawa and Jordan, 1968). Gametocytes, however, apparently lack cytostomes, although these usually are visible in that stage (Aikawa et al., 1969). Pigment granules "were conspicuous only in the meront residuum and in its food vacuole," their scarcity possibly due to age of the infection or processing of the material.

Garnia and Fallisia Ultrastructural evidence from their study of *Garnia gonatodi*, *G. uranoscodoni*, and *Fallisia effusa* was considered by Boulard et al. (1987) to support the familial distinction of those genera, as Garniidae, from the Plasmodiidae, Haemoproteidae, and Leucocytozoidae. Their basic structure demonstrates their haemosporidian identity. Absence of pigment from the erythrocytic *Garnia* species, and of course from the thrombocytic *Fallisia effusa*, as well as absence of a vacuolar digestive system showing phagocytic or pinocytic activity in both *Garnia* and *Fallisia* contrast to their presence in plasmodiids. Instead, a transfer of nutrients across the pellicular complex is suggested. In comparison to the location of centrosomes within the cytoplasm of the *Plasmodium*, *Haemoproteus*, *Parahaemoproteus*, and *Leucocytozoon* species that have been studied, caryokinesis arises from a centriolar plaque situated within a nuclear pore in *F. effusa* gametocytes and *G. gonatodi* schizonts (Boulard et al., 1987). Primary emphasis was placed by these authors on the absence of pigment and apparent absence of a vacuolar digestive system to justify the distinction of *Garnia* and *Fallisia* at the familial level. A subsequent study of the ultrastructure of *Garnia gonatodi* (Diniz-Jose et al., 2000) again confirmed the absence of hemozoin and the presence of ultrastructural features common to other genera of apicomplexans. Besides the presence of subpellicular microtubules, other microtubules were associated with the mitochondrion. A structure resembling the acidocalcisome of trypanosomatid flagellates was present, and the endoplasmic reticulum of the host erythrocyte, particularly the cisternae, was strongly associated with the parasitophorous vacuole. Although the previously described absence of a vacuolar digestive system in *G. gonatodi* (Boulard et al., 1987) was not mentioned in this study, it is interesting that a cytostome was present in trophozoites, and electron-dense material "similar to that found in the host erythrocyte cytoplasm" was present within it.

Silva et al. (2005) further studied the ultrastructure of *Fallisia effusa*, demonstrating that thrombocytes parasitized by gametocytes showed a distinctive "circumferential coil of microtubules," in comparison to uninfected thrombocytes in which microtubules were arranged as "bundles."

Macrogametocytes were bordered by "a four-layered pellicle composed of a plasma membrane and a three-layered inner complex, formed by three closely-apposed unit membranes." Microgametocytes, in contrast, were less electron dense than macrogametocytes and had a two-layer pellicle formed by "the plasma membrane and underlying membrane complex." Macrogametocytes also displayed invaginations of the pellicle that were sometimes deep, crossing large areas of the gametocyte. No cytostome was observed. The ultrastructure of meronts was similar to that of other plasmodiid species studied by electron microscopy. The authors believed that the ultrastructure of *F. effusa* had "characteristic features distinguishing it from other members of the Haemosporidian families," but the systematic significance of the described ultrastructural differences must await confirmation by study of other *Fallisia* species.

Haemocystidium The fine structure of gametocytes from five species of *Haemocystidium*, considered by the authors to be *Haemoproteus*, was described by Paperna and Boulard (2000). The species studied represented much of the geographic distribution of this saurian plasmodiid genus: Europe (*H. tarentolae*), the area of Cisjordan in western Asia (*H. edomensis, H. ptyodactyli*), and Australia (*H. oedurae, H. mackerrasi*). Except for *H. edomensis*, an agamid parasite, all are from gekkonid hosts. The ultrastructure of the gametocytes from these *Haemocystidium* species was consistent with that shown for the other reptilian plasmodiids studied, *Plasmodium floridense* (Aikawa and Jordan, 1968), *P. mexicanum* (Moore and Sinden, 1974), *P. tropiduri* (Scorza, 1971a), *P. gonatodi* and *Fallisia effusa* (Boulard et al., 1987), *F. copemani* (Paperna and Boulard, 1990), and *Haemoproteus "metchnikovi"* (Sterling, 1972).

All plasmodiid species of reptilian and avian hosts thus far studied have a gametocyte pellicle comprised of three layers. Much of the intraspecific variation appears to result from differences in gametocyte sex and age (juvenile, differentiated, "waiting", senile). The two species from Cisjordan, *H. edomensis* of agamids and *H. ptyodactyli* from gekkonids, were the most similar in structure. *Haemocystidium edomensis* was distinguished from the other species by presence of "An electron-dense, trapezoid body with a granular medium-electron-dense halo bordered by electron-dense droplets" (Paperna and Boulard, 2000). Paperna and Boulard suggested that it may represent a food vacuole, and is perhaps homologous to the "lipid-like" vesicle in *H. "metchnikovi"* (Sterling, 1972). TEM images of *H. tarentolae* gametocytes contained a "large electron-lucent central space," which may represent a "poorly stained zone" commonly seen in Giemsa-stained gametocytes of both sexes, reported as a "cisterna" by Paperna and Landau (1991).

Haemoproteus "metchnikovi" The ultrastructure of gametocytes and gametogenesis of *H. "metchnikovi"* were described by Sterling (1972) apparently from naturally infected turtles.

Gametocytes The structure of gametocytes is essentially similar to that of other species of reptilian and avian haemosporinids. The pellicle is trilaminate, with the inner membrane formed by two "closely apposed" membranes. Macrogametocytes have a more granular cytoplasm, resulting from the presence of numerous ribosomes; more numerous osmiophilic bodies in the cytoplasm; a more extensive, granular endoplasmic reticulum, often appearing as vesicular bundles that apparently store lipid material; more numerous mitochondria, which possess tubular cristae; and a more compact nucleus that contains a nucleolus. In microgametocytes, the cytoplasm is less granular; there are fewer osmiophilic cytoplasmic bodies and mitochondria; less-prominent vesicular bundles associated with endoplasmic reticulum; and a more poorly defined, although membrane-bound, nucleus that lacks a nucleolus. Gametocytes of both sexes contain food vacuoles surrounded by a single membrane, within which the digestion of hemoglobin leaves behind hemozoin granules. A single cytostome is found in gametocytes, surrounded by two electron-dense rings.

Gametogenesis Atypical centrioles are present in both young and mature gametocytes of both sexes and are comprised of nine peripheral tubules "with electron-dense fibers which merge at a single central tubule." The aggregation of electron-dense nuclear material near an atypical centriole signals the beginning of microgametogenesis. This aggregate is associated with microtubules within the nucleus that "end at the nuclear membrane at a dense plaque which may correspond to the atypical centriole." Basal bodies adjacent to atypical centrioles are attached to variably developed axonemes. Axonemes variably assembled are found in the cytoplasm before the gametocyte becomes extracellular. Their structure is comprised of "9 peripheral doublets of microtubules arranged around two central tubules" and appears the same in the microgametes. Development of many axonemes can be completed simultaneously near the nucleus periphery in the extracellular microgametocyte. At the periphery of the irregularly shaped nucleus, electron-dense nuclear material in aggregates apparently "leave the nucleus and assemble around the developing axonemes." After passing through the inner membranes of the microgametocyte, the nuclear buds become surrounded by the outer membrane, which possibly contributes to the membrane surrounding the individual microgamete. Microgametocytes contain "a single axoneme ... surrounded by a centrally located nucleus" and apparently lack mitochondria when mature. Overall, microgametogenesis of *H. "metchnikovi"* resembles that described for *Leucocytozoon simondi* and *Haemoproteus columbae* (Sterling, 1972), although timing of some events may differ in *H. "metchnikovi."* The direct budding of microgametes from the microgametocyte of *H. "metchnikovi"* is similar to that of *H. columbae* but differs from the microgametogenesis of *L. simondi*. There is some disparity in reports of axoneme numbers in microgametes, one or two in *L. simondi*, two in *H. columbae*, and one in *H. "metchnikovi,"* although a few microgametes of *H. "metchnikovi"* did contain more than one axoneme (Sterling, 1972).

Merogony Tissues from naturally infected turtles were examined by Sterling and DeGiusti (1972) to describe the merogony of *H. "metchnikovi."* Both merozoites and meronts were found only in tissues of the spleen. The pellicle of histotropic merozoites is comprised of three layers, beneath which lies a layer of subpellicular microtubules. These extend into a clear pellicular cavity in the anterior portion of the merozoite where the innermost membrane layer terminates. The nucleus, oval or round in shape, is located in the posterior region, and a large food vacuole containing membrane-bound "boluses" is present anterior to the nucleus. Within the boluses are particles that resemble ribosomes of the host cell cytoplasm. A cytostome may be present on the merozoite surface, near the food vacuole. Dedifferentiation of tissue merozoites into trophozoites is marked by disappearance of the thickened inner membrane and anterior-end complex, and the appearance of microvillus-like projections around the outer membrane. Cytoplasm of the trophozoite is surrounded by the middle membrane after disappearance of the inner membrane. Mitochondria, food vacuoles, and a "chromatinlike area" without a clear nuclear envelope are present in the trophozoite, and apparently as merogony approaches, microtubules radiate toward the chromatin-like area from a centriolar plaque.

As young meronts grow, nuclear division occurs, and the cytoplasm divides to form "discrete islands." Numerous ribosomes form a granular matrix within the islands, accompanied by less-granular areas at their periphery that contain chromatin-like material. Nuclear division is suggested by the presence of centriolar plaques and spindle fibers in the less-granular areas. Merozoite formation within mature megalomeronts results from "increased nuclear multiplication, cytoplasmic vacuolation, and differentiation within the islands" of the meront. Beneath the outer limiting membrane of the vacuolated islands, membrane thickenings occur in proximity to nuclei within the islands. Nuclear division and merozoite budding proceed at the same time, as indicated by centriolar plaques and spindle microtubules in association within nuclei. The conoids form from differentiation of the limiting membrane and thickened membranes at the free end of the merozoite, and "subpellicular microtubules extend posteriorly from the region of the polar rings." The limiting membranes of the multinucleated island give rise to the cytostome, which is incorporated into the merozoite pellicle. As budding merozoites grow outward, a thickened membrane complex develops and terminates where merozoites attach to the islands of the meront. As merozoite budding is completed, a mitochondrion and other elements within the cytoplasm become included within the body of the merozoite, and the inner membrane complex extends around the posterior portion of the merozoite bud. The completed merozoites are 2.5 × 1 μm in average size. Their structure is similar to that of the tissue merozoites at the beginning of dedifferentiation into trophozoites: an outer limiting membrane, two closely opposed membranes forming the inner membrane complex, and three polar rings above the clear pellicular cavity, on the top of which and to the exterior the inner membrane complex terminates. Microtubules, along one side only of the merozoite, originate near the pellicular cavity and extend at least one-half of the merozoite length. Elongate, tear-shaped paired organelles are present at the anterior end, associated with micronemes. At midbody of the merozoite, a mitochondrion with tubular cristae is usually bent around the associated spherical body. There is also a cytostome in the midregion of the merozoite. The oval or round nucleus is located in the posterior third of the merozoite, with chromatin material concentrated around its periphery. There are numerous ribosomes within the cytoplasm of the merozoite. When merozoites enter erythrocytes, they undergo a dedifferentiation similar to that of the histotropic merozoite when it becomes a trophozoite. The megalomeronts may contain both vacuolated islands and fully formed merozoites. Megalomeront margins often appear with irregular folds and have many uniformly spaced "microvilluslike projections" extending outward from the perimeter. These projections contain a central core. Megalomeronts are enveloped by extensive giant cell formation, and the perimeter of the granuloma within which the meront lies is comprised of giant cell fibers and fibroblast-like cells.

The pattern of exoerythrocytic merogony of *H. "metchnikovi"* resembles that demonstrated for *Plasmodium fallax* in vitro (Sterling and DeGiusti, 1972), that is, dedifferentiation, growth, and redifferentiation phases.

> Merozoite formation and structure ... is similar in most respects to in vivo ... descriptions of exoerythrocytic schizogony and merozoites in other malaria and malarialike parasites. Merozoite formation and structure in *H. "metchnikovi"* also show similarities with descriptions of erythrocytic schizogony in the Plasmodiidae. ... The characteristic structures within merozoites including the polar rings, subpellicular microtubules, paired organelles, micronemes, mitochondrion, spherical body, cytostome, and nucleus, are observed in *H. "metchnikovi"* merozoites. In addition, there is a clear pellicular cavity which extends around the anterior end of the merozoite and appears to be similar to the pellicular cavity described in *H. columbae* sporozoites. (Sterling and DeGiusti, 1972)

Taxonomic Characters

The descriptions of most reptilian haemosporinid species prior to the late 1960s were inadequate, primarily verbal, and contained little, if any, quantitative data that would permit comparisons with other species. In areas where the diversity of saurian *Plasmodium* species is great, notably the neotropics and East Africa, only morphometric comparisons could have effectively separated samples into species with distinctive and characteristic morphology. In a supplementary role, the effects of the parasite on host cells can be useful, but not primary. And, with the methods of genomic identification available today, species can be readily separated from each other, although the common presence of mixed infections can cause complications. Nevertheless, morphology visible under light microscopy is essential in the definition and description of species, whatever relevance it may possess to genetic relationships among a fauna. A species description must be based on the measurement of adequate series of mature sexual and asexual stages, that is, gametocytes and meronts. Telford (1974) listed both direct and indirect characters useful in making taxonomic decisions. The direct characters include those derived from measurement or obtained from calculations of those measurements: the length and maximum width of gametocytes and meronts in micrometers, the size or LW value (Length × Maximum width) stated as square micrometers, the shape or L/W ratio, and the merozoite numbers present in mature or nearly mature meronts. These quantitative data should be expressed as the mean ± 1 SD (standard deviation) and range. Shape can be verbally described as round, oval, elongate, lentiform, or bulky, and the arrangement of merozoites in meronts as a fan, rosette, morulum, cruciform, or amorphous. Describe the presence and relative quantity of pigment, its distribution as scattered granules or clumped as masses, and pigment color. Relative characters are host cell type; position of the parasite within the host cell usefully described as lateral, lateropolar, polar, halteridial, or filling most of the host cell; and the relative size of the gametocyte and meront to the host cell nucleus (LW/HNLW) and to the nuclei of uninfected erythrocytes (LW/NNLW). Indirect characters are hypertrophy or hypotrophy of the host cell, distortion of the host cell and its nucleus, and displacement of the nucleus.

Species Accounts

HAEMOSPORIDIA, Plasmodiidae

***PLASMODIUM* SPECIES OF AFRICAN LIZARDS**

AFRICAN *SAURAMOEBA* SPECIES

Plasmodium michikoa Telford 1988 (**Plate 1**)

Diagnosis A *Plasmodium* (*Sauramoeba*) species with variably shaped meronts 6–15 × 4–8 µm, and LW 28–78 µm^2 that produce 12–32 merozoites. Meronts are usually erythrocytic, but when proerythrocytes are parasitized, there is no difference in meront dimensions or number of merozoites produced. Meront size relative to host cell nucleus averages 1.91, and to normal erythrocyte nuclei is 1.56. Gametocytes are usually elongate, 6–14 × 4–8 µm, with LW 36–80 µm^2 and L/W 1.0–3.5. Gametocyte size relative to host cell nucleus is 1.81, and to normal erythrocyte nuclei is 1.75. The golden pigment granules are usually dispersed within mature meronts and macrogametocytes and tend to be marginal in microgametocytes. Both meronts and gametocytes cause hypotrophy of host erythrocytes and their nuclei. Gametocytes are not sexually dimorphic in size, but macrogametocytes are more elongate than microgametocytes.

Type Host *Bradypodion oxyrhinum* Klaver and Böhme (Sauria: Chamaeleonidae) (syn. *Chamaeleo tenuis* of Telford, 1988b).

Type Locality Eastern Udzungwa Mountains above Sanje, Kilombero District, Morogoro Region, Tanzania.

Other Hosts None known.

Other Localities None known.

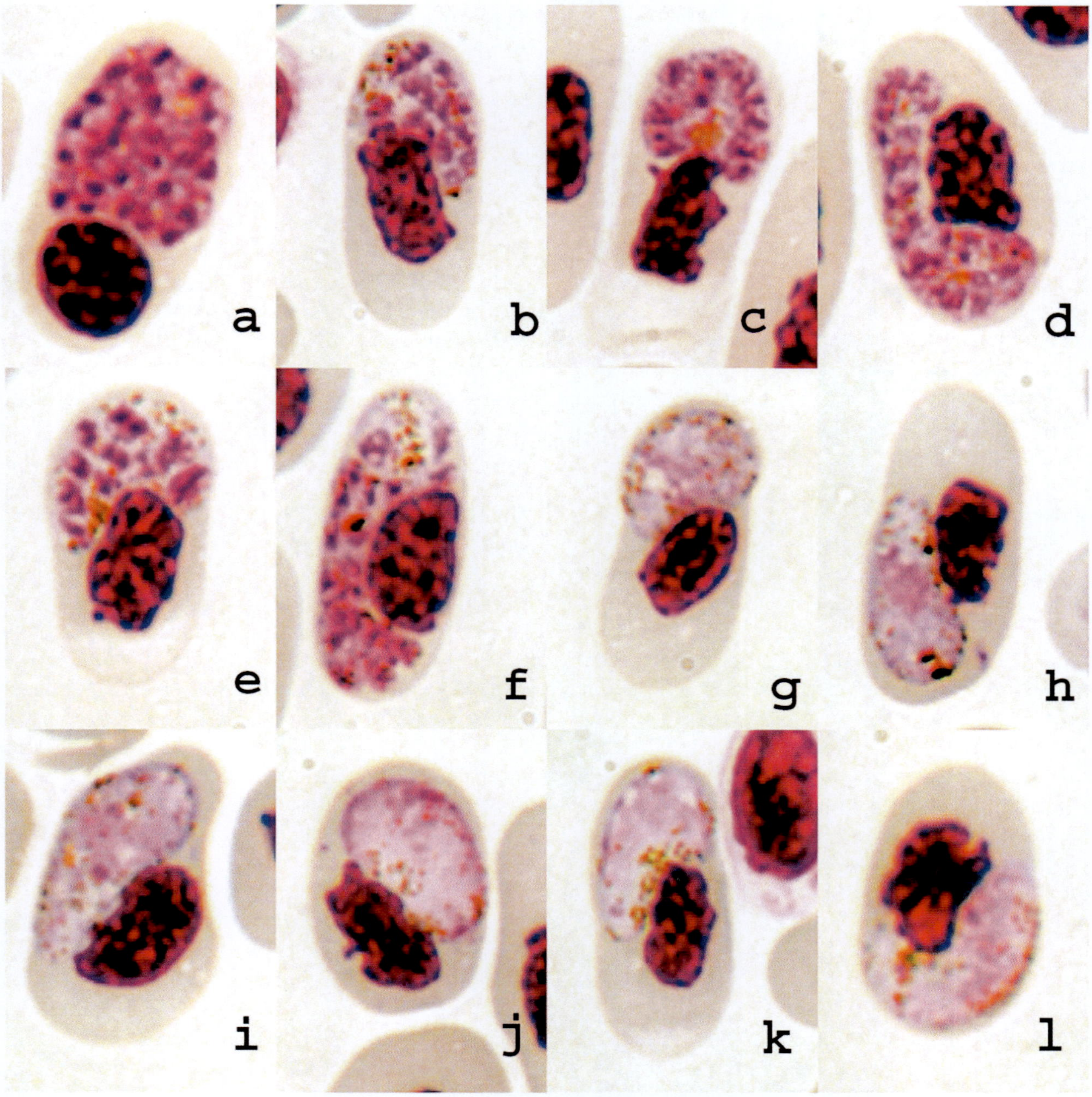

Plate 1 *Plasmodium michikoa* from *Bradypodion oxyrhinum* of Tanzania. Meronts, **a–f**; macrogametocytes, **g–i**; microgametocytes, **j–l**.

Prevalence The only *B. oxyrhinum* examined by Telford (1988b) was host to *P. michikoa*.

Morphological Variation Meronts are variably shaped, 8.4 ± 1.9 × 6.2 ± 1.0 μm (6–15 × 4–8, N = 75), with LW 51.2 ± 10.8 μm² (28–78), and contain 19.6 ± 3.8 (12–32) merozoites. Meront size relative to host cell nucleus is 1.91 ± 0.66 (0.8–4.0, N = 61), and to normal erythrocyte nuclei is 1.56 ± 0.40 (0.70–2.6). Meronts usually parasitize erythrocytes, but when proerythrocytic, neither meront dimensions nor number of merozoites produced differs. Golden pigment granules or small clumps are usually dispersed among merozoites in the meronts, sometimes clustered. Gametocytes are 9.9 ± 1.6 × 5.7 ± 1.0 μm (6–14 × 4–8, N = 105), with LW 56.1 ± 9.2 μm² (36–80) and L/W 1.82 ± 0.56 (1.0–3.5). Gametocyte size relative to host cell nucleus averages 1.88 ± 0.47 (1.0–3.2, N = 77), and to normal erythrocyte nuclei is 1.75 ± 0.38 (0.88–2.57, N = 105). Gametocytes do not differ sexually in size, but macrogametocytes are more elongate than microgametocytes: 10.3 ± 1.7 × 5.6 ± 1.0 μm (6–14 × 4–8, N = 57), LW 56.4 ± 9.8 μm² (36–80), and L/W 1.93 ± 0.59 (1.0–3.5) versus 9.5 ± 1.4 × 5.9 ± 1.0 μm

(7–12 × 4–8, N = 48), LW 55.6 ± 8.6 μm^2 (40–77), and L/W 1.68 ± 0.49 (1.0–3.0), respectively. Small, dark golden pigment granules are dispersed in macrogametocytes and usually lie along microgametocyte margins.

Exoerythrocytic Merogony Large phanerozoites were present in a fulminating mixed infection of *P. michikoa* and *P. gologoloense* but cannot be assigned to either species with confidence. Similarly, meronts present in thrombocytes might belong to either species. Free EE meronts appeared in circulating blood 5 days before the host died and infected polymorphonuclear leukocytes 2 days prior to death, while cardiac blood at death contained many heavily infected, large macrophages (Telford, 1988b), but examination of liver, lung, and spleen sections and smears showed no parasites in fixed cells.

Sporogony Unknown.

Effects on Host Erythrocytes host to both mature meronts and gametocytes, and their nuclei, are hypotrophic in size. Host cells of meronts and their nuclei are often distorted and nuclei displaced. Erythrocytes infected by gametocytes are seldom distorted, and their nuclei rarely so, but the latter were usually displaced (Telford, 1988b).

Remarks The rarely encountered host species has a disjunct range in the Uluguru and Udzungwa mountains of Tanzania, at a height of 1400–1900 m (Spawls et al., 2002). The type host specimen was collected in montane forest at 2100 m.

Plasmodium giganteum Theiler 1930 (**Plate 2**)

Diagnosis A *Plasmodium* (*Sauramoeba*) species characterized by rounded or ovoid meronts 9–18 × 4–11 μm, with LW 52–165 μm^2, that produce 28–74 merozoites, occasionally nearly 100. Meront size relative to host cell nucleus size is 2.0–5.9 and to normal erythrocyte nucleus size is 1.9–8.0. Pigment usually forms from small yellowish-brown granules into a large, blackish-brown mass. Gametocytes are nearly round to elongate or bulky, 9–22 × 4–10 μm, with LW 45–145 μm^2 and L/W 1.1–5.0. Gametocyte size relative to host cell nucleus size is 1.2–5.3, and to normal erythrocyte nuclei is 1.6–6.3. Pigment is dispersed in both gametocyte sexes, and forms as yellowish-brown granules. In older active and in chronic infections, gametocytes are dimorphic in dimensions, with macrogametocytes averaging larger in width and LW, sometimes in length, than microgametocytes.

Type Host *Agama agama* (Linnaeus) (syn. *A. colonorum*) (Sauria: Agamidae).

Type Locality Gbanga, Liberia, West Africa.

Other Hosts *Agama mossambica* (Telford), *A. cyanogaster* (Southgate, 1970).

Other Localities Nigeria (Macfie, 1914); Sierra Leone (Adler, 1924, as *P. agamae*; Garnham, 1966), Freetown (Garnham, 1966) and 22 other localities of Sierra Leone (Schall and Bromwich, 1994); Mali (Rousselot, 1953, as *P. agamae*); Harbel, Liberia (Bray, 1959); 12 km west southwest of Kinshasa, Kinsuka, on Congo River, Republic of the Congo (Telford); Marimonte (Ball, 1967a) and Pole, Kacheliba (Mutinga [Telford, 1994]), Bondo, Siakago, Kiambere, and Kiboku, Kenya (Southgate, 1970); Gonja, Pare Mountains (Pringle, vide Ball, 1967a), and Morogoro, Tanzania (Telford).

Prevalence In *A. agama*, the prevalence is as follows: Liberia, 2 of 30 (6.7%; Theiler, 1930), and 6 of 21 (28.6%; Bray, 1959); Charlesville, 36 of 139 (36%); Harbel, 2 of 28 (7%); Uoinjama, 1 of 12 (8%; Baker, 1961); Congo, 3 of 21 (14.3%; Telford); and Kenya, 3 of 48 (6.3%; Ball), and 2 of 77 (2.6%; Mutinga Collection [Telford, 1994]). In *A. cyanogaster*, the prevalence in Kenya, 6 of 125 overall, 6 of 57 (10.5%) in four positive localities (Southgate, 1970). In *A. mossambica*, prevalence is 25 of 69 (36.2%; Telford) in Tanzania.

Morphological Variation No dimensional data were provided by Theiler (1930), Bray (1959), or Ball (1967a). Garnham (1966) described macrogametocytes of *P. giganteum* as 16 to 18 μm in length and 6 or 7 μm in width, and microgametocytes as 12 × 8 μm. Meronts were said to reach 12 μm in diameter and produce 24–96 merozoites; the latter number was reported by Bray (1959). In active infection of *A. agama* from Sierra Leone, gametocytes were 15.8 ± 1.5 × 7.0 ± 0.9 μm (12–19 × 6–10, N = 50), with LW 109.3 ± 15.2 μm^2 (84–144), and L/W 2.31 ± 0.39 (1.3–3.2). Gametocytes in active infection of *A. agama* in Congo were 14.6 ± 2.0 × 5.7 ± 1.2 μm (9–20 × 4–9, N = 50), with LW 81.7 ± 17.4 μm^2 (45–104) and L/W 2.69 ± 0.71 (1.1–5.0). In a chronic-phase *A. agama* from Kenya, gametocytes were 17.9 ± 1.9 × 7.0 ± 1.0 μm (13–21 × 5–10, N = 25), with LW 125.1 ± 21.0 μm^2 (90–184.5) and L/W 2.62 ± 0.49 (1.4–3.7). Active-phase gametocytes in *A. mossambica* of Tanzania were 14.9 ± 1.9 × 5.8 ± 0.8 μm (10–22 × 4–8, N = 75), with LW 85.4 ± 14.5 μm^2 (52–132) and L/W 2.65 ± 0.58 (1.4–4.5). Macrogametocyte width and LW were greater than in

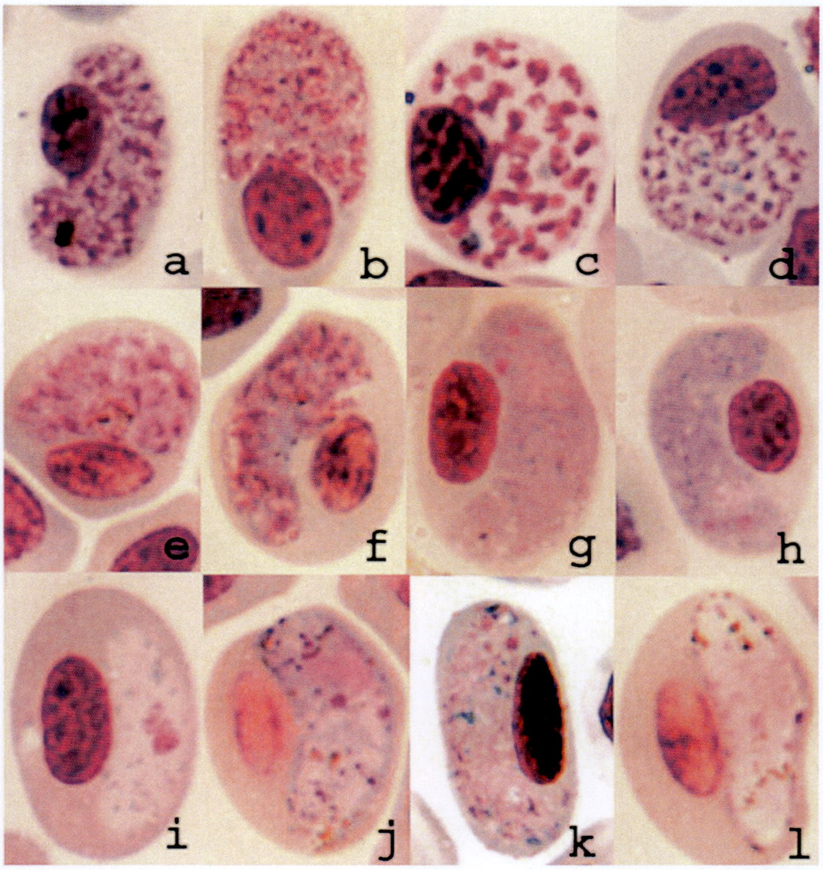

(A)

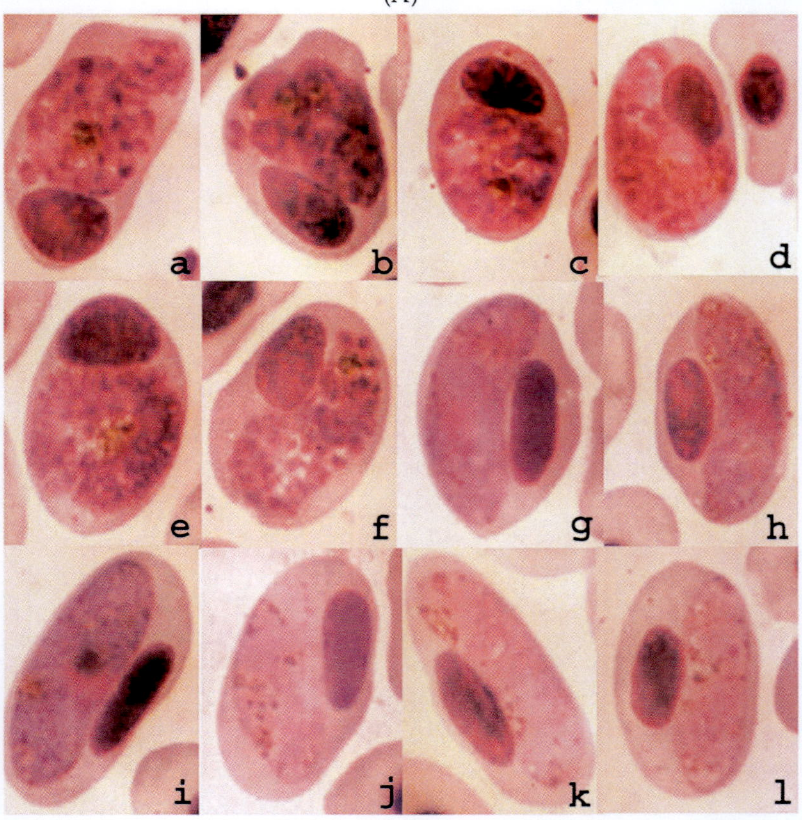

(B)

Plate 2 (**A**) *Plasmodium giganteum* from *Agama agama* and *A. mossambica*. Meronts, **a–f**; macrogametocytes, **g**, **h**, **j**; microgametocytes, **i**, **k**, **l**. Origin: *A. agama* from **a**, **b**, **g**, **h** Nigeria, and **c**, **d**, **i**, **k** Congo; *A. mossambica* from Tanzania **e**, **f**, **j**, **l**. (**B**) *Plasmodium* sp. *cf. giganteum* from *Mabuya striata* of Kenya. Meronts, **a–f**; macrogametocytes, **g–i**; microgametocytes, **j–l**.

microgametocytes in the samples from Sierra Leone (16.2 ± 1.8 μm length, 7.4 ± 1.0 μm width, and 118.0 ± 14.8 μm^2 LW vs. 15.4 ± 1.1 μm, 6.6 ± 0.5 μm, 100.7 ± 10.0 μm^2, respectively) and in Kenya were greater in length and LW but not width (18.7 ± 1.1 μm, 7.1 ± 1.0 μm, 133.2 ± 22.0 μm^2 vs. 16.9 ± 2.3 μm, 6.9 ± 1.1 μm, 114.8 ± 14.9 μm^2, respectively). Macrogametocytes from *A. mossambica* had greater average values in length, width, and LW (15.5 ± 2.0 μm, 6.0 ± 0.8 μm, 97.5 ± 13.6 μm^2) than did microgametocytes (14.3 ± 1.6 μm, 5.5 ± 0.8 μm, 78.2 ± 11.7 μm^2, respectively). In two apparently younger active infections of *A. agamae* from Congo, there were no dimensional differences in any characters: Macrogametocytes were 14.6 ± 1.7 μm, 5.7 ± 0.9 μm, and 83.0 ± 11.9 μm^2 versus 14.5 ± 2.4 μm, 5.6 ± 1.4 μm, and 80.4 ± 21.7 μm^2 in microgametocytes. Regardless of infection phase, host, or locality, gametocyte shape as indicated by L/W ratio did not differ by sex. Schall (1989) also reported macrogametocytes of *P. giganteum* as larger than microgametocytes in size (presumably LW).

Meronts averaged 11.2 ± 0.9 μm × 9.4 ± 0.7 μm (10–14 × 8–10, N = 25), with LW 105.5 ± 9.1 in Sierra Leone, and 10.8 ± 1.4 × 8.5 ± 1.3 μm^2 (9–16 × 4–11, N = 47), LW 90.6 ± 15.3 μm^2 in Congo and produced 48.3 ± 5.6 (39–61) and 42.2 ± 7.5 (28–66) merozoites, respectively. In Tanzanian *A. mossambica*, meronts were 13.1 ± 2.1 × 7.6 ± 1.4 μm^2 (9–18 × 5–11, N = 50), with LW 98.9 ± 21.2 μm^2 (72–165), and contained 41.0 ± 8.2 (28–74) merozoites. Relative to host cell nuclei, meront averages are 3.0–3.4, and to normal erythrocyte nuclei are 3.2–4.7, while gametocyte averages are 2.6–3.7 and 3.3–4.7, respectively.

Exoerythrocytic Merogony Bray (1959) reported two EE meronts from *A. agama* in Liberia. One, vermicular in shape, occupied a endothelium cell in brain capillary and contained 48 nuclei. The other, in a fixed macrophage cell of the liver, had 16 nuclei. Tissue smears of liver from *Agama mossambica* had both large and small meronts in abundance (**Plate 11B**), but these could not be identified as *P. giganteum* because of the presence of *P. "agamae."*

Sporogony Baker (1961), in Liberia, fed four *Aedes* species (*aegypti, simpsoni, apicoargenteus, africanus*) on *A. agama* infected with either *P. giganteum* or *P. agamae*. In a single *A. aegypti*, male and female gametes were seen in a smear of stomach contents about 15 minutes after ingesting *P. giganteum*. The macrogamete was oval, 10.3 × 7.4 μm, and an adjacent microgamete was about 15 μm in length. Ookinetes were found in six *A. aegypti* examined 1 day postfeeding (PF) and in four at 2 days. Ookinetes were 16.3 × 3.4 μm^2 (14.3–20.3 × 2.3–4.6, N = 10) and contained "a relatively large amount of pigment."

Effects on Host Schall (1990b) compared hematological parameters and oxygen consumption for *A. agama* from Sierra Leone infected with *P. giganteum* alone, mixed infections with *P. agamae*, and uninfected lizards. Uninfected lizards had 0.574% immature erythrocytes in their blood, on average, but in lizards infected with *P. giganteum* alone, the percentage of immature blood cells was 4.85%; in mixed infections, it was 5.91%. Normal hematocrit values were little different in these comparisons, as were hemoglobin concentrations. There were slight but significant increases in oxygen consumption for both categories of infected lizards. Running stamina was not affected by *P. giganteum* alone, but in mixed infection there was a significant reduction in capacity. Infection in either category did not affect ovary or testis mass or the prevalence of broken or regenerated tails, as an indication of predator evasion. In all samples of *P. giganteum*, gametocytes distorted 96–100% of host cells and displaced 93–100% of their nuclei. Erythrocytes of *A. agama* host to gametocytes were 2–13% larger than uninfected cells, but hypertrophy was greater in *A. mossambica* infected by *P. giganteum*, with 25–28% enlargement. Erythrocyte nuclei were distorted in 92–100% of cells in samples from *A. agama* but in only 16% of cells from *A. mossambica*. In both host species, the nuclei of infected erythrocytes were hypertrophied, by 30–36% in A. agama and by 26–52% in *A. mossambica*. Meronts in all samples of *P. giganteum* distorted 93–100% of host cells and displaced their nuclei, distorting nuclei in 96% and 98% in Sierra Leone and Congo, respectively, but only in 80% int *A. mossambica*. Erythrocytes of *A. agama* in one infection from Congo, parasitized by meronts, were enlarged by 11% and their nuclei by 28%. Erythrocytes host to *P. giganteum* meronts in *A. mossambica* were hypertophied by 23–26% and their nuclei by 35–44% over normal size. In *A. agama* from Sierra Leone and Congo, 32% and 69% of meronts, respectively, parasitized proerythrocytes, but only 4% of cells containing meronts in *A. mossambica* were immature. In the Congo samples, 35% of gametocytes occupied immature host cells, but all cells host to gametocytes in the Sierra Leone sample were erythrocytes only. Only 5% of gametocytes were proerythrocytic in the chronic infection from Kenya, and all gametocytes occupied erythrocytes in *A. mossambica*.

Remarks Schall and Bromwich (1994) examined 4772 blood smears from 2870 *A. agama* in Sierra Leone and apparently found an overall prevalence of 12.1% of *P. giganteum* (577 infections, their Table 2). Although not clearly stated, the authors presumably examined multiple slides of some lizards until they were satisfied about species present. In an analysis of ten single-species infections each of *P. giganteum* and *P. agamae*, they concluded that their data showed that *P. giganteum* primarily used immature host cells, and *P. agamae* occupied mature erythrocytes, leading to the conclusion that the presence of *P. agamae* infection in a lizard facilitated establishment of infection by *P. giganteum*, thus demonstrating that "*Plasmodium* species form interactive assemblages" (Schall and Bromwich, 1994). This conclusion is not strongly supported by the infections of *A. agama* from Sierra Leone and Congo reported above and not at all by infections of *P. giganteum* in *Agama mossambica*, where *P. giganteum* seldom to rarely infected proerythrocytes: 36 of 69 *A. mossambica* were infected by *Plasmodium* species, with single infections by *P. giganteum* present in 11, only *P. "agamae"* infections were in 13, and 12 lizards had mixed infections of the two species.

Plasmodium heischi Garnham and Telford 1984 (**Plate 3A**)

Diagnosis A *Plasmodium* (*Sauramoeba*) species with large, spindle-shaped gametocytes in which the nuclei usually occupy subterminal positions. Gametocytes are 8–12 × 4–9 μm, with LW 60–120 μm^2 and L/W 1.5–4.8. Their size relative to host cell nucleus size is 2.1–6.3, and to normal erythrocyte nuclei is 3.1–6.9. Meronts are elongate, often nearly halteridial, 8–18 × 6–11 μm, with LW 48–144 μm^2, and produce 20–65 merozoites. Meront size relative to host cell nucleus size is 1.8–5.3, and to normal erythrocyte nucleus size is 2.8–8.3. Although width does not differ, gametocytes are sexually dimorphic in length and LW, with macrogametocytes larger than microgametocytes, with the latter less elongated in shape. Pigment is dispersed as dark granules in gametocytes but tends to form small clumps in meronts.

Type Host *Mabuya striata* (Peters) (Sauria: Scincidae).

Type Locality Nairobi, Kenya.

Other Hosts None known.

Other Localities None known.

Prevalence *P. heischi* was identified in 11 of 60 *M. striata*.

Morphological Variation Gametocytes are 16.1 ± 1.9 × 5.7 ± 0.3 μm (12–20 × 4–9, N = 100), with LW 91.4 ± 12.9 μm^2 (60–126) and L/W 2.90 ± 0.60 (1.5–4.8). Gametocyte size relative to host cell nucleus size is 3.51 ± 0.76 (N = 100), and to normal erythrocyte nuclei is 4.48 ± 0.93. Gametocytes are sexually dimorphic in length, LW, and L/W: Macrogametocytes are 17.1 ± 1.6 × 5.7 ± 0.9 μm (13–20 × 4–9, N = 50), with LW 96.7 ± 11.7 μm^2 (68–126) and L/W 3.10 ± 0.65 (1.6–4.8); microgametocytes are 15.1 ± 1.7 × 5.7 ± 0.8 μm (12–18 × 5–8, N = 50), LW 86.1 ± 11.8 μm^2 (60–108), and L/W 2.70 ± 0.49 (1.5–3.6). Chronic-phase gametocytes differ sexually only in LW, with macrogametocytes averaging 97.4 ± 10.5 μm^2 (N = 25) and microgametocytes 90.4 ± 10.0 μm^2 (N = 25). Within the same sex, microgametocytes differ only in LW, 81.8 ± 12.0 μm^2 in active phase, versus 90.4 ± 10.0 μm^2 in chronic phase; while macrogametocytes differ by infection phase in length, width, and L/W, but not in LW: active 18.2 ± 1.0 × 5.3 ± 0.5 μm, LW 95.6 ± 12.5 μm^2, and L/W 3.18 ± 0.45 versus 16.1 ± 1.5 × 6.1 ± 1.0 μm, LW 97.4 ± 10.5 μm^2, and L/W 2.72 ± 0.60. Gametocytes have distinctly pointed ends in active infection and usually when chronic, although some in the latter phase of infection have more broadly rounded, nonpointed ends. Meronts are usually elongate and may curve incompletely around the host cell nucleus, but occasionally take an elongated ovoid shape, nearly filling the host cell. Meronts are 14.3 ± 2.8 × 7.7 ± 1.3 μm^2 (8–18 × 6–11, N = 25), with an LW of 109.4 ± 23.3 μm^2 (48–144). Size relative to host cell nucleus size is 4.11 ± 0.81, and to normal erythrocyte nuclei is 6.32 ± 1.34 (N = 25).

Exoerythrocytic Merogony Unknown.

Sporogony Unknown.

Effects on Host Immature parasites usually occur in proerythrocytes, which mature by the time the parasite matures. Both meronts and gametocytes produce significant hypertrophy of host erythrocytes, with host cells about 25% larger when parasitized by meronts and 40% larger when infected by gametocytes, and their nuclei are two-thirds larger than normal nuclei. Almost all host erythrocytes and their nuclei are distorted, and the latter are displaced, usually laterally, occasionally in a polar direction.

Remarks The infected series of *M. striata* were collected between 1949 and 1965 by R. B. Heisch in the vicinity of the Medical Research Laboratory in Nairobi, from rocky gullies on hillsides above the Athi Plains. In this habitat, *M. striata* and *Agama agama* coexist in boulder piles and termite nests. A single infection of *Plasmodium* in an *M. striata*, collected in 1949, appears to be *Plasmodium giganteum* instead of *P. heischi* and possibly represents one of the very few cross infections of a saurian *Plasmodium* species between host families (**Plate 2B**). Meronts are rounded or ovoid, usually filling the host cells, nearly half of which are proerythrocytes. Meronts average 11.2 ± 1.4 × 8.9 ± 0.8 μm (10–15 × 7–10, N = 25), with LW 99.2 ± 13.0 μm^2 (80–135), and contain 38.7 ± 10.5 (22–68) merozoites. Gametocytes are elongate with broad, rounded ends, not pointed as in *P. heischi*. Most gametocyte nuclei are large and central, never subterminal, the usual position of macrogametocyte

Plate 3 (**A**) *Plasmodium heischi* from *Mabuya striata* of Kenya. Meronts, **a–f**; macrogametocytes, **g–i**; microgametocytes, **j–l**. (Figures **a** and **b** from Garnham, P. C. C., and Telford, S. R., Jr., *J. Protozool.*, 31, 518, 1984, with permission, Blackwell Publishing.) (**B**) *Plasmodium robinsoni* from *Chamaeleo parsoni* of Madagascar. Meronts, **a–e**; macrogametocytes, **f–h**; microgametocytes, **i–l**.

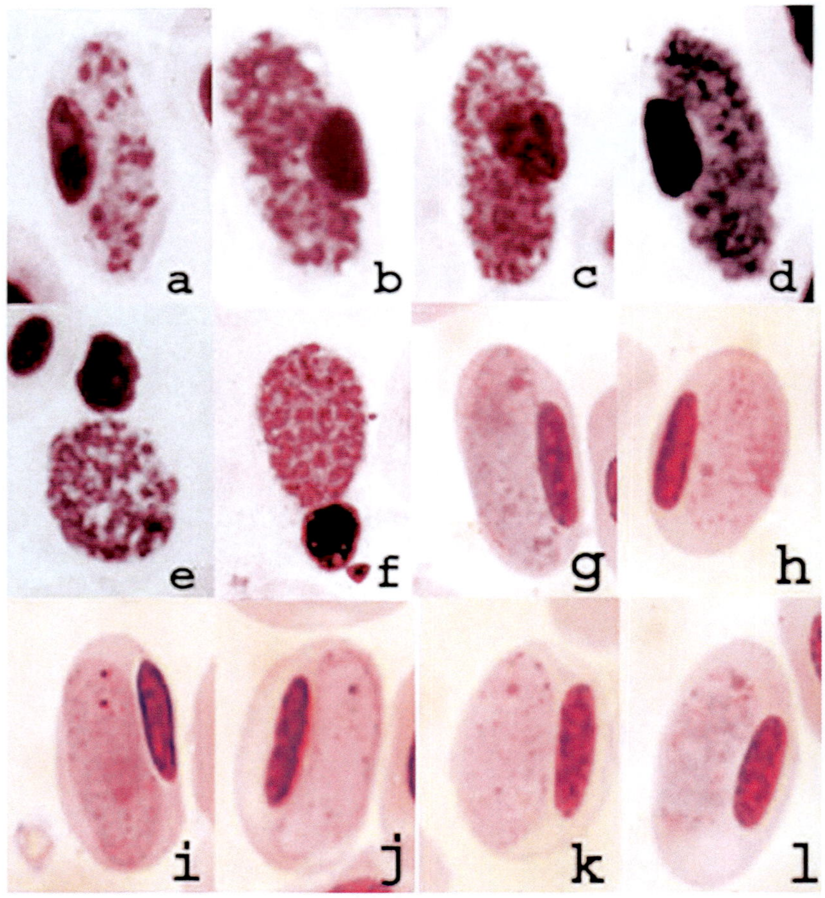

(A)

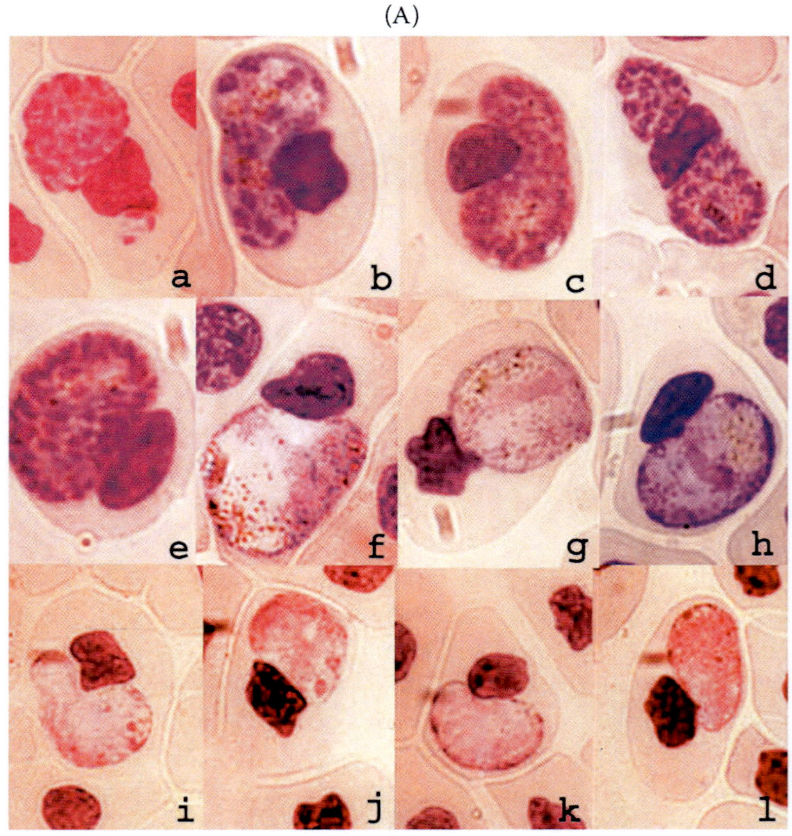

(B)

nuclei in *P. heischi*. Gametocytes are 15.4 ± 1.5 × 6.1 ± 0.9 µm (11–18 × 5–8, N = 50), with LW 92.8 ± 12.3 µm² and L/W 2.59 ± 0.48 (1.4–3.4). Their size relative to host cell nucleus averages 3.26 ± 0.64, and to normal erythrocyte nuclei is 4.20 ± 0.55. As in *P. giganteum* and *P. heischi*, macrogametocytes are longer, with greater LW, and more slender than microgametocytes but are similar in width. Macrogametocytes are 16.1 ± 0.5 × 6.1 ± 0.8 µm (13–18 × 5–8, N = 25), with LW 97.4 ± 9.5 µm² (75–112) and L/W 2.71 ± 0.47 (1.6–3.4); microgametocytes are 14.6 ± 1.4 × 6.1 ± 1.0 µm (11–17 × 5–8, N = 25), with LW 88.3 ± 13.2 µm² (65–128) and L/W 2.47 ± 0.47 (1.4–3.4). The slide of this infection is no. 998 in the Garnham Collection.

Plasmodium robinsoni (Brygoo) 1962 Telford and Landau 1987 (**Plate 3**)

Diagnosis A *Plasmodium* (*Sauramoeba*) species with large round, oval, or elongate meronts that approximate gametocytes in size. Meronts are 11–23 × 7–11 µm, with LW 90–184 µm², and contain 40–74 merozoites. Gametocytes are oval to elongate or bulky, 9–20 × 5–13 µm, with LW 72–221 µm² and L/W 1.0–3.4. Meront size relative to host cell nucleus size is 2.0–4.3, and to normal erythrocyte nuclei is 1.9–4.3. Gametocyte size relative to host cell nucleus size is 1.4–5.4, and to normal erythrocyte nuclei is 2.1–4.7. Pigment forms a light golden mass in meronts, often centered among nuclei, and is dispersed as several loose clumps of dark granules in gametocytes. Gametocytes are sexually dimorphic in dimensions, with microgametocytes longer and more slender than macrogametocytes.

Type Host *Chamaeleo brevicornis* Gunther.

Type Locality Fiherenana, Moramanga Subprefecture, Madagascar.

Other Hosts *Chamaeleo parsoni crucifer*.

Other Localities Périnet, Moramanga Subprefecture, Madagascar.

Prevalence Three *C. brevicornis* from the type locality and 2 of 47 (4.3%) collected at Périnet were infected by *P. robinsoni* (Brygoo, 1962).

Morphological Variation The type infection of *P. robinsoni* was in chronic phase at a parasitemia of less than 0.1% (Telford and Landau, 1987). Brygoo (1962) provided no dimensional data for either gametocytes or meronts but did comment that mature meronts contained 40–70 nuclei, and that gametocytes were variably shaped, elongate, ellipsoidal, sometimes with a projection that resembled a tennis racquet, sometimes curving around the erythrocyte nucleus in a halteridial form. Garnham (1966) described their form as a tennis racquet, which is not typical of the species. Examination of the type slide found a single mature meront, 10 × 9 µm, that contained 47 nuclei. In an active infection of *P. robinsoni* in *C. parsoni*, meronts are 16.1 ± 3.9 × 8.8 ± 1.2 µm (11–23 × 7–11, N = 17), with LW 138.8 ± 25.2 µm² (99–184) and 56.6 ± 12.0 (40–74) merozoites. Meront size relative to host cell nucleus is 3.13 ± 0.07 (2.0–4.3, N = 15), and to normal erythrocyte nuclei is 3.05 ± 0.70 (1.9–4.3, N = 16). Meronts are usually round or ovoid, but when elongate are halteridial around the erythrocyte nucleus or even occasionally divided into two portions by the latter, only one of which would contain the pigment mass. Pigment granules are clustered into a large, light golden mass. In *C. brevicornis*, gametocytes are 15.0 ± 1.5 × 7.4 ± 1.8 µm (12–18 × 5–12, N = 50), with LW 110.5 ± 24.5 µm² (75–170) and L/W 2.14 ± 0.57 (1.1–3.4). Their size relative to host cell nucleus is 3.10 ± 0.97 (1.4–5.4), and to normal erythrocyte nuclei is 3.04 ± 0.67 (2.1–4.7). Gametocytes in *C. parsoni* are 13.9 ± 2.4 × 9.4 ± 1.6 µm (9–20 × 6–13, N = 70), with LW 130.2 ± 32.6 µm² (72–221) and L/W 1.54 ± 0.42 (1.0–2.8). Gametocyte size relative to host cell nucleus size is 3.12 ± 0.79 (1.6–5.0, N = 25), and to normal erythrocyte nuclei is 3.35 ± 0.84 (1.9–5.7, N = 70). Gametocytes from *C. brevicornis* are longer and narrower than those from *C. parsoni*, with smaller LW, and are more elongate in shape, with a greater L/W ratio. Sexual dimorphism is present among gametocytes from both host species, varying between the sexes in exactly the same pattern: Macrogametocytes are shorter and wider, have greater LW values and lower L/W ratios (i.e., are more rounded) than in microgametocytes. In *C. brevicornis*, microgametocytes average 15.3 ± 2.2 × 6.4 ± 1.2 µm (12–18 × 5–9, N = 8), with LW 95.4 ± 9.1 µm² (84–108) and L/W 2.49 ± 0.67 (1.3–3.4) versus, respectively, in macrogametocytes, 14.9 ± 1.3 × 7.6 ± 1.8 µm (12–17 × 5–12, N = 42), 113.4 ± 25.5 µm² (75–170), and 2.07 ± 0.54 (1.1–3.2). Microgametocytes in *C. parsoni* are 14.7 ± 2.2 × 8.5 ± 1.4 µm (10–20 × 6–11, N = 25), with LW 124.8 ± 30.1 µm² (90–200) and L/W 1.78 ± 0.38 (1.0–2.5) versus, respectively, in macrogametocytes, 13.5 ± 2.4 × 9.9 ± 1.6 µm (9–20 × 6–13, N = 45), 133.2 ± 33.9 µm² (72–221), and 1.40 ± 0.38 (1.0–2.8). The dark pigment granules are distributed in several variably discrete groups in both gametocyte sexes.

Exoerythrocytic Merogony Unknown. Brygoo (1962) did not find EE meronts in sections of liver and spleen from two heavily parasitized chameleons.

Sporogony Brygoo (1962) fed *Culex fatigans* on a gametocyte-rich infection in *C. brevicornis*, but sporogonic development was not observed in 30 mosquitoes dissected between day 5 and 21 PF.

Effects on Host Infected erythrocytes were described as hypertrophied when host to meronts or mature gametocytes by Brygoo (1962). Host cells utilized by *P. robinsoni* are predominantly erythrocytes. Meronts and gametocytes are most commonly polar or lateropolar in *C. parsoni*; gametocytes are more commonly lateral or lateropolar in *C. brevicornis*. "In both hosts cells infected with meronts or gametocytes enlarged and distorted, with nuclei always displaced and often distorted; only meronts produced nuclear hypertrophy" (Telford and Landau, 1987).

Remarks Brygoo (1962) inoculated infected blood containing *P. robinsoni* subcutaneously into two *Chamaeleo lateralis* and one *C. verrucosus*, but neither species became infected. He also attempted transmission from a heavily infected *C. brevicornis* by subcutaneous inoculation of a broth prepared from liver and spleen into another *C. verrucosus* and two more *C. lateralis*, without success.

Plasmodium acuminatum Pringle 1960

Diagnosis A *Plasmodium* (*Sauramoeba*) species in which young asexual stages and gametocytes have prominent, pointed cytoplasmic projections at each end. Dark pigment granules tend to clump together at an extremity or along one side in gametocytes. Largest meronts observed contained six to nine nuclei, but mature meronts are undescribed. Immature meronts can equal host cell nuclei in size. Immature gametocytes are elongate and acuminate at both ends. Mature gametocytes, with irregularly rounded or bluntly tapered extremities, can occupy nearly the entire host erythrocyte, curving around the nucleus.

Type Host *Chamaeleo f. fischeri* (Reichenow) (Sauria: Chaemaeleonidae).

Type Locality Amani, Eastern Usambara Mountains, Tanga Region, Tanzania.

Other Hosts None known.

Other Localities None known.

Prevalence One of 42 *C. fischeri* collected at Amani was infected by *P. acuminatum* (Ball, 1967a).

Morphological Variation No dimensional data are available for this species.

Exoerythrocytic Merogony Unknown.

Sporogony Unknown.

Effects on Host Meronts "partially embraced the swollen and displaced red cell nucleus" while gametocytes (immature) "caused only a minor distortion of the host cell" (Pringle, 1960) and did not displace the nucleus.

Remarks This *Plasmodium* species has not been reported since its description by Pringle (1960). He did not describe mature meronts and gametocytes, but in an addendum to the description article, reported a third infection in which

> the blood contained scanty gametocytes apparently in a more advanced stage of development … the largest forms, which occupy almost the entire red cell, displace and partially embrace the red cell nucleus; they have a less characteristic shape, with irregularly rounded or bluntly tapered extremities … the macrogametocyte, usually a larger and more irregularly shaped parasite [than the microgametocyte]. … Rarely the shape of the microgametocyte approximates to the rounded bean shape. (Pringle, 1960)

This species is probably a *Sauramoeba* species based on having very large gametocytes that are sexually dimorphic in size and shape. The very large immature meronts figured by Pringle indicate that mature meronts would also be large, as with other *Sauramoeba* species. The host species was identified by Pringle (1960) as *Chamaeleo fischeri tavetanus*, which is incorrect.

AFRICAN *LACERTAMOEBA* AND *CARINAMOEBA* SPECIES

Plasmodium brygooi Telford and Landau 1987

Diagnosis A *Plasmodium* (*Lacertamoeba*) species with oval, oblong, or lentiform meronts, 6–9 × 5–8 μm, with LW 36–64 μm^2, that produce 10–16 merozoites. Meront size relative to host cell nucleus size is 0.5–1.2, and to normal erythrocyte nuclei is 0.5–1.1. Pigment in meronts is usually formed as three or four dark clusters, rarely coalesced into a single golden mass. Gametocytes are usually oval or elongate, 9–15 × 5–10 μm, with LW 66–126 μm^2 and L/W 1.1–3.0. Size of gametocytes relative to host cell nucleus is 1.0–3.4, and to normal erythrocyte nuclei is 1.2–2.3. Dark pigment granules in gametocytes are not widely dispersed but are somewhat localized.

Type Host *Chamaeleo brevicornis* Gunther (Sauria: Chamaeleonidae).

Type Locality Périnet, Madagascar.

Other Hosts None known.

Other Localities None known.

Prevalence Unknown.

Morphological Variation Meronts are variably shaped, oval, oblong, or lentiform, with merozoites arranged along the periphery as a rosette. They average 7.9 ± 0.9 × 6.3 ± 0.8 µm (6–9 × 5–8, N = 16), with LW 49.0 ± 6.7 µm² (36–64). Merozoites number 13.8 ± 2.1 (10–16). Meront size relative to host cell nucleus is 0.79 ± 0.17 (0.5–1.2), and to normal erythrocyte nuclei is 0.88 ± 0.12 (0.6–1.1). Pigment usually forms three or four prominent dark clusters, rarely coalescing as a golden mass. Gametocytes are usually oval or elongate, 11.5 ± 1.4 × 8.1 ± 1.3 µm (9–15 × 5–10, N = 48), with LW 92.5 ± 16.0 µm² (66–126) and L/W 1.47 ± 0.38 (1.1–3.0). Gametocyte size relative to host cell nucleus size is 1.74 ± 0.50 (1.0–3.4), and to normal erythrocyte nuclei is 1.65 ± 0.29 (1.2–2.3). There appears to be no sexual dimorphism in gametocyte dimensions.

Exoerythrocytic Merogony Unknown.

Sporogony Unknown.

Effects on Host Both meronts and gametocytes usually occupy polar or lateropolar positions in host erythrocytes, which are hypertrophied and distorted, with nuclei displaced and often distorted. Only meronts cause hypertrophy of erythrocyte nuclei.

Remarks Although Brygoo (1962) examined 47 *C. brevicornis* from the type locality, Périnet, between 1954 and 1962, he found only *Plasmodium robinsoni* present in their blood. A single *C. brevicornis* collected in 1972 was host to *P. brygooi.*

Plasmodium holaspi Telford 1986 (**Plate 4**)

Diagnosis A *Plasmodium* (*Lacertamoeba*) species with young asexual stages that occupy marginal positions in erythrocytes. Meronts are usually oblong or formed as rosettes, 5–13 × 4–7 µm, with LW 25–66 µm², and contain 8–18 merozoites. Meront size relative to host cell nucleus is 1.26, and to normal erythrocyte nuclei is 1.50. Pigment forms a single, dark irregular mass variably located within the meront. The usually elongate gametocytes are 6–18 × 3–8 µm, with LW 28–98 µm² and L/W 1.13–4.67. Their size relative to host cell nucleus size is 2.13, and to normal erythrocyte nuclei is 2.25. Dimensions of gametocytes are not sexually dimorphic. Large masses of apparent chromatin that stain intensely reddish occur in both sexes of maturing gametocytes, more dispersed in microgametocytes, becoming less prominent in the larger gametocytes. Irregular dark pigment granules are conspicuous and dispersed in both sexes.

Type Host *Holaspis guentheri* Gray (Sauria: Lacertidae).

Type Locality Kimboza Forest, 1 km north of the Ruvu River below Kibungo Village, south side of Uluguru Mountains, Morogoro Region, Tanzania.

Other Hosts None known.

Other Localities None known.

Prevalence Two of seven *H. guentheri* from the type locality were infected by *P. holaspi* (Telford, 1986a).

Morphological Variation Meronts are 7.8 ± 1.6 × 5.4 ± 0.8 µm (5–13 × 4–7, N = 66), with LW 42.1 ± 9.4 µm² (25–66), and produce 12.3 ± 2.3 (8–18) merozoites. Meront size relative to host cell nucleus is 1.26 ± 0.32 (0.73–2.29, N = 62), and to normal erythrocyte nuclei is 1.50 ± 0.52 (0.78–2.77). Meront shape is most often oblong or with merozoites arranged in a rosette, rarely as a fan or other forms, with a single dark irregular mass of pigment granules, variably located within the meront. Gametocytes, usually elongate, are 11.8 ± 2.4 × 5.0 ± 1.0 µm (6–18 × 3–8, N = 85), with LW 58.9 ± 15.2 µm² (28–98) and L/W 2.47 ± 0.75 (1.13–4.67). Gametocyte size relative to host cell nucleus size is 2.13 ± 0.77 (0.7–4.3, N = 60), and to normal erythrocyte nuclei is 2.25 ± 0.75 (0.9–4.1, N = 85). Gametocytes are not sexually dimorphic in dimensions: Macrogametocytes are 12.2 ± 2.1 × 4.9 ± 0.8 µm (7–18 × 3–7, N = 44), with LW 59.8 ± 14.1 µm² (35–90) and L/W 2.54 ± 0.66 (1.2–4.5); microgametocytes are 11.5 ± 2.6 × 5.1 ± 1.1 µm (6–15 × 3–8, N = 41), with LW 58.0 ± 16.4 µm² (28–98) and L/W 2.39 ± 0.83 (1.1–4.7). The prominent pigment granules are dispersed and irregular in shape. Blocks of apparent chromatin that stain deep reddish appear in the larger immature gametocytes and nearly mature gametocytes that mask the cell nucleus, more dispersed in micro- than in macrogametocytes. In the larger gametocytes, the blocks are somewhat reduced in size.

Exoerythrocytic Merogony EE meronts were common in the lungs of two *H. guentheri* (**Plate 11A**, **m–p**). Ovoid to elongate in shape, dimensions varied, 9–27 × 4–16 µm (14.6 ± 3.7 × 9.6 ± 2.4, N = 25). Most meronts were stained too intensely for an accurate count of nuclei in a single focal plane, but in six meronts less heavily stained, nuclei numbered approximately 66, 69, 78, 85, 99, and 210.

Plate 4 (**A**) *Plasmodium holaspi* from *Holaspis guentheri* of Tanzania. Meronts, **a–f**; macrogametocytes, **g**, **h**; prematuration gametocyte, **i**; microgametocytes, **j–l**. (Figures **a** and **b** modified from Telford, S. R., Jr., *J. Parasitol.*, 72, 271, 1986, Figs. 6 and 11, with permission.) (**B**) *Plasmodium uluguruense* from *Hemidactylus platycephalus* of Tanzania. Meronts, **a–f**; macrogametocytes, **g**, **h**; microgametocytes, **i–k**.

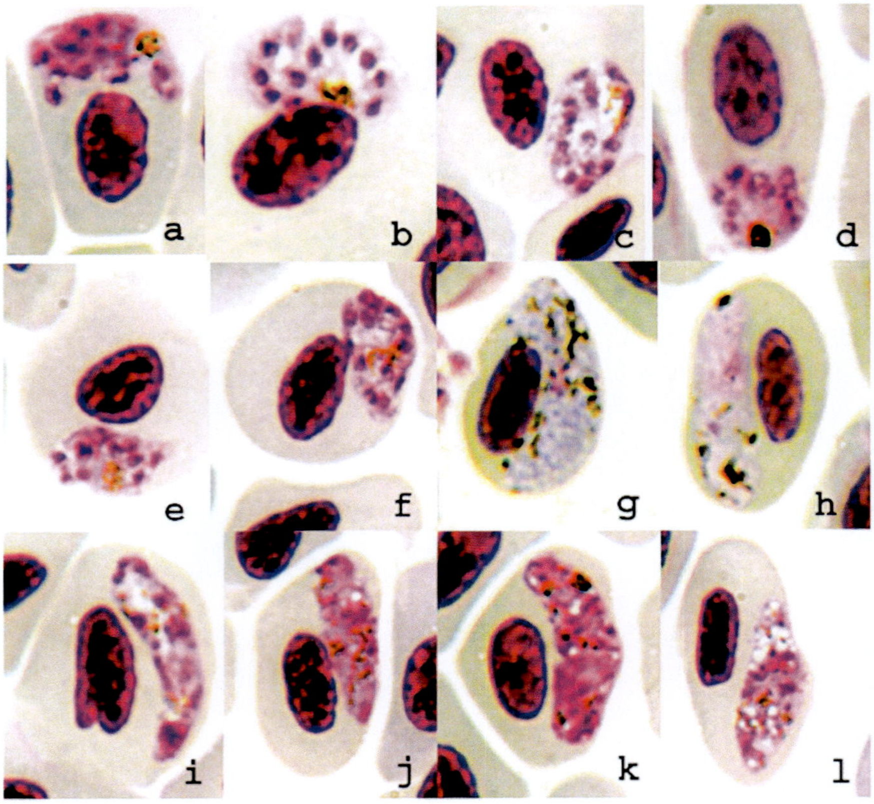

(A)

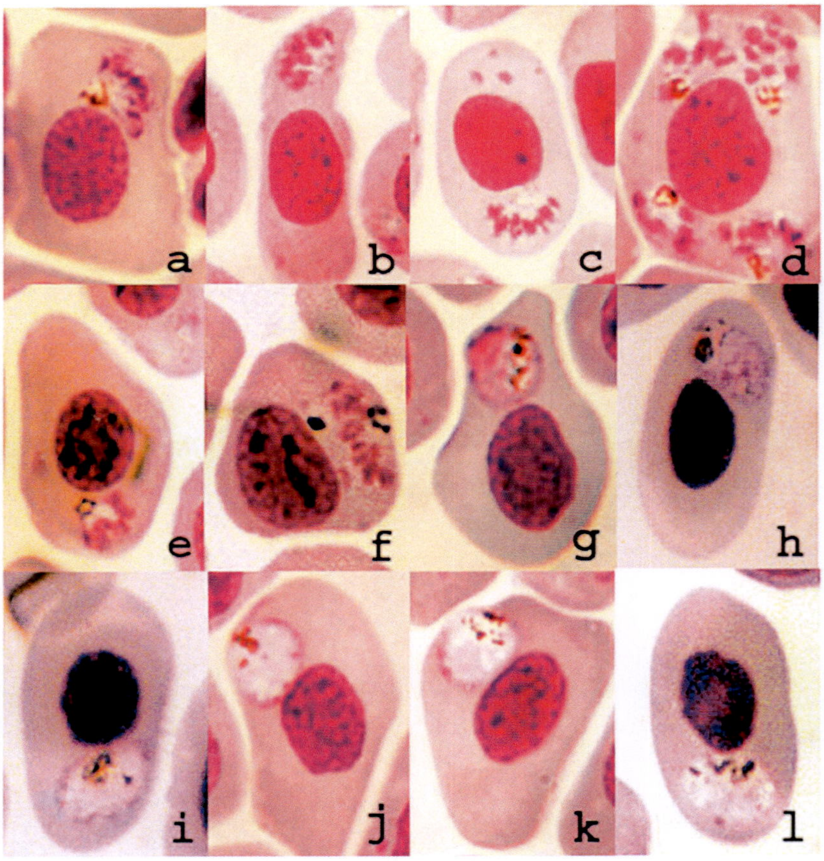

(B)

Sporogony Unknown.

Effects on Host Meronts often distort the host cell and displace its nucleus and can cause both the erythrocyte and its nucleus to become hypertrophied. Gametocytes commonly distort host cells and displace their nuclei, but both meronts and gametocytes only rarely distort the nuclei. In an infection in which hypertrophy of host cells and their nuclei appeared when meronts occupied erythrocytes; the same effect occurred with cells host to erythrocytes.

Remarks *Plasmodium holaspi* is distinguished from all other described malarial parasites of African lizards by the marginal position of young parasites and from all known *Plasmodium* species by the prominent blocks or masses of apparent chromatin present in maturing gametocytes (Telford, 1986a).

Plasmodium uluguruense Telford 1984 (**Plate 4**)

Diagnosis A *Plasmodium* (*Lacertamoeba*) species with asexual stages parasitic in both mature and immature erythrocytes. Meronts are 4–10 × 2–6 μm, with LW 12–54 μm², and contain 4–12 merozoites, usually arranged as a fan. Meront size relative to host cell nucleus averages 0.65, and to normal erythrocyte nuclei is 0.73. Proerythrocytic meronts produce more merozoites than erythrocytic meronts. Light golden pigment granules aggregate into a mass at the base of fans. The usually ovoid gametocytes are 5–10 × 4–7 μm, with LW 20–63 μm² and L/W 1.00–2.50. Gametocyte size relative to host cell nucleus averages 0.97, and to normal erythrocyte nuclei is 1.07. Dark greenish-yellow to black pigment granules are not dispersed in either sex of gametocyte but tend to aggregate in a single focus near the gametocyte margin. Microgametocytes exceed macrogametocytes in length and size but do not differ in shape.

Type Host *Hemidactylus platycephalus* Peters (Sauria: Gekkonidae).

Type Locality North slope of the Uluguru Mountains at Morogoro, Morogoro Region, Tanzania.

Other Hosts None known.

Other Localities Mindu Mountain about 5 km northwest of Morogoro; Kimboza Forest, 1 km north of the Ruvu River below Kibungo Village, south side of Uluguru Mountains, Morogoro Region; Mgeta, about 30 km southwest of Morogoro.

Prevalence *P. uluguruense* parasitized 46 of 71 (64.8%) *H. platycephalus* in the four localities in which it was found: 13 of 21 (61.9%) at the type locality, 26 of 41 (63.4%) in Kimboza Forest, 6 of 8 on Mindu Mountain, and the only gecko taken at Mgeta.

Morphological Variation Meronts, usually fan shaped, are 6.1 ± 1.0 × 4.2 ± 0.8 μm (4–10 × 2–6, N = 133), with LW 25.8 ± 6.4 μm² (12–54), and produce 7.2 ± 1.4 (4–12, N = 135) merozoites. Meront size relative to host cell nucleus is 0.65 ± 0.14 (0.63–1.00, N = 51), and to normal erythrocyte nuclei is 0.73 ± 0.16 (0.34–1.42, N = 133). Proerythrocytic meronts are similar in size to erythrocytic, 6.2 ± 1.1 × 4.1 ± 0.8 μm (4–10 × 2–6, N = 82), LW 25.7 ± 7.1 μm² (12–54), versus 5.9 ± 0.8 × 4.4 ± 0.7 μm (4–8 × 3–5, N = 51), LW 25.8 ± 5.2 μm², respectively, but produce more merozoites, 7.4 ± 1.5 (4–12, N = 84) versus 6.9 ± 1.2 (4–9, N = 51). Aggregations of the light golden pigment granules form a mass at the base of the fan or at one end of elongated meronts. Gametocytes are ovoid, 7.1 ± 1.1 × 5.2 ± 0.6 μm (5–10 × 4–7, N = 150), with LW 37.1 ± 7.2 μm² (20–63) and L/W 1.39 ± 0.28 (1.00–2.50). Gametocyte size relative to host cell nucleus is 0.97 ± 0.24 (0.58–2.29, N = 141), and to normal erythrocyte nuclei is 1.07 ± 0.19 (0.66–1.74, N = 150). Microgametocytes are longer and larger in size than macrogametocytes but do not differ in L/W ratio, 7.4 ± 1.2 × 5.4 ± 0.6 μm (6–10 × 4–7, N = 82), LW 39.5 ± 6.7 μm² (30–50), and L/W 1.40 ± 0.30 (1.00–2.50) versus 6.8 ± 0.9 × 5.0 ± 0.6 μm (5–10 × 4–7, N = 68), LW 34.2 ± 6.9 μm² (20–63), and L/W 1.38 ± 0.26 (1.00–2.50), respectively. Pigment is not dispersed in gametocytes but forms as an aggregate of dark greenish-yellow granules, usually situated near the cell margin in both sexes.

Exoerythrocytic Merogony Up to seven nuclei were present in thrombocytic meronts of one infection (Telford, 1984a).

Sporogony Unknown.

Effects on Host In one of three active infections, cells host to meronts were hypertrophied and occasionally distorted but were normal in size and less often distorted in chronic and relapse infections. Host cell nuclei were hypertrophied in two of three active infections, seldom distorted but usually displaced (Telford, 1984a). Nuclei were not distorted but were usually displaced in the chronic and relapse infections. In erythrocytes parasitized by gametocytes, hypertrophy of the cell was present in one active infection, and distortion of the cells was common in three of five infections. Erythrocyte nuclei were enlarged in all active infections of gametocytes but were normal in size in relapse and chronic infections. Nucleus distortion was uncommon, but nuclei were usually displaced.

Remarks *Plasmodium uluguruense*-parasitized geckoes in the lush lowland Kimboza Forest at 250 m elevation, in second growth savanna woodland at 600–800 m, and on a hillside with sparse remnants of savanna woodland, but were not found in peridomestic habitats in Morogoro or along the beach north of Dar-es-Salaam.

Plasmodium fischeri Ball and Pringle 1965

Diagnosis A *Plasmodium* (*Lacertamoeba*) species in which small asexual stages are rounded with a large central vacuole and no filiform cytoplasmic projections. Mature meronts have estimated dimensions about 9 × 6 µm and LW of 50 µm^2 and contain 21–25 merozoites arranged around the periphery as a broad fan. Large dark pigment granules are concentrated near the fan base. Gametocytes are oblong to elongate, with estimated dimensions 8–11 × 5–8 µm, LW 41–87 µm^2, and L/W 1.4–2.6. The dark pigment forms as large, irregular granules in microgametocytes and as smaller, dispersed granules in macrogametocytes. There is no evidence of sexual dimorphism in gametocyte dimensions.

Type Host *Chamaeleo f. fischeri* (Reichenow) (Sauria: Chamaeleonidae).

Type Locality Amani, Eastern Usambara Mountains, Tanga Region, Tanzania.

Other Hosts None known.

Other Localities None known.

Prevalence One of 42 *C. fischeri* was infected by *P. fischeri* at the type locality (Ball and Pringle, 1965; Ball, 1967a).

Morphological Variation No dimensional data were published for *P. fischeri*. Approximate dimensions were estimated by comparison of published illustrations (Ball and Pringle, 1965) with the average size of uninfected erythrocyte nuclei. The single mature meront illustrated is about 9.1 × 5.5 µm, with an estimated LW of 50.1 µm^2. Meronts produce 21–25 merozoites arranged along the margin of a broad fan, in which large, dark pigment granules are clustered at the base. Estimated gametocyte dimensions are 7.9–11.1 × 4.7–7.1 µm, LW 49.9–87.1 µm^2, and L/W 1.4–2.6. Size of both gametocytes and the single illustrated meront suggests that both are equal to or slightly larger than either infected or normal erythrocyte nuclei. Microgametocyte pigment is distributed as large, irregular dark granules, and that of macrogametocytes is dispersed as smaller granules. There is no evidence of sexual dimorphism in gametocyte dimensions.

Exoerythrocytic Merogony Unknown.

Sporogony Unknown.

Effects on Host Ball and Pringle (1965) stated, "There is little or no hypertrophy of cells infected with meronts of *P. fischeri*," although nuclear displacement and cell distortion is shown for both the mature meront and gametocytes.

Remarks Minute parasites that are probably *Plasmodium* trophozoites were present in 1 of 15 *C. fischeri* from Amani and 1 of 3 collected in the Western Usambara Mountains from 1981 to 1985, but no specific identification was possible. As described elsewhere (Telford, 1988b), the type slides of both *P. fischeri* and *P. acuminatum* are in poor condition or show very few parasites, and more precise descriptions are not possible.

Plasmodium tanzaniae Telford 1988 (**Plate 5**)

Diagnosis A *Plasmodium* (*Lacertamoeba*) species with usually fan-shaped or oblong meronts, 6–12 × 4–7 µm, with LW 28–70 µm^2, that produce 8–22 merozoites. Meront size relative to host cell nucleus averages 0.91, and to normal erythrocyte nuclei is 1.31. Meronts are predominantly proerythrocytic, no different in size from erythrocytic meronts, but with more merozoites. Pigment is dispersed as large, dark golden granules among merozoites of the meront or clumped at the base of fans. Gametocytes are usually elongate, 8–19 × 4–9 µm, with LW 48–112 µm^2 and L/W 1.0–4.3. Gametocyte size relative to host cell nucleus averages 1.89, and to normal erythrocyte nuclei is 2.07. Macrogametocytes are larger and slightly more elongate than microgametocytes. Some gametocytes have a nearly circular, light bluish-green area in the cytoplasm, apparently where pigment granules were previously aggregated. Pigment is dispersed as small dark golden granules or occasionally as loose clumps of granules among the smaller granules. Gametocytes are almost entirely erythrocytic.

Type Host *Chamaeleo werneri* Tornier (Sauria: Chamaeleonidae).

Type Locality Mufindi, Western Usambara Mountains, Iringa Region, Tanzania.

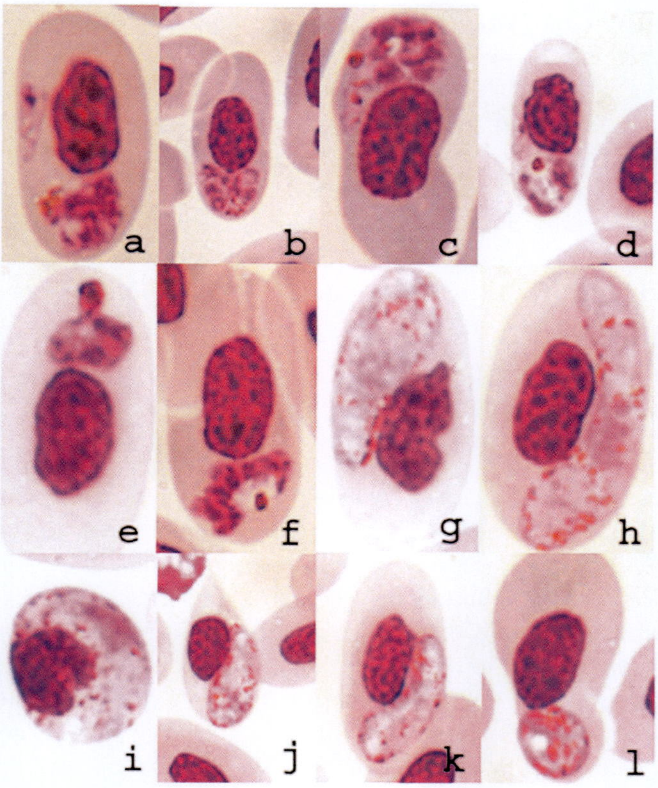

(A)

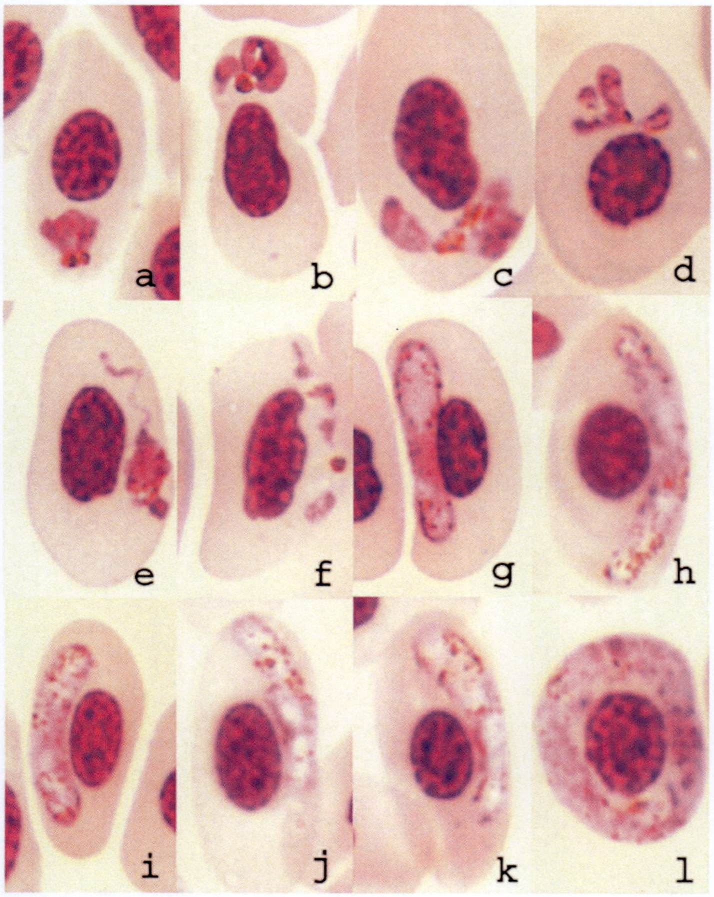

(B)

Plate 5 (**A**) *Plasmodium tanzaniae* from *Chamaeleo werneri* of Tanzania. Meronts, **a–f**; macrogametocytes, **g–i**; microgametocytes, **j–l**. (**B**) *Plasmodium arachniformis* from *Chamaeleo werneri* of Tanzania. Meronts, **a–f**; macrogametocytes, **g–i**; microgametocytes, **j**, **k**, double infection of macro- and microgametocytes.

Other Hosts None known.

Other Localities None known.

Prevalence *Plasmodium tanzaniae* parasitized 5 of 21 (23.8%) *C. werneri* at the type locality.

Morphological Variation Meronts are usually fan shaped or oblong, 7.6 ± 1.2 × 5.5 ± 0.7 µm (6–12 × 4–7, N = 55), with LW 41.5 ± 8.1 µm² (28–70), and contain 14.5 ± 3.1 (8–22) merozoites. Meront size relative to host cell nucleus is 0.91 ± 0.21 (0.67–1.20, N = 8), and to normal erythrocyte nuclei is 1.31 ± 0.25 (0.90–2.02, N = 55). Most meronts are proerythrocytic, 7.5 ± 1.2 × 5.5 ± 0.7 µm (6–12 × 4–7, N = 47), LW 41.0 ± 7.3 µm², with more merozoites, 15.0 ± 2.9 (10–22), than the similar size erythrocytic meronts, 7.9 ± 1.1 × 5.5 ± 0.8 µm (7–10 × 5–7, N = 8), LW 43.9 ± 12.3 µm², which contain 11.3 ± 2.3 (8–14) merozoites. Gametocytes are almost entirely erythrocytic, 12.9 ± 2.4 × 5.8 ± 1.1 µm (8–19 × 4–9, N = 125), with LW 73.8 ± 12.9 µm² (48–112) and L/W 2.37 ± 0.77 (1.00–4.24). Gametocyte size relative to host cell nucleus is 1.89 ± 0.41 (1.08–2.97, N = 74), and to normal erythrocyte nuclei is 2.07 ± 0.36 (1.23–3.08, N = 125). Macrogametocytes are 13.7 ± 2.2 × 5.8 ± 1.1 µm (8–18 × 4–9, N = 73), with LW 78.1 ± 12.9 µm² (48–112) and L/W 2.48 ± 0.75 (1.11–4.25), larger and slightly more elongate than microgametocytes, which are 12.0 ± 2.2 × 5.8 ± 1.1 µm (8–19 × 4–8, N = 52), with LW 67.7 ± 10.2 µm² (52–95) and L/W 2.21 ± 0.77 (1.00–4.00). Pigment is dark golden in both meronts and gametocytes, dispersed among merozoites or clustered at bases of fans, and dispersed in gametocytes as small granules, with occasional clumps of several granules. Some gametocytes have a clear, almost circular, bluish-staining area that apparently held pigment aggregates before dispersal.

Exoerythrocytic Merogony No EE meronts could be definitely identified as those of *P. tanzaniae.*

Sporogony Unknown.

Effects on Host Both host cells and their nuclei are hypertrophied when meronts are present, but neither cells nor their nuclei are distorted, although the latter are commonly displaced. Erythrocytes host to gametocytes show no hypertrophy but are usually distorted, nuclei are occasionally distorted and usually displaced. In active infections, erythrocyte nuclei are usually hypertrophied but are of normal size when infections are chronic.

Remarks Although thrombocytic and lymphocytic meronts were present in mixed infections with either *P. arachniformis* or *P. uzungwiense,* there was little probability that these represented *P. tanzaniae,* usually present in light active or chronic infection, but rather they probably belonged to the dominant species in the mixed infection. The infected *C. werneri* were collected where mature second-growth forest bordered hillside tea plantations of considerable age at Mufindi. No *Plasmodium* infections were found in 39 *C. tempeli* collected from the same area and often the same shrubs utilized by *C. werneri,* host to three *Plasmodium* species.

Plasmodium arachniformis Telford 1988 (**Plate 5**)

Diagnosis A *Plasmodium* (*Lacertamoeba*) species in which young asexual stages assume bizarre forms. Meronts are polymorphic, but most commonly form fans prior to segmentation. At segmentation, merozoites may become arranged linearly, in groups or singly, attached to others by narrow cytoplasmic connections. Mature meronts are 4–12 × 2–7 µm, with LW 12–49 µm², and produce 4–12 merozoites. Meronts occur in both mature and immature erythrocytes. Meront size relative to host cell nucleus averages 0.69, and to normal erythrocyte nuclei is 0.83. Although meront size is similar, proerythrocytic meronts contain more nuclei than do erythrocytic meronts. Prominent dark golden pigment granules form a clump at the base of fans, and in other formed meronts may be dispersed as individual granules or small clumps among nuclei. Gametocytes are typically elongate and thin, 6–17 × 3–8 µm, with LW 30–75 µm² and L/W 1.0–5.3. Gametocyte size relative to host cell nucleus averages 1.54, and to normal erythrocyte nuclei is 1.58. Macrogametocytes are longer and more slender than microgametocytes but are similar in size. Proerythrocytes are commonly parasitized by gametocytes, which are shorter and less slender than in erythrocytes, although size is similar. Small dark golden pigment granules are dispersed in both sexes but may occasionally aggregate in microgametocytes.

Type Host *Chamaeleo werneri* Tornier (Sauria: Chamaeleonidae).

Type Locality Mufindi, Western Usambara Mountains, Iringa Region, Tanzania.

Other Hosts None known.

Other Localities None known.

Prevalence *Plasmodium arachniformis* parasitized 4 of 21 (19.0%) *C. werneri* at the type locality.

Morphological Variation The polymorphic meronts most commonly are fan-shaped. Oddly configured meronts often become more condensed as segmentation approaches, assuming a peripheral, fan-like arrangement of nuclei, but at segmentation of merozoites again may become oddly configured (Telford, 1988b). Merozoites may become linearly arranged in groups or individually, connected to others by narrow cytoplasm bridges, and often appear arachnoid. Meronts are 5.9 ± 1.1 × 4.4 ± 0.7 μm (4–12 × 2–7, N = 125), with LW 25.9 ± 6.0 μm^2 (12–49), and contain 6.4 ± 2.0 (4–12) merozoites. Meront size relative to host cell nucleus is 0.69 ± 0.19 (0.33–1.25, N = 53), and to normal erythrocyte nuclei is 0.83 ± 0.19 (0.42–1.45, N = 125). Proerythrocytic meronts are similar in size to erythrocytic but produce more merozoites, with dimensions 5.8 ± 0.8 × 4.3 ± 0.7 μm (4–8 × 3–6, N = 72), LW 25.4 ± 5.5 μm^2 (15–40), and nuclei 6.8 ± 2.3 (4–12), versus 6.1 ± 1.3 × 4.4 ± 0.9 μm (4–12 × 2–7, N = 53), LW 26.5 ± 6.6 μm^2 (12–49), with 5.9 ± 1.4 (4–9) merozoites. Pigment in fans is clumped as a dark golden mass at the base of the fan, but granules are dispersed among nuclei in other meront forms. Gametocytes are typically elongate and slender with rounded ends but rarely may be ovoid or rounded. Their dimensions are 12.4 ± 2.1 × 4.2 ± 0.8 μm (6–17 × 3–8, N = 200), with LW 50.6 ± 8.9 μm^2 (30–75) and L/W 3.13 ± 0.99 (1.00–5.33). Gametocyte size relative to host cell nucleus is 1.54 ± 0.42 (0.88–2.67, N = 97), and to normal erythrocyte nuclei is 1.58 ± 0.35 (0.82–2.64, N = 200). Erythrocytic gametocytes are longer and more slender than those in proerythrocytes, 12.9 ± 2.2 × 4.0 ± 0.8 μm (6–17 × 3–6, N = 97), LW 51.2 ± 8.6 μm^2 (30–68), and L/W 3.37 ± 1.01 (1.20–5.33) versus 11.8 ± 2.0 × 4.3 ± 0.8 μm (7–16 × 3.3–8, N = 103), LW 50.1 ± 9.2 μm^2 (32–75), and L/W 2.90 ± 0.91 (1.00–5.33), respectively. Gametocytes do not differ in size, but macrogametocytes are longer and more slender than microgametocytes, 13.1 ± 2.0 × 4.0 ± 0.8 μm (8–17 × 3–8, N = 107), LW 51.0 ± 9.0 μm^2 (33–75), and L/W 3.47 ± 0.99 (1.00–5.33) versus 11.5 ± 2.0 × 4.4 ± 0.8 μm (6–16 × 3–6, N = 93), LW 50.2 ± 8.9 μm^2 (30–70), and L/W 2.73 ± 0.53 (1.20–5.33), respectively. Pigment in gametocytes is dispersed as small dark golden granules that occasionally may appear as variable-size aggregations in microgametocytes.

Exoerythrocytic Merogony In infections mixed with *P. tanzaniae*, in which *P. arachniformis* was the dominant species, thrombocytic and lymphocytic meronts were common and probably belonged to the latter species, although not certainly. Sections and smears of tissues were negative for EE stages.

Sporogony Unknown.

Effects on Host Erythrocytes host to either meronts or gametocytes are not hypertrophied and are seldom distorted. Meronts cause enlargement of host cell nuclei but rarely distort or displace them. Nuclei of erythrocytes infected by gametocytes are hypertrophied in some infections, not in others, often displaced but rarely distorted.

Remarks Although the tiny meronts of *Plasmodium* (*Asiamoeba*) *lionatum* in Southeast Asian flying geckoes may have its four to six merozoites lined up in linear arrangement, no other saurian *Plasmodium* species has merozoites at segmentation lining up in linear distribution, connected by thin links of cytoplasm that can appear arachnoid in form.

Plasmodium gologoloense Telford 1988 (**Plate 6**)

Diagnosis A *Plasmodium* (*Lacertamoeba*) species with usually oval or round meronts, 5–7 × 4–6 μm, and LW 20–42 μm^2, that produce 6–14 merozoites. Meront size relative to host cell nucleus averages 1.29, and to normal erythrocyte nuclei is 0.99. Meronts most commonly parasitize proerythrocytes, but erythrocytic meronts are larger and contain more merozoites. Pigment in meronts is formed as one or two prominent, often squarish, light greenish-gold granules. Gametocytes are usually ovoid or round, 5–11 × 4–6 μm, with LW 20–54 μm^2 and L/W 1.0–2.8. Gametocytes size relative to host cell nucleus averages 1.15, and to normal erythrocyte nuclei is 1.19. Microgametocytes are larger and more elongate than macrogametocytes. Gametocytes are usually erythrocytic but are commonly proerythrocytic as well. Erythrocytes host to meronts, and their nuclei, are hypotrophic but when occupied by gametocytes are normal in dimensions. Pigment is dispersed primarily along gametocyte margins as prominent dark gold granules.

Type Host *Bradypodion oxyrhinum* Klaver and Böhme (Sauria: Chamaeleonidae) (syn. *Chamaeleo tenuis* of Telford, 1988b).

Type Locality Eastern Udzungwa Mountains above Sanje, Kilombero District, Morogoro Region, Tanzania.

Other Hosts None known.

Other Localities None known.

Prevalence The only *B. oxyrhinum* examined by Telford (1988b) was host to *P. gologoloense*.

Morphological Variation Meronts are usually ovoid or round, 5.8 ± 0.6 × 5.1 ± 0.6 μm (5–7 × 4–6m N = 53), with

Plate 6 (**A**) *Plasmodium gologoloense* from *Bradypodion oxyrhinum* of Tanzania. Meronts, **a–f**; macrogametocytes, **g–i**; microgametocytes, **j–l**. (**B**) *Plasmodium uzungwiense* from *Chamaeleo werneri* of Tanzania. Meronts, **a–f**; macrogametocytes, **g–j**; microgametocytes, **k**, **l**.

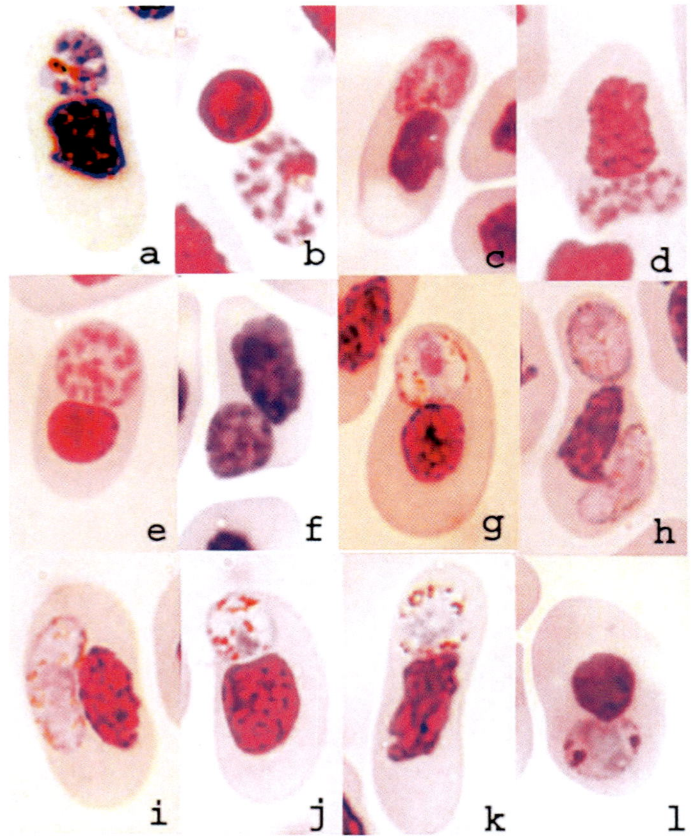

(A)

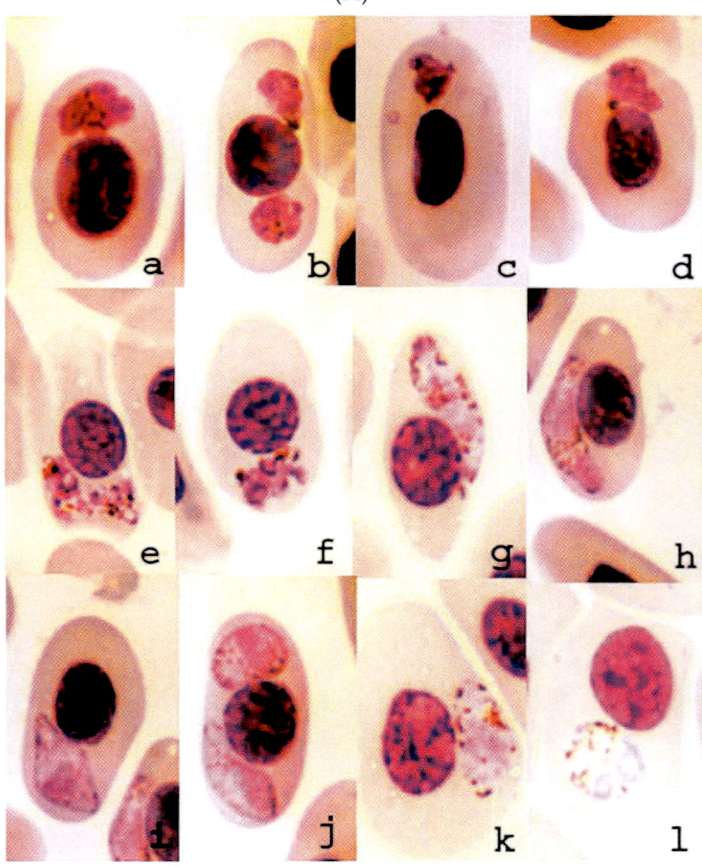

(B)

LW 29.8 ± 4.9 μm² (20–42), and contain 10.3 ± 2.6 (6–14, N = 54) merozoites. Meront size relative to host cell nucleus averages 1.29 ± 0.27 (0.8–1.8, N = 18), and to normal erythrocyte nuclei is 0.99 ± 0.16 (0.67–1.4, N = 53). Erythrocytic meronts are larger, 5.9 ± 0.5 × 5.4 ± 0.5 μm (N = 18), with LW 31.8 ± 4.7 μm², and produce more merozoites, 12.0 ± 2.2, than proerythrocytic meronts, 5.8 ± 0.7 × 5.0 ± 0.6 μm (N = 35), with LW 28.8 ± 4.7 μm², which contain 9.4 ± 2.5 (N = 36) merozoites. There are one or two large, greenish-gold pigment granules, often squarish in outline, in meronts. Gametocytes are round or ovoid usually, 7.3 ± 1.6 × 5.0 ± 0.7 μm (5–11 × 4–6, N = 75), with LW 35.7 ± 7.1 μm² (20–54), and L/W 1.50 ± 0.47 (1.00–2.75). Their size relative to host cell nucleus is 1.15 ± 0.25 (0.6–1.9, N = 38), and to normal erythrocyte nuclei is 1.19 ± 0.24 (0.7–1.8, N = 75). Microgametocytes are larger than macrogametocytes and more elongate, 7.6 ± 1.5 × 5.0 ± 0.7 μm (6–11 × 4–6, N = 37), LW 37.4 ± 6.1 μm² (30–54), and L/W 1.59 ± 0.50 (1.0–2.8) versus 6.9 ± 1.5 × 5.0 ± 0.6 μm (5–11 × 4–6, N = 38), with LW 34.0 ± 7.7 μm² (20–50), and L/W 1.41 ± 0.42 (1.0–2.8), respectively. Pigment in gametocytes forms as prominent dark gold granules that are dispersed, but tending to locate along gametocyte margins.

Exoerythrocytic Merogony Inasmuch as the infected chameleon was host to two type infections, *P. michikoa* and *P. gologoloense*, comments on the many EE meronts observed in circulating blood as host death approached are similarly appropriate for each species (see *P. michikoa* discussion).

Sporogony Unknown.

Effects on Host Erythrocytes parasitized by meronts, and their nuclei, are hypotrophic in size but do not differ from uninfected cells when occupied by gametocytes. Meronts rarely distort the host cell but commonly distort (by pressure) and displace their nuclei. Host cells and their nuclei are rarely distorted by gametocytes and seldom have altered shape, but the nuclei are often displaced. Meronts are usually proerythrocytic and gametocytes erythrocytic, but both can occur in either type of host cell.

Remarks Although both *P. michikoa* and *P. gologoloense* were found in mixed infection in a single *B. oxyrhinum*, their effects on host erythrocytes are both similar and different: Meronts and gametocytes of *P. michikoa* cause hypotrophy of the host erythrocyte and its nucleus, but only meronts of *P. gologoloense* similarly affect the erythrocyte. Cells host to gametocytes of the latter species are normal in dimensions. The name *gologoloense* is derived from another name for the Eastern Udzungwa Mountains, Gologolo Mountains.

Plasmodium uzungwiense Telford 1988 (**Plate 6**)

Diagnosis A *Plasmodium* (*Lacertamoeba*) species parasitic primarily in immature erythrocytes in all stages. Meronts are 4–8 × 3–6 μm, with LW 16–42 μm², and produce 4–12 merozoites, usually arranged as a fan or in oblong or rounded shape. Meronts are about one-half the size of host cell nuclei and, relative to the size of normal erythrocyte nuclei, average 0.9. Pigment is formed as large dark golden granules, clumped together at the base of the fan or in a single focus in other meront forms. Most gametocytes are elongate, 5–13 × 3–7 μm, with LW 24–63 μm² and L/W 1.0–4.3. Gametocyte size relative to the size of normal erythrocyte nuclei, averages 1.34. There is no sexual difference in gametocyte size, but microgametocytes are slightly shorter and are less elongate in shape than macrogametocytes. Pigment is dispersed as small dark golden granules, tending to be marginal in macrogametocytes and somewhat aggregated in microgametocytes.

Type Host *Chamaeleo werneri* Tornier (Sauria: Chamaeleonidae).

Type Locality Mufindi, Western Usambara Mountains, Iringa Region, Tanzania.

Other Hosts None known.

Other Localities None known.

Prevalence *Plasmodium uzungwiense* parasitized 2 of 21 (9.5%) *C. werneri* at the type locality.

Morphological Variation Meronts are usually fan shaped, oblong or round, 5.8 ± 1.0 × 4.5 ± 0.7 μm (4–8 × 3–6, N = 50), with LW 26.4 ± 7.3 μm² (16–42), and contain 6.7 ± 1.9 (4–12) merozoites. In their proerythrocytic host cells, meront size is about one-half of the cell nucleus, and relative to the size of normal erythrocyte nuclei is 0.88 ± 0.23 (0.56–1.46). Pigment forms as large dark golden granules, clumped together at the base when the meront is shaped like a fan or in a single focus in other meront forms. Gametocytes, almost entirely proerythrocytic, are typically elongate with blunt ends but rarely may be ovoid or rounded, 9.3 ± 1.7 × 4.3 ± 0.9 μm (5–13 × 3–7, N = 110), with LW 39.9 ± 8.7 μm² (24–63) and L/W 2.28 ± 0.78 (1.00–4.33). Gametocyte size relative to host cell nucleus is 1.09 ± 0.26 (0.80–1.43, N = 6), and to normal erythrocyte nuclei is 1.34 ± 0.28 (0.83–2.05, N = 110). Macrogametocytes are slightly longer than microgametocytes and more slender, 9.6 ± 1.7 × 4.1 ± 0.9 μm (6–13 × 3–6, N = 63), LW 39.2 ± 8.6 μm^2 (24–60), and L/W 2.45 ± 0.80 (1.00–4.33) versus 9.0 ± 1.6 × 4.6 ± 0.9 μm (5–13 × 3–7, N = 47), LW 40.9 ± 8.7 μm^2 (24–63), and L/W 2.06 ± 0.71 (1.00–4.33), respectively. Pigment, dispersed as small dark golden granules, tends to be marginal in macrogametocytes and may show aggregations as well as individual granules in microgametocytes.

Exoerythrocytic Merogony The many thrombocytic and lymphocytic meronts present at peak of infection by *P. uzungwiense* might belong to this species in view of the very light, chronic-phase infection by *P. tanzaniae*, also present.

Sporogony Unknown.

Effects on Host Neither asexual nor sexual stages produce hypertrophied host proerythrocytes. Meronts do not distort the host cell and its nucleus and seldom displace the nucleus. Gametocytes sometimes distort host cells and rarely their nuclei, usually displacing the latter, but do cause enlargement of the proerythrocyte nucleus.

Remarks *Plasmodium uzungwiense* is the only African saurian malaria species known that predominantly infects host proerythrocytes in all stages without regard to parasitemia or infection phase.

Plasmodium cnemaspi Telford 1984 (**Plate 7**)

Diagnosis A *Plasmodium* (*Lacertamoeba*) species with variably shaped meronts that parasitize both mature and immature red blood cells, 6–13 × 3–7 μm, with LW 24–91 μm^2, which contain 8–24 merozoites. Meront size relative to host cell nucleus averages 1.04, and to normal erythrocyte nuclei is 1.21. Proerythrocytic meronts are larger than erythrocytic and produce more merozoites. Dark, greenish-yellow pigment granules are variably located in loose aggregations within meronts. Gametocytes parasitize only erythrocytes and are usually elongate in active infection, ovoid or round in chronic infection, 7–14 × 3–9 μm, with LW 32–108 μm^2 and L/W 1.00–4.00. Gametocyte size relative to host cell nucleus averages 2.04, and to normal erythrocyte nuclei is 1.77. The dark pigment granules are dispersed in the cytoplasm of both sexes of gametocytes, which do not differ in dimensions.

Type Host *Cnemaspis barbouri* Perret (Sauria: Gekkonidae) (= *C. africana* of Telford, 1984a).

Type Locality Kimboza Forest, 1 km north of the Ruvu River below Kibungo Village, south side of Uluguru Mountains, Morogoro Region, Tanzania.

Other Hosts None known.

Other Localities None known.

Prevalence *Plasmodium cnemaspi* parasitized 10 of 22 (45.5%) *C. barbouri* at the type locality.

Morphological Variation Meronts are 7.4 ± 1.4 × 5.5 ± 0.9 μm (6–13 × 3–7, N = 100), with LW 40.8 ± 11.8 μm^2 (24–91), and produce 12.1 ± 3.0 (8–24) merozoites. Meront size relative to host cell nucleus is 1.04 ± 0.25 (0.69–2.14, N = 63), and to normal erythrocyte nuclei is 1.21 ± 0.34 (0.63–2.58, N = 100). Proerythrocytic meronts are larger, 7.8 ± 1.7 × 6.1 ± 0.7 μm (6–13 × 5–7, N = 37), LW 47.9 ± 13.7 μm^2 (30–91), versus 7.1 ± 1.2 × 5.2 ± 0.8 μm (6–12 × 3–7, N = 63), LW 36.8 ± 8.1 μm^2 (24–63), and produce more merozoites, 14.5 ± 2.9 (10–24), versus 10.6 ± 2.0 (8–16). Meronts are usually round, oval, or oblong, rarely fan-shaped or elongate, with dark greenish-yellow pigment granules, loosely aggregated and variably located within the meront. Gametocytes are 9.6 ± 1.6 × 6.6 ± 1.2 μm (7–14 × 3–9, N = 175), with LW 62.3 ± 12.0 μm^2 (32–108) and L/W 1.56 ± 0.59 (1.00–4.00). Gametocyte size relative to host cell nucleus is 2.04 ± 0.49 (0.80–3.86, N = 162), and to normal erythrocyte nuclei is 1.77 ± 0.31 (0.95–2.85, N = 175). There is no sexual dimorphism in gametocyte dimensions. Gametocytes from active infections are longer and narrower, with higher L/W ratio, but similar in LW, 10.0 ± 1.7 × 6.4 ± 1.3 μm (7–14 × 3–9, N = 125), LW 62.5 ± 13.4 μm^2 (32–108), and L/W 1.67 ± 0.65 (1.00–4.00), than gametocytes of chronic infections, 8.8 ± 1.0 × 7.1 ± 0.7 μm (7–12 × 3–9, N = 50), with LW 61.7 ± 7.7 μm^2 (42–80), and L/W 1.26 ± 0.26 (1.00–2.40). In chronic infections, more microgametocytes are round (92%) than macrogametocytes (50%) (Telford, 1984a). Dark pigment granules are dispersed throughout the cytoplasm in both sexes of gametocytes.

Exoerythrocytic Merogony A single EE meront containing 16 nuclei parasitized an apparent monocyte in circulating blood (Telford, 1984a). In a chronic infection of *C. barbouri*, five phanerozoites of *P. cnemaspi* were found in connective tissue of the lungs and gut wall (**Plate 11A, i–l**). Their dimensions averaged 13.8 × 9.8 μm (10 × 10–17 × 11).

Sporogony Unknown.

Effects on Host Cells host to meronts are not enlarged, but erythrocyte nuclei in one active infection were hypertrophied. Host cells are commonly distorted and their

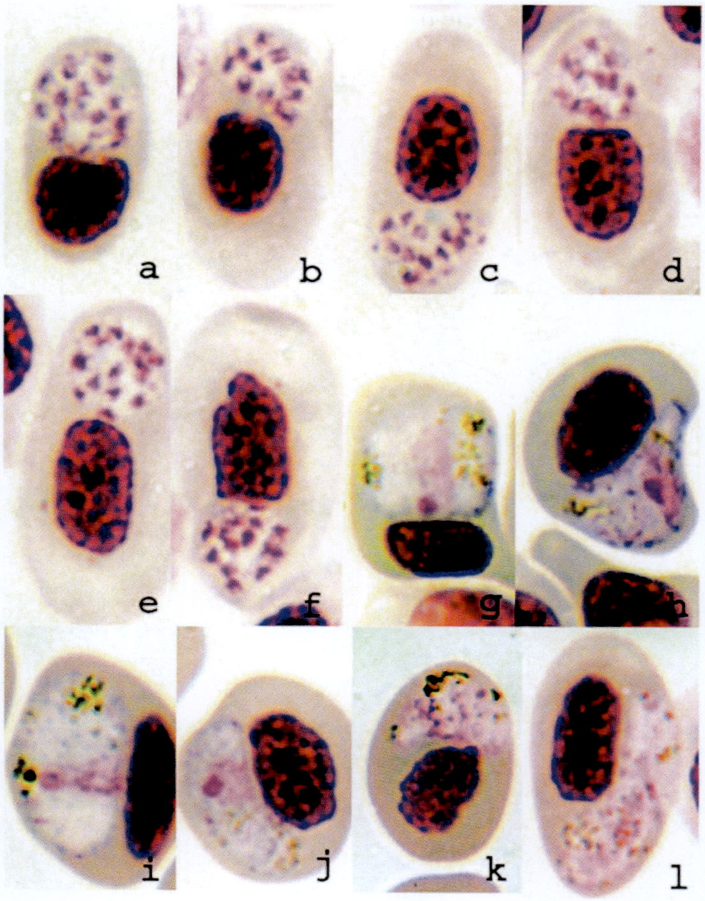

(A)

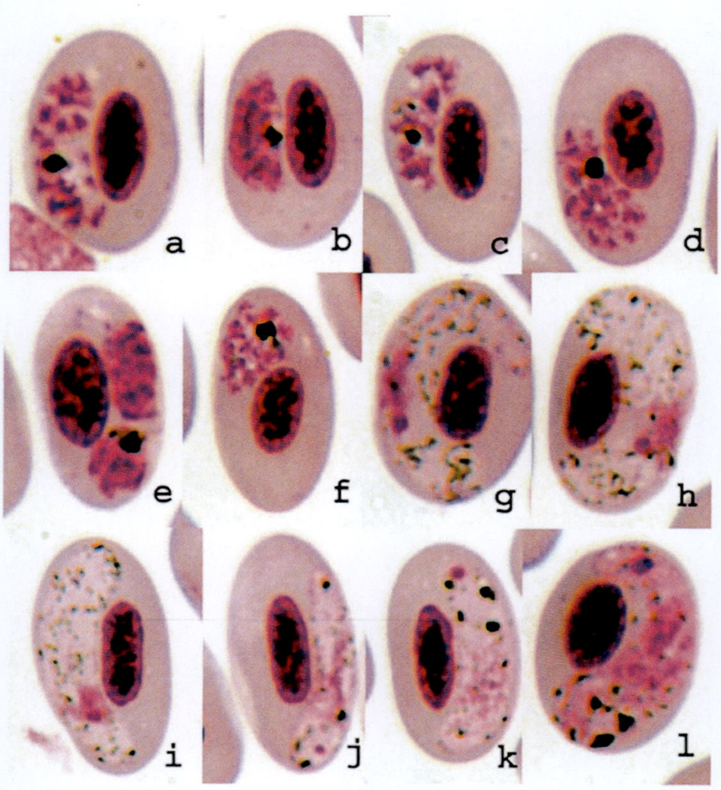

(B)

Plate 7 (**A**) *Plasmodium cnemaspi* from *Cnemaspis barbouri* of Tanzania. Meronts, **a–f**; macrogametocytes, **g–j**; microgametocytes, **k–l**. (**B**) *Plasmodium loveridgei* from *Lygodactylus luteopicturatus* of Tanzania. Meronts, **a–f**; macrogametocytes, **g–i**; microgametocytes, **j–l**.

nuclei displaced but seldom distorted. Erythrocytes host to gametocytes in active infections were hypertrophied in one of three samples, but nuclei were of normal size. Host cells were usually distorted and had displaced nuclei, but nuclear distortion was uncommon. In chronic infections, gametocytes caused striking hypotrophy of the host erythrocytes and their nuclei, with a reduction of 15–20% from their normal size (Telford, 1984a), usually producing distortion of both cell and its nucleus and always displacing the nucleus.

Remarks *Cnemaspis barbouri* infected by *P. cnemaspi* were collected from piles of boulders, deeply shaded within primary lowland rain forest at 250 m elevation, in association with *Hemidactylus platycephalus* infected by *P. uluguruense*. Although the two gecko species did not share their *Plasmodium* parasites, two other hemoparasites were held in common, *Trypanosoma cnemaspi* (Telford, 1995a) and a *Sauroleishmania* species (Telford, 1995b).

Plasmodium loveridgei Telford 1984 (**Plate 7**)

Diagnosis A *Plasmodium* (*Lacertamoeba*) species with polymorphic, usually elongate or fan-shaped meronts that parasitize both mature and immature erythrocytes. Meronts are 5–15 × 3–7 μm, LW 20–91 μm^2, and produce 6–26 merozoites. Meront size relative to host cell nucleus averages 1.45, and to normal erythrocyte nuclei is 1.81. Proerythrocytic meronts are larger and produce more merozoites than those in erythrocytes. Dark greenish-gold pigment granules, usually one to three in number, are dispersed among nuclei. Gametocytes are elongate, rarely rounded, 8–23 × 3–11 μm, with LW 48–176 μm^2 and L/W 1.00–6.00. Gametocyte size relative to host cell nucleus averages 3.19, and to normal erythrocyte nuclei is 2.89. There is no sexual dimorphism in gametocyte dimensions, and their size is not affected by infection phase. Pigment is dispersed as dark greenish-gold-to-black granules within the gametocyte cytoplasm.

Type Host *Lygodactylus l. luteopicturatus* Pasteur (Sauria: Gekkonidae) (= *L. picturatus* of Telford, 1984a).

Type Locality North slope of the Uluguru Mountains at the edge of University Campus, Morogoro, Morogoro Region, Tanzania.

Other Hosts *Lygodactylus capensis grotei.*

Other Localities None known.

Prevalence *Plasmodium loveridgei* infected 21 of 59 (35.6%) *L. luteopicturatus* and 2 of 50 (4%) *L. capensis grotei* at the type locality.

Morphological Variation In natural infections of *L. luteopicturatus*, meronts are 8.8 ± 1.8 × 5.2 ± 1.0 μm (5–15 × 3–7, N = 138), with LW 45.1 ± 11.4 μm^2 (20–91), and contain 16.9 ± 3.6 (6–26, N = 140) merozoites. Meront size relative to host cell nucleus is 1.43 ± 0.38 (0.67–3.00, N = 106), and to normal erythrocyte nuclei is 1.81 ± 0.61 (0.69–4.40, N = 138). Proerythrocytic meronts are larger, 9.5 ± 1.5 × 5.6 ± 0.9 μm (7–13 × 4–7, N = 32), LW 53.1 ± 11.3 μm^2 (35–91), and produce more merozoites, 18.9 ± 3.3 (14–26, N = 34), than erythrocytic meronts, which are 8.6 ± 1.8 × 5.0 ± 0.9 μm (5–15 × 3–7, N = 106), LW 42.7 ± 10.3 μm^2 (20–70), and contain 16.2 ± 3.5 (6–26) merozoites. Pigment is dispersed usually as one to three dark greenish-gold granules variably located within the meront. Gametocytes, usually elongate except when polar in position, are 13.5 ± 2.9 × 5.9 ± 1.6 μm (8–23 × 3–11, N = 200), with LW 77.5 ± 18.0 μm^2 (48–176) and L/W 2.52 ± 1.03 (1.00–6.00). Gametocyte size relative to host cell nucleus is 3.19 ± 0.91 (1.40–5.83), and to normal erythrocyte nuclei is 2.89 ± 0.71 (1.76–5.52. Gametocyte dimensions do not differ with sex or phase of infection. Dark greenish-gold-to-black pigment granules are dispersed within the cytoplasm of both gametocyte sexes. In natural infections of *Lygodactylus capensis grotei*, meronts are 7.2 ± 1.9 × 4.7 ± 0.9 μm (5–13 × 3–8, N = 49), LW 33.7 ± 11.1 μm^2 (18–66), and produce 12.1 ± 3.6 (7–22, N = 50) merozoites. Gametocytes are 14.1 ± 2.4 × 4.9 ± 0.9 μm (10–20 × 4–7, N = 50), with LW 68.2 ± 15.3 μm^2 (48–120) and L/W 3.00 ± 0.78 (1.67–4.50). In an experimental infection of *L. capensis grotei* derived from *L. luteopicturatus*, meronts are 6.3 ± 1.1 × 4.7 ± 0.6 μm (4–8 × 4–6, N = 24), LW 29.8 ± 5.9 μm^2 (16–42), and contain 10.0 ± 2.5 (6–18) merozoites. Gametocytes of the experimental infection are 14.8 ± 2.1 × 5.1 ± 0.7 μm (11–20 × 4–6, N = 25), with LW 76.2 ± 15.8 μm^2 (44–108) and L/W 2.96 ± 0.54 (2.17–4.00).

Exoerythrocytic Merogony Meronts with up to seven nuclei parasitized thrombocytes in one active infection. Phanerozoites found in connective tissue of the heart of *L. capensis grotei* averaged 22.2 × 16.0 μm (15 × 12–28 × 17) and contained approximately 52–126 nuclei (**Plate 11A, e–h**). Smaller phanerozoites, 7 × 6–12 × 9 μm, were present in connective tissue and endothelium of the liver.

Sporogony Unknown.

Effects on Host Erythrocytes host to meronts were hypertrophied in two of three acute infections, without nuclear enlargement, but in one infection host cells were of normal size but had enlarged nuclei (Telford, 1984a). Host cells are usually distorted with displaced nuclei, but distorted nuclei are less common. As with meronts, erythrocytes host to gametocytes were hypertrophied in two of three acute infections, but of normal size in the third infection. Most host cells are distorted and have displaced nuclei, but nuclei are seldom distorted. In a chronic infection, cells host to gametocytes are normal in size, but nuclei are distorted and hypotrophied.

Remarks Infected *L. luteopicturatus* and *L. capensis grotei* were collected from a small area of less-disturbed second-growth savanna woodland, especially from trunks of acacia trees, with 11 of 19 (58%) of the type host infected. Other habitats and localities from which these hosts were collected, 62 *L. luteopicturatus* and 25 *L. capensis grotei*, did not have geckoes infected by *P. loveridgei*.

Plasmodium maculilabre Schwetz 1931

Diagnosis A *Plasmodium* (*Lacertamoeba*) species with ovoid or round meronts that may nearly fill the host erythrocyte, estimated size about 10.0 × 6.9 μm, with LW approximately 69 μm², that produce 15–20 merozoites. Meront size relative to host cell nucleus and to normal erythrocyte nuclei is about 5.0. Gametocytes are 7–13 × 5–8 μm, with LW 42–91 μm² and L/W 1.13–2.60. Gametocyte size relative to host cell nucleus averages 2.62, and to normal erythrocyte nuclei is 3.61. Gametocytes are ovoid to elongate, sometimes filling the host cell; macrogametocytes are longer than microgametocytes, with greater LW and L/W ratio. Pigment appears to be sparse in meronts but is dispersed as round dark granules in gametocytes.

Type Host *Mabuya maculilabris* (Gray) (Sauria: Scincidae).

Type Locality Stanleyville (now Kisangani), Congo.

Other Hosts None known.

Other Localities None known.

Prevalence Schwetz (1931) reported 34 of 73 (46.6%) of *M. maculilabris* at the type locality were infected by *P. maculilabre*. In a series of ten *M. maculilabre* collected at Kisangani in March 1985, one skink was host to a chronic infection of *P. maculilabre*.

Morphological Variation Schwetz (1931) did not state dimensions of *P. maculilabre* meronts and gametocytes. Estimates derived from his figures indicate that the only mature meront shown was about 10.0 × 6.4 μm, with LW 69 μm². The number of merozoites produced by meronts is 15–20. Estimated size of this meront to its host cell nucleus and to normal erythrocyte nuclei is about 5.0 for both. The gametocyte dimensions calculated for those illustrated were 12.1 × 6.3 μm (10.3–15.1 × 4.6–8.0), with LW 83.1 μm² (58.0–104.2) and L/W 1.98 (1.36–2.74). Their size relative to normal erythrocyte nuclei averages 4.27. All of the estimated gametocyte dimensions are similar to the data obtained from the 1985 series of gametocytes, which are 10.6 ± 1.8 × 6.1 ± 0.9 μm (7–13 × 5–8, N = 20), with LW 63.7 ± 12.3 μm² (42–91) and L/W 1.79 ± 0.44 (1.13–2.60). Gametocyte size relative to host cell nucleus is 2.62 ± 0.53 (1.67–3.79, N = 19), and to normal erythrocyte nuclei is 3.62 ± 0.70 (2.39–5.17, N = 20). Both Schwetz (1931) and Garnham (1966) stated that microgametocytes tend to be larger than macrogametocytes. This is not confirmed either by Schwetz's figures or in dimensions derived from the 1985 gametocyte sample, in which macrogametocytes were longer, larger, and more elongate than microgametocytes. Macrogametocytes are 11.7 ± 1.3 × 6.1 ± 0.7 μm (10–13 × 5–7, N = 11), LW 71.1 ± 9.0 μm² (60–91), and L/W 1.96 ± 0.37 (1.43–2.60), versus 9.2 ± 1.4 × 6.0 ± 1.1 μm (7–11 × 5–8, N = 9), LW 54.7 ± 9.6 μm² (42–72), and L/W 1.60 ± 0.45 (1.13–2.20). Pigment appears to be sparse in meronts and is described as dispersed, round dark granules in gametocytes.

Exoerythrocytic Merogony Unknown.

Sporogony Unknown.

Effects on Host Erythrocytes parasitized by either meronts or gametocytes are hypertrophied and often distorted, with nuclei displaced and sometimes distorted.

Remarks This is another poorly described *Plasmodium* species for which new material and observations are much needed. *Plasmodium maculilabre* appears to be a Congo basin species similar in appearance, but different in effect on host cells, to *Plasmodium pitmani*, a Nile basin species, as suggested by Garnham (1966).

Plasmodium agamae (Wenyon) 1909 (**Plate 8**)

Diagnosis A *Plasmodium* (*Lacertamoeba*) species with usually fan-shaped meronts 4–11 × 3–6 μm, LW 12–55 μm², and merozoites 4–15, with meront size relative to host cell nucleus averaging 0.78–1.11, and to normal erythrocyte nuclei is 0.92–1.11. Dimensions of meronts and merozoite numbers are similar in mature and immature erythrocytes. In all localities, gametocytes are 6–19 × 3–8 μm, with LW

Plate 8 (**A**) *Plasmodium agamae* from *Agama agama*. Meronts, **a–f**; macrogametocytes, **g–i**; microgametocytes, **j–l**. Origin: **a**, **b**, **g**, **j** from Nigeria; **c–e**, **k** from Congo; **f**, **h**, **i**, **l** from Tanzania. (**B**) *Plasmodium mossambica* sp. nov. from *Agama mossambica* of Tanzania. Meronts, **a–d**; macrogametocytes, **e–h**; microgametocytes, **i–l**.

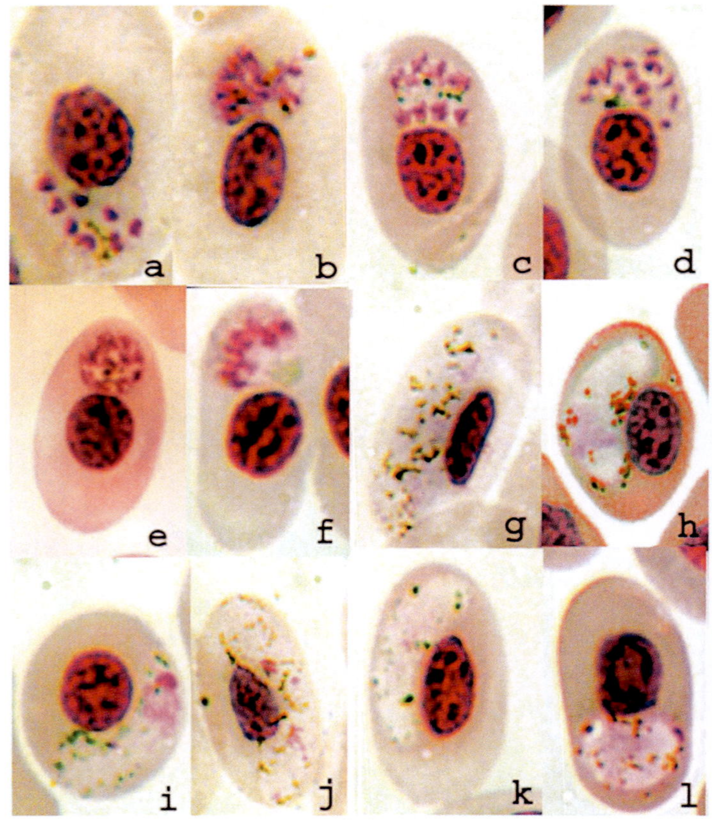

(A)

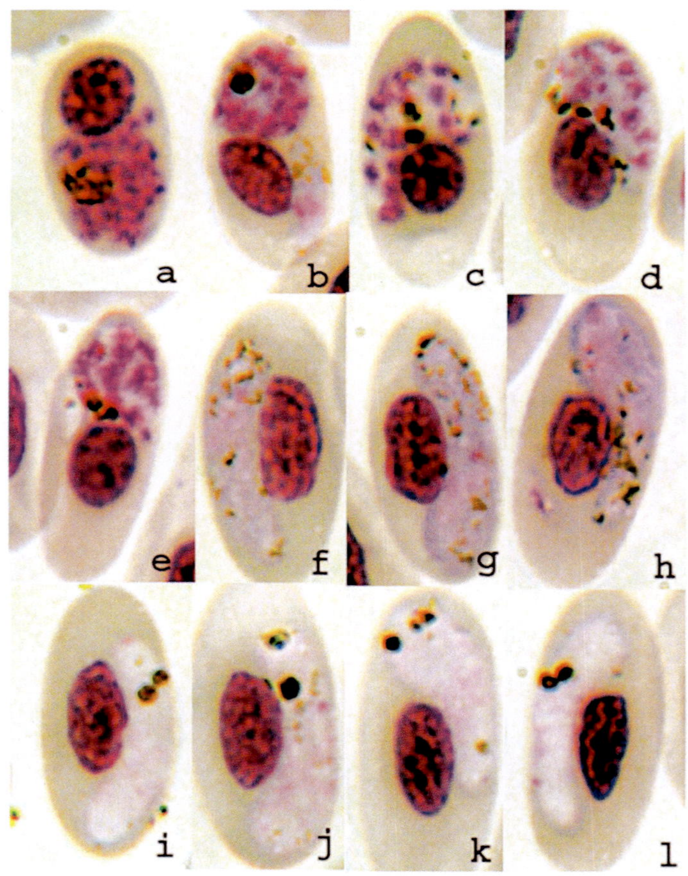

(B)

33–105 μm² and L/W 1.00–4.75. Gametocyte size relative to host cell nucleus averages 1.77–2.22, and to normal erythrocyte nuclei is 1.73–2.21. Gametocyte dimensions differ by sex and infection phase. Exoerythrocytic merogony occurs in hepatic and splenic macrophages and in monocytes in circulating blood. Pigment granules in meronts number two or three usually, are sometimes clumped, and are dark greenish-gold. Pigment in gametocytes is dispersed as dark greenish-gold-to-black granules in both sexes but sometimes tends to aggregate in one or more foci.

Type Host *Agama agama* (Linnaeus) (Sauria: Agamidae).

Type Locality Wau, Bahr-El-Ghazal Province, Sudan.

Other Hosts *Agama atricollis* (Pienaar, 1962) and *A. hispida aculeata* (Petit et al., 1983).

Other Localities Dem Besher Wood Station, Jur River, Sudan (Wenyon, 1909a); the Gambia (Garnham and Duke, 1953); Gambela, Ethiopia (Garnham and Duggan, 1986); and Nigeria (Garnham, 1952 unpublished, 1966). Liberia locations are Gbangu (Theiler, 1930), Charlesville (Baker, 1961), and Harbel (Bray, 1959; Baker, 1961). Other localities are 18 study sites in Sierra Leone (Schall and Bromwich, 1994); Congo site in Djugu, Lake Albert (Schwetz, 1931), 12 km west southwest of Kinshasa and Gemena, Equateur Province (Telford); Kenya sites in Marimont (Ball, 1967a) and Kacheliba, West Pokot District (Mutinga, ID by Telford); Tanzania in Morogoro Region, 5 km southwest of Turiani and Mikindo River, Nguru Mountains, Turiani District (Telford); South Africa, Northern Transvaal (Pienaar, 1962, needs confirmation); and Huimbo Region, Angola (Petit et al., 1983).

Prevalence In *A. agama*: Liberia, 3 of 21 (14.2%) by Bray (1959) and 3 of 28 (10.7%) by Baker (1961) at Harbel, 20 of 139 (14.4%) at Charlesville (Baker, 1961), 2 of 30 (6.7%) at Gbangu (Theiler, 1930); in Sierra Leone, *P. agamae* was present at 18 of 21 study sites, with an overall prevalence in 2986 *A. agama* of 18.7%, ranging from 5.0% to 71.9% with an average prevalence of 28.4% (Schall and Bromwich, 1994); in Congo, 7 of 14 (50%) at Djugu, Lake Albert (Schwetz, 1931), 2 of 13 (15.4% at 12 km southwest of Kinshasa) and 2 of 6 at Gemena, Equateur Province (Telford); in Kenya, 8 of 48 (16.7%) at Marimonti (Ball, 1967a) and 2 of 32 (6.3%) in Kacheliba, West Pokot District (Mutinga slides, ID by Telford); in Tanzania, 1 of 4 at 5 km southwest of Turiani and 1 of 5 from Mikindo River, Turiani District (Telford).

Morphological Variation Wenyon (1909a) described *P. agamae* gametocytes as 14 × 4 μ, with six merozoites in meronts. Bray (1959) commented that meronts contained four to eight merozoites, with two or three coarse pigment granules present, and in macrogametocytes there were numerous scattered coarse black pigment granules, while some granules appeared brown in microgametocytes. Garnham (1966) stated that the meront "at maturity … assumes a spherical shape, and the parasite rapidly divides into elongated merozoites in the form of a fan … with pigment … clumped at the base." In Gambian infections, merozoites numbered 10 or 12. Gametocytes "are sausage-shaped bodies about 10 μ long," while microgametocytes have "a less regularly elongate form and may be oval or club-shaped and up to 5 μ in width." There are apparently no other morphological data for *P. agamae* in the literature.

In four of the five populations of *P. agamae* compared in *A. agama*, meront length was shorter and LW less in the sample from Congo, and mean meront width and merozoite number were similar among all samples. Nigerian meronts are 6.4 ± 1.4 × 4.7 ± 0.7 μm (4–11 × 3–6, N = 49), with LW 30.4 ± 8.3 μm² (12–55), and merozoites are 9.5 ± 2.9 (4–15). Meront size relative to host cell nucleus is 0.89 ± 0.27 (0.33–1.72, N = 48), and to normal erythrocyte nuclei is 1.00 ± 0.28 (0.39–1.83, N = 49). Only one meront was proerythrocytic. In the sample from Sierra Leone, meronts are 6.3 ± 0.8 × 4.9 ± 0.7 μm (5–9 × 3–6, N = 48), with LW 30.4 ± 6.0 μm² (15–45), and merozoites are 9.1 ± 2.4 (6–14). Meront size relative to host cell nucleus is 0.77 ± 0.18 (0.46–1.20, N = 19), and to normal erythrocyte nuclei is 0.92 ± 0.18 (0.46–1.37, N = 48). About 60% of the meronts occupied proerythrocytes, but dimensions and merozoites produced are similar: Erythrocytic meronts are 6.0 ± 0.7 × 4.9 ± 0.6 μm (5–7 × 4–6, N = 19), LW 29.7 ± 5.8 μm² (24–42), and merozoites are 8.3 ± 2.2 (6–13) versus meronts in proerythrocytes, 6.4 ± 0.9 × 4.8 ± 0.7 μm (5–9 × 3–6, N = 29), LW 30.8 ± 6.2 μm² (15–45), and merozoite number 9.7 ± 2.3 (6–14). Congolese meronts are 5.6 ± 1.2 × 4.7 ± 0.8 μm (4–10 × 3–6, N = 30), with LW 26.3 ± 7.4 μm² (16–42), and merozoite number 9.4 ± 2.3 (6–14). Meront size relative to host cell nucleus is 0.90 ± 0.33 (0.46–1.50, N = 23), and to normal erythrocyte nuclei is 1.00 ± 0.29 (0.56–1.62, N = 30). About one-fourth of the meronts from Congolese *A. agama* were proerythrocytic, but their dimensions were similar to the erythrocytic meronts. Meronts from Tanzanian *A. agama*, all erythrocytic, are 6.5 ± 1.2 × 4.5 ± 0.7 μm (4–8 × 3–6, N = 25), with LW 28.8 ± 5.9 μm² (16–40), and merozoite number 8.2 ± 1.8 (4–10). Meront size relative to host cell nucleus is 1.11 ± 0.28 (0.67–1.75), and to normal erythrocyte nuclei is 1.11 ± 0.25 (0.67–1.63). Meront shape is most commonly a fan (27%) or a rosette (25%), less often elongate (15%) or oval (12%), and seldom round (8%). Pigment in meronts usually forms as two or three coarse greenish-gold granules, clumped at the base of fans or somewhat scattered in meronts of other shapes. Meronts from *A. agama* of Kenya averaged larger in each dimension than in the other four populations and produced more merozoites. The erythrocytic meronts are 7.5 ± 1.4 × 5.3 ± 0.7 μm (6–13 × 4–7, N = 25), with LW 39.6 ± 7.4 $μm^2$ (30–63), and contain 12.6 ± 2.4 (9–20) merozoites. Meront size relative to host cell nucleus is 1.81 ± 0.33 (1.29–2.60, N = 24), and to normal erythrocyte nuclei is 1.49 ± 0.28 (1.13–2.37, N = 25). Nearly mature meronts and some undergoing segmentation are highly amoeboid and often (36%) show cytoplasmic projections. Meront shape is most commonly oblong (48%) or elongate, rarely fans or rosettes.

Gametocytes of *P. agamae* are similar in size in all samples, and the length and width dimensions do not differ among samples from Sierra Leone, Congo, Kenya, and Tanzania, but length is shorter and width greater in the Nigerian sample when compared with the others. The L/W ratios are consistent with this pattern, lower in Nigerian *P. agamae* and higher among the other populations, which are similar among themselves. Nigerian gametocytes are 11.3 ± 2.2 × 5.2 ± 0.8 μm (6–16 × 4–8, N = 125), LW 57.4 ± 11.2 $μm^2$ (35–84), and L/W 2.26 ± 0.64 (1.00–3.75). Gametocyte size relative to host cell nucleus is 2.10 ± 0.51 (1.03–3.33, N = 75), and to normal erythrocyte nuclei is 1.81 ± 0.41 (1.01–2.79, N = 125). Gametocytes from Sierra Leone are 12.2 ± 1.0 × 4.7 ± 0.6 μm (10–14 × 4–6, N = 25), LW 57.0 ± 7.8 $μm^2$ (44–70), and L/W 2.64 ± 0.39 (1.67–3.50), with size relative to host cell nucleus 1.77 ± 0.33 (1.14–2.50) and to normal erythrocyte nuclei 1.73 ± 0.24 (1.34–2.13). The gametocytes from Congo are 11.8 ± 2.1 × 5.0 ± 1.2 μm (7–17 × 3–8, N = 100), LW 57.7 ± 14.3 $μm^2$ (33–105), and L/W 2.54 ± 0.81 (1.17–4.33), with size relative to host cell nucleus 2.22 ± 0.60 (1.18–4.00, N = 99) and to normal erythrocyte nuclei 2.11 ± 0.53 (1.22–3.70, N = 100). Gametocytes from Kenyan hosts are 11.8 ± 1.8 × 5.1 ± 1.0 μm (8–15 × 4–7, N = 25), with LW 59.4 ± 10.1 $μm^2$ (48–91) and L/W 2.43 ± 0.67 (1.2–3.3). Gametocyte size relative to host cell nucleus is 2.58 ± 0.60 (1.65–4.33), and to normal erythrocyte nuclei is 2.23 ± 0.38 (1.80–3.42). Pigment distribution is similar to other populations, but granules are smaller and less conspicuous. Tanzanian gametocytes are 11.8 ± 2.2 × 4.8 ± 0.9 μm (7–19 × 3–7, N = 75), LW 55.3 ± 9.1 $μm^2$ (35–76), and L/W 2.61 ± 0.80 (1.00–4.75), with size relative to host cell nucleus 2.16 ± 0.43 (1.20–3.61), and to normal erythrocyte nuclei 2.21 ± 0.43 (1.24 –3.79). In all samples of *P. agamae*, except those from Kenya where staining by sex was not consistent, macrogametocytes are longer on average than microgametocytes, with higher L/W ratios; width of macrogametocytes is less than microgametocytes except in the Tanzanian sample; and LW is greater in macrogametocytes except in the Sierra Leone sample, where it is the same. In three samples for which active-phase gametocytes could be compared with those from chronic infection (Tanzanian, Congo, Nigeria), gametocytes in active infections from Nigeria and Congo are longer and similar in length, with higher L/W ratios, but more narrow in width. Size (LW) is greater in active phase than in chronic phase in samples from Nigeria and Congo, but about the same in Tanzanian samples. In the samples from Nigeria for which comparisons by both sex and infection phase were feasible, the differences due to sex were exaggerated by infection phase: Mean dimensions for active-phase macrogametocytes are 12.7 ± 1.8 × 5.0 ± 0.6 μm (N = 49), LW 63.4 ± 9.0 $μm^2$, and L/W 2.57 ± 0.54 versus 12.1 ± 1.4 × 4.7 ± 0.7 μm (N = 26), LW 57.0 ± 8.8 $μm^2$, and L/W 2.62 ± 0.54 in microgametocytes. Chronic-phase gametocytes are 10.4 ± 1.4 × 5.3 ± 0.7 μm (N = 25), LW 55.8 ± 11.4 $μm^2$, and L/W 1.99 ± 0.34 in macrogametocytes versus 8.4 ± 1.2 × 5.7 ± 1.1 μm (N = 25), LW 47.4 ± 9.7 $μm^2$, and L/W 1.54 ± 0.41 in microgametocytes. In all samples of *P. agamae* gametocytes, pigment form and distribution are similar: fine-to-coarse, dark greenish-gold-to-black granules dispersed in both sexes of gametocytes, with an occasional tendency to loosely aggregate in one or more foci.

Exoerythrocytic Merogony Bray (1959) found three EE meronts of *P. agamae* in smears of spleen that contained 4, 9, and 12 nuclei. Another three meronts were seen in sections of spleen, which measured 8.5 × 5 μm· 9.5 × 5 μm, and a spherical meront 4 μm in diameter, with the respective numbers of nuclei 20, 35, and 8. Garnham and Duke (1953) reported EE meronts of *P. agamae* in Gambia in circulating monocytes, but later Garnham (1966) commented: "The host cell was not identified with certainty, but was probably an erythroblast or a monocyte, while on one occasion it seemed to be a thrombocyte." EE meronts "were fairly large bodies … up to 15 by 10 μ in size, lobulated, and with light blue cytoplasm, giving rise to about forty-five slightly elongated merozoites." In smears of the liver (primarily) and spleen of a Tanzanian *A. agama*, infected with a low parasitemia of *P. agamae*, there were hundreds of usually ellipsoidal cysts (**Plate 11C**). Most were opaque and white in color (unstained). A small series measured 26.9 ± 3.5 × 17.3 ± 1.9 μm (21–32 × 14.5–20.5, N = 7). Superficially, they resembled the dizoic cysts of *Hepatozoon* species, often seen in snake tissues, but there was no indication of zoites in those that were not opaque. Many cysts were not opaque and contained developing meronts. These cysts were similar in size, 27.9 ± 3.1 × 18.1 ± 1.5 μm (22–31 × 16–21, N = 9), with the contained meronts 23.1 ± 2.8 × 15.2 ± 1.4 μm (19.5–26.5 × 13.5–17.0, N = 9). Some nuclei were usually visible in the meront cytoplasm, but only two contained numbers indicating that merogony was advanced, with about 30 and 49 nuclei visible. Cyst walls were usually thick, but varied from 1 to about 5 μm in thickness. A few much smaller, ovoid cysts were seen, 12–12.5 × 8.5–9.5 μm, one of which contained two nuclei. Many rounded free meronts were present in smears, which presumably had lost their host cells during smear preparation. These were 17.5 ± 3.4 × 14.7 ± 2.3 μm (13–28 × 12–19, N = 16). In most, only a few nuclei could be discerned, but three contained 16, 30, and 41 nuclei. A few developing meronts were present in nonerythroid cells with bluish cytoplasm and single dark red nuclei, perhaps monocytes, which were 12.9 ± 3.7 × 9.8 ± 3.5 μm (8–23 × 7–19, N = 13). These appeared to have a capsule around them, which reduced clarity of the contained nuclei. Three contained nuclei sufficiently distinct for counting 22, 23, and 37 nuclei. Some host cells contained two to four of these meronts. One large ruptured meront was found, which had released 88 merozoites.

Sporogony Baker (1961, unpublished report) fed four species of Liberian *Aedes* on *P. agamae* infections in *A. agama*, all with negative results. Of four *Culicoides* sp. indet., one showed ookinetes of *P. agamae* at 1 day PF, but the other three were negative. Petit et al. (1983) obtained sporogony of *P. agamae* in laboratory-reared *Culicoides nubeculosus*. Oocysts were "extracellular but deeply embedded in the epithelium of the midguts." Mature oocysts were present at 7 days PF, in numbers up to 30 per midgut, and measured 20 μm in diameter. Oocysts contained about 100 sporozoites, 5.7 × 1.5 μm, with a central, round nucleus. Although oocysts were observed until 13 days PF, sporozoites did not leave the oocysts and enter the salivary glands. An ookinete, 18 × 5 μm, was found in the stomach of a *C. nubeculosus* dissected 24 hours PF.

Effects on Host In comparisons of *A. agama* infected by *P. agamae* with uninfected lizards, Schall (1990b) found that the average percentage of immature erythrocytes in the blood increased from 0.534 to 4.01, accompanied by a slight reduction in hematocrit values and blood hemoglobin concentration. In infected lizards, maximum oxygen consumption decreased by about 20%. Infection by *P. agamae* had no effect on egg mass in females, on testis mass in males, the ability of infected lizards to escape predators (as judged by damaged or regenerated tails), or their capacity running stamina. Host cells infected by gametocytes were hypotrophied by 7–20% in comparison to normal erythrocytes in a chronic infection from Congo and an active infection from Tanzanian *A. agama*. Meront infection produced 25% hypertrophy in host cell nuclei in *A. agama* from Sierra Leone. Infected erythrocyte nuclei were 20% smaller than normal in the chronic infection from Congo. Over the range of *P. agamae* from which samples were available, host cells were commonly distorted by gametocytes (54.8%, 28–80% range). Erythrocyte nuclei were seldom distorted (7.3%, 0–24%) but usually were displaced (82.9%, 74–100%). Meronts seldom distorted host erythrocytes (10.6%, 4–17%) or their nuclei (17.4%, 2–28%), but often displaced nuclei (37.0%, 16–80%).

Remarks Variation in gametocyte characters found in the samples from East, Central, and West Africa was probably influenced more by infection phase than by geography or gametocyte sex, although microgametocytes tended to be shorter, wider, and smaller in most samples. Within *A. agama*, meronts did not differ much among the geographic areas.

The presence of encysted EE meronts in *A. agama*, comparable to the encysted chronozoites of *Plasmodium sasai* in Japanese lacertids (Telford, 1989, 1998b), is the first report of these cysts in a tropical lizard, which appear to contribute to long-term maintenance of chronic *Plasmodium* infection.

Plasmodium mossambica sp. nov. (**Plate 8**)

Diagnosis A *Plasmodium* (*Lacertamoeba*) species with elongate or oblong meronts, 5–15 × 3–7 μm, LW 20–75 μm^2, that produce 6–34 merozoites. Meront size relative to host cell nucleus averages 1.83, and to normal erythrocyte nuclei is 1.79. Gametocytes are usually elongate, 6–17 × 3–8 μm, with LW 36–84 μm^2 and L/W 1.00–5.33. Gametocyte size relative to host cell nucleus averages 2.45, and to normal erythrocyte nuclei is 2.35. Gametocytes are not sexually dimorphic in dimensions but differ in pigment distribution. Macrogametocytes usually have an often-terminal aggregation of golden pigment clumps and numerous small black granules dispersed in the cytoplasm, while in microgametocytes, there are two or three large, black individual granules accompanied by small, black, dispersed granules.

Type Host *Agama mossambica* Peters (Sauria: Agamidae).

Type Locality Northern slope of the Uluguru Mountains at Morogoro, Morogoro Region, Tanzania.

Other Hosts None known.

Other Localities Kimboza Forest, south side of the Uluguru Mountains, 1 km north of the Ruvu River below Kibungo Village, Morogoro Region, Tanzania, and below Amani, Eastern Usambara Mountains, Tanga Region, Tanzania.

Prevalence *Plasmodium mossambica* infected 18 of 55 (32.7%) of *A. mossambica* at Morogoro, 1 of 6 in Kimboza Forest, and 1 of 3 below Amani.

Morphological Variation In *Agama mossambica* from Amani, meronts differed in length, width, and merozoite number from the Morogoro sample. The two samples of *P. mossambica* were similar only in meront LW; each sample, however, has greater numbers of merozoites than do meronts in all samples from *A. agama*, including those from the Tanzanian infections. In *A. mossambica* from Amani, meronts are 7.1 ± 1.6 × 4.3 ± 0.7 µm (5–11 × 3–6, N = 23), with LW 30.8 ± 9.0 µm^2 (20–54), and merozoites number 11.0 ± 4.0 (6–25, N = 25). Meront size relative to host cell nucleus is 0.90 ± 0.17 (0.69–1.29, N = 18), and to normal erythrocyte nuclei is 1.05 ± 0.31 (0.68–1.84, N = 23). About 20% of the meronts were proerythrocytic, similar in dimensions to the erythrocytic. Meronts from *A. mossambica* of Morogoro are larger than *P. agamae* and produce on average about twice the numbers of merozoites. Meronts are 8.5 ± 1.9 × 5.4 ± 0.8 µm (6–15 × 4–7, N = 61), with LW 45.7 ± 10.7 µm^2 (28–75), and merozoites number 18.1 ± 4.6 (8–34). Meront size relative to host cell nucleus is 2.12 ± 0.57 (1.17–4.17, N = 59), and to normal erythrocyte nuclei is 2.07 ± 0.40 (1.41–3.16, N = 61). Virtually all meronts from this sample are erythrocytic. Meront shape of *P. mossambica* is most commonly elongate (26%) or oblong (23%); less often fan shaped (11%), lentiform (12%), or oval (11%); seldom round (2%) or as a rosette (5%). Pigment, greenish-gold in color, is clumped at the base of fan-shaped meronts but occurs as small clumps or large granules variably distributed among the nuclei in other meront forms. Gametocytes of *P. mossambica* differed in dimensions between the two localities compared. Gametocytes from Morogoro are 12.3 ± 1.9 × 4.4 ± 0.8 µm (7–17 × 3–7, N = 125), LW 54.0 ± 9.8 µm^2 (36–84), and L/W 2.87 ± 0.76 (1.17–5.33), with size relative to host cell nucleus 2.55 ± 0.55 (1.71–4.67, N = 100) and to normal erythrocyte nuclei 2.52 ± 0.43 (1.57–3.93, N = 125). These values are similar to those of *P. agamae* in *A. agama* collected in Morogoro Region. Except in LW, all dimensions of *P. mossambica* gametocytes from *A. mossambica* of Amani in Tanga Region differ from the same host and from *A. agama* in Morogoro Region. Gametocytes are shorter and wider and have a lower L/W ratio in the Amani sample: 10.7 ± 1.7 × 5.4 ± 1.0 µm (6–14 × 4–8, N = 50), LW 56.7 ± 9.6 µm^2 (36–78), and L/W 2.06 ± 0.58, with size relative to host cell nucleus 2.07 ± 0.57 (1.25–3.71, N = 25) and to normal erythrocyte nuclei 1.94 ± 0.33 (1.23–2.66, N = 50). Micro- and macrogametocytes have similar dimensions.

Exoerythrocytic Merogony Secondary EE meronts were common in circulating blood of some infected hosts at Morogoro, usually occupying thrombocytes, and are 7–10 × 6–7 µm, containing 18–20 merozoites. Phanerozoites present in liver of *A. mossambica* (**Plate 11**) could represent *P. giganteum* or *P. mossambica*, as both species infected the hosts.

Sporogony Unknown.

Effects on Host Meronts in active infection distorted host erythrocytes sometimes (31%), and their nuclei (22%), and commonly displaced the nuclei (43%). In chronic infection, host cells were not distorted, but their nuclei were often distorted and displaced (55%). Gametocytes in active infection often distorted host cells (62%) and seldom (8%) distorted but usually (78%) displaced host cell nuclei. Host cells were rarely (4%) distorted in chronic infection, with their nuclei normal in appearance, but often (52%) were displaced.

Remarks *Plasmodium mossambica* is readily distinguished from *P. agamae* by its much larger, differently shaped meronts, which can produce twice as many merozoites on average than do the meronts of *P. agamae*. Gametocytes of the two species are similar in dimensions but differ in distribution of pigment, which is dispersed as fine-to-coarse greenish-gold-to-black granules in both sexes of *P. agamae*, but in *P. mossambica* aggregates in golden-yellow clumps accompanied by small dispersed black granules in macrogametocytes. Pigment forms two or three large black granules in addition to the dispersed smaller ones in microgametocytes. Eventually, genomic analysis may indicate sufficient similarity between *P. agamae* and *P. mossambica* to consider them to be subspecies, but the differences between their meronts suggests otherwise at present. Hapantotype blood films are deposited in the U.S. National Parasite Collection (USNPC), Beltsville, Maryland, nos. 100335–100336.

Plasmodium zonuriae Pienaar 1962, Telford 1987 (**Plate 9**)

Diagnosis A *Plasmodium* (*Lacertamoeba*) species with variably shaped meronts in which nuclei are usually arranged peripherally, 7–17 × 4–9 µm, with LW 36–120 µm^2, that produce 12–28 merozoites. Meront size relative to host cell nucleus averages 1.45, and to normal erythrocyte nuclei is 1.64. Dark golden-brown pigment granules are dispersed within meronts, rarely aggregating. Gametocytes are usually elongate, 7–20 × 4–10 µm, with LW 42–114 µm^2 and L/W 1.00–5.00. Gametocyte size relative to both host cell nucleus and to normal erythrocyte nuclei averages 1.69. Microgametocytes are more elongate than macrogametocytes, but other dimensions are similar between the sexes. The dark brown pigment granules are dispersed within gametocytes. Both meronts and gametocytes tend to curve around the host cell nucleus.

Type Host *Cordylus vittifer* (Reichenow) (Sauria: Cordylidae).

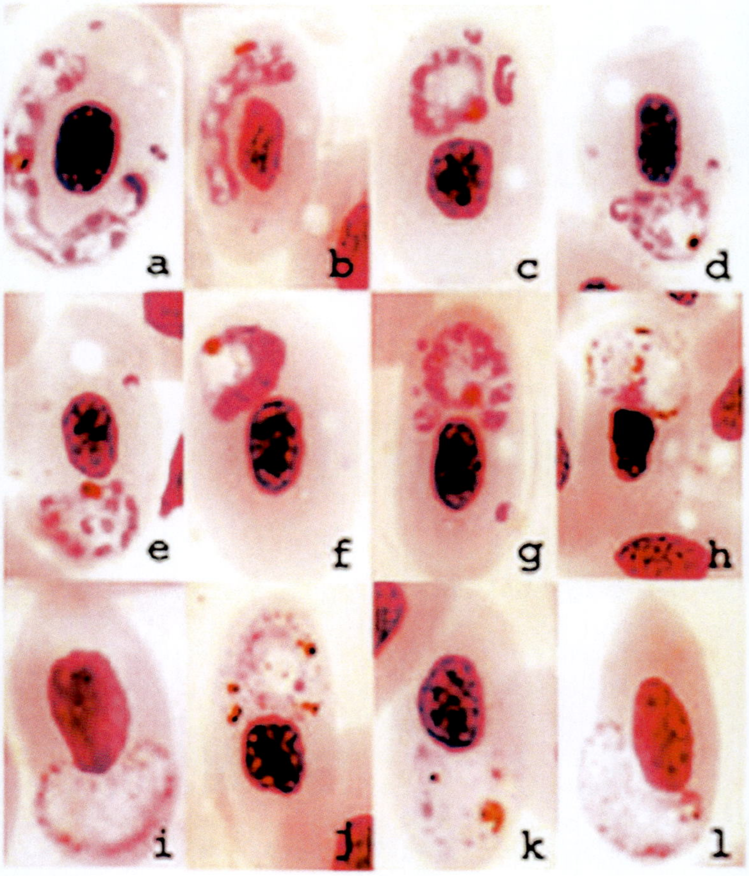

(A)

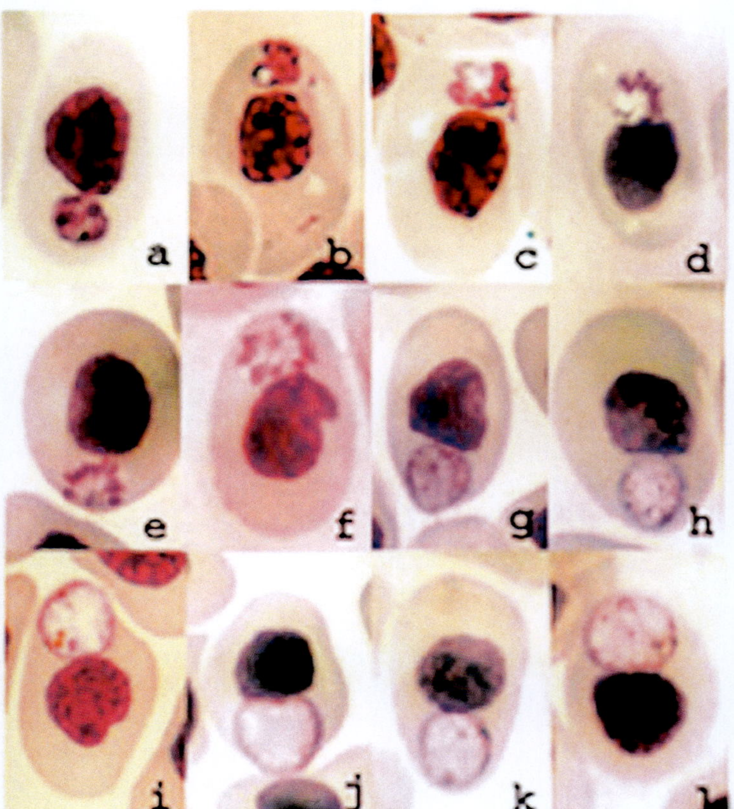

(B)

Plate 9 (**A**) *Plasmodium zonuriae* from *Pseudocordylus microlepidotus* of South Africa. Meronts, **a–g**; macrogametocytes, **h**, **i**; microgametocytes, **j–l**. (**B**) *Plasmodium cordyli* from *Cordylus tropidosternum* of Tanzania. Meronts, **a–f**; macrogametocytes, **g–i**; microgametocytes, **j–l**.

Type Locality Elandsfontein near Fochvill, between Johannesburg and Potschefstroom, southwest Transvaal, South Africa.

Other Hosts *Pseudocordylus microlepidotus melanotus.*

Other Localities In addition to Potschefstroom District, Die Berg, Pretoria, and Middelburg, southwest Transvaal, and vicinity of Capetown, South Africa.

Prevalence Pienaar (1962) examined 20 *C. vittifer* but did not state prevalence. Among the series of 20 *C. vittifer* examined in 1986 from southwest Transvaal (Telford), 12 of 13 from Potschefstroom District, 4 of 6 from Die Berg, 3 of 3 from Pretoria, and 1 of 3 from Middelburg were infected by *P. zonuriae.* Both *P. microlepidotus* from the vicinity of Capetown, examined in 1972, were infected.

Morphological Variation Pienaar (1962) described meronts of *P. zonuriae* as forming 18–24 merozoites, "the average being about 18." Microgametocytes measured 8.0–8.4 × 4.2 μm, and macrogametocytes were 8.0–8.4 × 4.6 μm, with the latter described as follows: "Pigment granules are fewer in number but much coarser and more regularly rounded than those of the microgametocytes." In the more recently examined topotypic material, *P. zonuriae* meronts are 10.3 ± 1.8 × 6.6 ± 1.2 μm (7–17 × 4–9, N = 102), with LW 67.6 ± 16.5 μm^2 (36–120), and contain 18.6 ± 3.5 (12–28) merozoites. Meront size relative to host cell nucleus is 1.45 ± 0.33 (0.75–2.75, N = 92), and to normal erythrocyte nuclei is 1.64 ± 0.39 (0.93–2.86, N = 102). Most meronts are erythrocytic; a very few occupy erythroblasts and proerythrocytes. Pigment granules are dark golden brown and are dispersed within meronts, rarely aggregating in a single focus. Meronts are polymorphic, assuming almost any shape from oval, round, or oblong to elongate, fan shaped, lentiform, or rosette, often curving around the host cell nucleus. Nuclei are usually arranged peripherally within the meront. Gametocytes are usually elongate, 10.9 ± 2.2 × 6.7 ± 1.1 μm (7–20 × 4–10, N = 125), with LW 72.2 ± 13.3 μm^2 (42–114) and L/W 1.70 ± 0.61 (1.00–5.00). Gametocyte size relative to host cell nucleus averages 1.69 ± 0.51 (0.67–2.84, N = 116), and to normal erythrocyte nuclei is 1.69 ± 0.39 (0.81–2.51, N = 125). Gametocyte dimensions are similar between the sexes except for the L/W ratios, which are 1.67 ± 0.5 (1.00–3.80, N = 78) in macrogametocytes and 1.75 ± 0.73 (1.00–5.00, N = 47) in microgametocytes, indicating a more elongate shape of the latter. Dark brown pigment granules are dispersed in both sexes of gametocyte. Meronts of *P. zonuriae* in *Pseudocordylus microlepidotus* are smaller than in *C. vittifer,* 8.1 ± 1.1 × 6.0 ± 0.8 μm (7–11 × 4–7, N = 25), with LW 48.4 ± 8.9 μm^2 (32–63), but the mean number of merozoites formed is almost identical, 18.2 ± 2.6 (12–23), versus 18.6 ± 3.5 (12–28), and the range in numbers is quite similar. Meront size relative to host cell nucleus is 1.43 ± 0.34 (1.00–2.25), and to normal erythrocyte nuclei is 1.42 ± 0.26 (0.94–1.85). Gametocytes are smaller and more elongate than in *C. vittifer,* 9.7 ± 2.9 × 5.9 ± 1.1 μm (7–18 × 4–8, N = 50), with LW 55.7 ± 12.2 μm^2 (35–88) and L/W 1.76 ± 0.86 (1.00–4.50). Gametocyte size relative to host cell nucleus is 1.89 ± 0.64 (1.20–4.44), and to normal erythrocyte nuclei is 1.77 ± 0.40 (1.20–3.01). In contrast to gametocytes of *P. zonuriae* in the type host, macrogametocytes are slightly larger and more elongate than microgametocytes, with L/W ratios of 2.08 ± 1.01 (1.14–4.50, N = 24) and 1.47 ± 0.56 (1.00–3.00, N = 26), respectively.

Exoerythrocytic Merogony Unknown.

Sporogony Unknown.

Effects on Host Pienaar (1962) commented that heavy infections of *P. zonuriae* "include marked debilitation and severe anemic changes in the peripheral blood." Telford (1987) reported that *P. zonuriae* meronts in *C. vittifer* caused hypertrophy of host cells in three of four infections and of their nuclei in two infections. Erythrocytes and nuclei were distorted and the latter displaced. All samples of gametocytes had hypertophied host cells, but in only one infection were the host cell nuclei enlarged. Erythrocytes were usually distorted, with their nuclei both displaced and distorted. In infected *P. microlepidotus,* meronts did not enlarge the host cells and caused little distortion of either cell or nucleus, only occasionally displacing the latter. One of two gametocyte samples showed host cell enlargement, but nuclei were not enlarged. Host cells were sometimes distorted by gametocytes, and nuclei were displaced but rarely distorted.

Remarks Pienaar (1962) thought prostigmatic mites, *Zonurobia circularis lotior,* might host sporogony of *P. zonuriae* given the close association of these mites with *C. vittifer* and the lizard's "formidable protective armor of hard, horny scales which appeared to be quite impervious" to bites of dipterans.

Plasmodium cordyli Telford 1987 (**Plate 9**)

Diagnosis A *Plasmodium* (*Carinamoeba*) species with polymorphic, but usually fan-shaped, meronts 4–7 × 3–6 μm, with LW 12–36 μm^2, that produce 4–11 merozoites. Meront size relative to host cell nucleus averages 0.49, and to normal erythrocyte nuclei is 0.53. Proerythrocytic meronts are larger than erythrocytic and produce more merozoites. Dark golden pigment granules are clumped at the base of fan-shaped meronts or aggregated in usually central foci in meronts of other forms. Meronts are strongly nucleophilic. Gametocytes are round or ovoid, 5–8 × 4–7 μm, with LW 20–49 μm^2 and L/W 1.00–1.16. Gametocyte size relative to host cell nucleus averages 0.74, and to normal erythrocyte nuclei is 0.89. There is no sexual dimorphism in gametocyte dimensions. Dark pigment granules are dispersed in macrogametocytes but form a single cluster in microgametocytes.

Type Host *Cordylus t. tropidosternum* (Cope) (Sauria: Cordylidae).

Type Locality Magrotto Mountain, Eastern Usambara Mountains, Tanga Region, Tanzania.

Other Hosts *Cordylus vittifer.*

Other Localities Tanga, Tanga District, and Rondo Forest, Lindi Region, Tanzania; four localities in southwest Transvaal, South Africa: Potchefstroom District, Die Berg, Pretoria, and Middelburg.

Prevalence In Tanzania, *P. cordyli* infected three of four *C. tropidosternum* at the type locality and one of one at Tanga and at Rondo Forest. In southwest Transvaal, South Africa, *C. vittifer* was infected by *P. cordyli* in 13 of 25 (52%) overall, 3 of 13 (23.1%) in Potchefstroom District, 4 of 6 at Die Berg, 3 of 3 at Middelburg, and 3 of 3 in Pretoria.

Morphological Variation In *Cordylus tropidosternum*, meronts are 5.0 ± 0.8 × 4.1 ± 0.6 μm (4–7 × 3–6, N = 88), with LW 20.9 ± 5.4 μm^2 (12–36), and produce 6.6 ± 1.6 (4–11) merozoites. Meront size relative to host cell nucleus is 0.49 ± 0.11 (0.2–0.83, N = 35), and to normal erythrocyte nuclei is 0.53 ± 0.14 (0.30–0.90, N = 88). Erythrocytic meronts are smaller than those in proerythrocytes, 4.9 ± 0.7 × 3.9 ± 0.4 μm (4–7 × 3–5, N = 35), LW 19.3 ± 3.8 μm^2 (12–28) versus 5.1 ± 0.8 × 4.3 ± 0.7 μm (4–7 × 3–6, N = 53), LW 22.0 ± 6.0 μm^2 (12–36), respectively, and produce fewer merozoites, 6.0 ± 1.4 (4–10) versus 7.0 ± 1.6 (4–11), respectively. Meronts are most commonly fan-shaped but may be round, oval, oblong, rosette, cruciform, or stellate. The dark golden pigment granules form a clump at the base of fan-shaped meronts or aggregate in a single, usually central, focus in meronts of other shapes. Gametocytes are round or ovoid, 6.3 ± 0.9 × 5.5 ± 0.7 μm (5–8 × 4–7, N = 50), with LW 35.1 ± 8.1 μm^2 (20–49), and L/W 1.15 ± 0.16 (1.00–1.60). Dimensions do not differ by sex. Gametocyte size relative to host cell nucleus is 0.74 ± 0.17 (0.43–1.17, N = 49), and to normal erythrocyte nuclei is 0.89 ± 0.21 (0.50–1.27). Pigment is dispersed as dark granules in macrogametocytes or is aggregated into a single dark golden focus in microgametocytes. In *Cordylus vittifer*, meronts are always proerythrocytic and larger than meronts in the type host, producing more merozoites. They also lack the tendency-to-nucleophily characteristic of the *C. tropidosternum* meronts. Meronts are 7.0 ± 0.7 × 5.6 ± 0.6 μm (6–9 × 5–7, N = 25), with LW 38.2 ± 4.4 μm^2 (30–49), and contain 11.9 ± 1.6 (8–14) merozoites arranged variably as those in the type host, but fewer are formed as fans. Meront size relative to host cell nucleus is 1.04 ± 0.19 (0.8–1.7), and to normal erythrocyte nuclei is 1.03 ± 0.12 (0.81–1.32). Gametocytes of *P. cordyli* are nearly identical in the two host species. In *C. vittifer*, gametocytes are 6.5 ± 0.9 × 5.3 ± 0.7 μm (5–9 × 4–7, N = 50), with LW 34.6 ± 7.3 μm^2 (20–49) and L/W 1.25 ± 0.20 (1.00–2.00). Gametocyte size relative to host cell nucleus is 0.85 ± 0.22 (0.36–1.40, N = 41), and to normal erythrocyte nuclei is 0.90 ± 0.20 (0.50–1.32).

Exoerythrocytic Merogony Phanerozoic meronts were common in endothelium and connective tissue of heart, lungs, and kidney (**Plate 11A, a–d**). None were observed in circulating leukocytes. Meronts in connective tissue of the heart were usually ovoid and averaged 14.4 ± 1.7 × 10.8 ± 1.4 μm (11–18 × 8–15, N = 20). Three meronts contained 55, 85, and 122 nuclei.

Sporogony Unknown.

Effects on Host Host cells infected by meronts, and their nuclei, were not enlarged, distortion of either was rare, but nuclei were sometimes displaced. Gametocytes caused hypertrophy and distortion of erythrocytes and their nuclei and commonly displaced nuclei.

Remarks Although meronts in the South African *C. vittifer* were larger and produced more merozoites than in the type host from Tanzania, the close similarity of gametocytes from both hosts is good reason to consider them as conspecific. In the forest, *C. tropidosternum* is very much a tree-hole inhabitant but will occupy fences and walls around settlements. The *C. tropidosternum* from Lindi was collected in August 1984, patent for *P. cordyli* at that time, and continuously showed asexual parasites whenever examined until it died in February 1988. Each year, in May and June, gametocytes appeared in the circulating blood. Tissues removed at death were heavily parasitized by phanerozoic meronts.

Plasmodium mabuiae Wenyon 1909, Telford 1983 (**Plate 10**)

Diagnosis A *Plasmodium* (*Carinamoeba*) species with usually fan-shaped meronts, 4–9 × 2–5 μm, with LW 10–30 μm^2, that produce 4–12 merozoites. Meront size

Plate 10 (**A**) *Plasmodium mabuiae* from *Mabuya striata* of Tanzania. Meronts, **a–f**; macrogametocytes, **g–i**; microgametocytes, **j–l.** (**B**) *Plasmodium pitmani* from *Mabuya striata* of Tanzania. Meronts, **a–f**; macrogametocytes, **g–i**; microgametocytes, **j–l**.

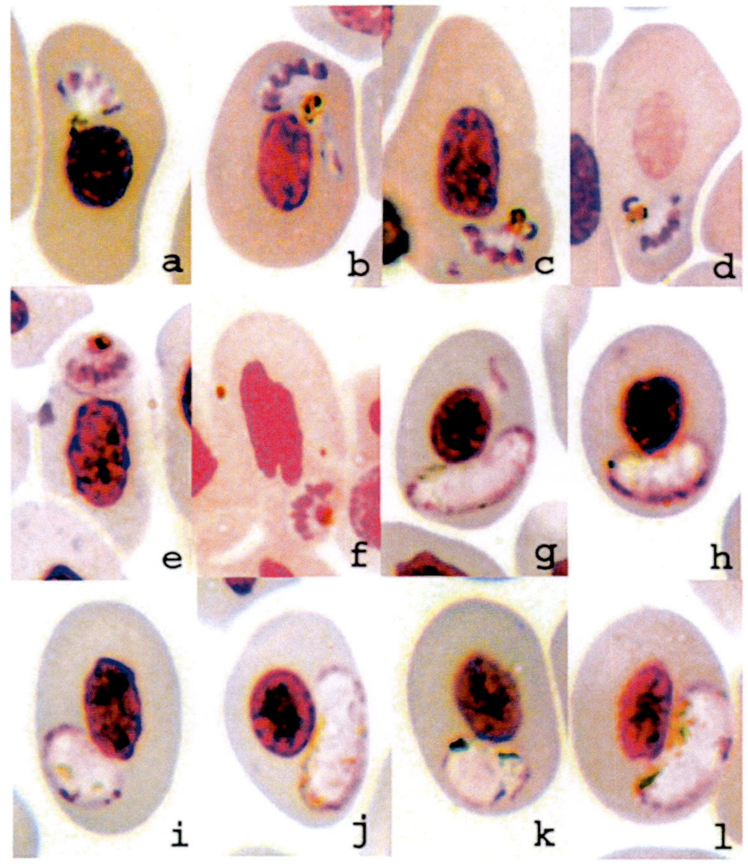

(A)

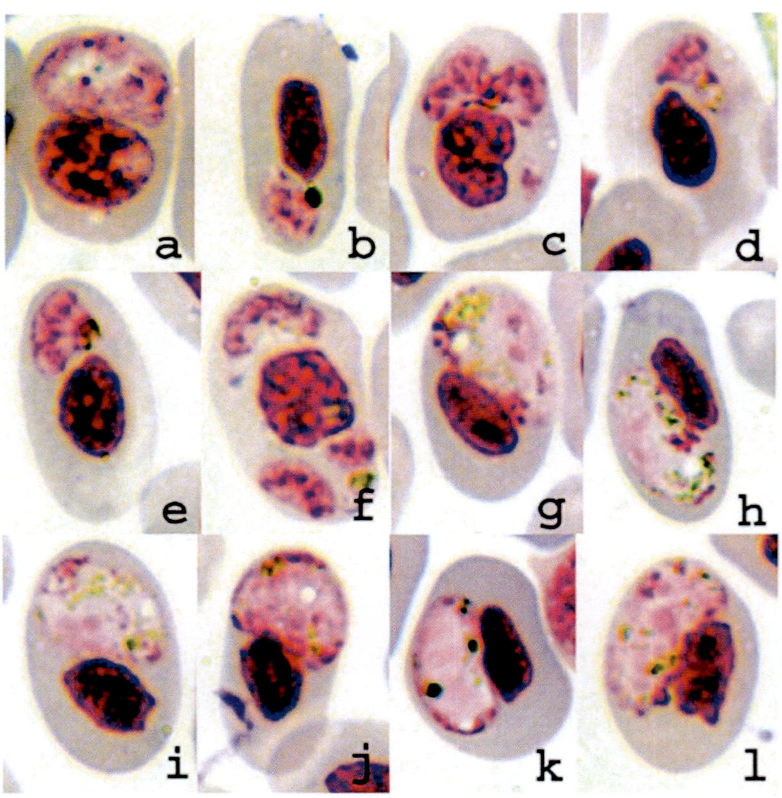

(B)

relative to host cell nucleus averages 0.77, and to normal erythrocyte nuclei is 0.86. Proerythrocytic meronts are larger and produce more merozoites than those in erythrocytes. Pigment is usually present as a dark yellow mass at the base of fan-shaped meronts. Young asexual stages are strongly nucleophilic, but immature gametocytes are less commonly in contact with the host cell nucleus. Gametocytes are predominantly elongate, rarely ovoid or round, and are 5–11 × 3–5 μm, with LW 18–44 μm² and L/W 1.20–3.67. Gametocyte size relative to host cell nucleus averages 1.45, and to normal erythrocyte nuclei is 1.38. Gametoctyes are not sexually dimorphic in dimensions but are smaller and more rounded in chronic infections. Although some dark pigment granules are dispersed, most remain clumped in one or two foci in most gametocytes. Microgametocytes are less heavily pigmented than macrogametocytes and usually show a thick reddish-staining area on one side or end.

Type Host *Mabuya quinquetaeniata* (Lichtenstein) (Sauria: Scincidae).

Type Locality Wau, Bahr-El-Ghazal Province, Sudan.

Other Hosts *Mabuya striata, M. maculilabris.*

Other Localities In Tanzania, northern slope of the Uluguru Mountains and Morogoro, Morogoro Region, Bahari Beach 21 km north of Dar-es-Salaam, and Amani, Eastern Usambara Mountains, Tanga Region; Nairobi, Pole, and Kacheliba, West Pokot District, Kenya; Lwiro, Kivu Province, and Kinshasa, Congo.

Prevalence Wenyon (1909a) commented that about half of the *M. quinquetaeniata* he examined at the type locality were positive for *P. mabuiae.* In Tanzania, *P. mabuiae* infected 15 of 144 (10.4%) *M. striata* at Morogoro and 2 of 9 at Bahari Beach and 1 of 11 (9.1%) *M. maculilabre* at Amani. In Kenya, 6 of 67 (9.0%) *M. striata* and 1 of 10 (10%) *M. maculilabre* examined from West Pokot District were positive for *P. mabuiae.* In Congo, 3 of 6 *M. maculilabre* taken at Lwiro and 2 of 26 (7.7%) at Kinshasa were infected by *P. mabuiae.*

Morphological Variation In Tanzanian *M. striata,* meronts are 5.1 ± 0.9 × 3.7 ± 0.6 μm (4–9 × 2–5, N = 72), with LW 18.4 ± 4.0 μm² (10–30), and contain 6.6 ± 1.2 (4–8, N = 75) merozoites. Meront size relative to host cell nucleus is 0.77 ± 0.18 (0.50–1.14, N = 53), and to normal erythrocyte nuclei is 0.86 ± 0.20 (0.48–1.44, N = 72). Erythrocytic meronts are smaller than those in proerythrocytes, 4.8 ± 0.8 × 3.6 ± 0.6 μm (4–8 × 2–5, N = 53), LW 17.3 ± 3.3 μm² (10–25) versus 5.7 ± 1.1 × 3.8 ± 0.5 μm (4–9 × 3–5, N = 19), LW 21.7 ± 4.2 μm² (15–30), respectively, and produce fewer merozoites, 6.1 ± 1.0 (4–8) versus 7.8 ± 0.7 (5–8), respectively. Gametocytes in *M. striata* are 7.9 ± 1.3 × 3.8 ± 0.5 μm (5–11 × 3–5, N = 100), with LW 29.9 ± 5.8 μm² (18–44) and L/W 2.15 ± 0.53 (1.20–3.67). Gametocyte size relative to host cell nucleus averages 1.45 ± 0.42 (0.60–2.67), and to normal erythrocyte nuclei is 1.38 ± 0.28 (0.83–1.91). There is no sexual difference in gametocyte dimensions, but gametocytes in chronic infection are smaller and more rounded, 6.6 ± 0.6 × 3.5 ± 0.5 μm (5–8 × 3–4, N = 25), LW 23.1 ± 3.1 μm² (18–28), and L/W 1.93 ± 0.40 (1.25–2.67), than in active infection, 8.4 ± 1.2 × 3.9 ± 0.5 μm (6–11 × 3–5, N = 75), LW 32.1 ± 4.5 μm² (21–44), and L/W 2.22 ± 0.54 (1.20–3.67). Some dark pigment granules are usually dispersed in gametocytes, but typically there are one or two foci of lighter colored, aggregated granules present in both sexes. Microgametocytes often have a thick, reddish-staining area resembling "a plaque of adherent stain" (Telford, 1983b) at one end or on one side. Chronic-phase gametocytes from an *M. striata* collected in Nairobi are 7.2 ± 0.7 × 3.8 ± 0.4 μm (6–9 × 3–4, N = 25), with LW 26.9 ± 4.4 μm² (18–36) and L/W 1.93 ± 0.30 (1.50–2.67). Early chronic-phase meronts in *M. maculilabre* collected at Lwiro, Congo, are 3.9 ± 0.7 × 2.9 ± 0.6 μm (2–5 × 2–4, N = 19), LW 11.3 ± 3.4 μm² (6–20), and contain 4.6 ± 1.9 (3–12) merozoites. Gametocytes are 6.0 ± 0.6 × 4.3 ± 0.6 μm (5–7 × 3–6, N = 24), with LW 25.3 ± 4.2 μm² (18–36) and L/W 1.43 ± 0.25 (1.00–2.00).

Exoerythrocytic Merogony Because of the presence of *Plasmodium pitmani* in most infections of *P. mabuiae,* specific allocation of leukocytic meronts is not possible (Telford, 1983b).

Sporogony Unknown.

Effects on Host Young asexual stages usually were strongly nucleophilic, young gametocytes less so, and often appeared to be touching host cell nuclei without apparent effect. Mature meronts that were erythrocytic also were in contact with host nuclei, but only rarely was there contact by proerythrocytic meronts (Telford, 1983b). Meronts rarely caused distortion of their host cells or their nuclei, but the latter were sometimes displaced. Gametocytes commonly appeared to be in contact with the nucleus, but this was probably less due to nucleophily than it was from their position, curving around the nucleus in a lateropolar or polar position. Gametocytes uncommonly caused distortion of the host cell and its nucleus but more commonly displaced the nucleus. Host cells of gametocytes were slightly hypertrophied.

Remarks *Plasmodium mabuiae* has a great geographic distribution in African *Mabuya* species, from Sudan south through Kenya and Tanzania, westward into Congo from Lwiro in the east to Kinshasa in the west. This suggests a relatively ancient relationship of the parasite with scincid hosts, in which there is greater variation in parasite morphology from infection phase than from host species or geography.

Plasmodium pitmani Hoare 1932 (**Plate 10**)

Diagnosis A *Plasmodium* (*Lacertamoeba*) species with meronts most commonly fan-shaped, a rosette, or elongate, 4–11 × 3–7 μm, with LW 12–66 μm^2, that produce 4–25 merozoites. Proerythrocytic meronts are larger than the erythrocytic and produce more merozoites. Pigment usually forms one or more pale yellow masses in meronts. Gametocytes are usually ovoid, 5–16 × 4–9 μm, with LW 25–91 μm^2 and L/W 1.00–3.25, without sexual dimorphism in size or shape. Gametocyte size relative to host cell nucleus averages 2.23, and to normal erythrocyte nuclei is 2.07. Pigment is dark greenish-gold, often loosely aggregated in one or two foci, with a few individual granules dispersed in the gametocyte cytoplasm.

Type Host *Mabuya striata* (Peters) (Sauria: Scincidae).

Type Locality Sese Islands on the west side of Lake Victoria, Uganda.

Other Hosts *Mabuya maculilabris, M. varia, M. quinquetaeniata.*

Other Localities Bugerere District, Uganda (Hoare, 1932); in Kenya, Kodera Valley, including Kavirondo 12 km inland from Lake Vistoria (Garnham, 1966), West Pokot District, and Nairobi; in Tanzania, northern slope of the Uluguru Mountains and Morogoro, Morogoro Region, 21 km north of Dar-es-Salaam, and Amani, Eastern Usambara Mountains, Tanga Region; Kinshasa, Congo.

Prevalence Hoare (1932) reported *P. pitmani* infections from six of six *M. striata,* and one of three *M. maculilabris* at the type locality and Bugerere District, Uganda. In West Pokot District, Kenya, 6 of 67 (9.0%) *M. striata* and 1 of 9 *M. quinquetaeniata* were infected by *P. pitmani.* In Tanzania, 32 of 144 (22.2%) *M. striata* and 2 of 13 (15.4%) *M. varia* were infected, as were 2 of 9 *M. striata* at Bahari Beach and 1 of 11 (9.1%) *M. maculilabris* at Amani. At Kinshasa, Congo, 1 of 26 (3.8%) *M. maculilabris* was host to *P. pitmani.*

Morphological Variation In *M. striata* at Morogoro, meronts of *P. pitmani* are 6.1 ± 1.4 × 4.6 ± 0.7 μm (4–11 × 3–6, N = 100), with LW 28.3 ± 9.0 μm^2 (12–66), and contain 9.1 ± 3.1 (4–18) merozoites. Meront size relative to host cell nucleus is 1.09 ± 0.38 (0.40–2.29, N = 75), and to normal erythrocyte nuclei is 1.18 ± 0.37 (0.51–2.69, N = 100). Meronts occupying erythrocytes are smaller, 5.7 ± 0.9 × 4.4 ± 0.6 μm (4–8 × 3–6, N = 75), LW 25.3 ± 5.7 μm^2 (12–42), than those in proerythrocytes, 7.6 ± 1.7 × 4.9 ± 0.7 μm (5–11 × 4–6, N = 25), LW 37.2 ± 11.2 μm^2 (20–66), and produce fewer merozoites, 8.0 ± 2.5 (4–16) versus 12.2 ± 2.6 (8–18), respectively. Pigment usually forms one or more pale golden masses in meronts. Gametocytes in active/acute infection are 8.1 ± 1.5 × 5.7 ± 0.7 μm (6–13 × 4–7, N = 100), with LW 46.3 ± 8.5 μm^2 (30–66) and L/W 1.46 ± 0.42 (1.00–3.25). Gametocyte size relative to host cell nucleus is 2.23 ± 0.54 (1.29–3.70, N = 97), and to normal erythrocyte nuclei is 2.07 ± 0.9 (1.35 –2.87, N = 100). Macrogametocytes and microgametocytes have similar dimensions, 8.2 ± 1.5 × 5.8 ± 0.8 μm (N = 51), LW 47.2 ± 8.8 μm^2, and L/W 1.45 ± 0.41 versus 8.1 ± 1.5 × 5.7 ± 0.7 μm (N = 49), LW 45.2 ± 8.1 μm^2, and L/W 1.46 ± 0.42, respectively. Pigment, dark greenish-gold, is usually loosely aggregated in one or more foci of granules, with a few individual granules dispersed. Chronic-phase gametocytes from *Mabuya varia* at Morogoro are 9.3 ± 1.9 × 6.0 ± 1.3 μm (5–16 × 4–9, N = 50), LW 56.3 ± 16.8 μm^2 (25–91), and L/W 1.60 ± 0.47 (1.00–3.20). Macrogametocytes in *M. varia* are more rounded than microgametocytes and slightly wider, 9.1 ± 1.5 × 6.5 ± 1.4 μm (5–11 × 4–9, N = 25), LW 59.6 ± 16.4 μm^2 (25–90), and L/W 1.47 ± 0.46 (1.00–2.75), versus 9.4 ± 2.3 × 5.6 ± 0.9 μm (6–16 × 4–7, N = 25), LW 53.0 ± 16.9 μm^2 (28–91), and L/W 1.73 ± 0.46 (1.00–3.20), respectively. In *M. striata* from Kenya, meronts are 6.5 ± 1.4 × 4.8 ± 0.8 μm (5–11 × 3–7, N = 49), with LW 30.9 ± 8.5 μm^2 (20–55), and contain 12.3 ± 3.9 (6–25) merozoites. Meronts are similar in dimensions and type of host cell but have more merozoites than the total Morogoro sample of meronts. Meront size relative to host cell nucleus is 1.03 ± 0.29 (0.60–2.04, N = 35), and to normal erythrocyte nuclei is 1.14 ± 0.30 (0.71–1.96, N = 49), similar values to those found in Morogoro. Meronts from *M. striata* from Kodera Forest were similar in dimensions to those in *M. striata* from Nairobi, 6.4 ± 1.6 × 5.7 ± 0.6 μm (N = 24), LW 33.1 ± 9.9 μm^2 versus 6.5 ± 1.0 × 4.4 ± 0.8 μm (N = 25), LW 28.8 ± 6.3 μm^2, and contained similar numbers of merozoites, 11.9 ± 3.8 (6–22) versus 12.8 ± 3.9 (6–25), respectively. Gametocytes from Kenyan *M. striata* were 9.7 ± 1.5 × 6.3 ± 1.0 μm (7–15 × 4–8, N = 76), with LW 61.1 ± 15.6 μm^2 (32–91) and L/W 1.59 ± 0.31 (1.00–2.50), larger and slightly more elongate than the Tanzanian gametocyte sample. The Kenyan gametocyte size relative to host cell nucleus is 2.29 ± 0.66 (1.03–4.29), and to normal erythrocyte nuclei is 2.22 ± 0.52 (1.23–3.25), similar values to those of Tanzanian gametocytes. Gametocytes from Kodera Forest were larger and less elongate

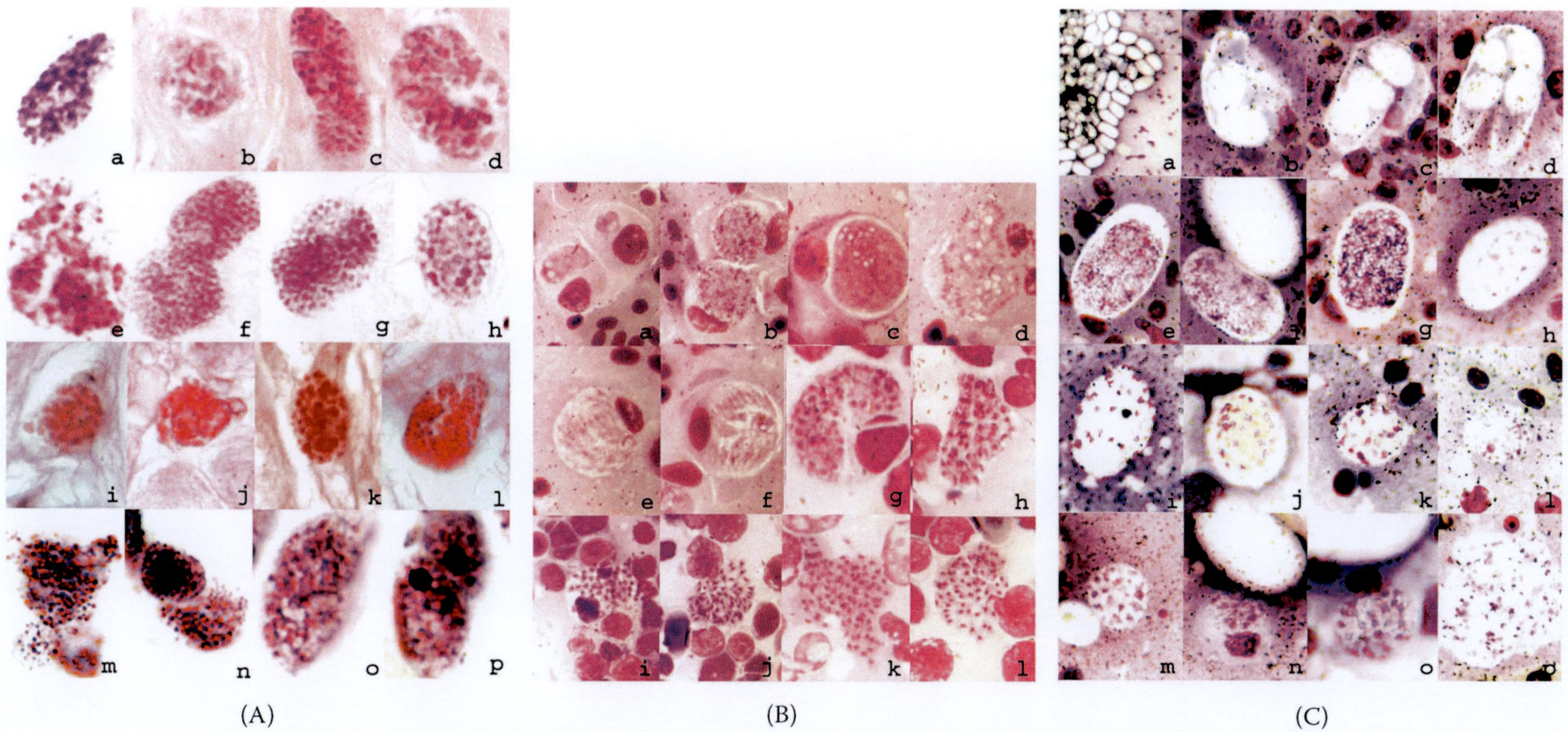

Plate 11 (**A**) Phanerozoites of four African *Plasmodium* species in cardiac tissues. **a–d**, *Plasmodium cordyli* of *Cordylus tropidosternum*; **e–h**, *Plasmodium loveridgei* of *Lygodactylus capensis grotei*; **i–l**, *Plasmodium cnemaspi* of *Cnemaspis barbouri*; **m–p**, *Plasmodium holaspi* of *Holaspis guentheri*. (**B**) Exo-erythrocytic stages in macrophages of *Agama mossambica* of Tanzania with mixed infection of *Plasmodium giganteum* and *P. mossambica*; **g**, **h**, **k**, **l** from spleen, remainder in liver. (**C**) Merogony of *Plasmodium agamae* in macrophages or within cysts in liver of *Agama agama*, Tanzania.

than those from Nairobi *M. striata*, 10.3 ± 1.4 × 6.7 ± 0.7 μm (N = 51), LW 69.3 ± 11.3 μm², and L/W 1.55 ± 0.29 versus 8.5 ± 0.9 × 5.2 ± 0.8 μm (N = 25), LW 44.2 ± 7.1 μm², and L/W 1.67 ± 0.35), respectively. No sexual dimorphism was present in the Kodera Forest gametocytes. Two gametocyte samples were available for *Mabuya maculilabris*, both from chronic infections, from Amani, Tanzania, and from Kinshasa, Congo. Both samples are similar in dimensions to active phase gametocytes from Tanzanian and Kenyan *M. striata* but are slightly more elongate in gametocyte shape. Gametocytes from Amani are 8.3 ± 1.2 × 5.2 ± 0.8 μm (6–11 × 4–7, N = 50), LW 43.1 ± 7.3 μm² (32–63), and L/W 1.63 ± 0.41 (1.00–2.75) versus 8.6 ± 1.8 × 5.1 ± 0.7 μm (6–14 × 4–7, N = 32), LW 44.2 ± 13.1 μm² (28–84), and L/W 1.71 ± 0.33 (1.20–2.80) in the Congo sample. Microgametocytes were more elongate, L/W 1.67 ± 0.39 (N = 25), than macrogametocytes, 1.58 ± 0.44 (N = 25) in *M. maculilabris* from Amani, but gametocyte shape was similar in microgametocytes (1.70) and macrogametocytes (1.72) in the Congo infection.

Exoerythrocytic Merogony Garnham (1966) described EE meronts of *P. pitmani* in smears of heart muscle and spleen of *M. maculilabris,* as well as in sections of liver and heart muscle. Phanerozoites in both fixed and circulating nonerythroid cells are evident in his figures. Garnham identified the circulating host cells as "mononuclear cells of the lymphoid-macrophage series," usually about 14 μm in diameter and containing 30–100 merozoites. Meronts in liver sections occupied "hypertrophied Kupfer cells," had a diameter of about 8 μm, and contained 50 or more merozoites. Telford (1994) illustrated phanerozoites in a spleen smear from *Mabuya striata.*

Sporogony Unknown.

Effects on Host In active infection, cells host to meronts were sometimes distorted (21%) with their nuclei displaced (43%). Distortion of the nuclei was similar (13–14%) for both meronts and gametocytes. The latter distorted erythrocytes more commonly (61%) and usually displaced their nuclei (88%). In active infection, erythrocytes host to meronts are hypotrophied by 20% and their nuclei by 18% less than uninfected cells. Gametocytes scarcely enlarge host cells and their nuclei, only 5% and 4%, respectively, over normal size.

Remarks Despite its broad geographic range, including Kenya, Uganda, and Tanzania to the west of Congo, dimensions of gametocytes of *Plasmodium pitmani* differed little by host species, infection stage, or geography. Meronts were similar in both Kenya and Tanzania in dimensions, but merozoite numbers varied by locality and infection phase in *M. striata.*

PLASMODIUM SPECIES OF NEOTROPICAL LIZARDS

NEOTROPICAL *SAURAMOEBA* SPECIES

Plasmodium diploglossi Aragão and Neiva 1909 (**Plate 12**)

Diagnosis A *Plasmodium* (*Sauramoeba*) species characterized by meronts 6–20 × 3–8 μm, with LW 30–136 μm², that produce 11–58 merozoites. Meronts are elongate in shape and usually partially encircle the host cell nucleus. Meront size relative to host cell nucleus is 1.1–4.0, and to normal erythrocyte nuclei is 1.5–5.8. Pigment is formed as a golden-yellow mass at one end of the meront. Gametocytes are ovoid to an elongate lentiform shape, 7–21 × 4–9 μm, with LW 35–120 μm² and L/W 1.0–4.2. Gametocyte size relative to host cell nucleus is 1.0–5.1, and to normal erythrocyte nuclei is 1.7–5.2. Gametocytes are seldom distinctly halteridial. A round vacuole may be present in about one-half of gametocytes. Pigment in both sexes is comprised of 20–30 scattered blackish granules. Sexual dimorphism is minimal.

Type Host *Diploglossus fasciatus* (Gray) (Sauria: Anguidae).

Type Locality Xerem, Rio de Janeiro State, Brazil.

Other Hosts *Mabuya mabouya* (Lacépède) (Sauria: Scincidae).

Other Localities Utinga Forest, Belem, Brazil (Lainson and Shaw, 1969b); Sasardi, San Blas Territory, Panama (Telford, 1970c); western Colombia (Ayala, 1978). Known only from *M. mabouya* in localities other than the type locality.

Prevalence Prevalence was 2 of 2 in *D. fasciatus* at the type locality (Aragão and Neiva, 1909), and in *M. mabouya,* was 6 of 20 (30%) in Belem (Lainson and Shaw, 1969b), 1 of 2 in Colombia (Ayala, 1978), and 4 of 11 (36%) in eastern Panama (Telford, 1970c).

Morphological Variation Overall, meronts average 12.3 ± 2.9 × 5.5 ± 1.5 μm (N = 103), with LW 66.0 ± 18.5 μm, and contain 28.3 ± 8.3 (N = 110) merozoites. Size relative to host cell nucleus 2.1 ± 0.7 (N = 78), and to normal erythrocyte nuclei is 3.0 ± 0.7 (N = 103). Mean values of characters in four infections were length 11.4–14.1, width 4.0–6.7, LW

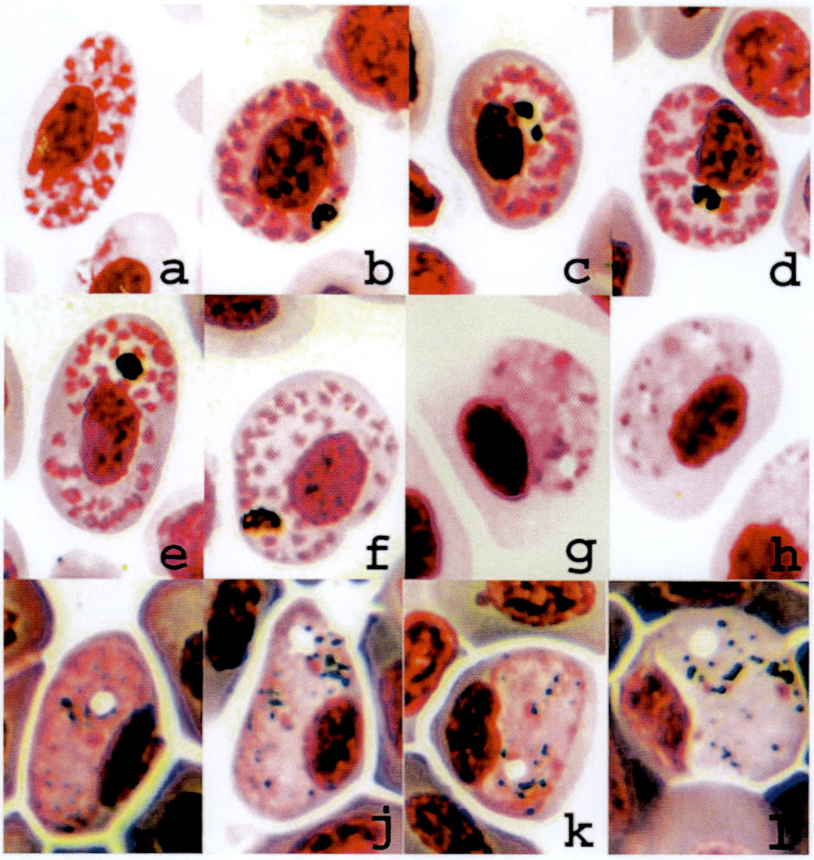

(A)

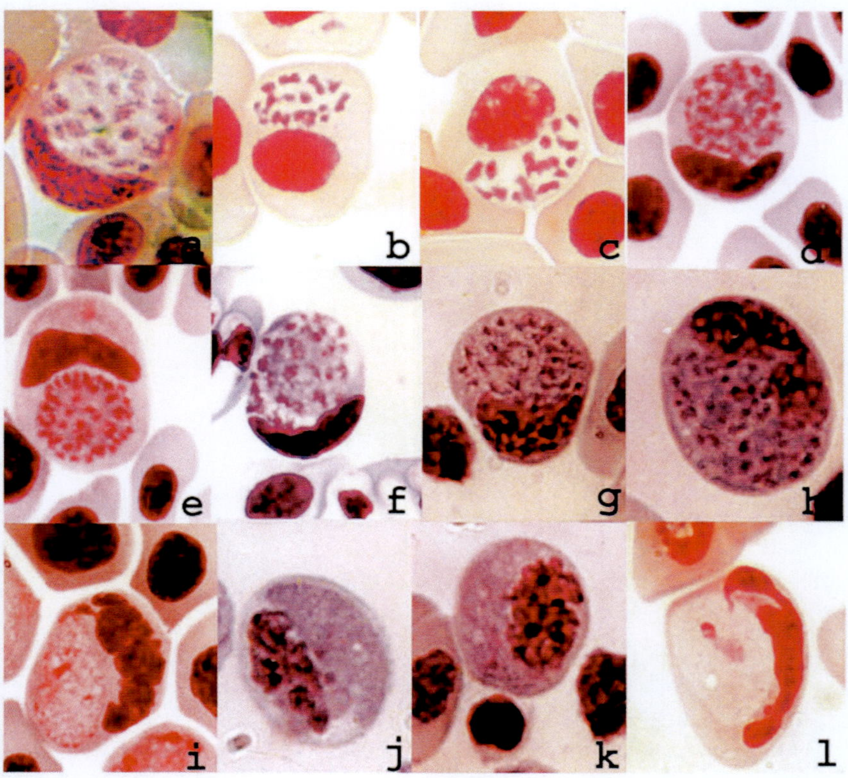

(B)

Plate 12 (**A**) *Plasmodium diploglossi* from *Mabuya mabouya*, Panama. Meronts **a–f**; macrogametocytes **g–i**; microgametocytes **j–l**. (**B**) *Plasmodium balli* from *Anolis lionotus*, **a–c**, **i**; *A. limifrons*, **d-h**, **j–l** from Panama. Meronts, **a–h**; macrogametocytes, **i–k**; microgametocytes, **l**. Pigment visible in **a**.

51.1–76.3, and merozoites 18.6–32.8. Gametocytes average 11.9 ± 2.1 × 6.0 ± 1.1 μm (N = 124), with LW 71.4 ± 18.9 μm^2 and L/W 2.06 ± 0.49. Size relative to the host cell nucleus is 2.6 ± 0.7 (N = 94), and to normal erythrocyte nuclei is 3.3 ± 0.7 (N = 124). Mean values of characters in three gametocyte infections were length 10.2–13.0, width 5.2–6.8, LW 60.4–87.3, and L/W 1.8–2.3. Macrogametocytes average 11.9 ± 1.8 × 6.1 ± 1.0 μm (7–16 × 4–9, N = 79), with LW 72.5 ± 17.6 μm^2 (35–120) and L/W 2.03 ± 0.43 (1.0–3.5). Microgametocytes average 12.0 ± 2.6 × 5.8 ± 1.2 μm (8–21 × 4–8, N = 44), with LW 69.8 ± 21.2 μm^2 (36–120) and LW 2.12 ± 0.59 (1.1–4.2). Meronts in Brazilian *M. mabouya* produced 40–50 merozoites (Lainson and Shaw, 1969b), and in *D. fasciatus*, over 40 (Aragão and Neiva 1909). Pigment was described as brownish by Aragão and Neiva in 1909 and green-yellow by Lainson and Shaw (1969b).

Exoerythrocytic Merogony Lainson and Shaw (1969b) reported possible EE meronts in monocytes and in cells of the liver and spleen, but cautioned that identification was uncertain due to the presence of *P. tropiduri* in all of the infected lizards.

Effects on Host Both asexual and sexual stages parasitize erythrocytes, only rarely being found in proerythrocytes (gametocytes, 1%). With growth of both meronts and gametocytes, host cells enlarge and may become rounded. Erythrocyte nuclei are often displaced to one side, are seldom enlarged, but are usually rounded (Telford, 1970c).

Sporogony Unknown.

Remarks *Plasmodium diploglossi* was the first *Plasmodium* species described from a saurian host (Aragão and Neiva 1909), followed by *P. tropiduri* in the same article. Slightly later, Wenyon (1909a) described *P. agamae* from the Sudan.

The *M. mabouya* hosts of *P. diploglossi* in Panama were collected in lowland moist tropical forest in December 1968 and February 1969. The lizard taken in early December showed only *Plasmodium morulum* infection at capture, but on 1 February *P. diploglossi* appeared in a massive infection that probably contributed, in combination with a high parasitemia of *P. morulum*, to the death of the host 10 days later. The other naturally infected *M. mabouya* had light infections present when captured, which remained at low parasitemias for more than 1 year following capture. Four infections induced in negative lizards by inoculation of infected blood became patent in 26–35 days; two of these reached peak at 41–42 days postinoculation, with maximum parasitemia of 64.3%.

Plasmodium balli Telford 1969 (**Plate 12**)

Diagnosis A *Plasmodium* (*Sauramoeba*) species that usually parasitizes immature erythrocytes, forming meronts that average 2.2–3.6 times normal erythrocyte nucleus size and produce 15–100 merozoites. Meronts are 7–19 × 5–13 μm with LW 45–255 μm^2. Gametocytes are typically elongate, 8–23 × 4–14 μm, with LW 40–196 μm^2, L/W 1.00–5.00, with average size relative to normal erythrocyte nucleus size of 2.06–2.79. Gametocytes do not differ in size or shape by sex in active infections, but in chronic phase, macrogametocytes may be larger than microgametocytes without difference in shape. Minute black dots of pigment are sometimes visible in erythrocytic meronts and gametocytes. Parasitized host cells in heavy infections are commonly enucleated; gametocytes often produce a lytic effect on host cell nuclei.

Type Host *Anolis lionotus* Cope (Sauria: Polychrotidae).

Type Locality Vicinity of Achiote, Colon Province, Panama (erroneously stated as 3 miles southeast of Achiote in original description instead of 0.3 miles).

Other Hosts *Anolis poecilopus, A. limifrons, A. lemurinus, A. humilis* (Telford, 1977); *A. frenatus* (Guerrero and Ayala, 1977); *A. fuscoauratus* (Guerrero et al., 1977).

Other Localities Localities in Panama are as follows: El Aguacate and Rio Madroño (Gaspar Sabanas and Madroño, 8 km north northwest of Chepo) in Panama Province; Santa Rita Ridge and Rio Agua Clara in Colon Province; Canal Zone, near Gamboa including Frijoles River, Quebrada Juan Grande, and Frijolito Creek, and Madden Forest; Sasardi, 5 km west of Mulatupo, San Blas Territory (Telford, 1977); Barro Colorado Island (Guerrero et al., 1977). Other localities are Costa Rica: Rio Frio, Heredia Province (Telford, 1977); Blancaneaux Lodge, 18 km north of Augustine, Belize (Telford, 1977); Rio Llullapichis, Huanuco about 300 km south of Pucallpa, Peru (Guerrero and Ayala, 1977).

Prevalence Panama: In *A. lionotus*, overall 94 of 249 (37.8%): Achiote 20 of 67 (29.9%), El Aguacate 72 of 175 (41.1%), Gaspar Sabanas 2 of 6. In *A. poecilopus*, overall 54 of 228 (23.7%): Santa Rita Ridge and Rio Agua Clara 4 of 16 (25.0%), Gaspar Sabanas and Madroño 9 of 64 (14.1%), Quebrada Juan Grande 11 of 44 (25.0%), Frijoles River 10 of 39 (25.6%), Frijolito Creek 20 of 60 (33.3%). In *A. limifrons*, overall 48 of 381 (19.6%): Achiote 13 of 112 (11.6%), Sasardi 7 of 107 (6.5%), Gaspar Sabanas 2 of 7, Quebrada Juan Grande 1 of 18 (5.6%), Frijoles River 1 of 7, Frijolito

Creek 19 of 73 (26.0%), Madden Forest 5 of 19 (26.3%) (Telford); Barro Colorado Island in *A. limifrons* overall, 326 of 735 (44.4%, Rand et al., 1983); 56 of 296 (18.9%), and in *A. frenatus* 2 of 26 (7.7%) (Guerrero et al., 1977). Costa Rica, Rio Frio: in *A. humilis* 2 of 67 (3.0%), *A. lemurinus* 4 of 31 (12.9%), *A. limifrons* 10 of 82 (12.2%), *A. lionotus* 2 of 39 (5.1%) (Telford, 1977). Belize: in *A. lemurinus* 1 of 4 (Telford, 1977).

Morphological Variation In the type host, *Anolis lionotus, P. balli* meronts are 12.1 ± 2.8 × 9.4 ± 2.2 μm (7–19 × 5–13, N = 109), with LW 118.3 ± 48.5 μm² (45–228), and contain 35.2 ± 10.0 (13–60, N = 111) merozoites. Meront size relative to host cell nucleus size averages 1.34 ± 0.67 (0.75–3.67, N = 22), and to normal erythrocyte nucleus size is 2.26 ± 0.90 (0.77–4.76, N = 109). Meronts in erythrocytes are smaller and contain fewer merozoites than those in immature host cells, respectively 9.8 ± 1.4 × 7.5 ± 1.2 μm (N = 22), LW 74.0 ± 17.5 μm², with merozoites 26.4 ± 8.3 (13–39), versus 12.7 ± 2.7 × 9.9 ± 2.1 μm (N = 87), LW 129.5 ± 47.5 μm², and 37.3 ± 9.1 (18–60, N = 89). Pigment is rarely seen in meronts (0.9%), but when present appears as a single, minute black dot to a large, squarish, brownish-black granule the size of a merozoite. The typically elongate gametocytes are 14.7 ± 2.8 × 6.6 ± 1.3 μm (9–23 × 4–11, N = 151), with LW 96.1 ± 20.9 μm² (44–165), and L/W 2.32 ± 0.73 (1.09–4.60). Gametocyte size relative to host cell nucleus size is 1.35 ± 0.31 (0.86–2.55, N = 56), and to normal erythrocyte nucleus size is 2.06 ± 0.57 (0.69–3.81, N = 151). There is no difference in gametocyte dimensions, size, or shape by sex or maturity of host cells. Pigment is usually absent but sometimes appears as one to occasionally several minute black dots in 9.3% of gametocytes, occasionally as a squarish larger granule, and usually when the host cell is a mature erythrocyte. Identity as hemozoin was verified under polarized light.

In *Anolis poecilopus*, meronts of *P. balli* are larger than in *A. lionotus*, the range in merozoite number is greater, but there is no difference in mean merozoite number. Meronts are 13.1 ± 2.1 × 10.5 ± 1.5 μm (9–17 × 7–15, N = 80), LW is 137.8 ± 35.6 μm² (72–255), and merozoites number 36.8 ± 16.3 (15–100, N = 82). Meront size relative to host cell nucleus size is 1.98 ± 0.67 (0.96–2.98, N = 10), and to normal erythrocyte nucleus size is 3.46 ± 0.87 (1.84–6.31, N = 80). Although average dimensions and size of meronts and merozoite number are slightly larger in proerythrocytes than in erythrocytes, there is no real difference. Meronts from *A. poecilopus* in the Canal Zone localities of Quebrada Juan Grande and Frijolito Creek produce more merozoites on average, 46.8 ± 17.0 (N = 42), than those from Rio Madroño in Panama Province, 26.4 ± 5.2 (N = 40); dimensions and size of meronts are similar. Gametocytes are larger than those of *P. balli* in *A. lionotus* and less elongate in shape, 14.1 ± 2.4 × 8.3 ± 1.8 μm (8–20 × 4–14, N = 229), with LW 115.9 ± 26.1 μm² (44–196) and L/W 1.81 ± 0.59 (1.00–4.00). Gametocyte size relative to host cell nucleus size is 1.70 ± 0.54 (0.69–4.17, N = 62), and to normal erythrocyte nucleus size is 2.79 ± 0.72 (1.09–5.01, N = 229). Pigment appears in 12.6% of gametocytes, usually those in mature host cells. Neither gametocyte sex nor maturity of host cell affects gametocyte size or shape. Gametocytes in chronic infection are larger and more rounded than those in active phase, respectively 13.5 ± 2.2 × 9.0 ± 1.7 μm (N = 150), LW 119.9 ± 25.7 μm², and L/W 1.57 ± 0.46 versus 15.4 ± 2.2 × 7.1 ± 1.4 μm (N = 79), LW 108.3 ± 25.4 μm², and L/W 2.26 ± 0.53.

In *Anolis limifrons, Plasmodium balli* parasitizes immature erythroid cells only, proerythrocytes, erythroblasts, and perhaps stem cells. Meronts are similar in size to those in *A. lionotus* and *A. poecilopus*, but fewer merozoites are produced, 16–45, in comparison to 13–60 in *A. lionotus* and 15–100 in *A. poecilopus*. In *A. limifrons*, meronts are 12.0 ± 1.3 × 10.0 ± 1.2 μm (10–16 × 8–13, N = 59), with LW 120.1 ± 22.9 μm² (88–176) and 28.8 ± 6.9 (16–45) merozoites. Meront size relative to normal erythrocyte nucleus size is 3.17 ± 0.67 (2.22–4.96). Meronts from eastern Panama (Sasardi) are of similar size to those from the central Panama localities (Canal Zone and Achiote) but contain more merozoites, 33.0 ± 5.8 (N = 25), than in the other localities, 25.7 ± 6.0 (N = 34). Overall, gametocytes of *P. balli* in *A. limifrons* are 13.8 ± 2.2 × 8.2 ± 1.8 μm (9–25 × 4–12, N = 149), with LW 113.0 ± 29.5 μm² (40–180) and L/W 1.78 ± 0.55 (1.00–5.00). Gametocyte size relative to normal erythrocyte nucleus size is 2.78 ± 0.74 (0.92–4.71). Gametocytes in active phase of infection are smaller and more elongate than chronic-phase gametocytes, respectively 13.4 ± 1.9 × 7.7 ± 1.6 μm (N = 109), LW 102.9 ± 26.2 μm², and L/W 1.82 ± 0.50 versus 15.1 ± 2.3 × 9.5 ± 1.5 μm (N = 40), LW 140.3 ± 19.0 μm², and L/W 1.68 ± 0.65. There is no sexual difference in dimensions and shape of active-phase gametocytes, but in chronic infection, although shape is similar, microgametocytes are smaller than macrogametocytes, respectively 14.7 ± 1.9 × 9.0 ± 1.2 μm (N = 20), LW 131.0 ± 15.4 μm², and L/W 1.68 ± 0.40 versus 15.5 ± 2.6 × 9.9 ± 1.7 μm (N = 20), LW 149.6 ± 18.0 μm², and L/W 1.68 ± 0.84. In active infections from Sasardi, gametocytes have a greater LW value, 127.2 ± 20.9 μm², than gametocytes in the central Panama localities, 96.2 ± 23.5 μm². Pigment is not visible in *P. balli* infections of *A. limifrons*.

Guerrero and Ayala (1977) reported *Plasmodium balli* from *Anolis fuscoauratus* of Peru. Mature meronts produced 27.8 ± 3.1 (24–34, N = 17) merozoites. Gametocytes were 14.2 ± 2.0 × 7.6 ± 1.1 μm (10.0–17.5 × 5.0–11.2, N = 80), from which LW is estimated at 107.9 μm² and L/W at 1.89, values well within the range of variation seen in *Anolis limifrons*.

Exoerythrocytic Merogony Unknown.

Sporogony Unknown.

Effects on Host In both *A. lionotus* and *A. poecilopus*, gametocytes of *P. balli* cause hypertrophy of host erythrocytes to the extent of 39% and 44%, respectively, and meronts by 21% and 28%, respectively. Host cell nuclei, when not distorted to an unmeasurable extent, are also hypertrophied, by 62% and 40%, respectively, for erythrocytes containing gametocytes; 47% and 65% in the case of meronts, respectively. *Plasmodium balli* in *A. lionotus, A. poecilopus*, and *A. limifrons* distorts 96–100% of cells host to gametocytes, and 84–100% of their nuclei, and displaces 80–100% of host cell nuclei. Meronts cause similar effects in the three host species, distorting 95–100% of host cells and 79–98% of their nuclei, and displacing 100% of the latter in all hosts. Enucleated host cells can be seen in both *A. lionotus* and *A. poecilopus* when parasitized by gametocytes, 1.3% and 4.6%, and by meronts, 5.5% and 6%, respectively. Enucleation is rare in *A. limifrons*, occurring in gametocyte-parasitized cells only (0.7%) and is not observed at all for meronts. Lysis of host cell nuclei can occur in all three hosts but is most common (43%) in *A. limifrons* erythrocytes parasitized by gametocytes. Distorted nuclei usually appear lobulated. All types of erythroid cells are parasitized by meronts: apparent stem cells, basophilic and polychromatophilic erythroblasts and proerythrocytes, and mature erythrocytes. Among these three hosts, 37–49% of gametocytes occupy erythrocytes, but in *A. limifrons*, only 3.4% to 20% of meronts occur in mature cells, on average. In some infections of *A. lionotus* and *A. poecilopus*, erythrocytes may be the predominant type of host cell infected, but phase of infection can alter host cell utilization.

In *A. fuscoauratus* of Peru, Guerrero and Ayala (1977) described a predilection by *P. balli* for immature host cells (polychromatophils), causing hypertrophy and distortion of cells and distortion of nuclei, with less nuclear distortion caused by meronts than by gametocytes. In their study of *Anolis limifrons* infected by *P. balli* and other species that certainly included *P. fairchildi, P. floridense*, and possibly *P. minasense anolisi*, Rand et al. (1984) found that *P. balli* had no greater effect on the proportion of immature red blood cells in its hosts than did the other smaller *Plasmodium* species (not identified to species), despite its greater effect on parasitized cells.

Ecology Telford (1977) described the annual fluctuations in prevalence and proportion of initial infections of *P. balli* as follows: Prevalences "did not differ during early [January-March] and late [April–June] dry seasons, and were at their minimum levels then." Prevalence increased significantly "in early wet season" (July–September) and rose even higher "to maximum rates in late wet season" (October–December), "followed by a sharp drop as dry season came on." Prevalences in this sequence of seasons were 15.6%, 11.3%, 19.4%, and 32.0%, respectively. The differences in prevalence "were accompanied by significant changes in proportions of initial vs. chronic infections. Two-thirds of infections found in late wet season were classified as initial, while those considered to be chronic were at their maximum (87%) immediately preceding, during early wet season," when these latter "provided maximum incidence of solely gametocytic infections (52%) … the rise evident in infection rate in early wet season perhaps indicated recrudescent infections as well." Although "some transmission clearly occurred throughout the year, it was heaviest in late wet season, October–December."

Both *Anolis lionotus* and *A. poecilopus* are semiaquatic species (Campbell, 1973) and occur only along intermittent or permanent streams, utilizing basking sites such as rocks, logs, tree trunks, and shrubs within 1 m of the stream surface, taking shelter beneath overhanging banks in holes or behind stones in air pockets. The vector of *P. balli* must be a species "closely associated with" the immediate vicinity of small streams flowing through primary or advanced secondary wet tropical forest. *Anolis limifrons* is usually "found within 2 m of the forest floor, on tree trunks, vines or shrubs" (Telford, 1977) and frequently utilizes stream margins or stream beds during foraging or escape from predators. The presence of infections occasionally in the arboreal *Anolis frenatus* (Guerrero et al., 1977) suggests a broader vertical distribution of the vector rather than strict restriction to the forest floor and stream bank.

Although Rand et al. (1984) did not differentiate among the smaller *Plasmodium* species in *Anolis limifrons* of Barro Colorado Island, where the prevalence of *P. balli* is 44.4%, they were unable to demonstrate any effect of malaria infections on the several parameters of *A. limifrons* ecology studied: physical condition of the hosts, food intake, time of activity of the lizards, reproduction, growth, predator pressure, or differential mortality.

Remarks At the time of its description, *P. balli* was compared with the *Plasmodium* (*Sauramoeba*) species known at the time, especially those that produced large numbers of merozoites (30–100+), that is, *P. cnemidophori, P. diploglossi, P. beltrani, P. giganteum, P. robinsoni*, and *P. egerniae*. All are prominently pigmented, irrespective of host cell maturity. Pigment is almost always visible in *P. balli* only when mature erythrocytes are host cells. Dimensions of *P. balli* meronts and gametocytes alone would indicate relationship to the *Sauramoeba* species. Telford (1988a) based inclusion of *P. balli* in the subgenus *Garnia* on the basis of variability in pigment presence. However, in *Garnia* the absence of pigment is independent of erythrocyte maturity, and for this reason *Plasmodium balli* should be classified as a *Plasmodium* (*Sauramoeba*) species. Genomic comparisons are necessary to elucidate its actual relationships as a species of *Sauramoeba* or *Garnia*.

Plasmodium cnemidophori Carini, 1941 (**Plate 13**)

Diagnosis A *Plasmodium* (*Sauramoeba*) species characterized by very large meronts, 10–19 × 7–15 μm, with LW 88–266 μm², that produce 42–127 merozoites, nearly filling the host cell. Meront size relative to the host cell nucleus is 2.2–12.0, and to normal erythrocyte nuclei is 3.8–13.9. Gametocytes are elongate and banana shaped to bulky, but not halteridial, 10–22 × 4–12 μm, with LW 60–240 μm² and L/W 1.0–4.0. Gametocyte size relative to host cell nucleus size is 1.1–6.0, and to normal erythrocyte nuclei is 3.0–10.6. Pigment is prominent, as dark greenish to golden masses, often squarish in meronts, and as dark dots dispersed in gametocytes from a golden mass in younger sexual stages. Gametocytes are sexually dimorphic in dimensions when in chronic phase of infection.

Type Host *Cnemidophorus l. lemniscatus* (Linnaeus) 1758 (Sauria: Teiidae).

Type Locality Goiás, Brazil.

Other Hosts *Ameiva a. ameiva* (Lainson and Shaw, 1969b; Telford, 1973b); *Ameiva ameiva praesignis* (Scorza, 1970; Telford, 1969, 1973a, 1980; Ayala, 1978); *Ameiva anomala, Ameiva bridgesi, Ameiva festiva,* (Ayala, 1978).

Other Localities Brazil: Belem, Para and Cachoera (Lainson and Shaw, 1969b); Venezuela: Aragua (Scorza, 1970); Municipios Ortiz in Guarico State, Araure, Piritu, and Nueva Florida, Portuguesa State (Telford, 1980); Vichada, Colombia (Ayala, 1978); vicinity of Georgetown, Guyana (Telford, 1973b); Madden Dam, Canal Zone, Panama (Telford, 1973a).

Prevalence In *C. lemniscatus*, two of four in Goiás, Brazil (Carini, 1941), and two of seven in Vichada, Colombia (Ayala, 1978). In *A. ameiva*, 29 of 66 (44%) in Pará, Brazil (Lainson and Shaw, 1969b); 2 of 12 (17%) at Madden Dam, Canal Zone (Telford, 1973a); 2 of 16 (13%) near Georgetown, Guyana (Telford, 1973b); and in Venezuela, 6 of 21 (29%) at Los Cumbitos, Municipio Ortiz; 4 of 65 (6%) in Araure, Municipio Araure, 2 of 3 at Santo Domingo, Municipio Nueva Florida (Telford, 1980); and 10 of 39 (26%) at El Limon, Aragua (Scorza, 1970).

Morphological Variation Carini (1941) reported mature meronts of *P. cnemidophori* in the type host to be oval in shape, 11 × 9 μm, producing 100 or more merozoites. The banana-shaped macrogametocytes were 12 × 4.5 μm, and microgametocytes were 9 × 4 μm, with these dimensions presumably averages. In Brazilian *A. ameiva*, Lainson and Shaw (1969b) described mature meronts as round or oval, containing 50–100 merozoites. They described the macrogametocytes as "about 5.0 × 6.0 μm in size," but this must be a typographical error as their figures indicate 15.0 × 6.0 μm as more likely. Mature meronts from two active infections (Santo Domingo, Venezuela, and Guyana, respectively) had these dimensions: 12–19 × 10–15 μm (15.1 ± 2.1 × 12.5 ± 1.5, N = 25), with LW 130–266 μm² (190.5 ± 40.1), containing 67–125 (96.8 ± 16.9) merozoites compared with 10–14 × 8–12 μm (12.3 ± 1.1 × 10.6 ± 1.2, N = 23), with LW 90–168 μm² (129.7 ± 19.8), containing 58–87 (70.4 ± 8.6, N = 25) merozoites. Gametocytes from the two active infections, in the same respective sequence, were 13–20 × 5–9 μm (15.7 ± 2.0 × 6.9 ± 1.3, N = 25), with LW 75–144 μm² (106.9 ± 16.9), and L/W 1.4–4.0 (2.40 ± 0.65) compared with 11–16 × 4–11 μm (13.9 ± 1.6 × 7.2 ± 1.8, N = 25), LW 60–135 μm² (98.8 ± 20.9), and L/W 1.0–3.8 (2.09 ± 0.70). Meronts from the chronic infections, Municipio Ortiz in Venezuela, and Canal Zone, Panama, respectively, were 11–15 × 9–11 μm (12.4 ± 1.1 × 10.4 ± 0.8, N = 10), with LW 108–165 μm² (129.1 ± 16.8), and contained 66–105 (79.2 ± 12.6) merozoites in comparison to 10–19 × 8–13 μm (13.0 ± 1.9 × 10.8 ± 1.4, N = 23), with LW 88–247 μm² (141.8 ± 35.4), containing 42–119 (66.0 ± 16.4, N = 25) merozoites. The chronic-phase gametocytes, respectively, were 12–22 × 7–12 μm (16.8 ± 2.1 × 9.1 ± 1.3, N = 50), LW 105–216 μm² (151.5 ± 24.8), and L/W 1.3–3.1 (1.90 ± 0.44) compared with 10–20 × 6–12 μm (15.1 ± 2.7 × 8.8 ± 1.4, N = 25), LW 88–240 μm² (133.4 ± 33.1), and L/W 1.0–2.7 (1.76 ± 0.46). Macrogametocyte and microgametocyte dimensions compared for active- (Santo Domingo) and chronic-phase (Municipio Ortiz) infections of *P. cnemidophori* found no difference between sexes in the active infection, but mean values of length, width, and LW were greater in macrogametocytes (17.5 ± 1.9 × 9.5 ± 1.4 μm, LW 164.7 ± 23.5 μm², N = 25) than in microgametocytes (16.0 ± 1.9 × 8.7 ± 1.2 μm, LW 138.2 ± 18.4 μm², N = 25) of the chronic infection. Gametocyte shape, however, did not differ between sexes (1.90 ± 0.43 versus 1.90 ± 0.45).

Exoerythrocytic Merogony Lainson and Shaw (1969b) reported meronts in lymphocytes and thrombocytes in lizards infected by *P. cnemidophori*. Up to 20 merozoites about 3.0 μm were produced.

Plate 13 (**A**) *Plasmodium cnemidophori* from *Ameiva ameiva*, Venezuela. Meronts, **a–f**; macrogametocytes, **g–i**; microgametocytes, **j–l**. (**B**) *Plasmodium guyannense* from *Plica plica*, Guyana. Meronts, **a–f**; macrogametocytes, **g–j**; microgametocytes, **k**, **l**.

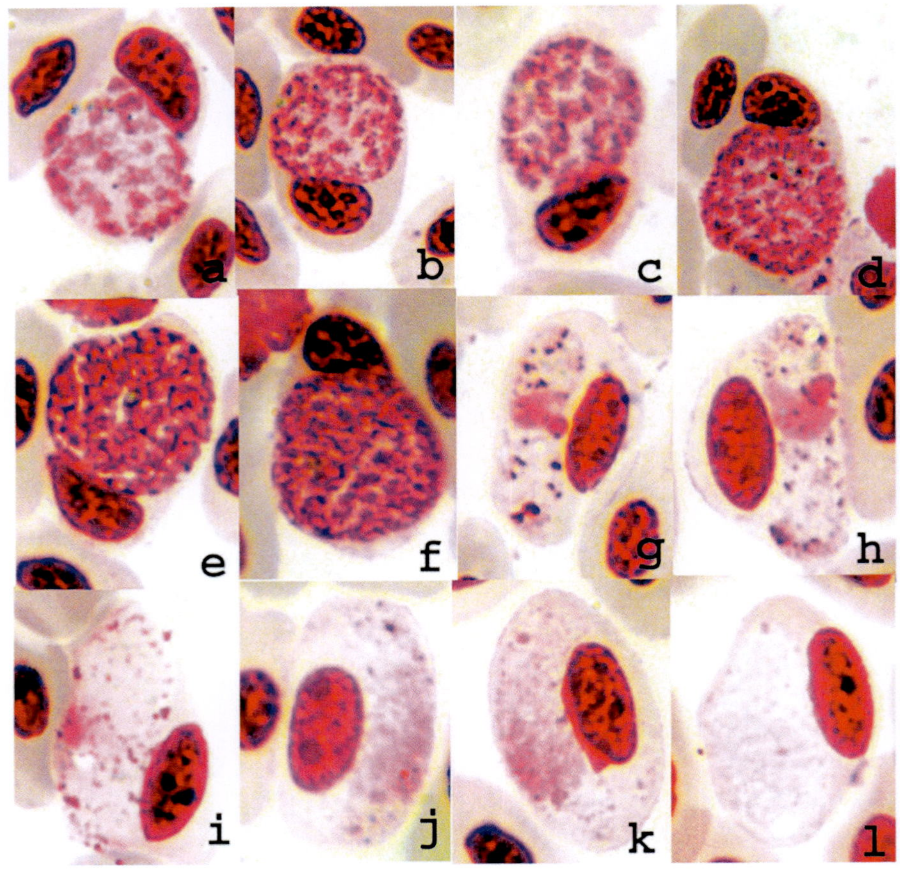

(A)

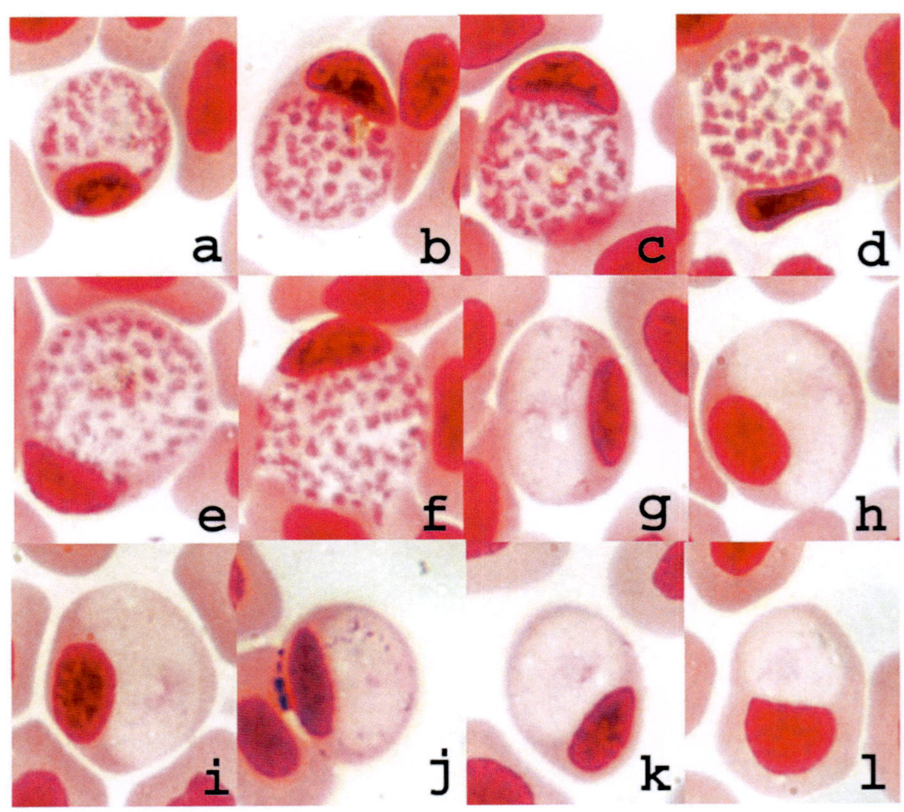

(B)

Effects on Host Carini (1941) mentioned changes in shape and size of erythrocytes host to *P. cnemidophori* meronts and gametocytes, with erythrocyte nuclei displaced and altered in appearance. Lainson and Shaw (1969b) also described displacement and enlargement of erythrocyte nuclei and some degree of degeneration. Although host cells were commonly deformed by meronts, gametocytes produced greater enlargement. Host cells to meronts and gametocytes in experimental and natural infections in *Ameiva a. ameiva* from Guyana were always erythrocytes (Telford, 1973b). However, infections in *Ameiva a. praesignis* in Panama showed uninucleate parasites present predominantly in proerythrocytes, with only one erythroblast and one erythrocyte found infected, while all meronts seen were in proerythrocytes, and both immature and mature gametocytes occupied erythrocytes (Telford, 1973a). In Venezuelan infections also, meronts were proerythrocytic, while gametocytes parasitized erythrocytes only. Host cells of both meronts and gametocytes of *P. cnemidophori* infections in Panama, Guyana, and Venezuela (Telford, 1980) were always distorted and hypertrophied and their nuclei distorted and displaced.

Sporogony Unknown.

Remarks One infection of *P. cnemidophori* in *Ameiva ameiva* from Guyana that resulted from inoculation of infected blood into a negative lizard became patent in 29 days, but never attained a parasitemia of 1%, and declined to occasional gametocytes only by 57 days postinoculation (Telford, 1980).

Plasmodium guyannense Telford, 1979 (**Plate 13**)

Diagnosis A *Plasmodium* (*Sauramoeba*) species characterized by round or oval meronts 11–16 × 6–12 μm, with LW 72–180 μm², that produce 40–74 merozoites, nearly filling the host cell. Size relative to host cell nucleus is 2.1–5.6, and to normal erythrocyte nuclei is 2.0–4.3. Prominent yellow pigment masses lie within vacuoles in trophozoites and younger asexual stages, becoming large, irregular masses of dark pigment usually centrally located in mature meronts. Gametocytes are spheroid to ovoid, seldom elongate, 8–17 × 5–10 μm, with LW 54–144 μm² and L/W 1.00–3.00. Size relative to host cell nucleus size is 0.9–3.9, and to normal erythrocyte nuclei is 1.5–3.9. Pigment is dispersed in both sexes as small, dark granules. Vacuoles are rarely present in gametocytes. Gametocytes are sexually dimorphic, with macrogametocytes longer and more slender than microgametocytes, resulting in greater LW values and L/W ratios.

Type Host *Plica plica* (Linnaeus) (Sauria: Iguanidae).

Type Locality Vicinity of Georgetown, Guyana.

Other Hosts None known.

Other Localities None known.

Prevalence One of ten *P. plica* (10%) was infected by *P. guyannense*.

Morphological Variation Meronts average 12.5 ± 1.3 × 10.2 ± 1.4 μm (N = 30), with LW 128.0 ± 22.8 μm², and contain 56.4 ± 9.1 merozoites. Size relative to host cell nucleus is 3.6 ± 0.8 (N = 30), and to normal erythrocyte nuclei is 3.4 ± 0.6 (N = 25). Gametocytes average 12.2 ± 2.3 × 7.1 ± 1.2 μm (N = 60), with LW 86.7 ± 21.8 μm² and L/W 1.77 ± 0.48. Size relative to host cell nucleus averages 1.91 ± 0.52 (N = 60), and to normal erythrocyte nuclei is 2.28 ± 0.56 (N = 50). Macrogametocytes are 13.5 ± 1.7 × 7.0 ± 1.3 μm (10–17 × 5–10, N = 35), with LW 94.1 ± 20.7 μm² (60–144) and L/W 2.01 ± 0.46 (1.2–3.0). Microgametocytes average 10.4 ± 1.6 × 7.3 ± 1.0 μm (8–15 × 5–9, N = 25), with LW 76.3 ± 19.1 μm² (54–135) and L/W 1.43 ± 0.26 (1.0–2.2).

Exoerythrocytic Merogony Not observed.

Effects on Host Meronts and gametocytes occupied erythrocytes, although trophozoites and young gametocytes more commonly were found in proerythrocytes, favoring polychromatophilic rather than basophilic host cells. A lateral position within the host cell by both meronts and gametocytes resulted in distortion and hypertrophy of the host erythrocytes, and displacement and commonly distortion of the host cell nucleus.

Sporogony Unknown.

Remarks *Plasmodium guyannense* is most similar to *P. cnemidophori*, especially in the appearance of mature meronts, and in fact was assigned to this species when first reported (Telford, 1973b). Meronts of *P. cnemidophori* are usually larger and contain more merozoites, however. The early stages of the two parasites are most distinctive: Trophozoites of *P. guyannense* contain a vacuole early on, in which a prominent pigment mass forms during growth, while comparable stages of *P. cnemidophori* lack prominent vacuoles, and pigment is never obvious as a large mass in them. Trophozoites and young meronts of the latter species favor younger host cells than do those of *P. guyannense*, usually occurring in basophilic proerythrocytes, and immature meronts of comparable size have more nuclei in *P. cnemidophori*, which perhaps suggests a faster rate of division (Telford, 1979d). Immature gametocytes of *P. cnemidophori* are elongate, never oval or round as in *P. guyannense*, and mature gametocytes commonly show a single vacuole, rarely observed in the latter species. Genomic comparison might indicate subspecific relationship of these two species that parasitize hosts of different families, Teiidae and Iguanidae.

Plasmodium achiotense Telford, 1972 (**Plate 14**)

Diagnosis A *Plasmodium* (*Sauramoeba*) species characterized by large, variably shaped meronts 10–18 × 9–13 µm, with LW 100–192 µm^2, that produce 36–56 merozoites. Meront size relative to host cell nucleus is 2.3–4.8, and to normal erythrocyte nuclei is 2.7–5.2. The spherical-to-ovoid gametocytes are also large, 9–13 × 8–12 µm, with LW 80–156 µm^2 and L/W 1.0–1.5. Gametocyte size relative to host cell nucleus is 1.8–4.0, and to normal erythrocyte nuclei is 2.1–4.2. Pigment granules are golden and prominent, forming an irregular central mass in meronts, and dispersed as 30 or more granules in gametocytes. Gametocytes do not show sexual dimorphism in dimensions.

Type Host *Basiliscus basiliscus* (Linnaeus) (Sauria: Corytophanidae).

Type Locality Achiote, Colon Province, Republic of Panama.

Other Hosts None known.

Other Localities El Aguacate, Panama Province, Republic of Panama, and Madden Forest, Canal Zone.

Prevalence Overall, 5 of 63 (7.9%) *B. basiliscus* were infected by *P. achiotense,* 1 of 14 (7%) at the type locality and 3 of 20 (15%) at El Aguacate.

Morphological Variation Meronts average 13.0 ± 1.8 × 10.9 ± 1.0 µm (N = 25), with LW 141.8 ± 22.0 µm^2, and contain 46.8 ± 5.8 merozoites. Size relative to host cell nucleus is 3.6 ± 0.8, and to normal erythrocyte nuclei is 3.9 ± 0.6. Gametocytes average 11.0 ± 0.4 × 9.8 ± 0.8 µm (N = 75), with LW 107.8 ± 13.0 and L/W 1.14 ± 0.13. Size relative to host cell nucleus is 2.67 ± 0.50, and to normal erythrocyte nuclei is 2.92 ± 0.34. Macrogametocytes average 10.9 ± 0.9 × 9.7 ± 0.7 µm (9–13 × 8–11, N = 35), with LW 105.7 ± 11.4 µm^2 and L/W 1.12 ± 0.14 (1.0–1.5). Microgametocytes are 11.2 ± 0.8 × 9.8 ± 0.9 µm (10–13 × 8–12, N = 40), with LW 109.7 ± 14.8 µm^2 (80–156) and L/W 1.15 ± 0.12 (1.0–1.5).

Exoerythrocytic Merogony Not observed.

Effects on Host Meronts and gametocytes parasitize erythrocytes, which are hypertrophied, especially in width, nearly always distorted, and with nuclei displaced. Nuclei are commonly distorted as well. Gametocytes are more often polar in position and meronts lateral to the erythrocyte nucleus.

Sporogony Unknown.

Remarks Gametocytes of *P. achiotense* are most likely to be confused with those of *P. basilisci* in mixed infections. Gametocytes of the latter species are only two-thirds the size of *P. achiotense* and have about half as many pigment granules, 10–20, instead of 30 or more. Basilisks host to *P. achiotense* were collected from secondary moist tropical forest areas in which cultivated crops or dwellings sparsely occurred. The parasite was not found in lizards collected along densely shaded forest streams, although these often were parasitized by *P. basilisci.*

Plasmodium beltrani Peláez and Pérez-Reyes, 1952 (**Plate 14**)

Diagnosis A *Plasmodium* (*Sauramoeba*) species characterized by variably shaped, often halteridial, meronts 9–15 × 5–11 µm that produce 23–48 merozoites, with LW 63–132 µm^2. Size relative to host cell nucleus is 1.6–3.8, and to normal erythrocyte nuclei is 2.7–5.5. Gametocytes are 6–17 × 3–9 µm, with LW 30–117 µm^2 and L/W 1.0–3.5, usually elongate or bulky, often halteridial, curving around the host cell nucleus. A prominent vacuole is present in most gametocytes of both sexes. The dark brown pigment granules, clustered in one or two sites in meronts, are dispersed as 30–40 fine granules in gametocytes. Gametocyte dimensions are not sexually dimorphic.

Type Host *Sceloporus v. variabilis* Smith (Sauria: Phrynosomatidae).

Type Locality Soyaltepec, Municipio Tuxtepec, Oaxaca, Mexico.

Other Hosts *Sceloporus teapensis*; *Sceloporus malachiticus taeniocnemis.*

Other Localities Tuxtlas region, Veracruz, Mexico; 4.3 km north of Rincon Chamula, Chiapas, Mexico.

Prevalence One of six *S. malachiticus* was parasitized by *P. beltrani.*

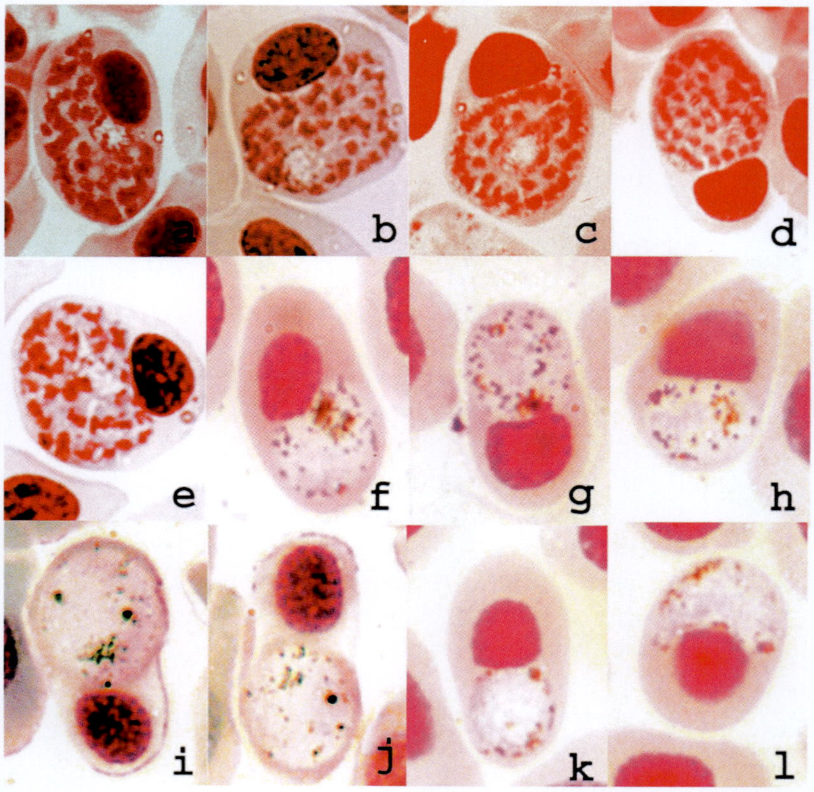

(A)

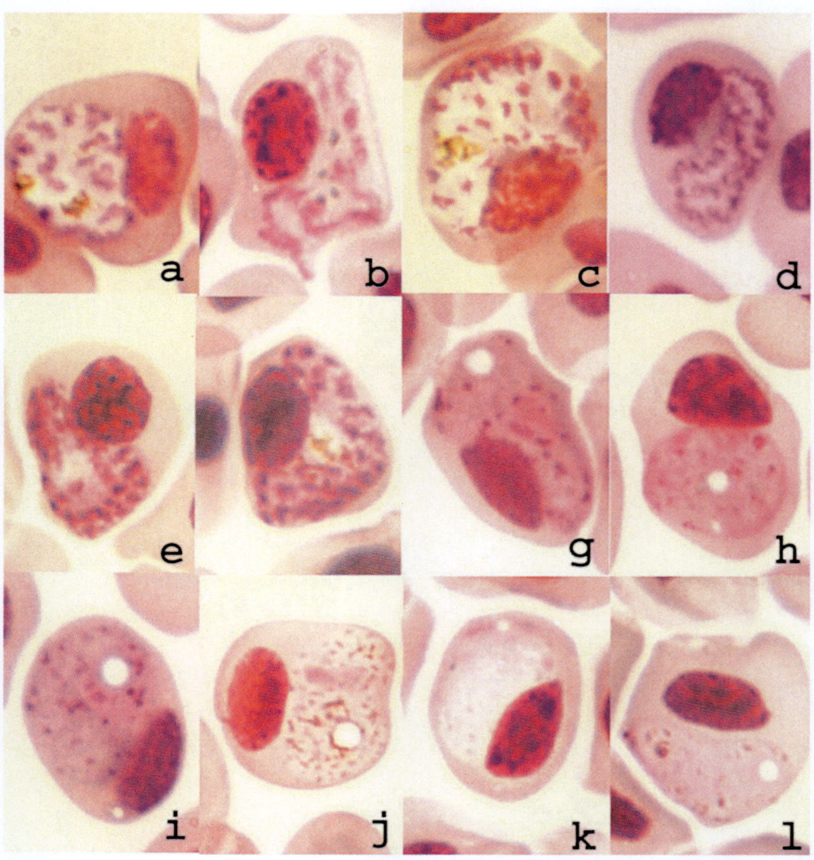

(B)

Plate 14 (**A**) *Plasmodium achiotense* from *Basiliscus basiliscus*, Panama. Meronts, **a–e**; macrogametocytes, **f–h**; microgametocytes, **i–l**. (Figures **a** and **c** from Telford, S. R., Jr., *J. Protozool.*, 19, 77, 1972, with permission, Blackwell Publishing.) (**B**) *Plasmodium beltrani* from *Sceloporus malachiticus*, Mexico. Meronts, **a–f**; macrogametocytes, **g–j**; microgametocytes, **k**, **l**.

Morphological Variation In *S. variabilis*, *P. beltrani* meronts produce 25.6 merozoites on average (Peláez and Pérez-Reyes, 1952), and in *S. malachiticus* 34.7 ± 6.6 (N = 15), with average meront dimensions in the latter host 11.9 ± 1.6 × 8.3 ± 1.7 μm and LW 98.4 ± 19.9 μm². Size relative to host cell nucleus is 2.71 ± 0.63, and to normal erythrocyte nuclei is 4.13 ± 0.81. A single meront seen in the chronic infection of *S. teapensis* was 13 × 7 μm and contained about 24 merozoites. Macrogametocytes in *S. variabilis* averaged 14.5 × 5.0 μm (12–17 × 3.5–5), and microgametocytes were 15.3 × 3.6 μm (13–17 × 3.0–4.5) (Peláez and Pérez-Reyes, 1952). In *S. teapensis*, macrogametocytes are 12.8 ± 1.3 × 6.1 ± 1.7 μm (10–15 × 5–9, N = 25), with LW 77.6 ± 14.5 μm² (60–108) and L/W 2.17 ± 0.46 (1.3–3.0). Microgametocytes in *S. teapensis* are 11.8 ± 1.6 × 4.6 ± 0.6 μm (6–14 × 4–6, N = 25), with LW 54.6 ± 11.5 μm² (30–84) and L/W 2.5 ± 0.48 (1.2–3.5). Both sexes together, relative to the host cell nucleus, are 2.31 ± 0.69 (1.0–4.4, N = 50), and to normal erythrocyte nuclei are 2.79 ± 0.73 (1.3–4.6). Macrogametocytes in *S. malachiticus* are 11.9 ± 1.4 × 7.2 ± 1.2 μm (9–15 × 5–9, N = 25), with LW 85.1 ± 12.4 μm² (65–117) and L/W 1.71 ± 0.41 (1.1–2.6). Microgametocytes in *S. malachiticus* are 10.7 ± 1.6 × 7.2 ± 1.1 μm (8–14 × 5–9, N = 25), with LW 76.6 ± 12.4 μm² (55–96) and L/W 1.53 ± 0.43 (1.0–2.8). Size of both sexes together, relative to host cell nucleus, is 2.42 ± 0.51 (1.4–4.0, N = 49), and to normal erythrocyte nuclei is 3.4 ± 0.58 (2.3–5.0, N = 50).

Exoerythrocytic Merogony Not observed.

Effects on Host Although not stated by Peláez and Pérez-Reyes (1952), *P. beltrani* appears to parasitize erythrocytes in *S. variabilis*, as it does in the other two host species. In *S. teapensis*, host erythrocytes and their nuclei are shorter than normal erythrocytes, but host cell width does not differ, and normal erythrocyte nucleus width is greater. Erythrocytes of *S. malachiticus* parasitized by *P. beltrani* are also shorter than uninfected cells but have greater width, and their nuclei are enlarged. In both host species, erythrocytes parasitized by gametocytes and their nuclei are distorted, and the latter are displaced. Meronts of *P. beltrani* affect host cells similarly to gametocytes, but shorten erythrocyte and nucleus length and cause hypertrophy in width of cell and nucleus dimensions. Both meronts and gametocytes usually lie lateral to the erythrocyte nucleus but are occasionally polar to it, curve around it, or even nearly fill the host cell.

Sporogony Unknown.

Remarks The slide of *P. beltrani* in *S. teapensis* was prepared by Professor D. Paláez and given to Professor J. V. Scorza, who gave it to me in 1974.

Plasmodium kentropyxi Lainson, Landau and Paperna 2001

Diagnosis A *Plasmodium* (*Sauramoeba*) species with large meronts, at least 14.5 × 6.5 μm, with estimated LW 94.3 μm², that produce 40–50 merozoites. Estimated meront size relative to host cell nucleus is 3.05, and to normal erythrocyte nuclei is 3.35. Meronts are semispherical to elongate and may curve around the host cell nucleus. Coarse greenish-black pigment granules are clustered in a distinct cytoplasmic vacuole. Elongate gametocytes contain a prominent vacuole at one end and are 9.6–15.0 × 3.6–9.8 μm, with L/W 2.2–5.0 and estimated LW 40–72 μm². Estimated gametocyte size relative to host cell nucleus is 2.17, and to normal erythrocyte nuclei is 1.90. Gametocytes do not differ in dimensions by sex. Fine greenish-black pigment granules are dispersed throughout gametocyte cytoplasm but tend to concentrate around the vacuole. Macrogametocytes are larger than microgametocytes, but the sexes are similar in shape.

Type Host *Kentropyx calcarata* Spix (Sauria: Teiidae).

Type Locality Outeiro Island, near Icoraci, Pará State, Brazil.

Other Hosts None known.

Other Localities Capanema and Belém, Pará State, Brazil (Lainson et al., 2001).

Prevalence Overall, *P. kentropyxi* infected 102 of 151 (67.5%) *K. calcarata*: in Outeiro, 31 of 39 (79.5%); in Capanema, 36 of 55 (65%); and in Belém, 35 of 57 (61.4%) (Lainson et al., 2001).

Morphological Variation Meronts are semispherical to elongate, usually "lateral in position, sometimes occupying almost the entire length of the erythrocyte and may curve round its nucleus" (Lainson et al., 2001). Size of meronts, estimated from a figure, is about 14.5 × 6.5 μm, with LW approximately 94.3 μm². "The largest meronts seen contained approximately 40–50 nuclei." Coarse, greenish-black pigment granules are clustered in a cytoplasmic vacuole. Estimated meront size relative to host cell nucleus is 3.05, and to normal erythrocyte nuclei is 3.35. Gametocytes are elongate and contain a prominent vacuole as large as 2–3 μm in diameter. Gametocytes are 9.6–15.0 × 3.6–4.8 μm, with L/W 2.9 (2.2–5.0) and estimated LW 40–72 μm². Estimated gametocyte size relative to host cell nucleus is 2.17, and to normal erythrocyte nuclei is 1.90. Gametocyte dimensions indicate that macrogametocytes are larger than microgametocytes; their respective dimensions are

13.5 × 4.5 µm (12.0 × 4.5–15.0 × 4.8m, N = 50), estimated LW 60.8 µm² (54–72) compared to 11.8 × 4.0 µm (9.6 × 4.2–13.2 × 3.6, N = 50), estimated LW 47.2 µm² (40.3–47.5). Shape is not sexually dimorphic. Fine greenish-black pigment granules are dispersed but tend to be concentrated around the large cytoplasmic vacuole of gametocytes.

Exoerythrocytic Merogony Unknown.

Sporogony Unknown.

Effects on Host Erythrocytes host to either meronts or gametocytes are scarcely enlarged, if at all. Meronts usually displace host cell nuclei but cause little distortion apart from slight indentation and rounding. The nuclei of cells host to gametocytes are displaced to one side and considerably flattened. Infected lizards showed no ill effects from being parasitized by *P. kentropyxi* (Lainson et al., 2001).

Remarks In four of the five *Plasmodium* species described or reported from *K. calcarata*, prominent cytoplasmic vacuoles are usually or always present in gametocytes—in *P. kentropyxi* and a *P. tropiduri*-like species (Lainson et al., 2001), and in *P. pifanoi* and *P. minasense calcaratae* (Telford and Telford, 2003). They are absent only in *P. lepidoptiformis*. This rather consistent presence of vacuoles is probably just coincidence but does suggest a possible influence by the host on parasite appearance. Regrettably, the otherwise useful description of *P. kentropyxi* does not furnish dimensional data for meronts.

NEOTROPICAL *CARINAMOEBA* SPECIES

Plasmodium attenuatum Telford 1973 (**Plate 15**)

Diagnosis A *Plasmodium* (*Carinamoeba*) species with meronts 3–6 × 2–4 µm, with LW 8–24 µm², containing three to eight merozoites arranged as a fan or rosette and elongate gametocytes 6–11 × 2–5 µm, LW 14–58 µm², and L/W 1.4–4.5. Meront size relative to host cell nucleus size averages 0.5–0.7, and to normal erythrocyte nucleus size is 0.5–0.8. Light golden pigment in meronts is scanty and loosely aggregated. Gametocyte size relative to host cell nucleus size is 0.8–1.1, and to normal erythrocyte nucleus size is 1.0–1.2. Pigment is dispersed in both gametocyte sexes as dark granules but when aggregated into a mass is golden yellow. Trophozoites and immature meronts are elongate. Gametocytes do not differ in dimensions by sex.

Type Host *Ameiva a. ameiva* (Linnaeus) (Sauria: Teiidae).

Type Locality Vicinity of Georgetown, Guyana.

Other Hosts *Ameiva ameiva praesignis.*

Other Localities Venezuelan localities are Portuguesa State: Municipios Piritu and Araure; Cojedes State, Municipio Cojedes; Guarico State, Municipio Ortiz (Telford, 1980); Aragua State, El Limon (Scorza, 1970, as *P. basilisci*).

Prevalence *P. attenuatum* was detected in 1 of 16 (6.3%) by blood inoculation from Guyana *A. ameiva* into an uninfected host (Telford, 1973b). Overall prevalence in Venezuelan *A. ameiva* was 28 of 131 (21.4%), with prevalence by municipio at 11 of 64 (17.2%) in Araure, 10 of 30 (33.3%) in Piritu, 2 of 13 (15.4%) in Cojedes, and 5 of 21 (23.8%) in Ortiz (Telford).

Morphological Variation In the type infection from Guyana, meronts are 4.3 ± 0.5 × 3.7 ± 0.5 µm (4–6 × 3–4, N = 34), with LW 15.4 ± 4.1 µm² (12–24), and merozoite number 6.0 ± 1.0 (4–8, N = 35). Merozoites are mostly arranged as a fan and occasionally as a rosette. Meront size relative to host cell nucleus size is 0.65 ± 0.19 (0.34–1.11, N = 32), and to normal erythrocyte nucleus size is 0.66 ± 0.19 (0.34–1.11, N = 33). Pigment forms a prominent, often round, light golden mass. Gametocytes in Guyana hosts are 7.5 ± 0.7 × 2.6 ± 0.7 µm (6–9 × 2–4, N = 25), with LW 19.8 ± 5.4 µm² (14–36) and L/W 3.05 ± 0.86 (1.75–4.50). Gametocyte size relative to host cell nucleus size is 0.82 ± 0.20 (0.57–1.20), and to normal erythrocyte nucleus size is 1.00 ± 0.27 (0.71–1.83). Most of the pigment is dispersed as small granules, but a small mass of golden pigment often is present as well. In Venezuelan *A. ameiva*, meronts are 4.3 ± 0.6 × 3.0 ± 0.5 µm (3–6 × 2–4, N = 34), with LW 13.0 ± 2.7 µm² (8–20), and contain 4.6 ± 1.0 (3–8) merozoites. Meront size relative to host cell nucleus size is 0.53 ± 0.15 (0.33–1.00, N = 30), and to normal erythrocyte nucleus size is 0.54 ± 0.13 (0.31–0.85, N = 34). Gametocytes are larger than in Guyana and less elongate, 8.2 ± 1.3 × 3.8 ± 0.8 µm (6–11 × 3–5, N = 20), with LW 31.4 ± 9.6 µm² (18–55) and L/W 2.20 ± 0.44 (1.40–3.00). Gametocyte size relative to host cell nucleus size is 1.14 ± 0.34 (0.60–1.83), and to normal erythrocyte nucleus size is 1.23 ± 0.36 (0.73–2.08). Pigment granules when dispersed are dark, but when adhering together are golden yellow. Gametocytes do not differ in dimensions or shape by sex. The infection phase was probably responsible for the observed differences in merozoite numbers and gametocyte size and shape between the samples from Guyana and Venezuela (Telford, 1980).

Exoerythrocytic Merogony Unknown.

Sporogony Unknown.

Plate 15 (**A**) *Plasmodium attenuatum* from *Ameiva ameiva*, Guyana, **a–d**, **i**, **j**, and Venezuela, **e–h**, **k**, **l**. Meronts, **a–h**; macrogametocytes, **i**, **j**; microgametocytes, **k**, **l**. (Figure **k** from Telford, S. R., Jr., *Int. J. Parasitol.,* 10, 365, 1980, with permission, Elsevier.) (**B**) *Plasmodium lepidoptiformis* from *Kentropyx calcarata*, Venezuela. Meronts, **a–f**; macrogametocytes, **g–i**; microgametocytes, **j–l**.

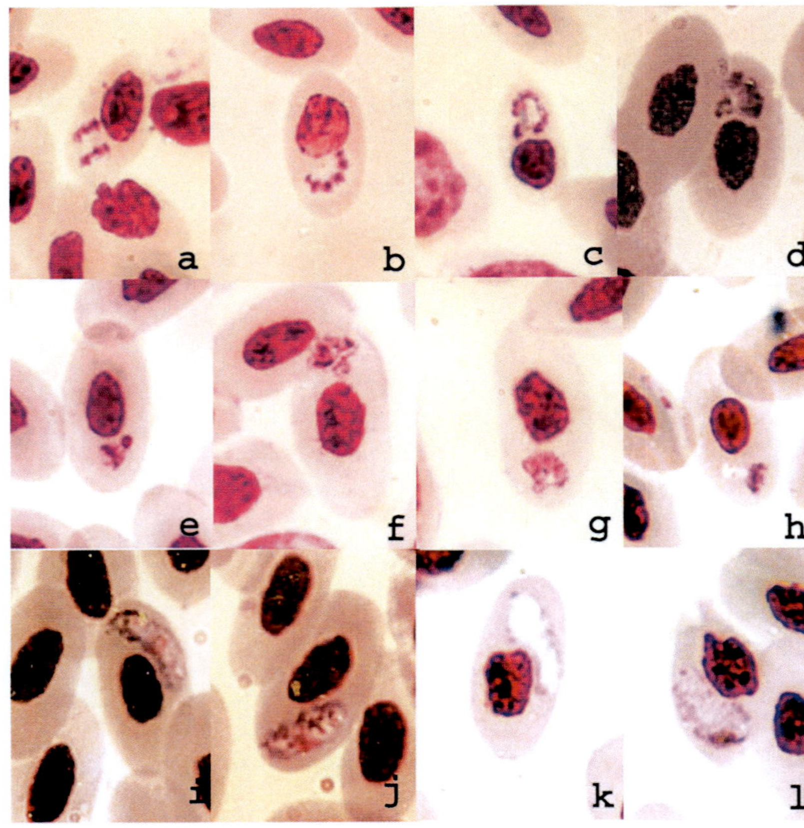

(A)

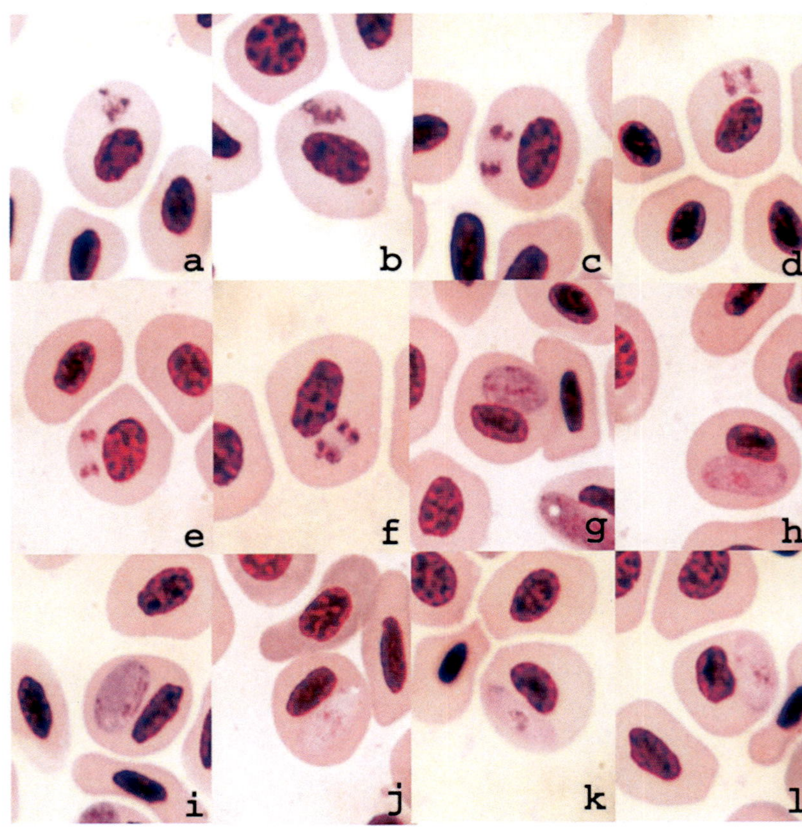

(B)

Effects on Host Erythrocytes host to meronts and gametocytes of both Guyanan and Venezuelan strains, and their nuclei, were not hypertrophied, and few were distorted, although nuclear displacement was occasionally seen.

Remarks The Guyana infection appeared in an experimental *A. ameiva* 35 days following inoculation with blood from a lizard supposedly infected only by *P. cnemidophori* and *P. telfordi* (Telford, 1973). *Plasmodium attenuatum* was the most common species found in Venezuelan *A. ameiva* (21.4%, overall), often in mixed infection with *P. cnemidophori* (5.3%), *P. telfordi* (3.8%), or both (1.5%). It also appears to be the species reported by Scorza (1970a) as *Plasmodium basilisci* (Telford, 1980).

Plasmodium lepidoptiformis Telford and Telford 2003 (Plate 15)

Diagnosis A *Plasmodium* (*Lacertamoeba*) species with tiny meronts 3–6 × 2.5–5 µm, LW 9–25 µm^2, that contain four to eight merozoites most often arranged as a butterfly, cruciform, or fan-shaped when few in number or as a rosette when more merozoites are present. Meront size relative to host cell nucleus size averages 0.51, and to normal erythrocyte nucleus size is 0.69. Gametocytes are 7–10 × 3–6 µm, LW 24–51 µm^2, and usually elongate, with L/W 1.27–3.33. Gametocyte size relative to host cell nucleus size averages 1.63, and to normal erythrocyte nucleus size is 1.76. Macrogametoycytes are longer than microgametocytes, but sexes are similar in other dimensions and L/W ratio. Pigment is sparse in all stages, forming as one or two barely visible granules in meronts and dispersed as several inconspicuous, dark grayish-golden granules in macrogametocytes, occasionally forming small golden clumps in microgametocytes.

Type Host *Kentropyx calcarata* Spix (Sauria: Teiidae).

Type Locality Mision Padomo, Territorio Federal de Amazonas, Venezuela.

Other Hosts Unknown.

Other Localities Unknown.

Prevalence One of four *K. calcarata* was infected by *P. lepidoptiformis*.

Morphological Variation The smaller meronts of *P. lepidoptiformis* usually show a cruciform arrangement of nuclei that most commonly become constricted in the middle to form a butterfly or bowtie shape. If three nuclear divisions occur, then the resulting merozoites usually form a rosette and rarely are elongate or round. Meronts are 4.6 ± 0.6 × 3.2 ± 0.6 µm (3–6 × 2.5–5, N = 27), with LW 14.9 ± 3.2 µm^2 (9–25), that contain 5.1 ± 1.4 (4–8) merozoites. Meront size relative to host cell nucleus size is 0.51 ± 0.11 (0.33–0.77), and to normal erythrocyte nucleus size is 0.69 ± 0.15 (0.41–1.15). Gametocytes are usually elongate, 9.0 ± 0.9 × 4.3 ± 0.6 µm (7–10 × 3–6, N = 50), with LW 38.3 ± 5.6 µm^2 (24–51) and L/W 2.16 ± 0.41 (1.27–3.33). Gametocyte size relative to host cell nucleus size is 1.63 ± 0.33 (1.14–2.16, N = 26), and to normal erythrocyte nucleus size is 1.76 ± 0.26 (1.12–2.34, N = 50). Macrogametocytes are longer than microgametocytes, but other dimensions are similar, respectively 9.2 ± 0.9 × 4.0 ± 0.5 µm, LW 36.6 ± 5.8 µm^2, and L/W 2.33 ± 0.41 versus 8.9 ± 0.8 × 4.5 ± 0.5 µm, LW 40.1 ± 4.6 µm^2, and L/W 1.98 ± 0.32 (N = 25 each). Pigment is sparse and inconspicuous in both meronts and gametocytes. There are usually one or two tiny golden granules in meronts, centrally located, and several dark grayish-golden granules dispersed in macrogametocytes. In microgametocytes, small clumps of golden granules may be present in addition to a few dispersed darker granules.

Exoerythrocytic Merogony Unknown.

Sporogony Unknown.

Effects on Host Erythrocytes host to meronts are not enlarged but may appear more rounded. Their nuclei are hypertrophied, noticeably in their width. Gametocytes always distort host erythrocytes and usually displace the nucleus but do not distort the latter. Host erythrocytes are shorter in length but greater in width, without affecting erythrocyte LW. Host cell nuclei are greater in width and LW but shorter in length than nuclei of uninfected cells.

Remarks Although the tiny butterfly-shaped meronts are more similar to *Carinamoeba* species in their size and merozoite production, gametocyte size suggests a likely relationship to *Lacertamoeba* species instead.

Plasmodium scelopori Telford 1977 (Plate 16)

Diagnosis A *Plasmodium* (*Carinamoeba*) species with meronts 4–8 × 3–6 µm, LW 12–48 µm^2, that contain four to ten merozoites usually arranged as a fan or rosette. Meront size relative to host cell nucleus size averages 0.7, and to normal erythrocyte nucleus size is 0.7. Gametocytes are usually rounded, 4–9 × 3–6 µm, with LW 16–45 µm^2 and L/W 1.00–2.25. Gametocyte size relative to host cell nucleus

Plate 16 (**A**) *Plasmodium scelopori* from *Sceloporus teapensis*, Belize. Meronts, **a–f**; macrogametocytes, **g**, **h**; microgametocytes, **i–l**. (**B**) *Plasmodium marginatum* from *Anolis frenatus*, Panama. Meronts, **a–f**; macrogametocytes, **g–j**; microgametocytes, **k**, **l**.

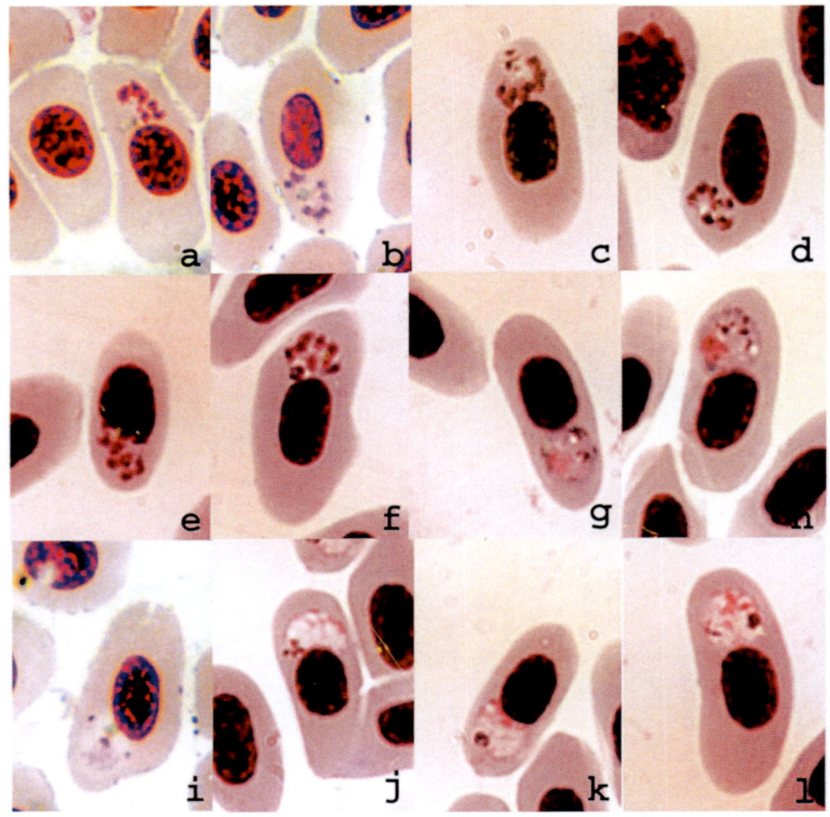

(A)

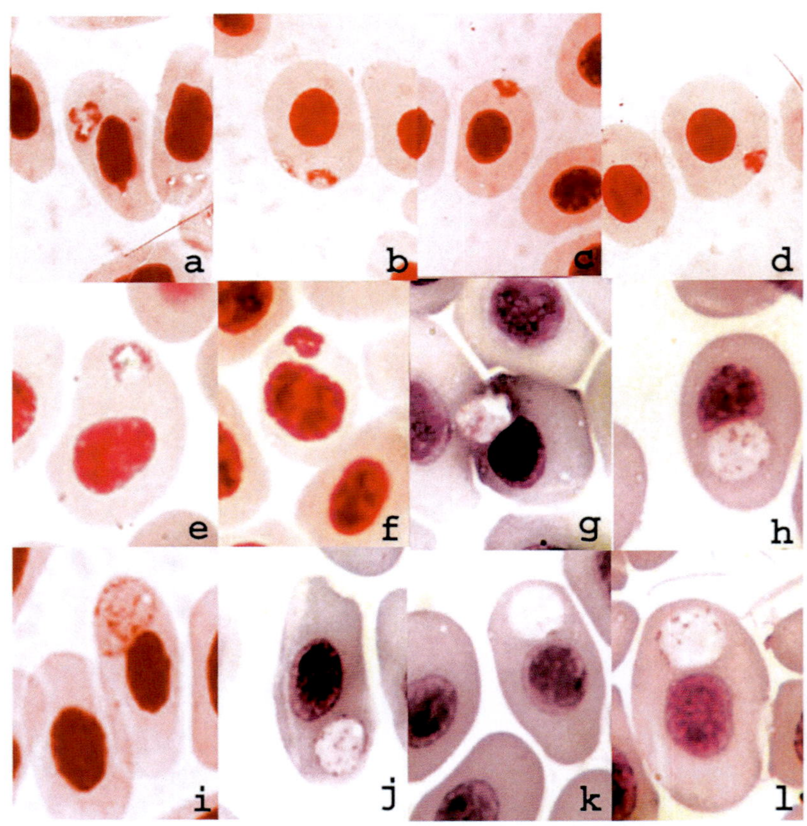

(B)

size averages 1.08, and to normal erythrocyte nucleus size is 1.06. Microgametocytes are larger than macrogametocytes in active infections. Pigment in meronts forms a greenish-yellow mass, central in rosettes and at the base of fans. In gametocytes, greenish-yellow pigment aggregates in one or two clumps, often with dispersed dark green granules around the clumps. Both meronts and gametocytes are nucleophilic within their host erythrocytes.

Type Host *Sceloporus teapensis* Günther (Sauria: Phrynosomatidae).

Type Locality Blancaneaux Lodge, 18 km north of Augustine, Belize.

Other Hosts *Sceloporus variabilis.*

Other Localities Gunacaste Province, Costa Rica, and Siguatepec, Honduras.

Prevalence At the type locality, 4 of 13 (30.8%) *S. teapensis* were infected. *Plasmodium scelopori* infected 4 of 44 (9.1%) *S. variabilis* from Costa Rica and 2 of 15 (13.3%) in Honduras.

Morphological Variation In *S. teapensis*, meronts are 4.9 ± 1.0 × 3.9 ± 1.6 μm (4–8 × 3–6, N = 40), with LW 19.6 ± 7.4 μm² (12–48), and produce 6.9 ± 1.6 (4–10) merozoites. Meront size relative to host cell nucleus size is 0.74 ± 0.33 (0.38–1.92, N = 37), and to normal erythrocyte nucleus size is 0.74 ± 0.33 (0.42–2.00, N = 40). Merozoites usually form as fans or rosettes, with a greenish-yellow pigment mass at the base of the fan or centrally located in rosettes. Gametocytes are 5.8 ± 1.1 × 4.8 ± 0.6 μm (4–9 × 3–6, N = 75), with LW 27.9 ± 6.8 μm² (16–45) and L/W 1.08 ± 0.29 (0.63–1.88). Gametocyte size relative to host cell nucleus size is 1.08 ± 0.29 (0.63–1.88), and to normal erythrocyte nucleus size is 1.06 ± 0.33 (0.55–1.88). In active infection, microgametocytes are larger than macrogametocytes, 5.5 ± 0.5 × 5.0 ± 0.5 μm (N = 25), LW 27.2 ± 4.1 μm², versus 5.0 ± 0.5 × 4.4 ± 0.5 μm (N = 25), LW 21.9 ± 4.7 μm², respectively, but in chronic infection phase, sexual difference in dimensions is not present. Gametocytes, however, are larger than in active phase: Microgametocyte LW is 34.3 ± 5.0 μm² (N = 12), and macrogametocyte LW is 35.7 ± 4.3 μm² (N = 13). Shape does not differ between sexes in either infection phase. Pigment forms in one or two greenish-yellow clumps, with small dark green granules dispersed around them. In chronic infections from Costa Rican *S. variabilis*, gametocytes are 6.3 ± 1.0 × 4.6 ± 0.7 μm (4–9 × 3–6, N = 50), with LW 28.7 ± 6.4 μm² (16–42) and L/W 1.41 ± 0.33 (1.00–2.33). Gametocyte size relative to host cell nucleus size is 0.97 ± 0.22 (0.56–1.39, N = 49), and to normal erythrocyte nucleus size is 0.97 ± 0.22 (0.56–1.39, N = 50). Microgametocytes are larger than macrogametocytes, 6.8 ± 2.3 × 4.8 ± 0.7 μm (N = 25), LW 32.2 ± 5.1 μm², and L/W 1.46 ± 0.33, versus 5.8 ± 0.8 × 4.4 ± 0.7 μm, LW 25.1 ± 5.6 μm², and L/W 1.36 ± 0.32, respectively. They are slightly more elongate than are gametocytes in *S. teapensis*. Microgametocytes in both hosts have rough or pitted surfaces, but the cell outline is smooth.

Exoerythrocytic Merogony Unknown.

Sporogony Unknown.

Effects on Host From the first nuclear division, meronts are nucleophilic and when mature are always polar in position. Host erythrocytes are neither hypertrophied nor distorted; their nuclei are sometimes displaced and rarely distorted by contact with the meront. Gametocytes affected host erythrocytes similarly, but more often displaced nuclei, especially in *S. variabilis,* in which lateral and lateropolar positions were assumed, in contrast to the polar position of gametocytes in *S. teapensis*. Nucleophily is characteristic for gametocytes in both host species.

Remarks Nucleophily is uncommon among neotropical *Plasmodium* species, recorded only from *P. sceloporí, P. minasense tegui, P. brumpti, P. diploglossi*, and meronts of *P. beltrani*. The rough, pitted surface of microgametocytes is also present in *P. brumpti*, but it has much larger meronts and gametocytes than *P. sceloporí*.

Plasmodium marginatum Telford 1979 (**Plate 16**)

Diagnosis A *Plasmodium* (*Carinamoeba*) species in which all stages lie along the cell margin. Immature asexual stages are highly amoeboid with prominent, elongate cytoplasmic processes. Meronts usually form as a flattened fan, 3–6 × 2–4 μm, with LW 6–24 μm², and contain three to eight merozoites. Meront size relative to host cell nucleus size averages 0.31, and to normal erythrocyte nucleus size is 0.32. Pigment is modest in quantity, light golden to dark when clumped. Gametocytes are round to oval, 5–9 × 4–7 μm, with LW 20–54 μm² and L/W 1.0–2.0, with no sexual difference in dimensions. Gametocyte size relative to host cell nucleus size averages 0.71, and to normal erythrocyte nucleus size is 0.67.

Type Host *Anolis frenatus* Cope (Sauria: Polychrotidae).

Type Locality Frijoles River, 5 km north of Gamboa, Canal Zone, Panama.

Other Hosts None known.

Other Localities Frijolito Creek, Quebrada Juan Grande, and Barro Colorado Island in Gatun Lake, all in the vicinity of the type locality.

Prevalence Overall, *P. marginatum* was present in 13 of 26 (50%) of *Anolis frenatus.*

Morphological Variation Larger trophozoites and all meronts are elongate, often with filiform cytoplasmic processes at one or both ends, and are almost all located along the cell margin. Meronts often appear to be flattened against the margin. Meronts are 4.4 ± 0.8 × 3.0 ± 0.6 μm (3–6 × 2–4, N = 96), with LW 13.0 ± 3.8 μm^2 (6–24), and produce 5.7 ± 1.5 (3–8, N = 99) merozoites. Meront size relative to host cell nucleus size is 0.31 ± 0.10 (0.17–0.69, N = 83), and to normal erythrocyte nucleus size is 0.32 ± 0.08 (0.17–0.52, N = 96). Meronts usually form fans but may appear as rosettes, a morulum, or in cruciform shape. Light golden-to-dark pigment clumps are obvious but modest in quantity. Gametocytes are 6.4 ± 0.8 × 4.7 ± 0.8 μm (5–9 × 4–7, N = 74), with LW 29.8 ± 6.7 μm^2 (20–54) and L/W 1.39 ± 0.27 (1.00–2.00). Gametocyte size relative to host cell nucleus size is 0.71 ± 0.18 (0.44–1.17, N = 68), and to normal erythrocyte nucleus size is 0.67 ± 0.17 (0.44–1.29, N = 74). Gametocytes lack sexual differences in dimensions. Small dark pigment granules are dispersed within their cytoplasm. Fewer gametocytes (32%) lie along the cell margin than do meronts (81%).

Exoerythrocytic Merogony Unknown.

Sporogony Unknown.

Effects on Host Both meronts and gametocytes cause hypertrophy of host cells, but their nuclei remain normal in size except in crisis-phase infections, when cells parasitized by meronts show hypertophied nuclei. Erythrocytes host to either stage commonly are distorted, their nuclei less often so, but the latter are sometimes displaced.

Remarks *Plasmodium marginatum* is the only neotropical *Plasmodium* species in which most asexual stages and about one-third of gametocytes are marginal in their position within cells. The host species, *Anolis frenatus*, is a canopy species but descends lower on the tree trunk in the late dry season.

Plasmodium rhadinurum Thompson and Huff 1944 (**Plate 17**)

Diagnosis A *Plasmodium* (*Carinamoeba*) species with cruciform or fan-shaped meronts, 3–5 × 3–5 μm, LW 9–25 μm^2, that contain four to six merozoites. Meront size relative to host cell nucleus size averages 0.77, and to normal erythrocyte nucleus size is 0.66. Trophozoites and young meronts usually have prominent and long filiform cytoplasmic projections, the frequency of which may vary geographically. Gametocytes are spherical or slightly ovoid, 5–8 × 4–7 μm, with LW 20–49 μm^2 and L/W 1.00–1.40. Gametocyte size relative to host cell nucleus size averages 1.47, and to normal erythrocyte nucleus size is 1.49. Pigment is sparse in meronts, usually a single brownish-gold granule, but forms as 35 or more small blackish granules dispersed in gametocytes. Gametocytes are not sexually dimorphic in size or shape.

Type Host *Iguana iguana rhinolopha* Wiegmann (Sauria: Iguanidae).

Type Locality Colima, Mexico.

Other Hosts *Iguana i. iguana* (Scorza, 1970; Telford, 1977, 1980); *Ctenosaura similis* (Garnham, 1966); *Sceloporus undulatus* as experimental host (Thompson and Huff, 1944b).

Other Localities Brazil: Goiás (Carini, 1945), Codajaz, Amazonas (Walliker, 1966); Venezuela: La Caimana, Guarico State (Scorza, 1970); Portuguesa State, Municipios Araure and Guanare (Telford, 1980); Panama: Sanbu River, Darien Province and Achiote, Colon Province (Telford, 1977); Choluteca, El Culito, Honduras (Telford); Belize (Garnham, 1966). In Mexico: Acapulco and Zihuatanejo, Guerrero; Apatzingán, Michoacan; Tuxtepec, Oaxaca; Nayarit (Peláez and Perez-Reyes, 1952).

Prevalence In Panama, prevalence of *P. rhadinurum* was 9 of 29 (31%) overall, including 4 of 5 from Sanbu River and 3 of 3 from Achiote (Telford, 1977); in Venezuela, 7 of 69 (10.1%) overall, 6 of 54 (11.1%) from Araure, 1 of 2 from Guanare (Telford), and 9 of 11 (81.8%) from Guarico (Scorza, 1970).

Morphological Variation In the description of *P. rhadinurum*, Thompson and Huff (1944b) reported the presence of "1 or 2 long cytoplasmic processes" in "a large proportion" of trophozoites and meronts with four or five merozoites "arranged in a cross or fan-shaped configuration." Gametocytes were 6.5–7.7 μm in diameter. Topotypic material that I obtained from Colima, Mexico, in 1965 from three *Iguana iguana* showed filamentous cytoplasmic processes in 75.2% and 60.0% in samples of 125 and 25 parasites, respectively. Meronts were 4.4 ± 0.6 × 3.9 ± 0.7 μm (3–5 × 3–5, N = 50), with LW 17.2 ± 4.5 μm^2 (9–25), and contained 4.2 ± 0.4 (4–6) merozoites. Meront size relative

Plate 17 (**A**) *Plasmodium rhadinurum* from *Iguana iguana*, Mexico, **a**, **c–e**, **i–l**, and Venezuela, **b**, **f–h**. Young meronts, **a**, **b**; meronts, **c–h**; macrogametocytes, **i**, **j**; microgametocytes, **k**, **l**. (**B**) *Plasmodium minasense carinii* from *Iguana iguana*, Colombia, **a–d**; and Trinidad, **e–l**. Meronts, **a–h**; gametocytes, **i–l**. (Figures **e–l** from C. M. Wenyon slide, 1915.)

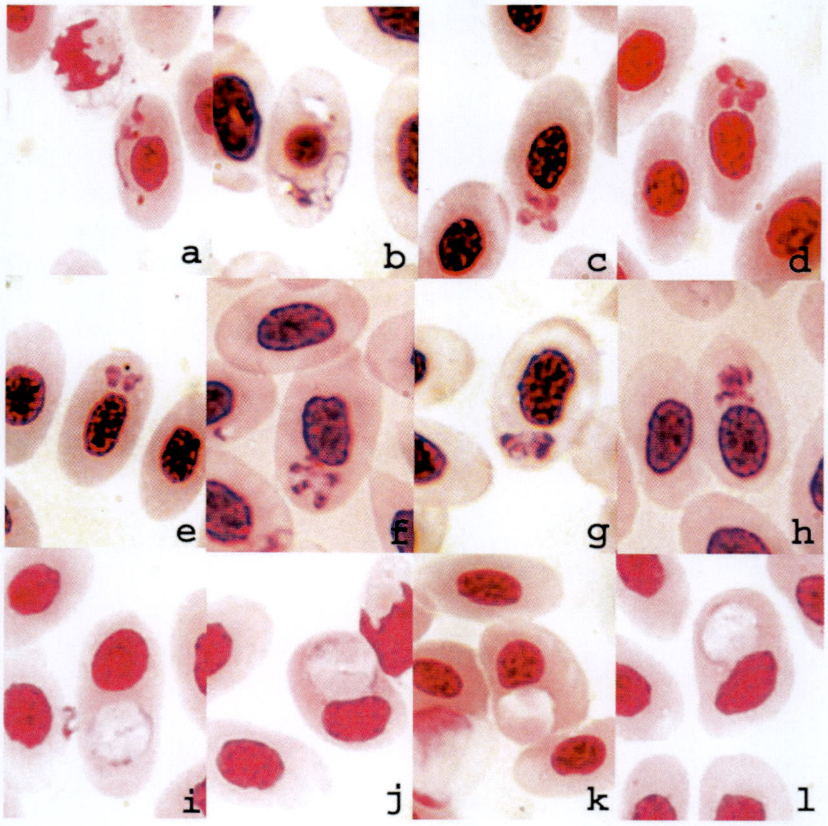

(A)

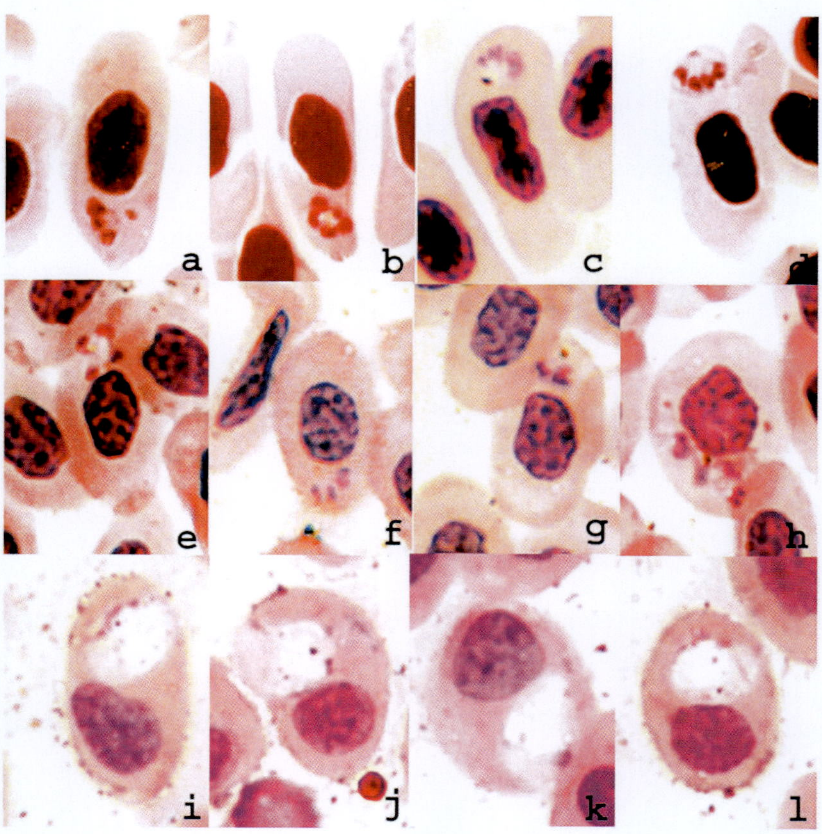

(B)

to host cell nucleus size was 0.77 ± 0.22 (0.38–1.39), and to normal erythrocyte nucleus size was 0.66 ± 0.17 (0.34–0.45). Merozoites were arranged in a cruciform manner (86%) or as fans (14%). Pigment formed as a single, brownish-gold granule at base of fans or centrally in cruciform meronts. Gametocytes were 6.7 ± 0.7 × 5.8 ± 0.6 μm (5–8 × 4–7, N = 34), with LW 39.0 ± 6.3 μm² (20–49) and L/W 1,17 ± 0.11 (1.00–1.40). Gametocyte size relative to host cell nucleus size was 1.47 ± 0.31 (1.00–2.10, N = 33), and to normal erythrocyte nucleus size was 1.49 ± 0.24 (0.76–1.87, N = 34). Pigment was dispersed as small dark granules in both sexes (numbering 35 or more according to Thompson and Huff, 1944).

In Venezuelan *I. iguana*, young asexual parasites showed much lower prevalence of the characteristic filopodia, 6–26% in three infections, overall 16.8% in samples ranging from 25 to 50 parasites. Scorza (1970a) described the filopodia in trophozoites or young meronts as "often ramifying, may attain four to five times the length of the body of the trophozoite or the young meront." Meronts in his material from Guarico were 4 × 4 or 4 × 3 μm, with usually four merozoites (71.4%), and, rarely, six (2.4%). Scorza described meront shape as a "square around a central pigment granule," that is, cruciform, or as fan-shaped. Gametocytes were 6–7.5 μm in diameter, abundant in one infection, but with fewer pigment granules than in the type description, only "10–15 heavy grains, uniformly dispersed." Meronts of *P. rhadinurum* in an iguana from Guanare are 3.8 ± 0.8 × 2.7 ± 0.5 μm (3–5 × 2–3, N = 25), LW 10.3 ± 2.5 μm² (6–15), and contain 4.3 ± 0.5 (4–6) merozoites. Meront size relative to host cell nucleus size is 0.38 ± 0.14 (0.20–0.75, N = 18), and to normal erythrocyte nucleus size is 0.40 ± 0.10 (0.23–0.58, N = 25). Merozoites were arranged primarily as fans (72%), in comparison to cruciform arrangement (24%). Gametocytes were rare in the Guanare infection, only five being found. These averaged 6.6 ± 0.5 × 5.8 ± 0.8 μm (6–7 × 5–6), with LW 37.2 ± 8.2 μm² (30–49) and L/W 1.19 ± 0.14 (1.00–1.40). Their size relative to normal erythrocyte nucleus size averaged 1.44 ± 0.32 (1.16–1.90).

Exoerythrocytic Merogony In the Colima infections, one meront was seen in a thrombocyte.

Sporogony Unknown.

Effects on Host Meronts of *P. rhadinurum* have no effect on host erythrocytes or their nuclei except for hypotrophy of the cell by about 16% in the Colima sample, but only 7% in the Venezuelan material. Most cells host to gametocytes are distorted and their nuclei displaced, with nearly 40% of the nuclei distorted as well. Cells host to gametocytes are not significantly reduced in size.

Remarks The conclusion by Ayala and Spain (1976) that *P. minasense* represents "crisis forms" of *P. tropiduri* was rejected by Telford (1979e), and the synonymy of *P. rhadinurum* with *P. minasense*, whatever its status, by Ayala (1978) was not accepted by Telford (1979e, 1980, 1994). Adequate samples of meronts from Mexican and Venezuelan *P. rhadinurum* and from Colombian *P. minasense carinii* demonstrate that the cruciform meronts of *P. rhadinurum* may exceed the proportion formed as fans, but both types are common. In *P. minasense carinii*, cruciform meronts are present, but a broad fan shape is the predominant form (about 90%). Meront dimensions are considerably influenced by phase of infection and are less useful than shape in distinguishing the two species. Although too few gametocytes of either Venezuelan *P. rhadinurum* or Colombian *P. minasense carinii* are available, the former are comparable to gametocytes of Mexican *P. rhadinurum*. The *P. minasense carinii* gametocytes are smaller. More adequate samples are necessary to prove the conclusion that *P. rhadinurum* has larger gametocytes.

Plasmodium minasense Carini and Rudolph 1912, Telford 1979

Diagnosis A small *Plasmodium* (*Carinamoeba*) species distinguished by meronts equal to or smaller than the nuclei of uninfected erythrocytes, and gametocytes that may slightly exceed these nuclei in size but are usually smaller. Meronts produce four to ten merozoites, with the mean number usually less than six. All stages past young meronts are pigmented. Gametocytes parasitize erythrocytes, appear to be common only immediately following the acute phase of infection, and are often apparently absent, with infection recognized by presence of asexual stages alone. Neither the host cell nor its nucleus is significantly distorted by the presence of the parasite.

Type Host *Mabuya agilis* (Raddi) (= *M. mabouya* Lacépède) (Sauria: Scincidae).

Type Locality River Paranahyba and its tributary Bagagem, Minas Gerais State, Brazil.

Other Hosts Species of the families Iguanidae, Polychrotidae, and Teiidae.

Other Localities Southern Middle America to southern Brazil, and Caribbean islands.

Remarks *Plasmodium minasense* is represented by seven subspecies, as recognized next.

Plasmodium minasense minasense Carini and Rudolph 1912, Telford 1979

Diagnosis A *Plasmodium* (*Carinamoeba*) species with meronts fan-shaped, approximately 5 × 4 µm, similar in size to nuclei of uninfected erythrocytes. Gametocytes are approximately 5–6 × 4–5 µm, with LW values of 23–30 µm^2 and L/W 1.2–1.4. Gametocyte size relative to host erythrocyte nucleus size is approximately 1.1–1.3.

Type Host *Mabuya mabouya* Lacépède (Sauria: Scincidae).

Type Locality River Paranahyba and its tributary Bagagem, Minas Gerais State, Brazil.

Other Hosts None known.

Other Localities None known.

Prevalence Unknown.

Morphological Variation The only data available are from the type description and are stated in the Diagnosis section.

Exoerythrocytic Merogony Unknown.

Sporogony Unknown.

Effects on Host Unknown.

Remarks This species is known only from the minimal description of Carini and Rudolph (1912) and has not been reported again. Size estimates are derived from comparison of the authors' figures with known erythrocyte nucleus average size from *M. mabouya*.

Plasmodium minasense carinii Leger and Mouzels 1917, Telford 1979 (**Plate 17**)

Diagnosis A *Plasmodium* (*Carinamoeba*) species with broad, fan-shaped, occasionally cruciform, meronts 3–5 × 2–4 µm, with LW 6–20 µm^2, that contain four to eight merozoites. Meront size relative to host cell nucleus size averages 0.46, and to normal erythrocyte nucleus size is 0.45. Pigment forms a small dark golden-brown mass at the base of fans, central when the meront is cruciform. Gametocytes, round to elongate, are rarely seen in circulating blood and are 4–7 × 3–4 µm, with LW 12–28 µm^2 and L/W 1.2–1.8. Gametocyte size relative to host cell nucleus size averages 0.83 and to normal erythrocyte nucleus size is 0.76. Dark brown pigment granules tend to aggregate in one or two sites in gametocytes. Immature meronts lack long, filamentous cytoplasmic processes but may occasionally show one or two short, digitiform processes.

Type Host *Iguana iguana* (Linnaeus) (Sauria: Iguanidae).

Type Locality Vicinity of Cayenne, French Guiana.

Other Hosts *Polychrus acutirostris*.

Other Localities Mathias Leme, Minas Gerais State, Brazil (da Silva Cordeiro, 1977); Salvador, Bahia State, Brazil (Telford, 1979e); Trinidad (Wenyon, 1915); vicinity of Barranquilla, Colombia (Telford, 1979e).

Prevalence Five of 16 (31.3%) *Iguana iguana*, collected presumably in the vicinity of Barranquilla, Colombia were infected with *P. minasense carinii*; 1 of 1 *Polychrus acutirostris* from Minas Gerais, Brazil was infected (da Silva Cordeiro, 1977).

Morphological Variation Estimated meront dimensions of *P. minasense carinii* from *I. sapidissima* (Wenyon, 1915) (= *I. iguana*, Ayala, 1978) are 3.3–4.5 × 2.2–3.4 µm, LW 9.5–14.6 µm^2, containing four merozoites. The two gametocytes figured by Wenyon are approximately 7 × 3.8 µm, LW 26.6 µm^2, and L/W 1.84, and 6.9 × 4.1 µm, LW 28.3 µm^2, and L/W 1.68, about the size of their host cell nuclei. Leger and Mouzels (1917) described *P. carinii* from *I. nudicollis* (= *I. iguana*, Ayala, 1978), providing little information other than a size not exceeding 5 µm for intracellular meronts, 5–7 µm when "extraglobulaire," with four merozoites present. Macrogametocytes were described as 3.5 to 5–6 µm in size, with microgametocytes a little smaller, but were not illustrated. In the sample from Colombia (Telford, 1979e), meronts are broadly fan-shaped, sometimes cruciform 3.9 ± 0.6 × 3.0 ± 0.3 µm (3–5 × 2–4, N = 22), with LW 11.8 ± 2.7 µm^2 (6–20), and contain 4.2 ± 0.8 (4–8, N = 25) merozoites. Meront size relative to host cell nucleus size is 0.46 ± 0.11 (0.25–0.67), and to normal erythrocyte nucleus size is 0.45 ± 0.10 (0.23–0.76). Only three gametocytes were found, 4–7 × 3–4 µm, with LW 12–28 µm^2 and L/W 1.2–1.8. Gametocyte size relative to host cell nucleus size is 0.50–1.17, and to normal erythrocyte nucleus size is 0.45–1.00. Pigment in meronts forms a small, dark golden-brown clump at the base of the fan shape, central in those cruciform, and in gametocytes appears to be localized in one or two sites. Immature meronts may occasionally show one or two short, digitiform processes but lack long, filiform cytoplasmic projections.

In *Polychrus acutirostris* (da Silva Cordeiro, 1977), meronts of *P. minasense* are 1.5–3.5 × 3.5–6.0 µm and form two to six merozoites. The ovoid gametocytes, with pigment in compact lateral masses, are 3.0–3.5 × 4.5–6.0 µm in macrogametocytes and 4.0–4.5 × 5.0–5.6 µm^2 in microgametocytes.

Exoerythrocytic Merogony Unknown.

Sporogony Unknown.

Effects on Host Neither meronts nor the few gametocytes seen in the sample from Colombia had any discernible effects on host erythrocytes or their nuclei.

Remarks The illustrations of da Silva Cordeiro (1977) closely resemble those of *P. minasense carinii* in Telford (1979e), and this appears to be the most reasonable assignment of the parasite in *Polychrus acutirostris* to *P. minasense* subspecies at present.

Plasmodium minasense anolisi Telford 1979 (Plate 18)

Diagnosis A *Plasmodium* (*Carinamoeba*) species with predominantly fan-shaped meronts 4–8 × 2–5 µm, LW 8–30 µm^2, that contain four to ten merozoites. Meront size relative to host cell nucleus size averages 0.48, and to normal erythrocyte nucleus size is 0.73. Gametocytes are round or oval usually, 4–9 × 3–6 µm, with LW 12–42 µm^2 and L/W 1.00–2.67. Gametocyte size relative to host cell nucleus size averages 0.61, and to normal erythrocyte nucleus size is 1.00. Pigment is present as a single golden mass usually located at the base of fan-shaped meronts and in gametocytes is usually clumped as three to six granules in one or two foci. Sexual difference in dimensions and shape may be present in Caribbean populations.

Type Host *Anolis limifrons* Cope (Sauria: Polychrotidae).

Type Locality Frijolito Creek, 4.8 km north of Gamboa, Canal Zone, Panama.

Other Hosts *Anolis cybotes, A. distichus*, and *A. coelestinus.*

Other Localities Sasardi, San Blas Territory, Panama; Port au Prince and Fond Verretes, Departement de l'Ouest, Haiti; Dominican Republic: Pedro Sanchez and Morro de Miches, El Seibo Province, Baharona, and 5 km north northeast of Polo, Baharona Province.

Prevalence Panama: 10 of 381 (2.6%) overall in *A. limifrons*; 7 of 107 (6.5%) at Sasardi, 2 of 74 (2.7%) at Frijolito Creek, and 1 of 18 (5.6%) at Quebrada Juan Grande; Hispaniola: 7 of 97 (7.2%) in *A. cybotes*, 2 of 27 (7.4%) in *A. distichus*, and 1 of 6 *A. coelestinus*; El Seibo Province, Dominican Republic: 3 of 15 (20%) in *A. cybotes*, 1 of 1 *A. distichus* at Pedro Sanchez and 2 of 3 *A. cybotes* at Morro de Miches; Baharona Province, 1 of 11 *A. cybotes* and 1 of 4 *A. coelestinus*; Haiti: 1 of 7 *A. cybotes* at Port au Prince and 1 of 15 *A. distichus* at Fond Verretes.

Morphological Variation In *Anolis limifrons* (**Plate 18A**), *P. minasense anolisi* meronts are 4.5 ± 0.68 × 3.5 ± 0.53 µm (4–8 × 2–4, N = 78), with LW 15.8 ± 3.2 µm^2 (8–24), and produce 5.8 ± 1.2 (4–8) merozoites. Meront size relative to host cell nucleus size is 0.48 ± 0.10 (0.27–0.75, N = 51), and to normal erythrocyte nucleus size is 0.45 ± 0.10 (0.24–0.73, N = 76). Pigment forms a single golden mass at the base of fan-shaped meronts. Erythrocytic meronts have slightly smaller dimensions and fewer merozoites on average than do those occupying proerythrocytes: 4.4 ± 0.6 × 3.4 ± 0.5 µm, LW 14.9 ± 3.0 µm^2, and 5.4 ± 1.1 merozoites (N = 51) versus 4.7 ± 0.8 × 3.7 ± 0.4 µm, LW 17.4 ± 3.0 µm^2, and 6.5 ± 1.0 merozoites in proerythrocytes (N = 27). Gametocytes are typically round or oval, 5.2 ± 0.3 × 4.4 ± 0.8 µm (4–8 × 3–6, N = 100), LW 22.9 ± 6.8 µm^2 (12–42), and L/W 1.21 ± 0.23 (1.00–2.67). Gametocyte size relative to host cell nucleus size is 0.61 ± 0.18 (0.30–1.07, N = 97), and to normal erythrocyte nucleus size is 0.58 ± 0.13 (0.33–0.93, N = 74).

In *Anolis cybotes* (**Plate 18B**), *P. minasense anolisi* meronts are 5.0 ± 0.6 × 3.8 ± 0.5 µm (4–6 × 3–5, N = 75), with LW 19.0 ± 3.5 µm^2 (12–30), and contain 5.9 ± 1.2 (4–8) merozoites. Meront size relative to host cell nucleus size is 0.63 ± 0.16 (0.34–1.00, N = 48), and to normal erythrocyte nucleus size is 0.64 ± 0.14 (0.46–1.17, N = 75). Gametocytes are 6.3 ± 0.7 × 4.7 ± 0.5 µm (5–8 × 4–6, N = 75), with LW 29.3 ± 4.0 µm^2 (20–30) and L/W 1.35 ± 0.26 (1.00–2.00). Gametocyte size relative to host cell nucleus size is 1.00 ± 0.22 (0.57–1.50, N = 49), and to normal erythrocyte nucleus size is 1.06 ± 0.18 (0.74–1.40, N = 75). In *Anolis distichus*, meronts are 5.4 ± 0.6 × 4.0 ± 0.6 µm (4–6 × 3–5, N = 50), with LW 21.9 ± 4.7 µm^2 (12–30), and contain 6.4 ± 1.4 (4–10) merozoites. Meront size relative to host cell nucleus size is 0.73 ± 0.20 (0.30–1.07, N = 25), and to normal erythrocyte nucleus size is 0.70 ± 0.15 (0.38–0.96, N = 50). Gametocytes are 6.6 ± 1.0 × 4.3 ± 0.6 µm (5–9 × 3–5, N = 50), with LW 28.2 ± 5.0 µm^2 (20–36) and L/W 1.56 ± 0.39 (1.0–2.67). Gametocyte size relative to host cell nucleus size is 0.97 ± 0.27 (0.57–1.50, N = 25), and to normal erythrocyte nucleus size is 0.90 ± 0.16 (0.64–1.15, N = 50).

In the type host *A. limifrons, P. minasense anolisi* has smaller meronts and gametocytes, and the latter are more

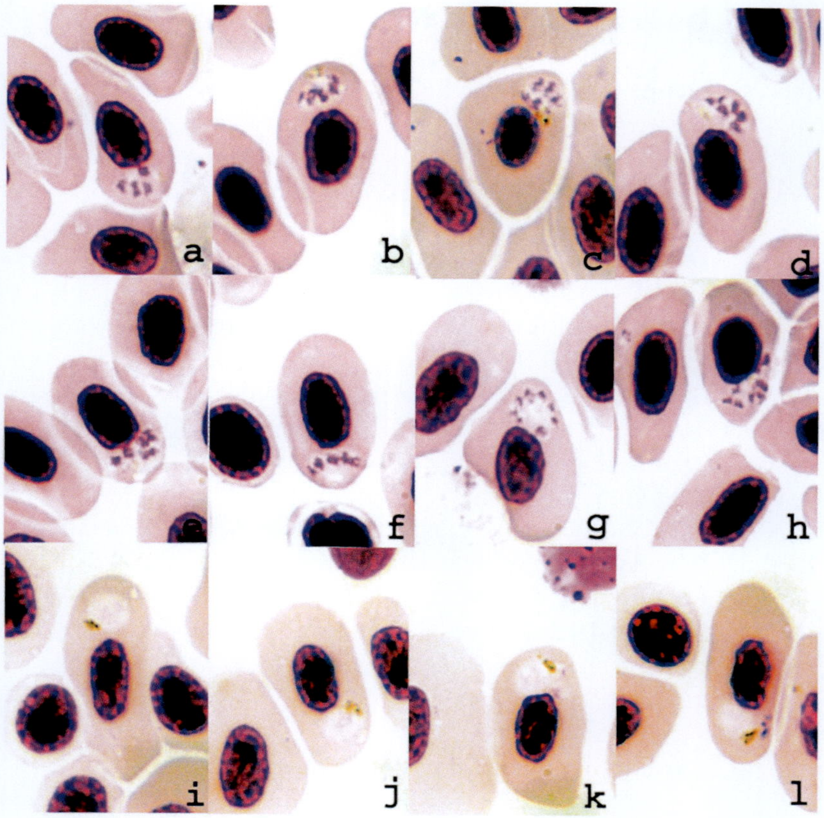

(A)

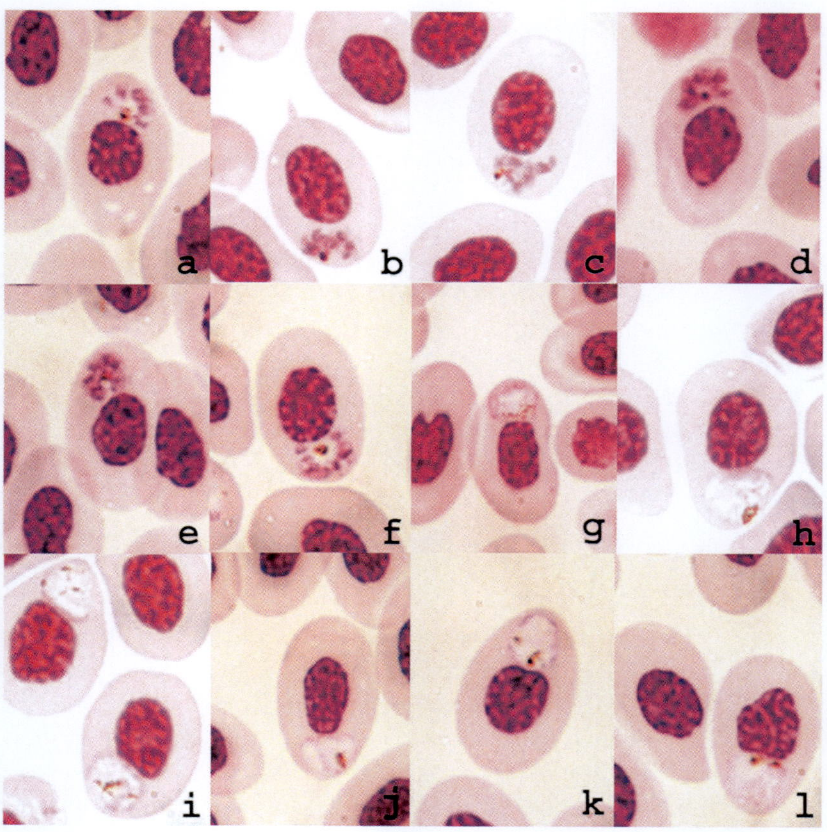

(B)

Plate 18 (**A**) *Plasmodium minasense anolisi* from *Anolis limifrons*, Panama. Meronts, **a–h**; macrogametocytes, **i**, **j**; microgametocytes, **k**, **l**. (**B**) *Plasmodium minasense anolisi* from *Anolis cybotes*, Hispaniola. Meronts, **a–f**; macrogametocytes, **g–i**; microgametocytes, **j–l**.

rounded than in *A. cybotes* and *A. distichus*. In the host species from Hispaniola, macrogametocytes are more elongate than microgametocytes (Telford et al., 1989). Values of each character, however, overlap considerably.

Exoerythrocytic Merogony Unknown.

Sporogony Unknown.

Effects on Host In *A. limifrons*, meronts have little visible effect on their host erythrocytes and their nuclei, and gametocytes do not alter the size of erythrocytes or their nuclei. The latter sometimes distort cells and displace the nuclei. In *A. cybotes* and *A. distichus*, the effects on host erythrocytes and their nuclei are similar with respect to distortion and displacement, but in *A. distichus*, the host cells but not their nuclei are hypotrophied (Telford et al., 1989).

Remarks In an article that appeared before the revision of *P. minasense* (Telford, 1979e), Guerrero et al. (1977) reported prevalence of three *Plasmodium* species from 296 *A. limifrons* of Barro Colorado Island but did not distinguish the very small meronts and gametocytes of *P. minasense* from *Plasmodium "tropiduri."* Ayala (1977), however, figured *P. minasense "sensu* Telford," apparently based on the inadequate information available to Telford (1974). *Plasmodium minasense anolisi* does occur on Barro Colorado Island in Panama, however, which is part of the same drainage system and only a few kilometers from the type locality of Frijolito Creek north of Gamboa. The latter is the type locality, rather than Frijoles River, stated by Telford (1979e).

Plasmodium minasense capitoi Telford 1979 (**Plate 19**)

Diagnosis A *Plasmodium* (*Carinamoeba*) species with variably shaped meronts 3–7 × 3–5 µm, LW 9–28 µm², that contain four to eight merozoites. Meront size relative to host cell nucleus size averages 0.46, and to normal erythrocyte nucleus size is 0.50. Gametocytes are 5–9 × 3–6 µm, with LW 21–48 µm² and L/W 1.00–2.33. Gametocyte size relative to host cell nucleus size averages 0.92, and to normal erythrocyte nucleus size is 0.97. Both meronts and gametocytes are heavily pigmented, as a large golden mass in meronts and large black granules dispersed in gametocytes. There are no sexual differences in gametocyte dimensions.

Type Host *Anolis capito* Peters (Sauria: Polychrotidae).

Type Locality Frijoles River, 4.8 km north of Gamboa, Canal Zone, Panama.

Other Hosts None known.

Other Localities None known.

Prevalence Overall, two of nine *A. capito* were infected by *P. minasense capitoi* and two of two at the type locality.

Morphological Variation The variably shaped meronts are 4.9 ± 0.8 × 3.6 ± 0.5 µm (3–7 × 3–5, N = 50), with LW 18.0 ± 4.5 µm² (9–28). Merozoites number 4.6 ± 1.1 (4–8). Meront size relative to host cell nucleus size is 0.46 ± 0.14 (0.27–0.78, N = 26), and to normal erythrocyte nucleus size is 0.50 ± 0.14 (0.23–0.84, N = 50). Pigment forms a prominent yellowish mass. Gametocytes are 7.1 ± 0.9 × 4.9 ± 0.8 µm (5–9 × 3–6, N = 50), with LW 34.9 ± 6.4 µm² (21–48) and L/W 1.50 ± 0.35 (1.00–2.33). Gametocyte size relative to host cell nucleus size averages 0.93 ± 0.23 (0.53–1.50), and to normal erythrocyte nucleus size is 0.97 ± 0.19 (0.54–1.35). Pigment is dispersed as large black granules in gametocytes. There are no sexual differences in dimensions of gametocytes.

Exoerythrocytic Merogony Unknown.

Sporogony Unknown.

Effects on Host About one-half of mature meronts parasitize proerythrocytes; all gametocytes are erythrocytic. Neither meronts nor gametocytes alter the size of host erythrocytes or their nuclei, seldom causing distortion of the cell or nucleus, and uncommonly displacing the latter.

Remarks Pigmentation of both meronts and gametocytes is more prominent than in other subspecies of *Plasmodium minasense*. The host species, *Anolis capito*, is uncommon and terrestrial, associated with the leaf litter of the forest floor, and very likely is exposed to a different community of hematophagous arthropods than most *Anolis* species that utilize tree trunks, shrubs, or the canopy of the mature moist tropical forest.

Plasmodium minasense plicae Telford 1979 (**Plate 19**)

Diagnosis A *Plasmodium* (*Carinamoeba*) species with small, usually fan-shaped meronts 3–6 × 2–4 µm, LW 6–18 µm², that contain four to eight merozoites. Meront size relative to host cell nucleus size averages 0.30, and to normal erythrocyte nucleus size is 0.35. Smaller mature meronts tend to be marginal in position. There is a prominent light golden pigment mass. Gametocytes are round or oval, 4–6 × 4–5 µm, with LW 16–30 µm² and L/W 1.0–1.5,

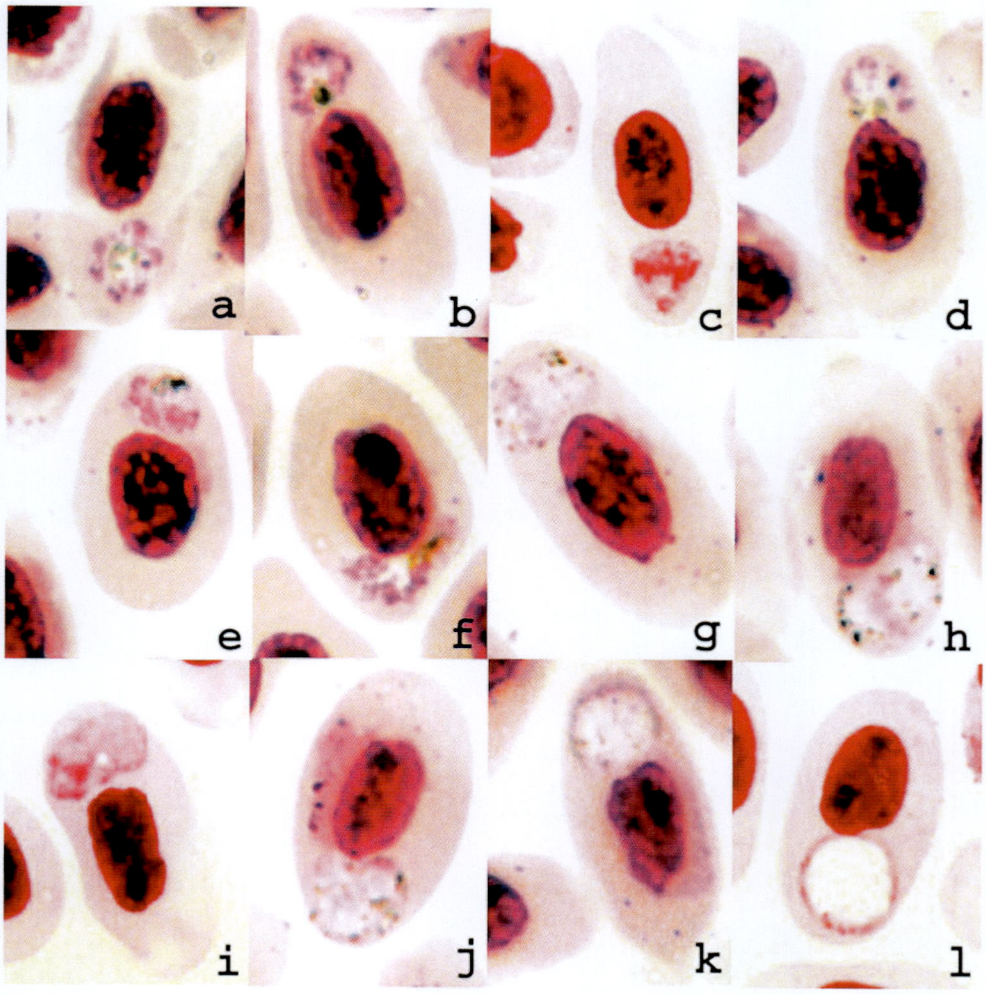

(A)

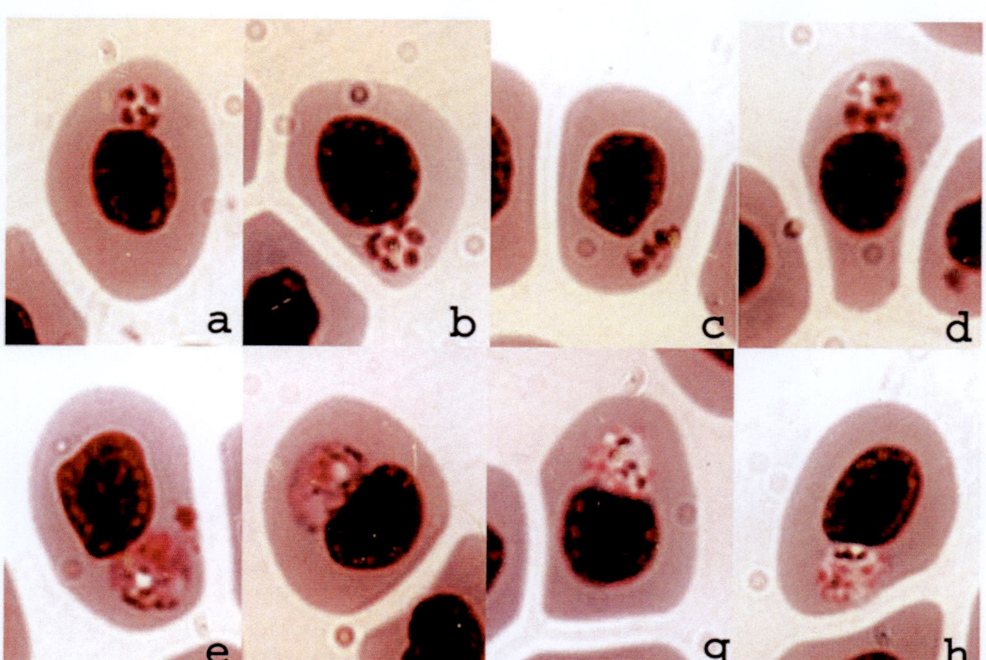

(B)

Plate 19 (**A**) *Plasmodium minasense capitoi* from *Anolis capito*, Panama. Meronts, **a–f**; macrogametocytes, **g–k**; microgametocyte, **l**. (**B**) *Plasmodium minasense plicae* from *Plica umbra*, Guyana. Meronts, **a–d**; macrogametocytes, **e**, **f**; microgametocytes, **g**, **h**.

and tend to be nucleophilic. Gametocyte size relative to host cell nucleus size averages 0.61, and to normal erythrocyte nucleus size is 0.71. Pigment, light golden when aggregated, is usually localized with the gametocyte but may disperse as dark greenish-gold granules. There is no sexual dimorphism in dimensions.

Type Host *Plica umbra* (Linnaeus) (Sauria: Iguanidae).

Type Locality Vicinity of Georgetown, Guyana.

Other Hosts *Plica plica.*

Other Localities None known.

Prevalence Two of seven *P. umbra* and one of ten *P. plica* were infected by *P. minasense plicae* (Telford, 1973b).

Morphological Variation Meronts are small and usually fan-shaped but occasionally resemble a morulum, 4.0 ± 1.0 × 2.8 ± 0.5 μm (3–6 × 2–4, N = 25), with LW 11.1 ± 3.6 μm^2 (6–18), and contain 5.0 ± 1.3 (4–8) nuclei. Meront size relative to host cell nucleus size is 0.30 ± 0.08 (0.17–0.45), and to normal erythrocyte nucleus size is 0.35 ± 0.12 (0.18–0.57). The smaller mature meronts are usually marginal in the erythrocyte. Pigment forms a prominent light golden mass. Gametocytes are round or oval, 5.3 ± 0.6 × 4.2 ± 0.4 μm (4–6 × 4–5, N = 25), with LW 22.2 ± 4.1 μm^2 (16–30) and L/W 1.28 ± 0.14 (1.00–1.50). Gametocyte size relative to host cell nucleus size is 0.61 ± 0.09 (0.46–0.83), and to normal erythrocyte nucleus size is 0.71 ± 0.13 (0.51–0.96). Gametocytes tend to be nucleophilic and show several prominent light golden pigment granules localized in one site, which may show some dispersal as dark greenish-gold-to-black granules. There is no sexual dimorphism in gametocyte dimensions.

Exoerythrocytic Merogony Unknown.

Sporogony Unknown.

Effects on Host Meronts have little effect on host erythrocytes except for occasional displacement of the nucleus. Gametocytes caused no hypertrophy, but sometimes distorted host cells, usually displacing their nuclei and producing a slight flattening of nuclei when in proximity.

Remarks This species was reported as *Plasmodium minasense* by Telford (1973b) and later recognized as a distinct subspecies (Telford, 1979e). The combination of often marginal meronts and usually nucleophilic gametocytes is distinctive among saurian *Plasmodium* species.

Plasmodium minasense tegui Telford 1979, 1980 (Plate 20)

Diagnosis A *Plasmodium* (*Carinamoeba*) species with small nucleophilic meronts and gametocytes. Meronts, fan-shaped or cruciform, are 3–6 × 3–5 μm, LW 9–25 μm^2, and contain four to six merozoites. Meront size relative to host cell nucleus size averages 0.64, and to normal erythrocyte nucleus size is 0.57. Gametocytes are nearly round, 3–5 × 2–4 μm, with LW 6–20 μm^2 and L/W 1.00–1.50. Gametocyte size relative to host cell nucleus size averages 0.59, and to normal erythrocyte nucleus size is 0.59. Pigment granules are small and dark greenish-yellow to black in both meronts and gametocytes, and in the latter form one or two large grains on the periphery. The nucleus of macrogametocytes forms as a thick band along one margin of the cell. Gametocyte dimensions are not sexually dimorphic.

Type Host *Tupinambis teguixin* (Linnaeus) (Sauria: Teiidae).

Type Locality San Jorge, Municipio Pirítú, Portuguesa State, Venezuela.

Other Hosts None known.

Other Localities Municipio Cojedes, Cojedes State, Venezuela. Reports in the literature from Pará, Brazil (Landau et al., 1973), French Guiana (Leger, 1919), and Colombia (Ayala et al., 1973) probably refer to *P. minasense tegui.*

Prevalence In Venezuela, 3 of 28 (10.7%) of tegu lizards from the type locality and 4 of 44 (9.1%) from Cojedes State were infected (Telford). Prevalences of a *Plasmodium* species from *Tupinambis* hosts were 5 of 6 in Pará, Brazil, 1 of 40 (2.5%) in Guiana, and 1 of 66 (1.5%) in Colombia.

Morphological Variation Trophozoites and immature meronts commonly show short cytoplasmic processes. Mature meronts are fan-shaped or cruciform, 4.2 ± 0.8 × 3.4 ± 0.5 μm (3–6 × 3–5, N = 49), with LW 14.8 ± 4.5 μm^2 (9–25), and contain 4.2 ± 0.6 (4–6, N = 50) merozoites. Meront size relative to host cell nucleus size is 0.64 ± 0.21 (0.32–1.33), and to normal erythrocyte nucleus size is 0.57 ± 0.15 (0.38–0.91). Pigment appears as small, dark greenish-yellow granules. Gametocytes are nearly round, 4.2 ± 0.7 × 3.6 ± 0.5 μm (3–5 × 2–4, N = 50), with LW 15.1 ± 4.0 μm^2 (6–20) and L/W 1.17 ± 0.15 (1.00–1.50). Gametocyte size relative to host cell nucleus size is 0.59 ± 0.17 (0.25–1.00), and to normal erythrocyte nucleus size is 0.59 ± 0.16 (0.25–0.84). In macrogametocytes, the nucleus is formed as a thick band along one margin, usually opposite one or two grains of greenish-black pigment, also marginal. There

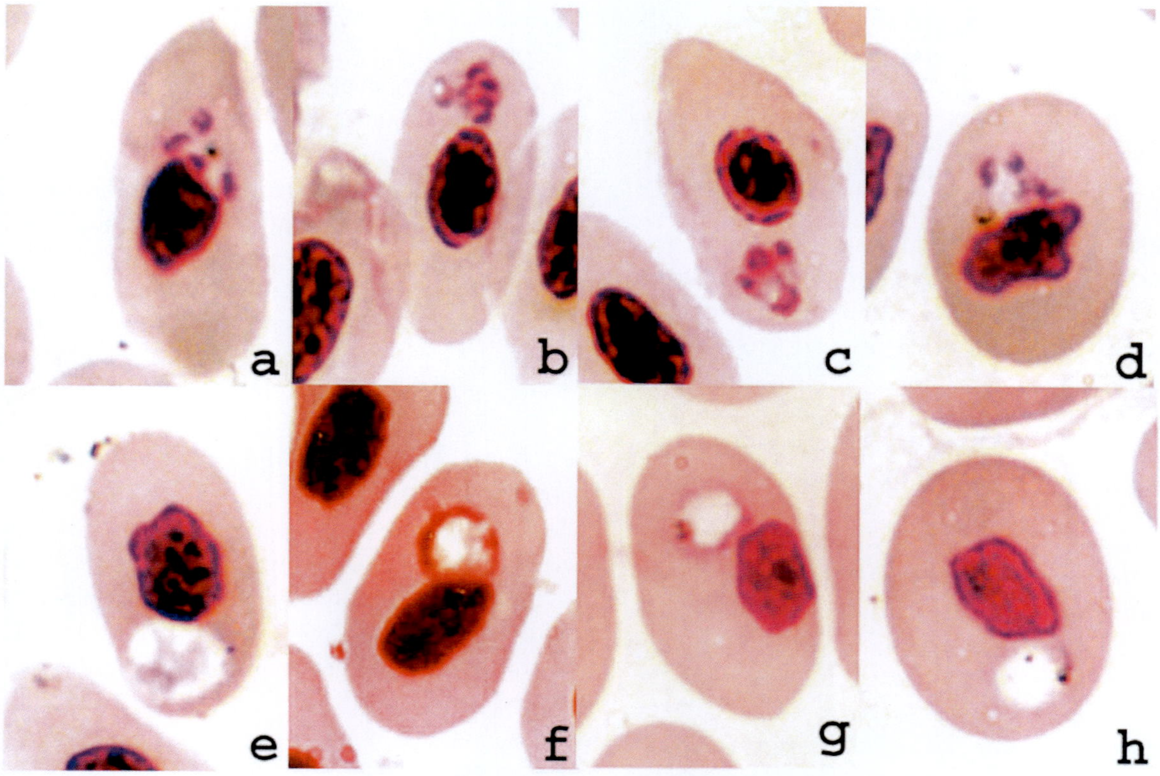

(A)

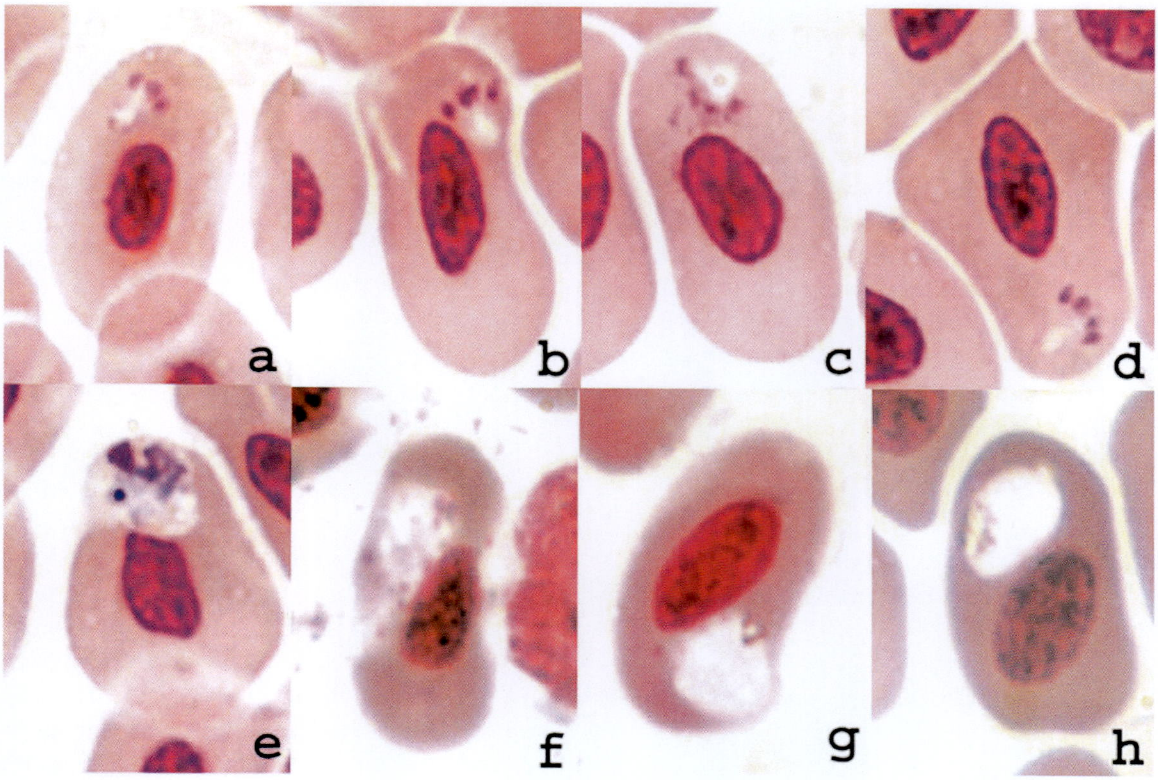

(B)

Plate 20 (**A**) *Plasmodium minasense tegui* from *Tupinambis teguixin*, Venezuela. Meronts, **a–d**; macrogametocytes, **e**, **f**; microgametocytes, **g**, **h**. (**B**) *Plasmodium minasense diminutivum* from *Ameiva ameiva*, Panama. Meronts, **a–d**; macrogametocytes, **e**, **f**; microgametocytes, **g**, **h**.

are no sexual differences in gametocyte dimensions. Both meronts and gametocytes are usually nucleophilic.

Exoerythrocytic Merogony Unknown.

Sporogony Unknown.

Effects on Host About one-half of the cells host to asexual stages and young gametocytes are polychromatophilic proerythrocytes, commonly in multiple infections of two to six parasites. Neither meronts nor gametocytes cause hypertrophy of host erythrocytes or their nuclei. Meronts sometimes distort the host cell and its nucleus and usually displace the latter. Gametocytes often distort the host erythrocyte but not its nucleus, which is usually displaced.

Remarks The type infection was isolated into a juvenile *T. teguixin* by inoculation of infected blood. This subspecies is distinguished from all others of *P. minasense* by nucleophily of both meronts and gametocytes, the peripheral band-like nucleus, and by having the least disparity in size between average LW values for meronts (14.8 μm^2) versus gametocytes (15.1 μm^2). The gametocytes are also the smallest among the six adequately described subspecies.

Plasmodium minasense diminutivum Telford 1973, 1979 (**Plate 20**)

Diagnosis A *Plasmodium* (*Carinamoeba*) species with meronts 3–5 × 3–4 μm, LW 9–20 μm^2, that contain four to six merozoites arranged as an elongate fan. Meront size relative to host cell nucleus size averages 0.57, and to normal erythrocyte nucleus size is 0.63. A prominent mass of grayish-yellow pigment forms the base of the fan-shaped meronts. Gametocytes are round to oval, 5–9 × 3–6 μm, with LW 18–42 μm^2 and L/W 1.00–2.25. Gametocyte size relative to host cell nucleus size averages 0.89, and to normal erythrocyte nucleus size is 1.25. Pigment is aggregated into a grayish-gold mass, accompanied by several darker dispersed granules. Gametocytes are not sexually dimorphic in dimensions.

Type Host *Ameiva ameiva praesignis* (Baird and Girard) (Sauria: Teiidae).

Type Locality Guayabalito Village on Rio Chagres, Colon Province, Panama.

Other Hosts None known.

Other Localities El Aguacate and Santa Rita Chorrera, Panama Province, and Madden Dam, Canal Zone, Panama.

Prevalence Overall, *P. minasense diminutivum* infected 20 of 92 (21.7%) of *A. ameiva* in Panama, 10 of 13 (79.9%) at El Aguacate, 4 of 34 (11.8%) at Santa Rita Chorrera, and 5 of 12 (41.7%) at Madden Dam (Telford).

Morphological Variation Meronts form as an elongate fan, with a mass of grayish-yellow pigment forming the handle. Meronts are 3.9 ± 0.4 × 3.1 ± 0.4 μm (3–5 × 3–4, N = 50), with LW 12.2 ± 2.2 μm^2 (9–20), and produce 4.2 ± 0.5 (4–6) merozoites. Meront size relative to host cell nucleus size is 0.51 ± 0.12 (0.32–0.89, N = 47), and to normal erythrocyte nucleus size is 0.63 ± 0.12 (0.45–1.08, N = 50). The round or oval gametocytes are 5.9 ± 1.0 × 4.4 ± 0.9 μm (5–9 × 3–6, N = 22), with LW 26.1 ± 7.3 μm^2 (18–42) and L/W 1.40 ± 0.38 (1.00–2.25). Gametocyte size relative to host cell nucleus size averages 0.89 ± 0.20 (0.64–1.50), and to normal erythrocyte nucleus size is 1.25 ± 0.36 (0.84–1.96). Gametocytes usually have grayish-gold pigment aggregated into a single mass, with several darker granules dispersed within the cytoplasm. Gametocyte dimensions do not differ between sexes.

Exoerythrocytic Merogony Unknown.

Sporogony Unknown.

Effects on Host Meronts have no effect on host erythrocytes or their nuclei. Gametocytes do not cause enlargement of the host cells or their nuclei but sometimes slightly distort the cell and commonly displace the nucleus.

Remarks Gametocytes of *P. minasense diminutivum* are uncommon; parasitemias of up to 13% usually appear to be comprised only of asexual stages.

Plasmodium minasense calcaratae Telford and Telford 2003

Diagnosis A *Plasmodium* (*Carinamoeba*) species with variably shaped meronts 2.5–4.5 × 2–3 μm, LW 5–12 μm^2, that contain three or four merozoites. Meront size relative to host cell nucleus size averages 0.38, and to normal erythrocyte nucleus size is 0.38. Pigment forms a single, rounded, light yellow mass. Gametocytes are usually round or oval, 4.5–9 × 3–7 μm, with LW 15.8–54 μm^2 and L/W 1.00–2.33. Gametocyte size relative to host cell nucleus size averages 1.42, and to normal erythrocyte nucleus size is 1.42. Pigment is occasionally formed as a prominent dark yellow mass or is dispersed as darker granules. There is no sexual dimorphism in gametocyte dimensions.

Type Host *Kentropyx calcarata* Spix (Sauria: Teiidae).

Type Locality Mision Padomo, Territorio Federal de Amazonas, Venezuela.

Other Hosts None known.

Other Localities None known.

Prevalence One of four *K. calcarata* was host to *P. minasense calcaratae*.

Morphological Variation Meronts are usually fan-shaped or cruciform when small, sometimes elongate, oval, or lepidoptiform. Meronts are 3.4 ± 0.6 × 2.6 ± 0.4 μm (2.5–4.5 × 2–3, N = 25), with LW 8.8 ± 2.1 μm^2 (5–12), and produce 3.9 ± 0.3 (3–4) merozoites. Meront size relative to host cell nucleus size is 0.38 ± 0.10 (0.24–0.57), and to normal erythrocyte nucleus size is 0.38 ± 0.09 (0.22–0.52). Pigment occurs as a single light yellow, rounded mass. Gametocytes are spherical or ovoid, rarely elongate, 6.7 ± 0.8 × 5.0 ± 0.8 μm (4.5–9 × 3–7, N = 75), with LW 33.7 ± 7.3 μm^2 (15.8–54) and L/W 1.38 ± 0.29 (1.00–2.33). Gametocyte size relative to host cell nucleus size is 1.42 ± 0.27 (1.03–2.00), and to normal erythrocyte nucleus size is 1.42 ± 0.26 (1.07–2.28). A prominent dark yellow pigment mass occasionally remains in mature gametocytes or disperses as darker granules. Gametocytes are not sexually dimorphic in dimensions.

Exoerythrocytic Merogony Unknown.

Sporogony Unknown.

Effects on Host Meronts have no effect on host erythrocytes or their nuclei. Gametocytes rarely distort the host cell or its nucleus but almost always displace the latter. There is no hypertrophy, but without affecting erythrocyte LW values, host cells may be shorter and broader than uninfected erythrocytes.

Remarks The disparity in size between meronts and gametocytes is greater than in other subspecies of *P. minasense*, with average LW 8.8:33.7 μm^2, respectively. Among the subspecies, meronts of *P. minasense calcaratae* average least in LW and second largest in gametocyte LW (33.7 μm^2) to those of *P. minasense capitoi* (34.9 μm^2).

NEOTROPICAL *LACERTAMOEBA* SPECIES

Plasmodium tropiduri Aragão and Neiva 1909 Telford 1979

Diagnosis A *Plasmodium* (*Lacertamoeba*) species characterized by mature meronts usually formed as a rosette or fan, which are typically the size of uninfected erythrocyte nuclei or smaller, and contain 4–24 merozoites. The mean number of nuclei in individual infections ranges from 5.3 to 17.3, and meront shape and size vary according to host species, stage of infection, and maturity of the red blood cell parasitized. Gametocytes are round to ovoid, seldom elongate except when lateral to host cell nucleus, with mean L/W in individual infections 1.1–1.6. Gametocyte size (LW) ranges from 16 to 80 μm^2, with means of individual infections 26–60 μm^2. Both size and shape vary with host species, stage of infection, and maturity of red blood cell. Gametocyte size relative to that of uninfected erythrocyte nuclei is 0.6–1.5, with ratio varying with host species. Pigment is always present in meronts and gametocytes unless the parasite occupies an immature erythrocyte, in which case its presence and quantity may be variable. Pigment masses in meronts and immature gametocytes are golden yellow; these become dispersed as minute dark granules in the cytoplasm of mature gametocytes. Trophozoites and meronts usually parasitize proerythrocytes, while gametocytes are more common in erythrocytes. Younger asexual stages are not highly amoeboid and lack prominent cytoplasmic processes. Meronts and gametocytes usually occupy polar or lateropolar positions in host cells. Host cells are distorted more commonly than their nuclei, which are frequently displaced. Both meronts and gametocytes can produce significant hypertrophy of host cell and nucleus, but this varies with host species and stage of infection.

Type Host *Tropidurus torquatus* (Wied) (Sauria: Tropiduridae).

Type Locality Bicudos, Minas Gerais State, Brazil.

Other Hosts Species of the families Tropiduridae, Polychrotidae, and Scincidae.

Geographic Range Southeastern Brazil to Costa Rica and in the Caribbean.

Remarks *Plasmodium tropiduri* is represented by four subspecies, as recognized next.

Plasmodium tropiduri tropiduri Aragão and Neiva 1909 Telford 1979 (**Plate 21**)

Diagnosis A *Plasmodium* (*Lacertamoeba*) species with usually rosette or fan-shaped meronts, sometimes elongate, 3–12 × 2–7 μm, LW 9–63 μm^2, that contain 4–24 merozoites. Meront size relative to host cell nucleus size averages 0.51–1.17, and to normal erythrocyte nuclei is 0.58–1.40,

Plate 21 (**A**) *Plasmodium t. tropiduri* from *Tropidurus hispidus*, Venezuela. Meronts, **a–f**; macrogametocytes, **g–i**; microgametocytes, **j–l**. (**B**) *Plasmodium tropiduri caribbense* from *Anolis cybotes,* Hispaniola. Meronts, **a–f**; macrogametocytes, **g–i**; microgametocytes, **j–l**.

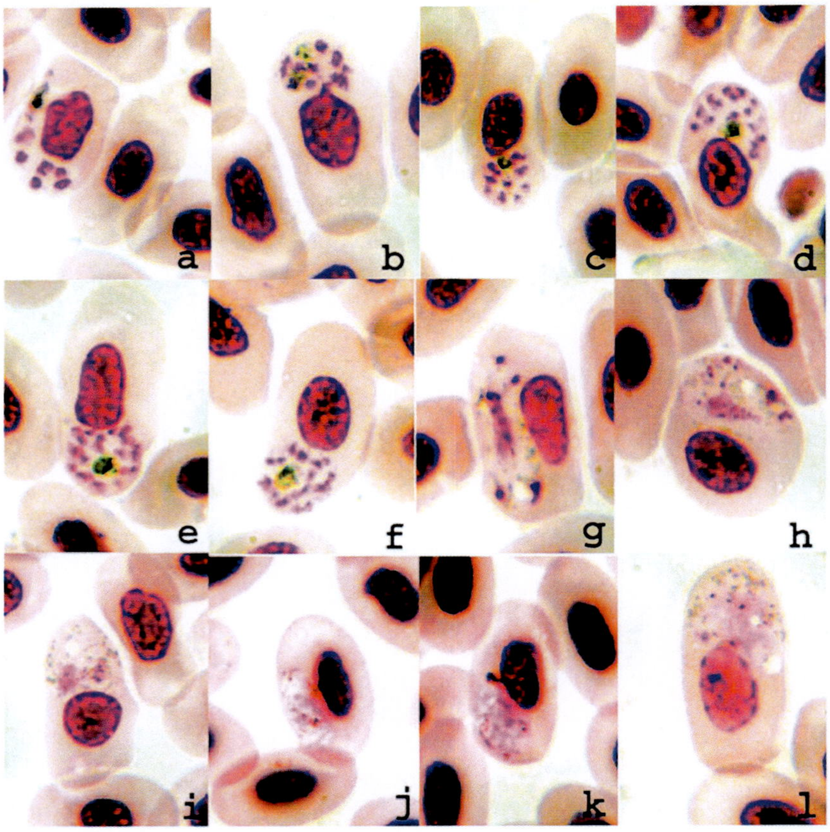

(A)

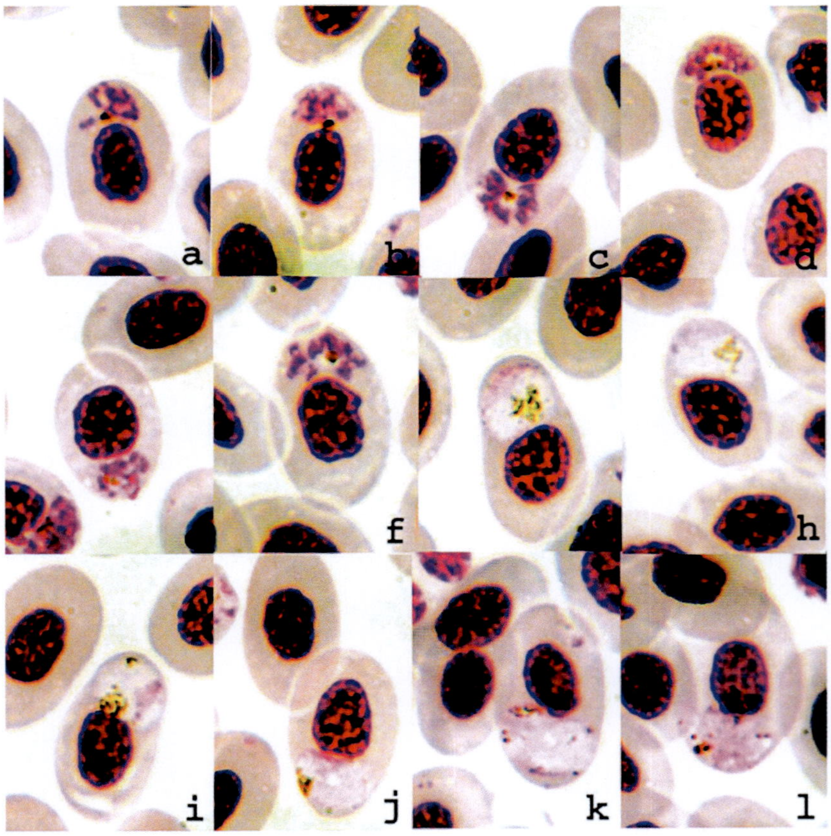

(B)

depending on infection phase, which influences meront size. Meronts in immature host cells produce more merozoites on average than erythrocytic meronts. Pigment masses are golden yellow in meronts and are usually central in rosettes or at the base of fans. Gametocytes are round to elongate, usually ovoid, 5–14 × 4–8 μm, with LW 20–80 μm^2 and L/W 1.00–3.50. Gametocyte size relative to host cell nucleus size averages 1.06–1.51, and to normal erythrocyte nuclei is 1.08–1.58. Pigment is dispersed as fine dark granules in gametocytes. Gametocytes do not differ in shape or dimensions by sex or phase of infection.

Type Host *Tropidurus torquatus* (Wied) (Sauria: Tropiduridae).

Type Locality Bicudos, Minas Gerais State, Brazil.

Other Hosts *Tropidurus hispidus* (Scorza, 1970; Ayala, 1978; Telford, 1980); *Mabuya mabouya* (Lainson and Shaw, 1969b).

Other Localities In *T. torquatus*, in Brazil: Porto Nacional, Goias State (Carini, 1941); Jacobina, Bahia State (Pessôa and Lopes, 1963); Bairro Cachoeirinha in Municipio Davinolândia, Bairro Boa Vistas in Municipio Pinhal, and Bairro Fonte Platina in Municipio Aguas da Prata, São Paulo State (Rocha e Silva and Rodrigues, 1974). In Guyana, *T. torquatus* in the vicinity of Georgetown (Telford, 1973). In *M. mabouya*, Utinga Forest, Belém, Pára State, Brazil (Lainson and Shaw, 1969b). In *T. hispidus*, in Venezuela: Ortiz, Dos Caminos, La Caimana, Peñas Negras, Paya, Veladero, Parapara, and San Juan Morros, Guarico State (Scorza, 1970); Municipios Guanare and Araure, Portuguesa State, and Municipio Cojedes, Cojedes State (Telford, 1980).

Prevalence In *T. torquatus* of Brazil, *P. tropiduri* parasitized 4 of 39 (10.3%) in Bahia State (Pessôa and Lopez, 1963); in São Paulo State, 10 of 51 (19.6%) in Municipio Davinolândia, 2 of 31 (6.5%) and 6 of 49 (12.2%) in Municipio Aguas da Prata (Rocha e Silva and Rodrigues, 1974), and 1 of 1 in Minas Gerais and Goias states (Aragão and Neiva, 1909; Carini, 1941). In *M. mabouya*, Lainson and Shaw (1969b) reported 8 of 20 (40%) infected in Utinga Forest and later (Lainson et al., 1974b) 9 of 63 (14.3%), presumably including those previously reported. Telford (1973b) found 1 of 12 (8.3%) *T. torquatus* in Guyana infected. In Venezuelan *T. hispidus*, Scorza (1970a) reported overall prevalence of 54 of 211 (25.6%) in Guarico State and in the seven localities where *P. tropiduri* was found, prevalences by locality varied from 5 of 29 (14.5%) in Dos Caminos to 14 of 26 (53.8%) in Veladera; Telford (1980) found an overall prevalence of 13 in 106 (12.3%), with the prevalence by municipio 10 of 30 (33.3%) in Guanare, 2 of 57 (3.5%) in Araure, and 1 of 1 in Cojedes.

Morphological Variation In the type description, Aragão and Neiva (1909) described meronts as 7–8 μm in diameter, with 12 merozoites, and macrogametocytes 6–8 μm and microgametocytes 9 μm in diameter. Rocha e Silva and Rodriques (1974) reported usually 8 merozoites, rarely 10 or 12 in meronts. Meronts described by Scorza (1970a) averaged 10.6 × 5.5 μm (7–15 × 3.5–6), containing 7–14 merozoites. Chronic infections averaged 10.4, active natural infections were 18.5 ± 1.1, and in active experimental infection, there were 14.5 ± 2.2 merozoites. Microgametocytes were 10.3 × 7.5 μm, and macrogametocytes, classed as spherical, ovoid, or elongate, were 7.5 × 7.5 μm, 9.0 × 6.0 μm, and 12.0 × 3.5 μm, respectively. In the skink *Mabuya mabouya*, Lainson and Shaw (1969b) described meronts as shaped primarily as rosettes, some as fans, 5–7 μm in diameter, with an average of 6.4 (6–12) merozoites. Gametocytes were spherical, 7.5 μm in diameter. In *T. torquatus* of Guyana, a chronic-to-recrudescent *P. tropiduri* infection was sampled four times in 6 weeks. Meronts were 4.6 ± 0.9 × 3.6 ± 0.6 μm (3–8 × 2–5, N = 42), LW 17.3 ± 5.6 μm^2 (9–35), with 7.1 ± 2.1 (4–12) merozoites. Meront size relative to host cell nucleus size was 0.52 ± 0.13 (0.23–0.75, N = 37), and to normal erythrocyte nuclei was 0.58 ± 0.18 (0.31–1.12, N = 42). Gametocytes were 6.2 ± 0.8 × 5.3 ± 0.7 μm (5–9 × 4–7, N = 50), with LW 32.8 ± 7.5 μm^2 (20–54). Gametocyte size relative to host cell nucleus size was 1.06 ± 0.23 (0.63–1.54), and to normal erythrocyte nuclei was 1.08 ± 0.26 (0.65–1.85). Gametocytes did not differ by sex in dimensions or shape.

In Venezuelan *T. hispidus*, meronts were 7.2 ± 1.8 × 5.0 ± 1.1 μm (3–12 × 3–7, N = 64), LW 36.9 ± 13.7 μm^2 (9–63), and contained 13.4 ± 4.7 (4–24, N = 75) merozoites. Meront size relative to host cell nucleus size was 1.17 ± 0.56 (0.32–2.50, N = 40), and to normal erythrocyte nuclei was 1.40 ± 0.53 (0.33–2.51, N = 64). Meront dimensions differed by infection phase and maturity of host cells. In active phase, meronts were larger and produced more than twice the number of merozoites, on average, than those from chronic infections, respectively, 7.7 ± 1.5 × 5.2 ± 1.0 μm (N = 55), LW 40.1 ± 11.6 μm^2, and 14.6 ± 4.0 merozoites versus 4.4 ± 1.2 × 3.8 ± 0.7 μm (N = 9), LW 17.3 ± 7.2 μm^2, and 6.7 ± 2.8 (N = 11) merozoites. Erythrocytic meronts were nearly similar in size but produced fewer merozoites on average than those in proerythrocytes, respectively, 7.1 ± 1.6 × 5.1 ± 1.1 μm (N = 31), LW 36.6 ± 12.6 μm^2, and 13.0 ± 4.0 merozoites versus 8.4 ± 0.9 × 5.3 ± 0.8 μm (N = 24), LW 44.7 ± 8.5 μm^2, and 17.3 ± 2.1 merozoites. Gametocytes were 7.9 ± 1.5 × 5.6 ± 1.0 μm (5–14 × 4–8, N = 132), with LW 44.4 ± 11.4 μm^2 and L/W 1.44 ± 0.39 (1.00–3.50). Gametocyte size relative to host cell nucleus size was 1.51 ± 0.46

(0.70–2.92, N = 128), and to normal erythrocyte nuclei was 1.58 ± 0.37 (0.73–2.54, N = 132). Neither phase of infection nor sex affected gametocyte dimensions or shape.

Exoerythrocytic Merogony Scorza (1971c) reported the presence of meronts and gametocytes in the thrombocytes of *T. torquatus* (= *T. hispidus*) in 4 of 54 *Plasmodium tropiduri* infections. Meronts were 6–9 μm in diameter and contained 10–18 nuclei, with 24–26 nuclei present in mature meronts. Microgametocytes were 9–10 × 7–8 μm, similar to pigmented microgametocytes in erythrocytes. Macrogametocytes were slightly smaller, 7–9 × 6–8 μm. Pigment was not present in thrombocytic meronts or gametocytes. Ultrastructural comparisons of erythrocytic and thrombocytic gametocytes found no significant differences except the absence of pigment in the latter gametocytes. The appearance of thrombocytic parasites in a lizard infected with *P. tropiduri* by blood inoculation, during a 4-month latent period following the induced active infection, furnished additional evidence to strengthen the conclusion that *P. tropiduri* can produce both merogonic and gametogonic cycles in thrombocytes as well as in erythrocytes. Two of 10 infections of *P. tropiduri* from Municipio Guanare had gametocytes present in thrombocytes as well (Telford). Thrombocytic meronts were reported by Rocha e Silva and Rodrigues (1974) in Brazilian infections of *P. tropiduri*.

Sporogony Unknown.

Effects on Host Anemia produced in *T. hispidus* by *P. tropiduri* infection was described by Scorza (1971b) as the normochromic and normocytic type, with hemoglobin concentrations below normal but with cell counts only slightly below those of uninfected lizards. Extreme anemia developed during the crisis phase of infection or shortly thereafter, with intense erythropoietic activity occurring at this time. Lizards with severe anemia sought cooler areas of their cages during this time. In active infections, both meronts and gametocytes may cause significant hypertrophy of host cells and their nuclei but seldom distort host cells and nuclei, with the latter often displaced. Meronts in chronic infection may produce hypertrophy of host cells also, but nuclei remain normal in size. Host cells of chronic-phase gametocytes are not hypertrophied, and nuclei are normal in size except in younger chronic infections, but older infections show greater cell and nucleus distortion than younger infections, with nuclei almost always displaced.

Remarks Scorza (1972) and Scorza et al. (1971, 1972) provided the only information on cytochemistry available for saurian malaria parasites. Although Scorza's (1971c) ultrastructural evidence is convincing that gametocytes of *Plasmodium tropiduri* may parasitize thrombocytes of *Tropidurus hispidus*, it is vital to use genomic comparison of infections with and without thrombocytic infections to provide the absolute evidence necessary to satisfy skeptics. If the identification of *P. tropiduri* in the skink *Mabuya mabouya* is confirmed by genomic comparison of infections in skinks and tropidurids, this would represent only the fifth example of a saurian *Plasmodium* species parasitizing in nature hosts of two different families, with the exception of the *Plasmodium minasense* complex.

Plasmodium tropiduri caribbense Telford, Johnson and Young 1989 (**Plate 21**)

Diagnosis A *Plasmodium* (*Lacertamoeba*) species with meronts 4–8 × 3–6 μm, LW 12–48 μm², that form elongate, thin cytoplasmic projections during maturation. Merozoites number 5–14, usually arranged as broad fans or, less often, rosettes. Meront size relative to host cell nucleus size averages 0.67–0.77, and to normal erythrocyte nuclei is 0.68–0.89. Dark brown pigment granules form an irregular mass in meronts. Gametocytes, usually round or ovoid, are 5–14 × 4–7 μm, with LW 25–84 μm² and L/W 1.00–3.25. Gametocyte size relative to host cell nucleus size averages 1.27–1.55, and to normal erythrocyte nuclei is 1.46–1.55. Dark pigment granules tend to be focused in their distribution, with one to three granules in microgametocytes and somewhat more in macrogametocytes. Host cells at all stages are usually erythrocytes.

Type Host *Anolis cybotes* Cope (Sauria: Polychrotidae).

Type Locality Morro de Miches, El Seibo Province, Dominican Republic.

Other Hosts *Anolis lineatopus*.

Other Localities Dominican Republic: Rio Seibo and Pedro Sanchez, in El Seibo Province, 5 km north northeast of Polo in Baharona Province. Haiti: Fond Verretes, Departement de l'Ouest, and Port au Prince. Jamaica: Montego Bay.

Prevalence In Dominican Republic: 3 of 3 at Morro de Miches, 3 of 3 at Rio Seibo, 9 of 15 (60%) at Pedro Sanchez, and 1 of 3 in Baharona Province. In Haiti, 3 of 22 (13.6%) at Fond Verretes and 1 of 7 at Port au Prince. In Jamaica, 1 of 78 (1.3%) overall, 1 of 6 at Montego Bay.

Morphological Variation In *A. cybotes*, as meronts near maturity, some have elongate, thin cytoplasmic projections that may extend up to the length of the meront. Meronts are usually broadly fan-shaped, less often formed as rosettes, with a few otherwise shaped. Meronts are 5.9 ± 0.8 × 4.4 ± 0.7 μm (4–8 × 3–6, N = 88), LW 25.8 ± 5.9 μm^2 (15–48), and contain 7.9 ± 2.2 (5–14) merozoites. Meront size relative to host cell nucleus size is 0.77 ± 0.26 (0.38–1.60, N = 72), and to normal erythrocyte nuclei is 1.89 ± 0.26 (0.49–1.87, N = 88). The few meronts found in proerythrocytes are similar in size to erythrocytic meronts and contain no more merozoites, on average (8.1 ± 1.6, N = 9). Gametocytes are usually round or ovoid, seldom elongate, 7.5 ± 1.2 × 5.6 ± 0.7 μm (5–11 × 4–7, N = 156), with LW 42.2 ± 8.2 (25–60) μm^2 and L/W 1.35 ± 0.32 (1.00–2.75). Gametocyte size relative to host cell nucleus size averages 1.27 ± 0.32 (0.60–2.08, N = 129), and to normal erythrocyte nuclei is 1.46 ± 0.35 (0.82–2.33, N = 156). Dimensions of gametocytes are similar for both sexes, but the average L/W ratio of macrogametocytes, 1.30 ± 0.30 (N = 77), in comparison to that of microgametocytes, 1.40 ± 0.34 (N = 79), suggests a slightly more rounded shape of macrogametocytes. The dark brown pigment forms an irregular mass in meronts, central in rosettes and at the base of those formed as fans. Pigment is sparse in microgametocytes, only one to three granules, but more granules are present in macrogametocytes. Rather than dispersed in a pattern, the granules in both gametocyte sexes tend to be focused in one or more sites. Meronts and gametocytes usually parasitize only erythrocytes, although a few meronts can be found in proerythrocytes. In *A. lineatopus* of Jamaica, meronts are very similar to those in the type host, 5.1 ± 0.8 × 3.7 ± 0.6 (4–7 × 3–5, N = 50), with LW 19.1 ± 4.3 (12–35), and contain 8.6 ± 1.9 (5–14) merozoites. Meront size relative to host cell nucleus size is 0.67 ± 0.17 (0.43–1.19, N = 47), and to normal erythrocyte nuclei is 0.68 ± 0.16 (0.44–1.28, N = 50). Gametocytes are more elongate than in *A. cybotes*, although their size is very similar: 8.6 ± 1.8 × 4.9 ± 0.8 (6–14 × 4–7, N = 50), with LW 41.4 ± 10.1 (28–84) and L/W 1.83 ± 0.53 (1.00–3.25). Gametocyte size relative to host cell nucleus size averages 1.55 ± 0.48 (0.94–3.50, N = 49), and to normal erythrocyte nuclei is 1.47 ± 0.37 (0.96–3.08, N = 50). There is no difference in gametocyte dimensions by sex.

Exoerythrocytic Merogony Unknown.

Sporogony Unknown.

Effects on Host Erythrocytes host to either meronts or gametocytes are not enlarged; gametocytes but not meronts occasionally cause some distortion of the host cell. Meronts rarely distort or displace host cell nuclei but may produce nuclear hypertrophy in some infections. Gametocytes commonly displace erythrocyte nuclei but seldom distort them and in some infections may cause nuclear hypertrophy.

Remarks All of the *Plasmodium tropiduri* subspecies look very similar to each other regarding gametocyte shape and size, and with merozoites produced by meronts most commonly formed as rosettes or fans, the characters of greatest similarity. They differ primarily in presence and degree of pigmentation and somewhat in relative size of meronts and gametocytes to the nuclei of uninfected erythrocytes.

Plasmodium tropiduri panamense Telford 1979 (**Plate 22**)

Diagnosis A *Plasmodium* (*Lacertamoeba*) species in which the presence of visible pigment is variable and usually absent when the host cell is immature. Meronts are 4–8 × 2–6 μm, LW 8–40 μm^2 and contain 4–18 merozoites, most commonly arranged as a rosette and seldom as fans. Proerythrocytic meronts produce more merozoites on average than do those in erythrocytes. Meront size relative to host cell nucleus size averages 0.40, and to normal erythrocyte nuclei is 0.56. Gametocytes, more commonly ovoid or round than elongate, are 5–9 × 3–7 μm, with LW 18–56 μm^2 and L/W 1.00–2.67. Gametocyte size relative to host cell nucleus size averages 0.83, and to normal erythrocyte nuclei is 0.84. There is no sexual difference in gametocyte dimensions or shape. Meronts predominantly parasitize immature host cells and seldom show visible pigment. Gametocytes are more commonly erythrocytic, with pigment granules sparse when present and more often visible as infections become chronic.

Type Host *Anolis biporcatus* (Wiegmann) (Sauria: Polychrotidae).

Type Locality El Aguacate, Panama Province, Panama.

Other Hosts *Anolis pentaprion.*

Other Localities Achiote and Rio Gatuncillo, Colon Province, and Capira Caimito, Panama Province, Panama.

Prevalence Overall, 54 of 99 (54.5%) *A. biporcatus* were infected by *P. tropiduri panamense*, 51 of 86 (59.3%) at El Aguacate, 1 of 2 each at Rio Gatuncillo and Capira Caimito, and 1 of 3 at Achiote. Overall, 2 of 19 (10.5%) *A. pentaprion* and 2 of 10 (20%) from El Aguacate were found infected.

Morphological Variation Meronts are 5.4 ± 1.0 × 3.8 ± 0.7 μm (4–8 × 2–6, N = 162), with LW 20.7 ± 6.2 μm^2 (8–40), and contain 8.6 ± 2.9 (4–18, N = 163) merozoites. Meront size relative to host cell nucleus size is 0.40 ± 0.14

Plate 22 (**A**) *Plasmodium tropiduri panamense* from *Anolis biporcatus*, Panama. Meronts, **a–f**; macrogametocytes, **g–i**; microgametocytes, **j–l**. (**B**) *Plasmodium tropiduri aquaticum* from *Anolis lionotus*, Panama. Meronts, **a–f**; macrogametocytes, **g–i**; microgametocytes, **j–l**.

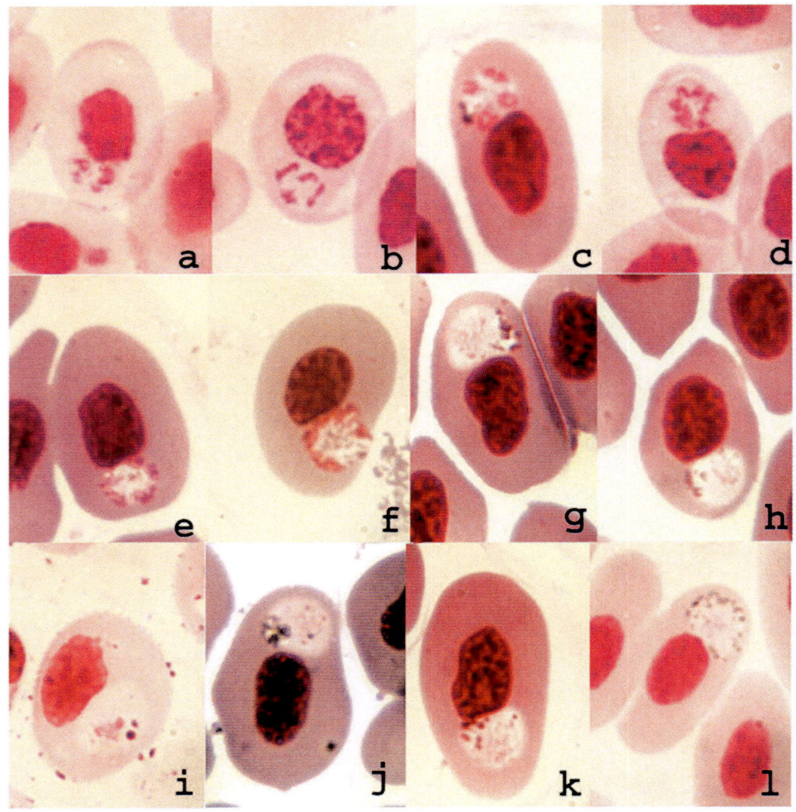

(A)

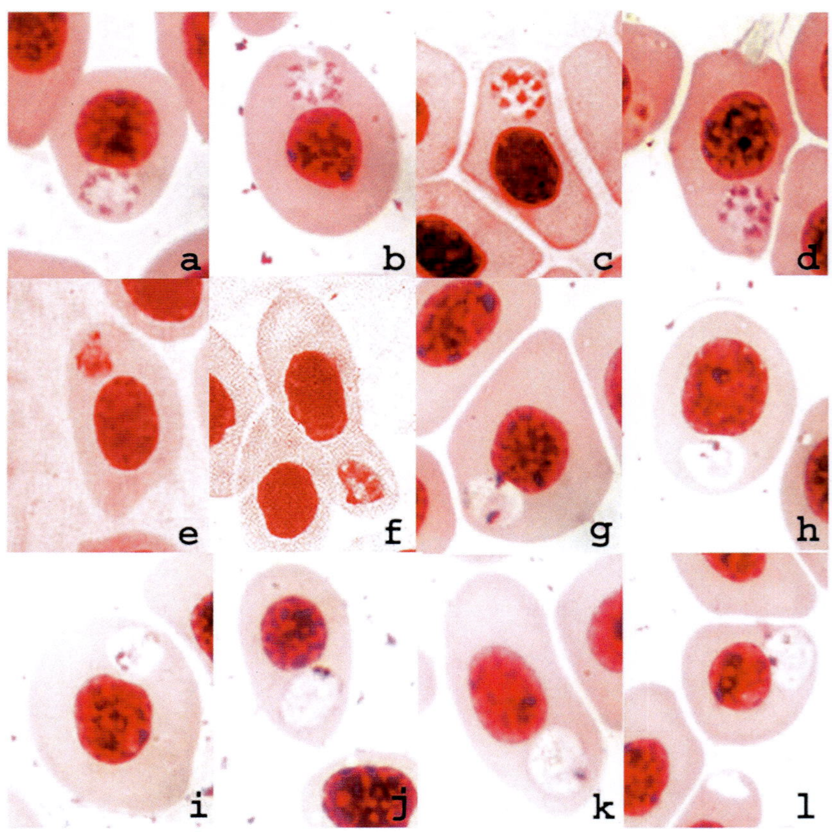

(B)

(0.23–0.69, N = 16), and to normal erythrocyte nuclei is 0.56 ± 0.17 (0.22–1.01, N = 162). Meronts are larger in active-phase infections than in chronic phase and produce more merozoites, respectively 5.5 ± 1.0 × 3.9 ± 0.7 μm (N = 142), LW 21.6 ± 6.1 μm^2, and 9.0 ± 2.9 (N = 143) merozoites versus 4.5 ± 0.8 × 3.3 ± 0.4 μm (N = 20), LW 14.6 ± 3.0 μm^2, and 6.3 ± 1.0 merozoites. In active infections, erythrocytic meronts are smaller than proerythrocytic, respectively 4.8 ± 0.9 × 3.4 ± 0.7 μm (N = 14), LW 16.2 ± 5.3 μm^2, and 6.4 ± 2.0 merozoites versus 5.6 ± 1.0 × 4.0 ± 0.7 μm (N = 128), LW 22.2 ± 5.9 μm^2, and 9.2 ± 2.8 (N = 129) merozoites. Most meronts are proerythrocytic (97.6%), and most lack visible pigment (96.5%). Merozoites usually form as a rosette or are variable, seldom appearing as a fan. Gametocytes are usually ovoid or round, seldom elongate, 6.5 ± 0.9 × 4.8 ± 0.8 μm (5–9 × 3–7, N = 154), with LW 31.3 ± 7.3 μm^2 (18–56) and L/W 1.39 ± 0.31 (1.00–2.67). Gametocyte size relative to host cell nucleus size is 0.83 ± 0.19 (0.50–1.60, N = 110), and to normal erythrocyte nuclei is 0.84 ± 0.20 (0.48–1.35, N = 154). Gametocytes do not differ in dimensions or shape by sex, but active-phase macrogametocytes and microgametocytes average larger and less rounded than the respective sexes in chronic infection: Macrogametocytes are 6.7 ± 1.0 × 4.9 ± 0.8 μm, LW 32.9 ± 7.8 μm^2, and L/W 1.40 ± 0.33 (N = 85) versus 6.0 ± 0.4 × 4.8 ± 0.6 μm, LW 28.6 ± 4.7 μm^2, and 1.28 ± 0.16 (N = 12); microgametocytes are 6.5 ± 0.8 × 4.7 ± 0.8 μm, LW 30.4 ± 6.2 μm^2, and L/W 1.42 ± 0.31 (N = 44), versus 5.8 ± 0.9 × 4.4 ± 0.5 μm, LW 25.5 ± 4.2 μm^2, and L/W 1.35 ± 0.29 (N = 13). Within active-phase infections, erythrocytic gametocytes average slightly larger and are more rounded than gametocytes occupying proerythrocytes, respectively 6.7 ± 0.9 × 5.0 ± 0.8 μm, LW 33.8 ± 7.2 μm^2, and L/W 1.37 ± 0.32 (N = 94) versus 6.3 ± 0.8 × 4.3 ± 0.7 μm, LW 27.4 ± 5.7 μm^2, and L/W 1.49 ± 0.30 (N = 35). Gametocytes usually parasitized erythrocytes (69%), and more had visible pigment (14.5%) than when proerythrocytes were occupied and pigment was rarely discernible (2.9%). In two chronic infections of *A. pentaprion*, meronts were 4.4 ± 0.5 × 3.2 ± 0.4 μm (4–5 × 3–4, N = 10), LW 14.0 ± 1.8 μm^2 (12–16), and contained 7.4 ± 2.1 (4–12) merozoites. Meront size relative to host cell nucleus size was 0.37 ± 0.05 (0.30–0.46, N = 8), and to normal erythrocyte nuclei was 0.40 ± 0.05 (0.34–0.46, N = 10). All but two meronts (80%) were erythrocytic, and all were very lightly pigmented. Gametocytes were 5.7 ± 0.6 × 4.7 ± 0.5 μm (5–7 × 4–6, N = 50), with LW 26.7 ± 3.9 μm^2 (20–36) and L/W 1.23 ± 0.22 (1.00–1.75). Gametocyte size relative to host cell nucleus size was 0.69 ± 0.12 (0.51–0.92, N = 49), and to normal erythrocyte nuclei was 0.72 ± 0.11 (0.51–0.92, N = 50). One gametocyte was proerythrocytic; all were lightly pigmented.

Exoerythrocytic Merogony A single meront with 12 nuclei was seen in a thrombocyte, but identity is uncertain due to the presence also of *Plasmodium floridense* in the host.

Sporogony Unknown.

Effects on Host In active infections, cells host to gametocytes, and their nuclei, were hypertrophied, but both cells and nuclei were of normal size when host to meronts. Erythrocytes, however, were distorted and their nuclei displaced by both meronts and gametocytes, and some nuclei were distorted by the latter. In infections at crisis, both cells and their nuclei were hypotrophied by both sexual and asexual parasites. Gametocytes, but not meronts, caused host cell and nuclei distortion, and both stages displaced nuclei. In chronic infections, cells host to meronts were smaller than normal, but their nuclei were enlarged. Gametocytes also caused nuclear hypertrophy, but host cells were normal in size. Both meronts and gametocytes distorted host cells, distorted their nuclei, and displaced them, although distortion was greater when meronts were present.

Remarks Two infections of *P. tropiduri* were found in *Anolis pentaprion* at El Aguacate. Both were chronic infections; one consisted of gametocytes only, and the other had a few meronts present, along with gametocytes. Comparison of dimensions with those of *P. tropiduri panamense* chronic-phase meronts and gametocytes revealed close similarity in dimensions of both stages and in merozoite numbers. The only real difference was distinct pigment in all gametocytes and sparse but visible pigment in all meronts of the *A. pentaprion* samples. Both hosts *A. biporcatus* and *A. pentaprion* inhabit secondary forest at El Aguacate; the former is a canopy species, and *A. pentaprion* utilizes a foraging zone closer to the ground. Despite the uniform presence of pigment in the *A. pentaprion* gametocytes and meronts, the two hosts are probably infected by the same parasite species, *P. tropiduri panamense.*

Plasmodium tropiduri aquaticum Telford 1979 (**Plate 22**)

Diagnosis A *Plasmodium* (*Lacertamoeba*) species in which both meronts and gametocytes average smaller in size than nuclei of uninfected erythrocytes. Meronts, usually shaped as rosettes, seldom as fans, are 4–8 × 3–5 μm, with LW 12–35 μm^2, and contain 4–14 merozoites. Meront size relative to host cell nucleus size averages 0.42, and to normal erythrocyte nuclei is 0.50. Gametocytes, round or ovoid usually, are 4–10 × 3–6 μm, with LW 16–50 μm^2 and L/W 1.00–2.00. Gametocyte size relative to host cell nucleus size averages 0.66, and to normal erythrocyte nuclei is 0.65. Although meronts predominantly parasitize proerythrocytes, all stages are prominently pigmented.

Type Host *Anolis lionotus* Cope (Sauria: Polychrotidae).

Type Locality Achiote, Colon Province, Panama.

Other Hosts *Anolis poecilopus.*

Other Localities In *A. lionotus* in Panama: El Aguacate and Rio Madroño at Gaspar Sabanas, about 8 km northwest of Chepo, Panama Province. In *A. poecilopus* in Panama: Quebrada Juan Grande, Frijoles River, and Frijolito Creek, about 3 km northwest of Gamboa, Canal Zone; Rio Gatuncillo, Santa Rita Ridge, Colon Province; Cerro Azul and Rio Madroño at Gaspar Sabanas; Sasardi, San Blas Territory. In *A. lionotus,* Costa Rica: Rio Frio, Heredia Province.

Prevalence Overall in Panama, 118 of 228 (51.8%) *A. lionotus* were infected by *P. tropiduri aquaticum*: 39 of 67 (58.2%) at Achiote, 69 of 175 (39.4%) at El Aguacate, and 1 of 6 from Rio Madroño. In *A. poecilopus*, overall, 118 of 228 (51.8%): 65 of 143 (45.5%) in the Canal Zone, at Frijoles River, 20 of 39 (51.3%), 31 of 60 (51.7%) at Frijolito Creek, and at Quebrada Juan Grande, 14 of 44 (31.8%); 37 of 63 (58.7%) from Rio Madroño, and 2 of 2 at Cerro Azul; 7 of 9 from Rio Agua Clara and 2 of 3 at Sasardi. In Costa Rica, 15 of 39 (38.5%) *A. lionotus* were infected by *P. tropiduri aquaticum.*

Morphological Variation In active infection from *A. lionotus*, meronts are 5.3 ± 0.8 × 4.1 ± 0.6 µm (4–8 × 3–5, N = 120), LW 21.9 ± 4.7 µm^2 (12–35), and contain 8.3 ± 1.6 (4–14) merozoites. Meront size relative to host cell nucleus size is 0.42 ± 0.08 (0.29–0.63, N = 40), and to normal erythrocyte nuclei is 0.50 ± 0.11 (0.24–0.79, N = 120). Nuclei are usually arranged as a rosette, seldom as fans or otherwise; meronts are always pigmented and typically occupy immature erythrocytes. Meronts in erythrocytes and proerythrocytes are similar in size, but the latter produce slightly more merozoites, on average 8.7 ± 1.6 (N = 80), than do those in erythrocytes, 7.5 ± 1.5 (N = 40). Gametocytes are 6.1 ± 1.0 × 4.8 ± 0.6 µm (4–10 × 3–6, N = 103), with LW 29.0 ± 5.8 µm^2 (16–50) and L/W 1.31 ± 0.32 (1.00–2.50). Gametocyte size relative to host cell nucleus size is 0.66 ± 0.16 (0.42–1.20, N = 73), and to normal erythrocyte nuclei is 0.65 ± 0.15 (0.36–1.24, N = 103). Gametocytes more commonly parasitize erythrocytes and average slightly larger in dimensions, with a more rounded shape than when proerythrocytic, respectively, 6.2 ± 0.9 × 4.9 ± 0.6 µm (N = 73), LW 30.4 ± 5.4 µm^2, and L/W 1.28 ± 0.27 versus 5.8 ± 1.1 × 4.4 ± 0.7 µm (N = 30), LW 25.7 ± 5.3 µm^2, and L/W 1.36 ± 0.41. Macrogametocytes are slightly smaller and more rounded than microgametocytes, respectively 5.7 ± 0.7 × 4.7 ± 0.6 µm (N = 56), LW 26.7 ± 5.0 µm^2, and L/W 1.26 ± 0.29 versus 6.5 ± 1.1 × 4.9 ± 0.6 µm (N = 47), LW 31.8 ± 5.5 µm^2, and L/W 1.36 ± 0.34.

In *A. poecilopus*, meronts are 5.0 ± 0.9 × 4.0 ± 0.8 µm (3–7 × 2–6, N = 84), LW 20.3 ± 6.3 µm^2 (6–36), and contain 6.9 ± 2.0 (4–12). Merozoites are more commonly arranged as fans than in *A. lionotus*, about one-half as often as rosettes are formed. Proerythrocytes are more commonly host to meronts than are erythrocytes. Erythrocytic meronts average slightly smaller than proerythrocytic and produce fewer merozoites, 6.1 ± 1.2 (N = 22), versus 7.3 ± 1.7 (N = 62). The few meronts available from crisis or chronic infections are smaller than those from active infections and produce fewer merozoites, respectively 3.9 ± 0.3 × 3.0 ± 0.5 µm (N = 9), LW 11.8 ± 2.5 µm^2, and 5.3 ± 1.0 merozoites versus 5.1 ± 0.9 × 4.1 ± 0.7 µm (N = 75), LW 21.3 ± 5.8 µm^2, and 7.1 ± 2.0 merozoites. Gametocytes are usually ovoid or round, 6.2 ± 0.9 × 5.1 ± 0.7 µm (5–9 × 4–7, N = 83), with LW 39.5 ± 7.7 µm^2 (20–54), and L/W 1.23 ± 0.22 (1.00–2.00). Gametocyte size relative to host cell nucleus size is 0.62 ± 0.14 (0.33–1.36, N = 59), and to normal erythrocyte nuclei is 0.69 ± 0.24 (0.37–1.44, N = 83). Gametocytes more commonly parasitize erythrocytes than proerythrocytes; dimensions and shape are not affected by host cell maturity. The only gametocyte character apparently influenced by infection phase is shape, for which the L/W ratio of crisis-chronic phase gametocytes is 1.12 ± 0.13 (N = 22), in comparison to 1.27 ± 0.23 (N = 61) in active-phase infections. There is little difference in dimensions and shape by sex: Macrogametocytes are 5.9 ± 0.7 × 4.9 ± 0.7 µm (N = 47), LW 29.0 ± 6.4 µm^2, and L/W 1.24 ± 0.21 versus 6.4 ± 1.1 × 5.4 ± 0.7 µm (N = 36), LW 34.7 ± 8.1 µm^2, and L/W 1.22 ± 0.24 in microgametocytes.

Exoerythrocytic Merogony A single EE meront has been seen in *A. poecilopus,* in a lymphocyte of a chronic infection at very low parasitemia (0.1%).

Sporogony Unknown.

Effects on Host Erythrocytes host to either meronts or gametocytes in active or chronic phase of infection are not hypertrophied. Host cell nuclei are significantly hypertrophied by meronts in active infection and much less so by gametocytes. Host cells and nuclei are seldom distorted by either stage, but nuclei are displaced by both. At crisis, cells host to either stage are distorted, and nuclei are displaced. Meronts do not distort or enlarge host cell nuclei, but gametocytes both distorted them and caused some hypertrophy. In chronic phase, host cells and nuclei are of normal size; neither is distorted by either meronts or gametocytes, but both stages displace nuclei (Telford, 1979d).

Remarks *Plasmodium tropiduri aquaticum* is a riverine species and only parasitizes the two semiaquatic *Anolis* species, *A. lionotus* and *A. poecilopus*, apparently wherever they occur in Panama, and in fairly high prevalence. *Anolis poecilopus* inhabits streams that drain into the Caribbean, and *A. lionotus* is found along streams of the Pacific drainage. The only known site where the two hosts are sympatric is the Rio Madroño, about 8 km northwest of Chepo in Panama Province. The host species are similar in appearance and in their ecology (Campbell, 1973).

Plasmodium torrealbai Scorza and Dagert 1957

Diagnosis A *Plasmodium* (*Lacertamoeba*) species characterized by irregularly shaped trophozoites and meronts with prolonged filiform cytoplasmic projections. At maturity, meronts form elongate fans 8–10 × 4–6 μm, estimated LW 42–60 μm^2, and contain 8–20 elongate merozoites. Pigment forms a gray or black mass at the base of the fan. Meront size relative to host cell nucleus size is estimated to be 1.8–2.7, and to normal erythrocyte nuclei is 1.6–2.4. Gametocytes are ovoid to elongate, 7–22 × 3–6 μm, with estimated LW 42–105 μm^2 and L/W 1.4–6.2. Gametocyte size relative to host cell nucleus size is an estimated 2.2–4.8, and to normal erythrocyte nuclei is 1.7–4.3. Macrogametocytes are larger and more elongate than microgametocytes, with a different distribution of pigment by sex, dispersed as dark bacilliform granules in macrogametocytes and as fine granules aggregated at one site in microgametocytes.

Type Host *Anolis sp. indet.* (Sauria: Polychrotidae).

Type Locality Guayaraca, Región del Auyantepui, Bolivar State, Venezuela.

Other Hosts None known.

Other Localities None known.

Prevalence *Plasmodium torrealbai* infected two of three *Anolis* sp. at the type locality (Scorza and Dagert, 1957).

Morphological Variation Young asexual stages were described by Scorza and Dagert (1957) as irregular and elongate, with long filiform cytoplasmic projections. These became compacted at maturity into fan-shaped meronts that contained eight to ten elongated merozoites. Apparently, up to 20 merozoites can be produced according to Walliker (1966), who examined a slide from the type material. No dimensions were stated in the description, but as estimated from the figures by Scorza and Dagert, meronts are 8–10 × 4–6 μm, with LW 42–60 μm^2. Meront size relative to host cell nucleus size is estimated to be 1.8–2.7, and to normal erythrocyte nuclei is 1.6–2.4. Pigment forms a gray-to-black prominent mass at the base of the fan. Stated average (?) dimensions for the elongate macrogametocytes are 8.5 × 3.6 μm, and for ovoid microgametocytes are 6.5 × 3.0 μm. Dimensions estimated from the type figures suggest ranges of 7–22 × 3–6 μm, in ovoid to elongate-shaped gametocytes, with LW 42–105 μm^2 and L/W 1.4–6.2. Estimated gametocyte size relative to host cell nucleus size is 2.2–4.8, and to normal erythrocyte nuclei is 1.7–4.3. Gametocytes are apparently sexually dimorphic in size, shape, and distribution of pigment. Up to 30 bacilliform dark granules are dispersed in the cytoplasm of macrogametocytes, and fine granules aggregate at one site in microgametocytes.

Exoerythrocytic Merogony Unknown.

Sporogony Unknown.

Effects on Host Meronts apparently broaden the host erythrocyte at the ends of the cells where they are situated. The figures in the description do not show nuclear hypertrophy, distortion, or displacement in cells host to meronts. Gametocytes cause hypertrophy of the erythrocyte but apparently do not displace the nucleus or distort it.

Remarks *Plasmodium torrealbai* has not been reported since its description except for the brief comment by Walliker (1966), who examined the type slide. Unfortunately, the *Anolis* host was not identified to species.

Plasmodium colombiense Ayala and Spain 1976 (**Plate 23**)

Diagnosis A *Plasmodium* (*Lacertamoeba*) species with fan- or rosette-shaped meronts 4–11 × 3–7 μm, LW 15–66 μm^2, that contain 4–14 merozoites. Meront size relative to host cell nucleus size averages 0.68, and to normal erythrocyte nuclei is 0.68. Golden-to-dark-brown pigment forms a compact mass at the base of fans or centrally in rosettes. Gametocytes are 4.4–8.9 × 2.5–7.6 μm, with LW 10–62 μm^2 and L/W 1.00–1.75. Gametocyte size relative to host cell nucleus size averages 0.94–1.05, and to normal erythrocyte nuclei is 1.04. Gametocytes are round or lentiform to, usually, elongate and are not sexually dimorphic in dimensions or shape. Pigment is dispersed as brown-to-black granules in macrogametocytes but forms a rounded compact peripheral mass in microgametocytes.

Type Host *Anolis auratus* Daudin (Sauria: Polychrotidae).

Plate 23 (**A**) *Plasmodium colombiense* from *Anolis auratus*, Venezuela, **a–g**, **i–l**, and Colombia, **h**. Meronts, **a–f**; macrogametocytes, **g–i**; microgametocytes, **j–l**. (**B**) *Plasmodium fairchildi* from *Anolis limifrons*, Panama. Meronts, **a–f**; macrogametocytes, **g–i**; microgametocytes, **j–l**.

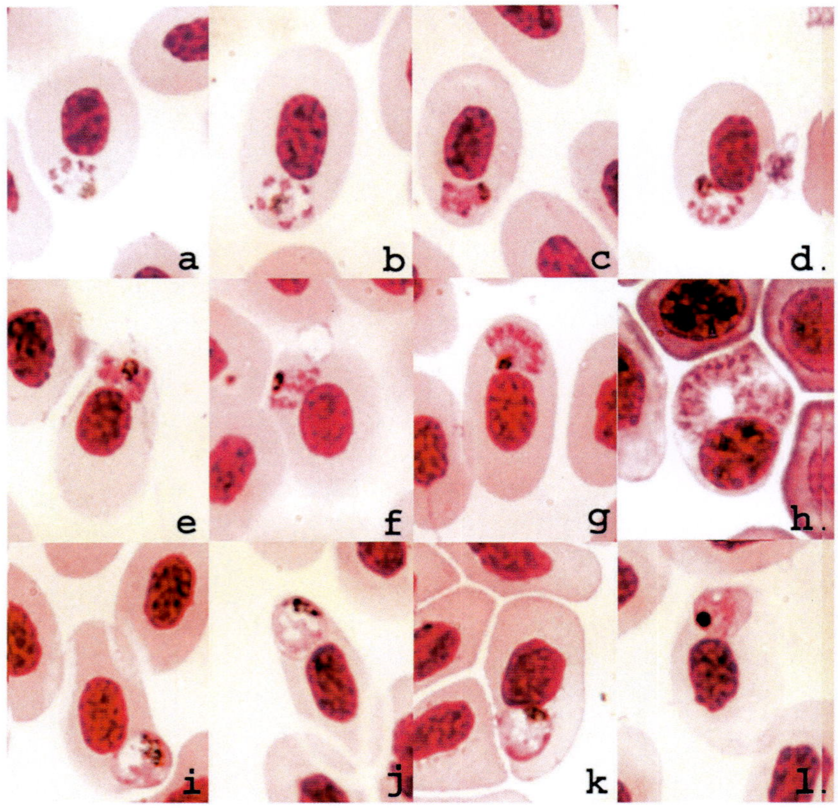

(A)

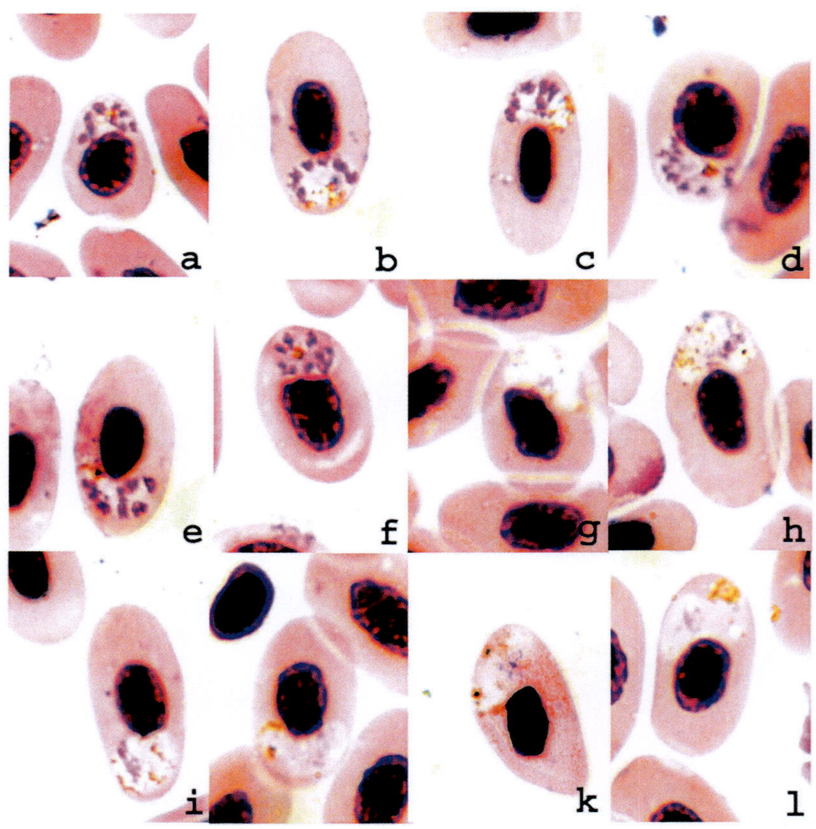

(B)

Type Locality Vicinity of Cali, Colombia.

Other Hosts None known.

Other Localities Fundo Vega Honda, 25 km north of Guanare, Municipio Guanare, Portuguesa State, Venezuela.

Prevalence In Colombia, 274 of 1131 (24.2%) *Anolis auratus* were infected by *Plasmodium colombiense* (Ayala and Spain, 1976). Overall, 1 of 32 (3.1%) Venezuelan *A. auratus* and 1 of 17 (5.9%) from Municipio Guanare were positive (Telford, 1980).

Morphological Variation Meronts of *P. colombiense* were described by Ayala and Spain (1976) as 5–6 μm "across," "approximately same size as mature gametocytes," with merozoites 3–14, "usually 8, 10, or 12, average 8.4 (N = 614), or 7.9 (N = 585) depending on definition of segmenter." Shape was described as "either rosette or fan-shaped or a distorted form of one or the other." Macrogametocytes were 6.1 × 4.7 μm (4.4–7.6 × 2.5–6.3, N = 557), with LW 28.4 μm^2 (10–48.4). Microgametocytes were 6.2 × 4.9 μm (5.1–8.9 × 2.5–7.6, N = 353), with LW 30.3 μm^2 (16.1–62). With sexes combined, the L/W ratio was 1.26, and LW was 29.1, "scarcely larger than normal host cell nucleus; L × W/HNCL × W = 1.05." On a slide from Colombian *A. auratus*, sent to me by Dr. Ayala, meronts are all proerythrocytic, 6.6 ± 1.3 × 5.1 ± 0.8 μm (5–11 × 4–7, N = 25), LW 34.1 ± 10.4 μm^2 (20–66), and contain 9.4 ± 1.7 (6–13) merozoites. Gametocytes of this infection are 6.0 ± 0.6 × 4.8 ± 0.6 μm (5–7 × 4–6, N = 25), with LW 29.0 ± 4.9 μm^2 (20–42) and L/W 1.27 ± 0.20 (1.00–1.75). Gametocyte size relative to host cell nucleus size is 0.94 ± 0.19 (0.57–1.25, N = 23), and to normal erythrocyte nuclei is 1.07 ± 0.18 (0.74–1.54, N = 25). In the infection from Venezuela, most meronts are erythrocytic, 5.4 ± 0.9 × 4.2 ± 0.6 μm (4–8 × 3–6, N = 49), with LW 22.5 ± 4.6 μm^2 (15–36), and contain 6.4 ± 1.7 (4–12, N = 50) merozoites. Meront size relative to host cell nucleus size is 0.68 ± 0.17 (0.38–1.04, N = 43), and to normal erythrocyte nuclei is 0.68 ± 0.14 (0.45–1.08, N = 49). Gametocytes are 5.6 ± 0.5 × 5.1 ± 0.3 μm (5–7 × 4–6, N = 50), with LW 28.2 ± 3.7 μm^2 (20–36) and L/W 1.10 ± 0.11 (1.00–1.40). Gametocyte size relative to host cell nucleus size is 0.94 ± 0.17 (0.57–1.29, N = 48), and to normal erythrocyte nuclei is 1.04 ± 0.13 (0.74–1.32, N = 50). Pigment in meronts and gametocytes is similar in both Colombian and Venezuelan samples: a single, compact, usually round, golden-to-brown mass at the base of fans or centrally in rosette-shaped meronts, dispersed as prominent black granules in macrogametocytes, and formed as a marginal clump in microgametocytes. Dimensions and shape are similar in macro- and microgametocytes.

Exoerythrocytic Merogony Ayala and Spain (1976) reported the presence of a meront with six nuclei in a heterophil and commented that "no obvious exoerythrocytic schizonts were seen in the other 273 blood films, organ impression smears of five heavily infected lizards, or in tissue sections of three lizards." Presumably, this statement indicates they were observed, but the authors did not provide further information.

Sporogony Unknown.

Effects on Host Erythrocytes host to either meronts or gametocytes are not hypertrophied. Meronts do not distort host cells or their nuclei, except when very large, and seldom displace nuclei. Gametocytes sometimes distort erythrocytes and slightly distort nuclei, but often displace the latter.

Parasitemia averaged 2.7% in 255 infections, ranging from barely detectable to 61% (Ayala and Spain, 1976). Significant anemia and a distinct erythropoietic response were noted in infections, with the blood picture returning to normal during chronic phase of infection. During active infection, *P. colombiense* meronts predominantly parasitize immature erythrocytes, from erythroblasts to polychromatophilic erythrocytes, and gametocytes also occasionally occupy immature cells. Ayala and Spain suggested that "nearly complete replacement of the original erythrocytes had occurred" in many infections. They considered *P. colombiense* to be of "intermediate pathogenicity," perhaps equally pathogenic as *Plasmodium mexicanum* [the most pathogenic saurian *Plasmodium* species yet known] in view of the "small size and total blood volume of *A. auratus*."

Ecology No "regular seasonal alternation of active and latent infections" was found by Ayala and Spain (1976). Prevalence increased with host age and size, with 81% of the parasite population found in "middle age lizards," in females 45–49 mm snout-vent length (SVL), and in males 40–49 mm SVL. Prevalence appears to be affected more in the short term by variation in host population structure than by monthly variation in precipitation. Infection is uncommon before attainment of maturation size, 38–40 mm SVL. Two of the four infections found in lizards 34 mm or less were overwhelming and potentially fatal, and "the others were acute with high parasitemias." Oogenesis by female hosts did not affect infection by *P. colombiense*. Although local epidemics of *P. colombiense* occur, there is "little overall seasonal synchronization," a sharp contrast to the temperate zone *P. mexicanum*, "where the regional infection pattern is seen in all local populations" (Ayala and Spain, 1976). "Temperate Zone species show 'spring relapses'," coincident with host reproductive season, "while

P. colombiense shows active infections all year around" in its continuously reproductive tropical host.

Remarks Ayala and Spain (1976) presented a thorough study on *P. colombiense* from examination of blood films alone, without following natural or induced infections in the laboratory. In addition to the presentation of much information about the species, Dr. Ayala's color figures rival and perhaps exceed the quality of the color plates drawn by Esther Bohlman in Thompson and Huff (1944a).

Plasmodium fairchildi fairchildi Telford, Johnson and Young 1989 (**Plate 23**)

Diagnosis A *Plasmodium* (*Lacertamoeba*) species with typically elongate gametocytes and rosette or fan-shaped meronts. Meronts are 4–10 × 4–7 μm, LW 16–70 μm², that contain 6–12 merozoites. Meront size relative to host cell nucleus size averages 0.71, and to normal erythrocyte nuclei is 0.95. Pigment forms a golden-yellow mass in meronts, located centrally in rosettes or at the base of fan-shaped meronts. Gametocytes are 5–16 × 4–7 μm, with LW 25–75 μm² and L/W 1.00–4.00. Gametocyte size relative to host cell nucleus size averages 1.34, and to normal erythrocyte nuclei is 1.30. Dark brown pigment granules are dispersed in gametocytes. There are no sexual differences in gametocyte dimensions or shape.

Type Host *Anolis limifrons* Cope (Sauria: Polychrotidae).

Type Locality Pipeline road between Gamboa and Frijoles, Canal Zone, Panama.

Other Hosts *Anolis cupreus* (Perkins and Schall, 2002), *A. fuscoauratus* (Guerrero and Ayala, 1977).

Other Localities In Panama: Barro Colorado Island (Guerrero et al., 1977); Sasardi, San Blas Territory, Gaspar Sabanas-Madroño, 8 km north of Chepo, Panama Province; Quebrada Juan Grande and Frijolito Creek, between Gamboa and Frijoles, Canal Zone (Telford, 1977). Costa Rica. Region of Llullapichis, Huanuco, 300 km south of Pucallpa, Peru (Guerrero and Ayala, 1977).

Prevalence In *A. fuscoauratus*, 3 of 16 (18.8%, Guerrero and Ayala, 1977); in *A. limifrons*, overall 26 of 371 (7.0%); 2 of 107 (1.9%) at Sasardi, 2 of 7 at Gaspar Sabanas-Madroño, 6 of 18 (33.3%) at Quebrada Juan Grande, 10 of 73 (13.7%) at Frijolito Creek (Telford); and at Barro Colorado Island, 36 of 296 (12.2%, Guerrero et al., 1977).

Morphological Variation Meronts form as rosettes or fans, 6.8 ± 1.3 × 5.3 ± 0.9 μm (4–10 × 4–7, N = 44), with LW 36.5 ± 11.6 μm² (16–70), and contain 8.7 ± 1.5 (6–12) merozoites. Meront size relative to host cell nucleus size is 0.71 ± 0.15 (0.46–0.94, N = 10), and to normal erythrocyte nuclei is 0.95 ± 0.32 (0.40–1.85, N = 44). Erythrocytic meronts are smaller than those in proerythrocytes and produce fewer merozoites, respectively 5.3 ± 0.8 × 4.2 ± 0.4 μm (N = 10), LW 22.3 ± 4.3 μm², with 7.3 ± 0.8 merozoites versus 7.3 ± 1.1 × 5.6 ± 0.8 μm (N = 34), LW 40.6 ± 9.6 μm², and 9.1 ± 1.4 merozoites. Meronts in an entirely erythrocytic, blood-induced experimental infection of *A. limifrons* were nearly identical with erythrocytic meronts in natural infections: 5.5 ± 0.9 × 4.2 ± 0.5 μm (5–8 × 3–5, N = 25), LW 23.0 ± 4.8 μm² (18–40), with 7.8 ± 0.8 (6–10) merozoites. The golden-yellow pigment granules accumulate as a mass, centrally located in rosettes and at the base of fans. Gametocytes, usually elongate, are 10.3 ± 2.1 × 4.8 ± 0.7 μm (5–16 × 4–7, N = 51), with LW 49.2 ± 9.9 μm² (25–75) and L/W 2.17 ± 0.60 (1.00–4.00). Gametocyte size relative to host cell nucleus size is 1.34 ± 0.44 (0.52–2.67,N = 34), and to normal erythrocyte nuclei is 1.30 ± 0.26 (0.63–1.98, N = 51). In experimental infection of *A. limifrons*, gametocytes were smaller and more ovoid, 7.5 ± 1.5 × 4.1 ± 0.9 μm (5–10 × 3–7, N = 25), LW 31.3 ± 9.9 μm² (18–50), and L/W 1.88 ± 0.41 (1.00–2.50). Pigment is dispersed as dark brown granules in both sexes, which do not differ in dimensions or shape. Gametocytes commonly have irregular margins. In *Anolis cupreus*, meronts were 6.0 ± 0.6 × 4.7 ± 0.6 μm (5–7 × 4–6, N = 25), LW 28.4 ± 4.9 μm² (22.5–42), and contained 8.9 ± 1.9 (4–13) merozoites. Gametocytes were 11.1 ± 2.2 × 4.8 ± 0.5 μm (7–15 × 4–6, N = 15), with LW 52.7 ± 9.4 μm² (35–67.5) and L/W 2.37 ± 0.59 (1.40–3.33). Guerrero and Ayala (1977) reported *P. fairchildi* of *A. fuscoauratus* in Peru as forming rosette or fan-shaped, round, or lentiform meronts containing 8.8 ± 1.2 (6–12, N = 20) merozoites.

Exoerythrocytic Merogony Unknown.

Sporogony Unknown.

Effects on Host Both meronts and gametocytes commonly parasitize proerythrocytes, the latter about one-third as often as the meronts. Both stages cause host cell hypotrophy, and erythrocytes host to meronts have smaller nuclei than uninfected cells. Nuclei of host cells are often displaced but only rarely does either stage distort host cell or nucleus.

Remarks The "*P. tropiduri*-like" parasites reported by Telford (1974) from four Panamanian *Anolis* species were described by Telford (1979d) as *P. tropiduri panamense* from *A. biporcatus*, *P. tropiduri aquaticum* from *A. lionotus* and *A. poecilopus*, and *P. marginatum* from *A. frenatus*. During the intervening years from 1974 until 1989, the

"*P. tropiduri*-like" species in *A. limifrons* was referred to thus by Guerrero and Ayala (1977), Guerrero et al. (1977), and Ayala (1978). When adequate material for description became available, the species was designated *Plasmodium fairchildi fairchildi* (Telford et al., 1989).

Plasmodium fairchildi hispaniolae Telford, Johnson and Young 1989

Diagnosis A *Plasmodium* (*Lacertamoeba*) species with ovoid-to-elongate gametocytes and usually rosette or fan-shaped meronts, 5–9 × 3–5 μm, LW 24–45 μm², that contain 8–14 merozoites. Meront size relative to host cell nucleus size averages 1.01, and to normal erythrocyte nuclei is 1.01. Dark greenish-yellow-to-black pigment forms as one or more clusters in meronts. Gametocytes are 7–15 × 4–7 μm, with LW 35–66 μm² and L/W 1.14–3.75. Gametocyte size relative to host cell nucleus size averages 1.77, and to normal erythrocyte nuclei is 1.59. Although similar in size, macrogametocytes are more elongate than microgametocytes. Pigment granules in macrogametocytes are dispersed and more uniform in size than in microgametocytes, where the greenish-black granules are irregular in size and tend to be marginal in distribution.

Type Host *Anolis distichus* Cope (Sauria: Polychrotidae).

Type Locality Pedro Sanchez, El Seibo Province, Dominican Republic.

Other Hosts None known.

Other Localities None known.

Prevalence One of 19 (5.3%) of Hispaniolan *A. distichus* was infected by *P. fairchildi hispaniolae* (Telford et al., 1989).

Morphological Variation Meronts are most commonly rosettes or fans but may be oblong, oval, or elongated, 6.7 ± 1.0 × 4.7 ± 0.5 μm (5–9 × 3–5, N = 25), with LW 31.6 ± 5.4 μm² (24–45), and contain 10.7 ± 1.7 (8–14) merozoites. Meront size relative to host cell nucleus size is 1.01 ± 0.18 (0.71–1.46, N = 23), and to normal erythrocyte nuclei is 1.01 ± 0.17 (0.77–1.44, N = 25). Pigment forms one or more dark greenish-yellow-to-black clusters, usually central in rosettes or at the base of fans. Gametocytes are ovoid, somewhat elongated, 8.8 ± 1.4 × 5.7 ± 0.7 μm (7–15 × 4–7, N = 52), with LW 49.7 ± 8.3 μm² (35–66) and L/W 1.60 ± 0.41 (1.14–3.75). Gametocyte size relative to host cell nucleus size is 1.77 ± 0.31 (1.29–2.50, N = 25), and to normal erythrocyte nuclei is 1.59 ± 0.27 (1.12–2.12, N = 52). Gametocyte size is similar in both sexes, but macrogametocytes are more elongate than microgametocytes, with respective dimensions 9.5 ± 1.5 × 5.4 ± 0.7 μm (N = 27), LW 51.0 ± 9.0 μm², and L/W 1.81 ± 0.46 versus 8.1 ± 0.6 × 6.0 ± 0.6 μm (N = 25), LW 48.3 ± 7.4 μm², and L/W 1.36 ± 0.14. Greenish-black pigment granules are more uniform in size and better dispersed in macrogametocytes than in microgametocytes, where granules variable in size tend to be marginal in distribution.

Exoerythrocytic Merogony Unknown.

Sporogony Unknown.

Effects on Host Host cells are primarily erythrocytes, but occasionally proerythrocytes are host to meronts. Cells parasitized by either meronts or gametocytes are hypotrophied but never distorted by meronts and only rarely so by gametocytes. Erythrocyte nuclei are rarely distorted by either stage, are normal in size, but are sometimes displaced by meronts and usually displaced by gametocytes.

Remarks *Plasmodium fairchildi hispaniolae* resembles the nominate subspecies in appearance and morphometrically and was considered to be a subspecies on the latter grounds. Genomic comparison is necessary to confirm the relationship apparent from morphology.

Plasmodium iguanae Telford 1980 (**Plate 24**)

Diagnosis A *Plasmodium* (*Lacertamoeba*) species with meronts 4–9 × 3–7 μm, LW 12–49 μm², that contain 6–25 merozoites. Meront size relative to host cell nucleus size averages 0.95, and to normal erythrocyte nucleus size is 1.21. Erythrocytic meronts, usually fan-shaped, are smaller, more heavily pigmented, and produce fewer merozoites than meronts in proerythrocytes, which are frequently unpigmented and formed usually as rosettes. Pigment, when present in meronts, forms a large, golden mass. Gametocytes are round or ovoid, 5–9 × 5–7 μm, with LW 25–56 μm² and L/W 1.00–1.60, and more often occur in proerythrocytes than in mature host cells. When erythrocytic, gametocyte size relative to host cell nucleus size averages 1.66, and to normal erythrocyte nucleus size is 1.74. Despite their presence in immature host cells, gametocytes are always pigmented, with small black granules dispersed within the cytoplasm. Gametocytes are not sexually dimorphic in size or shape.

Type Host *Iguana iguana* (Linnaeus) (Sauria: Iguanidae).

Type Locality Fundo Vega Honda, 25 km north of Guanare, Portuguesa State, Venezuela.

Plate 24 (**A**) *Plasmodium iguanae* from *Iguana iguana*, Venezuela. Meronts, **a–h**; macrogametocytes, **i**, **j**; microgametocytes, **k–l**. (**B**) *Plasmodium uncinatum* from *Plica plica*, Guyana. Meronts, **a–d**; macrogametocyte, **e**; microgametocytes, **f–h**. (Figures **a**, **d** from Telford, S. R., Jr., *Int. J. Parasitol.,* 3, 829, 1973, with permission, Elsevier.)

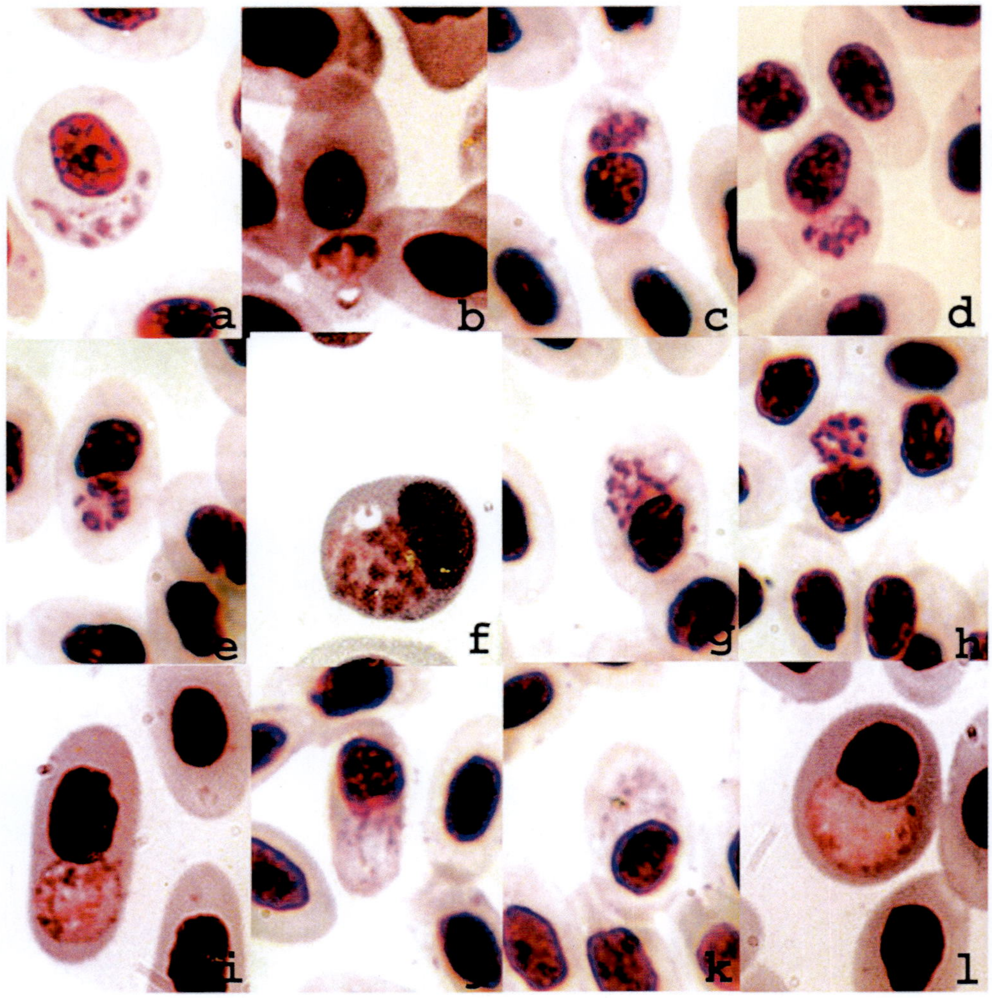

(A)

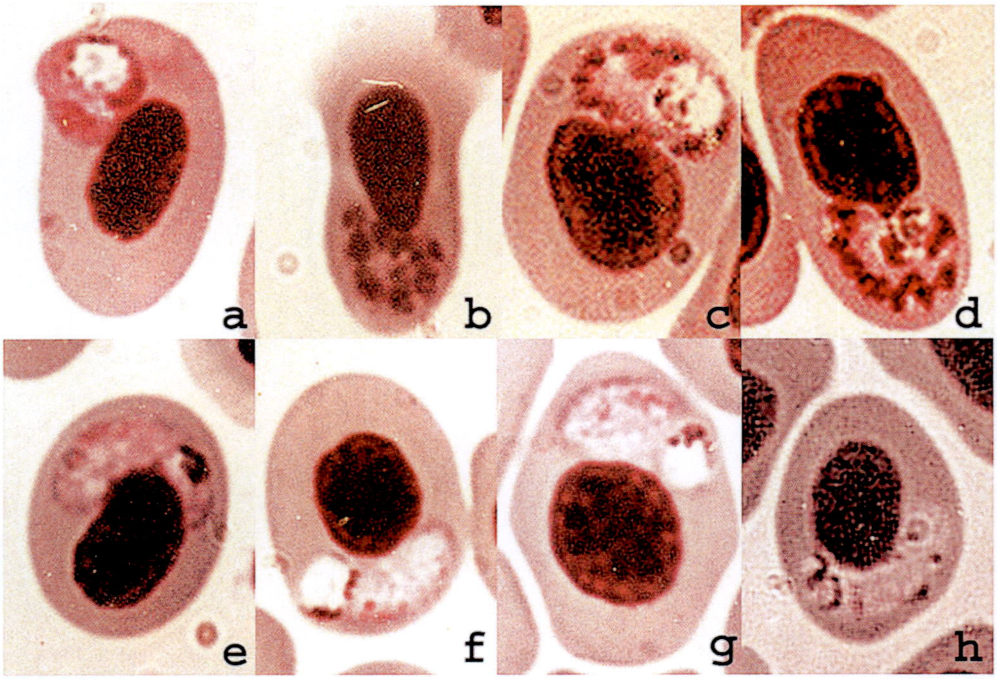

(B)

Other Hosts None known.

Other Localities None known.

Prevalence *Plasmodium iguanae* was present in 1 of 70 *Iguana iguana* examined in Venezuela and in 1 of 2 from the type locality (Telford, 1980).

Morphological Variation Meronts are 6.0 ± 1.3 × 4.9 ± 0.9 μm (4–9 × 3–7, N = 49), with LW 30.1 ± 10.3 μm^2 (12–49), and contain 13.7 ± 6.0 (6–25) merozoites. Meront size relative to host cell nucleus size is 0.95 ± 0.40 (0.43–2.00, N = 30), and to normal erythrocyte nucleus size is 1.21 ± 0.41 (0.48–2.02). Erythrocytic meronts usually form as fans, are smaller than those from immature host cells, and produce fewer merozoites: 5.6 ± 1.3 × 4.6 ± 0.9 μm (4–9 × 3–6, N = 30), LW 26.8 ± 10.2 μm^2 (12–48), with 11.4 ± 5.0 (6–21) merozoites versus 6.9 ± 1.1 × 5.3 ± 0.8 μm (5–9 × 4–7, N = 19), LW 35.3 ± 8.3 μm^2 (24–49), with 17.3 ± 5.7 (10–25) merozoites. Meronts in proerythrocytes or erythroblasts are usually formed as rosettes, and nearly half (44%) lack visible pigment. Pigment forms a prominent golden mass at the base of the fan in erythrocytic meronts. Gametocytes are 7.1 ± 1.0 × 6.0 ± 0.7 μm (5–9 × 5–7, N = 25), with LW 42.8 ± 9.4 μm^2 (25–56) and L/W 1.20 ± 0.18 (1.00–1.60). Gametocyte size relative to host cell nucleus size is 1.66 ± 0.40 (1.20–2.24, N = 6), and to normal erythrocyte nucleus size is 1.74 ± 0.37 (1.00–2.31, N = 25). Gametocytes are more common in proerythrocytic host cells (76%) than in erythrocytes, but all are pigmented, with small dark granules dispersed within the cytoplasm. There is no difference by sex in dimensions or shape of gametocytes.

Exoerythrocytic Merogony A single meront with ten nuclei, presumably of *P. iguanae*, was found in a monocyte or lymphocyte (Telford, 1980).

Sporogony Unknown.

Effects on Host Meronts usually distort their host cells and displace their nuclei but seldom distort the latter. Erythrocytes are not significantly hypertrophied, but their nuclei show an increase of 19% in size. Cells host to gametocytes are always distorted and have displaced nuclei. Proerythrocyte nuclei are not distorted, but those of host erythrocytes usually differ in shape from normal. The few erythrocytes seen parasitized by gametocytes were mildly hypotrophied, about 6% smaller than normal cells, but their nuclei were 29% larger than in normal cells.

Remarks Carini (1942) reported a *Plasmodium* species in *I. iguana* from Porto Nacional, Goiás, Brazil, the gametocytes of which were irregularly oval, 7–8 μm in diameter, with fine pigment granules. A single meront was found and figured by Carini, about 5.6 × 3.9 μm (estimated from scale bar), that contained 12 nuclei arranged as a rosette around a large central clump of pigment granules. The meront clearly resembles those of *P. iguanae* and suggests that the species occurs in Brazil as well as in Venezuela.

The gametocytes of *P. iguanae* are unusual among *Plasmodium* species in their apparent predilection for immature erythrocytes. The type infection was followed for 90 days, and this host cell preference did not change with time.

Plasmodium uncinatum Telford 1973 (**Plate 24**)

Diagnosis A *Plasmodium* (*Lacertamoeba*) species with meronts 6–12 × 3–7 μm, LW 24–63 μm^2, that contain 8–18 merozoites, usually fan-shaped or arranged as a rosette, sometimes elongate. Meront size relative to host cell nucleus size averages 1.13, and to normal erythrocyte nucleus size is 1.14. Gametocytes are elongate, 8–15 × 3–6 μm, LW 27–90 μm^2, and L/W 2.00–3.50. Gametocyte size relative to host cell nucleus size averages 1.23, and to normal erythrocyte nucleus size is 1.29. Pigment forms a large golden-yellow mass in meronts, comprising at least one-third of the meront size before maturity, then diminishing in extent. Pigment is similarly colored and forms a prominent mass at one end of immature gametocytes from which, at maturity, individual darker granules disperse within the cytoplasm. There are no sexual differences in dimensions or shape of gametocytes.

Type Host *Plica plica* (Linnaeus) (Sauria: Iguanidae).

Type Locality Vicinity of Georgetown, Guyana.

Other Hosts None known.

Other Localities None known.

Prevalence One of ten (10%) *Plica plica* were infected by *Plasmodium uncinatum* (Telford, 1973).

Morphological Variation Immature meronts are heavily pigmented with a golden-yellow mass at least one-third of their size, and often larger. As segmentation approaches, the pigment mass contracts somewhat but remains prominent. Meronts are 8.1 ± 1.5 × 4.9 ± 0.9 μm (6–12 × 3–7, N = 23), with LW 39.7 ± 9.5 μm^2 (24–63), and produce 13.5 ± 3.1 (8–18) merozoites. Meront size relative to host cell nucleus size is 1.13 ± 0.28 (0.60–1.79), and to normal erythrocyte nucleus size is 1.14 ± 0.27 (0.69–1.81. Their shape is variable, usually with merozoites arranged as a fan or a rosette, but sometimes in elongate form. Gametocytes when immature have prominent golden pigment formed as a large mass, usually at one end; but with maturity, darker granules disperse into the cytoplasm, leaving behind a still prominent golden clump of granules. Gametocytes are 10.6 ± 1.7 × 4.2 ± 0.7 µm (8–15 × 3–6, N = 29), with LW 44.7 ± 14.1 µm² (27–90) and L/W 2.55 ± 0.35 (2.00–3.50). Gametocytes are always elongate regardless of maturity and strongly tend to curve around the erythrocyte nucleus. Gametocyte size relative to host cell nucleus size is 1.23 ± 0.24 (0.77–1.75, N = 28), and to normal erythrocyte nucleus size is 1.29 ± 0.41 (0.78–2.59, N = 29). Gametocytes do not differ in dimensions or shape by sex.

Exoerythrocytic Merogony Unknown.

Sporogony Unknown.

Effects on Host Both meronts and gametocytes parasitize only erythrocytes, without distorting but commonly displacing the host cell nucleus, but distorting the cells. Erythrocytes host to meronts are hypertrophied by 22%, and with gametocytes, 24%, but host cell nuclei are normal in size.

Remarks *Plasmodium uncinatum* is known only from the type infection. The presence of consistently elongate gametocytes that usually curve closely around erythrocyte nuclei and the very prominent golden-yellow pigment masses in both meronts and gametocytes easily distinguish the species.

Plasmodium vacuolatum Lainson, Shaw and Landau 1975

Diagnosis A *Plasmodium* (*Lacertamoeba*) species characterized by the presence of prominent vacuoles in both asexual and sexual stages, which contain a large mass of light yellow-to-golden pigment granules. The variably shaped meronts are 4–7 × 3–5 µm, LW 16–30 µm², and contain 8–20 merozoites. Meront size relative to host cell nucleus size averages 0.80, and to normal erythrocyte nucleus size is 0.73. Gametocytes are 4–7 × 3–6 µm, with LW 16–42 µm² and L/W 1.00–2.00. Gametocyte size relative to host cell nucleus size averages 0.93 and to normal erythrocyte nucleus size is 0.81. Vacuoles tend to become less distinct with maturity of either meronts or gametocytes, but the pigment mass contained within remains localized with little or no dispersal, and in the case of gametocytes, tends to become flattened against one cell margin.

Type Host *Plica umbra* (Linnaeus) (Sauria: Iguanidae).

Type Locality Vicinity of Belém, Pará State, Brazil.

Other Hosts None known.

Other Localities Vicinity of Georgetown, Guyana (Telford, 1973).

Prevalence *Plasmodium vacuolatum* was identified in 31 of 235 (13.2%) *Plica umbra* in Brazil (Lainson et al., 1975), and in 3 of 7 *P. umbra* from Guyana (Telford, 1973).

Morphological Variation In Brazil, Lainson et al. (1975) described mature or nearly mature meronts as "bean-shaped, oval, rounded or fan-shaped", 4.0 × 3.0 to 6.0 × 4.0, "average 4.8 × 3.3". The data suggest LW values of 12–24 µm², averaging 15.8 µm². Merozoite number varied from 8 to 20, averaging 16. Meronts from Guyana are usually formed as a rosette or fan, occasionally as a morulum, or are elongate. Their dimensions are 5.5 ± 0.9 × 4.2 ± 0.5 µm (4–7 × 3–5, N = 24), with LW 22.7 ± 4.3 µm² (16–30), and they contain 11.7 ± 1.9 (8–14) merozoites. Meront size relative to host cell nucleus size is 0.80 ± 0.22 (0.46–1.40), and to normal erythrocyte nucleus size is 0.73 ± 0.14 (0.51–0.96). Lainson et al. (1975) reported dimensions of *P. vacuolatum* gametocytes in Brazilian *P. umbra* as "4.0 × 3.5 to 4.5 × 4.0, average 4.3 × 3.5" for macrogametocytes and microgametocytes as "3.5 × 3.0 to 4.5 × 4.5, average 4.0 × 3.5," oval or rounded in shape. These data suggest average LW values of 15.1 µm² and 14.0 µm² for the respective sexes, with L/W ratios 1.23 and 1.14. In Guyanan *P. umbra*, gametocytes are 5.4 ± 0.7 × 4.6 ± 0.8 µm (4–7 × 3–6, N = 25), with LW 25.4 ± 6.3 µm² (16–42) and L/W 1.20 ± 0.24 (1.00–2.00). Gametocyte size relative to host cell nucleus size is 0.93 ± 0.25 (0.63–1.75), and to normal erythrocyte nucleus size is 0.81 ± 0.20 (0.51–1.35). Gametocytes are apparently somewhat larger in the material from Guyana. The large vacuoles containing light yellow-to-golden pigment granules are at maximum size just prior to maturation, after which they become less conspicuous, although the pigment remains localized in a clump of granules in both meronts and gametocytes. In gametocytes, the pigment-containing vacuole appears smaller and is pressed tightly against the gametocyte margin.

Exoerythrocytic Merogony Unknown.

Sporogony Unknown.

Effects on Host Lainson et al. (1975) reported "no noticeable increase in the size of the infected erythrocyte," but there was sometimes mild displacement or indentation of the host cell nucleus. Around one-third of the infected erythrocytes in Guyanan *P. umbra* showed distortion of host cell and nucleus displacement but no nuclear distortion. Neither meronts nor gametocytes cause significant change in size of infected cells or their nuclei.

Remarks *Plasmodium vacuolatum* in *Plica umbra* of Guyana was reported as *Plasmodium tropiduri* by Telford (1973b), but differences in dimensions and merozoite numbers from those of *P. tropiduri* in *Tropidurus torquatus* were noted.

Plasmodium vautieri Pessôa and de Biasi 1973

Diagnosis A *Plasmodium* (*Lacertamoeba*) species with rounded meronts, and 10–20 merozoites usually arranged as rosettes, that apparently exceed gametocytes in size. Estimated meront dimensions are 4.8–7.5 × 3.5–6.0 µm with LW 18.2–31.5 µm^2. Estimated meront size relative to host cell nucleus size averages 0.81, and to normal erythrocyte nuclei is 0.86. Dark pigment granules aggregate as a mass centrally in the meront or at one of its margins. Gametocytes of both sexes are ovoid, estimated at 5.0–7.3 × 3.0–4.0 µm, with LW 17.1–25.1 µm^2 and L/W 1.45–2.43. Estimated gametocyte size relative to host cell nucleus size averages 1.02, and to normal erythrocyte nuclei is 0.64. Small pigment granules are dispersed in the cytoplasm. There is apparently no sexual difference in size or shape of gametocytes.

Type Host *Urostrophus vautieri* Duméril and Bibron (Sauria: Iguanidae).

Type Locality São Paulo State, Brazil.

Other Hosts None known.

Other Localities None known.

Prevalence Unknown.

Morphological Variation Meronts were described by Pessôa and de Biasi (1973) as rounded, 7–10 µm in diameter, and producing 10–20 merozoites. Dark pigment granules accumulate in the center or on one border of the meront, as a mass. Gametocytes are rounded or ovoid in both sexes, with microgametocytes 6–8 µm and macrogametocytes 9–10 µm. Pigment granules are small and (presumably) dispersed. Calculations from the photomicrographs of Pessôa and de Biasi (1973), at magnification of ×2000, provided the following morphometric dimensions and ratios. Merozoites are arranged as rosettes or large fans. Meronts are estimated as 6.0 × 4.6 µm (4.8–7.5 × 3.5–6), with LW 27.5 µm^2 (18.2–31.5). Their estimated size relative to host cell nucleus size is 0.81 (0.67–1.05), and to normal erythrocyte nuclei is 0.86 (0.61–1.05). Estimated gametocyte size is 5.7 × 3.4 µm (5.0–7.3 × 3.0–4.0), with LW 20.4 µm^2 (17.1–25.5) and L/W 1.70 (1.20–2.43). Estimated gametocyte size relative to host cell nucleus size is 1.02 (0.59–1.25), and to normal erythrocyte nuclei is 0.64 (0.50–0.73).

Exoerythrocytic Merogony Pessôa and de Biasi (1973) reported finding two EE meronts, both in monocytes of circulating blood.

Sporogony Unknown.

Effects on Host Erythrocytes host to meronts and gametocytes are apparently normal in size and appearance.

Remarks This *Plasmodium* species has not been reported since its inadequate description in 1973.

Plasmodium basilisci Peláez and Péres-Reyes 1959 (**Plate 25**)

Diagnosis A *Plasmodium* (*Lacertamoeba*) species with typically fan-shaped meronts in erythrocytes or formed as a rosette in immature host cells, 3–8 × 3–6 µm, LW 9–40 µm^2, containing 4–14 merozoites. Proerythrocytic meronts are larger and produce more merozoites than meronts in mature erythrocytes. Meront size relative to host cell nucleus size averages 0.35, and to normal erythrocyte nuclei is 0.56. Golden pigment granules form a single mass at the base of fan-shaped meronts or are usually central in rosettes. Gametocytes are ovoid or round, occasionally elongate, 6–10 × 5–8 µm, with LW 36–72 µm^2 and L/W 1.00–1.60. Gametocyte size relative to host cell nucleus size averages 1.47, and to normal erythrocyte nuclei is 1.38. Gametocyte dimensions and shape do not differ by sex. Prominent dark brown-to-black pigment granules are usually dispersed in both gametocyte sexes but may form loose aggregates in some gametocytes.

Type Host *Basiliscus vittatus* Wiegmann (Sauria: Corytophanidae).

Type Locality Laguna Encantada, San Andres Tuxtla, Veracruz State, Mexico.

Other Hosts *Basiliscus basiliscus* (Telford, 1972b), *B. galerita* (Ayala, 1978), *B. plumifrons* (Gorgas Memorial

Plate 25 (**A**) *Plasmodium basilisci* from *Basiliscus basiliscus*, Panama. Meronts, **a–f**; macrogametocytes, **g–i**; microgametocytes, **j–l**. (**B**) *Plasmodium telfordi* from *Ameiva ameiva*, Guyana, **b–g**, **i**, **l**, and Venezuela, **a**, **h**, **j**, **k**. Meronts, **a–h**; macrogametocytes, **i–k**; microgametocyte, **l**. (Figure **d** from Telford, S. R., Jr., *Int. J. Parasitol.,* 3, 829, 1973, with permission, Elsevier.)

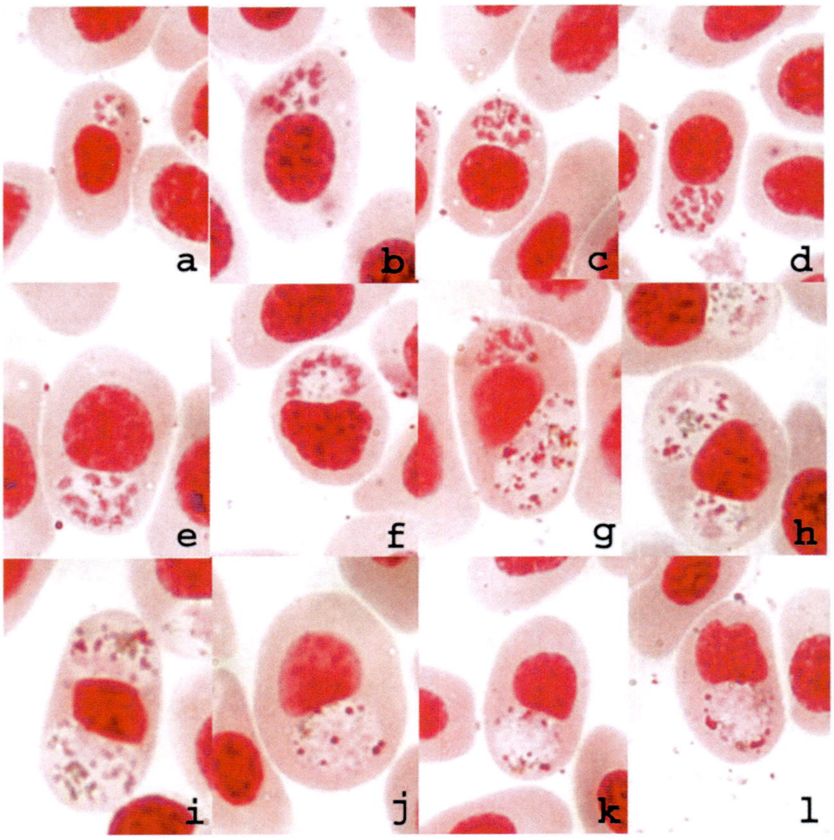

(A)

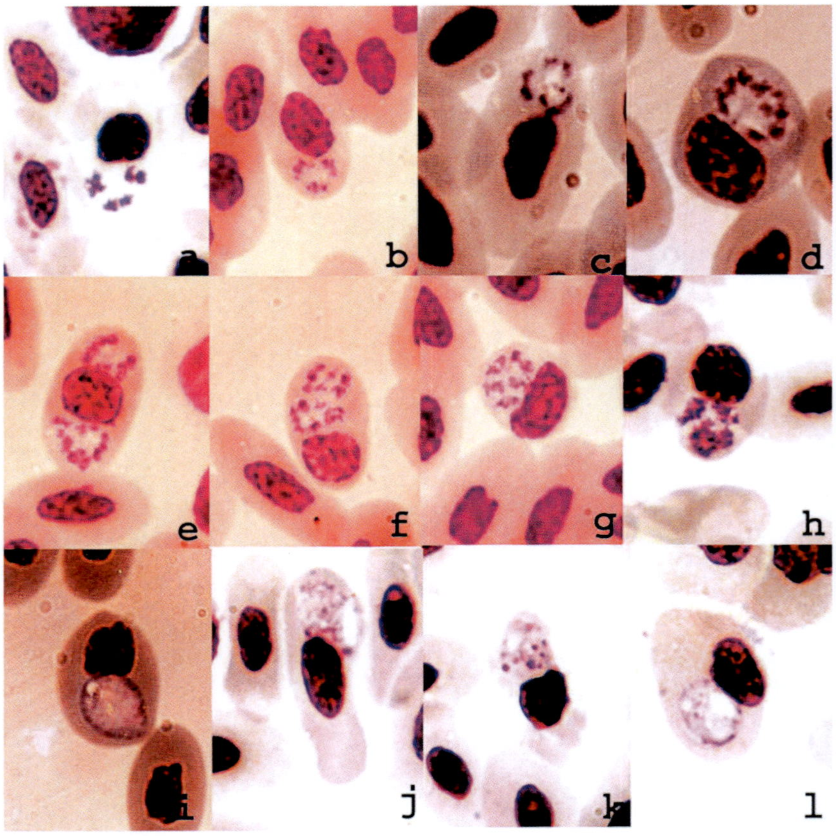

(B)

Laboratory, 1964), *Iguana iguana* (Herban and Coatney, 1969).

Other Localities Western Colombia (Ayala, 1978); El Salvador (Herban and Coatney, 1969); Belize (Garnham, 1966); Cortez, Charancaco, Honduras (Telford); Panama: Almirante (Gorgas Memorial Laboratory, 1964); Achiote, Colon Province; Sasardi, San Blas Territory; El Aguacate, Panama Province; Madden Forest, Canal Zone (Telford).

Prevalence In *B. vittatus*: 2 of 4 at San Andres Tuxtla (Peláez and Pérez-Reyes, 1959); 1 of 2 at Cortez, Honduras (Telford); 11 of 54 (20.4%) in Belize (Garnham, 1966); 34 of 215 (15.8%) at Almirante, Panama (Gorgas Memorial Laboratory, 1964). In *B. plumifrons*, 2 of 2 at Almirante, Panama. In *B. basiliscus*, Panama overall, 20 of 63 (31.7%); Achiote, 6 of 14 (42.9%); Sasardi, 5 of 11 (45.5%); El Aguacate, 8 of 20 (40%); Madden Forest, 1 of 2 (Telford). In *Iguana iguana*, in El Salvador, 2 of 2 (Herban and Coatney, 1969).

Morphological Variation Peláez and Peréz-Reyes (1959) described *P. basilisci* meronts in *B. vittatus* as 3.5–4.5 × 3–3.5 μm, usually formed as a fan, and containing five or six, rarely eight, merozoites. Young meronts rarely showed filamentous cytoplasmic projections. Gametocytes were ovoid, macrogametocytes 6–7 × 4–5 μm, and microgametocytes 6 × 5 μm. Pigment distribution differed between sexes, with fine brown granules dispersed in the cytoplasm of macrogametocytes or accumulated granules forming masses. In microgametocytes, pigment aggregated to form a single cluster. Herban and Coatney (1969) described meronts in *Iguana iguana* as 3 × 6 μm, with six to nine (average seven) merozoites, elongate macrogametocytes 10 × 5 μm, and irregularly shaped microgametocytes 4 × 8 μm. Garnham (1966) described meronts of *P. basilisci* from *B. vittatus* in Belize as often fan-shaped, 4 μm in diameter, with four or five, rarely eight, merozoites and black pigment at the base of the fan. Microgametocytes were about 6 μm and macrogametocytes up to 8 μm in length. In *Basiliscus basiliscus* of Panama, meronts are usually fan-shaped, occasionally elongate, a rosette, or a morulum. Meronts are 5.2 ± 1.3 × 3.8 ± 0.8 μm (3–8 × 3–6, N = 95), with LW 20.3 ± 8.3 μm^2 (9–40), and contain 7.2 ± 2.7 (4–14, N = 100) merozoites. Meront size relative to host cell nucleus size is 0.35 ± 0.06 (0.22–0.52, N = 49), and to normal erythrocyte nuclei is 0.56 ± 0.27 (0.19–1.24, N = 95). Erythrocytic meronts are smaller and produce fewer merozoites, 4.3 ± 0.9 × 3.4 ± 0.6 μm (N = 49), LW 15.1 ± 4.4 μm^2, with 5.5 ± 1.8 (4–10, N = 51) merozoites, than those occupying proerythrocytes, 6.0 ± 1.2 × 4.2 ± 0.8 μm (3–8 × 3–6, N = 46), LW 25.9 ± 7.9 μm^2 (9–40), with 9.0 ± 2.3 (4–14, N = 49) merozoites. Gametocytes are round to elongate, usually ovoid, 8.0 ± 0.9 × 6.8 ± 0.6 μm (6–10 × 5–8, N = 50), with LW 53.0 ± 9.8 μm^2 (36–72) and L/W 1.21 ± 0.13 (1.00–1.60). Gametocyte size relative to host cell nucleus size is 1.47 ± 0.37 (0.86–2.52, N = 34), and to normal erythrocyte nuclei is 1.38 ± 0.35 (0.91–2.22, N = 50). Gametocytes do not differ in dimensions by sex, but microgametocytes more commonly appear to be elongate than do macrogametocytes. Pigment in meronts forms a deep golden mass at the base of fans but is less conspicuous in proerythrocytic meronts, which usually form as a rosette. Dark pigment granules are dispersed in both sexes.

Exoerythrocytic Merogony Telford (1972b) reported that thrombocytic meronts containing 10–16 nuclei were commonly seen in all infections studied, averaging 12.8 ± 1.2 nuclei in a series of 11. In blood-induced experimental infections, "EE-schizonts appeared early following patency and persisted throughout the course of infections." Most were thrombocytic, but occasional lymphocytic meronts were observed.

Sporogony Unknown.

Effects on Host Infected cells, either mature or immature, were normal in size in comparison to uninfected cells, with nuclei also of normal size. Proerythrocytes hosted either meronts or gametocytes, and their nuclei were usually distorted and the nuclei often displaced. Nuclei usually appeared somewhat "disorganized" when distorted (Telford, 1972b). In erythrocytic infections, distortion of cell and nucleus and nucleus displacement occurred about one-half as often as when parasites occupied proerythrocytes.

Remarks Scorza (1970a) reported *P. basilisci* from two Venezuelan hosts, *Tropidurus hispidus* and *Ameiva ameiva*. Slides from the latter host (Telford, 1980) showed infection by *Plasmodium attenuatum*, which was undescribed until 1973, and not *P. basilisci*. Identity of the parasites reported as *P. basilisci* from *T. hispidus* is less clear. Meronts were 3.8 × 4–4 × 4.5 μm and contained seven or eight merozoites (Scorza, 1970). Microgametocytes were round, 5.5–6 to 7 × 4.5 μm, and macrogametocytes smaller, 5–5.5 μm in diameter, round or irregular in shape. These could represent an undescribed species. Scorza's comment that "Infections by *P. basilisci* were always chronic with a low parasitemia" tends to support the hypothesis that these infections were chronic-phase *P. tropiduri*. Telford (1979d) described crisis stages of *P. tropiduri* in both *Tropidurus torquatus* and *T. hispidus,* with both meronts and gametocytes smaller in dimensions than in active or fully chronic infections and with fewer merozoites. In *T. hispidus* and *T. torquatus*, *P. tropiduri* meronts were 3–5 × 2–4 μm, with respective means 4.1 × 3.6 μm and 4.4 × 3.7 μm, LW values 14.5–16.5 μm^2 (9–20), and 4–10 nuclei, averaging 6.0 and

6.4–7.8, respectively. Gametocytes in both species at crisis were 5–9 × 4–6 μm, means 5.6–6.7 × 4.7–5.4 μm, and LW 26.4–34.6 μm^2 (20–54). Given the similarity in size of crisis-stage meronts and gametocytes to "*P. basilisci*" of *Tropidurus hispidus*, it is likely that these were crisis phase infections of *P. tropiduri*.

Plasmodium telfordi (Lainson, Landau and Shaw) 1971 (**Plate 25**)

Diagnosis A *Plasmodium* (*Lacertamoeba*) species in which meronts and gametocytes within a single infection may entirely lack pigment, or be fully pigmented, or a mixture of both unpigmented and pigmented stages. The presence of pigment is dependent on maturity of host erythrocytes, with absence of pigment typical of infections in immature host cells, but unpigmented meronts and gametocytes also occur in mature erythrocytes. Nearly mature meronts often show nuclei in lobes, resembling "cauliflower heads." Meronts shaped as fans or rosettes are 3–9 × 3–7 μm, LW 9–49 μm^2, and contain 5–24 merozoites. Proerythrocytic meronts are larger and produce more merozoites than those occupying erythrocytes. Meront size relative to host cell nucleus size averages 0.62–0.70, and to normal erythrocyte nuclei is 0.80–1.52. Pigment, when present, forms as a golden-yellow mass at the base of fans or centrally in rosettes. Gametocytes are round to elongate, usually ovoid, and are 5–10 × 3–8 μm, with LW 20–64 μm^2 and L/W 1.00–2.67. Gametocyte size relative to host cell nucleus size averages 1.31–1.48, and to normal erythrocyte nuclei is 1.55–1.66. Dark pigment granules are dispersed in gametocytes, when present. There is no sexual dimorphism in gametocyte size.

Type Host *Ameiva ameiva ameiva* (Linnaeus) (Sauria: Teiidae).

Type Locality Kilometer 260 on Xavantina-Cachimbo road, Mato Grosso State, Brazil.

Other Hosts *Ameiva ameiva praesignis, A. ameiva vögli.*

Other Localities Vicinity of Georgetown, Guyana (Telford, 1973). In Venezuela: Portuguesa State, Araure, Hoja Blanca, and Los Tanques in Municipio Araure and San Jorge in Municipio Pirítú; Los Cumbitos, Municipio Ortiz, Guarico State (Telford, 1980).

Prevalence In Guyana, 1 of 16 (6.3%) (Telford, 1973). In Venezuela, 11 of 131 (8.4%) overall; 6 of 64 (9.4%) in Municipio Araure, 3 of 30 (10%) in Municipio Pirítú, and 2 of 21 (9.5%) in Muncipio Ortiz.

Morphological Variation In the type infection (Lainson et al., 1971), all parasites were unpigmented and occupied mature erythrocytes. Meronts averaged 5.9 × 4.9 μm (4.8–6.8), with estimated LW 28.9 μm^2, and contained 6–12 (usually 8 or 10) merozoites. Meronts were commonly lobulated, resembling cauliflower heads. Gametocytes were ovoid or lentiform. Macrogametocytes averaged 7.0 × 4.5 μm (6.3 × 3.8–7.6 × 6.3, N = 50), with estimated LW 31.5 μm^2 (23.9–47.9) and L/W 1.56 (1.21–1.66), and microgametocytes were 6.5 × 4.5 μm (6.3 × 3.8–7.5 × 5.0), with estimated LW 29.3 μm^2 (23.9–37.5) and L/W 1.44 (1.50–1.66). In a natural infection of *Ameiva a. ameiva* from Guyana, meronts were formed either as rosettes (mostly in proerythrocytes and unpigmented) or as fans (mostly erythrocytic and pigmented). Cauliflower-shaped meronts were commonly seen but not measured. Twice as many erythrocytic meronts are fan-shaped than are rosettes. Meronts are 5.6 ± 1.2 × 4.6 ± 0.9 μm (4–8 × 3–6, N = 30), with LW 26.7 ± 9.1 μm^2 (12–48), and contain 10.1 ± 2.8 (5–16) merozoites. Meront size relative to host cell nucleus size is 0.70 ± 0.12 (0.57–0.86, N = 6), and to normal erythrocyte nuclei is 1.11 ± 0.39 (0.58–1.97, N = 30). Erythrocytic meronts are smaller and produce fewer merozoites than those in proerythrocytes, respectively 5.1 ± 1.6 × 4.2 ± 1.3 μm (N = 12), LW 21.6 ± 7.1 μm^2, with 7.7 ± 2.4 merozoites versus 6.1 ± 1.6 × 4.9 ± 1.2 μm (N = 18), LW 30.1 ± 7.8 μm^2, with 11.7 ± 3.0 merozoites. Gametocytes, mostly erythrocytic and all but one pigmented, are 6.4 ± 0.7 × 5.0 ± 0.7 μm (5–8 × 3–6, N = 29), with LW 32.2 ± 6.2 μm^2 (20–48) and L/W 1.31 ± 0.35 (1.00–2.67). Gametocyte size relative to host cell nucleus size is 1.40 ± 0.37 (0.71–2.33, N = 25), and to normal erythrocyte nuclei is 1.55 ± 0.28 (0.99–2.08, N = 29).

In *A. ameiva* ssp. in Venezuela, cauliflower-shaped meronts were present in both pigmented and unpigmented infections. Erythrocytic meronts are 4.6 ± 0.8 × 3.5 ± 0.6 μm (3–6 × 3–5, N = 25), with LW 16.5 ± 4.4 μm^2 (9–25), and contain 8.8 ± 2.0 (6–12) merozoites. Meront size relative to host cell nucleus size is 0.62 ± 0.18 (0.38–1.04, N = 20) and to normal erythrocyte nuclei is 0.80 ± 0.21 (0.44–1.22, N = 25). Proerythrocytic meronts are mostly irregular in shape, with very few describable as fans or rosettes, and are larger, with nearly twice as many merozoites on average than those from erythrocytic infection. Meronts are 6.8 ± 0.8 × 5.1 ± 0.8 μm (6–9 × 4–7, N = 17), with LW 34.4 ± 7.7 μm^2 (24–49), and contain 15.2 ± 3.6 (10–24) merozoites. Meront size relative to normal erythrocyte nuclei is 1.52 ± 0.38 (1.10–2.55). In pigmented meronts, golden-yellow pigment granules formed a conspicuous mass at the base of fans or center of rosettes. Pigmented gametocytes are 7.0 ± 1.0 × 4.8 ± 0.7 μm (5–9 × 4–6, N = 25), with LW 33.6 ± 5.7 μm^2 (25–54) and L/W 1.49 ± 0.35 (1.00–2.25). Gametocyte size relative to host cell nucleus size is 1.31 ± 0.41 (0.71–2.70, N = 24), and to normal erythrocyte nuclei is

1.64 ± 0.28 (1.22–2.63, N = 25). Pigment is dispersed as dark granules in both gametocyte sexes. There is no difference in gametocyte dimensions by sex, but microgametocytes average greater in L/W ratio, 1.61, than do macrogametocytes, 1.44. In the unpigmented infection, 3 of 75 gametocytes are lightly pigmented, 68% of gametocytes occupy proerythrocytes, and 32% erythrocytes. Gametocytes are 7.5 ± 1.1 × 5.1 ± 0.9 μm (5–10 × 4–8, N = 75), with LW 38.3 ± 8.2 μm² (24–64) and L/W 1.52 ± 0.36 (1.00–2.50). Gametocyte size relative to host cell nucleus size is 1.48 ± 0.45 (0.86–2.70, N = 24), and to normal erythrocyte nuclei is 1.66 ± 0.39 (0.94–2.92, N = 75). In this infection, macrogametocytes have greater L/W ratios, 1.70, in comparison to 1.41 in microgametocytes, indicating a more elongate shape, on average, but dimensions are similar.

Exoerythrocytic Merogony In an experimental infection induced by blood inoculation from the naturally infected *Ameiva ameiva* from Guyana (Telford, 1973), meronts appeared in thrombocytes and lymphocytes on day 7 of patency. Their dimensions were 6.1 × 4.2 μm (5–9 × 3–7), and they contained 11.4 nuclei (8–20). They disappeared from circulating blood after day 24 and did not reappear before death of the host on day 101, when it was apparently negative.

Sporogony Unknown.

Effects on Host Neither meronts nor gametocytes noticeably affect dimensions of host cells or their nuclei. Meronts more often distort host cells than do gametocytes, nuclei are seldom distorted, but both stages commonly displace erythrocyte nuclei (Telford, 1973). In Venezuelan infections (Telford, 1980), pigmented meronts seldom distorted cells or their nuclei but occasionally displaced the latter. Unpigmented meronts caused slightly more distortion of host cell nuclei and almost always displaced them. Pigmented gametocytes caused slightly more distortion of host cells and their nuclei than did unpigmented gametocytes, but the latter displaced nuclei slightly more.

Remarks The presence or absence of pigment in meronts and gametocytes of *P. telfordi* is apparently related to host cell maturity in some infections (Telford, 1973, 1980). Unpigmented parasites were produced in experimental infection from a natural infection comprised of both unpigmented and pigmented parasites (Telford, 1973). Certainly, the type infection consisted of unpigmented parasites in mature erythrocytes (Lainson et al., 1971). Without genomic comparison of pigmented versus unpigmented infections, the significance of pigment in systematic arrangements will remain controversial.

Plasmodium brumpti Peláez and Pérez-Reyes 1952

Diagnosis A *Plasmodium* (*Lacertamoeba*) species with variably shaped meronts 7–13 × 4–7 μm, LW 28–65 μm², that contain 12–22 merozoites. Meront size relative to host cell nucleus size averages 2.25, and to normal erythrocyte nuclei is 2.16. Golden pigment granules aggregate in one or more clumps variably located in meronts. Gametocytes are ovoid when polar in position to usually elongate, 8–14 × 4–7 μm, with LW 32–84 μm² and L/W 1.14–3.50. Gametocyte size relative to host cell nucleus size averages 2.53, and to normal erythrocyte nuclei is 2.33. Pigment is dispersed as fine golden granules in both sexes of gametocytes, which are not sexually dimorphic in dimensions, although microgametocytes tend to be more ovoid in shape than macrogametocytes.

Type Host *Sceloporus h. horridus* Wiegmann (Sauria: Phrynosomatidae).

Type Locality Central region of Morelos State, Mexico, here restricted to Coatetelco.

Other Hosts None known.

Other Localities Alpuyeca and Puente de Ixtla, Morelos State (Peláez and Pérez-Reyes, 1952).

Prevalence *Plasmodium brumpti* infected one of two *S. horridus* from Alpuyeca, three of seven from Coatetelco, and one of three from Puente de Ixtla.

Morphological Variation Peláez and Pérez-Reyes (1952) described mature meronts of *P. brumpti* as 8–12 × 4.5–6 μm, containing 15–20 merozoites. In polar positions, meronts were ovoid or reniform, but when lateral, formed as a fan. Macrogametocytes in polar position were 7.5–10 × 5–8.5 μm, and when lateral, 10–12 × 4–5 μm. Microgametocytes in the same respective positions were 7–9 × 5–6.5 μm and 10–12 × 5 μm, respectively. On a topotypic slide from Coatetelco, meronts are 9.3 ± 1.9 × 5.3 ± 1.1 μm (7–13 × 4–7, N = 23), with LW 48.3 ± 10.5 μm² (28–65), and contain 17.4 ± 3.1 (12–22) merozoites. Meront size relative to host cell nucleus size is 2.25 ± 0.56 (1.33–3.33, N = 15), and to normal erythrocyte nuclei is 2.16 ± 0.47 (1.25–2.90, N = 23). Meronts rarely form as fans but usually are elongate or with merozoites arranged as a rosette. Pigment forms one or more clumps of golden granules. Gametocytes are 10.7 ± 1.7 × 4.9 ± 0.9 μm (8–14 × 4–7, N = 25), with LW 52.1 ± 12.5 μm² (32–84) and L/W 2.25 ± 0.52 (1.14–3.50). Gametocyte size relative to host cell nucleus size is 2.54 ± 0.76 (1.1–4.0), and to normal erythrocyte nuclei is 2.33 ± 0.56

(1.43–3.75). Gametocytes are similar in dimensions by sex: Macrogametocytes are 10.9 × 4.8 μm, LW 51.9 μm^2, and microgametocytes are 10.2 × 5.1 μm, LW 52.4 μm^2. Microgametocytes, however, are less elongate than macrogametocytes, with respective L/W ratios of 2.01 ± 0.28 (N = 9) versus 2.39 ± 0.60 (N = 16). Fine golden pigment granules are dispersed in gametocyte cytoplasm in both sexes.

Exoerythrocytic Merogony Unknown.

Sporogony Unknown.

Effects on Host Host erythrocytes are seldom distorted by either meronts or gametocytes, and their nuclei are never distorted but are often displaced. Infected erythrocytes and their nuclei are of normal size.

Remarks Although Morelos State lies within the neotropical region, the type host, *Sceloporus horridus,* is nearctic in its relationships, and *Plasmodium brumpti* may be better considered as a North American species. The topotypic slide I examined was given to me by Professor J. V. Scorza, who had obtained it from Professor D. Peláez. *Plasmodium brumpti* has not been reported since its description.

Plasmodium pelaezi Malagón and Salmeron 1988

Diagnosis A *Plasmodium* (*Lacertamoeba*) species with variably shaped, usually round, meronts 5.4–9.5 × 4.8–8.2 μm, LW 25.9–77.7 μm^2, that produce 16 merozoites. Meront size relative to host cell nucleus size and to normal erythrocyte nuclei averages 2.15. Pigment forms a single, black mass located marginally in the meront. Gametocytes are round or ovoid to broadly lentiform in shape, 4.1–9.5 × 2.7–6.8 μm, with LW 13.1–64.7 μm^2 and L/W 1.00–2.00. Dimensions are similar between sexes, but macrogametocytes are more ovoid than microgametocytes. Gametocyte size relative to host cell nucleus size averages about 1.36, and to normal erythrocyte nuclei is 1.41. Golden-brown pigment granules are dispersed or form several small masses within the cytoplasm.

Type Host *Urosaurus b. bicarinatus* (Dumeril) (Sauria: Phrynosomatidae).

Type Locality Chila de la Sal, southern Puebla State, Mexico.

Other Hosts None known.

Other Localities None known.

Prevalence Two of 12 (16.7%) *U. bicarinatus* were infected by *P. pelaezi* (Malagón and Salmeron, 1988).

Morphological Variation Meronts of *P. pelaezi* were described by Malagón and Salmeron (1988) as "round, with a tendency for some to become oval or elongate when they get older." Large vacuoles containing the pigment form in some meronts but disappear before segmentation.

Meronts are 7.8 ± 1.2 × 6.0 ± 0.8 μm (5.4–9.5 × 4.8–8.2, N = 25), with LW 47.6 ± 11.7 μm^2 (25.9–77.7). Meronts are said to contain "exactly 16 merozoites" when mature. Estimated meront size relative to host cell nucleus size and to normal erythrocyte nuclei averages 2.15 (1.17–3.50). A single black pigment mass is located at one margin of the meront. Gametocytes are 4.1–9.5 × 2.7–6.8 μm, with LW 13.1–64.7 μm^2 and L/W 1.00–2.00. Estimated gametocyte size relative to host cell nucleus size averages 1.36 (0.57–2.91), and to normal erythrocyte nuclei is 1.41 (0.59–2.92). Dimensions of macrogametocytes average 6.5 ± 1.3 × 4.6 ± 1.0 μm (4.6–9.5 × 2.8–6.8, N = 25), LW 30.7 ± 11.6 μm^2 (14.8–64.7), and L/W 1.44 ± 0.22 (1.33–2.00), and microgametocytes are 6.4 ± 1.1 × 4.8 ± 0.9 μm (4.1–8.2 × 2.7–6.1, N = 25), LW 31.8 ± 10.1 μm^2 (13.1–55.5), and L/W 1.31 ± 0.20 (1.00–2.00). The difference in L/W ratio indicates that macrogametocytes are more ovoid on average than are microgametocytes. Pigment is golden brown, "present in three forms: as individual granules; as several small masses; and in a mixture of small masses and isolated granules: in all cases it was dispersed in the cytoplasm" (Malagón and Salmeron, 1988).

Exoerythrocytic Merogony Unknown.

Sporogony Unknown.

Effects on Host Meronts usually parasitize erythrocytes but may occur in immature host cells as well. Meronts do not affect host erythrocyte size except to slightly lengthen the cell and reduce its width. Nucleus size is not increased by the presence of meronts, but they become more rounded and are displaced. Erythrocytes host to gametocytes are hypotrophied and more rounded. Nuclei are displaced and normal in size but more rounded in shape than uninfected erythrocyte nuclei.

Remarks Malagón and Salmeron (1988) presented all of the dimensional data necessary for objective comparison of *P. pelaezi* with other well-described parasites, an exception to the usually inadequate taxonomic descriptions of saurian *Plasmodium* species. If *P. pelaezi* truly produces a precise number of merozoites (16), then it is unique among reptilian *Plasmodium* parasites, which always vary to some degree in merozoite number.

Plasmodium josephinae Peláez 1967

Diagnosis A *Plasmodium* (*Lacertamoeba*) species with variable, but usually fan-shaped meronts preceded by elongated fusiform, often very large, immature meronts that tend to nearly encircle the host cell nucleus. Meronts are 6–16.5 × 4–5.3 µm, with estimated LW 24.0–87.5 µm², and produce 7–22 merozoites. Estimated meront size relative to host cell nucleus size averages 1.72, and to normal erythrocyte nuclei is 2.01. Immature gametocytes are elongate with acuminate ends when young and form rounded to ovoid macrogametocytes and ovoid or ellipsoidal microgametocytes. Gametocytes are 7.5–8.7 × 5.5–6.5 µm with estimated LW 41.2–56.6 µm² and L/W 1.33–1.36. Estimated gametocyte size relative to host cell nucleus size averages 2.76, and to normal erythrocyte nuclei is 2.81. Dark brown pigment granules form a central or peripheral mass in meronts. Brownish-black pigment granules are prominent, abundant, and dispersed in gametocytes. Dimensions of gametocytes are similar between sexes.

Type Host *Ameiva undulata amphigramma* Smith and Laufe (Sauria: Teiidae).

Type Locality San Andrés Tuxtla, Veracruz State, Mexico.

Other Hosts None known.

Other Localities None known.

Prevalence Five of 16 (31.3%) *Ameiva undulata* were infected by *P. josephinae* at the type locality (Peláez, 1967).

Morphological Variation Young meronts of *P. josephinae* are elongate and amoeboid, with pointed ends. As they increase in size, they may become almost fusiform and tend to curve around the erythrocyte nucleus, nearly encircling it in some cases. Mature meronts usually form large fans or, less often, a rosette, or are oval to elongate in shape. Mature meronts and segmenters are 6–16.5 × 4–5.5 µm, with estimated LW 24.0–87.5 µm², and contain 7–22 merozoites, averaging 14.1. As estimated from the figures, meront size relative to host cell nucleus size averages 1.72 (1.19–2.45), and to normal erythrocyte nuclei is 2.01 (1.61–2.34). Gametocytes are 7.5–8.7 × 5.5–6.5 µm, with estimated LW 41.2–56.6 µm² and L/W 1.33–1.36. Estimated gametocyte size relative to host cell nucleus size averages 2.76 (1.89–4.26), and to normal erythrocyte nuclei is 2.81 (1.63–4.01). Dark brown pigment granules form a prominent clump at the base of fan-shaped meronts, central in rosettes, or variably situated in elongate forms. Macrogametocytes are 7.5–8.7 × 6–6.5 µm, and microgametocytes are 8.2–8.7 × 5.5–6.5 µm, "aproximadamente el mismo tamano que los macrogametocitos" (Peláez, 1967). Macrogametocytes are ovoid to almost round, microgametocytes ovoid or ellipsoidal. Mature gametocytes are preceded by "pregametocitos," which can be distinguished by sex from their staining reaction. These are 9.5–13.5 × 3.5–4.5 µm in females and 9.5–12.5 × 4–4.2 µm in males. They are elongate or narrowly reniform in both sexes, with dispersed pigment granules, and occupy lateral positions in the cell. Other observers would probably consider them to be mature in addition to the more rounded forms, polar in host cells, that are described as mature gametocytes. Microgametocytes are slightly less heavily pigmented than macrogametocytes, with 34–45 dark brown, dispersed granules in comparison to 50–60 in macrogametocytes.

Exoerythrocytic Merogony Unknown.

Sporogony Unknown.

Effects on Host Peláez (1967) provided little information on host cell effects from being parasitized except to comment that gametocytes cause "very little hipertrophy [sic] and distortion of host cell and nuclear displacement." Meronts and segmenters may cause a slight hypertrophy of the cell, with some increase in its length, but usually without displacing the nuclei. Nuclei appear to be normal in size and shape as figured by Peláez.

Remarks Peláez (1967) briefly described exflagellation of microgametocytes and illustrated the cells, showing two to five flagella extending from the two-to-five nucleate body of the exflagellate. His illustrations of *P. josephinae* are among the best available of saurian *Plasmodium* species.

Plasmodium pifanoi Scorza and Dagert 1956, Telford and Telford 2003 (**Plate 26**)

Diagnosis A *Plasmodium* (*Paraplasmodium*) species with meronts 4–8 × 3–6 µm that contain 7–16 merozoites arranged variably within, with LW 16–42 µm². Meront size relative to host erythrocyte nucleus is 0.51–1.56, and to normal erythrocyte nuclei is 0.73–1.93. Gametocytes are usually elongate, 8–16 × 4–10 µm, with LW 52–112 µm² and L/W 1.05–3.25. Gametocyte size relative to infected erythrocyte nucleus size is 2.07–5.48, and to normal erythrocyte nuclei is 2.39–4.36. Large round vacuoles are always present in gametocytes. Gametocytes are sexually dimorphic in size, macrogametocytes are longer than microgametocytes with greater LW, but width and L/W ratio do not differ.

Type Host *Ameiva ameiva ameiva* (Linnaeus) (Sauria: Teiidae).

Plate 26 (**A**) *Plasmodium pifanoi* from *Kentropyx calcarata*, Venezuela. Meronts, **a–f**; macrogametocytes, **g–i**; microgametocytes, **j–l**. (Figures **d**, **e** modified from Telford, S. R., Jr., and Telford, S. R., III, *J. Parasitol.,* 89, 362, 2003, Figures 2 and 4, with permission.) (**B**) Exoerythrocytic meronts of three New World *Plasmodium* species: *P. floridense* from *Anolis carolinensis*, Florida, **a–d**; *P. aurulentum* from *Thecadactylus rapicaudus*, Panama, **e**, **f**; meronts in spleen of *Gonatodes albogularis* inoculated with blood of *Mabuya mabouya*, Panama, heavily infected with *P. diploglossi* and *P. morulum* in mixed infection.

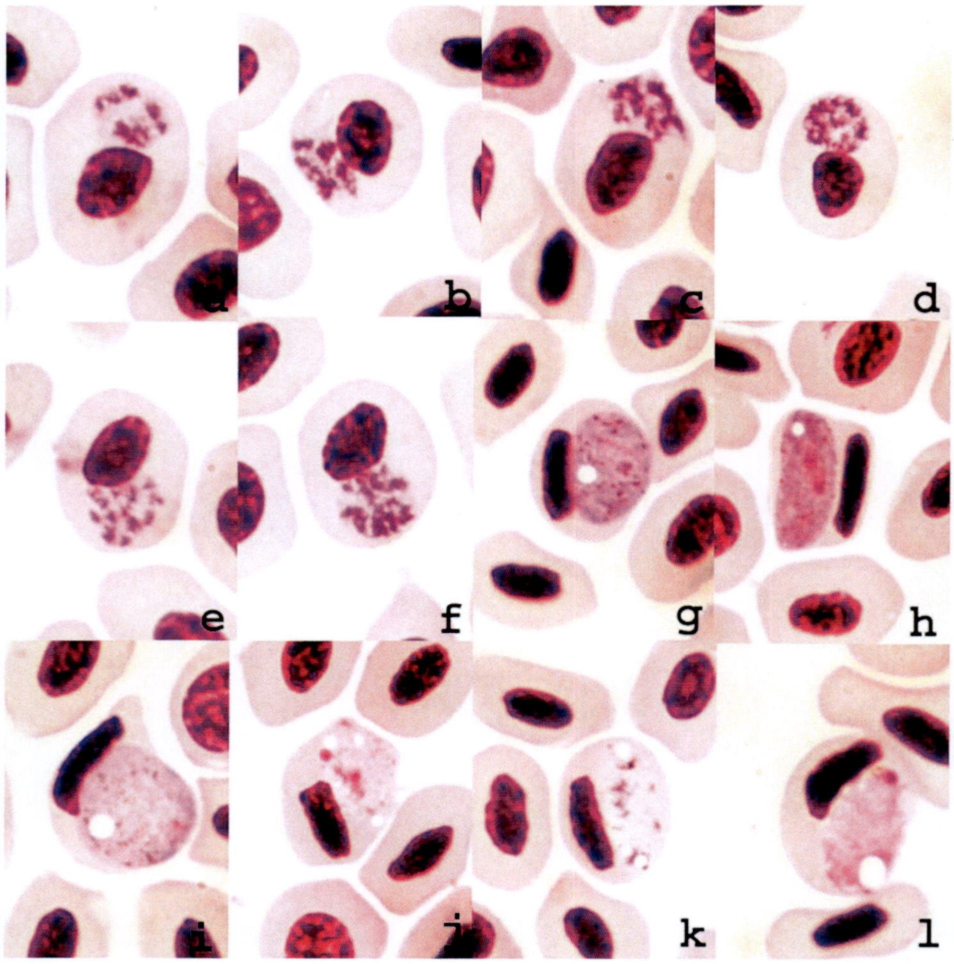

(A)

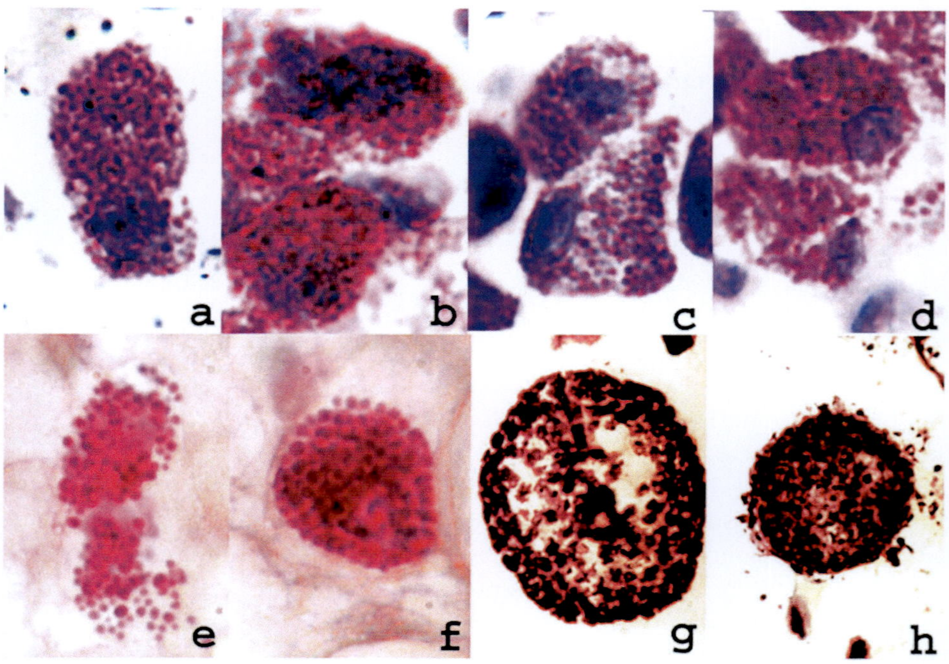

(B)

Type Locality Venezuela, Territorio Federal de Amazonas, Departamento de Atures, Coramoto.

Other Hosts *Kentropyx calcarata* Spix (Sauria: Teiidae).

Other Localities Venezuela, Territorio Federal de Amazonas, Mision Padomo.

Prevalence Three of four *K. calcarata* were infected by *P. pifanoi* (Telford and Telford, 2003).

Morphological Variation In the type host *Ameiva ameiva*, (Scorza and Dagert, 1956) meronts produce 6–12 merozoites; dimensions are not given, but their figures suggest a size of about 27 μm^2. Macrogametocytes averaged 15.4 × 8.4 μm, and microgametocytes averaged 17.6 × 6.0 μm, with estimated LW values of 129.4 and 105.6 μm^2, respectively. Vacuoles are described as polar, very conspicuous, and well-defined. In *Kentropyx calcarata*, mature or nearly mature meronts are 6.2 ± 0.9 × 4.5 ± 0.6 μm (4–8 × 3–6, n = 25), with LW 28.0 ± 5.5 μm^2 (16–42), and contain 7–16 (11.9 ± 3.2) merozoites arranged most commonly in an oblong meront (40%), as a rosette (20%), a fan (8%), or otherwise. Meront size relative to infected erythrocyte nucleus is 1.00 (0.51–1.56), and to normal erythrocyte nuclei is 1.28 (0.73–1.93). Mature gametocytes are usually elongate, 12.4 ± 1.6 × 6.0 ± 1.2 μm (8–16 × 4–10, N = 50), with LW 72.9 ± 11.1 μm^2 (52–112) and L/W 2.18 ± 0.59 (1.05–3.25). Gametocyte size relative to infected erythrocyte nucleus size is 3.11 (2.07–5.48), and to normal erythrocyte nuclei is 3.32 (2.39–4.36). Large round vacuoles are always present, one in macrogametocytes and one to four, usually two or three (68%), in microgametocytes. Gametocytes are sexually dimorphic in size, and macrogametocytes are longer than microgametocytes with greater LW. Their width and L/W ratio do not differ. Macrogametocyte dimensions are 12.9 ± 1.6 × 6.1 ± 1.2 μm (N = 25), LW 77.2 ± 11.6 μm^2, and L/W 2.24 ± 0.59, and microgametocytes 11.9 ± 1.4 × 5.9 ± 1.1 μm (N = 25), LW 68.6 ± 9.0 μm^2, and L/W 2.12 ± 0.60. Chronic-phase gametocytes are 11.2 ± 1.5 × 5.8 ± 1.1 μm (9–14 × 4–9, n = 25), with LW 64.0 ± 7.5 μm^2 (52–94) and L/W 2.06 ± 0.65 (1.12–3.50), and did not differ in mean dimensions by sex. Active-phase microgametocytes differed from those in chronic phase only by greater LW, and chronic-phase macrogametocytes are shorter than those in active phase, with lesser LW values. A single vacuole is present in chronic-phase macrogametocytes, and one to five vacuoles, most commonly three (42%), were seen in microgametocytes, as in active-phase gametocytes.

Exoerythrocytic Merogony Unknown.

Sporogony Unknown.

Effects on Host Meronts parasitized only erythrocytes, causing distortion of the host cell (96%), but rarely the nucleus (8%), seldom displacing it (28%). Host erythrocytes and their nuclei are hypertophied, with widths of both cell and nucleus increased but with no change in lengths of either. Gametocytes always distorted the host cell and displaced its nucleus and usually distorted the nucleus (92%). Host erythrocytes and nucleus length are enlarged, but other dimensions do not differ from uninfected cells. In chronic phase, host erythrocytes do not differ in dimensions from uninfected cells, but their nucleus length is greater and width less than that of uninfected cells, with no difference in LW of either cell or nucleus.

Remarks Scorza and Dagert (1956) described *Plasmodium pifanoi* from six infections in *Ameiva a. ameiva* collected at Coromoto, Departamento de Atures, Territorio Federal de Amazonas, Venezuela, and it was not reported again until it was redescribed from *Kentropyx calcarata* (Telford and Telford, 2003). It was not found in an extensive survey of Venezuelan lizards by the latter authors in 1973–74 (Telford, 1980). The lizards were from the upper llanos of Venezuela rather than in Amazonas, which probably explains the absence of *P. pifanoi* if it occurs only in the southern portion of the country. It is perhaps relevant to mention that adult *Kentropyx calcarata* resemble somewhat juvenile *A. ameiva*.

As stated by Telford and Telford (2003), "Scorza and Dagert (1956) considered *P. pifanoi* to 'exhibe una gran afinidad' with *Plasmodium mexicanum* Thompson and Huff, from which they distinguished it by the types of host cells invaded, pigment characteristics in trophozoites and gametocytes, the numbers of merozoites formed, gametocyte shape, and the presence of a vacuole in *P. pifanoi*." Using only the scanty information in the description by Scorza and Dagert and their figures, Telford (1988a) tentatively placed *P. pifanoi* in the subgenus *Paraplasmodium* with *P. mexicanum* and *Plasmodium chiricahuae* Telford. This subgeneric assignment is supported on morphological grounds by the present material from *K. calcarata*. *Plasmodium* species in the subgenus *Paraplasmodium* "have schizonts 0.5–2.2 times the size of uninfected erythrocyte nuclei that produce 4–30 merozoites, a mean gametocyte size 3–6 times that of uninfected erythrocyte nuclei and of meronts, and sexually dimorphic gametocytes in which macrogametocytes exceed microgametocytes in size." The morphology of *Plasmodium pifanoi* is consistent with this definition, but this relationship needs comparison by genome analysis with *P. mexicanum* and *P. chiricahuae*.

Plasmodium beebei Telford 1978 (**Plate 27**)

Diagnosis A *Plasmodium* (*Lacertamoeba*) species with variably shaped meronts 5–15 × 3–7 μm, LW 18–70 μm^2, that contain 8–20 merozoites. Meront size relative to host cell nucleus size averages 0.72 and to normal erythrocyte nucleus size is 0.98. Gametocytes, round to elongate in shape, are preceded by prematuration stages that are elongate with irregular margins and tend to encircle the host cell nucleus. Gametocytes are 8–21 × 5–11 μm, with LW 54–168 μm^2 and L/W 1.00–2.83. Gametocyte size relative to host cell nucleus size averages 2.17, and to normal erythrocyte nucleus size is 2.16. Pigment in meronts forms a dark yellowish-brown mass and a dark brown mass in prematuration gametocytes, which disperses as prominent granules in mature gametocytes. Macrogametocytes are larger than microgametocytes and more elongate, on average.

Type Host *Gonatodes taniae* Roze (Sauria: Gekkonidae).

Type Locality Parque Nacional Henri Pittier (Rancho Grande), Aragua State, Venezuela.

Other Hosts None known.

Other Localities None known.

Prevalence Two of three *G. taniae* were infected with *Plasmodium beebei* (Telford, 1978d).

Morphological Variation Trophozoites have long filiform cytoplasmic projections that disappear when nuclear division begins. Meronts usually form as rosettes, sometimes as fans or in elongate shape, often appearing triangular when immature. Meronts are 8.6 ± 2.0 × 4.7 ± 1.1 μm (5–15 × 3–7, N = 44), with LW 39.7 ± 12.0 μm^2 (18–70), and contain 13.9 ± 2.8 (8–20) merozoites. Meront size relative to host cell nucleus size is 0.72 ± 0.24 (0.38–1.46, N = 37), and to normal erythrocyte nucleus size is 0.98 ± 0.30 (0.47–1.81, N = 44). Elongate immature gametocytes are fusiform with approaching maturity, then become variable in form with onset of differential staining by sex, with margins frequently crenulated and irregular and one or both ends attenuated, curving out and encircling the erythrocyte nucleus. These prematuration gametocytes assume regular, smooth margins at maturity, becoming round, oval, or broadly elongate. Gametocytes are 12.7 ± 3.2 × 7.3 ± 1.2 μm (8–21 × 5–11, N = 50), with LW 93.3 ± 28.2 μm^2 (54–168) and L/W 1.78 ± 0.54 (1.00–2.83). Gametocyte size relative to host cell nucleus size is 2.17 ± 0.67 (1.19–3.69), and to normal erythrocyte nucleus size is 2.16 ± 0.65 (1.25–3.89). Macrogametocytes are larger and more elongate than microgametocytes, 15.1 ± 2.11 × 7.6 ± 1.3 μm, LW 113.6 ± 24.5 μm^2, and L/W 2.06 ± 0.50 versus 10.4 ± 1.9 × 7.1 ± 0.9 μm, LW 73.1 ± 13.3 μm^2, and L/W 1.50 ± 0.43 (N = 25 each), respectively. Pigment forms a dark yellowish-brown mass in meronts and a prominent dark brown mass in one of the extended ends of prematuration gametocytes, this latter dispersing as prominent dark granules in mature gametocytes.

Exoerythrocytic Merogony Two meronts were seen in thrombocytes, one 13 × 5 μm with 12 nuclei, and the other 8 × 4 μm with 5 nuclei (Telford, 1978d).

Sporogony Unknown.

Effects on Host Meronts, almost entirely erythrocytic, always distort host cells and their nuclei and displace the latter, causing enlargement of 30% in nucleus size but only 12% in size of host cells. Gametocytes always distort host cells and displace their nuclei but neither distort nor enlarge nuclei. Host erythrocytes, however, are hypertrophied 54% over normal size.

Remarks The only other neotropical *Plasmodium* species with prematuration gametocytes is an unpigmented species, *Plasmodium* (*Garnia*) *gonatodi* (Telford, 1970c; Lainson et al., 1971). It also parasitizes species of *Gonatodes, G. albogularis* and *G. humeralis*, in Panama and Brazil, respectively. The presence of such a rarely found life history stage (prematuration gametocytes) in two *Plasmodium* species that parasitize three species of the same gekkonid genus, *Gonatodes*, may be relevant to the phyletic significance of pigment in saurian plasmodiid parasites.

Plasmodium aurulentum Telford 1971 (**Plate 27**)

Diagnosis A *Plasmodium* (*Lacertamoeba*) species with crudely fan-shaped meronts 4–10 × 4–7 μm, LW 16–70 μm^2, that contain 6–22 merozoites. Meront size relative to host cell nucleus size averages 0.72, and to normal erythrocyte nuclei is 0.95. A prominent pinkish-red staining mass appears in older meronts but usually disappears before segmentation. Gametocytes are round to lentiform, 6–11 × 4–9 μm, LW 28–81 μm^2, and L/W 1.00–2.20. Gametocyte size relative to host cell nucleus size averages 1.09, and to normal erythrocyte nuclei is 1.18. There is no sexual dimorphism in gametocyte dimensions. Pigment forms a prominent pale golden mass in meronts and an elongate, narrow mass in gametocytes, from which smaller clumps or individual black granules sometimes disperse into the cytoplasm.

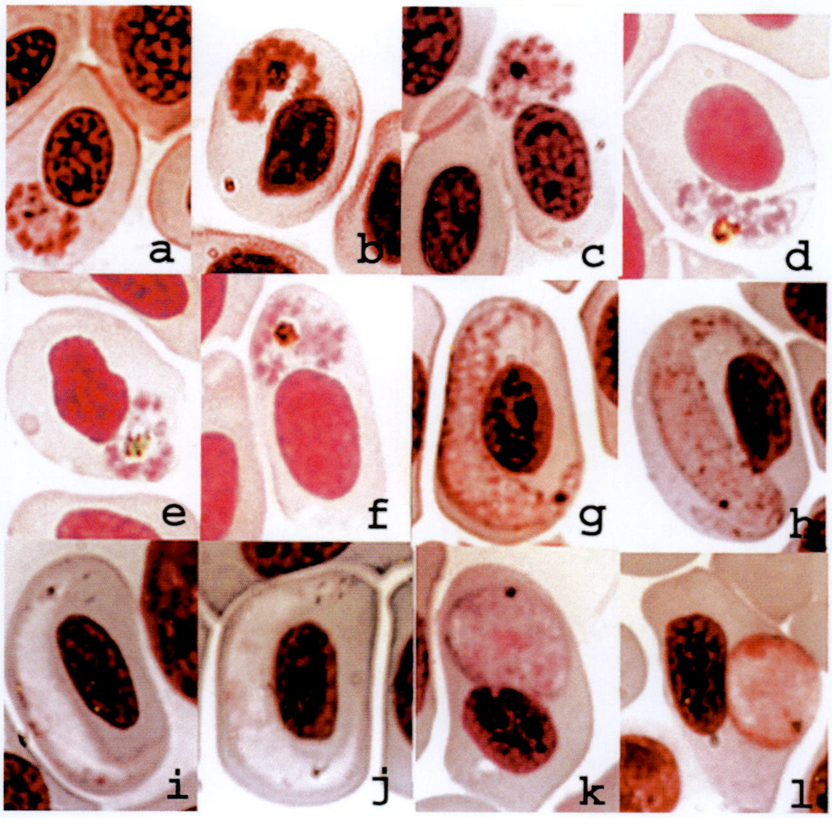

(A)

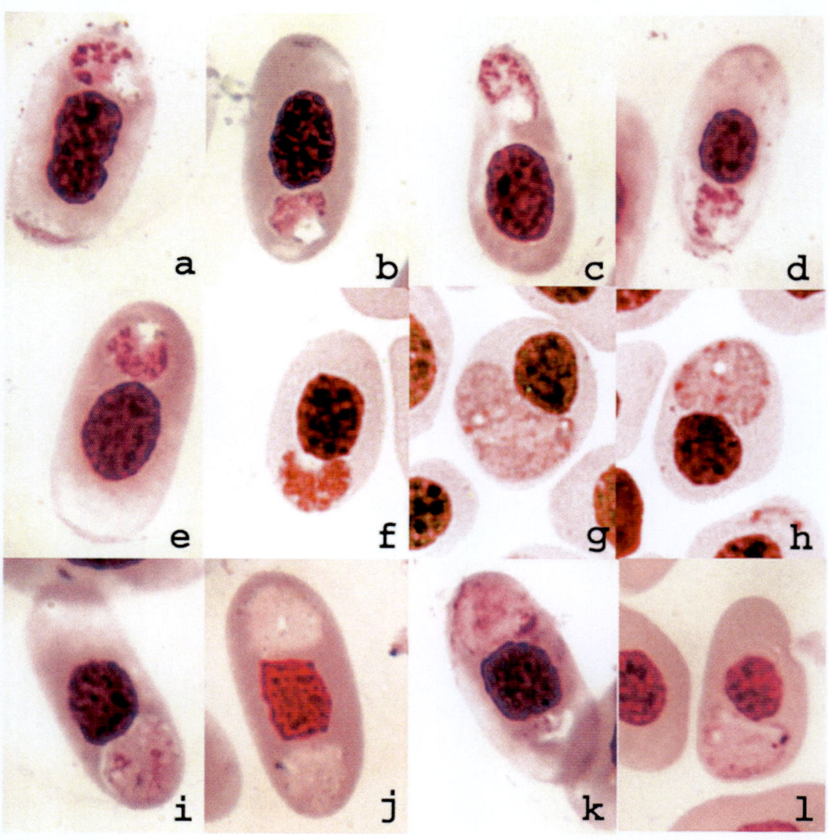

(B)

Plate 27 (**A**) *Plasmodium beebei* from *Gonatodes taniae*, Venezuela. Meronts, **a–f**; prematuration gametocytes, **g–j**; macrogametocyte, **k**; microgametocyte, **l**. (**B**) *Plasmodium aurulentum* from *Thecadactylus rapicaudus*, Panama. Meronts, **a–f**; macrogametocytes, **g–i**; microgametocytes, **j–l**.

Type Host *Thecadactylus rapicaudus* (Houttuyn) (Sauria: Gekkonidae).

Type Locality Boy Scout Camp Chagres, Madden Lake, Canal Zone, Panama.

Other Hosts None known.

Other Localities Quebrada Bonita, Colon Province, Sasardi, San Blas Territory, and Gaspar Sabanas, 8 km northwest of Chepo, Panama Province, Panama. In Venezuela, Araure, Municipio Araure, Portuguesa State, and Tierra Caliente, Municipio Manrique, Cojedes State.

Prevalence Overall, *P. aurulentum* infected 9 of 25 (36%) *T. rapicaudus* in Panama, 4 of 7 at the type locality, 2 of 7 at Quebrada Bonita, 2 of 4 at Sasardi, and 1 of 3 at Gaspar Sabana-Madroño; 3 of 22 (13.6%) overall in Venezuela, 2 of 18 (11.1%) at Araure, and 1 of 4 at Tierra Caliente.

Morphological Variation In Panamanian *T. rapicaudus*, meronts are 6.7 ± 1.3 × 5.2 ± 0.9 μm (4–10 × 4–7, N = 52), with LW 35.5 ± 11.5 μm^2 (16–70), and contain 13.1 ± 4.5 (6–22, N = 78) merozoites. Meront size relative to host cell nucleus size is 0.72 ± 0.20 (0.48–1.02, N = 8) and to normal erythrocyte nuclei is 0.95 ± 0.31 (0.44–1.87, N = 52). The few erythrocytic meronts measured are similar in dimensions to those in proerythrocytes, respectively 6.6 ± 1.2 × 5.1 ± 1.0 μm (N = 8), LW 34.1 ± 9.4 μm^2 versus 6.7 ± 1.4 × 5.3 ± 0.8 μm (N = 44), LW 35.8 ± 11.9 μm^2, but proerythrocytic meronts produce more merozoites, 15.4 ± 4.2 (6–22) versus 10.4 ± 2.5 (8–15). Six erythrocytic meronts only were found in a chronic infection of *P. aurulentum* in Venezuela: 6.3 ± 1.0 × 4.2 ± 0.4 μm (5–8 × 4–5), LW 26.3 ± 4.5 μm^2 (20–32), with 8.3 ± 2.2 (6–12) merozoites. Panamanian gametocytes are 7.7 ± 1.1 × 6.1 ± 0.9 μm (6–11 × 4–9, N = 80), with LW 46.4 ± 10.9 μm^2 (28–81) and L/W 1.29 ± 0.25 (1.00–2.20). Gametocyte size relative to host cell nucleus size is 1.09 ± 0.28 (0.67–1.93, N = 64), and to normal erythrocyte nuclei is 1.18 ± 0.28 (0.77–1.96, N = 80). Gametocyte dimensions by sex are similar, but macrogametocytes are slightly more rounded than microgametocytes, with respective dimensions 7.6 ± 1.0 × 6.4 ± 0.9 μm (N = 42), LW 48.8 ± 11.2 μm^2, and L/W 1.21 ± 0.19 versus 7.8 ± 1.3 × 5.8 ± 0.7 μm (N = 38), LW 44.8 ± 9.6 μm^2, and L/W 1.37 ± 0.28. Gametocytes in active infection average larger than those in chronic phase, which are slightly more rounded; respective dimensions are 8.1 ± 1.1 × 6.2 ± 1.0 μm (N = 50), LW 50.5 ± 11.9 μm^2, and L/W 1.33 ± 0.26 versus 7.0 ± 0.7 × 5.9 ± 0.6 μm (N = 30), LW 41.0 ± 5.1 μm^2, and L/W 1.21 ± 0.21. Gametocytes in Venezuelan *T. rapicaudus* are 8.3 ± 1.5 × 6.0 ± 0.8 μm (6–12 × 5–7, N = 25), with LW 50.4 ± 13.5 μm^2 (30–84) and L/W 1.39 ± 0.25 (1.00–2.00). Gametocyte size relative to host cell nucleus size is 1.29 ± 0.39 (0.71–2.33), and to normal erythrocyte nuclei is 1.26 ± 0.34 (0.75–2.10). Pigment forms as a large, pale golden mass at the narrower end of meronts, sometimes resembling a handle to the fan. There is a prominent mass of pinkish-red-staining material, less intensely stained than nuclei, in meronts with eight or more nuclei, which diminishes in size as segmentation nears, usually disappearing in the largest meronts. In gametocytes, pigment forms a very light golden, elongate, compact, narrow mass, usually central in the gametocyte. Occasionally, smaller clumps of individual granules do not aggregate but are dispersed within the cytoplasm, often black in appearance.

Exoerythrocytic Merogony Due to mixed infections with *Fallisia thecadactyli*, identification of meronts in thrombocytes and lymnphocytes as EE meronts of *P. aurulentum* is unwise, as may be that of phanerozoites observed in endothelium and connective tissue of the various organs (**Plate 26B**).

Sporogony Unknown.

Effects on Host Cells host to meronts and their nuclei are slightly hypertrophic, mostly in their breadth, and distortion of host cell and nucleus commonly occurs, but nuclear displacement is less common. Gametocytes cause no hypertrophy to either cell or nucleus but often distort cells and nuclei and displace the latter.

Remarks Pigment in this species is often so light in hue that it is easily overlooked or at first considered to be a vacuole.

Plasmodium floridense Thompson and Huff 1944

Diagnosis A *Plasmodium* (*Lacertamoeba*) species characterized by mature meronts typically formed as rosettes, less often as fans or in elongate shapes, which are usually the size of uninfected erythrocyte nuclei or larger, with ratios ranging from 0.64 to 1.71, varying with host species. Meronts contain 4–32 merozoites, with mean numbers among host species ranging from 9.1 to 21.7. Meront size and merozoite number are influenced by phase of infection but not by maturity of host cell. Gametocytes are more commonly elongate or ovoid than spherical except when occupying a polar position within the host erythrocyte, with mean L/W ratios varying from 1.52 to 2.29 among host species and from 1.16 to 2.44 among individual infections. Gametocyte size (LW) ranges from 20 to 128 μm^2, with means of individual infections 38.3–70.9 µm². Gametocyte size and shape may vary by sex in some host species and are affected by phase of infection. Gametocyte size relative to that of normal erythrocyte nuclei is 1.22–1.95, with ratio varying by host species. Pigment is always present, forming a dark golden mass in meronts and dispersed as dark gold-to-black granules in gametocytes. Young asexual stages are variably shaped, often elongate, and lack prominent cytoplasmic projections. Meronts more commonly occupy polar positions in erythrocytes than do gametocytes, which tend to be lateropolar to lateral in position. Cells host to either meronts or gametocytes are often distorted, with displaced nuclei, but nuclear distortion is uncommon, and hypertrophy of either cell or nucleus rarely occurs.

Type Host *Sceloporus undulatus* (Latreille) (Sauria: Phrynosomatidae).

Type Locality Vicinity of Silver Springs, Marion County, Florida, USA.

Other Hosts Species of the families Polychrotidae and Phrynosomatidae.

Geographic Range Eastern Panama from the Colombian border north throughout Middle America in the Caribbean versant to Las Tuxtlas, Veracruz State, Mexico, throughout most of the Caribbean on islands of the Greater and Lesser Antilles, San Andres and Cayman Islands, and the Bahamas, and peninsular Florida west to the Appalachicola River, northward into southeastern Georgia and its offshore islands.

Remarks *Plasmodium floridense* has been reported from at least 29 species of *Anolis* and *Sceloporus*, yet there is no objective evidence to support recognition of subspecies. Because of the extended range both geographically and among host species, *P. floridense* is considered next in separate accounts for Middle America, the Caribbean, and the southeastern United States.

Plasmodium floridense in Middle American Lizards (**Plate 28**)

Hosts In Panama: *Sceloporus malachiticus* (Huff and Marchbank, 1953); *Anolis limifrons, A. biporcatus, A. pentaprion, A. frenatus* (Telford, 1973, 1977). In Costa Rica: *A. limifrons, A. humilis, A. lionotus* (Telford, 1977). In Honduras: *A. tropidonotus* (Telford, 1977). In Belize: *A. lemurinus* (Telford, 1977). In Mexico: *A. tropidonotus, Sceloporus variabilis* (Lowichik et al., 1988).

Localities and Prevalences Panama overall, 58 of 381 (15.2%) *A. limifrons*; 33 of 99 (33.3%) *A. biporcatus*; 5 of 26 (19.2%) *A. frenatus*; 1 of 19 (5.3%) *A. pentaprion* (Telford, 1977); 14 of 244 (5.7%) *S. malachiticus* (Huff and Marchbank, 1953). Panama by locality: Achiote, Colon Province, 22 of 112 (9.8%) *A. limifrons*, 1 of 3 *A. biporcatus*, 2 of 3 *A. frenatus*; Sasardi, San Blas Territory, 14 of 107 (13.1%) *A. limifrons*; Cerro Pirre, Darien Province, 1 of 1 *A. limifrons*; Panama Province, 1 of 4 *A. limifrons*, 1 of 3 *A. biporcatus*, and 1 of 1 *A. frenatus* at Gaspar Sabanas, 8 km northwest of Chepo; 1 of 1 *A. biporcatus* on Chiva Chiva road and 1 of 2 at Capira Caimito; 28 of 86 (32.6%) *A. biporcatus*, 1 of 3 *A. frenatus*, and 1 of 10 (10%) *A. pentaprion* at El Aguacate; Canal Zone localities about 5.4 km north of Gamboa: Quebrada Juan Grande, 3 of 18 (16.7%) *A. limifrons*, 2 of 12 (16.7%) *A. frenatus*; Frijolito Creek, 12 of 73 (16.4%) *A. limifrons* and 1 of 7 *A. frenatus*; Madden Forest, 4 of 19 (21.1%) *A. limifrons*; and Barro Colorado Island in Gatun Lake, 71 of 296 (24.0%) *A. limifrons* (Guerrero et al., 1977); Chiriqui Province, 7 of 105 (6.7%) *S. malachiticus* at Jurado's Mill, Llano, and 7 of 139 (5.0%) at Palo Santo (Huff and Marchbank, 1953). Costa Rica, at Rio Frio, Heredia Province: 2 of 67 (3.0%) *A. humilis*, 5 of 82 (6.1%) *A. limifrons*, and 1 of 39 (2.6%) *A. lionotus* (Telford, 1977). Honduras at Siguatepec, 3 of 34 (8.8%) *A. tropidonotus* (Telford, 1977). Belize, 18 km north of Augustine, two of four *A. lemurinus* (Telford, 1977). Mexico, Las Tuxtlas in Veracruz State: 103 of 281 (36.7%) *A. tropidonotus* and 2 of 37 (3.5%) *Sceloporus variabilis* (Lowichik et al., 1988).

Morphological Variation In active infections of *Anolis limifrons* in Panama, meronts of *P. floridense* are 7.7 ± 1.4 × 5.8 ± 0.9 µm (5–15 × 4–8, N = 240), with LW 44.5 ± 10.9 µm² (20–90), and contain 12.4 ± 3.4 (6–24, N = 244) merozoites. Meront size relative to host cell nucleus size is 1.21 ± 0.30 (0.57–2.00, N = 188), and to normal erythrocyte nucleus size is 1.11 ± 0.29 (0.56–2.44, N = 240). In shape, meronts most commonly form as rosettes (52%) and are about equally elongate (19%) or fan-shaped (16%), seldom oblong, oval, or round. Golden-yellow pigment granules form a prominent mass, central in rosettes, at the base of fans, or variably placed in other meront shapes. Erythrocytic meronts are typical and do not differ in dimensions from those that less often occupy proerythrocytes, although on average, slightly more merozoites (12.8 ± 3.3, N = 188) are produced in mature host cells than in immature (11.0 ± 3.0, N = 56). Gametocytes are 9.2 ± 1.6 × 6.3 ± 1.1 µm (6–15 × 4–9, N = 316), with LW 57.4 ± 12.6 µm² (30–99) and L/W 1.53 ± 0.43 (1.00–3.75). Gametocyte size relative to host cell nucleus size is 1.54 ± 0.47 (0.67–3.09, N = 308), and to normal erythrocyte nucleus size is 1.44 ± 0.37 (0.72–2.70,

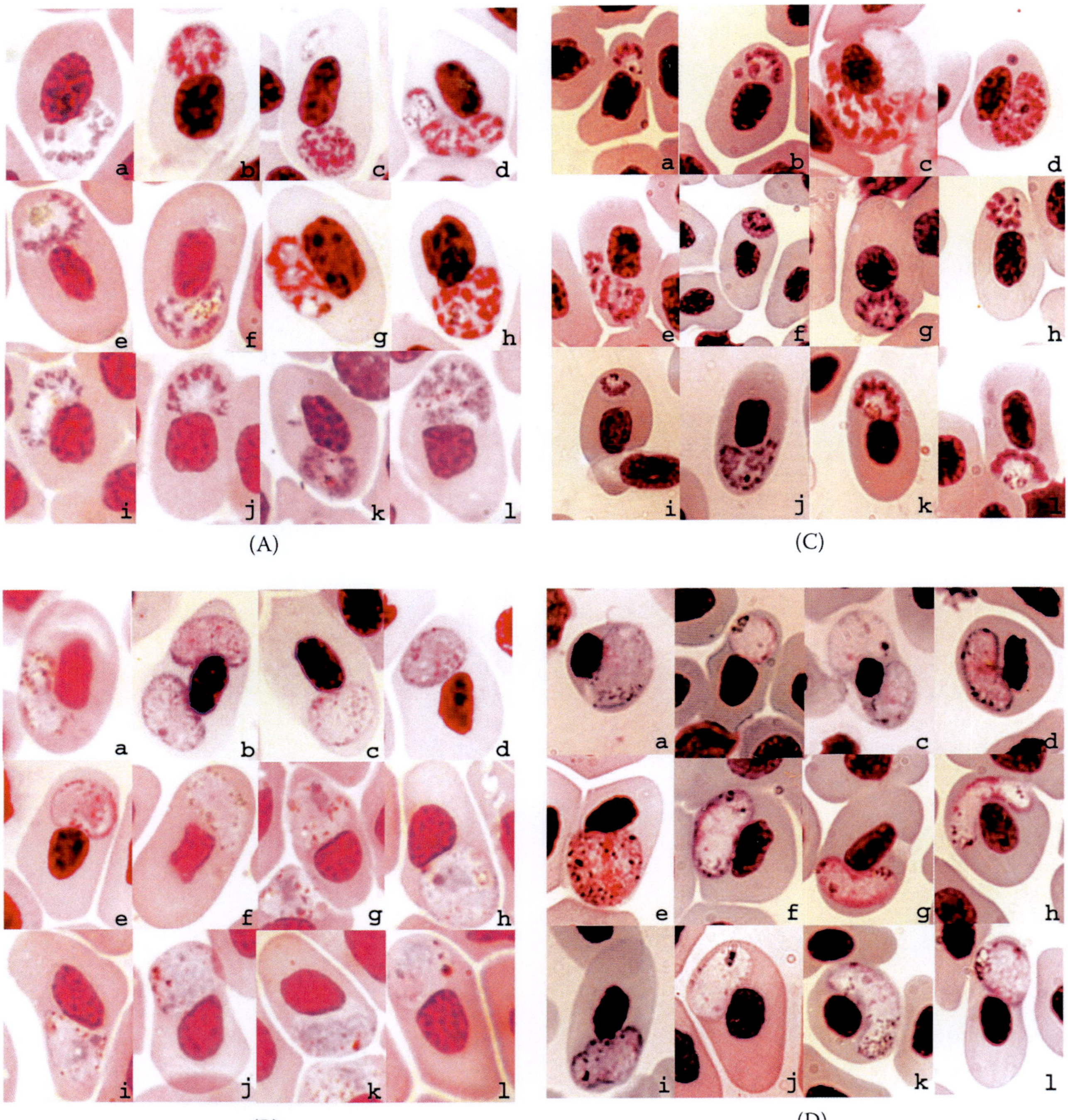

Plate 28 (**A**) and (**B**) *Plasmodium floridense* from Middle American hosts. (**A**) Meronts: **a–c**, *Anolis limifrons*; **d–f**, *A. biporcatus*; **g**, **h**, *A. pentaprion*; **k**, **l**, *Sceloporus malachiticus*, all Panamanian; **i**, **j**, *Anolis lemurinus*, Belize. (**B**) Gametocytes: **a–c**, *A. limifrons*; **d–f**, *A. biporcatus*; **g–l**, *S. malachiticus*. Macrogametocytes are **b** (above) and **g–i**, **k** (below). Others are microgametocytes. (**C**) and (**D**) *Plasmodium floridense* from Caribbean *Anolis* hosts. (**C**) Meronts: **a–c**, *A. cybotes*, Hispaniola; **d**, **e**, *A. pulchellus*, Puerto Rico; **f–h**, *A. garmani*, Jamaica; **i**, **j**, *A. conspersus*, Grand Cayman; **k**, **l**, *A. sagrei*, Bimini. (**D**) Gametocytes: **a–c**, *A. cybotes*; **d**, **e**, *A. pulchellus*; **f–h**, *A. garmani*; **i**, **j**, *A. conspersus*; **k**, **l**, *A. sagrei*. Macrogametocytes are **a**, **c** (below), **d**, **f**, **i**, **k**. Others are microgametocytes.

N = 316). Dimensions of gametocytes do not differ by sex, but macrogametocytes are more elongate on average, with greater L/W ratio (1.60 ± 0.43, N = 153) than microgametocytes (1.46 ± 0.43, N = 163). Dark pigment granules are dispersed in both gametocyte sexes. In a comparison of samples from the Canal Zone localities (Madden Forest, Quebrada Juan Grande, Frijolito Creek), Achiote in Colon Province, and eastern Panama (Sasardi, San Blas Territory and Cerro Pirre, Darien Province), little difference is present in meronts and gametocytes among the three areas.

In *Anolis biporcatus*, meronts are 8.4 ± 2.1 × 5.9 ± 1.1 µm (6–13 × 4–9, N = 80), with LW 50.5 ± 17.3 µm² (24–104), and contain 19.5 ± 4.7 (12–32, N = 82) merozoites. Meront size relative to host cell nucleus size is 1.43 ± 0.49 (0.60–2.89, N = 71), and to normal erythrocyte nucleus size is 1.52 ± 0.44 (0.76–2.88, N = 80). Meronts from infections in active phase are larger than in chronic phase and produce more merozoites, respectively 9.3 ± 2.0 × 6.1 ± 1.2 µm (N = 54), LW 56.7 ± 17.4 µm², and merozoites 21.5 ± 4.0 versus 6.6 ± 0.8 × 5.7 ± 0.8 µm (N = 26), LW 37.7 ± 6.9 µm², and 15.2 ± 2.8 merozoites. Gametocytes are 10.1 ± 1.7 × 5.5 ± 0.9 µm (5–14 × 4–9, N = 126), with LW 55.8 ± 13.8 µm² (20–91) and L/W 1.89 ± 0.41 (1.11–3.00). Gametocyte size relative to host cell nucleus size is 1.68 ± 0.46 (0.57–2.60, N = 116), and to normal erythrocyte nucleus size is 1.73 ± 0.43 (0.67–2.84, N = 128). Active-phase gametocytes are larger and more elongate than those in chronic phase, respectively 10.5 ± 1.4 × 5.6 ± 0.9 µm (N = 100), LW 59.0 ± 12.6 µm², and L/W 1.93 ± 0.40 versus 8.7 ± 1.8 × 5.0 ± 0.7 µm (N = 26), LW 43.2 ± 11.1 µm², and L/W 1.76 ± 0.39. Neither size nor shape differ between sexes in active infection. Meronts typically are shaped as rosettes and seldom occupy immature host cells.

Only meronts are available from *Anolis frenatus* and *Anolis pentaprion*. These more closely resemble those from *A. biporcatus* in size and merozoite production than they do meronts from *A. limifrons*. In *A. frenatus*, meronts in active phase are 9.2 ± 1.1 × 6.4 ± 1.0 µm (7–12 × 5–9, N = 20), LW 58.6 ± 11.5 µm² (42–81), and contain 21.7 ± 4.3 (12–28) merozoites, with size relative to host cell nucleus size 1.37 ± 0.34 (0.92–2.00, N = 16), and to normal erythrocyte nucleus size 1.34 ± 0.26 (0.96–1.85, N = 20). In acute infection of *A. pentaprion*, meronts are 8.6 ± 1.2 × 5.7 ± 0.9 µm (7–12 × 4–7, N = 25), LW 49.1 ± 11.7 µm² (28–84), and contain 16.2 ± 3.1 (8–22) merozoites. Their size relative to host cell nucleus size is 1.19 ± 0.31 (0.58–1.80), and to normal erythrocyte nucleus size is 1.25 ± 0.30 (0.71–2.14).

An acute-phase infection of *P. floridense* in *Sceloporus malachiticus* from Chiriqui Province (from C. G. Huff) has meronts more similar to those in *A. limifrons* than to the other Panamanian *Anolis* hosts, but gametocytes are consistent in size with both *A. limifrons* and *A. biporcatus* samples. Meronts are 8.9 ± 1.5 × 6.0 ± 1.1 µm (7–12 × 3–8, N = 25), LW 52.2 ± 10.2 µm² (35–80), with 14.1 ± 4.0 (6–23) merozoites. Meront size relative to host cell nucleus size is 1.65 ± 0.36 (1.00–2.67, N = 20), and to normal erythrocyte nucleus size is 1.71 ± 0.33 (1.14–2.61, N = 25). Gametocytes are 9.6 ± 1.7 × 6.3 ± 1.1 µm (6–16 × 4–8, N = 40), with LW 59.7 ± 15.1 µm² (35–128) and L/W 1.58 ± 0.46 (1.00–3.25). Gametocyte size relative to host cell nucleus size is 1.86 ± 0.46 (1.00–3.20, N = 39), and to normal erythrocyte nucleus size is 1.95 ± 0.49 (1.14–4.18, N = 40). There is no sexual difference in size, but microgametocytes are more elongate than macrogametocytes, with respective L/W ratios 1.75 ± 0.54 (1.00–3.25, N = 20) versus 1.42 ± 0.31 (1.00–2.00, N = 20).

In *Anolis lemurinus* from Belize, meronts of *P. floridense* are 7.7 ± 1.1 × 4.7 ± 0.7 µm (5–10 × 4–6, N = 50), with LW 35.7 ± 7.1 µm² (20–50), and contain 11.7 ± 2.4 (6–17) merozoites. Meront size relative to host cell nucleus size is 0.95 ± 0.22 (0.58–1.40, N = 48), and to normal erythrocyte nucleus size is 1.00 ± 0.23 (0.50–1.54, N = 50). Meronts are formed as rosettes usually (62%) or as fans (36%), and are rarely elongate in shape. Gametocytes are usually elongate, 9.5 ± 1.5 × 4.3 ± 0.6 µm (6–13 × 3–5, N = 50), with LW 40.3 ± 7.7 µm² (21–65) and L/W 2.29 ± 0.55 (1.20–3.67). Gametocyte size relative to host cell nucleus size is 1.16 ± 0.28 (1.20–3.67), and to normal erythrocyte nucleus size is 1.13 ± 0.28 (0.65–2.00). Macro- and microgametocytes are similar in size, but the latter are less elongate on average, L/W ratios 2.08 ± 0.32 (1.68–2.60, N = 19), than are macrogametocytes, LW 2.42 ± 0.63 (1.20–3.67, N = 31).

In *Anolis tropidonotus* from Honduras, *P. floridense* meronts are 6.0 ± 1.0 × 4.5 ± 0.7 µm (5–10 × 3–6, N = 24), with LW 26.8 ± 4.3 µm² (20–36), and contain 9.1 ± 1.7 (6–14, N = 25) merozoites. Meront size relative to host cell nucleus size is 0.70 ± 0.13 (0.44–0.94), and to normal erythrocyte nucleus size is 0.64 ± 0.10 (0.48–0.86). Gametocytes are 9.4 ± 1.6 × 5.5 ± 0.9 µm (7–15 × 4–7, N = 33), with LW 50.8 ± 8.6 µm² (40–66) and L/W 1.80 ± 0.56 (1.14–3.75). Gametocyte size relative to host cell nucleus size is 1.35 ± 0.26 (0.89–1.88, N = 32), and to normal erythrocyte nucleus size is 1.27 ± 0.25 (0.96–1.85, N = 33). In Mexican *A. tropidonotus*, meront and gametocyte dimensions averaged slightly larger, and merozoite numbers were greater. The differences can be attributed to phase of infection: The Honduran samples came from two infections nearing chronic phase, while the Mexican infections were both acute. Meronts from the Mexican *A. tropidonotus* are 8.1 ± 2.1 × 5.7 ± 1.1 µm (5–15 × 4–9, N = 50), with LW 46.7 ± 16.8 µm² (24–81), and contain 14.9 ± 5.3 (4–26) merozoites. Meront size relative to host cell nucleus size is 1.06 ± 0.36 (0.50–1.93, N = 33), and to normal erythrocyte nucleus size is 1.39 ± 0.45 (0.76–2.33, N = 50). Meronts in immature erythrocytes are similar in size to those in mature cells but average slightly higher merozoite counts, respectively 16.9 ± 6.5 (4–26, N = 17) versus 13.9 ± 4.3 (6–23, N = 33). Gametocytes are 10.7 ± 2.4 × 5.6 ± 0.8 µm (7–18 × 4–7, N = 80), with LW 59.2 ± 13.7 µm² (40–108) and L/W 1.96 ± 0.57 (1.00–3.60). Gametocyte size relative to host cell nucleus size is 1.53 ± 0.50 (0.79–3.50, N = 50), and to normal erythrocyte nucleus size is 1.77 ± 0.45 (1.15–3.41, N = 80). Macrogametocytes average slightly larger in size with higher L/W ratios than microgametocytes, with respective dimensions $11.6 \pm 2.0 \times 5.4 \pm 0.7$ µm (N = 43), LW 62.7 ± 15.0 µm^2, and L/W 2.20 ± 0.46 versus $9.5 \pm 2.3 \times 5.8 \pm 0.8$ µm (N = 36), LW 54.7 ± 10.6 µm^2, and L/W 1.69 ± 0.57.

Exoerythrocytic Merogony Unknown.

Sporogony Huff and Marchbank (1953) did not obtain sporogony by feeding *Culex pipiens, Aedes aegypti, A. albopictus,* and *Anopheles quadrimaculatus* on an infected *Sceloporus malachiticus.*

Effects on Host In most hosts, infected cells and their nuclei differed by no more than about 5% in size (LW) from that of uninfected erythrocytes and nuclei. In the acute infections of Mexican *A. tropidonotus,* infected erythrocyte nuclei were hypertrophied 28% by meronts and 23% by gametocytes. *Anolis frenatus* erythrocytes host to meronts were 15% larger than normal, and erythrocytes infected with gametocytes in *A. tropidonotus* from Honduras were 8% larger than normal. In all hosts, meronts commonly distorted host erythrocytes (35–70%) and displaced cell nuclei (16–60%), but seldom distorted nuclei (0–28%) except in *A. frenatus* and *A. pentaprion,* for which 50% and 60%, respectively, were distorted. Gametocytes commonly distorted host cells (32–94%) and displaced nuclei (19–64%) but caused little nuclear distortion (0–14%).

Ecology Host lizards infected by *Plasmodium floridense* show the entire range of diversity in perch sites, from the terrestrial, leaf litter-inhabiting *Anolis humilis* to the canopy species *A. frenatus* and *A. biporcatus,* the former usually associated with the tree trunks and the latter with branches. The other host species perch on tree trunks or buttresses near the ground or shrubs within or along the edge of forest. Telford (1977) suggested that *P. floridense* is transmitted by a vector that occurs from the canopy to within a meter or so of the ground but is not necessarily associated with tree margins, given the absence of *P. floridense* infections in the semiaquatic anoles, *Anolis lionotus* and *A. poecilopus* in Panama, and its rarity in the former species (2.6%) in Costa Rica. Habitats in which *P. floridense* occurs, where known, are mature primary or secondary moist tropical forest. There are no records of *P. floridense* from the more arid Pacific coast of Middle America and Mexico, only from the Caribbean versant. The prevalence of *P. floridense* in Panamanian anoles is at minimum in the late wet season, increases significantly in early dry season, remaining at the same level through late dry season, then again increasing significantly to its maximum in early wet season (Telford, 1977). However, continuous transmission is indicated by similar proportions of initial versus chronic infections throughout the year. Sampling bias probably influenced analysis of the Panamanian data.

Remarks On morphological grounds, *P. floridense* and *P. tropiduri* are very similar and probably are derived from a common stock in anoles that was separated by the opening of the Panamanian Portal in the Eocene (Telford, 1977). This hypothesis needs evaluation by genome analysis.

Plasmodium floridense in Lizards of the Caribbean (**Plate 28**)

Hosts *Anolis cybotes, A. distichus, A. coelestinus, A. pulchellus, A. garmani, A. lineatopus, A. grahami, A. opalinus, A. sagrei, A. conspersus* (Telford, 1975; Telford et al., 1989, and unpublished); *A concolor* (Ayala, 1975); *A. gundlachi* (Guerrero and Pickering, 1984; Schall and Vogt, 1993; Schall et al., 2000); *A. gingivinus, A. sabanus, A. bimaculatus, A. lividus* (Staats and Schall, 1996a); and *A. oculatus* (Perkins and Schall, 2002).

Localities and Prevalences Bahamas: in *A. sagrei,* North Bimini, 1 of 17 (5.9%) (Telford, 1975); Cat Cay (Telford et al., 1989). Hispaniola: Haiti, in *A. cybotes* overall, 3 of 51 (5.9%), 2 of 22 (9.0%) at Fond Verettes and 1 of 5 at Petionville, Departement de l'Ouest (Telford, 1975); Dominican Republic, in *A. cybotes,* 13 of 46 (28.3%) overall, Rio Seibo 3 of 3, Pedro Sanchez 8 of 15 (53.3%), Morro, 10 km south of Miches 1 of 3, all El Seibo Province, and 1 of 3 from 5 km north northeast of Polo, Baharona Province (Telford et al., 1989, and unpublished); in *A. distichus,* 1 of 8 at 2.8 km north of Las Cuchillas, El Seibo Province, and in 1 of 4 *A. coelestinus* at 5 km north northeast of Polo, Baharona Province (Telford et al., 1989, and unpublished). Jamaica: in *A. grahami* overall, 1 of 38 (2.6%), and at Grange, 1 of 4; in *A. garmani,* 11 of 32 (34.3%) overall, Lacovia with 4 of 12 (33.3%), 4 of 6 at Somerset, 3 of 4 at Worthy Park; in *A. opalinus,* 4 of 48 (8.3%) overall, 1 of 7 at Lacovia, 1 of 5 at Magotty, 1 of 7 at Mandeville, 1 of 10 at Worthy Park; in *A. lineatopus,* overall 10 of 78 (12.8%), 1 of 2 at Bog Walk, 2 of 5 at Portland Cottage, 3 of 6 at Salt River, 1 of 7 at Somerset, and 3 of 4 at Worthy Park (Telford, 1975, and unpublished). Puerto Rico: 3 of 5 (5.8%) overall in *A. pulchellus,* 2 of 15 (13.3%) at Rio Grande, 1 of 23 (4.3%) at Rio Piedras (Telford, 1975, and unpublished); in *A. gundlachi,* 3 of 58 (?) (Guerrero and Pickering, 1984), and 11 of 406 (2.7%) (Schall and Vogt, 1993). Cayman Islands: 6 of 42 (14.3%) in *A. conspersus,* Grand Cayman; 1 of 4 in *A. maynardi,* Little Cayman (Telford, 1975; Telford et al., 1989). San Andres Island in *A. concolor,* 1 of 98 (1.0%) (Ayala, 1975). Lesser Antilles: *A. gingivinus* on Anguilla; 377 of 1762 (21.4%) in *A. sabanus* of Saba; *A. bimaculatus* on St. Kitts; *A. lividus* on Monserrat (Staats and Schall, 1996a); *A. oculatus* on Dominica (Perkins and Schall, 2002).

Morphological Variation Meronts of *P. floridense* in *Anolis cybotes* of Hispaniola are 6.0 ± 1.0 × 4.5 ± 0.7 μm (4–8 × 3–6, N = 81), with LW 25.6 ± 6.3 μm² (12–42), and contain 10.0 ± 2.7 (4–16, N = 87) merozoites. Meront size relative to host cell nucleus size is 0.80 ± 0.27 (0.43–1.60, N = 60), and to normal erythrocyte nucleus size is 0.88 ± 0.22 (0.42–1.47, N = 81). Meronts from the two areas sampled, Fond Verretes, Haiti, and Rio Seibo, Dominican Republic, were similar in meront dimensions and merozoite numbers. Erythrocytic meronts do not differ from those occupying proerythrocytes in dimensions and merozoite numbers. Gametocytes in Hispaniolan *A. cybotes* are 9.5 ± 1.0 × 5.7 ± 0.9 μm (6–15 × 4–8, N = 100), with LW 54.1 ± 10.7 μm² (35–88) and L/W 1.72 ± 0.46 (1.00–3.25). Gametocyte size relative to host cell nucleus size is 1.71 ± 0.47 (0.83–3.21, N = 97), and to normal erythrocyte nucleus size is 1.86 ± 0.39 (1.18–3.22, N = 100). Gametocyte dimensions and shape are similar in the two areas sampled. Dimensions are also similar by sex, but microgametocytes have a higher average L/W ratio and are therefore more elongate than are macrogametocytes, with respective dimensions 9.6 ± 1.3 × 5.3 ± 0.8 μm (N = 41), LW 50.6 ± 9.3 μm², and L/W 1.87 ± 0.41 versus 9.4 ± 1.8 × 6.0 ± 0.9 μm (N = 59), LW 56.5 ± 10.9 μm², and L/W 1.61 ± 0.47. Meronts in *Anolis distichus* from El Seibo Province, Dominican Republic, are slightly larger than *P. floridense* meronts in *A. cybotes*, 7.3 ± 1.0 × 5.4 ± 0.6 μm (6–10 × 4–7, N = 50), with LW 39.6 ± 7.8 (28–60) μm², and contain 12.3 ± 2.3 (6–16, N = 51) merozoites. Meront size relative to host cell nucleus size is 1.37 ± 0.28 (1.00–2.22, N = 25), and to normal erythrocyte nucleus size is 1.37 ± 0.27 (0.97–2.07, N = 50). Gametocytes are larger and more elongate than in *A. cybotes*, 12.1 ± 2.8 × 5.7 ± 1.1 μm (6–20 × 4–8, N = 52), with LW 68.0 ± 16.0 μm² (30–105) and L/W 2.23 ± 0.80 (1.00–4.00). Gametocyte size relative to host cell nucleus size is 2.74 ± 0.71 (1.50–4.67, N = 25), and to normal erythrocyte nucleus size is 2.34 ± 0.55 (1.03–3.62, N = 52). Gametocytes do not differ in size or shape by sex.

In *Anolis garmani* of Jamaica, meronts are 6.7 ± 1.2 × 4.7 ± 0.7 μm (5–10 × 3–6, N = 56), with LW 31.1 ± 7.3 μm² (15–50), and contain 9.8 ± 2.6 (6–18) merozoites. Meront size relative to host cell nucleus size is 0.95 ± 0.22 (0.63–1.50, N = 53), and to normal erythrocyte nucleus size is 1.06 ± 0.26 (0.50–1.71, N = 56). Gametocytes are 9.9 ± 2.2 × 5.7 ± 1.2 μm (6–15 × 3–9, N = 100), with LW 55.2 ± 14.0 μm² (28–90) and L/W 1.86 ± 0.70 (1.00–4.33). Gametocyte size relative to host cell nucleus size is 1.92 ± 0.56 (0.80–3.67, N = 98), and to normal erythrocyte nucleus size is 1.87 ± 0.48 (0.96–2.99, N = 100). Microgametocyte dimensions are similar to those of macrogametocytes, but their shape is more elongate, respectively 10.3 ± 2.4 × 5.1 ± 1.0 μm (N = 51), LW 51.8 ± 12.3 μm², and L/W 2.16 ± 0.81 versus 9.4 ± 1.8 × 6.2 ± 1.0 μm (N = 49), LW 58.8 ± 14.9 μm², and L/W 1.56 ± 0.32. In chronic-phase infections of *Anolis lineatopus*, the few meronts available are 7.0 ± 1.5 × 4.7 ± 1.1 μm (5–10 × 4–7, N = 7), LW 34.0 ± 16.3 μm² (24–70), and contain 11.0 ± 3.6 (7–16) merozoites. Their size relative to host cell nucleus size is 1.00 ± 0.31 (0.70–1.67), and to normal erythrocyte nucleus size is 1.10 ± 0.58 (0.78–2.36), parameters very similar to active-phase meronts in *A. garmani*. Gametocytes are also similar in dimensions to those in *A. garmani*, 10.5 ± 1.9 × 5.4 ± 1.1 μm (7–16 × 4–8, N = 50), LW 55.9 ± 12.5 μm² (32–96), and L/W 2.05 ± 0.65 (1.00–3.75). Gametocyte size relative to host cell nucleus size is 1.93 ± 0.49 (1.14–3.43), and to normal erythrocyte nucleus size is 1.83 ± 0.42 (1.06–3.24). These chronic-phase gametocytes do not differ by sex in size or shape. In chronic-phase infections of *Anolis opalinus*, gametocytes are 11.7 ± 2.0 × 5.7 ± 0.8 μm (8–16 × 4–8, N = 50), with LW 67.2 ± 15.8 μm² (40–105) and L/W 2.08 ± 0.43 (1.33–3.20). Gametocyte size relative to host cell nucleus size is 3.05 ± 0.75 (1.67–5.06), and to normal erythrocyte nucleus size is 3.17 ± 0.75 (1.89–4.95). Gametocyte dimensions and shape are similar by sex. The chronic-phase gametocytes in *A. opalinus* are larger and more elongate than active-phase gametocytes in *A. garmani* and larger on average, but similar in L/W ratio to the chronic-phase gametocytes in *A. lineatopus*.

Plasmodium floridense meronts in active infections of *Anolis pulchellus* of Puerto Rico are 9.0 ± 1.8 × 5.9 ± 1.0 μm (6–14 × 4–8, N = 49), with LW 52.1 ± 9.4 μm² (35–72), and contain 13.3 ± 2.5 (9–22, N = 50) merozoites. Their size relative to host cell nucleus size is 1.95 ± 0.47 (1.25–3.43, N = 47), and to normal erythrocyte nucleus size is 2.02 ± 0.42 (1.39–3.14, N = 49). Gametocytes are 11.3 ± 1.6 × 7.1 ± 1.4 μm (8–15 × 4–11, N = 100), with LW 80.0 ± 18.6 μm² (40–143) and L/W 1.65 ± 0.47 (1.00–3.50). Gametocyte size relative to host cell nucleus size is 2.92 ± 0.81 (1.43–5.96, N = 98), and to normal erythrocyte nucleus size is 3.02 ± 0.74 (1.39–5.77, N = 100). Gametocytes are similar in dimensions by sex, but microgametocytes are more elongate than macrogametocytes, respectively 11.4 ± 1.4 × 6.3 ± 1.3 μm (N = 30), LW 71.1 ± 14.3 μm², and L/W 1.91 ± 0.52 versus 11.2 ± 1.7 × 7.5 ± 1.3 μm (N = 70), LW 83.8 ± 19.0 μm², and L/W 1.54 ± 0.41.

Meronts of *Plasmodium floridense* in active infections of *Anolis conspersus* of Grand Cayman Island are 6.7 ± 1.2 × 4.7 ± 0.6 μm (5–13 × 3–6, N = 75), with LW 34.1 ± 6.8 μm² (20–65), and contain 12.0 ± 1.9 (8–16) merozoites. Meront size relative to host cell nucleus size is 1.00 ± 0.27 (0.44–1.75, N = 67), and to normal erythrocyte nucleus size is 1.04 ± 0.26 (0.61–2.35, N = 75). Gametocytes are 8.8 ±

1.3 × 5.3 ± 0.8 μm (6–11 × 3–7, N = 65), with LW 46.5 ± 9.0 μm^2 (21–66) and L/W 1.70 ± 0.37 (1.00–2.75). Gametocyte size relative to host cell nucleus size is 1.56 ± 0.44 (0.75–2.50, N = 61), and to normal erythrocyte nucleus size is 1.55 ± 0.33 (0.76–2.38, N = 65). Gametocytes do not differ in dimensions or shape by sex.

In *Anolis sagrei* of the Bahamas, meronts of *P. floridense* are 7.8 ± 1.4 × 4.7 ± 0.8 μm (6–14 × 3–7, N = 58), with LW 36.3 ± 7.5 μm^2 (24–63), and contain 15.1 ± 4.4 (7–32) merozoites. Meront size relative to host cell nucleus size is 1.26 ± 0.30 (0.62–2.10, N = 41), and to normal erythrocyte nucleus size is 1.35 ± 0.29 (0.88–2.39, N = 58). Samples of meronts from *A. sagrei* of two islands, North Bimini and Cat Cay, were similar in dimensions and merozoite production, respectively 7.7 ± 1.2 × 5.2 ± 0.7 μm (N = 23), LW 39.4 ± 7.3 μm^2, and 16.9 ± 2.9 (12–22) merozoites versus 7.9 ± 1.6 × 4.4 ± 0.7 μm (N = 35), LW 34.2 ± 7.1 μm^2, and merozoites 13.9 ± 4.8 (7–32). Meronts from Cat Cay averaged more merozoites in proerythrocytes (16.4 ± 5.7, 11–32, N = 16) than in erythrocytes (11.8 ± 2.5, 7–16, N = 19), but dimensions were similar, respectively 8.7 ± 2.0 × 4.4 ± 0.9 μm, LW 37.8 ± 8.4 μm^2 versus 7.3 ± 0.8 × 4.3 ± 0.5 μm, LW 31.2 ± 3.8 μm^2. Overall, gametocytes in *A. sagrei* are 11.3 ± 2.5 × 5.2 ± 1.1 μm (6–18 × 3–9, N = 75), LW 58.6 ± 16.9 μm^2 (30–108), and L/W 2.27 ± 0.80 (1.00–4.67). Gametocyte size relative to host cell nucleus size is 1.95 ± 0.68 (1.03–4.00, N = 49), and to normal erythrocyte nucleus size is 2.17 ± 0.63 (1.14–4.09, N = 75). Gametocytes from Cat Cay average longer than those from North Bimini and have a higher L/W ratio, with respective dimensions 12.0 ± 2.1 × 5.0 ± 1.0 μm (N = 50), LW 59.1 ± 14.8 μm^2, and L/W 2.54 ± 0.78 versus 9.8 ± 2.7 × 5.8 ± 1.1 μm, LW 57.0 ± 20.8 μm^2, and L/W 1.73 ± 0.53. Dimensions of micro- and macrogametocytes in the Cat Cay sample are similar, but the latter are more elongate, on average, with respective dimensions 12.3 ± 2.3 × 4.7 ± 0.7 μm (N = 25), LW 57.5 ± 15.5 μm^2, and L/W 2.71 ± 0.77 versus 11.7 ± 1.9 × 5.3 ± 1.2 μm (N = 25), LW 60.7 ± 14.2 μm^2, and L/W 2.37 ± 0.77 for microgametocytes.

Exoerythrocytic Merogony Unknown.

Sporogony Unknown.

Effects on Host In most of the Caribbean hosts, erythrocytes infected by *P. floridense* meronts are approximately normal in size in comparison to uninfected cells, ± 3%. Hypertrophy of 8–12% was present in *A. sagrei* of North Bimini and *A. distichus*, while erythrocytes host to meronts in *A. sagrei* of Cat Cay were 10% smaller than normal. Only gametocyte-infected erythrocytes in *A. lineatopus* and *A. opalinus*, both with chronic infections, and in *A. conspersus* were less than 5% larger than uninfected cells. Erythrocytes of *A. garmani, A. pulchellus*, and *A. sagrei* of North Bimini were hypertrophied by 7.5–12%, *A. distichus* by 17%, while gametocyte-infected erythrocytes of *A. cybotes* were 22% smaller than uninfected cells. The nuclei of meront-infected erythrocytes were hypertrophied by 10–13% in *A. sagrei* of North Bimini, *A. cybotes, A. pulchellus*, and *A. garmani*; and in cells host to gametocytes, nuclei were 10–14% larger than normal in *A. sagrei* of North Bimini and Cat Cay, *A. cybotes*, and *A. pulchellus*. Meronts sometimes distorted erythrocytes (2–48%) in most host species, but in *A. pulchellus* 86% of host cells were distorted. Host cell nuclei were rarely distorted (0–20%) but often were displaced (17–60%) by meronts. Erythrocytes were commonly distorted by *P. floridense* gametocytes (28–56%), but in three hosts, *A. sagrei* of North Bimini, *A. garmani*, and *A. pulchellus*, 88–99% of erythrocytes were distorted. Host cell nuclei were seldom distorted by gametocytes (0–20%) but were commonly displaced (32–73%), and most nuclei were displaced in *A. sagrei* of North Bimini (80%) and *A. pulchellus* (87%).

Ecology *Anolis* species known to host *Plasmodium floridense* in the Greater Antilles are represented by five of the six anole ecomorphs defined by Williams (1983). On Hispaniola, three species are recorded hosts, each a different ecomorph: *A. cybotes* is a trunk-ground species, *A. coelestinus* is a trunk-crown type, and *A. distichus* occupies the middle of this spectrum as a trunk species. Neither crown-giant nor bush-grass species are so far known hosts to *P. floridense*, perhaps as a result of inadequate sampling of these categories. The localities positive for *P. floridense* all appear to be mesic habitats. On Jamaica, *A. lineatopus* is a trunk-ground species, *A. grahami* and *A. opalinus* are trunk-crown, and *A. garmani* is a crown-giant. In the case of Puerto Rico, only two anole species are known hosts, the bush-grass species *A. pulchellus,* and *A. gundlachi*, a trunk-ground species. This host spectrum suggests that the vector of *P. floridense* in the Greater Antilles probably forages from ground to forest crown in mesic habitat.

In a wet-evergreen tropical forest in Puerto Rico, El Verde, during a 9-year period, Schall et al. (2000) found that the prevalence of *P. floridense* in *Anolis gundlachi* was lowest (~13%) in 1990, in the aftermath of Hurricane Hugo, and reached maximum prevalence (~40%) in the winter period of 1997, remained at a stable level of about 28–30% in winter and summer of 1998, and dropped again in winter 1999 to about 17%. The relative prevalence of *P. floridense* to that of *Plasmodium azurophilum* remained constant during the study period, but because data for the two species were combined as "malaria," other aspects of the ecology of *P. floridense* cannot be assessed from the published information. In *Anolis sabanus* of Saba, Staats and Schall (1996b) found a higher prevalence of *P. floridense* and *P. azurophilum*, considered together, in moist areas of

Saba and a lower prevalence in arid, wind-blown sites or the wet mountain top, but without elevational effects on prevalence. In male lizards, but not females, *P. floridense* prevalence was greater among the larger (therefore, older) lizards, and *P. azurophilum* infected more of the smaller, younger *A. sabanus*. Little difference was present in prevalence of *P. floridense* by sex of its hosts. Parasitemias were highest in *P. floridense* infections, apparently at a little over 1% of erythrocytes. Prevalence differed considerably among sites, suggesting that "vector density or feeding behavior varies among closely situated sites" in the complex topography of Saba. *Anolis sabanus* because of its small size is probably short-lived, indicating that the transmission rate of the *Plasmodium* parasites "must be very high." Because Saba has no "real dry season," transmission is possible year round.

Remarks Present knowledge indicates that *Plasmodium floridense* is distributed from the Greater Antilles southward through the northern Lesser Antilles to Dominica (Staats and Schall, 1996a; Perkins and Schall, 2002) in *Anolis* species of the *A. bimaculatus* group. The *bimaculatus* group possibly has an origin in the Greater Antilles (Staats and Schall, 1996a). Anoles of the *Anolis roquet* group, of South American origin, occupy the Lesser Antilles south of Dominica, from Martinique to Grenada, and apparently are not hosts to *P. floridense* but are parasitized by *P. azurophilum*. This distribution pattern supports the suggestion by Telford (1994) that *P. floridense* entered the Caribbean with "beta-anoles" from Middle America, which colonized and speciated into the diverse anole fauna of the Greater Antilles. The distinctive morphology of *P. floridense* seen in the Middle American hosts has been retained in its Caribbean hosts. Future genome studies will undoubtedly clarify this postulated history of *P. floridense* in the Caribbean and must relate it to the genetic variation of *P. floridense* in its Middle American hosts.

Plasmodium floridense in Lizards of the Southeastern United States (**Plate 29**)

Hosts *Sceloporus undulatus* (Thompson and Huff, 1944b; Goodwin, 1951; Jordan, 1964; Telford, 1978c); *Anolis carolinensis* (Thompson and Huff, 1944b; Jordan, 1964; Telford, 1978c); and *Anolis sagrei* (Wozniak et al., 1996). Experimental hosts are *Crotaphytus collaris* and *Gambelia wislizenii* (Thompson and Huff, 1944b), *Sceloporus olivaceous* (Thompson, 1944), and *Sceloporus woodi* (Telford, 1974).

Localities and Prevalence In the description of *P. floridense*, Thompson and Huff (1944b) reported its presence in 44 of 135 (32.6%) *S. undulatus* and 32 of 52 (61.5%) *A. carolinensis* at Ocala, Marion County, Florida. In Baker County, Georgia, Goodwin (1951) found 50 of 331 (15.1%) *S. undulatus* infected. Also in Georgia, Jordan (1964) reported *P. floridense* prevalence in *S. undulatus* to be 40 in 771 (5.2%) from Lee County, 104 of 2232 (4.7%) in Clinch and Echols counties, and 22 of 554 (4.0%) from unspecified areas. In *A. carolinensis*, prevalence was 32 of 276 (11.6%) in Lee County, 510 of 1502 (34.0%) in Clinch and Echols counties, 13 of 101 (12.9%) from unspecified areas, 19 of 157 (12.1%) on Jekyll Island, and 42 of 312 (13.5%) on Sapelo Island. Overall prevalence in Georgia was 166 of 3557 (4.7%) in *S. undulatus* and 616 of 2348 (26.2%) in *A. carolinensis*. In a later study, Jordan and Friend (1971), including the data from the 1964 study, reported that annual prevalence of *P. floridense* varied from 10% to 52%, with an average of 35% in *A. carolinensis*, in comparison to 2% to 10% in *S. undulatus*, averaging 5% in the Okefenokee Swamp-Fargo area, with 11,420 lizards of both species examined during 13 years. *Plasmodium floridense* is known from 18 Georgia counties (Appling, Atkinson, Baker, Berrien, Brantley, Camden, Charlton, Clinch, Coffee, Dougherty, Early, Echols, Lanier, Lee, Lowndes, McIntosh, Ware, and Wayne) and from three coastal islands (Jekyll, Sapelo, and Cumberland). All known Georgia localities extend south or southwest from the vicinity of the Altamaha River and lie east of the Appalachicola River. Prevalence of *P. floridense* in Florida was reported by Telford (1978c) as 12.7% in 973 *A. carolinensis* and 3.1% in 163 *S. undulatus*, with prevalence 24.0% from Hardee County southward, and 8.6% from Marion county northward. Additional records from Florida lizards collected from 1978 to 2005 confirm the greater prevalence of *P. floridense* in *A. carolinensis* in southern Florida in comparison to northern Florida. In eight counties of southern Florida (Dade, Monroe, Collier, Palm Beach, Lee, De Soto, Charlotte, and Hendry), 67 of 278 (24.1%) *A. carolinensis* were infected; in Polk and Hillsborough counties of central Florida, 7 of 86 (8.1%), and in 5 counties of northern Florida (Alachua, Marion, Gilchrist, Putnam, and Nassau), 85 of 1146 (7.4%) *A. carolinensis* were host to *P. floridense*. Infected *S. undulatus* were found in Alachua and Marion counties only, in 12 of 219 (5.5%). Klein et al. (1987a) reported prevalence of *P. floridense* in *A. carolinensis* as 36 of 133 (27.1%) in Alachua County and 2 of 3 in Levy County, and prevalence in *S. undulatus* as 6 of 114 (5.3%) in Alachua County. No infections were found in 103 *A. sagrei* as reported in 1978, but additional material collected since found *P. floridense* to be common in *A. sagrei* in Palm Beach County, 25 of 62 (40.3%), and present but less common in Collier (2 of 21, 9.5%) and Monroe counties (1 of 17, 5.9%), all in southern Florida. A single infection has been found in Hillsborough County in central Florida (1 of 103), but none were found in 33 *A. sagrei* examined from Alachua and Marion counties of north-central Florida.

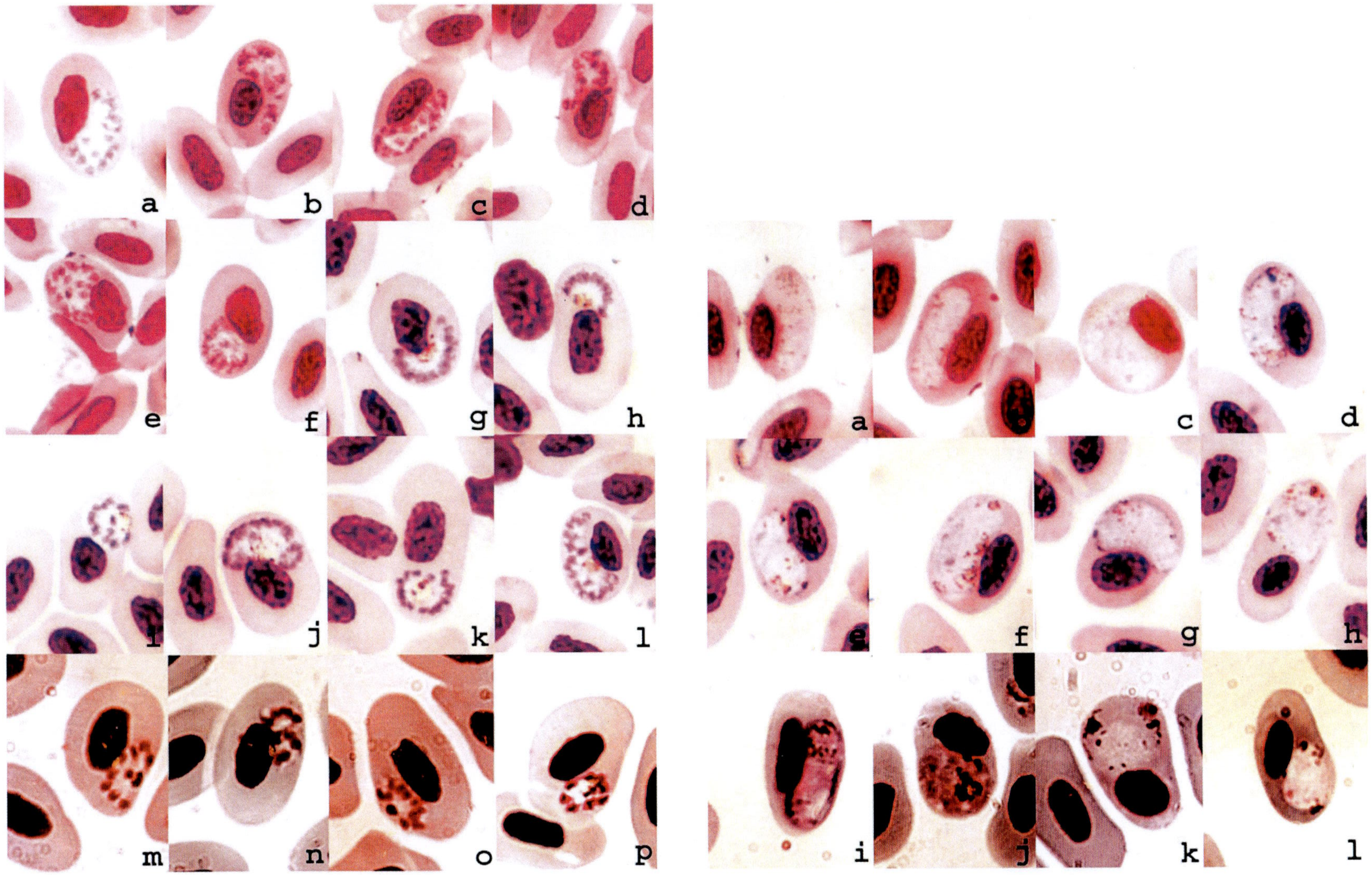

(A) (B)

Plate 29 (**A**) and (**B**) *Plasmodium floridense* from Florida hosts. (**A**) Meronts: **a–f**, *Anolis carolinensis*; **g–l**, *A. sagrei;* **k**, **l**, **m–p**, *Sceloporus undulatus.* (**B**) Gametocytes: **a–d**, *A. carolinensis*; **e–h**, *A. sagrei;* **i–l**, *S. undulatus.* Macrogametocytes are **a**, **b**, **e**, **f**, **i**, and **j**. Others are microgametocytes.

Morphological Variation In *Anolis carolinensis* of Florida and Georgia, meronts usually form as rosettes or fans: Among 235 meronts, 58% are fan-shaped, 30% are rosettes, 9% elongate, and the remainder can be described as a morulum, round, oval, oblong, or of indefinite form. Meronts are 7.7 ± 1.5 × 5.2 ± 1.0 μm (4–13 × 3–8, N = 320) and contain 12.7 ± 4.6 (4–30, N = 412) merozoites. Meront LW is 40.8 ± 12.4 μm² (15–84), and relative to host cell nucleus size is 1.38 ± 0.44 (0.53–2.86, N = 254), and to normal erythrocyte nucleus size is 1.50 ± 0.46 (0.54–3.27, N = 297). In southern Georgia (Baker County and the Fargo-Okefenokee area), meronts are 7.9 ± 1.7 × 5.0 ± 0.9 μm (N = 41), LW 39.1 ± 10.8 μm², and contain 13.6 ± 5.0 (6–27) merozoites. In north-central Florida (Alachua, Gilchrist, Marion counties), meronts are 7.5 ± 1.5 × 5.3 ± 1.0 μm (N = 173), LW 40.4 ± 13.3 μm², with 12.0 ± 4.4 (4–30, N = 263) merozoites. Meronts in *A. carolinensis* from south Florida (De Soto, Lee, Collier, Dade counties) are 8.0 ± 1.5 × 5.2 ± 0.9 μm (N = 106), LW 42.1 ± 11.6 μm², and produce 14.2 ± 4.6 (4–28, N = 108) merozoites. These data demonstrate remarkably little variation in meront characteristics throughout the range of *P. floridense* in *A. carolinensis* of the southeastern United States. The greatest variation in morphology of meronts results from maturity of the host cells. In Alachua County infections, erythrocytic meronts are 7.2 ± 1.5 × 5.1 ± 0.9 μm (N = 77), LW 37.4 ± 12.4 μm², with 11.9 ± 4.1 (4–22) merozoites, but in proerythrocytes, meronts are larger with more merozoites, 9.6 ± 1.0 × 5.9 ± 1.0 μm (N = 15), LW 56.5 ± 12.5 μm², and 19.7 ± 2.5 (16–26) merozoites. Similar differences are present in a Gilchrist County infection, with 25 erythrocytic meronts 6.8 ± 1.1 × 4.8 ± 0.7 μm (N = 25), LW 32.5 ± 7.2 μm², and merozoites 7.9 ± 2.1 (4–13) versus 25 proerythrocytic meronts, 7.8 ± 0.9 × 5.3 ± 0.9 μm, LW 41.3 ± 8.5 μm², and merozoites 12.5 ± 2.8 (8–20). The latter sample was taken immediately postcrisis, while the Alachua County samples were in active phase, preceding crisis. Jordan (1975) found a distinct effect of infection phase on merozoite production by natural infections of *P. floridense* in *A. carolinensis*, merozoite number declining from 13.44 ± 0.15 during the acute rise in parasitemia to 10.28 ± 0.09 during the period of declining parasitemia. Overall, gametocyte dimensions of *P. floridense* in *A. carolinensis* from Georgia and Florida are 11.8 ± 2.6 × 5.9 ± 1.2 μm (7–22 × 3–10, N = 360), with LW 68.5 ± 17.7 μm² (32–132), and L/W 2.12 ± 0.76 (1.00–5.25). Gametocyte size relative to host cell nucleus size is 2.66 ± 0.90 (1.14–8.80, N = 297), and to normal erythrocyte nuclei is 2.58 ± 0.67 (1.16–4.98, N = 235). In southern Georgia gametocytes are 12.7 ± 3.0 × 5.0 ± 0.9 μm (N = 42), LW 62.8 ± 15.8 μm², and L/W 2.65 ± 0.92. Gametocytes in north-central Florida anoles are 11.3 ± 2.2 × 6.1 ± 1.3 μm (N = 175), LW 67.7 ± 17.7 μm², and L/W 1.96 ± 0.66. Gametocytes in south Florida are 12.1 ± 2.7 × 5.9 ± 1.2 μm (N = 143), LW 71.0 ± 18.0 μm², and L/W 2.15 ± 0.76. As with meronts, gametocyte morphology varies little over the geographic range of *P. floridense* in *A. carolinensis*. Average dimensions of *P. floridense* macrogametocytes are greater in width and LW, with a lower L/W ratio than microgametocytes, 11.6 ± 2.5 × 6.3 ± 1.3 μm (N = 208), LW 72.0 ± 19.3 μm², and L/W 1.94 ± 0.68 versus 12.1 ± 2.7 × 5.4 ± 1.0 μm (N = 152), LW 63.7 ± 14.0 μm², and L/W 2.36 ± 0.81, respectively. Gametocyte dimensions in active infections are similar to those in chronic phase, and in the comparisons between the same sex by infection phase, there was little difference.

In the type host, *Sceloporus undulatus*, meronts are 7.6 ± 1.9 × 4.9 ± 1.0 μm (5–15 × 3–7, N = 128), with LW 37.1 ± 10.6 μm² (15–72), and contain 10.3 ± 3.2 (5–21, N = 132) merozoites. Meront size relative to host cell nucleus size is 1.35 ± 0.44 (0.43–2.61, N = 106), and to normal erythrocyte nucleus size is 1.34 ± 0.40 (0.58–2.88, N = 128). Samples from southern Georgia, Baker County, and the Fargo-Okefenokee area, are similar in dimensions to those from north-central Florida (Alachua County). Jordan (1964) found a pattern of reduction in merozoite numbers during change in phase of infection from 10.6 ± 0.11 during acute rise to 8.55 ± 0.16 during decline, similar to that observed in *A. carolinensis* from the same areas. Gametocyte dimensions, overall, of *P. floridense* in *S. undulatus* from southern Georgia to north-central Florida are 11.0 ± 2.9 × 6.0 ± 1.5 μm (6–21 × 3–10, N = 200), with LW 63.7 ± 16.1 μm² (35–114), and L/W 2.06 ± 1.10 (1.00–5.25). Gametocyte size relative to host cell nucleus size is 2.49 ± 0.75 (0.90–6.00, N = 171), and to normal erythrocyte nucleus size is 2.36 ± 0.75 (1.03–4.52, N = 200). Gametocytes in the Georgia samples in comparison to those from north-central Florida are of similar size but are more elongate and narrower, with higher L/W ratio, 12.3 ± 2.9 × 5.2 ± 1.4 μm (N = 102), LW 61.7 ± 16.4 μm², and L/W 2.62 ± 1.06 versus 9.7 ± 2.2 × 6.8 ± 1.1 μm (N = 98), LW 65.7 ± 15.7 μm², and L/W 1.48 ± 0.50, respectively. In contrast to *P. floridense* in *A. carolinensis*, macro- and microgametocytes are similar in dimensions and shape, respectively 11.5 ± 2.8 × 5.9 ± 1.6 μm (N = 93), LW 65.6 ± 18.7 μm², and L/W 2.20 ± 1.00 versus 10.7 ± 2.9 × 6.0 ± 1.4 μm (N = 107), LW 62.0 ± 13.4 μm², and L/W 1.94 ± 0.95. In an experimental infection of *Sceloporus woodi*, which does not host *P. floridense* in nature, derived from a natural infection in north-central Florida *S. undulatus*, meronts were smaller in *S. woodi* but produced similar numbers of merozoites to those in *S. undulatus*: 6.0 ± 0.9 × 4.4 ± 0.7 μm (N = 25), LW 26.4 ± 6.2 μm², and 9.1 ± 2.3 merozoites. Gametocyte dimensions and shape were similar to those in *S. undulatus*, 9.0 ± 1.6 × 6.1 ± 1.0 μm (N = 25), LW 55.0 ± 14.2 μm², and L/W 1.51 ± 0.36.

In the third natural host species of *P. floridense* in Florida, *Anolis sagrei* of south Florida, meronts are similar in size and merozoite number to those in *A. carolinensis* of the same area. Meronts are 7.4 ± 1.9 × 5.0 ± 1.3 μm 5.5–10 × 3–7.5, N = 25), LW 37.3 ± 11.6 μm² (21–68), and contain 13.0 ± 4.1 (7–24) merozoites. Their size relative to host cell nucleus size is 1.23 ± 0.36 (0.66–2.05), and to normal erythrocyte nucleus size is 1.30 ± 0.40 (0.73–2.34). Gametocytes are also similar in dimensions to those in *A. carolinensis*, but shape tends to be a little less elongate with a lower L/W ratio. Gametocytes are 10.7 ± 1.5 × 6.3 ± 1.1 μm (8–14 × 4.5–9.5, N = 60), with LW 67.1 ± 13.6 μm² (45–114), and L/W 1.75 ± 0.47 (1.13–3.11). Gametocyte size relative to host cell nucleus size is 2.47 ± 0.47 (1.67–3.42, N = 25), and to normal erythrocyte nucleus size is 2.32 ± 0.47 (1.56–3.96, N = 60).

Exoerythrocytic Merogony No one has reported the presence of EE meronts in circulating white blood cells of lizards host to *P. floridense*. In an *A. carolinensis* from north-central Florida chronically infected with *P. floridense*, phanerozoic meronts were abundant in endothelium and connective tissue of the heart, liver, lungs, and kidneys (**Plate 26B**). None were seen in macrophages, monocytes, granulocytes, lymphocytes, or thrombocytes within these organs. Although some of the host cells with early infection resembled erythrocytes in being elongate with an elongated ovoid nucleus, their size appeared smaller, and none were present within the lumen of capillaries or blood sinuses. Some meronts in endothelium were especially elongated and narrow. Meronts averaged smaller in lung, 13.5 ± 2.1 × 8.6 ± 1.2 μm (10–17 × 7–10, N = 9), and heart, 14.8 ± 4.4 × 9.2 ± 2.4 μm (10–26 × 6–15, N = 14), than in liver, 16.2 ± 4.0 × 8.0 ± 1.8 μm (11–21 × 6–15, N = 14), and kidney, 15.9 ± 3.3 × 10.3 ± 2.6 μm (10–20 × 7–14, N = 9). These were not compared statistically because their selection for measurement was not random. Nuclei counted in single focal planes of 11 meronts that were not too heavily stained to permit counts varied from 40 to 57 in smaller meronts, 10–14 × 7–10 μm, and from 74 to 132 at meront sizes of 17–20 × 7–15 μm.

Sporogony Huff (1941) reported the presence of a single "nearly mature" oocyst on the stomach of an *Aedes aegypti* fed on *Sceloporus undulatus* infected with malaria, later described as *Plasmodium floridense* (Thompson and Huff, 1944b). Jordan (1964) attempted to feed 16 local mosquito species of the genera *Aedes, Mansonia, Culex, Anopheles*, and *Psorophora* on *A. carolinensis* and *S. undulatus*, noting that when exposed at the same time, mosquitoes preferred to feed on *Sceloporus*. Eight of the species tested fed on the lizard hosts. Oocysts were found in only four species, and none had infected salivary glands. One of 80 *A. aegypti* dissected had a single oocyst, 2 of 150 *Culex quinquefasciatus* showed 1 and 3 oocysts, 4 of 70 *Culex territans* were host to 23, 1, 16, and 2 oocysts, while 1 of 3 *Culex* sp. had 70 oocysts present. This last mosquito species was not identified by Jordan; Klein et al. (1987a) suggested that it may have been *Culex erraticus*. Klein (1985) succeeded in obtaining complete sporogony, including invasion of salivary glands, in *C. erraticus* of Alachua County, Florida, and accomplished transmission by bite of infected mosquitoes to *A. carolinensis* but not to *S. undulatus* (Klein et al., 1987a, 1988c). Although in addition to *C. erraticus*, blood-fed *Culex territans, C. salinarius/nigripalpis*, and *C. perturbans* were collected from lizard-baited traps, only *C. erraticus* supported sporogony of *P. floridense*. Sporogonic development was completed in oocysts as early as 9 days PF, but development was "highly asynchronous" (Klein et al., 1988c). Mean oocyst size by days 10 and 11 PF averaged 41.8 μm and 42.7 μm, respectively, with oocysts varying in size as 23.4–54.6 μm and 28.6–59.8 μm on those days, respectively. Sporozoites appeared in salivary glands as early as 13–14 days PF and averaged 15.1 ± 0.5 × 1.03 ± 0.02 μm (8–22 × 0.8–1.5, N = 51). When phlebotomine sand flies, *Lutzomyia vexator*, fed on *P. floridense*-infected hosts, they readily developed sporulated oocysts on the midgut, but sporozoites did not invade salivary glands (Klein et al., 1987a, 1988c).

Effects on Host In the type host, *S. undulatus*, neither gametocytes nor meronts affect host cell or nucleus size. Erythrocytes were distorted by both meronts (29%) and gametocytes (75%) and their nuclei displaced, respectively, 33% and 67%, but nucleus distortion was rare (7–10%). In *A. carolinensis*, erythrocytes host to gametocytes were slightly enlarged (12%), but those with meronts were only 7% larger. Host cell nuclei were of normal size in gametocyte infections, but averaged slightly larger (7%) when meronts were present. Host erythrocytes were commonly distorted by meronts (52%) and usually were by gametocytes (94%), with little effect on nucleus shape by either stage (4%). Nuclei were usually displaced by both meronts (71%) and gametocytes (86%). In a natural infection of *A. sagrei*, erythrocytes infected by gametocytes were hypertrophied 16.5% over normal size, but only slightly by meronts (7.5%). Nuclei were similar in size compared to those of uninfected cells. Distortion of host cells was greater by gametocytes (96%) than by meronts (32%), nuclear distortion was uncommon (8% and 12%, respectively), but nuclei were always displaced by gametocytes, less often by meronts (68%). In an experimental infection of *Sceloporus woodi* from an *S. undulatus* donor, erythrocytes were hypotrophic when parasitized by either stage, with erythrocytes host to gametocytes 7.6% smaller and those containing meronts 17.5% smaller than uninfected cells. Host cell nuclei were similar in size to those of normal cells. Host cells were seldom distorted by meronts (8%) but commonly were (64%)

by gametocytes. Nuclei were often displaced by meronts (36%) and usually were by gametocytes (84%).

In experimental infections, Thompson (1944) found that *P. floridense* could produce higher parasitemias in *S. undulatus* than in *A. carolinensis*. Goodwin and Stapleton (1952) reported peak parasitemias of 25,384 and 35,192 parasites per 10^4 erythrocytes in natural infections of *S. undulatus*. Jordan (1975) reported the average peak of parasitemia in *A. carolinensis* to be 1600 parasites per 10^4 erythrocytes, but in *S. undulatus* it was 11,600 per 10^4 erythrocytes, ranging from 5000 to 22,000. Not only was parasitemia higher in *S. undulatus*, but duration of infection was twice as long, 150 (100–253) days, than in *A. carolinensis*, in which parasitemia lasted 71 (60–90) days. This was attributed by Jordan to the development of acquired immunity to *P. floridense*, with twice as long required on average in *S. undulatus* than in *A. carolinensis*. *Sceloporus undulatus* that developed infections that reached a peak of 12,900 died 4 days after the peak, while *A. carolinensis* recovered from infections that lasted 81 days.

Ecology Goodwin (1951), in Baker County, Georgia, found infected juvenile *Sceloporus undulatus* in November (5.3%) and January (14.3%). The highest prevalence in adults was also in January (68.9%), with lowest prevalence seen from May to October. Jordan (1964) reported that the prevalence of initial and acute infections in *Anolis carolinensis* began to rise in August and continued into November, when maximum monthly prevalence occurred. Overall prevalence, also, reached a peak in October during the 1959–63 study. Between 1959 and 1963, Jordan (1964) found a steady increase in prevalence in *A. carolinensis* by month, from a little over 30% in July–August, rising in September to about 40% in October, thereafter declining again until February, when it rose to slightly exceed 40%, again declining to a minimum of around 25% in May, then rising sharply to a monthly peak in June of 45%. Prevalence in 1960–61 was higher than for the overall 5-year period, with peaks of 50% in June and about 51% in November, with the curve of monthly prevalence mirroring well that for the entire period, but somewhat higher, especially in October and November. Most important, the seasonal percentage of new infections rose sharply from a level of about 5% from January to July, through August (30%), September (47%), and October (55%), to the peak of about 58% in November, after which the percentage of new infections declined sharply to 25% in December. This acquisition of new infections in the anole population clearly demonstrates that in southern Georgia, *P. floridense* is seasonally transmitted from late summer (August) into November.

During the 13-year period (1958–70) of Jordan's study of *P. floridense* in the Fargo-Okefenokee Swamp area (Jordan and Friend, 1971), in which 11,240 lizards, comprised of about equal numbers of *A. carolinensis* and *S. undulatus*, were examined, annual prevalence of *P. floridense* varied from a high of 52% in 1958 in anoles to a low of 10% in 1964 and then recovered to nearly the same initial level, about 46%, by 1970. Overall prevalence of *P. floridense* in *A. carolinensis* during the 13 years was 35%. In *S. undulatus* from the same locality, prevalence varied from 2% to 10%, with overall prevalence of 5%. The greatly lower prevalence in *S. undulatus* provided only a weakly similar pattern of fluctuation in prevalence of *P. floridense* over the 13 years to that seen in *A. carolinensis*. Jordan and Friend (1971) also commented that their observations supported a "spotty" distribution of saurian malaria parasites within endemic areas, suggested by Thompson and Huff (1944b). Within the 25-mile radius of the Okefenokee Swamp study area, prevalence in samples of 50 or more *A. carolinensis* could be as high as 67%, but the overall prevalence was 40%. The authors attributed this variation in distribution to "(a) optimum climatic conditions conducive to the breeding of the vector in that particular spot and (b) to the proximity and restricted range of a dense population of lizards." Jordan (1986) summarized 30 years of field research on *P. floridense* (1954–84), which involved the capture and examination of 15,144 *S. undulatus* and *A. carolinensis*. Overall prevalence in *S. undulatus* was 9%, and 31% in *A. carolinensis* in the Fargo-Okefenokee Swamp study area. In 7068 lizards from Cumberland island examined from 1976 to 1984, 10% of *S. undulatus* and 31% of *A. carolinensis* were infected by *P. floridense*. Jordan characterized the incidence of infection over 30 years as "amazingly uniform."

In a much smaller mark-and-release study involving 839 *A. carolinensis* from an ecologically very different environment, conducted in 1970–73 and 1985–98 in two suburban residential yards of Gainesville, Florida (Telford, unpublished), prevalence of *P. floridense* was at a minimum in July–August (1.4%), rising to 4.3% in September–October, then to 7.9% in November, declining in December to 4.6%, then dropping to 2.8% for January–February, reaching higher prevalence in March (9.6%), April (12.9%), May (7.3%), and June (12.1%). Infections in juvenile anoles appeared in October and November, when they comprised 20% and 31%, respectively, of the total infections found in those months. In the Gainesville area, Klein et al. (1987a) reported peak abundance of *Culex territans* to occur in June, and that of *Culex erraticus* in August, with light traps showing maximum activity of *C. erraticus* in late July to early August. Klein (1985) reported that most new infections of *P. floridense* were observed in October and November, subsequent to the peak activity period of *C. erraticus*.

Remarks To date, there are no records of *Plasmodium floridense* north of McIntosh County, Georgia, on the north side of the Altamaha River, or west of the Appalachicola River in Georgia and Florida, although this might be an artifact of collecting. Certainly, the vector *Culex erraticus* extends well north and west of these limits. *Plasmodium floridense* occurs throughout peninsular Florida and at least as far south as Big Pine Key in the Florida Keys. It has been suggested (Telford, 1994) that *P. floridense* entered the southeastern United States with *Anolis carolinensis*, a species of Cuban origin (Williams, 1983) that successfully invaded both the Bahamas and Florida. *Sceloporus* possibly reached Florida twice from its western North American origin, the first invasion resulting in the isolation of a progenitor of *Sceloporus woodi* on islands of the Pleistocene interglacial periods, today represented by the relatively xeric habitats of scrub and sand hills to which *S. woodi* is restricted (Clark et al., 1999). A later invasion brought *S. undulatus* into contact with the long-established *A. carolinensis,* which occupied the less-xeric mesic habitats in which *S. undulatus* occurs today. Transfer of *P. floridense* to the new host, *S. undulatus*, may have occurred along the wetter and interdigitating margins of mesic forest where suitable breeding habitat for *Culex erraticus* is present. In contrast to the widespread distribution of *P. floridense* in much greater prevalence in *A. carolinensis*, which has lower parasitemias and a shorter duration of infection (Jordan, 1975) with little evidence of parasite-induced mortality, *P. floridense* in *S. undulatus* is characterized by much lower prevalence, a duration of infection twice as long, and commonly massive parasitemias following which mortality can result. This indicates a more recently established host-parasite association between *P. floridense* and *Sceloporus undulatus*. Some of the lower prevalence is likely due to behavioral differences between *S. undulatus*, which spends nights sheltered beneath bark or in tree holes and crevices, in comparison to *A. carolinensis,* which is commonly found on vegetation exposed to the mosquito vector at night. This difference in vector exposure is undoubtedly responsible for the absence of *Plasmodium* infections in the experimentally suitable host, *Sceloporus woodi,* which has a disjunct distribution in the most xeric habitats of Florida, where it disappears into the loose sand in late afternoon, not to reemerge until next midmorning, when mosquito activity is at its minimum.

THE *GARNIA* SPECIES OF NEOTROPICAL LIZARDS

Plasmodium gonatodi Telford 1970 (Plate 30)

Diagnosis A species of *Plasmodium* (*Garnia*) with polymorphic meronts 8–19 × 4–9 μm, LW 40–114 μm², that produce 12–50 merozoites. Meront dimensions and productivity are not affected by host cell maturity. Meront size relative to host cell nucleus size averages 1.78, and to normal erythrocyte nucleus size is 1.96. Gametocytes are typically elongate and slender, banana-shaped, 7–21 × 4–10 μm, with LW 42–114 μm² and L/W 1.00–5.25. Gametocyte size relative to host cell nucleus size averages 2.45, and to normal erythrocyte nucleus size is 2.38. Microgametocytes average slightly smaller in size, longer but more slender than macrogametocytes, with greater L/W ratios. Prematuration stages, with irregular margins, often bizarre shapes, and larger, with one or both ends commonly extended, precede mature gametocytes. All stages lack visible pigment.

Type Host *Gonatodes albogularis fuscus* (Hallowell) (Sauria: Gekkonidae).

Type Locality Sasardi, about 5 km west of Mulatupo, San Blas Territory, Panama.

Other Hosts *Gonatodes humeralis* (Lainson et al., 1971), *G. vittatus* (Ayala, 1978).

Other Localities Vicinity of Belém, Pará State, Brazil (Lainson et al., 1971), and northern Colombia (Ayala, 1978).

Prevalence Overall, *Plasmodium gonatodi* parasitized 12 of 172 (7.0%) *Gonatodes albogularis* from Panama and 12 of 47 (25.5%) at the type locality. In Brazil, 26 of 52 (50%) *Gonatodes humeralis* were infected.

Morphological Variation All stages lack visible pigment. Meront shape is highly variable, round or often elongate and curving around the host cell nucleus. Rarely, merozoites may be arranged as rosettes or broad fans. Occasional elongate meronts apparently break into two portions when the cytoplasm becomes narrowed, resembling a multiply infected cell. Meronts are 11.8 ± 2.1 × 6.1 ± 1.0 μm (8–19 × 4–9, N = 70), with LW 71.8 ± 13.7 μm² (40–114), and contain 28.0 ± 7.9 (12–47) merozoites. Meront size relative to host cell nucleus size is 1.76 ± 0.38 (1.12–2.75, N = 40), and to normal erythrocyte nucleus size is 1.96 ± 0.38 (1.03–2.81, N = 70). Meront dimensions and merozoite numbers are not affected by host cell maturity. Gametocytes are preceded by prematuration stages, often enlarged and bizarrely shaped, with irregular margins and one or both ends narrowed, even attenuated into blunt points. Gametocytes are characteristically elongate and thin, banana-shaped, but occasionally may be rounded or ovoid, 15.5 ± 3.3 × 5.6 ± 1.3 μm (7–21 × 4–10, N = 125), with LW 84.3 ± 14.6 μm² (42–114) and L/W 2.95 ± 1.03 (1.00–5.25). Gametocyte size relative to host cell nucleus size is 2.45 ± 0.57 (1.20–4.15), and to normal erythrocyte nucleus size is 2.38 ± 0.47 (1.15–3.60, N = 100). Macrogametocytes are

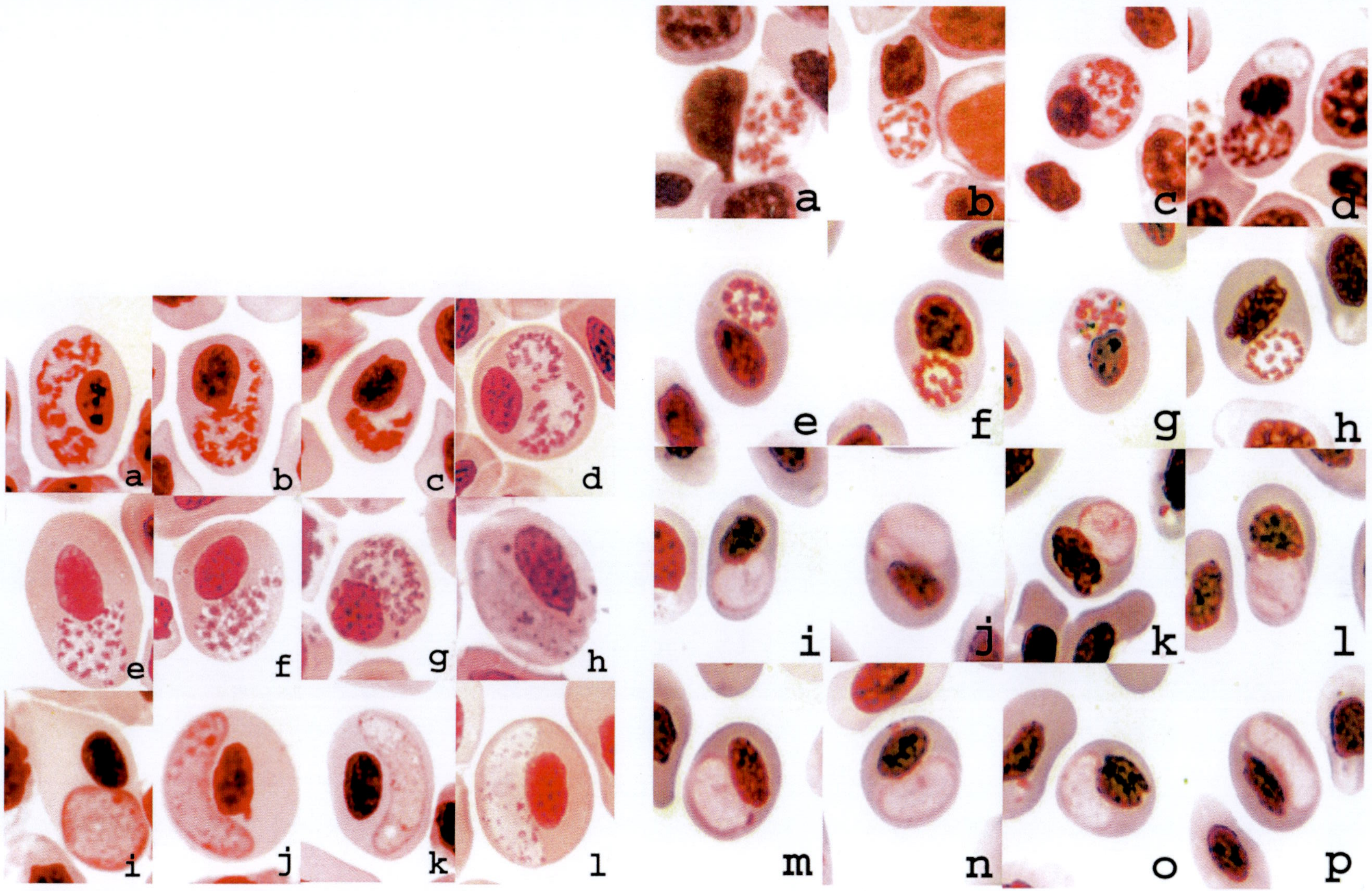

(A) (B)

Plate 30 (**A**) *Plasmodium gonatodi* from *Gonatodes albogularis*, Panama. Meronts, **a–g**; prematuration gametocyte, **h**; macrogametocytes, **i**, **j**; microgametocytes, **k**, **l**. (Figure **a** from Telford, S.R., Jr., *J. Protozool.,* 17, 566, 1970, with permission, Blackwell Publishing.) (**B**) *Plasmodium morulum* from *Mabuya mabouya*, Panama. Lymphocytic meront, **a**; erythrocytic meronts, **b–h**; macrogametocytes, **i–l**; microgametocytes, **m–p**.

slightly shorter but wider than microgametocytes, larger in average LW and less elongate, respectively 15.2 ± 3.5 × 5.9 ± 1.4 μm (N = 85), LW 86.3 ± 14.9 μm^2, and L/W 2.79 ± 1.06 versus 16.1 ± 2.9 × 5.1 ± 0.9 μm (N = 40), LW 80.0 ± 13.0 μm^2, and L/W 3.3 ± 0.87. Gametocyte dimensions and shape are not affected by infection phase. In *Gonatodes humeralis* of Brazil (Lainson et al., 1971), parasite dimensions were similar. Meronts were 10–17.5 × 5.6–12.5 μm, estimated LW 56–219 μm^2, with 14–50 merozoites (average 28). Microgametocytes were 13 × 4–20 × 4 μm, average 15.78 × 4.1 μm, estimated LW 64.7 μm^2 (52–80), and L/W 3.85 (3.25–5.00), and macrogametocytes were 13 × 4–20.5 × 6 μm, average 16.67 × 4.1 μm^2, with estimated LW 68.3 μm^2 (52–123), and L/W 4.07 (3.25–3.42), in a total sample of 100. Oval or spherical forms varied as 7 × 6–10 × 10 μm. Prematuration gametocytes similar to those in Panamanian infections were present in all infections studied.

Exoerythrocytic Merogony One meront, containing 20 nuclei, was seen in a lymphocyte (Telford, 1970c), and Lainson et al. (1971) reported a "small, immature form" in a lymphocyte.

Sporogony Unknown.

Effects on Host Host erythrocytes are hypertophied by 24% when parasitized by gametocytes of *P. gonatodi* and by 14% when meronts are present. Host cell nuclei are enlarged 16% by meronts, but their size is not affected by gametocytes. Meronts always distort host cells, and most cells (90%) show displaced nuclei, but nuclei are not distorted. Host cells are also always distorted by gametocytes, but nucleus shape appears unaffected except for often being displaced (51%). Distorted gametocytes usually were broadened, often rounded in shape.

Ecology Most of the active infections of *P. gonatodi* from Panama were collected in June, at the end of the dry season, while chronic infections were characteristic of February, in early dry season. All infected hosts were taken along the edge of advanced secondary wet tropical forest.

Remarks The samples of *Plasmodium gonatodi* from Panamanian *Gonatodes albogularis* (Telford, 1970c) and Brazilian *Gonatodes humeralis* (Lainson et al., 1971) differ little in morphological parameters despite the distance between their respective localities. *Plasmodium gonatodi* is the type species of the subgenus *Garnia*.

Plasmodium morulum Telford 1970 (**Plate 30**)

Diagnosis A *Plasmodium* (*Garnia*) species with rounded meronts containing merozoites arranged around the periphery similarly to rosettes or in a globular mass within the meronts, resembling a morulum, accompanied by ovoid or round-to-lentiform gametocytes. Meronts, predominantly proerythrocytic, are 5–8 × 3–6 μm, LW 18–48 μm^2, and contain 9–40 merozoites. Meront size relative to host cell nucleus size averages 1.28, and to normal erythrocyte nucleus size is 1.19. Gametocytes are 5–10 × 4–7 μm, with LW 20–60 μm^2 and L/W 1.00–2.25. Gametocyte size relative to host cell nucleus size averages 1.30, and to normal erythrocyte nucleus size is 1.34. Gametocytes are less restricted to immature host cells and do not differ in dimensions or shape by sex. Visible pigment is absent from all stages.

Type Host *Mabuya mabouya* (Lacépède) (Sauria: Scincidae).

Type Locality Sasardi, about 5 km west of Mulatupo, San Blas Territory, Panama.

Other Hosts None known.

Other Localities Vicinity of Belém, Pará State, Brazil (Lainson et al., 1971).

Prevalence In *Mabuya mabouya* of Panama, prevalence was 1 of 63 (1.6%), and 1 of 13 (7.7%) at the type locality was infected by *P. morulum*; in Brazil (Lainson et al., 1974b), 18 of 63 (28.6%) *M. mabouya* were infected.

Morphological Variation Meronts of *P. morulum* are 6.0 ± 0.8 × 4.9 ± 0.8 μm (5–8 × 3–6, N = 95), with LW 29.7 ± 7.4 μm^2 (18–48), and contain 17.2 ± 3.4 (9–28) merozoites. In the very small sample of meronts from mature erythrocytes, meront size relative to host cell nucleus size is 1.28 ± 0.31 (1.00–1.75, N = 5), and to normal erythrocyte nucleus size is 1.19 ± 0.23 (0.71–1.71, N = 95). Meronts may be formed as a rosette with merozoites arranged along the periphery of the round or oval meront or as a morula, a globular mass of merozoites within the meront. Gametocytes are ovoid, round, or lentiform, 6.7 ± 1.1 × 5.1 ± 0.6 μm (5–10 × 4–7, N = 100), with LW 34.3 ± 6.8 μm^2 (20–60) and L/W 1.34 ± 0.29 (1.00–2.25). Gametocyte size relative to host cell nucleus size is 1.30 ± 0.31 (0.67–2.10, N = 35), and to normal erythrocyte nucleus size is 1.34 ± 0.30 (0.85–2.54, N = 100). Erythrocytic gametocytes do not differ in size or shape from those in proerythrocytes or between the sexes.

In Brazilian *M. mabouya* (Lainson et al., 1974b), meronts appear to be slightly smaller than in Panama and produce on average fewer merozoites. Meronts were 3.5–4.5 μm for rounded parasites with 6–8 merozoites, and with up to 30 merozoites were 5 × 5 to 6 × 5 μm. Estimated LW of the latter meronts is 25–30 μm², more comparable to meronts of Panamanian infections. Gametocytes are 5 × 4.5 to 7.5 × 4.5 μm, average 5.5 × 4.5 μm. Estimated LW values would be 22.5–33.8 μm², average 24.8 μm², within the range but smaller on average in LW than in Panamanian skinks. Similarly, L/W ratios are estimated at 1.11–1.67, average 1.22, within the range of the Panamanian strain but somewhat more rounded in shape.

Exoerythrocytic Merogony Meronts containing 7–22 nuclei were commonly seen in lymphocytes (16%) and thrombocytes (84%) in the one Panamanian natural infection of *P. morulum,* and in experimental infections appeared early and persisted throughout the course of infection (Telford, 1970c). In one of the experimental infections, only lymphocytes were parasitized, and meronts contained 18–24 nuclei. Lainson et al. (1974b) reported that EE meronts were uncommon in *P. morulum* infections, lymphocytes and thrombocytes were parasitized, and nuclei ranged in numbers from 6 to 40.

Sporogony After feeding *Culex pipiens fatigans* on skinks infected with *Plasmodium morulum* (Lainson et al., 1974b), ookinetes were found 2 hours PF. These were about 15.0 μm in length, with prominent nuclei, and had one end "more swollen and pointed than the other." Azurophilic granules "mostly concentrated at the broader end." Rounded stages, possibly young oocysts but perhaps degenerating ookinetes, were also found. Similar stages were present at 12 and 24 hours PF, but no further development occurred, and mosquitoes dissected from 48 hours to 4 days PF were negative.

Effects on Host In a very small sample (five only), meronts caused hypertrophy of erythrocytes by 11% and their nuclei by 21%. Erythrocytes host to gametocytes were not enlarged, but their nuclei were 10% larger than those of uninfected erythrocytes. Meronts distorted mature and immature host cells (86%) and sometimes their nuclei, almost always displacing the nuclei (92%). Gametocytes often (81%) distorted host cells and displaced their nuclei (90%) but rarely distorted nuclei (4%).

Remarks Tiny dark granules present in a few meronts and some of the gametocytes of *P. morulum* as described (Telford, 1970c) did not give the polarized light test for hemozoin. In a later report (Telford, 1973), refractility under polarized light was observed in some gametocytes in experimental infections derived from the type infection but was not seen in meronts. Lainson et al. (1974b) referred to what may be the same type of inclusions in Brazilian *P. morulum* as "azurophilic granules" that did not refract under polarized light and suggested that the 1973 observations may have resulted from mixed-species infections in the experimental lizards. This is unlikely due to complete absence of refractile granules in asexual stages of the infections. The pigmented gametocytes likely would not have been present without pigmented asexual stages if there were two species present. It is more likely that occasional gametocytes can produce pigment granules through some physiological malfunction from the normal steps in digestion of hemoglobin, leaving a visible hemozoin residue.

Plasmodium scorzai Telford 1978 (**Plate 31**)

Diagnosis A *Plasmodium* (*Garnia*) species with meronts formed most commonly as rosettes, often elongate in form, rarely as fans, accompanied by elongate gametocytes. Meronts are 5–12 × 4–8 μm, with LW 25–88 μm², and contain 11–35 merozoites. Meront size relative to host cell nucleus size averages 1.12, and to normal erythrocyte nucleus size is 1.50. Proerythrocytic meronts are slightly larger than erythrocytic and produce more merozoites. The elongate, mature gametocytes are preceded by prematuration stages with irregular margins and one or both ends sharply pointed. Gametocytes are 9–16 × 4–10 μm, with LW 40–120 μm² and L/W 1.00–3.50. Gametocyte size relative to host cell nucleus size averages 1.93, and to normal erythrocyte nucleus size is 2.16. Macrogametocyte dimensions are slightly larger than microgametocytes, with little difference in shape. Neither asexual nor sexual stages produce visible pigment.

Type Host *Phylodactylus ventralis* O'Shaughnessy (Sauria: Gekkonidae).

Type Locality Fundo Vega Honda, 25 km north of Guanare, Municipio Guanare, Portuguesa State, Venezuela.

Other Hosts Unknown.

Other Localities Unknown.

Prevalence *Plasmodium scorzai* parasitized 6 of 86 (7.0%) Venezuelan *Phyllodactylus ventralis* and 6 of 15 (40%) at the type locality.

Morphological Variation Merozoites most commonly are arranged as rosettes or are in elongate meronts, seldom as fans, most of the fans occupying mature erythrocytes. Meronts are 8.5 ± 1.5 × 5.5 ± 0.9 μm (5–12 × 4–8, N = 69), with LW 46.8 ± 10.4 μm² (25–88), and contain 19.5 ± 4.2 (11–35) merozoites. Meronts in erythrocytes are slightly smaller and produce fewer merozoites than proerythrocytic meronts, respectively 8.2 ± 1.4 × 5.4 ± 0.8 μm (N = 44), LW 44.2 ± 8.3 μm², and 18.0 ± 3.7 (11–26) merozoites

Plate 31 (**A**) *Plasmodium scorzai* from *Phyllodactylus ventralis*, Venezuela. Meronts, **a–h**; macrogametocytes, **i**, **j**; microgametocytes, **k**, **l**. (**B**) *Plasmodium lainsoni* from *Phyllodactylus ventralis*, Venezuela. Meronts, **a–h**; macrogametocytes, **i–k**; microgametocytes, **l**.

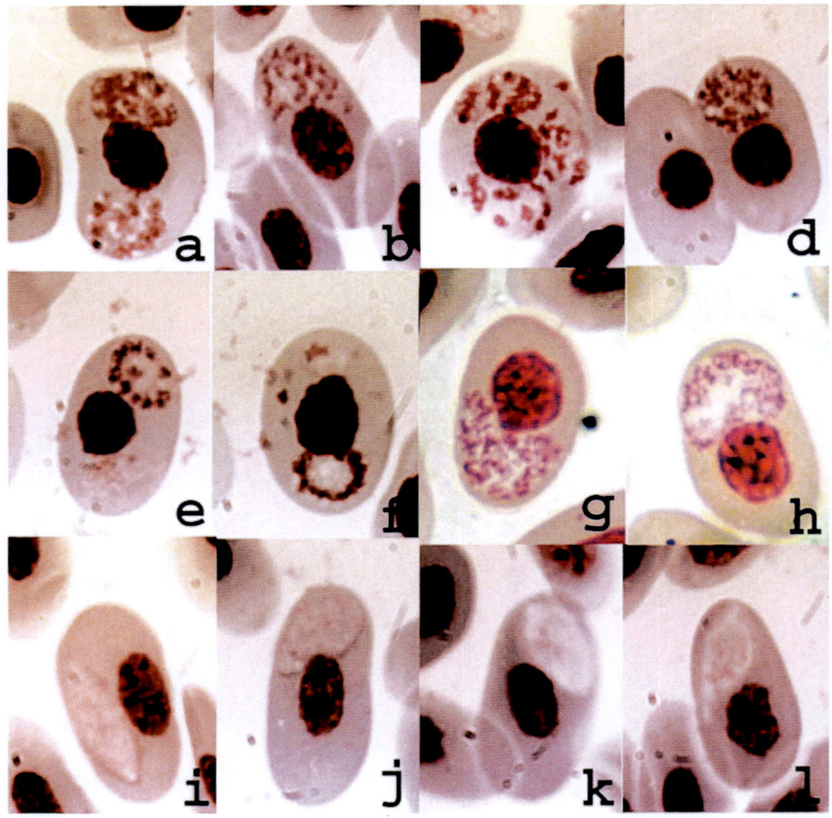

(A)

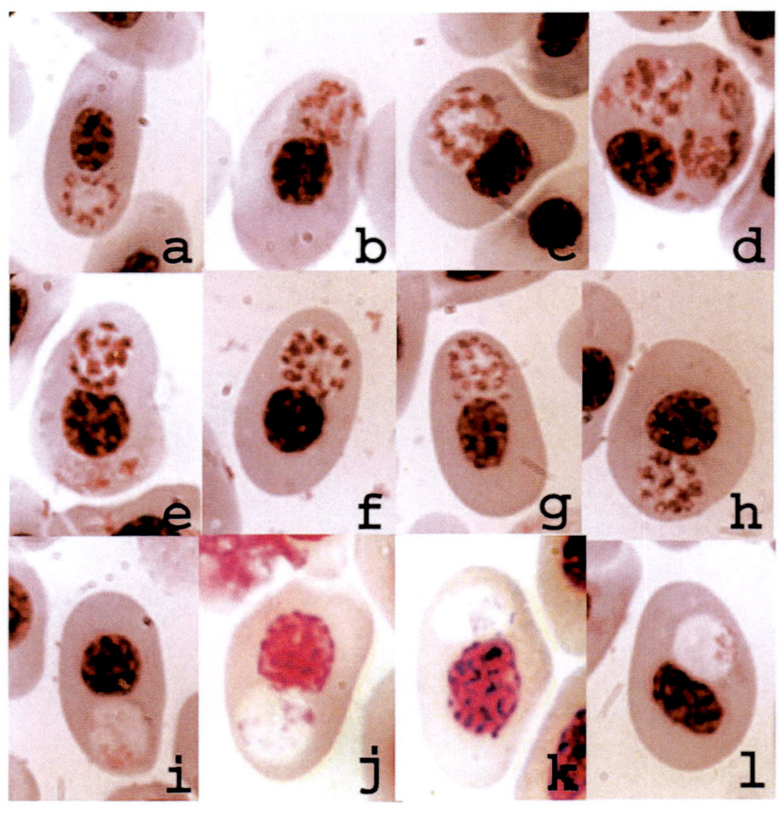

(B)

versus 9.0 ± 1.6 × 5.7 ± 0.9 µm (N = 25), LW 51.4 ± 12.2 µm², and 22.1 ± 3.9 (15–35) merozoites. Meront size relative to host cell nucleus size is 1.12 ± 0.25 (0.82–2.00, N = 44), and to normal erythrocyte nucleus size is 1.50 ± 0.41 (0.82–3.14, N = 69). The elongate gametocytes are preceded by prematuration forms that have highly irregular margins and one or both ends pointed or attenuated. Mature gametocytes are 11.9 ± 1.5 × 5.5 ± 1.0 µm (9–16 × 4–10, N = 101), with LW 65.1 ± 11.6 µm² (40–120) and L/W 2.24 ± 0.48 (1.00–3.50). Gametocyte size relative to host cell nucleus size is 1.93 ± 0.46 (1.14–4.00, N = 99), and to normal erythrocyte nucleus size is 2.16 ± 0.43 (1.34–4.21, N = 101). Macrogametocytes average slightly larger in size but are similar in shape to microgametocytes, respectively 12.0 ± 1.6 × 5.7 ± 1.1 µm (N = 60), LW 67.3 ± 11.9 µm², and L/W 2.21 ± 0.49 versus 11.8 ± 1.4 × 5.3 ± 0.8 µm (N = 41), LW 61.9 ± 10.4 µm², and L/W 2.30 ± 0.47. Visible pigment is not present in any stage of *P. scorzai*.

Exoerythrocytic Merogony A lymphocytic meront containing 30 nuclei was seen in a mixed infection of *P. scorzai* and *Plasmodium lainsoni* and is therefore of uncertain specific identity.

Sporogony Unknown.

Effects on Host Gametocytes and meronts cause hypertrophy of host erythrocytes by 11% and 5%, respectively, and of host cell nuclei by 14% and 32%, respectively. Gametocytes, because of their usual diagonal position in erythrocytes, almost always distort host cells, sometimes causing rounding of shape, and often displace nuclei but seldom distort them. Meronts sometimes cause rounding of host cells, almost always distorting them and displacing but seldom distorting nuclei.

Remarks *Plasmodium scorzai* is easily distinguished from *P. lainsoni* in mixed infections by the elongate gametocytes that lie diagonally in host cells and by the presence of prematuration gametocytes and meronts usually formed as rosettes. Prematuration gametocytes occur in very few plasmodiid species, notably *Plasmodium beebei* and *P. holaspi*, but are known also from *Plasmodium gonatodi* and *P. azurophilum*. The host species, *Phyllodactylus ventralis*, is probably the most common gecko that inhabits houses in Venezuela but is far less commonly seen in sylvatic or agricultural situations. The apparent rarity of *P. scorzai* and its restriction to the type locality certainly is an artifact of collecting. Active infections were found in August and September, while those from May were chronic or approaching that infection phase.

Plasmodium lainsoni Telford 1978 (**Plate 31**)

Diagnosis A *Plasmodium* (*Garnia*) species with rounded meronts containing 14–28 merozoites irregularly positioned within the meront, and ovoid gametocytes without prematuration stages. Visible pigment is absent. Meronts are 6–9 × 5–7 µm, with LW 30–63 µm², and their size relative to normal erythrocyte nucleus size averages 1.61. Meronts are larger and produce more merozoites in immature host cells than in erythrocytes. Gametocytes are 6–12 × 4–9 µm, with LW 30–90 µm² and L/W 1.00–2.50. Gametocyte size relative to host cell nucleus size averages 1.57, and to normal erythrocyte nucleus size is 1.73. Microgametocytes are smaller and more elongate than macrogametocytes.

Type Host *Phylodactylus ventralis* O'Shaughnessy (Sauria: Gekkonidae).

Type Locality Fundo Vega Honda, 25 km north of Guanare, Municipio Guanare, Portuguesa State, Venezuela.

Other Hosts Unknown.

Other Localities Unknown.

Prevalence *Plasmodium lainsoni* parasitized 2 of 86 (2.3%) Venezuelan *Phyllodactylus ventralis* and 2 of 15 (13.3%) at the type locality.

Morphological Variation Meronts are rounded with merozoites arranged in no particular manner, more or less filling the meront. Meronts are 7.2 ± 0.7 × 6.0 ± 0.8 µm (6–9 × 5–7, N = 45), LW 42.7 ± 7.6 µm² (30–63), and contain 20.7 ± 3.3 (14–28, N = 46) merozoites. Meront size relative to normal erythrocyte nucleus size is 1.61 ± 0.37 (1.01–2.42, N = 46). Proerythrocytic meronts are larger than those in mature erythrocytes and produce more merozoites, respectively 7.5 ± 0.8 × 6.2 ± 0.7 µm (N = 20), LW 46.3 ± 7.7 µm², with 23.2 ± 2.2 (16–28, N = 21) merozoites versus 6.9 ± 0.5 × 5.8 ± 0.8 µm (N = 25), LW 39.9 ± 6.2 µm², with 18.5 ± 2.5 (14–22, N = 25) merozoites. Gametocytes are 8.6 ± 1.4 × 6.0 ± 0.9 µm (6–12 × 4–9, N = 100), with LW 51.3 ± 12.2 µm² (30–90), and L/W 1.47 ± 0.32 (1.00–2.50). Gametocyte size relative to host cell nucleus size is 1.57 ± 0.46 (0.86–3.33, N = 91) and to normal erythrocyte nucleus size is 1.73 ± 0.40 (0.99–3.08, N = 100). Microgametocytes are smaller and more elongate than macrogametocytes, respectively 8.3 ± 1.4 × 5.5 ± 0.7 µm (N = 43), LW 45.5 ± 8.3 µm², and L/W 1.54 ± 0.37 versus 8.8 ± 1.4 × 6.3 ± 0.9 µm (N = 57), LW 55.7 ± 12.9 µm² (35–90), and L/W 1.42 ± 0.27. All stages of *P. lainsoni* lack visible pigment.

Exoerythrocytic Merogony Unknown.

Sporogony Unknown.

Effects on Host Both meronts and gametocytes cause about 5% hypertrophy of host cells, and about one-half of the cells containing meronts are rounded. Nuclei of cells infected by gametocytes are enlarged by 25%, and meronts cause similar hypertrophy. Meronts always distort host cells and their nuclei and displace nuclei. Gametocytes often distort host erythrocytes (47%), occasionally causing a rounding effect, frequently displacing nuclei (61%) but seldom distorting them (19%).

Remarks In mixed infection, *P. lainsoni* is easily distinguished from *P. scorzai* by its ovoid gametocytes, absence of prematuration forms, and meronts that are rounded with little tendency for merozoites to arrange along the periphery as a rosette, instead more or less filling the meront. Gametocytes also do not stain as intensely as do those of *P. scorzai*, with the macrogametocytes pale blue and microgametocytes white, seldom in the least pinkish.

Plasmodium azurophilum Telford 1975 (Plate 32)

Diagnosis A *Plasmodium* (*Garnia*) species with variably shaped meronts 7–13 × 4–10 μm, LW 30–121 μm², that contain 8–46 merozoites with no particular arrangement within the meronts. Meront size relative to host cell nucleus size averages 2.15, and to normal erythrocyte nucleus size is 2.35. Meronts lack visible pigment. Proerythrocytic meronts are slightly larger than erythrocytic and produce more merozoites on average. Gametocytes are usually ovoid, 7–14 × 5–10 μm, with LW 45–108 μm² and L/W 1.00–2.80. Gametocyte size relative to host cell nucleus size averages 2.53, and to normal erythrocyte nucleus size is 2.55. Although similar in size, microgametocytes are slightly more elongate than macrogametocytes. Gametocytes are preceded by prematuration stages that have irregular or lobulated margins and may be oddly shaped. Visible pigment is very rarely present in gametocytes.

Type Host *Anolis cybotes* Cope (Sauria: Polychrotidae).

Type Locality Fond Verretes, northern slope of Massif de la Selle, Departement de l'Ouest, Haiti.

Other Hosts *Anolis distichus, A. chlorocyanus, A. coelestinus, A. baharocoensis, A. lineatopus, A. grahami, A. krugi, A. acutus* (Telford, 1975; Telford et al., 1989, and unpublished); *A. gundlachi, A. evermanni, A. stratulus, A. cristatellus* (Schall and Vogt, 1993, as *Plasmodium azurophilum s.l.*; Schall et al., 2000); *A. schwartzei, A. oculatus, A. sabanus, A. marmoratus, A. bimaculatus, A. trinitatis, A. richardi* (Perkins, 2000a); and *A. roquet* (Ayala and Hertz, 1981).

Other Localities Puerto Rico, Jamaica, Dominican Republic, St. Croix (Telford, 1975; Telford et al., 1989, and unpublished); St. Marten, Saba, St. Kitts, Guadeloupe, Dominica, St. Vincent, and Grenada (Perkins, 2000a); Martinique (Ayala and Hertz, 1981).

Prevalence Hispaniola: Dominican Republic, El Seibo Province: *A. cybotes* 5 of 23 (21.7%), *A. distichus* 1 of 8; Baharona Province: *A. baharocoensis* 1 of 4, and *A. coelestinus* 2 of 4. Hispaniola: Haiti, Departement de l'Ouest: *A. cybotes* 10 of 22 (45.5%) at Fond Verretes, 3 of 5 at Petionville, 2 of 7 at Port au Prince; and *A. chlorocyanus* 1 of 6 (Telford, 1975, and unpublished). Jamaica, various localities: *A. lineatopus* 4 of 78 (5.1%), and *A. grahami*, 1 of 38 (2.6%). Puerto Rico, El Verde: *A. krugi* 1 of 18 (5.6%) (Telford, 1975), 1 of 47 (2.1%) (Schall and Vogt, 1993); *A. cristatellus* 1 of 216 (0.5%), *A. evermani* 1 of 386 (0.3%), *A. stratulus* 1 of 256 (0.4%), and *A. gundlachi* 70 of 406 (17.2%) (Schall and Vogt, 1993). St. Croix 2 of 13 (15.4%) *A. acutus* (Telford et al., 1989, as *A. wattsi*, in error). Saba, 371 of 1762 (21.1%) *A. sabanus* (Staats and Schall, 1996b).

Morphological Variation In erythroid cells of the type host, *Anolis cybotes* of Hispaniola, *Plasmodium azurophilum* meronts are of variable shape and contain merozoites in no particular arrangement. Meronts are 9.3 ± 1.1 × 7.6 ± 0.9 μm (7–13 × 6–10, N = 72), with LW 70.4 ± 14.2 μm² (42–104), and contain 28.5 ± 5.8 (20–46) merozoites. Meront size relative to host cell nucleus size is 2.15 ± 0.48 (1.37–3.33, N = 41), and to normal erythrocyte nucleus size is 2.35 ± 0.48 (1.47–3.70, N = 72). Although proerythrocytic meronts are larger than erythrocytic, there is no difference in average merozoite number, respectively 9.7 ± 1.1 × 7.9 ± 1.0 μm (N = 31), LW 76.8 ± 13.9 μm², 30.5 ± 6.2 (N = 37) merozoites versus 8.9 ± 1.0 × 7.3 ± 0.8 μm (N = 43), LW 65.5 ± 12.6 μm², and 26.9 ± 4.8 (N = 46) merozoites. Gametocytes are spherical to elongate, usually ovoid in shape, 10.1 ± 1.2 × 7.6 ± 0.9 μm (7–13 × 6–10, N = 85), with LW 76.4 ± 13.1 μm² (49–108) and L/W 1.35 ± 0.25 (1.00–2.17). Gametocyte size relative to host cell nucleus size is 2.53 ± 0.56 (1.50–4.00, N = 44), and to normal erythrocyte nucleus size is 2.55 ± 0.47 (1.81–4.00, N = 85). Microgametocytes are slightly more elongate than macrogametocytes, but there is no difference in size, respectively 10.6 ± 1.2 × 7.4 ± 0.9 μm (N = 24), LW 78.2 ± 11.0 μm², and L/W 1.46 ± 0.30 versus 9.9 ± 1.1 × 7.6 ± 0.9 μm (N = 61), LW 75.7 ± 13.8 μm², and L/W 1.31 ± 0.22. In *Anolis distichus*, meronts are 9.6 ± 1.3 ×

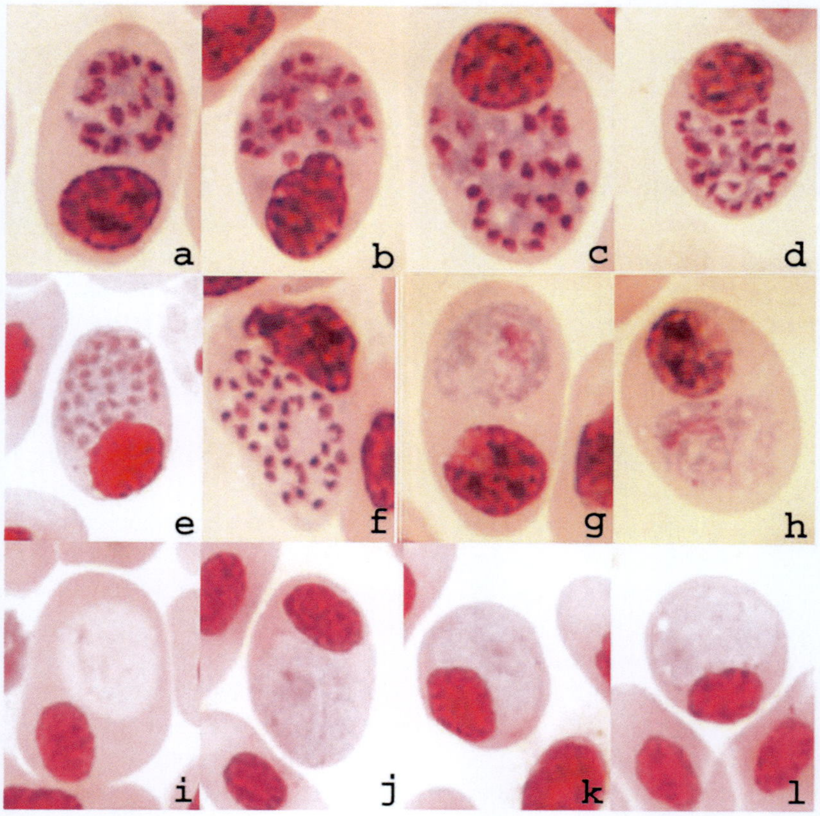

(A)

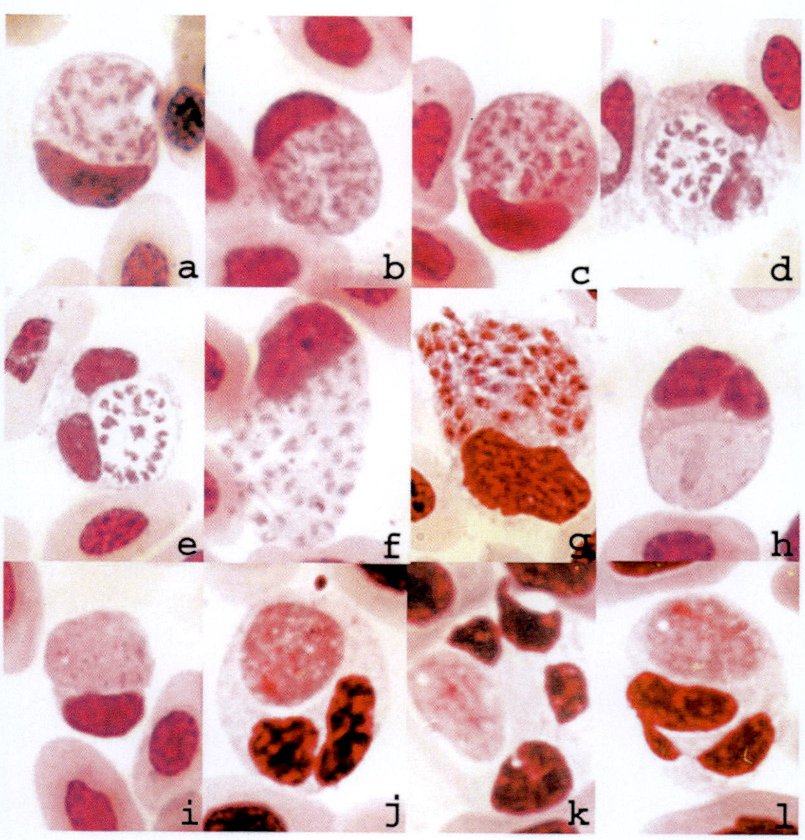

(B)

Plate 32 (**A**) *Plasmodium azurophilum* from *Anolis cybotes*, Hispaniola, **a–c**, **f–k**, and *A. lineatopus*, Jamaica, **d**, **e**, **l**. Meronts, **a–g**; macrogametocytes, **h–j**, **l**; microgametocyte, **k**. (**B**) *Plasmodium leucocytica* from *Anolis cybotes*, Hispaniola, **a–h**, **j–l**, and *A. krugi*, Puerto Rico. Meronts, **a–f**; macrogametocytes, **g–i**; microgametocytes, **j–l**.

7.5 ± 0.66 μm (7–13 × 7–9 , N = 24), LW 71.9 ± 11.4 μm² (49–91), and contain 29.1 ± 4.1 (22–36, N = 25) merozoites. Meront size relative to host cell nucleus size is 1.97 ± 0.35 (1.17–2.60), and to normal erythrocyte nucleus size is 2.31 ± 0.36 (1.57–2.92). Gametocytes are 10.0 ± 0.9 × 7.6 ± 0.9 μm (9–13 × 6–9, N = 25), with LW 76.2 ± 12.3 μm² (54–104) and L/W 1.33 ± 0.17 (1.11–1.67). Gametocyte size relative to host cell nucleus size is 2.01 ± 0.32 (1.43–2.75), and to normal erythrocyte nucleus size is 2.44 ± 0.40 (1.73–3.33).

In chronic infections of Jamaican hosts, meronts in *Anolis lineatopus* are 10.8 ± 0.5 × 7.5 ± 2.9 μm (10–11 × 4–11, N = 4), LW 80.8 ± 32.3 μm² (44–121), and contain 25.5 ± 3.7 (23–31) merozoites. Gametocytes are 9.9 ± 1.6 × 6.6 ± 1.1 μm (7–14 × 5–8, N = 20), with LW 64.1 ± 13.0 μm² (45–91) and L/W 1.56 ± 0.43 (1.00–2.8). In *A. grahami*, meronts are 10.3 ± 1.5 × 8.0 ± 1.0 μm (9–12 × 7–9, N = 3), LW 83.3 ± 21.4 μm² (70–108), and contain 30.0 ± 5.3 (26–36) merozoites. Gametocytes are 9.5 ± 1.2 × 7.3 ± 0.7 μm (8–11 × 6–8, N = 10), with LW 69.1 ± 9.1 μm² (56–88), and L/W 1.32 ± 0.25 (1.00–1.83). In active infection of *Anolis krugi* from Puerto Rico, meronts are 9.5 ± 1.8 × 7.5 ± 1.3 μm (6–12 × 5–10, N = 22), with LW 73.4 ± 23.4 μm² (30–110), and contain 19.3 ± 5.4 (8–28, N = 25) merozoites. Meront size relative to host cell nucleus size is 1.97 ± 0.35 (1.17–2.60), and to normal erythrocyte nucleus size is 2.32 ± 0.74 (0.95–3.48). Gametocytes are 9.8 ± 1.0 × 8.1 ± 0.8 μm (7–13 × 6–10, N = 50), with LW 79.6 ± 11.6 μm² (49–108) and L/W 1.23 ± 0.19 (1.00–1.86). Gametocyte size relative to host cell nucleus size is 2.73 ± 0.67 (1.60–5.50, N = 47), and to normal erythrocyte nucleus size is 2.67 ± 0.45 (1.55–3.82, N = 50).

Pigment, verified under polarized light, was present in 2 of 582 (0.3%) gametocytes from *A. cybotes* and, unverified, 1 microgametocyte in a much smaller sample from *A. lineatopus* (Telford, 1975).

Exoerythrocytic Merogony Although common, EE meronts found in *A. cybotes* and *A. lineatopus* could not be identified as secondary EE meronts of *P. azurophilum* due to the presence of a similar *Plasmodium* species in white blood cells in mixed infections of both hosts.

Sporogony Unknown.

Effects on Host In *Anolis cybotes*, neither meronts nor gametocytes of *P. azurophilum* causes significant enlargement of host erythrocytes or their nuclei. Meronts and gametocytes always distort host cells and displace their nuclei but rarely distort nuclei. In the description of *Plasmodium azurophilum* (Telford, 1975), it was stated that "Both schizonts and gametocytes produced significant hypertrophy of erythrocytic host cells and their nuclei, with infected cells commonly appearing round." *Hypertrophy* as used referred to increased width of erythrocytes when parasites were situated lateral to the nucleus and not to a hypertrophy of area (LW), which is not characteristic of infection.

Ecology In Puerto Rico, over a 9-year period, *P. azurophilum* infecting erythroid cells accounted for 60–80% of the malaria infections found, indicating far greater prevalence than that of *P. floridense* and the leukocytic *Plasmodium* species previously thought to be *P. azurophilum* (Telford, 1975), which together accounted for 10–30% of malarial infections (Schall et al., 2000). The lack of significant annual fluctuations in prevalence of *P. azurophilum,* in years following the initial 1990 sample in which its proportion of malaria infections was at the observed maximum, exceeding 80%, indicated that there was no correlation between parasite prevalence and either temperature or rainfall, although both factors affected fitness of the host, *Anolis gundlachi*. Other conclusions may not be warranted because the three *Plasmodium* species were considered as "malaria" without analysis by species. Ecological conclusions drawn from studies conducted prior to the genomic differentiation of the erythrocytic and leukocytic species (Perkins, 2000b), earlier thought to be the same, have to be reexamined by specific data analysis differentiating between the two species by host cell type.

Remarks The presence of asexual and sexual cycles in leukocytic cells and the similarity of those stages to the unpigmented meronts and gametocytes in erythroid cells of the host, *Anolis cybotes*, seemed at that time to be best explained as a possible mechanism to ensure survival of the parasite against immunological defense directed at erythrocytic parasites. One of the five experimental infections described (Telford, 1975) actually demonstrated that the leukocytic cycle developed without the appearance of erythrocytic parasites from an inoculum that contained leukocytic stages only of *P. azurophilum*, *P. floridense*, and *Schellackia cf. golvani*. Ayala and Hertz (1981) suggested that *Plasmodium azurophilum* as described could represent a composite species, one in erythroid cells and the other in leukocytes. The association of erythrocytic and leukocytic "*P. azurophilum*" on most of the Lesser Antilles was reported by Schall and his associates in several studies (Schall, 1992; Staats and Schall, 1996a, 1996b; Perkins 2000a, 2000b), and the suspicion grew that "*P. azurophilum*" red and white strains were in fact separate species. Perkins (2000a, 2000b) and Perkins and Schall (2002) finally provided conclusive evidence that *P. azurophilum* as described by Telford (1975) is actually comprised of two genetically distinct species and suggested that the leukocytic species is derived from the species utilizing erythroid cells (Perkins, 2000b).

Reconsideration of the original morphometric data from *Anolis cybotes*, and inclusion of more recently acquired samples from this host, has demonstrated that in the type host, at least, the two species can be distinguished by significant differences in the morphological characters of meront and gametocyte length, width, and LW. There is no difference, however, in mean merozoite number or gametocyte L/W ratio between the two species in *A. cybotes*. Because description of the erythrocytic stages preceded that of the leukocytic stages (Telford, 1975), although inappropriate, the name *azurophilum* has to be applied to the *Plasmodium* species inhabiting erythroid cells, and the leukocytic species requires a different specific name.

As stated in the description (Telford, 1975), pigment was visible in 2 of 582 (0.3%) gametocytes from *A. cybotes*, and its identity as hemozoin was verified under polarized light by Professor Garnham when I visited his laboratory at Imperial College in June 1973.

Plasmodium leucocytica sp. nov. (**Plate 32**)

Diagnosis A *Plasmodium* (*Garnia*) species parasitic exclusively within nonerythroid cells, lacking an erythrocytic cycle. Meronts in nonlymphocytic cells are 8–15 × 6–12 μm, LW 56–165 μm², and produce 14–52 merozoites. Meronts in lymphocytes are 10–24 × 10–21 μm, LW 100–504 μm², and contain 43–80 merozoites. Gametocytes are 6–15 × 4–10 μm, with LW 32–110 μm² and L/W 1.00–2.50. Macrogametocytes are larger than microgametocytes but do not differ in shape.

Type Host *Anolis cybotes* Cope (Sauria: Polychrotidae).

Type Locality Fond Verretes, northern slope of Massif de la Selle, Departement de l'Ouest, Haiti.

Other Hosts Jamaica, *Anolis lineatopus* (Telford, 1975); Puerto Rico, *A. gundlachi* (Schall et al., 2000); Saba, *A. sabanus* (Staats and Schall, 1996b); Guadeloupe, *A. marmoratus*; Dominica, *A. oculatus*; St. Vincent, *A. trinitatus*; St. Kitts, *A. bimaculatus* (Perkins, 2000a); St. Maarten, *A. gingivinus* (Schall, 1992); and Martinique, *A. roquet* (Ayala and Hertz, 1981).

Other Localities Dominican Republic, El Seibo Province (Telford); Jamaica, Montego Bay (Telford, 1975); Puerto Rico, El Verde (Schall et al., 2000); Martinique, Morne Calebasse (Ayala and Hertz, 1981); Saba, St. Kitts, Guadeloupe, Dominica, St. Vincent (Perkins, 2000a); St. Maarten (Schall, 1992).

Prevalence In *A. cybotes* of Haiti, 6 of 34 (17.6%); Jamaica, 3 of 78 (3.8%) *A. lineatopus*; 4 of 207 (1.9%) *A. bimaculatus* in St. Kitts, 2 of 162 (1.2%) *A. marmoratus* in Guadeloupe, 4 of 318 (1.2%) *A. oculatus* in Dominica, 8 of 312 (2.6%) *A. trinitatus* in St. Vincent, 1 of 89 (1.1%, Ayala and Hertz, 1981) and 2 of 275 (0.7%, Perkins, 2000a) *A. roquet* in Martinique; and approximately 12–26% *A. gundlachi* between 1990 and 1999 in Puerto Rico (Schall et al., 2000).

Morphological Variation *Plasmodium leucocytica* in its type host, *Anolis cybotes*, parasitizes at least 7 types of white blood cells: neutrophils (polymorphonuclear amphophilic special leukocytes of Pienaar, 1962), lymphocytes, monocytoid and lymphocytoid azurophil granulocytes, monocytes, and apparent lymphocytoid stem cells (Telford, 1975). Overall, leukocytic meronts are 11.3 ± 2.7 × 9.5 ± 2.4 μm (8–24 × 6–21, N = 72), with LW 112.4 ± 64.8 μm² (56–504), and contain 32.2 ± 14.9 (14–80, N = 75) merozoites. Uncommon lymphocytic meronts, which possibly represent secondary exoerythrocytic merogony of either *P. leucocytica* or *P. azurophilum*, are larger than those in other types of white cells, 16.0 ± 3.8 × 13.4 ± 3.3 μm (10–24 × 10–21, N = 10), with LW 224.0 ± 113.0 μm² (100–504), and produce twice as many merozoites than other leukocytic cell types, 58.3 ± 13.5 (43–80). Meronts in azurophils are 11.4 ± 1.4 × 10.1 ± 0.8 μm (9–15 × 9–12, N = 25), with LW 115.2 ± 19.1 μm² (81–165), and contain 37.4 ± 6.7 (24–52) merozoites. Meronts in neutrophils of *A. cybotes* are smaller, 9.9 ± 1.3 × 7.9 ± 1.0 μm (8–13 × 6–10, N = 35), LW 78.4 ± 16.3 μm² (56–120), with 20.6 ± 4.8 (14–38, N = 36) merozoites. Meronts in neutrophils of the Jamaican *A. lineatopus* are similar in size, 10.7 ± 1.9 × 8.5 ± 1.1 μm (9–18 × 7–11, N = 24), LW 91.6 ± 27.2 μm² (70–198), and produce nearly the same number of merozoites, 22.4 ± 6.0 (14–38, N = 25). Gametocytes in leukocytes of *A. cybotes* are 9.5 ± 1.4 × 7.0 ± 1.0 μm (6–15 × 4–10, N = 100), with LW 67.4 ± 15.3 μm² (32–110) and L/W 1.38 ± 0.26 (1.00–2.50). Macrogametocytes are larger than microgametocytes but are similar in shape, respectively 9.7 ± 1.4 × 7.3 ± 1.1 μm (N = 64), LW 70.5 ± 15.7 μm², and L/W 1.36 ± 0.27 versus 9.3 ± 1.3 × 6.7 ± 0.9 μm (N = 36), LW 61.9 ± 12.9 μm², and L/W 1.41 ± 0.25. Gametocytes in lymphocytes and azurophils are similar in size and shape, respectively 9.8 ± 0.8 × 7.6 ± 1.4 μm (9–11 × 6–10, N = 19), LW 74.5 ± 17.0 μm² (54–110), and L/W 1.33 ± 0.24 (1.00–1.83) versus 10.3 ± 1.1 × 7.3 ± 0.8 μm (8–12 × 6–9, N = 26), LW 75.0 ± 11.6 μm² (54–108), and L/W 1.43 ± 0.21 (1.00–1.83). Gametocytes in neutrophils are smaller but of similar shape, 9.4 ± 1.5 × 6.6 ± 0.8 μm (7–15 × 5–8, N = 32), with LW 62.4 ± 11.1 μm² (40–90) and L/W 1.44 ± 0.31 (1.00–2.50). Gametocytes in lymphoid stem cells are smaller and more rounded than in other host cell types, 8.5 ± 1.2 × 6.9 ± 1.1 μm (6–11 × 4–9, N = 20), LW 58.9 ± 16.0 μm² (32–99), and L/W 1.26 ± 0.23 (1.00–2.00). In neutrophils of *A. lineatopus*, gametocytes are larger than in neutrophils of *A. cybotes* but are similar in shape, 10.3 ± 1.7 × 7.4 ± 1.1 µm (7–14 × 6–10, N = 25), with LW 76.2 ± 16.1 µm² (49–104) and L/W 1.43 ± 0.35 (1.00–2.33).

Exoerythrocytic Merogony The very large meronts present in lymphocytes of *A. cybotes* may belong to a secondary EE cycle, but because of mixed infections their specific identity is uncertain.

Sporogony Unknown.

Effects on Host Some white cells host to *P. leucocytica* may be larger than uninfected cells, but comparison is difficult due to the variability present in the different types of leukocytes. In *Anolis gingivinus* of St. Maarten, as parasitemia by *P. leucocytica* increased, production of the white blood cell types used by the parasite also increased, along with the abundance of leukocytes relative to erythrocytes (Schall, 1992). Production of acid phosphatase by monocytes and neutrophils was reduced, and perhaps their function as components of the host immune response thereby loses effectiveness. This increase in the leukocyte proportions of total blood cells as parasitemia rose was also seen in experimental infections of *A. cybotes* (Telford, 1975).

Ecology *Plasmodium azurophilum* (*P. leucocytica*) infecting leukocytic cells of *Anolis gundlachi* in Puerto Rico, over a 9-year period, maintained proportions of 22–25% of total malaria infections, considerably less than that of *P. azurophilum* in erythrocytes (Schall et al., 2000). This absence of significant annual fluctuations in prevalence of *P. leucocytica* indicated that parasite prevalence and either temperature or rainfall differences were not correlated.

Remarks Genome analysis has clearly demonstrated that *P. azurophilum* populations infecting erythroid and leukocytic blood cells (Telford, 1975) represent distinct species (Perkins, 2000a, 2000b) and suggests that the leukocytic species is derivative from the erythrocytic species. It is necessary, therefore, to recognize this distinction by the use of *Plasmodium azurophilum* for the *Garnia* species utilizing erythroid cells, and to provide the leukocytic species with the taxonomic distinction *Plasmodium leucocytica*. Comparison by one-way analysis of variance (ANOVA) of the samples from *Anolis cybotes* demonstrates that the meronts of *P. leucocytica* are larger in length, width, and LW than those of *P. azurophilum*, but the mean number of merozoites is similar between the two species. Gametocytes of *P. leucocytica*, however, are smaller in length, width, and LW than are those of *P. azurophilum* with no difference in L/W ratio; that is, they are similar in shape. The conclusion that *P. leucocytica* is derived from *P. azurophilum* complicates their placement into the subgenera of *Plasmodium*, as discussed elsewhere. Hapantotype blood films are deposited in the USNPC, Beltsville, Maryland, nos. 100329, 100327, 100331–100334.

Plasmodium karyolytica (Lainson and Naiff) 1999

Diagnosis A *Plasmodium* (*Garnia*) species with rounded meronts up to 12.0 × 10.0 µm, LW 120 µm², that contain 20–28 merozoites. Gametocytes are spherical to usually elongate, 12–31 × 5–16 µm, LW 75–294 µm², and L/W 1.08–5.17. Gametocyte size relative to host cell nucleus size is 1.91–5.57 and to normal erythrocyte nuclei is 4.01–4.58. Macrogametocytes are larger than microgametocytes, but their shape is similar. All stages lack visible pigment. Nuclei of the host erythrocytes characteristically show some degree of lysis.

Type Host *Thecadactylus rapicaudus* (Houttuyn) (Sauria: Gekkonidae).

Type Locality Nova Repartmento near Tucuruí, Pará State, Brazil.

Other Hosts None known.

Other Localities Sasardi, about 5 km west of Mulatupo, San Blas Territory, Panama; Tierra Caliente, Municipio Manrique, Cojedes State, Venezuela.

Prevalence *Plasmodium karyolytica* was found in 1 of 4 geckoes from the type locality (Lainson and Naiff, 1999), 1 of 25 (4%) *T. rapicaudus* from Panama, and in 1 of 22 (4.5%) in Venezuela (Telford, 1978d).

Morphological Variation *Plasmodium karyolytica* in Brazil has rounded meronts that may be 12.0 × 10.0 µm and contain 20–28 merozoites and perhaps more (Lainson and Naiff, 1999). Estimated LW is 120 µm². Prior to segmentation, cytoplasm of the meronts sometimes divides "into separate clumps" with peripheral nuclei. Both sexes of gametocytes are predominantly elongate, macrogametocytes average 16.6 × 6.3 µm (13.3–21.4 × 4.4–8.1, N = 50), with L/W 2.6 (1.8–4.0), and microgametocytes are 15.25 × 6.24 µm (12.6–18.5 × 4.4–8.1, N = 50), L/W 2.4 (1.6–3.3). Estimated averages for LW are 104.6 µm² and 95.2 µm², respectively. Some mature gametocytes of both sexes were "round to broadly ovoid." Macrogametocytes were 9.5 × 8.0 µm (7.4–11.1 × 6.6–9.6, N = 13) with L/W 1.2 (1.0–1.5), and microgametocytes are 9.4 × 8.4 µm (7.4–11.8 × 7.4–9.6, N = 16), L/W 1.1 (1.0–1.4). Estimated averages for LW values of these rounded gametocytes are 76 μm^2 and 79 μm^2, respectively. Elongated gametocytes of both sexes were present in the type infection and had pointed extremities or irregular outlines. These probably represented prematuration gametocytes. Gametocytes in the Panamanian infection are 19.0 ± 1.3 × 12.0 ± 1.7 µm (17–23 × 9–16, N = 25), with LW 228.3 ± 32.8 μm^2 (162–294) and L/W 1.61 ± 0.27 (1.13–2.30). Host erythrocyte nuclei were unaltered in only 2 of 25 infected cells, and in these ratios of gametocyte size relative to host cell nucleus size were 3.48 and 3.67. Gametocyte size relative to normal erythrocyte nuclei is 4.58 ± 0.66 (3.25–5.90). Macrogametocytes are larger than microgametocytes, but their shape is similar, respectively 19.3 ± 1.5 × 12.7 ± 1.9 µm (N = 13), LW 245.6 ± 34.7 μm^2, and L/W 1.55 ± 0.31 versus 18.7 ± 1.0 × 11.3 ± 1.1 µm (N = 12), LW 209.6 ± 17.3 μm^2, and L/W 1.68 ± 0.21. In the Venezuelan infection, gametocytes are smaller and more elongate in both sexes. Gametocytes are 16.7 ± 3.7 × 9.7 ± 2.1 µm (12–31 × 5–13, N = 25), with LW 160.8 ± 42.6 μm^2 (75–273) and L/W 1.84 ± 0.80 (1.08–5.17). Gametocyte size relative to host cell nucleus size is 3.32 ± 0.87 (1.91–5.57, N = 18), and to normal erythrocyte nuclei is 4.01 ± 1.06 (1.87–6.81, N = 25). Macrogametocytes are larger than microgametocytes but similarly shaped, respectively 17.2 ± 4.1 × 9.9 ± 2.0 µm (N = 17), LW 169.5 ± 43.4 μm^2, and L/W 1.85 ± 0.90 versus 15.6 ± 2.2 × 9.1 ± 2.1 µm (N = 8), LW 142.4 ± 36.5 μm^2, and L/W 1.82 ± 0.58. Two-thirds of the Panamanian gametocytes contained one to five dark dots, some as large as 1–1.5 µm in diameter, that resembled pigment, but these did not refract under polarized light. These dots were not observed in Venezuelan gametocytes.

Exoerythrocytic Merogony Unknown.

Sporogony Unknown.

Effects on Host In the type infection, there is a "progressively lytic effect on the host cell nucleus, particularly the mature gametocytes, which enlarge and deform the erythrocyte" (Lainson and Naiff, 1999). Host cells tended to become more rounded and often were unusually shaped. In the Panamanian strain, gametocytes always distorted the host cell and almost always (92%) its nucleus, with lysis of the nucleus commonly evident. Nuclei were always displaced by gametocytes. Lysis was not seen in the Venezuelan strain, although nuclei were commonly distorted (56%). Host cells were always distorted and their nuclei displaced. About one-fourth of the Venezuelan host cells were proerythrocytes, and all of the Panamanian infections were in mature cells. Host erythrocytes were hypertrophied in size by 45% in the Panamanian sample and 36% in the Venezuelan, and erythrocyte nuclei were enlarged by 27% in the latter material.

Ecology The host of *P. karyolytica*, *Thecadactylus rapicaudus*, is a large gecko, closely and typically associated with large buttresses and tree holes in wet primary or secondary tropical forest and occasionally with outbuildings in the vicinity of forest. Buttresses and tree holes are characteristic microhabitat for phlebotomine sand flies, favored by Lainson and Naiff (1999) as possible vectors. Kimsey (1992) found no evidence of malaria parasite transmission by sand flies in Panama and no malaria infections in 36 *T. rapicaudus*, but his study areas were in the central part of the country, well west of the single known locality for *P. karyolytica* in Panama.

Remarks This parasite was first seen in Panama in 1968, but despite repeated samples over several weeks, asexual stages were never found. The Venezuelan infection was also chronic, and no definite meronts were identified (Telford, 1978d). Some gametocytes occupied white cells, which were possibly monocytes and macrophages.

Plasmodium utingensis (Lainson, Landau and Shaw) 1971

Diagnosis A *Plasmodium* (*Garnia*) species with delicate, elongate, and small developing meronts that often have a conspicuous cleft along the midline and trophozoites with filiform pseudopodia. Fan-shaped to oval mature meronts average 7.25 × 5.0 µm, estimated LW 36.3 μm^2, and produce 8–12 merozoites. Estimated meront size relative to host cell nucleus size is 0.65–0.79, and to normal erythrocyte nucleus size is 0.63–0.89. Gametocytes are spherical to oval, 6.3–8.8 × 5.6–6.3 µm, with microgametocytes slightly smaller than macrogametocytes. Estimated LW averages 42.5–47.5 μm^2, with L/W 1.22 in both sexes. Estimated gametocyte size relative to host cell nucleus size is 0.77–1.14, and to normal erythrocyte nucleus size is 0.83–1.24. There is no visible pigment at any stage.

Type Host *Anolis punctatus* Daudin (Sauria: Polychrotidae).

Type Locality Utinga Forest, near Belèm, Parà State, Brazil.

Other Hosts None known.

Other Localities None known.

Prevalence *Plasmodium utingensis* parasitized two of six *A. punctatus* from the Utinga Forest (Lainson et al., 1971).

Morphological Variation Trophozoites form filiform pseudopodia, and young meronts are elongate, often showing a cleft along the midline. Meronts are fan-shaped or oval, approximately 7.25 × 5.0 μm, with estimated LW 36.3 μm², and contain 8–12 merozoites. Their LW relative to host cell nucleus size, estimated from illustrations, is 0.65–0.79, and to normal erythrocyte nucleus size is 0.63–0.89. Immature gametocytes are "markedly crescentic or wedge-shaped, often with a finely crenated border" (Lainson et al., 1971). Macrogametocytes average 7.6 × 6.25 μm (6.25 × 6.25–8.75 × 6.25, N = 50), with estimated LW 47.5 μm² (39.1–54.7), slightly larger than microgametocytes, which average 7.2 × 5.9 μm (6.25 × 5.6–8.75 × 6.25), estimated LW 42.5 μm² (35–54.7), with respective L/W ratios estimated at 1.22 (1.00–1.40) and 1.22 (1.11–1.40). Neither azurophilic granules nor pigment is visible in any stage of *P. utingensis*.

Exoerythrocytic Merogony Possible monocytes are host to EE meronts, 8.1 × 5.5 μm (7.5 × 3.1–13.75 × 7.5), and larger meronts that are oval or elongate and contain 4–20 merozoites. No mature forms were observed.

Sporogony Unknown.

Effects on Host Host cells are not distorted but their nuclei are sometimes displaced by gametocytes. Illustrations suggest that neither the erythrocyte nor its nucleus is hypertrophied by either asexual or sexual parasites.

Remarks The irregular margins of immature but sexually differentially stained gametocytes suggest the presence of prematuration gametocytes seen in other *Plasmodium* (*Garnia*) species.

Plasmodium multiformis (Lainson, Shaw and Landau) 1975

Diagnosis A *Plasmodium* (*Garnia*) species in which merogony following infection begins in erythroblasts. As the infection progresses, meronts sequentially infect proerythrocytes and ultimately mature red blood cells, with merozoite numbers fewer as host cell maturity increases. Erythroblastic meronts are 8.0 × 6.4 μm with 24 merozoites to 17 × 13 μm containing about 150 nuclei. In proerythrocytes, meronts are 5.2 × 3.3–9.1 × 7.8 μm and contain 6–40 merozoites, and in erythrocytes meronts are 4.0 × 3.5–7.8 × 5.2 μm, with 4–14 merozoites. Gametocytes may develop in erythroblasts and proerythrocytes but reach maturity in erythrocytes. Before maturity, gametocytes are round or oval, becoming elongate with irregular margins when mature, at dimensions of 10.0–21.0 × 2.5–4.0 μm, with size not affected by sex. Estimated LW values average 37.8–43.1 μm² and L/W 3.51–4.20. Estimated gametocyte size relative to host cell nucleus size is 0.50–1.25, and to normal erythrocyte nuclei is 1.46–2.08. Azurophilic granules are prominent in gametocytes but are less often present in meronts. All stages lack visible pigment.

Type Host *Plica umbra* (Linnaeus) (Sauria: Iguanidae).

Type Locality Secondary tropical rain forest near Belèm, Parà State, Brazil.

Other Hosts None known.

Other Localities None known.

Prevalence *Plasmodium multiformis* was identified with certainty in 41 of 235 (17.5%) *Plica umbra* and, including probable asexual infections, in 64 of 235 (27.2%) (Lainson et al., 1975).

Morphological Variation Erythroblastic meronts in early infection were variable in size and shape and, depending on merozoite number present, usually oval to spherical. Meronts with 24 merozoites averaged about 8.0 × 6.5 μm; those with about 150 nuclei were about 17 × 13 μm (Lainson et al., 1975). Ruptured meronts revealed cytoplasm formed as "a network of finger-like protrusions from which the merozoites are budded off, leaving behind a bulky residual body." Calculated estimates of LW indicated a range of 52–221 μm². Proerythrocytic meronts had merozoites distributed peripherally, scattered within the cytoplasm or in a fan shape. These were 5.2 × 3.3–9.1 × 7.8 μm, with an average of 7.0 × 5.2 μm. Calculated estimates of meront LW averaged 36.4 μm² (17.2–71.0). Merozoite numbers averaged 16.0 (6–40). In erythrocytes, meronts were 4.0 × 3.5–7.8 × 5.2 μm, average 5.5 × 3.8 μm, and produced still fewer merozoites, on average 8.0 (4–14). Estimated average LW of meronts in mature erythrocytes is 36.4 μm². As estimated from photographs, erythrocytic meront size relative to host cell nucleus size is 0.52–0.80, and to normal erythrocyte nuclei is 0.65–1.20. Although immature gametocytes occurred in erythroblasts and proerythrocytes, almost all mature gametocytes were erythrocytic (Lainson et al., 1975). Before maturity, but showing sexually different staining, gametocytes were oval or round, with smooth margins. At maturity, gametocytes became elongated with uneven crenulated margins. Microgametocytes averaged 12.3 × 3.5 μm (10.0 × 4.0–21.0 × 3.0), and macrogametocytes were 12.6 × 3.0 µm (10.5 × 2.5–20.0 × 3.0). Their estimated LW values are respectively 43.1 µm² (40–63) and 37.8 µm² (26.3–60.0), with L/W ratios 3.51 (2.50–7.00) and 4.20 (4.20–6.67). Estimated gametocyte size relative to host cell nucleus size is 0.50–1.25, and to normal erythrocyte nuclei is 1.46–2.08. Azurophilic granules were abundant and prominent in gametocytes and less commonly seen in meronts. Pigment is absent from all stages of *P. multiformis*.

Exoerythrocytic Merogony A few meronts were observed in monocytes of one lizard examined (Lainson et al., 1975).

Sporogony Unknown.

Effects on Host Hypertrophy of erythroblastic host cells was minor except when multiple infections were present. Nuclei were displaced and "variably indented," sometimes forming a "band around the cell margin" (Lainson et al., 1975). Proerythrocytes, when multiply infected, were enlarged, but "displacement and indentation of the host-cell nucleus" was the main effect by both meronts and gametocytes. The small sizes of erythrocytic meronts caused little alteration of host cells and their nuclei, and gametocytes had little effect on host cell size but did displace host erythrocyte nuclei.

Remarks Had Lainson et al. not carefully studied the host cells used and maturation of the parasites, the irregularly formed mature gametocytes might have been considered to be "prematuration gametocytes" and the round or ovoid forms as mature gametocytes. As suggested by these authors, following infection by sporozoites, and (in my opinion) a cryptozoic and possibly metacryptozoic merogony in hepatic parenchymal cells, merogony proliferated in the erythroblast series of bone marrow. This became sufficiently intense to "stimulate a marked marrow hyperplasia," with the subsequent invasion of circulating blood in a variety of infected immature blood cells, followed by a diminishing hyperplasia as proerythrocytes and erythrocytes were utilized as host cells by *P. multiformis*.

Plasmodium uranoscodoni (Lainson, Shaw and Landau) 1975

Diagnosis A *Plasmodium* (*Garnia*) species with very amoeboid young asexual stages that form long filopodia. Meronts are round, oval, or fan-shaped, 4.6 × 4.0–11.1 × 6.5 µm, and produce 6–16 merozoites. Estimated LW values range from 18.4 to 72.2 µm²; estimated size relative to host cell nucleus size is 0.41, and to normal erythrocyte nuclei is 0.67. Young gametocytes are amoeboid with irregular edges, becoming oval or round with smooth margins at maturity. Gametocytes are 7.8–11.7 × 5.2–7.8 µm, with macrogametocytes slightly larger than microgametocytes. Estimated LW is 45.2–91.3 µm², L/W is 1.34–1.75, with size relative to host cell nucleus size 0.85–2.00, and to normal erythrocyte nuclei 0.85–1.32. A large, irregular vacuole may be present, most commonly in macrogametocytes. All stages lack visible pigment.

Type Host *Uranoscodon superciliosa* (Linnaeus) (Sauria: Iguanidae).

Type Locality Secondary tropical rain forest near Belèm, Parà State, Brazil.

Other Hosts Unknown.

Other Localities Unknown.

Prevalence *Plasmodium uranoscodoni* infected 46 of 167 (27.6%) *U. superciliosa* (Lainson et al., 1975).

Morphological Variation Young asexual stages are amoeboid and often form long filopodia. In round, oval, or fan-shaped meronts 6.3 × 4.9 µm (4.6 × 4.0–11.1 × 6.5), an average of 8.0 (6–16) merozoites are arranged "often in a rosette or half-circle" (Lainson et al., 1975). Estimated LW is 30.9 µm² (18.9–72.2). Calculated from illustrations, meront size relative to host cell nucleus size is 0.41, and to normal erythrocyte nuclei is 0.67. Gametocytes when young may form amoeboid projections and remain somewhat irregularly shaped after appearance of sexually different staining but then round up into ovoid shapes with smooth margins. Macrogametocytes average slightly larger than microgametocytes, respectively 10.4 × 6.5 µm (9.1 × 5.2–11.7 × 7.8) versus 8.7 × 7.0 µm (7.8 × 5.8–10.4 × 6.5). Estimated LW values are, respectively, 67.6 µm² (47.3–91.3) and 60.9 µm² (45.2–67.6), and respective L/W ratios are 1.60 (1.50–1.75) and 1.24 (?1.34–1.60). Calculated from illustrations, gametocyte size relative to host cell nucleus size is 0.85–2.00, and to normal erythrocyte nuclei is 0.85–1.32. Azurophilic granules are tiny or fine, rare in meronts but in increased numbers, although inconspicuous, in gametocytes. No pigment is visible in *P. uranoscodoni* at any stage.

Exoerythrocytic Merogony Thrombocytes are more commonly host to EE meronts than are lymphocytes. Thrombocytic meronts average 7.8 × 5.8 µm, producing on average 8.0 (4–14) merozoites. The larger lymphocytic meronts are at least 10.0 × 7.0 µm and contain up to 12 merozoites (Lainson et al., 1975).

Sporogony Unknown.

Effects on Host Rarely, erythrocytes host to the larger meronts and gametocytes were slightly enlarged, and their nuclei were "variably displaced or deformed" (Lainson et al., 1975). Thrombocytic meronts were sometimes slightly enlarged and their nuclei indented or displaced. Nuclei of infected lymphocytes were "invariably pressed to one side … reducing it to a mere peripheral band."

Remarks This is yet another *Plasmodium* (*Garnia*) species in which the maturing gametocytes, showing sexually different staining reactions, are irregular in form, becoming rounded or ovoid with smooth margins, at least suggesting a brief appearance of "prematuration gametocytes," which appear in *P. gonatodi, P. azurophilum,* and *P. scorzai*, prior to gametocyte maturation.

Plasmodium audaciosa (Lainson, Shaw and Landau) 1975

Diagnosis A *Plasmodium* (*Garnia*) species with merogony and gametogony exclusively in neutrophil granulocytes of circulating blood. Spherical or oval meronts are 13 × 11–18 × 18 µm, estimated LW 143–324 µm², and produce 50–150 merozoites. Gametocytes, oval or round, are 7.8 × 6.5–10.4 × 7.8 µm. Estimated LW is 50.7–81.1 µm², and L/W is 1.20–1.33. Microgametocytes are slightly smaller than macrogametocytes. Azurophilic granules are common in gametocytes, rare or absent in meronts.

Type Host *Plica umbra* (Linnaeus) (Sauria: Iguanidae).

Type Locality Secondary tropical rain forest near Belém, Pará State, Brazil.

Other Hosts Unknown.

Other Localities Unknown.

Prevalence *Plasmodium audaciosa* was present in 4 of 235 (1.7%) *P. umbra* (Lainson et al., 1975).

Morphological Variation Meronts are spherical or ovoid, often filling the host cell, 13.0 × 11.0 µm with about 100 merozoites to 18 × 18 µm containing about 150 merozoites. Merozoite numbers varied from 50 to 150. Estimated LW is 143–324 µm². Azurophilic granules were rarely seen in meronts. Gametocytes are rounded to oval, with azurophilic granules more prominent in macrogametocytes than in microgametocytes. Macrogametocytes are slightly larger than microgametocytes, respectively 9.5 × 7.5 µm (7.8 × 7.8–10.4 × 7.5) versus 9.0 × 7.0 µm (7.8 × 6.5–9.1 × 7.8). Estimated LW values and L/W values are respectively 71.3 µm² (60.8–78.0) versus 63.0 µm² (50.7–71.0) and 1.27 (1.00–1.39) versus 1.29 (1.17–1.20).

Exoerythrocytic Merogony Tissue smears of "spleen, liver, lung, kidney, bone-marrow and brain revealed no other division process" (Lainson et al., 1975).

Sporogony Unknown.

Effects on Host Neutrophils host to maturing meronts "became enlarged and its nucleus forced to the periphery" (Lainson et al., 1975). Nuclei commonly were fragmented and degenerate. Neutrophils host to one or two gametocytes differed little from normal except for slight enlargement of the cells and displacement of the nuclei. Hypertrophy of cells was greater in multiple infections.

Remarks This species was described as a species of *Fallisia*, a genus in which all other known species parasitize only thrombocytes and lymphocytes. This placement became illogical and untenable when genome analysis indicated that the morphologically similar *Plasmodium* (*Garnia*) *leucocytica*, also a parasite of neutrophils and other granulocytes, is derived from the erythrocytic *Plasmodium* (*Garnia*) *azurophilum*. Gametocytes of *P. leucocytica* are nearly identical in size and shape to those of *P. audaciosa*. Meronts of *P. leucocytica*, however, are somewhat smaller in dimensions and produce far fewer merozoites, 14–80, in comparison to 50–150 in *P. audaciosa*.

PLASMODIUM SPECIES OF NORTH AMERICAN LIZARDS

Plasmodium mexicanum Thompson and Huff 1944 (**Plate 33**)

Diagnosis A *Plasmodium* (*Paraplasmodium*) species with round, elongate, or usually fan-shaped meronts, 3–22 × 3–7 µm, LW 9–88 µm², that produce 3–26 merozoites. Meront size relative to host cell nucleus size averages 1.05–1.28, and to normal erythrocyte nucleus size is 1.17–1.32. Gametocytes are round to elongate, usually ovoid, 8–23 × 4–14 µm, with LW 56–216 µm² and L/W 1.00–4.60. Gametocyte size relative to host cell nucleus size averages 3.15–5.79, and to normal erythrocyte nucleus size is 3.94–5.31. Gametocytes in chronic phase of infection are larger and more rounded than those in active infections. Macrogametocytes are larger than microgametocytes but differ little in shape. Pigment is dark brown to almost black, focused at the base of fan-shaped meronts, and dispersed as small, bacilloid granules in gametocytes. Exoerythrocytic merogony may

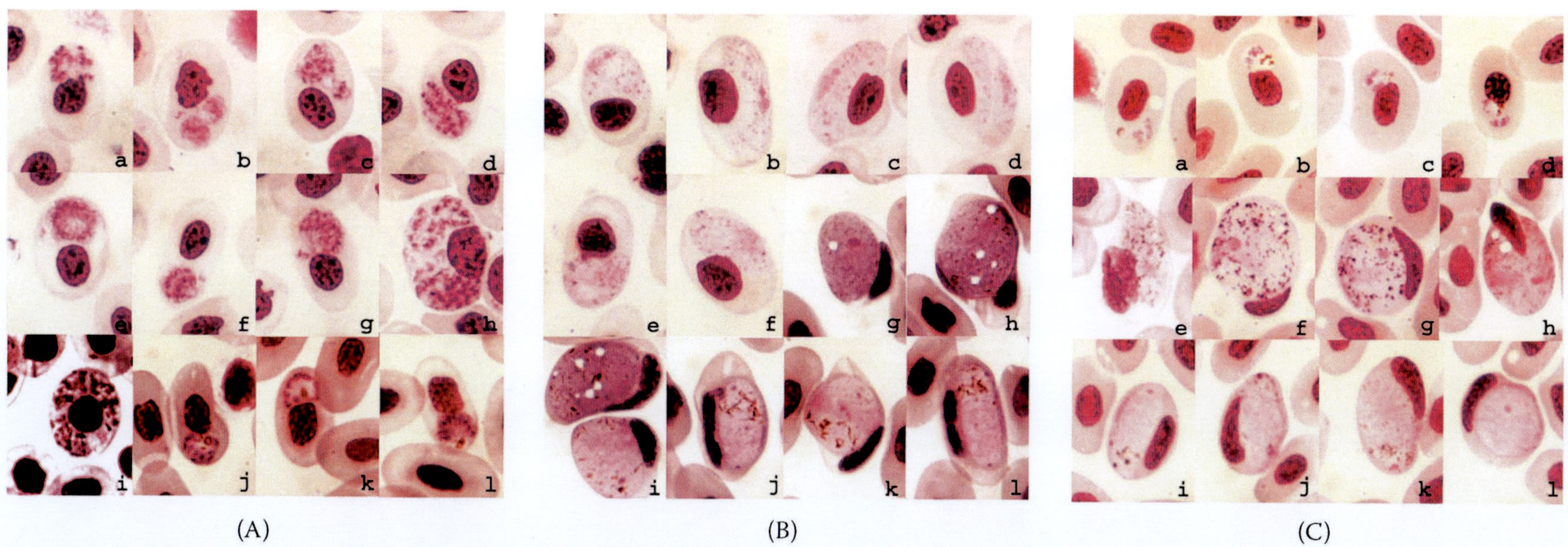

Plate 33 (**A**) and (**B**) *Plasmodium mexicanum* from *Sceloporus occidentalis*, California, and *Sceloporus torquatus*, Mexico. (**A**) Meronts: **a–i**, *S. occidentalis*; **j–l**, *S. torquatus*. (**B**) Gametocytes: **a–f**, *S. occidentalis*; **g–l**, *S. torquatus*. Macrogametocytes are **a–d**, **g–l**. Others are microgametocytes. (**C**) *Plasmodium chiricahuae* from *Sceloporus jarrovii*, Arizona (**a–d**, **f–l**), and *Sceloporus poinsetti*, Texas (**e**). Meronts, **a–d**; macrogametocytes, **e–h**; microgametocytes, **i–l**.

be profuse in both circulating, nonerythroid blood cells, especially lymphocytes and monocytes, and in fixed tissue cells, particularly in vascular endothelium and connective tissue of most organs. The natural vectors are the psychodid fly *Lutzomyia vexator* and related species.

Type Host *Sceloporus torquatus* Wiegmann (syn. *S. ferrariperezi*) (Sauria: Phrynosomatidae).

Type Locality Michoacan, Mexico.

Other Hosts *Sceloporus occidentalis, S. graciosus* (Wood and Wood, 1936; Ayala, 1970b); *S. undulatus* (Greiner and Daggett, 1973); *S. microlepidotus* (Peláez et al., 1948); *Gerrhonotus multicarinatus* (Ayala, 1973); and *Ctenosaura hemophila* (?) (Mahrt, 1979). Experimental hosts: *Sceloporus olivaceous, S. undulatus, Crotaphytus collaris, Phrynosoma cornutum, P. asio* (Thompson and Huff, 1944a); *Sceloporus spinosus* (Peláez et al., 1948); *Sceloporus jarrovi, S. magister, Phrynosoma coronatum, P. platyrhinos, G. multicarinatus* (Ayala, 1970b); and *Sceloporus woodi* (Telford, 1984b).

Other Localities Mexico: Distrito Federal, Lomas de Chapultepec; Michoacan State, 7.2 km northeast of Cotija; Jalisco State, 20.2 km northeast Union de la Tula; Mexico State, 9.4 km east turnoff from Highway 15 to Via Victoria (Telford). California, by counties: Berkeley (Wood and Wood, 1936; Jordan, 1970), Alameda; Concord (Telford), Mt. Diablo (Jordan, 1970; Ayala, 1970b), Contra Costa; El Dorado (Jordan, 1970), El Dorado; Kern (Ayala, 1973); Hobergs (Jordan, 1970), Middleton (Ayala, 1970b), Lake; Coarsegold (Telford), San Joaquin Experimental Range (Jordan, 1970), Madera; Petaluma (Jordan, 1970), Marin; Ukiah (Telford), Hopland Valley and Hopland Field Station (Ayala, 1970b; Schall and Marghoob, 1995; Telford), Mendocino; Napa and Pope Valley (Jordan, 1970), Napa; Riverside (Ayala, 1973); Scott Road and El Dorado Hills (Telford), Sacramento; Santa Clara (Ayala, 1973); Lake Herman (Ayala, 1970b), Rakvill Park (Telford), Solano; Skagg Springs (Jordan, 1970; Telford), Sonoma; Sutter Buttes (Jordan, 1970), Gualala River (Wood and Wood, 1936), Sutter; and Yolo (Klein et al., 1987b).

Prevalence Mexico: in Lomas de Chapultepec, Distrito Federal, 45 of 50 (90%) *Sceloporus torquatus* and 2 of 28 (7.1%) *S. microlepidotus* were infected (Peláez et al., 1948); *P. mexicanum* infected 3 of 5 *S. torquatus* in Michoacan, 2 of 7 in Jalisco, and 1 of 6 in Mexico State (Telford).

Plasmodium mexicanum was reported from one of four *Sceloporus graciosus* from Gualala River, Sonoma County, and in *Sceloporus occidentalis* in 17 California counties. Prevalence data obtained by Wood and Wood (1936), Jordan (1970), Ayala (1970b, 1973), and Telford (unpublished) have been combined for each county to simplify presentation: Alameda, 22 of 235 (9.4%); Contra Costa, 15 of 79 (19.0%); El Dorado, 2 of 3; Lake, 10 of 30 (33.3%); Madera, 13 of 28 (35.7%); Marin, 1 of 26 (3.8%); Mendocino, 119 of 542 (22.0%) excluding Schall and Marghoob (1995), below; Napa, 33 of 144 (22.9%); Sacramento, 6 of 56 (10.7%); Solano, 8 of 37 (21.6%); Sonoma, 24 of 161 (34.3%); and Sutter, 4 of 8. Ayala (1970) reported that (W.) Harmon found *P. mexicanum* in Fresno County. In a 13-year study of *P. mexicanum* in *S. occidentalis* at Hopland Field Station, Mendocino County, Schall and Marghoob (1995) reported overall prevalence of 1808 infections in 9179 lizards, 19.7%, which is very similar to the pooled prevalence of 22.0% at that locality. Although Ayala (1973) reported the presence of *P. mexicanum* in Riverside County, no infections were found in 116 *S. occidentalis* and 71 *S. graciosus* in the San Jacinto Mountains of Riverside County (Telford, 1970d). Ayala (1973) found two natural infections of *P. mexicanum* in *Gerrhonotus multicarinatus* (Anguidae) collected in Mendocino and Sonoma counties but gave no host sample size for this species at these sites. Natural infections of *P. mexicanum* were present in 4 of 33 (12.1%) *Sceloporus undulatus* collected in Albany County, Wyoming (Greiner and Daggett, 1973).

Morphological Variation The experimental host *Sceloporus olivaceous* provided the sample used for description of *P. mexicanum*, isolated from *S. torquatus* by blood inoculated into *Crotaphytus collaris* and subsequently into *S. olivaceous* (Thompson and Huff, 1944a). Mature meronts were round to usually elongate and contained 17–20 merozoites in erythrocytes, normoblasts, and polychromatophil erythroblasts. Gametocytes were elliptical to oval, 12–16 × 6–7.7 μm, suggesting LW values of 72–123 μm^2 and L/W ratios of 2.00–2.08. Pigment was dispersed as "15–30 or more coarse, round or bacilloid black granules." Peláez et al. (1948) described meronts of *P. mexicanum* in *S. torquatus* from Distrito Federal as fan-shaped or almost circular and illustrated an elongate meront from a lysed cell. The largest meronts measured 8.46 × 6.40 μm and produced 10–17 merozoites. At the base of fans, eight to ten grains of dark brown to almost black pigment were aggregated. Meronts from *S. microlepidotus* appeared to be identical. The ovoid mature macrogametocytes in *S. torquatus* were 15.4 × 8.3 μm (12.6–18.1 × 6.3–10.2), suggesting estimated LW values of 127.8 μm^2 (79.4–184.6) and L/W ratios of 1.86 (1.77–2.00). In *S. microlepidotus*, macrogametocytes were 12.6 × 10.5 μm (10.8–17.2 × 8.4–12.8), with estimated LW 132.2 μm^2 (90.7–220.2). Microgametocytes in *S. torquatus* were 13.8 × 7.1 μm (11.0–16.5 × 5.5–8.6), estimated LW 98.0 μm^2 (61.1–141.9) and L/W 1.94 (1.92–2.00). In *S. microlepidotus*, microgametocytes were 12.0 × 9.3 μm

(10.2–14.0 × 7.3–10.9), estimated LW 111.6 μm^2 (74.5–152.6) and L/W 1.29 (1.28–1.40).

Recent study of slides from *S. torquatus* of Distrito Federal (Peláez material) and more recently collected *S. torquatus* from the states of Mexico and Jalisco made possible a more thorough description of *P. mexicanum* in the type host. Meronts are 5.9 ± 1.0 × 4.8 ± 0.8 μm (3–8 × 3–6, N = 36), with LW 28.7 ± 8.2 μm^2 (9–48), and contain 9.3 ± 2.7 (3–15) merozoites. Meront size relative to host cell nucleus size is 1.05 ± 0.26 (0.32–1.60, N = 31), and to normal erythrocyte nucleus size is 1.17 ± 0.38 (0.31–2.06, N = 36). Meronts from chronic infections had smaller average dimensions and slightly fewer merozoites than in active infection, respectively 5.5 ± 1.1 × 4.4 ± 0.8 μm (N = 11), LW 24.5 ± 6.8 μm^2, with 7.4 ± 2.4 (3–10) merozoites versus 6.1 ± 1.0 × 5.0 ± 0.8 μm (N = 25), LW 30.6 ± 8.1 μm^2, and 10.2 ± 2.4 (6–16) merozoites. Gametocytes are 15.6 ± 1.9 × 9.4 ± 1.7 μm (10–20 × 6–14, N = 125), with LW 147.1 ± 32.6 μm^2 (84–216) and L/W 1.71 ± 0.35 (1.00–2.71). Gametocyte size relative to host cell nucleus size is 5.79 ± 2.48 (2.40–15.11, N = 122), and to normal erythrocyte nucleus size is 5.31 ± 1.13 (3.00–7.65, N = 125). Gametocytes in active phase of infection are smaller and more elongate than in chronic phase, respectively 14.6 ± 1.4 × 7.9 ± 1.2 μm (12–17 × 6–11, N = 25), LW 115.6 ± 22.3 μm^2 (84–165), and L/W 1.89 ± 0.30 (1.33–2.50) versus 15.9 ± 1.9 × 9.8 ± 1.6 μm (10–20 × 6–14, N = 100), LW 155 ± 29.9 μm^2 (93.5–216), and L/W 1.67 ± 0.35 (1.00–2.71). In active phase, microgametocytes are smaller but shaped similarly to macrogametocytes, respectively 14.0 ± 1.3 × 7.5 ± 0.9 μm (N = 13), LW 104.2 ± 14.2 μm^2, and L/W 1.90 ± 0.30 versus 15.3 ± 1.2 × 8.3 ± 1.3 μm (N = 12), LW 128 ± 23.4 μm^2, and L/W 1.88 ± 0.30. In chronic phase, however, microgametocytes are both smaller and more elongate than macrogametocytes, respectively 15.2 ± 1.8 × 8.9 ± 1.1 μm (N = 45), LW 135.4 ± 19.5 μm^2, and L/W 1.74 ± 0.36 versus 16.4 ± 1.9 × 10.5 ± 1.5 μm (N = 55), LW 171.0 ± 27.4 μm^2, and L/W 1.61 ± 0.34.

In California, dimensions of *P. mexicanum* in *Sceloporus occidentalis* were not stated by Jordan (1970), but merozoites in erythroid cells averaged 16, ranging from 8 to 26. Ayala (1970) described meronts of *P. mexicanum* in this host (and *S. graciosus*?) as round, fan-shaped, or reniform, 8 × 5 μm, producing 8–20 merozoites, usually 12–16. Macrogametocytes were 18.4 × 7.9 μm (15.0–20.6 × 6.4–9.1, N = 50), and microgametocytes were 19.4 × 7.0 μm (15.6–22.6 × 5.0–7.6, N = 50). These data suggest estimated LW values and L/W ratios of 145.4 μm^2 (96–187.5) and 2.33 (2.26–2.34), respectively, for macrogametocytes and 135.8 μm^2 (78–171.8) and 2.77 (1.97–3.12), respectively, for microgametocytes. Ayala commented that average length of 100 gametocytes from ten captive lizards continued to grow through the winter, averaging 15.8 μm in length in May, 18.4 μm in October, with several in January "over 25 μm." His measured series were from September.

Meronts of *P. mexicanum* in erythroid cells of *S. occidentalis* are 7.5 ± 1.9 × 5.1 ± 0.9 μm (5–22 × 3–7, N = 97), with LW 38.2 ± 9.7 μm^2 (20–88), and contain 13.1 ± 3.4 (7–21, N = 100) merozoites. Meront size relative to host cell nucleus size is 1.28 ± 0.29 (0.75–2.00, N = 67), and to normal erythrocyte nucleus size is 1.32 ± 0.34 (0.75–3.17, N = 97). Proerythrocytic meronts average slightly smaller in size than erythrocytic meronts, although their range in size (20–88) is greater than those in mature cells (27–54). The average number of merozoites is the same, and merozoite range is similar, 7.6 ± 2.6 × 4.7 ± 0.9 μm (N = 40), LW 35.5 ± 12.2 μm^2, and 13.1 ± 3.7 (8–20, N = 43) merozoites in proerythrocytes versus 7.5 ± 1.2 × 5.4 ± 0.9 μm (N = 57), LW 40.0 ± 6.9 μm^2, and 13.1 ± 3.1 (7–21). Gametocytes are 16.6 ± 2.3 × 7.3 ± 1.2 μm (8–23 × 4–10, N = 175), with LW 120.9 ± 24.1 μm^2 (56–184) and L/W 2.35 ± 0.56 (1.14–4.60). Gametocyte size relative to host cell nucleus size is 3.15 ± 0.80 (1.50–6.33, N = 149), and to normal erythrocyte nucleus size is 3.94 ± 0.77 (2.01–6.09, N = 175). Gametocytes in active phase of infection average smaller and are more elongate than chronic-phase gametocytes, respectively 16.3 ± 2.3 × 7.0 ± 1.2 μm (N = 100), LW 113.4 ± 24.2 μm^2 (56–162), and L/W 2.41 ± 0.60 versus 17.1 ± 2.3 × 7.7 ± 1.0 μm (N = 75), LW 130.9 ± 20.1 μm^2 (96–184), and L/W 2.27 ± 0.50. Their size relative to host cell nucleus size and to normal erythrocyte nucleus size, however, does not differ by phase of infection, respectively 3.16 ± 0.77 (N = 99) and 3.92 ± 0.80 (N = 100) in active infection versus 3.13 ± 0.86 (N = 50) and 3.98 ± 0.73 (N = 75) in chronic phase. Microgametocytes are smaller than macrogametocytes but do not differ in shape despite difference in infection phase. In active-phase infection, microgametocytes average 15.4 ± 2.3 × 6.5 ± 1.0 μm (N = 41), LW 100.4 ± 23.4 μm^2, and L/W 2.43 ± 0.54 versus 16.9 ± 2.0 × 7.3 ± 1.2 μm (N = 59), LW 122.4 ± 20.6 μm^2, and L/W 2.40 ± 0.64 in macrogametocytes, while the respective average dimensions in chronic phase are 16.4 ± 2.2 × 7.5 ± 1.0 μm (N = 34), LW 121.5 ± 16.7 μm^2, and L/W 2.26 ± 0.54 versus 17.6 ± 2.2 × 7.9 ± 1.0 μm (N = 41), LW 138.7 ± 19.5 μm^2, and L/W 2.28 ± 0.47.

Greiner and Daggett (1973) reported natural infections of *P. mexicanum* in *Sceloporus undulatus* in Wyoming, in which meronts, mostly fan-shaped, produced 8–17 merozoites. Macrogametocytes were 16.7 × 5.8 μm (11.8–19.6 × 3.9–9.8, N = 33), and microgametocytes measured 16.3 × 5.9 μm (11.8–18.6 × 3.9–8.8, N = 28). Estimated LW and L/W, respectively, are 96.9 μm^2 (46.0–192.1) and 2.88 (2.00–3.03) in macrogametocytes and 96.2 μm^2 (46.0–163.7) and 2.76 (2.11–3.03) in microgametocytes.

Exoerythrocytic Merogony Exoerythrocytic stages of *P. mexicanum* were described in detail from both fixed and circulating nonerythroid cells by Thompson and Huff (1944a), the first demonstration of them for a saurian *Plasmodium* species. These were present in experimental infections of unnatural host species, established by inoculation of infected blood from the natural host, *Sceloporus torquatus*, into *Crotaphytus collaris* and, several passages later, into other host species, in all of which both erythrocytic and EE stages occurred. In circulating cells, in addition to all types of erythroid cells (erythrocytes, normoblasts, polychromatophilic and basophilic erythroblasts), asexual parasites were observed in granulocytes, myelocytes, lymphocytes, monocytes, thrombocytes, and free macrophages. In fixed cells of tissues, EE stages were found in "reticular cells, littoral cells of the spleen and liver, hemocytoblasts, tissue macrophages, true endothelial cells of capillaries, of other small blood vessels, and of the interstices between the trabeculae of the cardiac muscle" (Thompson and Huff, 1944a). Both endothelium and fixed connective tissue cells were heavily parasitized, but their presence, number, and distribution varied somewhat by host species. Thompson and Huff (1944a) considered that the type of EE merogony found in *P. mexicanum* corresponded to both the *elongatum* and *gallinaceum* types of avian *Plasmodium* EE merogony.

In *Sceloporus occidentalis*, Jordan (1970) reported that monocytes and lymphocytes comprised the "greatest percentage of white blood cells … infected by *P. mexicanum*" and that "in large monocytes segmenters produced as many as 200 merozoites." Ayala (1970) found EE stages of *P. mexicanum* in *S. occidentalis* to occur in "(1) vascular endothelial cells of the brain, (2) in small, medium-sized, and large mononuclear leukocytes, and (3) free in the blood." Meronts in endothelium of the brain produced 70–120 nuclei, while those occupying small and medium-size leukocytes produced 15–30 merozoites within a vacuole and were common in smears of heart, spleen, liver, and femur. "Large, distended leukocytes in liver and heart smears often contained several meronts with a total of 50 to 200 chromatin masses" (Ayala, 1970b). Free EE meronts contained up to 50, usually 15–30, merozoites, apparently surrounded by a membrane. These were seen in juvenile lizards near death from massive parasitemias. The large meronts found in capillary endothelium of the brain were typical of fatal experimental infections in two *Sceloporus* species from Florida, *S. undulatus* and *S. woodi* (Telford, 1984b), and it was suggested that these indicate pathology by the parasites in "severe infections, which overwhelm the host's defense mechanisms, and represent a deviation from the normal course of infection." Klein et al. (1987b) also reported the presence of EE meronts in endothelium of the capillaries in the brain of lizards that died from vector-transmitted experimental infections of *S. undulatus*.

Sporogony Sporogony of *P. mexicanum* in the phlebotomine sand flies *Lutzomyia vexator* and *L. stewarti* was described by Ayala and Lee (1970) and Ayala (1971a). Experimental infections by inoculation of sporozoites into hatchling lizards were reported by Ayala (1971a). Ookinetes appeared within the sand fly blood meal 10–18 hours PF and then formed oocysts on the hemocoel surface of the sand fly stomach. After 10 days, mature oocysts 22–38 μm in diameter were observed, containing active sporozoites. Several hundred oocysts covering the entire stomach were sometimes seen. Sporozoites were found in hemolymph following oocyst rupture when oviposition by the sand fly occurred. Fresh sporozoites 7–9 μm long entered salivary glands, seldom exceeding 80–100 in number, and were also found in the lumen of the sand fly cardia. Inoculation of sporozoites from the hemocoel into 12 laboratory hatched *S. occidentalis* and 10 negative yearling lizards resulted in infections of two of the hatchlings and one yearling lizard, detected at 22 days postinoculation.

Klein (1985) fed *L. vexator* of North Florida origin on *S. occidentalis* collected in Yolo County, California, and infected by *P. mexicanum* and observed oocysts on day 2 PF, often in large numbers. Development proceeded more rapidly at 27°C than at 24°C. At 27°C, sporozoites were present in the hemocoel on day 6 PF and at day 6.5 PF had entered the salivary glands. At 24°C, sporozoites were not seen until 8.5–9.0 days PF. Oocysts were not found in sand flies maintained at 19°C and dissected on days 4 and 5 PF (Klein et al., 1987b). Oocysts reached maximum average dimensions of 33.5 μm at 27°C in 6.5 days and 34.2 μm at 24°C on day 8.5. Sporozoites were highly infective at 8–10 days PF (Klein et al., 1988b). *Sceloporus u. undulatus* from Alachua County, Florida, negative for *Plasmodium floridense* for 30 days, were fed on for a second blood meal by *L. vexator* infected with *P. mexicanum* 7–10 days following their infected meal from *S. occidentalis*. One to three sand flies were fed on each of 13 *S. undulatus*, and all of the sand flies, dissected following their second blood meal, had sporozoites in their salivary glands. Nine of the 13 lizards showed patent infection by *P. mexicanum* on average 28.6 (23–40) days postinfection. *Culex erraticus* and *C. territans* from North Florida and *Culex apicalis* from California were fed on the infected *S. occidentalis*, and none supported sporogony of *P. mexicanum*. All *L. vexator* dissected had an average of 22 (9–54) oocysts present. Oral transmission by ingestion of infected sand flies was unsuccessful (Klein et al., 1987b).

Fialho and Schall (1995) found that "only about 2% [of oviposited] sand flies lived long enough to take a second blood meal." For transmission of *P. mexicanum* during a second blood meal "*P. mexicanum* must have a development period that is shorter than the time needed for the insect to produce and oviposit its clutch of eggs and feed again." The presence of oocysts by 2 days PF and infected salivary glands by day 6.5 (Klein, 1985) evidently meets this requirement. Fialho and Schall (1995) found that temperatures between 16°C and 32°C affected the rate of parasite development, the rate of maturation of sand fly eggs, and the probability of sand flies becoming infected but did not affect longevity of female sand flies: Increase in temperature between 22°C and 32°C shortened parasite development but did not reduce the time between first and second blood meals.

Effects on Host

Cellular effects In the type host, *S. torquatus*, erythrocytes host to active-phase gametocytes of *P. mexicanum* are enlarged by 16% and in chronic infection by 42%, with nucleus size increased by 21% and 23%, respectively. Erythrocytes infected by meronts and their nuclei are similar in size to uninfected cells and nuclei. In active infection, gametocytes always distort host cells and displace their nuclei, which are usually (64%) distorted. Meronts seldom distort host cells (16%) or displace nuclei (24%), but commonly distort nuclei (44%). In chronic infections, erythrocytes host to gametocytes are distorted, as are their nuclei, and the latter are displaced. The few meronts found in chronic infections had little effect on host cells. In active infections of *P. mexicanum* in *S. occidentalis*, erythrocytes infected by gametocytes were enlarged by 35% and their nuclei by 20%. Host cells were always distorted and nuclei displaced (94%), but overall fewer nuclei were distorted (66%). Nucleus distortion varied among infections from 12% to 100%, perhaps related to age of infection. Erythrocytes host to meronts and their nuclei showed little enlargement and were variably distorted, from 0% to 68% among infections. Their nuclei were seldom distorted (21%) but more often displaced (61%). In chronic infections, host erythrocytes were hypertrophied by 40% and their nuclei by 42%, the cells were distorted and their nuclei displaced and usually (74%) distorted.

Hematological parameters Schall et al. (1982, reviewed by Schall, 1990b), found infection by *P. mexicanum* caused an increase in the percentage of immature erythrocytes from 2.6% in uninfected lizards to 9.5%, and a decrease in hemoglobin concentration in the blood from 7.3 g/100 ml to 5.5 g/100 ml in infected *S. occidentalis*. However, hematocrit values were similar between infected lizards, 33.4% and uninfected, 32.3%, as were the numbers of erythrocytes × 10^3 per mm^2 of blood, 971.1 and 843.4, respectively. There was no difference in oxygen consumption of resting lizards, 0.59 ml/(g.h.) and 0.54 ml/(g.h.). Maximum oxygen consumption, however, was lower in infected lizards, 1.30 ml/(g.h.) than in uninfected, 1.53 ml/(g.h.), with a consequent decrease in aerobic scope, .71 versus 1.00. Schall et al. suggested that the difference in maximal oxygen consumption was due to differences in blood hemoglobin concentration. The effect of these differences in hematological parameters was visible in comparison of running stamina where stamina was reduced in infected lizards. There was, however, no difference in burst speed, important in escape and avoidance of predators.

Effects on fat storage During a 5-year study on the effects of infection by *P. mexicanum* on reproductive success of *S. occidentalis* (Schall, 1983), clutch size of infected females was decreased by 1–2 eggs, an average of 9.2 eggs versus 10.6 in uninfected females. Infected females "stored only 68% as much fat as non-infected animals in 1978, and 80% in 1979," with the decrement in fat stored equal to the equivalent of 1.46 eggs in 1978 and 1.02 eggs in 1979. Although average clutch size varied somewhat among years, significantly so for uninfected females, "Infected females consistently produced smaller clutches." Apparently, the presence of infection lowers fat storage "by the end of the activity season and this results in smaller clutches of eggs the next spring." During a 2-year sampling period, testis size of male lizards infected by *P. mexicanum* was significantly smaller in late summer than for uninfected males (Schall, 1983), decreasing by 38% in 1978 and 37% in 1979. As with females, stored fat in males was 45% less in 1978 in infected lizards in comparison to uninfected males, and 22% less in 1979.

Hormonal levels The effects of infection by *P. mexicanum* on hormonal levels of *S. occidentalis* were studied by Dunlap and Schall (1995) in wild-caught and experimental lizards. Infected lizards had similar levels of basal corticosterone, 7.8 ng/ml, to that of uninfected lizards, 11.5 ng/ml, but lower levels of basal testosterone, 23.5 ng/ml to 37.5 ng/ml, respectively, which was measured immediately after capture. There was no difference in the stress levels of testosterone, measured 1 hour after capture. Stress levels of corticosterone were higher in infected lizards. Steroid concentration did not vary with infection age, parasitemia, time of day, or lizard size. There were lower basal plasma glucose levels in infected lizards but no difference in stress levels of glucose. Parasitemia was correlated negatively with basal glucose levels. Male lizards with implanted exogenous corticosterone showed pathologies similar to those produced by malarial infection: The concentrations of basal corticosterone increased five- to sixfold, testosterone levels were lowered, testis mass reduced, plasma glucose levels increased, and hematocrit lowered in comparison to control lizards. Fat body mass was decreased to a level of borderline significance. Host testosterone apparently does not affect the timing of gametocyte production (Eisen and DeNardo, 2000).

Behavioral effects Schall and Sarni (1987), in a field study, compared the behavioral time budget of *S. occidentalis* that were infected by *P. mexicanum* with uninfected lizards. Both groups utilized the microhabitat similarly, except that infected lizards more often perched in shade. Behavioral patterns were similar, but uninfected adult males engaged in social behaviors more often than did infected males. Their running activity in terms of burst runs and length of runs was similar between infected and uninfected lizards. Earlier, Schall et al. (1982) had found no difference in the frequency of damaged tails, a useful index of predator attacks, between infected and uninfected lizards. Male competition for females was compared by Schall and Dearing (1987) between males infected with *P. mexicanum* and uninfected males. The latter showed greater display activity to females and to other males than did infected males. Infected males that apparently "won" encounters with uninfected males were those with lower parasitemias by *P. mexicanum*. In a further study of the behavioral effects of being parasitized by *P. mexicanum*, Ressel and Schall (1987) compared the attraction of female lizards to males of varying ventral color patterns. Ventral color intensity was not related to *P. mexicanum* infections, but infected males tended to belong to the darkest color class. However, there was little support generated for the hypothesis that females using male ventral coloration would tend to choose uninfected over infected males. Schall and Houle (1992) found in another field study that although uninfected male *S. occidentalis* were more often sighted and more active in chasing other lizards than males infected with *P. mexicanum*, the presence of infection did not affect either minimum or maximum home range size. In every case, males considered to be dominant in their behavior were uninfected.

Ecology New infections of *P. mexicanum* in adult and yearling *Sceloporus occidentalis* apparently can appear at any time during the warmer months, following emergence from hibernation in March (Ayala, 1970b; Bromwich and Schall, 1986). Apparent initial infections were found as early as 4 March by Jordan (1970), and she suggested that infection could be acquired in early spring. Jordan thought that gametocytemia persisted through hibernation, reporting presence of a 6% parasitemia comprised only of gametocytes in a hibernating lizard, which would support transmission beginning in early spring. Ayala (1970) called this early appearance of gametocytes in spring a "spring relapse," timed to precede emergence from hibernation of the highly susceptible lizard hatchlings of the preceding summer. He thought that most transmission at Hopland Field Station took place between April and June, with a "lower level of transmission continuing through the summer." Hatchlings appeared after the peak period of transmission. Ayala described the presence of infections, not of overwhelming parasitemia, but with "high levels of asexual and sexual forms" in four infected hatchlings collected in October. He observed fatal results of massive parasitemias in yearling lizards collected in April and July, which apparently convinced him that "in the spring and early summer, infection undoubtedly proves fatal to a percentage of yearling lizards." Captive lizards maintained a largely gametocytic infection from September until the following spring, and Ayala reported that the first infection seen in adult lizards in late March "consisted of mature gametocytes and asexual forms," with merogony "most prevalent from April to early June." Although Bromwich and Schall (1986) claimed that "over the winter gametocytes are cleared from the lizards blood and parasitemia" diminished, their basis for this disappearance of gametocytes during hibernation is unclear because there is no evidence that collections were made at Hopland from early March through April or after the first week of September during their several years of sampling. This probably accounts for the discrepancy between the observations of Ayala (1970) and of Jordan (1970) and those of Bromwich and Schall (1986) concerning the overwintering of gametocytemia as an important factor in the early springtime infection of sand flies.

Schall and Marghoob (1995) studied 51 sites at Hopland Field Station; the sites varied from 244 to 854 m elevation. The habitat is described as hilly oak savanna at lower elevation and chaparral at higher elevation. A total of 10,546 *S. occidentalis* were examined for *P. mexicanum* during the 13-year project. Male lizards were more often parasitized, 22.4%, in comparison to females, 16.3%, which the authors thought could be a result of different nocturnal resting sites or possibly the effect of elevated testosterone levels on the efficiency of male immune systems. Prevalence increased with body size: The larger lizards were older and therefore had a greater chance of exposure to infected sand flies than younger lizards. Lizards were divided into two groups, those less than 60 mm SVL and those larger than 60 mm (adults). In the smaller group of females, 7.8% were infected versus 19.1% of adults, and in males, 12.6% of younger lizards were infected compared with 25.6% of adult males. During the 13 years studied, there were significant differences in prevalence in *P. mexicanum* among years: Among the 3523 adult males, prevalence varied from 17.2% to 39.4%, and among 2884 females from 10.0% to 29.7%. The younger and smaller lizards maintained low infection levels until maturity. Schall and

Marghoob suggested that lizards may not develop immunity to malaria but are infected for life.

Among the sites sampled, prevalence was usually less than 50%, which could result from a low transmission rate and high mortality among younger lizards. Environmental temperature appears to be important in determining the distribution and abundance of *P. mexicanum* and other saurian *Plasmodium* species. *Lutzomyia vexator* may be an inefficient vector because its preferred body temperature is close to the minimum needed for completion of sporogony before the next blood meal (Fialho and Schall, 1995). Temperature is the only environmental variable with a significant effect on activity of the vector, which is minimal at temperatures below 16°C. Ground squirrel burrows are the preferred resting and reproductive site of *L. vexator*, yet malaria may be found in lizards hundreds of meters distant from burrows. Prevalence of *P. mexicanum* is not related to apparent vector abundance. Sand flies apparently can fly considerable distances. In the opinion of Schall and Marghoob (1995), "Only knowledge of the complexities of sand fly flight, host seeking and feeding behavior will allow an understanding of the number of blood meals taken by a sand fly during its lifetime and consequently the distribution of *P. mexicanum* among sites." Because *S. occidentalis* has a short life span (1–2 years) and the lizards typically remain at one location for entire warm seasons, annual variation in prevalence "must represent changes in the parasite's transmission and maintenance in the host population." Although one would expect environmental factors like temperature and precipitation, which have an effect on vector abundance and feeding behavior, would be correlated with the prevalence of *P. mexicanum*, this is not the case. The prevalence of *P. mexicanum* was not correlated with temperature, rainfall, or herbage produced per hectare (a measure of environmental quality used by Schall and Marghoob). The prevalence data over the 13-year period suggests a cycle of about 10 years in *P. mexicanum* prevalence, similar to that found by Jordan and Friend (1971) for *Plasmodium floridense* in Georgia over 13 years.

Remarks All of Professor Schall's prolific research with his students on *Plasmodium mexicanum* in *Sceloporus occidentalis* has been conducted to elucidate the ecology of this host-parasite system and to relate it to certain ecological theory. The many articles are laden with theory, and it has been challenging to abstract only the data, separate from theory, which is beyond the scope of this work. It is hoped that there has been little missed of significant importance to this species account.

Because wild-caught sand flies can safely be used to transmit *P. mexicanum* to clean lizards and no insect colony need be established, it is regrettable that no one has attempted to obtain the primary merogonic cycle of this parasite in the lizard host. No experimental study has demonstrated the postsporozoite stages of a saurian malaria parasite, and it is a glaring gap in our knowledge.

Plasmodium chiricahuae Telford 1970 (**Plate 33**)

Diagnosis A *Plasmodium* (*Paraplasmodium*) species characterized by large elongate-to-round gametocytes and small meronts formed as rosettes or fans. Meronts are 4–8 × 3–6 µm, LW 12–48 µm², and contain four to ten merozoites. Meront size relative to host cell nucleus size averages 0.69, and to normal erythrocyte nucleus size is 0.80. Gametocytes in active phase of infection are 11–19 × 5–14 µm, with LW 70–210 µm² and L/W 1.07–3.20, their size relative to host cell nucleus size averages 3.08, and to normal erythrocyte nucleus size is 4.33. Macrogametocytes are larger than microgametocytes, but shape is similar between sexes. All erythrocytic stages past young trophozoites show prominent dark brown or black pigment granules.

Type Host *Sceloporus jarrovi* Cope (Sauria: Phrynosomatidae).

Type Locality In the United States, Cochise County, Arizona, in the Chiricahua Mountains 3.2 km above Onion Saddle at 2342 m elevation.

Other Hosts *Sceloporus clarki, S. magister* (Mahrt, 1979); *S. poinsetti*, and *Urosaurus inornatus* (Telford, 1978b).

Other Localities Arizona: Graham County, Pinaleño Mountains; Pima County (Telford, 1970b); Graham County, Pinaleño Mountains, Mt. Graham and Stockton Pass; Cochise County, Huachuca Mountains, Carr Canyon; Cochise County, Chiricahua Mountains, South Fork, Herb Martyr, Turkey Creek; Pima County, Santa Rita Mountains, Madera Canyon; Bahoquivari Mountains, Kitt Peak; Redington Pass east of Tucson (Mahrt, 1979, 1987, 1989). Texas: Llano County, Enchanted Rock (Telford, 1978b).

Prevalence Prevalence of *P. chiricahuae* in *S. jarrovi* in the Chiricahua and Pinaleño Mountains varied altitudinally from 6% to 54% (Telford, 1970b; see below), and in the same host from eight localities in five mountain ranges of Cochise, Graham, and Pima counties from 1979 to 1986 prevalence ranged between 32% and 58% (Mahrt, 1989). In *S. clarki* and *S. magister*, respectively, 6 of 24 (25%) and 1 of 13 (7.7%) were infected by *P. chiricahuae* (Mahrt, 1979). In Texas, 7 of 20 (35%) *S. poinsetti* and 1 of 38 (3%) *U. inornatus* were parasitized by *P. chiricahuae*.

Morphological Variation In the type host, *S. jarrovi*, meronts usually form as a fan or crude rosette, 5.4 ± 1.1 × 3.9 ± 0.7 μm (4–8 × 3–6, N = 71), with LW 20.8 ± 6.2 μm^2 (12–48), and produce 6.3 ± 1.6 (4–10) merozoites. Meront size relative to host cell nucleus size is 0.69 ± 0.22 (0.43–1.56, N = 53), and to normal erythrocyte nucleus size is 0.80 ± 0.26 (0.45–1.88, N = 71). The few meronts found in proerythrocytes do not differ in dimensions or merozoite numbers from erythrocytic meronts. Dark brown-to-black pigment granules form one or two masses at the base of fans or are variably positioned in rosette-shaped meronts. In active infections, the elongate-to-round gametocytes are 14.8 ± 1.7 × 8.0 ± 1.7 μm (11–19 × 5–14, N = 185), with LW 119.4 ± 29.0 μm^2 (70–210) and L/W 1.92 ± 0.44 (1.07–3.20). Gametocyte size relative to host cell nucleus size is 3.08 ± 0.84 (1.70–7.08, N = 119), and to normal erythrocyte nucleus size is 4.33 ± 1.19 (2.51–7.17, N = 185). Macrogametocytes average larger than microgametocytes, respectively 15.3 ± 1.5 × 8.5 ± 1.8 μm (N = 101), LW 129.6 ± 30.2 μm^2, and L/W 1.90 ± 0.46 versus 14.2 ± 1.7 × 7.5 ± 1.3 μm (N = 84), LW 107.2 ± 22.1 μm^2, and L/W 1.94 ± 0.42. There is no difference in shape between sexes. Chronic-phase gametocytes are larger and more rounded than those from active-phase infections, 16.2 ± 2.0 × 11.3 ± 2.0 μm (11–22 × 7–17, N = 124), with LW 184.7 ± 46.8 μm^2 (108–330) and L/W 1.47 ± 0.26 (1.00–2.43). Their size relative to host cell nucleus size is 5.10 ± 1.26 (3.00–8.25, N = 50), and to normal erythrocyte nucleus size is 6.28 ± 1.40 (3.91–10.2, N = 124). Macrogametocytes in chronic phase are also larger than microgametocytes, respectively 17.2 ± 1.8 × 12.3 ± 1.9 μm (N = 67), LW 212.0 ± 43.6 μm^2, and L/W 1.43 ± 0.24 versus 15.1 ± 1.6 × 10.1 ± 1.4 μm (11–19 × 7–14, N = 57), LW 152.7 ± 25.1 μm^2, and L/W 1.52 ± 0.28. There is no difference in shape between sexes. Newly mature gametocytes are elongate with blunt or slightly pointed ends. The golden-brown pigment of immature gametocytes forms one or two masses, often at the ends of the parasites. In mature gametocytes, pigment is dispersed as minute bacilliform granules, 20–50 or more, which sometimes aggregate to form larger masses, especially in microgametocytes.

In *S. poinsetti* of Texas, *P. chiricahuae* meronts, very few of which have been seen, are similar in shape, size, and merozoite number to those in the type host. Meronts are 5.3 ± 0.5 × 4.7 ± 0.5 μm (5–6 × 4–5, N = 6), LW 25.0 ± 4.5 μm^2 (20–30), and contain 8.3 ± 1.4 (6–10) merozoites. Their size relative to host cell nucleus size is 0.64 ± 0.10 (0.57–0.83), and to normal erythrocyte nucleus size is 0.86 ± 0.20 (0.64–1.09). In infections that appeared to be early chronic phase, gametocytes are 15.7 ± 1.8 × 9.5 ± 1.4 μm (12–20 × 6–13, N = 150), with LW 147.3 ± 21.8 μm^2 (96–209) and L/W 1.70 ± 0.41 (1.00–3.00). Gametocyte size relative to host cell nucleus size is 3.30 ± 0.85 (1.56–5.36, N = 100), and to normal erythrocyte nucleus size is 4.64 ± 1.03 (2.86–8.03, N = 150). Macro- and microgametocytes are similar in dimensions and shape, respectively 15.9 ± 1.9 × 9.5 ± 1.4 μm (13–20 × 7–13, N = 87), LW 150.1 ± 20.8 μm^2 (104–209), and L/W 1.73 ± 0.43 versus 15.3 ± 1.5 × 9.4 ± 1.5 μm (12–19 × 6–13, N = 63), LW 143.5 ± 22.8 μm^2 (96–192), and L/W 1.67 ± 0.38. A single infection of *P. chiricahuae* in *Urosaurus inornatus* collected at the same site as *S. poinsetti* was similar in range of gametocyte dimensions to those from *S. poinsetti*, but averaged slightly smaller in LW values. Gametocytes are 14.6 ± 2.1 × 8.8 ± 1.2 μm (12–19 × 7–11, N = 25), LW 126.6 ± 19.5 μm^2 (91–165), and L/W 1.71 ± 0.40 (1.18–2.71). Gametocyte size relative to host cell nucleus size is 3.51 ± 0.87 (2.02–5.67), and to normal erythrocyte nucleus size is 4.00 ± 0.63 (2.89–5.24). Only one erythrocytic meront could be found, which was 7 × 4 μm and contained 14 merozoites.

Exoerythrocytic Merogony Although immature parasites were seen commonly in several types of nonerythroid cells in the type host—thrombocytes, "leukocytes" (some of which were lymphocytes), and eosinophils (Telford, 1970b)—EE meronts were identified only in thrombocytes and probable lymphocytes. Thrombocytic meronts contained 4–6 nuclei, and those in cells of uncertain identity had 10 and 18 nuclei. In *S. poinsetti*, no EE stages were observed, but in *U. inornatus* several parasitized thrombocytes were found, as was one infected lymphocyte. Thrombocytic meronts were 7.1 ± 1.0 × 5.1 ± 0.4 μm (6–9 × 5–6, N = 8), with LW 36.4 ± 4.4 μm^2 (30–45), and contained 16.8 ± 2.1 (14–20) nuclei.

Sporogony Unknown. Given the genomic relationship of *P. chiricahuae* to *Plasmodium mexicanum* (Perkins and Schall, 2002), it is reasonable to suspect *Lutzomyia vexator* or similar species as the vectors.

Effects on Host In *S. jarrovi*, host erythrocytes infected by active-phase gametocytes were hypertrophied by 32% and even more, 47%, by those gametocytes in chronic phase. Nuclei also were enlarged by 38% and 20%, respectively. Meronts, however, produced little enlargement, 6% only, of either infected erythrocytes or their nuclei. Erythrocytes and their nuclei, parasitized by gametocytes, were distorted and nuclei were displaced. Meronts had little effect on either host cell or nucleus apart from occasional displacement of the latter. The same effects were seen in *S. poinsetti* erythrocytes infected by probably chronic-phase gametocytes, 32% hypertrophy of the host cell and 48% enlargement of its nucleus, again with distortion of cell and nucleus, and displacement of nuclei. In the probably active infection of *U. inornatus*, hypertrophy of both cell and nucleus occurred, but with less enlargement, 15% and 17%, respectively.

Ecology

Altitudinal distribution *Plasmodium chiricahuae* was reported to occur in greatest prevalence in the evergreen oak-ponderosa pine communities in the Chiricahua and Pinaleño Mountains between 1829 and 2408 m, 54% of 57 *S. jarrovi*, declining to 26% of 35 lizards at 2438–2713 m, and to only 6% of 36 lizards between 2743 and 3018 m. Samples of 22 lizards from 2865 m and 21 from 3261 m were negative (Telford, 1970b). The maximum elevation at which an infected *S. jarrovi* was collected was 2804 m. Mahrt (1979) found 1 of 13 (7.7%) *Sceloporus magister* from a creosote bush community at 610 m infected by *P. chiricahuae* and 6 of 24 (25%) *Sceloporus clarki* infected in a paloverde-saguaro community at 914–1158 m. His *S. jarrovi* sample, 10 of 35 (28.6%) positive, came from 1829–2438 m, in the same community and elevation of maximum prevalence reported by Telford (1970b).

Geographic distribution Mahrt (1987, 1989) described the distribution of *P. chiricahuae* in *S. jarrovi* as an example of island biogeography among five isolated mountain ranges in southern Arizona: Chiricahua, Huachuca, Pinaleño, Santa Rita, and Baboquivari, thought to be isolated for 8000–12,000 years as lowland vegetation became increasingly desertic (Mahrt, 1989). The presence of *P. chiricahuae* in *Sceloporus poinsetti* at Enchanted Rock, Llano County, Texas (Telford, 1978b), is consistent with Mahrt's description of the range of host and parasite. Both *S. jarrovi* and *S. poinsetti* belong to the *Sceloporus torquatus* group of species that ranges from Arizona and Texas south to Guatemala. In the central and western portions of the range of this species group, *S. torquatus* (*S. ferrariperezi* of Thompson and Huff, 1944a, and Peláez et al., 1948) is host to the related *Plasmodium mexicanum*. *Sceloporus jarrovi* ranges southward into central Mexico in isolated montane populations, but there are no records of *P. chiricahuae* in populations south of Arizona and Texas.

Seasonality and age of host Telford (1970b) reported that prevalence of *P. chiricahuae* in samples of *S. jarrovi* from the Pinaleño Mountains taken at 1828–2134 m in June (N = 32) and October–November (N = 16) did not differ, and there was no difference in the proportion of asexual:sexual parasites in samples that might indicate seasonality of transmission. Mahrt (1989) found little difference in prevalence between yearling lizards and adults during his 8-year study from 1979 to 1986 in eight different areas of the five mountain ranges. In April–June, prevalence of *P. chiricahuae* was 37% in yearlings and 56% in adults, and from July–November, the respective prevalences were 42% and 48%. *Sceloporus jarrovi* is an ovoviviparous species that mates in the fall and gives birth in May–June (Ballinger, 1973). Neonate lizards apparently can be infected almost immediately after birth. Mahrt (1989) found patent infections in 1 of 3 juveniles in July, 3 of 32 in August, with a sharp rise to 70% (16 of 23) by September at three different sites. There is no evidence of congenital transmission from infected females to their offspring. Only gametocytes were seen in lizards from March to May and occasional meronts in late spring and summer. Mahrt suggested that "long parasitemia of gametocytes ensuring availability for transmission" contributed to the success of the host-parasite system. Parasitemia occurs and is maintained at low levels, comprised almost entirely of gametocytes (Telford, 1970b), with gametocytes produced early in infection, almost simultaneously with appearance of asexual stages. The maximum parasitemia seen was 219 parasites at 10^4 erythrocytes (2.2%), in great contrast to the related species, *P. mexicanum*.

Remarks Although the taxonomic significance of morphological similarity may eventually be shown to be dubious, the relationship of *P. chiricahuae* to *P. mexicanum* appeared obvious to me on morphological grounds alone, despite skepticism of their relationship expressed to me by Professor Garnham. Perkins and Schall (2002) have demonstrated a closer relationship through their analyses of cytochrome-b sequences of these two species than to the other North American saurian *Plasmodium* species, *P. floridense*. Genomic comparison of the isolated montane populations of *P. chiricahuae* among themselves and with the disjunct populations of *P. mexicanum* in central and western Mexico and California can be expected to provide valuable insight into speciation from their common ancestor, as well as clarification of their zoogeographic history.

PLASMODIUM SPECIES OF ASIAN LIZARDS

Plasmodium sasai Telford and Ball 1969 (**Plates 34 and 35**)

Diagnosis A species of *Plasmodium* (*Lacertamoeba*) with meronts 3–10 × 2–8 μm and LW 6–56 μm² that produce 2–14 merozoites, usually arranged as a fan or, in the largest meronts, a rosette. Meront size relative to host cell nucleus is 0.69, and to normal erythrocyte nuclei is 0.82. Pigment forms as a single, prominent dark mass. The spherical-to-elongate gametocytes are 5–9 × 3–7 μm, with LW 18–49 μm² and L/W 1.0–2.7. Gametocyte size relative to host cell nucleus is 1.28, and to normal erythrocyte nuclei is 1.38. Pigment in both sexes of gametocyte usually comprises a prominent single dark granule, rarely two or three granules. Sexual dimorphism in dimensions is present in gametocyte shape only, with macrogametocytes less rounded than microgametocytes. Phanerozoic

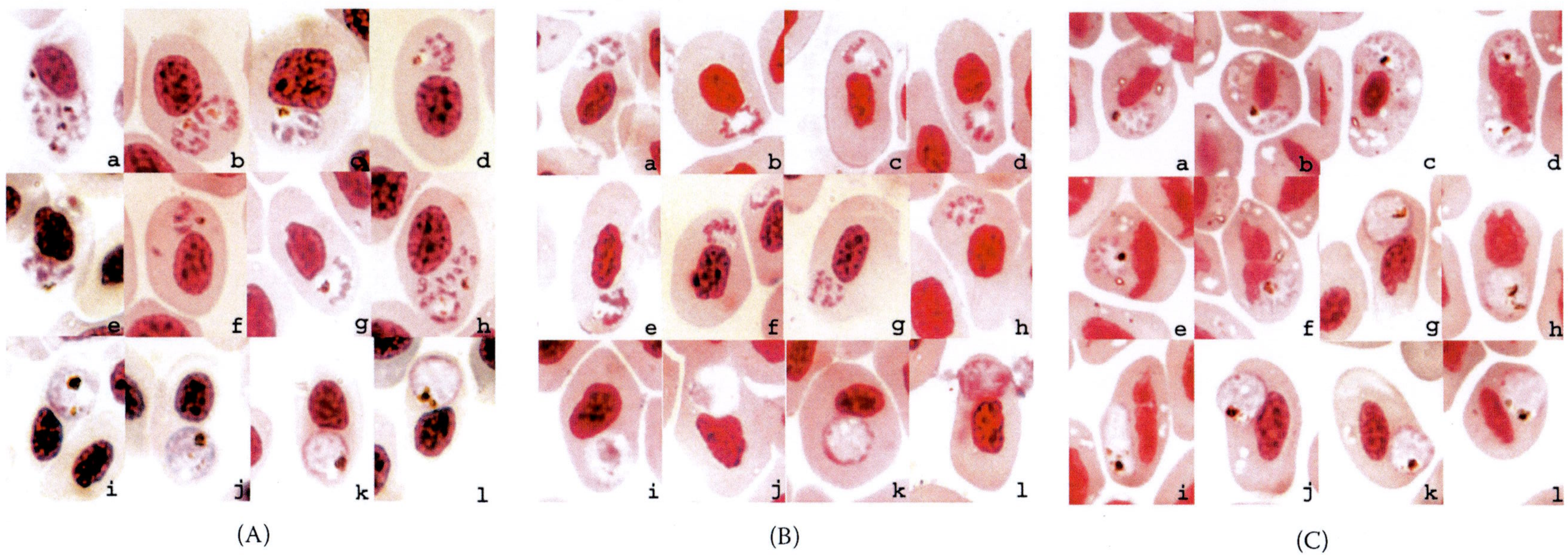

Plate 34 (**A**)–(**C**) *Plasmodium sasai* from *Takydromus tachydromoides*, Honshu, Japan (**A**), *T. smaragdinus*, Ryukyu Islands (**B**); and *T. sexlineatus* (**C**), Thailand. Meronts in (**A**) and (**B**), **a–h**, and in (**C**), **a–e**; macrogametocytes in (**A**) and (**B**), **i**, **j**, and in (**C**), **g**, **h**; microgametocytes in (**A**) and (**B**), **k**, **l**, and in (**C**), **h–l**.

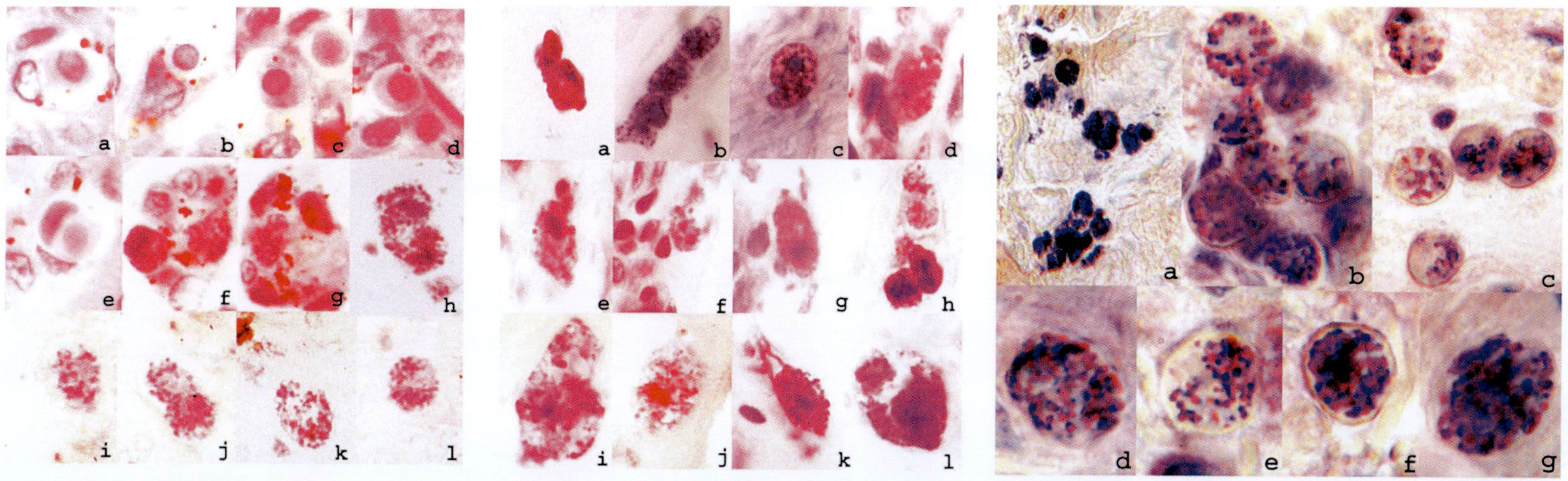

(A) (B) (C)

Plate 35 Preerythrocytic stages and exoerythrocytic meronts of *Plasmodium sasai* in *Takydromus tachydromoides*. (**A**) Preerythrocytic stages in liver parenchymal cells (**a–d**, **i–l**) and hepatic macrophages (**e–h**): apparent hypnozoites, **a–d**; uninucleate parasite, **e**, and meronts, **f–h**; **i–l**, meronts in parenchymal cells. Natural infection, posthibernation. (**B**) Phanerozoic meronts of *P. sasai* in various cells of experimentally infected host: phanerozoites in liver (**a**), brain (**b**), and lung (**c–g**); chronozoites in connective tissue between spleen and pancreas (**h**), heart (**i**), femoral muscle (**j**), brain (**k**), and lung (**l**). Cyst walls not distinct. (Figure **c** modified from Telford, S. R., Jr., *J. Parasitol.,* 82, 226, 1996, and Figures **b** and **c** modified from Telford, S. R., Jr., 1997, *Ecology of a Symbiotic Community*, vol. 2, ch. 15, Figures 1b and 1c, Krieger Publishing Company, Malabar, FL, with permission.) (**C**) **a**, phanerozoites in heart of *P. sasai* in experimental infection of juvenile *T. tachydromoides* at 144 hours postinoculation; **b–g**, experimental infection of *P. sasai* from *Takydromus smaragdinus* into *T. tachydromoides* at 71 days postinoculation; **d**, phanerozoites; remainder are encysted chronozoites in connective tissue of heart.

meronts parasitize endothelium and connective tissues of most organs, especially those of the heart.

Type Host *Takydromus tachydromoides* Schlegel (Sauria: Lacertidae).

Type Locality Hanno, Saitama Prefecture, Honshu, Japan.

Other Hosts *Takydromus smaragdinus, T. sexlineatus.*

Other Localities In *T. tachydromoides*: Mt. Yahiko, Niigata Prefecture; Tokorozawa, Saitama Prefecture; Tsurukawa, Hiyoshi, and Tsunashima, Kanagawa Prefecture; urban Tokyo. In *T. smaragdinus*, Koniya, Amami Island, Ryukyu Islands, Japan. In *T. sexlineatus*, near Ramintra, Bangkok, Thailand.

Prevalence As detected by routine screening of blood films from *T. tachydromoides* at the type locality, only five infections of *P. sasai* were found in 1132 lizards examined between 1965 and 1967 (0.4%). Corresponding prevalences for the sampling periods during which these infections were detected are 1.7% in May–June 1966, 0.8% in September 1966, and 1.7% in September 1967 (Telford, 1997b). When hearts from 424 lizards collected from 1965 to 1967 were examined histologically, overall prevalence of *P. sasai*, indicated by presence of phanerozoic meronts, was 88.9%, 85.4% in 206 first-year lizards and 92.2% in 218 adults, with no difference between males (88.5%, N = 252) and females (89.5%, N = 172) (Telford, 1996, 1997b). In six other localities, prevalences determined by histology of the heart were Tokorozawa, 100%; Niigata, 40%; Tsurukawa, 50%; Hiyoshi, 66.7%; Tsunashima, 80%; and in Tokyo, 50% (Telford, 1997b). In the other host species, *P. sasai* was found by blood film examination only in 7.4% of 81 *Takydromus smaragdinus* and 24% of 42 *Takydromus sexlineatus* (Telford, 1982a).

Morphological Variation In the type host, meronts are 4.9 ± 0.9 × 4.0 ± 0.7 μm (3–10 × 2–8, N = 525), with LW 19.6 ± 6.9 μm² (6–56). Merozoite number averages 6.8 ± 2.3 (2–14, N = 527). Meront size relative to host cell nucleus is 0.69 ± 0.21 (0.26–1.39, N = 304), and to normal erythrocyte nuclei is 0.82 ± 0.28 (0.26–2.52, N = 446). The few meronts measured from a barely detectable, chronic, natural infection were 4.0 ± 0.2 × 3.1 ± 0.5 μm (3–5 × 2–4, N = 11), with LW 12.5 ± 8.7 μm² (8–20), and contained 4.3 ± 1.6 (3–8) merozoites. Most meronts (91%) are fan-shaped; the remainder form as rosettes. Pigment forms as a single, prominent clump of dark granules at the fan base or in the center of the rosette. Gametocytes are 6.5 ± 0.9 × 4.9 ± 0.8 (5–9 × 3–7 μm, N = 125), with LW 31.9 ± 6.4 μm² (18–49) and L/W 1.37 ± 0.34 (1.00–2.67). Gametocyte size relative to host cell nucleus is 1.28 ± 0.28 (0.63–1.94, N = 105), and to normal erythrocyte nuclei is 1.38 ± 0.28 (0.72–2.11, N = 125). Macrogametocytes are more elongate than microgametocytes, but other dimensions are similar, respectively 6.7 ± 1.0 × 4.9 ± 0.8 μm (5–9 × 3–7, N = 50), LW 32.3 ± 6.3 μm² (24–49), and L/W 1.42 ± 0.37 (1.00–2.67) versus 6.4 ± 0.9 × 5.1 ± 0.8 μm (5–9 × 4–7, N = 50), LW 32.8 ± 6.8 μm² (20–49), L/W 1.30 ± 0.30 (1.00–2.25). Pigment occurs as a large, irregular dark granule in gametocytes, rarely as two or three granules clustered together.

In *T. smaragdinus* from Amami Island, meronts are 5.1 ± 0.8 × 3.9 ± 0.6 μm (4–7 × 2–5, N = 53), with LW 19.9 ± 4.9 μm² (8–30), and produce 8.3 ± 1.9 (5–13) merozoites. Meront size relative to host cell nucleus is 0.86 ± 0.25 (0.40–1.50, N = 48), and to normal erythrocyte nuclei is 0.85 ± 0.23 (0.27–1.42, N = 53). Merozoites arranged as rosettes are more common (24%) than in the type host. Gametocytes are 6.1 ± 0.8 × 5.3 ± 0.6 μm (5–9 × 4–7, N = 95), with LW 32.5 ± 7.1 μm² (20–63) and L/W 1.16 ± 0.17 (1.00–1.75). Their size relative to host cell nucleus is 1.38 ± 0.38 (0.83–3.15, N = 94), and to normal erythrocyte nuclei is 1.42 ± 0.28 (0.94–2.55, N = 95). Compared to *P. sasai* from *T. tachydromoides*, both sexes of gametocytes are more rounded: Macrogametocytes are 6.0 ± 0.6 × 5.2 ± 0.6 μm (5–8 × 4–7, N = 50), LW 31.6 ± 6.0 μm² (20–49), and L/W 1.16 ± 0.16 (1.00–1.56), and microgametocytes are 6.2 ± 0.9 × 5.4 ± 0.6 μm (5.9 × 4–7, N = 45), LW 33.5 ± 8.0 μm² (24–63), and L/W 1.15 ± 0.17 (1.00–1.75).

In the Thai host, *T. sexlineatus*, meronts again more commonly form as rosettes (40%) than in *T. tachydromoides*. Meronts are 5.6 ± 0.8 × 4.2 ± 0.8 μm (4–8 × 2–6, N = 50), with LW 23.8 ± 5.9 μm² (10–40), and contain 8.4 ± 2.7 (4–14) merozoites. Their size relative to host cell nucleus is 1.19 ± 0.37 (0.57–1.87, N = 23), and to normal erythrocyte nuclei is 1.01 ± 0.26 (0.45–1.79, N = 50). Gametocytes are 6.5 ± 1.0 × 5.5 ± 0.7 μm (5–10 × 3–7, N = 75), with LW 35.8 ± 7.6 μm² (20–56) and L/W 1.22 ± 0.29 (1.00–2.50). Gametocyte size relative to host cell nucleus is 2.01 ± 0.68 (0.71–3.86, N = 67), and to normal erythrocyte nuclei is 1.50 ± 0.31 (0.85–2.28, N = 75). As with gametocytes from *T. smaragdinus*, gametocytes of both sexes are similar in L/W but are more rounded than in *T. tachydromoides* and slightly more elongated than those from *T. smaragdinus*: Macrogametocytes are 6.5 ± 1.0 × 5.4 ± 0.7 μm (5–10 × 4–7, N = 45), LW 35.4 ± 6.9 μm² (20–54), and L/W 1.22 ± 0.27 (1.00–2.50), and microgametocytes are 6.6 ± 1.1 × 5.5 ± 0.9 μm (5–9 × 3–7, N = 30), LW 36.4 ± 8.7 μm² (21–56), and L/W 1.22 ± 0.34 (1.00–2.33). Pigment appearance is the same in all three hosts of *P. sasai*.

Exoerythrocytic Merogony Telford (1989) found apparent preerythrocytic stages of *P. sasai* in *T. tachydromoides* by examination of tissue sections from two naturally infected hosts, one collected immediately after leaving hibernation and the other later in the spring (**Plate 35**). Uninucleate parasites that averaged 6.5 × 5.4 µm occurred in hepatic parenchymal cells within parasitophorous vacuoles. Other uninucleate parasites, 6.4 × 5.1 µm, occupied macrophages in liver sinuses, and larger, more elongate uninucleate stages (7.6 × 5.4 µm) were seen in pulmonary endothelium. Binucleate meronts also occurred in the same sites, slightly larger in size in the lung endothelium, 12.3 × 9 µm, than in the hepatic cells, 10 × 6 µm. Multinucleate meronts in liver parenchyma were smaller, 9 × 6.8 µm, than those from macrophages, 13.5 × 11.5 µm. Segmenters 6–7 × 5–6 µm found in parenchymal cells contained 14–16 elongate merozoites. Swollen macrophages also contained large meronts with 30–40 or more nuclei. Variably shaped, usually elongate, meronts in capillary endothelium of lungs averaged 10.8 × 7.3 µm and contained a minimum of 22–47 nuclei. Two lizards with natural chronic infections, collected in early fall, had meronts in tissues of nearly every organ examined, in connective tissue of liver, lung, spleen, pancreas, brain, heart, kidney, intestine, testis, and femoral muscle, and in endothelium of lung, brain, kidney, and femoral muscle. In femoral muscle, some meronts also were intra- or intermuscular. Many meronts had a hyaline cyst wall surrounding them and were generally similar among the various tissues in their dimensions, averaging 13.5–21.9 × 9.5–14.4 µm, with maxima of length and width 34 µm and 19 µm, respectively. Experimental infections of the type host, induced by inoculation of infected blood, demonstrated that erythrocytic parasites are able to establish the phanerozoic portion of the life cycle in fixed endothelium and connective tissue (Telford, 1998b). This verified the identification of phanerozoites found in natural infections as *P. sasai* (**Plate 35**). Dimensions of phanerozoic meronts induced by blood inoculation averaged 11–14 × 6.3–10.4 µm. Phanerozoites were detected 48 hours postinoculation, at which time some in femoral muscle were surrounded by hyaline cyst walls as chronozoites. Chronozoites comprised 60–100% of phanerozoites by 296 days postinfection, in heart (75%), lung (60%), kidney (100%), and femur (100%). Chronozoites were identified in experimental lizards on postinoculation days 2, 4, 75, 78, 121, 200, and 296. Their dimensions and shape differed among the various tissues. In every infection, phanerozoites were most plentiful in the heart, often occurring in clusters exceeding 30 meronts. Phanerozoites apparently required 4 days for maturation, at which time their rupture occurred, and merozoites were released to begin another cycle. As early as day 4, a large, ruptured phanerozoite in capillary endothelium of the brain released at least 181 merozoites (Telford, 1998b). There was little difference in phanerozoite production and morphology among two strains of *P. sasai* from the type locality and one strain isolated into *T. tachydromoides* from a natural infection in *T. smaragdinus* of the Ryukyu Islands.

Sporogony Unknown. No infections resulted when *Culex pipiens, Aedes togoi, Ophionyssus natricis, Ornithodoros talaje*, and *Ixodes nipponensis* were fed on *T. tachydromoides* infected by *P. sasai* (Telford, 1997b).

Course of Infection Considering the factors of month of host collection, stage, and intensity of infection, the course of pre-erythrocytic infection was postulated by Telford (1989) as follows: Sporozoites, perhaps inoculated late in the transmission period before host hibernation, entered liver parenchymal cells and rounded up within parasitophorous vacuoles into a resting stage, analogous to the hypnozoites found in primate *Plasmodium* species (Krotowski et al., 1981, 1982a, 1982b). On emergence from hibernation, these began merogony, resulting in small segmenters with 14–16 merozoites, representing cryptozoic meronts. Cryptozoites produced by them formed metacryptozoic meronts in hepatic macrophages or possibly in hepatic parenchyma. Their progeny, metacryptozoites, then entered capillary endothelium, forming phanerozoic meronts, the merozoites from them entering erythrocytes. Some phanerozoites and possibly metacryptozoites as well entered the endothelium, connective tissue, and skeletal muscle, where many encysted as the chronozoite stage (Telford, 1989), which perhaps ensures the survival of infection through time. Sporozoites that were transmitted early in the transmission period could have entered hepatic cells and, following phanerozoite merogony, initiated the erythrocytic cycle prior to hibernation. Telford (1972a) studied the course of blood-induced infections of *P. sasai* in *Takydromus tachydromoides* and *T. smaragdinus*. No strain differences were found in the prepatent period, onset of the acute phase of infection, or level of parasite-induced erythropoiesis in homologous infections, in which the parasite strain originated from the same host species as the experimental host. There was no effect on course of infection by age of host (i.e., adult or juvenile). Prepatent periods ranged from 6 to 50 days, with the interval dependent on the numbers of parasites in the inoculum. The Hanno strain of *P. sasai* reached peak of infection 25-96 days postinoculation, with mean parasitemia at peak 4792 parasites per 10^4 erythrocytes. The maximum parasitemia observed was 17,850 parasites per 10^4 erythrocytes. *Plasmodium sasai* multiply infected erythrocytes, with up to 13 parasites in a single cell. After peak, infections declined slowly over 50–75 days to chronic infections of less than 500 parasites per 10^4 red blood cells. During chronic phase, multiply infected cells disappeared. Relapse of infections was seen on days 142 and 213 postinoculation in two lizards; when multiply infected cells reappeared, parasitemias rose to 742 and 2220, respectively, and were followed again by decline in parasitemia. Another relapse occurred in the second infection on day 277 and again declined. Infected hosts survived up to 291 days postinoculation, never becoming negative, but dying from causes other than malaria.

Effects on Host In experimental infections (Telford, 1972a), no deaths could be attributed to *P. sasai* alone. Fatalities were observed only when lizard erythrocytic virus (LEV) infection was also present. The only demonstrable effects were increased erythropoiesis of 11–34% in adults and 27–37% in juveniles. Maximum percentages of immature red blood cells were found 10–16 days (mean 7.0%) following peak parasitemia in adults and 13–28 days (mean 17.0%) in juveniles, with the blood picture in both adults and juveniles quickly returning to near-normal levels. Telford and Ball (1969) reported that *P. sasai* often has little obvious effect on infected erythrocytes, but when multiple infections, which may include 11 or more parasites, or mature gametocytes are present, the host cell may be hypertrophied and distorted. The nucleus is often hypertrophied as well.

Ecology At the type locality, transmission of *Plasmodium sasai* to *Takydromus tachydromoides* occurs primarily during late summer and fall, the period in which hatchlings appear, at an average size at hatching of 22.5 mm SVL, varying from 21 to 24 mm (Telford, 1969). Most clutches hatch in September and October. The average daily growth of hatchlings is 0.21 mm/day prior to hibernation (Telford, 1997a). A hatchling lizard of 23 mm SVL contained phanerozoites in the heart, which indicates an immediate posthatching infection (Telford, 1998b). Between hatching and the time the first-year cohort of the host population enters hibernation in mid-to-late November, the prevalence of *P. sasai* can reach 100%. The capture of two yearling lizards with acute infections in May and June 1966 indicates that some transmission in spring or early summer may occur, and this is supported by the 100% prevalence in May 1966, an increase over the posthibernation decline seen in March and April of that year. These could have been relapsed infections. Infections were not lost during hibernation. Chronozoites were more commonly found following hibernation than before. Blood films of 46 lizards apparently negative from routine screening but proven positive for *P. sasai* by histological examination of hearts were reexamined (Telford, 1997b), and trophozoites or very small meronts were found on 20 slides (43%), at barely detectable parasitemias of 0.01–0.06%. The lizards were captured in fall 1966 (10), spring 1967 (16), and fall 1967 (20). Erythrocytic parasitemias were found in one-half of the phanerozoite-positive lizards in both fall samples and in 31% of the spring sample, again indicating that transmission of *P. sasai* occurs primarily in the late summer and fall. The duration of active infections of *P. sasai* in lizards in nature may be much shorter than that found in the experimental laboratory infections (Telford, 1972a) and occur at barely detectable parasitemias in peripheral blood. Many more infective parasites are probably introduced by inoculation of infected blood than during the blood meal of a sporozoite-infected vector, thus establishing a more severe infection (Telford, 1997b, 1998b). *Plasmodium sasai* is probably transmitted to *T. tachydromoides* in central Honshu by the bite of vectors infected from chronically infected adult lizards with low parasitemias. In a given year, most of the hatchling cohort is infected following hatching during August–October until hibernation in November. The infections occur at very low parasitemias and persist as phanerozoites or encysted chronozoites through hibernation from November to March. Chronozoites possibly occur in greater proportion than phanerozoites when hibernation is prolonged because of a severe winter. Following hibernation, active parasitemias in spring could result from relapses produced by the exoerythrocytic meronts, activation of hypnozoites (Telford, 1989), or from persistent patent infections. As vector populations expand following winter, transmission to uninfected lizards may occur in late spring and early summer but probably represents only 10–15% of transmission by vectors.

Remarks *Plasmodium sasai* is a component of a host-parasite complex that may have been present in a common ancestor of the three *Takydromus* species that it is known to parasitize. This progenitor must have extended southward in eastern Asia during the Pliocene prior to geological activity that gave rise to the Ryukyu archipelago (Telford, 1982a, 1997b). Variation in the taxonomically important characters of meront size (LW) and merozoite number and gametocyte LW and shape (L/W) differs as much within the type population in Honshu as among the three host species in four populations over a distance of 4600 km, from Niigata in Honshu to Bangkok, Thailand. In turn, this relative stability of morphological characters suggests that:

> It seems unlikely that such a low degree of morphological variability would persist if a variety of vector types is utilized. Rather, a long established association with a single type of vector … mosquito, phlebotomine or ceratopogonid is suggested … also unlikely that a single vector species remains sympatric with *Takydromus* species through the … climatic differences extending from tropical Southeast Asia well into cool temperate Honshu along the subtropical Ryukyu Archipelago. (Telford, 1982a)

Plasmodium lionatum Telford 1982 (**Plate 36**)

Diagnosis A *Plasmodium* (*Asiamoeba*) species with elongate, thin gametocytes, 5–12 × 2–6 µm, with LW 18–45 µm² and L/W 1.0–4.5. Meronts are usually tiny, 3–6 × 1.5–3 µm, with LW 4.5–18 µm², and contain four to six merozoites. Merozoites usually are arranged linearly or as a morulum and rarely are fan-shaped or cruciform. Meronts are situated on the margin of the erythrocyte. Macrogametocytes are more slender, with higher L/W ratio, than microgametocytes. Pigment is present as a single, small dark dot variably located in meronts and difficult to discern but is dispersed as dark granules in the cytoplasm of gametocytes.

Type Host *Ptychozoon lionatum* Annandale (Sauria: Gekkonidae).

Type Locality Peninsular Thailand, no precise locality.

Other Hosts None known.

Other Localities Malaysia, specific locality unknown.

Prevalence Overall, 16 of 44 (36.4%) *Ptychozoon lionatum* examined from 1976 to 1980 were infected by *Plasmodium lionatum*, with annual prevalences 7 of 15 in 1976, 6 of 15 in 1977, 2 of 10 in 1979, and 1 of 4 in 1980.

Morphological Variation Meronts average 3.7 ± 0.8 × 2.1 ± 0.4 µm (3–6 × 1.5–3, N = 50), with LW 8.0 ± 3.0 µm² (4.5–18), and produce 4.5 ± 0.7 (4–6) merozoites. Merozoites are arranged most commonly in linear form (38%) or as a morulum (48%), rarely as a fan (8%), a rosette (2%), or in cruciform pattern (4%). Little cytoplasm is visible, and they are most commonly marginal (52%) in position. A single, variably placed small dot of dark pigment, difficult to discern, is present in most meronts, but 12% of meronts appear to lack pigment. Pigment is dispersed as dark granules in gametocytes. Gametocytes average 8.9 ± 1.7 × 3.6 ± 0.7 µm (5–12 × 2–6, N = 100), with LW 31.4 ± 6.4 µm² (18–45) and L/W 2.62 ± 0.80 (1.0–4.5). Relative to the host cell nucleus, gametocyte LW averages 0.97 ± 0.23, and to normal erythrocyte nuclei is 0.87 ± 0.17. Macrogametocytes are 9.2 ± 1.7 × 3.3 ± 0.7 µm (5–12 × 2–6, N = 50), with LW 30.0 ± 5.7 µm² (18–44) and L/W 2.88 ± 0.79 (1.0–4.5), more slender and with higher L/W ratio than microgametocytes, with their respective dimensions 8.7 ± 1.7 × 3.8 ± 0.6 µm (5–11 × 3–5, N = 50), 32.7 ± 6.8 µm² (20–45), and 2.37 ± 0.73 (1.3–3.7). Chronic-phase microgametocytes are narrower with smaller LW than those in active infection: 8.4 ± 1.8 × 3.5 ± 0.5 µm, LW 28.9 ± 5.6 µm², and L/W 2.42 ± 0.79 (N = 25) versus 9.0 ± 1.6 × 4.1 ± 0.6 µm, LW 36.6 ± 5.7 µm², and L/W 2.27 ± 0.67 (N = 25), respectively. Chronic-phase macrogametocytes do not differ in width, but are shorter, with lower LW and L/W than in active phase: 8.5 ± 2.0 × 3.4 ± 0.8 µm, LW 28.5 ± 6.3 µm², and L/W 2.65 ± 0.94 (N = 25) versus 9.8 ± 1.1 × 3.2 ± 0.4 µm, LW 31.6 ± 4.7 µm², and L/W 3.10 ± 0.53 (N = 25), respectively.

Exoerythrocytic Merogony Phanerozoites were present in the heart and, more commonly, lungs in two chronically infected *Ptychozoon lionatum* (**Plate 41, a and b**). Meronts in the lungs were usually elongate and slender, occasionally ovoid. Dimensions averaged 11.5 ± 3.1 × 5.5 ± 1.6 µm (7–18 × 2.5–9, N = 18). Nuclei counted in one meront numbered 83.

Sporogony Unknown.

Effects on Host Gametocytes utilize erythrocytes as host cells, as do most meronts, but occasionally the latter occupy proerythrocytes (4%). Neither host cells nor their nuclei are hypertrophied, and nuclei are not distorted. Meronts rarely slightly distort host cells (4%) or displace their nuclei (4%). Erythrocytes are commonly distorted by gametocytes (30%), and their nuclei are displaced (33%).

Remarks Most parasitemias were very light, nearly undetectable. One infection, observed from June until the following February, remained active throughout that period, with the asexual:sexual parasite ratios changing only from 69:31 to 43:57 from 4 July to 12 December. A slide from *Ptychozoon lionatum* shown to me at the Institute for Medical Research in Kuala Lumpur in June 1976 demonstrated the presence of *Plasmodium lionatum* in Malaysia as well as peninsular Thailand.

Plasmodium saurocaudatum Telford 1983 (**Plate 36**)

Diagnosis A *Plasmodium* (*Asiamoeba*) species distinguished by the greatly disproportionate sizes of meronts and gametocytes in active infection, with average macrogametocyte size 16.8 times that of meronts and average microgametocyte size 14.3 times average meront size. Gametocytes are 8–19 × 5–14 µm, with LW 60–252 µm² and L/W 1.0–2.8. Size of gametocytes relative to host cell nucleus size is 1.8–14.6, and to normal erythrocyte nuclei is 2.4–10.1. Meronts, usually located in the margin of host cells, are 2–4 × 1–3 µm, with LW 3.0–9.0 µm²; they produce three to six merozoites, usually arranged in fan shape. Meront size relative to host cell nucleus size and to normal erythrocyte nuclei is 0.1–0.4. Gametocytes are sexually dimorphic in dimensions, with macrogametocytes larger in length, width, and LW than microgametocytes.

Plate 36 (**A**) *Plasmodium lionatum* from *Ptychozoon lionatum*, Thailand. Meronts **a–g**; macrogametocytes, **h–j**; microgametocytes, **k**, **l**. (Figures **e** and **f** from Telford, S. R., Jr., *J. Parasitol.*, 81, 53, 1995, with permission.) (**B**) *Plasmodium saurocaudatum* from *Mabuya multifasciata*, Singapore (**a**, **f**, **j**) and Thailand. Meronts, **a–e**; macrogametocytes from active (**f–h**) and chronic infections (**i**, **j**); microgametocytes from active (**j**, **k**) and chronic (**l**) infections.

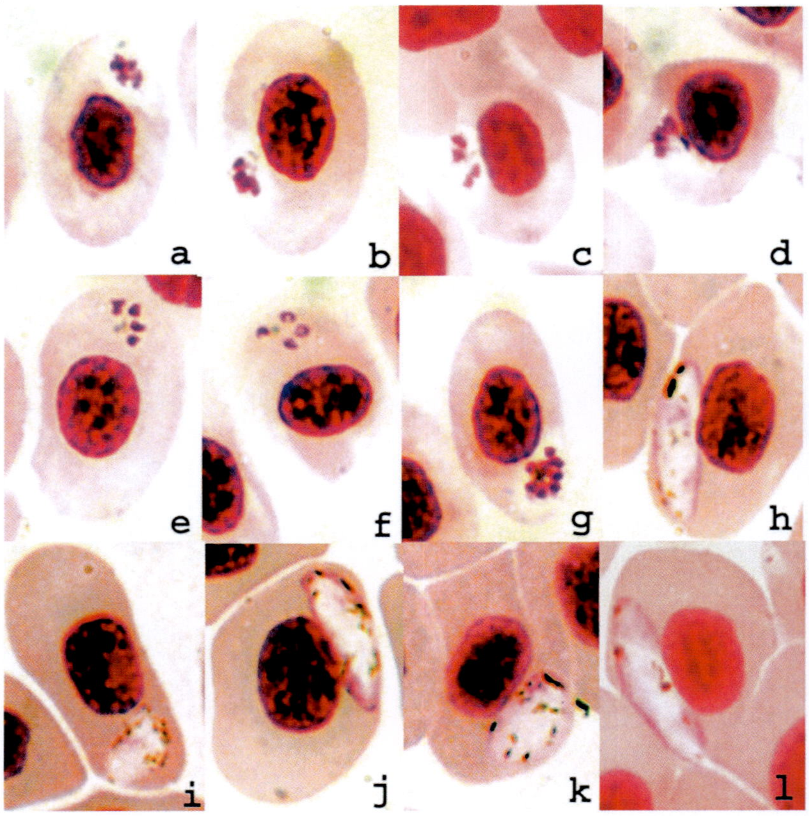

(A)

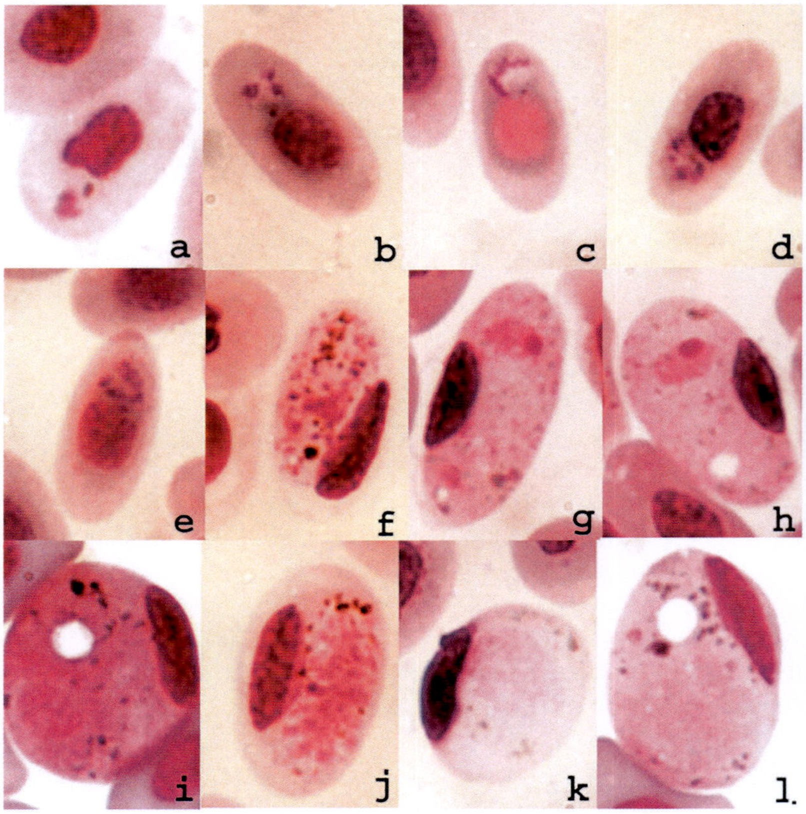

(B)

Type Host *Mabuya multifasciata* (Kuhl) (Sauria: Scincidae).

Type Locality MacRitchie Reservoir, Singapore.

Other Hosts None known.

Other Localities Southern Thailand, no precise localities.

Prevalence *P. saurocaudatum* infected 1 of 6 *M. multifasciata* from Singapore and 7 of 123 (5.7%) from Thailand. Annual prevalences in Thai samples were 0/5 in June 1976, 0/48 in December 1976, 1/13 in May 1977, 0/28 in April 1979, and 6/29 in April 1980 (Telford, 1983c).

Morphological Variation Meronts are 2.73 ± 0.5 × 2.0 ± 0.3 μm (2–4 × 1–3, N = 26), with LW 5.6 ± 1.5 μm^2 (3–9). Merozoites average 3.8 ± 0.4 (3–6, N = 27). Overall, gametocytes are 13.9 ± 2.4 × 8.9 ± 2.1 μm (8–19 × 5–14, N = 301), with LW 125.9 ± 46.0 μm^2 (60–252) and L/W 1.62 ± 0.36 (1.0–2.8). Gametocyte LW relative to host cell nucleus averages 4.90 ± 2.37 (1.8–14.6, N = 200), and to normal erythrocyte nuclei is 5.04 ± 1.70 (2.4–10.1, N = 301). Macrogametocytes, overall, are 14.2 ± 2.3 × 9.2 ± 2.2 μm^2 (8–19 × 5–14, N = 149), with LW 138.4 ± 47.2 μm^2 (60–252) and L/W 1.60 ± 0.38 (1.0–2.8), and microgametocytes are 13.5 ± 2.5 × 8.6 ± 2.0 μm (8–14 × 5–14, N = 152), with LW 118.5 ± 43.7 μm^2 (60–238) and L/W 1.63 ± 0.35 (1.0–2.7). Only the L/W ratio does not differ. Macrogametocytes from active infections are 12.7 ± 1.9 × 7.7 ± 1.1 μm (8–16 × 5–10, N = 63), LW 94.0 ± 14.5 μm^2 (60–126), and L/W 1.75 ± 0.45 (1.0–2.8) and from chronic infections are 16.3 ± 1.7 × 11.8 ± 1.1 μm (11–19 × 10–14, N = 36), LW 192.4 ± 31.6 μm^2 (110–252), and L/W 1.39 ± 0.15 (1.1–1.7); all values are significantly different. Active-phase microgametocytes are 11.6 ± 1.4 × 6.9 ± 0.9 μm (8–15 × 5–9, N = 62), with LW 79.8 ± 10.1 μm^2 (60–104) and L/W 1.71 ± 0.34 (1.0–2.5), all values significantly different from those of chronic phase, 16.1 ± 1.8 × 11.1 ± 1.0 μm (12–19 × 9–14, N = 39), LW 178.1 ± 24.6 μm^2 (140–238), and L/W 1.47 ± 0.23 (1.0–2.0). Length, width, and LW are significantly different by gametocyte sex and phase of infection. Size (LW) of gametocytes relative to those of host cell nucleus and normal erythrocyte nuclei reflects both sexual and phase differences of gametocytes; these respective ratios are 5.18 ± 2.54 (N = 100) and 5.35 ± 1.76 (N = 149) for all macrogametocytes; 4.61 ± 2.17 (N = 100) and 4.7 ± 1.59 (N = 152) for all microgametocytes; 2.94 ± 0.52 (N = 38) and 4.00 ± 0.63 (N = 63) in active-phase macrogametocytes; 7.84 ± 2.39 (N = 25) and 7.48 ± 1.33 (N = 36) in chronic-phase macrogametocytes; 2.65 ± 0.53 (N = 37) and 3.40 ± 0.43 (N = 62) in active phase microgametocytes; and 7.36 ± 1.62 (N = 25) and 6.91 ± 1.03 (N = 39) in chronic-phase microgametocytes.

Exoerythrocytic Merogony Unknown.

Sporogony Unknown.

Effects on Host Only erythrocytes have been found parasitized. Meronts are almost always situated at the margin of the host cell, usually lateral or lateropolar to the nucleus, rarely polar, and cause no discernible effect on the cell or its nucleus (Telford, 1983c) but occasionally occupy a polar, almost nucleophilic position. Host cells in either active or chronic phase of infection are hypertrophied, often grossly, to the point of near doubling of erythrocyte size. In active infections but not in chronic phase, erythrocyte nuclei are also hypertrophied. Chronic-phase gametocytes nearly fill host cells, compressing their nuclei into elongate, thin shapes (Telford, 1983c).

Remarks *Plasmodium saurocaudatum* has the largest mean values for gametocyte size among described saurian *Plasmodium* species and the smallest mean meront size. As stated previously (Telford, 1983c), "It appears that gametocyte production begins early in the course of infection, overshadowing an inconspicuous erythrocytic schizogony of short duration, a trend which might perhaps eventually result in the loss of the asexual cycle in circulating blood cells."

Plasmodium clelandi Manawadu 1972

Diagnosis A *Plasmodium* (*Asiamoeba*) species with elongate gametocytes, 18–20 × 2–4 μm, estimated LW 40–72 μm^2 and L/W 4.5–10.0. Meronts are tiny, approximately 2.4–3.2 × 2.1 μm, estimated LW 5.0–6.7 μm^2, and produce four to eight merozoites arranged as a fan. Meronts are polar in position within the host cell. Gametocytes tend to encircle the erythrocyte nucleus. Blackish-brown pigment granules are scanty in meronts and are dispersed as 6–8 granules in microgametocytes and 12–16 in macrogametocytes.

Type Host *Varanus bengalensis* (Daudin) (syn. *V. cepedianus*) (Sauria: Varanidae).

Type Locality Kal-Aru, Mannar District, Ceylon.

Other Hosts Unknown.

Other Localities Unknown.

Prevalence *P. clelandi* infected two of three *V. bengalensis* at the type locality (Manawadu, 1972).

Morphological Variation The dimensional data provided by Manawadu (1972) are meager. Average measurements of ten macrogametocytes and ten microgametocytes are 18 × 4 µm and 20 × 2 µm, respectively, which provide estimates of 72 µm^2 and 40 µm^2 in LW and 4.5 and 10.0 as L/W ratios. Gametocytes, described as elongate with irregular borders, "tended to encircle the host cell nucleus" (Manawadu, 1972). Macrogametocytes contain 12–16 pigment granules, twice as many as the 6–8 present in microgametocytes. Pigment forms as coarse, black-brown granules, dispersed in the cytoplasm. Data on meronts is even more limited. Measurements from the two mature meronts in Manawadu's figures indicate dimensions of 2.4 × 2.1 µm and 3.2 × 2.1 µm, suggesting LW of 5.0–6.7 µm^2, respectively. Mature meronts usually produce four merozoites, but occasionally twice that number, and are roughly fan-shaped, with scanty pigment.

Exoerythrocytic Merogony Unknown.

Sporogony Unknown.

Effects on Host Erythrocytes containing meronts are "unaltered in size, shape, and staining characteristics" (Manawadu, 1972). Erythrocytes host to gametocytes tend to "round off," with the nucleus "encircled or displaced laterally." Some host cells may be enlarged.

Remarks *P. clelandi* is the only *Plasmodium* species described from varanid lizards until now and has not been reported again since its description. *Plasmodium* species, however, parasitize *Varanus grayi* on Luzon (described below) and *V. prasinus* in West Irian, New Guinea (Telford).

Plasmodium draconis Telford 1995 (**Plate 37**)

Diagnosis A species of *Plasmodium* (*Lacertamoeba*) with variably shaped meronts slightly smaller than gametocytes, 5–9 × 3–6 µm, with LW 15–48 µm^2 that produce 4–18 merozoites. Gametocytes are 6–13 × 3.5–8 µm, with LW 30–88 µm^2, and L/W 1.0–3.3. Pigment is scanty in meronts, 1–4 grayish-green granules usually clustered at a single locus, but dispersed as prominent dark brown granules in gametocytes. Meront size relative to host cell nucleus is 1.15, and to normal erythrocyte nuclei, 1.11. Gametocyte size relative to host cell nucleus is 1.69, and to normal erythrocyte nuclei is 1.71.

Type Host *Draco volans* Linnaeus (Sauria: Agamidae).

Type Locality Pamauyay Falls on Pamauyay River, about 3 km from Port Barton, Palawan Island, Philippines.

Other Hosts None known.

Other Localities Gunung Mulu National Park, Sarawak, Malaysia.

Prevalence One of 2 *D. volans* was infected at each known locality in Palawan and Sarawak.

Morphological Variation Meronts in the type infection are 6.5 ± 1.1 × 5.0 ± 0.9 µm (5–9 × 3–6, N = 17), with LW 32.4 ± 7.8 µm^2 (21–48), and contain 8.4 ± 3.8 (4–18) merozoites. Meronts are highly variable in shape, oblong, oval, squarish, dumbbell-shaped, cruciform, or with nuclei arranged as a rosette or fan. Meront size relative to host cell nucleus size averages 1.15 (0.18–1.6, N = 16), and to normal erythrocyte nuclei is 1.11 (0.7–1.6, N = 17). In the small sample from Sarawak, meronts are 5.3 ± 0.4 × 3.3 ± 0.3 µm (5–6 × 3–3.5, N = 5), with LW 17.5 ± 2.2 µm^2 (15–21), and contain 7.4 ± 1.1 (6–9) merozoites. Their size relative to host cell nucleus is 0.44 (N = 1), and to normal erythrocyte nuclei is 0.67 ± 0.08 (0.6–0.8, N = 5). Pigment in both samples is represented by 1–4 greenish-gray granules in a cluster. Gametocytes in the Palawan sample are 8.0 ± 1.1 × 6.2 ± 0.9 µm (7–13 × 4–8, N = 95), with LW 49.8 ± 11.3 µm^2 (32–88), and L/W 1.31 ± 0.29 (1.0–3.3). Gametocyte size relative to host cell nucleus averages 1.69 ± 0.47 (0.8–3.5), and to normal erythrocyte nuclei is 1.71 ± 0.39 (1.1–3.0). Macrogametocytes from Palawan are smaller than microgametocytes, 7.7 ± 1.0 × 6.0 ± 0.8 µm (7–11 × 5–8, N = 43), with LW 46.0 ± 10.0 µm^2 (35–70), versus microgametocytes, 8.3 ± 1.2 × 6.4 ± 1.0 µm (7–13 × 4–8, N = 52), and LW 53.0 ± 11.5 µm^2 (32–88). The L/W ratio, however, is similar, respectively 1.30 ± 0.20 (1.0–2.2) versus 1.33 ± 0.35 (1.0–3.3). Dark brown pigment granules are prominent and dispersed throughout the cytoplasm. Gametocytes of the Sarawak sample are 9.0 ± 1.2 × 5.0 ± 0.9 µm (6–12 × 3.5–7, N = 74), with LW 45.0 ± 9.4 µm^2 (30–65), and L/W 1.85 ± 0.41 (1.0–3.1). The Sarawak gametocytes are longer and more narrow than are those from Palawan, clearly elongate in shape rather than oval as in the type infection. Gametocyte sizes relative to host cell nucleus are 1.69 ± 0.38 (1.0–2.8), and to normal erythrocyte nuclei is 1.72 ± 0.36 (1.1–2.5), and are virtually identical to the sample from Palawan. In contrast to the type sample, macrogametocytes do not differ in length or shape ratio from microgametocytes, but are wider and slightly larger in LW, respectively 9.0 ± 1.2 × 5.3 ± 1.0 µm (6–11 × 3.5–7, N = 35),

Plate 37 (**A**) *Plasmodium draconis* from *Draco volans*, Palawan (**a**, **b**, **g**, **h**, **k**) and Sarawak (**i**, **j**, **l**). Meronts, **a–e**; segmenter, **f**; macrogametocytes, **g–j**; microgametocytes, **k–l**. (Figures **e** and **f** modified from Telford, S. R., Jr., *J. Parasitol.,* 81, 53, 1995, Figures 7 and 8, with permission.) (**B**) *Plasmodium volans* from *Draco volans*, Palawan (**d–l**) and Sarawak (**a–c**). Meronts, **a–f**; macrogametocytes, **g–i**; microgametocytes, **j–l**.

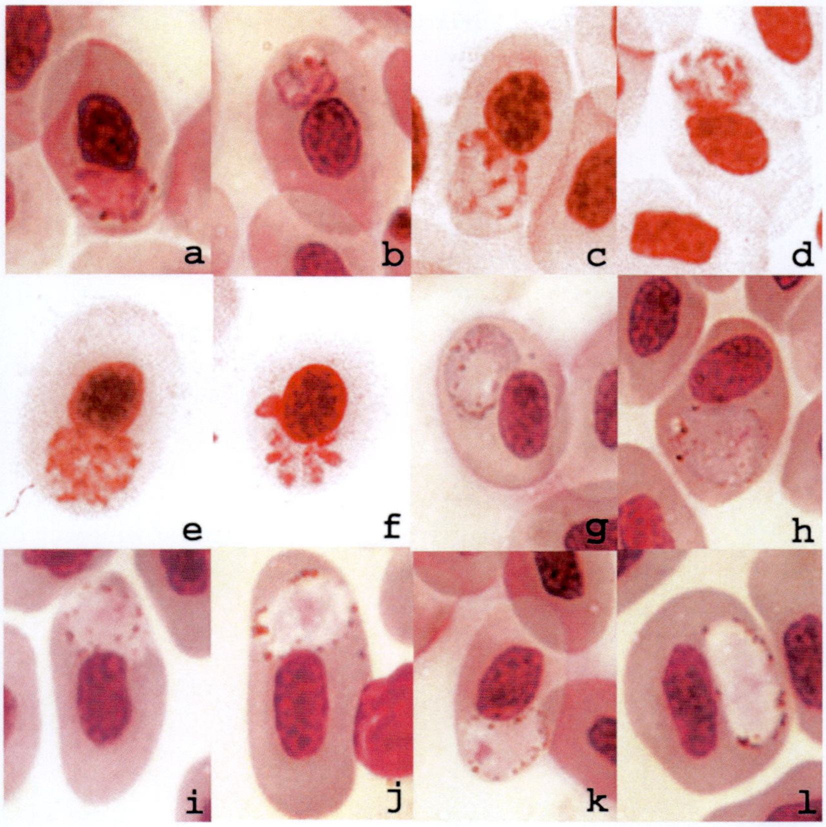

(A)

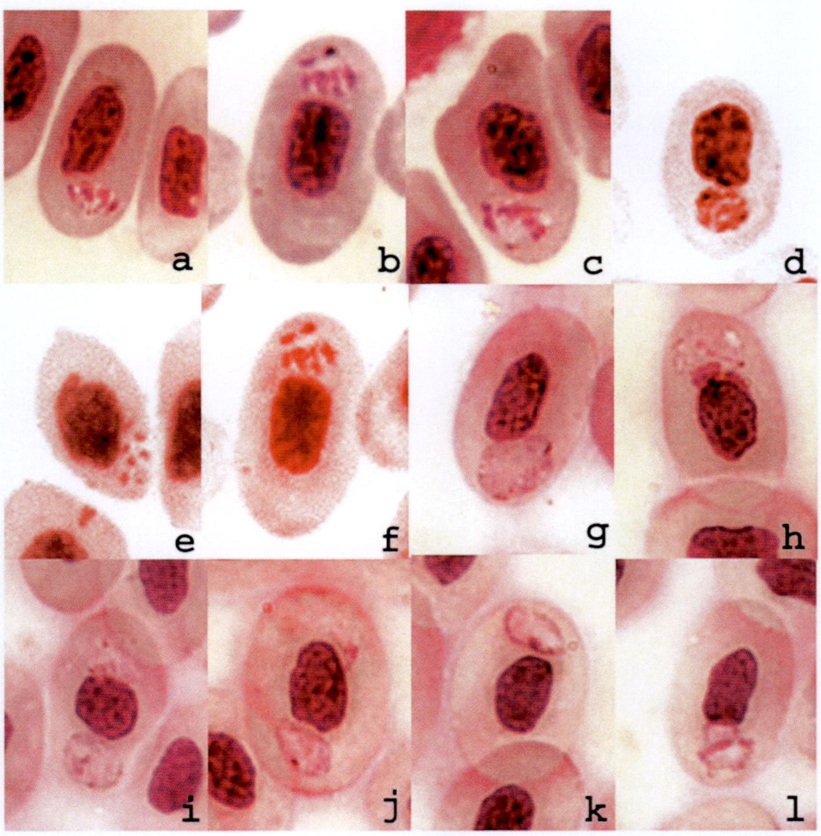

(B)

LW 47.3 ± 10.0 μm^2 (32–65), L/W 1.81 ± 0.46 (1.0–3.1) versus 8.9 ± 1.2 × 4.8 ± 0.7 μm (7–12 × 3.5–6.5, N = 39), LW 42.9 ± 8.3 μm^2 (30–57), and L/W 1.88 ± 0.36 (1.2–2.7).

Exoerythrocytic Merogony Unknown.

Sporogony Unknown.

Effects on Host In the type sample from Palawan, neither gametocytes nor meronts cause hypertrophy of the host erythrocyte or its nucleus. Nuclei are sometimes altered in shape and commonly displaced by both stages, but only gametocytes occasionally distort the host cell. Erythrocytes host to meronts in the Sarawak sample occasionally showed displaced nuclei only, but cells host to gametocytes are hypertrophied by 12% (Telford, 1995d), and often are distorted, with displaced but not distorted erythrocyte nuclei.

Remarks The infected *D. volans* from Palawan was infected also by *Plasmodium vastator*. Although only a single gametocyte was found, the characteristic nucleophilic and amoeboid asexual parasites were more common. These contained red-staining material as previously reported by Laird (1960) which occurred occasionally also in young asexual parasites thought to be *P. draconis* (Telford, 1995d).

Plasmodium vastator Laird 1960

Diagnosis A *Plasmodium* (*Asiamoeba*) species characterized by large gametocytes, reniform to elongate or bulky, 12.4–22.1 × 3.7–8.1 μm, with estimated LW 45.9–179 μm^2, and L/W 2.0–3.4. Meronts are nucleophilic, elongate and amoeboid when immature, becoming fan-shaped at segmentation, estimated to be 5.3 × 3.2 μm with LW 17.0 μm^2, that produce 4–8 merozoites. Macrogametocyte dimensions exceed those of microgametocytes. Pigment forms as discrete golden-brown granules.

Type Host *Draco volans* Gray (Sauria: Agamidae).

Type Locality Ulu Langat, Selangor, Malaysia.

Other Hosts None known.

Other Localities Pamauyay Falls on Pamauyay River, about 3 km from Port Barton, Palawan Island, Philippines (Telford, 1995d).

Prevalence At the type locality, 1 of 1 *D. volans* was infected by *P. vastator* (Laird, 1960); 1 of 2 *D. volans* from Palawan was infected (Telford, 1995d).

Morphological Variation Scanty information on dimensions of *P. vastator* was provided by Laird (1960). Meronts are amoeboid and nucleophilic as they develop, becoming fan-shaped at segmentation, with dimensions, estimated from Laird's Figure 12, approximately 5.3 × 3.2 μm, LW 17.0 μm^2. Laird found 12 segmenting parasites to produce 4–8 merozoites, with an average of his merozoite count, 6.2. Meronts contained up to 15 discrete golden-brown granules. Some trophozoites are highly amoeboid with long to very long filopodia. Microgametocytes were 12.6–15.5 × 5.1–7.6 μm, which indicates LW values 64–118 μm^2 and L/W ratios 2.0–2.5. The more plentiful macrogametocytes averaged 15.3 × 5.6 μm (12.4–22.1 × 3.7–8.1), with estimated LW 46–179 μm^2 and L/W 2.1–3.4. Macrogametocytes show a definite pale staining area on the side adjacent to the erythrocyte nucleus. The dispersed golden-brown pigment granules average 27 (9–43). Gametocytes are reniform to elongate "quadrangular" (bulky), with the latter nearly filling the host cell.

Exoerythrocytic Merogony Unknown.

Sporogony Unknown.

Effects on Host Erythrocytes infected by asexual stages may be rounded in shape, and nuclear displacement and rounding is evident in Laird's figures. Gametocytes cause hypertrophy of their host cells, which are enlarged by approximately one-third over uninfected erythrocyte size. Host cell nuclei are "swollen and strongly displaced" (Laird, 1960), often "irregular outline and prone to destruction" (i.e., more fragile). Both infected cells and their nuclei show paler staining reactions than uninfected erythrocytes.

Remarks A single gametocyte of *P. vastator*, 15 × 10 μm, was seen on slides from one *D. volans* taken on Palawan Island, in mixed infection with two other *Plasmodium* species (Telford, 1995d).

Plasmodium volans Telford 1995 (**Plate 37**)

Diagnosis A *Plasmodium* (*Carinamoeba*) species with meronts smaller than gametocytes, 3.5–6.0 × 2.5–5.0 μm, LW 11.2–25.0 μm^2, that produce 4–6 merozoites arranged cruciform or fan-shaped at segmentation. Meront size relative to host cell nucleus is 0.69, and to normal erythrocyte nuclei is 0.66. Pigment granules are tiny and dark. Gametocytes 4.5–9.0 × 2.5–5.5 μm, with LW 16–33 μm^2 and L/W 1.0–3.6, usually polar to the host cell nucleus. Pigment is dark and dispersed as dust-like granules. Gametocyte size relative to host cell nucleus is 0.88, and to normal erythrocyte nuclei is 0.89.

Type Host *Draco volans* Linnaeus (Sauria: Agamidae).

Type Locality Pamauyay Falls on Pamauyay River, about 3 km from Port Barton, Palawan Island, Philippines.

Other Hosts None known.

Other Localities Gunung Mulu National Park, Sarawak, Malaysia.

Prevalence One of 2 *D. volans* was infected at each known locality in Palawan and Sarawak.

Morphological Variation Meronts in the type infection from Palawan are 4.8 ± 0.5 × 3.9 ± 0.6 μm (4–6 × 3–5, N = 23), with LW 19.0 ± 3.2 μm^2 (13.5–25.0), and contain 4.8 ± 0.9 (4–6) merozoites at segmentation in cruciform or fan-shape arrangement. Their size relative to host cell nucleus is 0.69 ± 0.13 (0.5–1.0, N = 21), and to normal erythrocytic nuclei, 0.66 ± 0.11 (0.5–0.9, N = 23). The dark pigment granules are tiny and central to the nuclei. The few meronts available from Sarawak are 4.0 ± 0.4 × 3.1 ± 0.5 μm (3.5–4.5 × 2.5–3.5, N = 4), with LW 12.4 ± 1.2 μm^2 (11.2–14.0), and contain 4.5 ± 1.0 (4–6) merozoites. Their size relative to host cell nucleus averages 0.49 ± 0.06 (0.4–0.5, N = 3), and to normal erythrocyte nuclei is 0.47 ± 0.04 (0.4–0.5, N = 4). Gametocytes from Palawan are 5.7 ± 0.5 × 4.4 ± 0.5 μm (4.5–7.0 × 3.0–5.5, N = 96), with LW 25.2 ± 4.2 μm^2 (16–33), and L/W 1.29 ± 0.18 (1.0–2.0). The dark, dispersed pigment granules are dust-like in appearance. Relative to host cell nucleus size, gametocytes are 0.84 ± 0.17 (0.5–1.3, N = 94), and to normal erythrocyte nuclei are 0.87 ± 0.14 (0.6–1.1). Macrogametocytes are slightly wider than microgametocytes, with greater LW, 5.7 ± 0.5 × 4.5 ± 0.5 μm (4.5–6.5 × 4.0–5.5, N = 48), LW 26.0 ± 4.0 μm^2 (18–33), and L/W 1.27 ± 0.15 (1.0–1.5) versus 5.6 ± 0.5 × 4.3 ± 0.6 μm (5–7 × 3.5–5.5, N = 48), LW 24.4 ± 4.3 μm^2 (16–30), and L/W 1.31 ± 0.20 (1.0–2.0), but are similar in length and L/W. Gametocytes from Sarawak are more elongate and narrower than in the type infection, but LW values are the same, 7.1 ± 0.8 × 3.6 ± 0.6 μm (5.5–9.0, N = 37), LW 25.2 ± 3.5 μm^2 (19–32), and L/W 2.02 ± 0.52 (1.2–3.6). Their size relative to host cell nucleus size is 0.93 ± 0.19 (0.7–1.3), and to normal erythrocyte nuclei 0.96 ± 0.13 (0.7–1.2). Macrogametocytes are slightly shorter than microgametocytes, but other dimensions are similar, respectively 6.7 ± 0.6 × 3.6 ± 0.6 μm (5.5–8.0, N = 17), LW 24.4 ± 4.1 μm^2 (19–32), and L/W 1.90 ± 0.39 (1.3–2.3), versus 7.3 ± 0.9 × 3.6 ± 0.6 μm (6–9 × 2.5–5, N = 20), LW 25.9 ± 2.8 μm^2 (21.0–31.5), and L/W 2.13 ± 0.61 (1.2–3.6).

Exoerythrocytic Merogony Unknown.

Sporogony Unknown.

Effects on Host Neither erythrocytes nor their nuclei are hypertrophied when host to meronts, occasional nuclei may be slightly altered in shape and they are commonly displaced, but most host cells are not distorted. Gametocytes produce 7–12% hypertrophy in host erythrocyte size, with nuclei enlarged to the same extent in the Sarawak sample, but not in those from Palawan. Nuclei are often displaced in both samples.

Remarks Infection phase probably accounts for differences in the parasites and their effects upon host cells. The mixed infections of both *P. volans* and *P. draconis* in *D. volans* from Sarawak were chronic, while both species were in active phase in the Palawan host.

Plasmodium "minasense" Laird 1960

Diagnosis A *Plasmodium* (*Carinamoeba*) species with meronts 2.4–3.7 × 1.2–3.2 μm and estimated LW 7.4 μm^2 that produce 3–4 merozoites arranged cruciform or as a fan. Pigment is brownish in an apical or central small mass. Gametocytes are 6.1–8.8 × 4.7–7.3 μm, estimated LW 28.7–64.2 μm^2, and L/W 1.21–1.30. Gametocytes do not differ sexually in size, but macrogametocytes are more rounded than the more commonly oval or reniform microgametocytes. Brownish-black pigment granules are scanty and dispersed in both gametocyte sexes.

Type Host *Goniocephalus borneensis* (Schlegel) (Sauria: Agamidae).

Type Locality Kepong, Selangor, Malaysia.

Other Hosts None known.

Other Localities Bukit Lagong, Selangor, Malaysia.

Prevalence Unknown.

Morphological Variation Meronts average 3.1 × 2.4 μm (2.4–3.7 × 1.2–3.2) with estimated average LW 7.4 μm^2. Merozoites average 3.98 (3–4), in oval, cruciform, or fan-shaped arrangement. Pigment is brownish, in a small mass apical in fans or central when meronts are cruciform. Gametocytes are 6.1–8.8 × 4.7–7.3 μm. Microgametocytes are more commonly oval or reniform, 8.0 × 5.8 μm (7.1–8.8 × 4.7–6.9) with estimated LW 46.4 μm^2 and L/W 1.38. Macrogametocytes are more rounded, 7.4 × 6.4 μm (6.1–8.7 × 5.4–7.3) with estimated LW 47.4 μm^2 and L/W 1.16. Pigment is brownish-black, dispersed as 1–11 dust-like granules in microgametocytes and 1–8 slightly coarser granules in macrogametocytes.

Exoerythrocytic Merogony Unknown.

Sporogony Unknown.

Effects on Host Neither infected erythrocyte nor its nucleus is altered by meront presence. Erythrocytes infected by gametocytes are "often slightly hypertrophied and distorted, commonly somewhat rounded; their nuclei displaced, a little enlarged, of paler staining reaction than normal" (Laird, 1960).

Remarks It is unlikely that the parasite described by Laird (1960) from Malaysian agamid lizards is conspecific with the *Plasmodium minasense* complex in Neotropical scincids, teiids, and polychrotids. Without material at hand, I am reluctant to provide a specific name for it, but this would be justified by geographical distribution alone. Genome comparison with *P. minasense* ssp. would be interesting.

Plasmodium auffenbergi sp. nov. (**Plate 38**)

Diagnosis A *Plasmodium* (*Carinamoeba*) species with tiny variably shaped meronts and very small, sometimes nucleophilic, spherical-to-ovoid gametocytes. Meronts are 2–5 × 1.5–3 µm, with LW 4–12 µm², and produce two to five merozoites. Meront size relative to host cell nucleus averages 0.28, and to normal erythrocyte nuclei is 0.27. Gametocytes are 3–7 × 2.5–6 µm, with LW 7.5–42 µm² and L/W 1.00–1.50. Gametocyte size relative to host cell nucleus is 0.78, and to normal erythrocyte nuclei is 0.80. There is no sexual difference in gametocyte dimensions or shape. Microgametocytes contain a large central vacuole. All stages show a single dot of dark brown pigment.

Type Host *Varanus grayi* Boulenger (Sauria: Varanidae).

Type Locality Caputatan, Caramoan Municipio, Caramoan Peninsula, Luzon, Philippines.

Other Hosts None known.

Other Localities Known only from sites within Caramoan Municipio, Caramoan Peninsula, Luzon.

Prevalence All 15 *V. grayi* collected in 1982 from Caramoan Municipio were infected by *P. auffenbergi*. An additional seven of seven *V. grayi* in zoological park collections were also positive for this parasite.

Morphological Variation Most meronts are cruciform (47%) or fan-shaped (40%), seldom elongate, oblong, or triangular. Meronts are 3.0 ± 0.6 × 2.2 ± 0.3 µm (2–5 × 1.5–3.0, N = 53), with LW 6.5 ± 1.7 µm² (4–12), and contain 3.7 ± 0.8 (2–5, N = 56) merozoites. Merozoites are comprised mostly of nucleus with little cytoplasm discernible. Meront size relative to host cell nucleus is 0.30 ± 0.10 (0.17–0.60, N = 29), and to normal erythrocyte nuclei is 0.28 ± 0.07 (0.18–0.53, N = 53). The spherical-to-ovoid gametocytes are 4.7 ± 1.3 × 3.9 ± 1.1 µm (3–7 × 2.5–6, N = 40), with LW 19.6 ± 10.4 µm² (7.5–42) and L/W 1.20 ± 0.15 (1.00–1.50). Gametocyte size relative to host cell nucleus is 0.78 ± 0.37 (0.31–1.50, N = 39), and to normal erythrocyte nuclei is 0.80 ± 0.42 (0.32–1.69, N = 40). Gametocytes do not differ by sex in dimensions or shape. A large vacuole is usually centrally located in microgametocytes. Nucleophilic gametocytes are commonly seen. Both meronts and gametocytes are most often polar in position, meronts are sometimes lateral to the host cell nucleus, but gametocytes are rarely so. All stages contain a single dot of dark brown pigment, verified under polarized light.

Exoerythrocytic Merogony Unknown.

Sporogony Unknown.

Effects on Host Both erythrocytes and proerythrocytes are parasitized by asexual stages; gametocytes have been seen only in mature cells. Asexual stages multiply infect host cells, with usually three parasites per cell and in massive parasitemias. A single infection, typical of all seen, had trophozoites or meronts (usually) in 99 of 102 erythrocytes, with a total of 55 trophozoites and 99 meronts, indicating a parasitemia of 15,400 parasites in 10^4 erythroid cells. Cells parasitized by either meronts or gametocytes do not differ in appearance of cells or nuclei in comparison to the rare uninfected cells, but host cell average sizes (LW) for both stages show 9–11% hypotrophy, with no change in nucleus dimensions.

Remarks When first seen, the erythrocytes appeared to contain minute inclusions of chromatin similar to *Sauroplasma*. But when placed under polarized light, a single dot of refractive pigment appeared in both asexual and sexual stages. The infections are always dominated by asexual stages, and gametocytes are rare and difficult to find, similar to some other *Carinamoeba* species. Gametocyte size, while tiny, 3–7 × 2.5–6 µm with LW 7.5–42 µm², is comparable to that of the neotropical species *Plasmodium minasense tegui,* which has gametocytes 3–5 × 2–4 µm, with LW of 6–20 µm². The species name recognizes the contributions to our knowledge of monitor lizard ecology by the collector, the late Walter Auffenberg, my close friend for a half-century. Hapantotype blood films are deposited in the USNPC, Beltsville, Maryland, nos. 100324–100327.

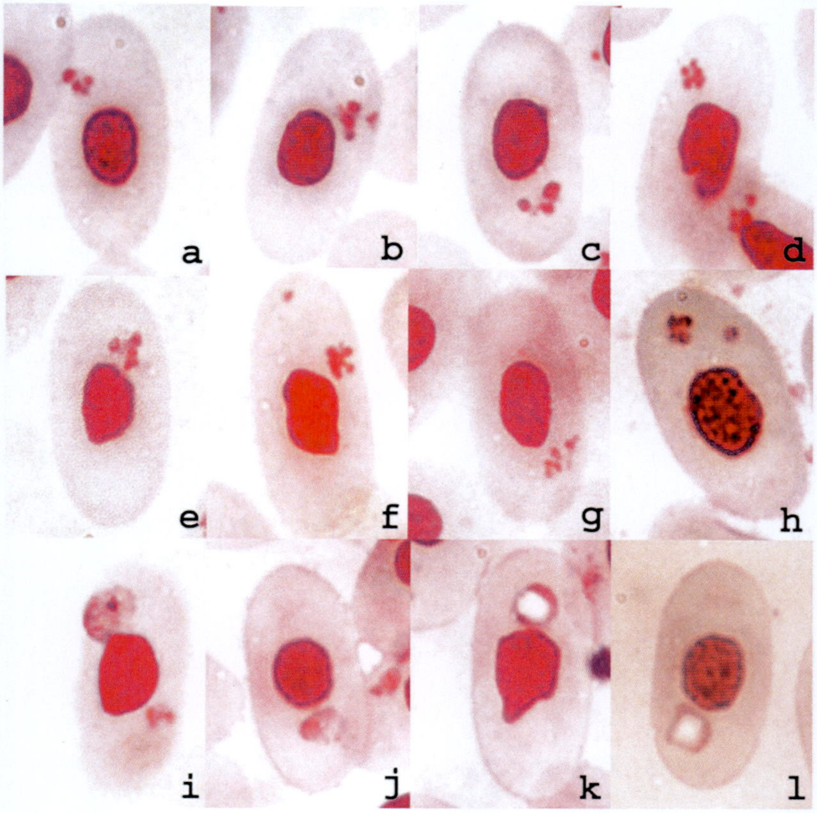

(A)

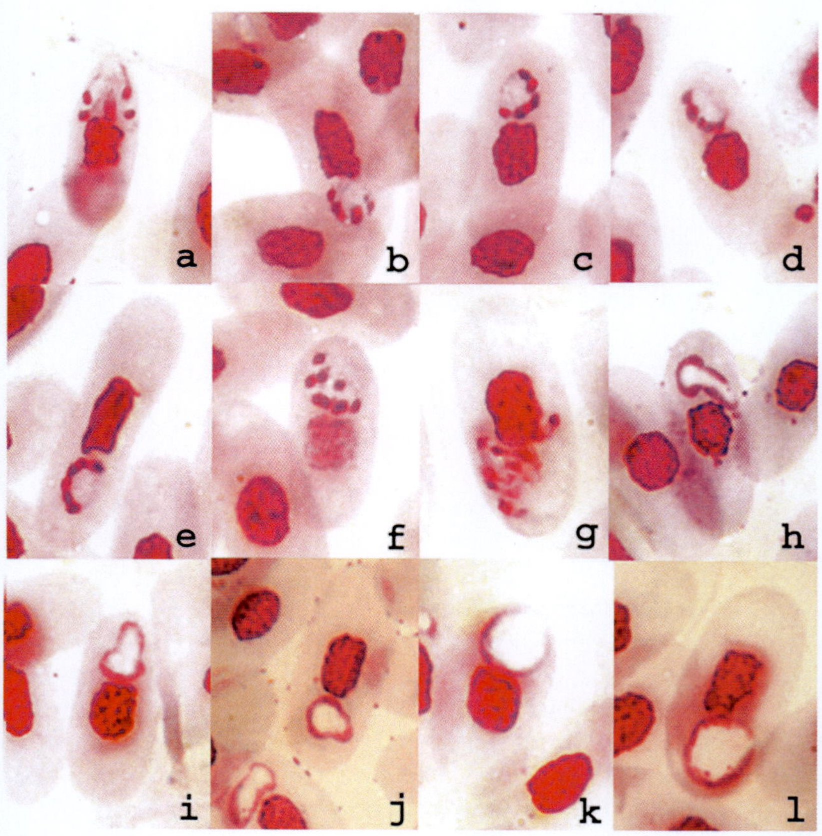

(B)

Plate 38 (**A**) *Plasmodium auffenbergi* sp. nov. from *Varanus grayi*, Philippines. Meronts, **a–h**; macrogametocytes, **i**, **j**; microgametocytes, **k–l**. (**B**) *Plasmodium rhacodactyli* sp. nov. from *Rhacodactylus leachianus*, New Caledonia. Meronts, **a–g**; gametocytes, sex not evident, **h–l**.

PLASMODIUM SPECIES OF AUSTRALASIAN LIZARDS

Plasmodium australis Garnham 1966

Diagnosis A *Plasmodium* (*Sauramoeba*) species with elongate, nearly club-shaped, to ovoid meronts approximately 11 × 9–14 to 14 × 7 μm, with LW about 99 μm^2, that contain 22–49 merozoites. Dark pigment granules form a prominent cluster. Gametocytes are approximately 10–13.5 × 4.7–9.3 μm, with estimated LW 47–112 μm^2 and L/W 1.3–2.4. Pigment is variable, dispersed as scanty fine granules to abundant round black granules.

Type Host *Amphibolurus barbatus* (Cuvier) (Sauria: Agamidae).

Type Locality Wacul, Deception Bay, Eidsvold, South Queensland, Australia.

Other Hosts None known.

Other Localities None known.

Prevalence *P. australis* was present in 3 of 72 (4.2%) *A. barbatus* from various localities in South Queensland (Mackerras, 1961a).

Morphological Variation Neither Mackerras (1961a) nor Garnham (1966) reported dimensional data for this species, although Mackerras mentioned two mature meronts with 22 and 49 merozoites. Measurements of parasites in photographs by Mackerras (in 1961a, Plate 10, Figures 1–6) provide estimates of 14 × 7 μm and 11 × 9 μm for meronts, indicating LW values of 98 μm^2 and 99 μm^2, respectively. Macrogametocytes illustrated were 13.3 × 5.3 μm, LW 70.5 μm^2, and 12 × 9.3 μm, LW 111.6 μm^2, and the single microgametocyte shown was 10 × 4.7 μm, LW 47 μm^2, with corresponding L/W ratios 1.29 and 2.41 for macrogametocytes and 2.13 for the microgametocyte. Mackerras stated that nuclei of gametocytes "were sometimes dispersed in several masses," which is not especially evident in the photographs, although Garnham's (1966) figures, drawn from the photographs, show a divided nucleus in a macrogametocyte.

Exoerythrocytic Merogony No EE meronts were found in sections of liver, spleen, and lung of one infected lizard (Mackerras, 1961a).

Sporogony Unknown.

Effects on Host *P. australis* parasitized mature erythrocytes, producing hypertrophied cells 14.5–18 × 8–11 μm when host to gametocytes and 15–20 × 7–12 μm when parasitized by meronts, in comparison to normal erythrocytes that measured 11.5–17 × 7–9.5 μm. Nuclei were displaced but not distorted in parasitized cells.

Remarks Mackerras (1961a) tentatively assigned this parasite to *Plasmodium giganteum*, stating that it agreed well with Theiler's (1930) description of this species, but distinguished it from *P. giganteum* as a parasite of mature erythrocytes, in contrast to Bray's (1959) contention that host cells were "never mature erythrocytes." Certainly *P. giganteum* of East African *Agama* spp. does parasitize mature cells preferentially, while those from Central and West Africa commonly utilize proerythrocytes. Mackerras thought host cell type to be inadequate for species distinction. Ayala (1977, 1978) and Telford (1984b) followed Garnham's trinomial for this parasite, but later Telford (1988a, 1994) recognized the Australian parasite as a distinct species, *P. australis*.

Plasmodium egerniae Mackerras 1961

Diagnosis A *Plasmodium* (*Sauramoeba*) species with meronts that fill the host erythrocytes, about 14 × 10.7 μm, with estimated LW of 150 μm^2, which produce at least 40–50 merozoites. Gametocytes also fill the host cell, flattening the nucleus against the cell membrane. Gametocytes are approximately 14–18 × 8–10.7 μm, with estimated LW 122–157 μm^2 and L/W ratio 1.3–2.1. Pigment granules appear to be dispersed in macrogametocytes and perhaps form a polar cluster in microgametocytes.

Type Host *Egernia m. major* (Gray) (Sauria: Scincidae).

Type Locality Innisfail, Northern Queensland, Australia.

Other Hosts None known.

Other Localities None known.

Prevalence Infections of *P. egerniae* were found in three of an unspecified number of *E. major* collected in rain forest.

Morphological Variation Mackerras (1961a) did not provide dimensional data for meronts and gametocytes of *P. egerniae*, but stated that meronts, in which segmentation into merozoites was still incomplete, contained "between 40 and 50 discrete masses of chromatin." The nearly mature meront of Mackerras's (1961a) Plate 10, Figure 9, is approximately 14 × 10.7 μm, which indicates an estimated LW of about 150 μm^2. Gametocytes in Figures 10–12 of Plate 10 are approximately 18 × 8.7 μm and 14 × 10.7 μm (macrogametocytes) and 15.3 × 8 µm (microgametocyte), indicating LW values of 156.6 µm², 149.8 µm², and 122.4 µm² and L/W ratios of 2.07, 1.31, and 1.91, respectively. Their shape can be described as elongate to bulky. Macrogametocytes have dispersed pigment and may show vacuoles, and pigment of microgametocytes was described by Mackerras (1961a) as polar, appearing in her figure as a cluster of granules.

Exoerythrocytic Merogony Unknown.

Sporogony Unknown.

Effects on Host Both meronts and gametocytes displace erythrocyte nuclei to the cell margin, where, in the case of gametocytes, "nuclei were drawn out into band-like or wedge-shaped objects … up to 20 in length by 1–3 in width" (Mackerras, 1961a). Uninfected erythrocytes were 14.5–17 × 7–9 µm, and those containing gametocytes were 19–22 × 9–12 µm.

Remarks This *Plasmodium* species may be restricted in distribution to the rain forest of Queensland. It was not found in hundreds of slides from *Egernia stokesii* in South Australia examined by J. Stein in a 3-year field study of this species.

Plasmodium lygosomae Laird 1951

Diagnosis A *Plasmodium* (*Asiamoeba*) species characterized by very small meronts 3.2–3.7 × 2.3–3.1 µm that usually contain four merozoites arranged in fan shape. Macrogametocytes may be round, 6.0–8.5 µm in diameter, but are most commonly reniform in shape, 11.0–16.5 × 4.0–6.7 µm, with estimated LW 44–111 µm² and L/W 1.0–2.8. Microgametocytes are always elongated, 8.1–16.7 × 3.3–4.8 µm, with estimated LW 27–55 µm² and L/W 2.5–3.5. Pigment in asexual stages consists of a few brownish granules and in gametocytes forms up to 30–35 dispersed black granules.

Type Host *Oligosoma nigriplantare* (Peters) (Sauria: Scincidae) [Laird, 1951, identified the type host as *Lygosoma moco* (Gray), but in the opinion of Allison and Desser, 1982, following Hardy, 1977, this was probably *Leiolopisma* (= *Oligosoma*) *nigriplantare*].

Type Locality Two miles south of Makara Stream, Wellington, New Zealand.

Other Hosts None known.

Other Localities None known.

Prevalence Two of nine *O. nigriplantare* collected at the type locality were infected by *P. lygosomae* (Laird, 1951).

Morphological Variation Laird (1951) described in detail the trophozoites of *P. lygosomae* but provided little information about erythrocytic meronts, commenting that they were "rarely encountered." Dimensions "averaging" 3.2–3.7 × 2.1–3.1 µm indicate an LW of 7.4–11.5 µm². Meronts were described as "usually fan-shaped … with several (usually four) chromatin masses at the broader extremity." Macrogametocytes were "most commonly reniform in shape, although circular … forms also occur." Circular forms were 6.0–8.5 µm in diameter, providing estimated LW of 36–72 µm², and reniform gametocytes measured 11.0–16.7 × 4.0–6.7 µm, with estimated LW 44–109 µm² and L/W 2.4–2.8. Up to 35 dispersed black pigment granules were seen in macrogametocytes, with the "heavier ones often being massed at the polar regions." All microgametocytes were elongate, 8.1–16.7 × 3.3–4.8 µm, with estimated LW 27–55 µm² and L/W 2.5–3.5. They contained up to 30 pigment granules.

Exoerythrocytic Merogony Laird (1951) found EE meronts within "hypertrophied and distorted thrombocytes" and one in a lymphocyte. EE meronts free from host cells in the circulating blood stained deep blue and contained 4–20 nuclei. They were 4.2–10.0 × 3.3–5.4 µm in size and unpigmented.

Sporogony Unknown.

Effects on Host Meronts do not cause hypertrophy of the host cell or displace its nucleus. Rounded macrogametocytes are polar within erythrocytes, and reniform gametocytes lie "alongside" the host cell nucleus, displacing it laterally. Erythrocytes host to macrogametocytes are often distorted and somewhat hypertrophied. The elongate microgametocytes also lie alongside the host cell nucleus, slightly displacing it laterally, and cause only slight hypertrophy.

Remarks Laird's clear description of EE meronts demonstrated the presence of secondary exoerythrocytic merogony of *P. lygosomae* in circulating leukocytes and thrombocytes.

Plasmodium nucleoversans Garnham 1966

Diagnosis A *Plasmodium* (*Asiamoeba*) species with elongate gametocytes, 13.4–20.6 × 5.4–6.9 µm in estimated dimensions, with LW 99–142 µm² and L/W 3.0–3.5, situated lateral to the erythrocyte nucleus and curving slightly around it. Asexual stages are nucleophilic, with estimated dimensions of meronts 3.7–5.1 × 2.9–4.0 μm, LW 11–16 μm^2. They produce four to six merozoites, arranged as a fan or rounded form with nuclei at the periphery. Pigment granules are black, often inconspicuous in meronts, prominent and dispersed in gametocytes.

Type Host *Emoia cyanura* (Lesson) (Sauria: Scincidae).

Type Locality Guadalcanal, Solomon Islands.

Other Hosts *Prasinohaema virens anolis*, "*Lygosoma guppyi*."

Other Localities Florida Island, Solomons.

Prevalence Sample size not stated. Found in 2 *E. cyanura* on Guadalcanal, and in 2 *P. virens anolis*, 2 "*L. guppyi*", and 1 *E. cyanura* of Florida Island (Garnham, 1966).

Morphological Variation Dimensions given here were calculated from Garnham's (1966) figures and his statement, "Final division results in a spherical body, 4 μ in diameter and with the four nuclei lying in the periphery." This corresponds to his Plate LXV, Figure 25, which serves as a scale for calculations. If all figures are to the same scale, and erythrocyte sizes support this, then his comment that the macrogametocyte "is about 10–12 μm in length" is in error because calculations suggest the four gametocytes illustrated are 17.4–20.6 × 5.4–6.9 μm, with LW 99–142 μm^2 and L/W 3.0–3.5. Similarly, the three meronts shown are 3.7–5.1 × 2.9–4.0 μm, with LW 11–16 μm^2. Merozoite number is given as "four (or rarely six)" (Garnham, 1966). Asexual stages are nucleophilic. Pigment occurs as black granules, inconspicuous in asexual stages, and prominently dispersed in gametocytes.

Exoerythrocytic Merogony Unknown.

Sporogony Unknown.

Effects on Host There are probably no effects on host by asexual stages. Garnham (1966) stated, "Neither type of gametocyte causes any enlargement of the host cell."

Remarks This species description is typical of most written before 1969 in the absence of dimensional data adequate for comparisons with other species. The catalogue of Garnham's collection (Garnham and Duggan, 1986) lists three slides (nos. 934–936), which may represent the material collected by R. E. Kuntz on Florida Island in 1944. The slides should be reexamined. This *Plasmodium* is considered here as *P. nucleoversans* rather than *P. lygosomae nucleoversans* as Garnham named it on the grounds of nucleophilic asexual stages, presence of elongate gametocytes only, and the great oceanic distance between the Solomons and New Zealand. The identification of one of the lizard species found infected on Florida Island is dubious: The only lizard occurring there with the specific name "*guppyi*" is a gecko, *Lepidodactylus guppyi*, and it is doubtful that skinks and geckoes share the same *Plasmodium* species.

Plasmodium rhacodactyli sp. nov. (**Plate 38B**)

Diagnosis A *Plasmodium* (*Lacertamoeba*) species with meronts 4–7 × 3.5–6 μm, LW 14–42 μm^2, that contain four to ten merozoites, usually arranged as a fan. Meront size relative to that of the host cell nucleus is 0.98, and to the size of normal erythrocyte nuclei is 0.90. The variably shaped gametocytes are 5–8.5 × 4–6.5 μm, with LW 22.5–52 μm^2 and L/W 1.08–1.78. Gametocyte size relative to that of the host cell nucleus is 1.31, and to the size of normal erythrocyte nuclei is 1.26. Pigment is scanty and dispersed in both meronts and gametocytes. Gametocytes are not sexually dimorphic in dimensions.

Type Host *Rhacodactylus leachianus* (Linnaeus) (Sauria: Gekkonidae).

Type Locality New Caledonia Island.

Other Hosts None known.

Other Localities None known.

Prevalence Unknown.

Morphological Variation Meronts are usually fan-shaped, 5.7 ± 0.9 × 4.7 ± 0.8 μm (4–7 × 3.5–6, N = 25), with LW 26.4 ± 7.6 μm^2 (14–42), and contain 6.5 ± 1.9 (4–10) merozoites. Their size relative to that of the host cell nucleus is 0.98 ± 0.26 (0.44–1.48), and to the size of normal erythrocyte nuclei is 0.90 ± 0.25 (0.47–1.40). Gametocytes are variably shaped, lentiform, nearly spherical to elongate, or as a fat, irregular comma. Gametocytes are 7.3 ± 0.8 × 5.1 ± 0.7 μm (5–8.5 × 4–6.5, N = 20), with LW 37.6 ± 7.4 μm^2 (22.5–52) and L/W 1.45 ± 0.22 (1.08–1.78). Gametocyte size relative to that of the host cell nucleus is 1.31 ± 0.31 (0.64–1.82), and to the size of normal erythrocyte nuclei is 1.26 ± 0.25 (0.75–1.74). There is no sexual dimorphism in gametocyte dimensions. Pigment is sparse, difficult to discern, and dispersed as tiny granules in both sexual and asexual stages. Verification of its identity could not be established under polarized light.

Exoerythrocytic Merogony Unknown.

Sporogony Unknown.

Effects on Host Neither meronts nor gametocytes have marked effect on host cells, except that nuclei of erythrocytes parasitized by meronts are shorter than those of uninfected cells, perhaps because of pressure from meronts in polar position.

Remarks *Plasmodium rhacodactyli* was reported from *Rhacodactylus leachianus* but not named by Dollahon et al. (1996), following my examination of the slides for confirmation of generic identity. Some meronts superficially resemble *Dactylosoma* species of anurans. When additional material is obtained, it is desirable that the apparent pigment be verified under polarized light. Hapantotype blood film is deposited in the USNPC, Beltsville, Maryland, no. 100337.

Plasmodium mackerrasae Telford 1979, Telford and Stein 2000 (**Plates 39 and 41**)

Diagnosis A *Plasmodium* (*Lacertamoeba*) species with meronts most commonly oblong, elongate, or oval, 3–7 × 2.5–6.0 μm, with LW 7.5–42 μm², that produce 4–14 merozoites. Meront size relative to host cell nucleus is 0.75, and to normal erythrocyte nuclei is 0.75. Pigment is scanty, present as a few dark dots. Immature gametocytes, elongate with terminal nucleus, may occur in multiple infections of six or more parasites. Gametocytes, usually rounded, are 3.5–9.0 × 3.0–6.0 μm, with LW 10.5–42.0 μm² and L/W 1.0–2.3. Gametocyte size relative to host cell nucleus is 0.98, and to normal erythrocyte nuclei is 1.03. Pigment occurs as one to several small dark dispersed granules.

Type Host *Egernia cunninghami* (Gray) (Sauria: Scincidae).

Type Locality Southern Queensland, Australia, no precise locality.

Other Hosts *Egernia stokesii, E. striolata* and *E. whitei* (experimental host).

Other Localities South Australia, at any location where the host species occur between 31°42′S and 32°15′S.

Prevalence Variable, 5–54% among localities (Telford and Stein, 2000).

Morphological Variation As redescribed from *E. stokesii* (Telford and Stein, 2000), meronts are 5.1 ± 1.1 × 3.7 ± 1.1 μm (3–7 × 2.5–6.0, N = 25), with LW 19.8 ± 9.3 μm² (7.5–42). Merozoites number 4–14 (7.2 ± 2.8, N = 25), arranged variably, most commonly in oblong or elongate meronts, rarely as rosettes, fans, or in cruciform or oval configurations. The scanty dark dots of pigment may be difficult to discern and their presence dependent on host cell maturity. Meront size relative to host cell nucleus is 0.75 ± 0.35 (0.29–1.75), and to normal erythrocyte nuclei is 0.75 ± 0.36 (0.28–1.62). Immature gametocytes are 5–7 × 1.5–2.5 μm, elongate, with a terminal nucleus and a small cluster of pigment granules, often producing multiple infections in host cells of six or more parasites. Gametocytes usually are rounded, 5.8 ± 0.9 × 4.6 ± 0.7 μm (3.5–9.0 × 3.0–6.0, N = 75), with LW 26.7 ± 6.6 μm² (10.5–42.0) and L/W 1.29 ± 0.25 (1.0–2.3). Chronic-phase gametocytes are 6.1 ± 0.9 × 4.6 ± 0.5 μm (5.0–9.0 × 3.5–6.0, N = 25), with LW 27.9 ± 4.5 μm² (20–36) and L/W 1.35 ± 0.3 (1.0–2.3). Many gametocytes are irregular in shape, from almost rectangular to slightly lobate, in contrast to more regular rounded or oval shapes in active infections. Chronic-phase gametocytes differ in mean dimensions of length and LW by sex. Macrogametocytes are greater in length, 6.4 ± 1.0 μm and LW, 29.3 ± 4.4 μm², than microgametocytes, 5.8 ± 0.7 μm, 26.3 ± 4.3 μm². Gametocytes in active infections are not sexually dimorphic in size or shape. Gametocyte size relative to host cell nucleus is 0.98 ± 0.26 (0.4–1.6), and to normal erythrocyte nuclei is 1.03 ± 0.25 (0.4–1.6). One to several small dark granules of pigment are dispersed. Vacuoles are rarely present.

In the experimental host *Egernia whitei*, into which the parasite was isolated by inoculation of infected blood from *E. cunninghami* (Telford, 1979a), meronts are 4.7 ± 0.90 × 3.8 ± 0.9 μm (3–6 × 2–5, N = 55), with LW 18.6 ± 7.1 μm². Meronts contain 10.1 ± 2.3 (6–12, N = 66) merozoites. Their size relative to host cell nucleus is 0.47 ± 0.11 (0.3–0.8, N = 24), and to normal erythrocyte nuclei is 0.86 ± 0.27 (0.4–1.3, N = 55). Gametocytes are 5.2 ± 0.7 × 4.6 ± 0.7 μm (4–7 × 3–6, N = 78), with LW 24.3 ± 6.4 μm² (12–36) and L/W 1.16 ± 0.14 (1.0–1.5). Gametocyte size relative to host cell nucleus is 0.78 ± 0.28 (0.4–1.3, N = 23), and to normal erythrocyte nuclei is 1.15 ± 0.30 (0.6–1.9, N = 78). There is no sexual dimorphism in dimensions. In a third host, *Egernia striolata*, only gametocytes, apparently those of *P. mackerrasae* (Telford, 1979a), were present. Gametocytes are 6.5 ± 0.9 × 5.6 ± 1.0 μm (5–8 × 4–7, N = 25), with LW 37.0 ± 11.0 μm² (20–56) and L/W 1.17 ± 0.12 (1.0–1.5). Size of gametocytes relative to host cell nucleus is 1.32 ± 0.46 (0.6–2.3, N = 25), and to normal erythrocyte nuclei is 1.29 ± 0.38 (0.7–2.0). In this chronic infection, macrogametocytes were larger in every dimension and more rounded in shape than microgametocytes, with respective dimensions 7.2 ± 0.7 × 6.4 ± 0.5 μm (6–8 × 6–7, N = 13), LW 45.8 ± 6.4 μm² (36–56), and L/W 1.12 ± 0.12 (1.0–1.3) versus

Plate 39 (**A**) *Plasmodium mackerrasae* from *Egernia* spp., Australia: *E. stokesii* meronts, **a–e**; macrogametocytes, **g**, **h**; microgametocyte, **j**; *E. cunninghami* meront from experimental infection in *E. whitei*, **f**; *E. striolata* macrogametocyte, **i**, and microgametocytes, **k**, **l**. (**B**) *Plasmodium circularis* from *Egernia stokesii*, Australia: young meronts, **a–e**; meronts, **f–h**; macrogametocytes, **i–k**; microgametocyte, **l**. (Figures **g**, **f**, **h**, and **k** modified from Telford, S. R., Jr., and Stein, J., *J. Parasitol.,* 86, 395, 2000, Figures 27, 37, 40, and 42, with permission.)

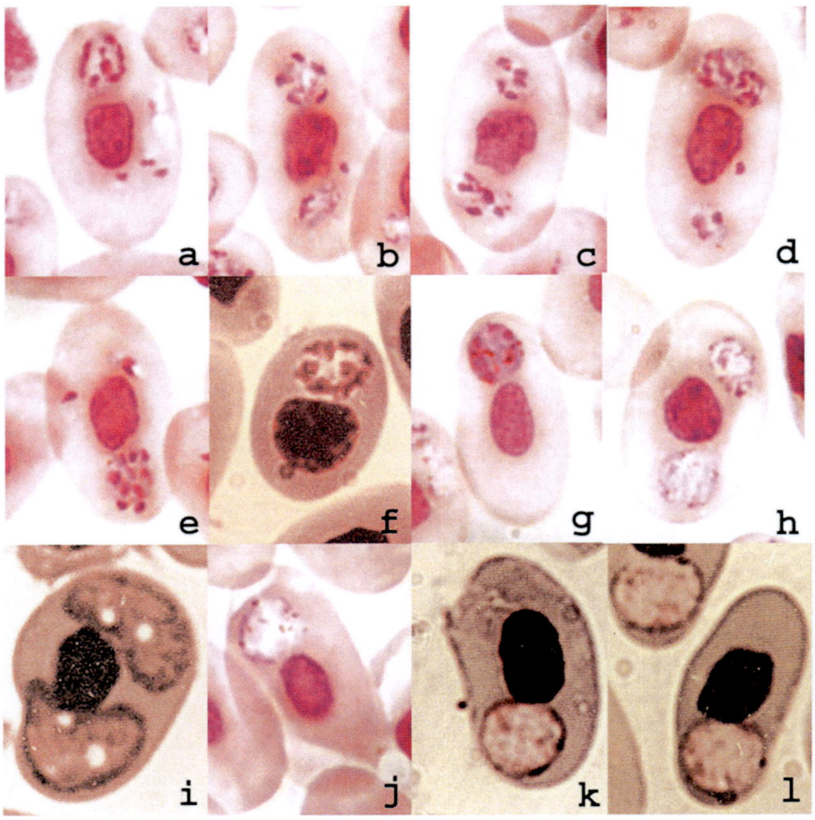

(A)

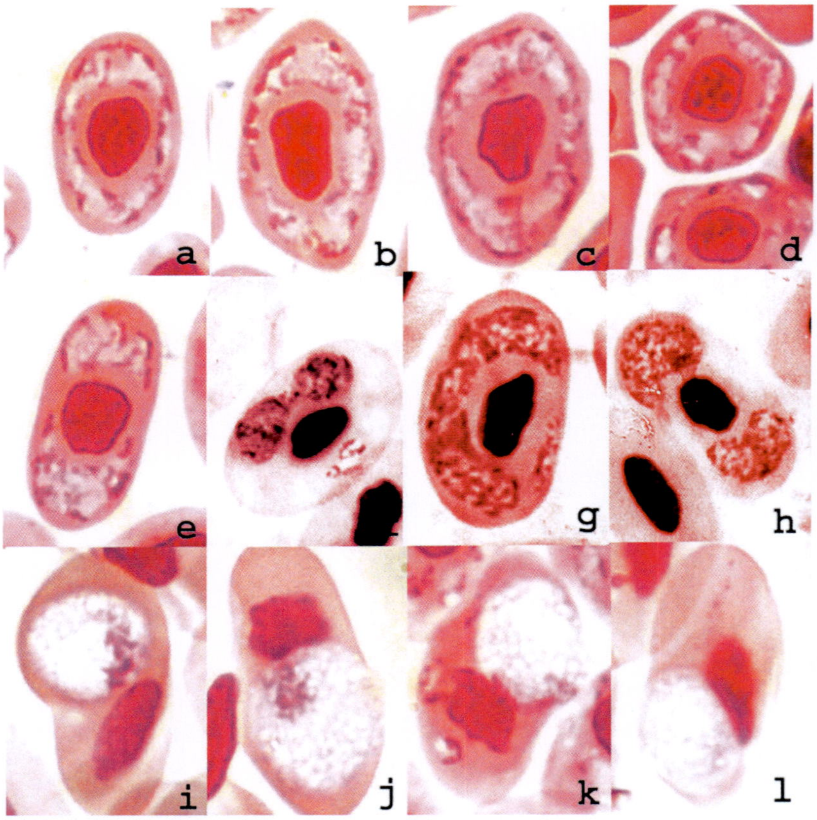

(B)

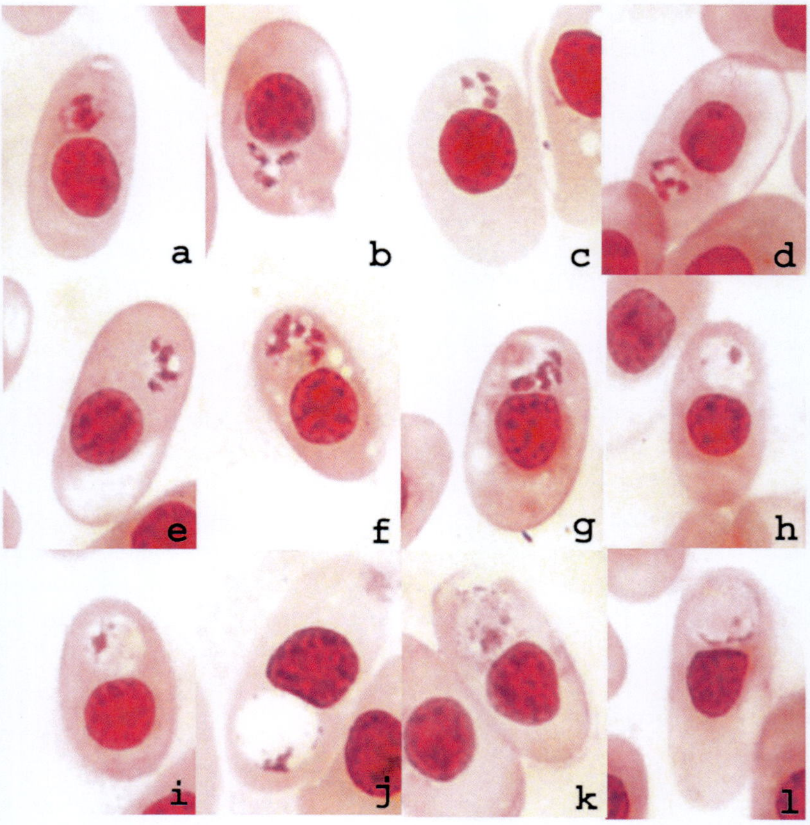

(A)

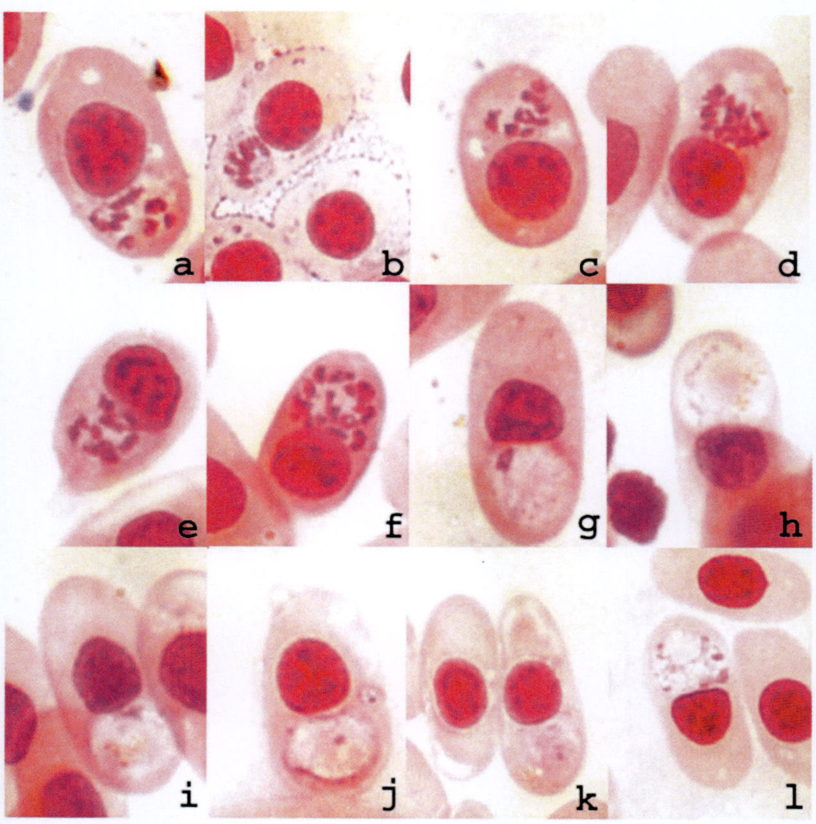

(B)

Plate 40 (**A**) *Plasmodium gracilis* from *Tribolonotus gracilis*, West Irian, Indonesia: meronts, **a–g**; macrogametocytes, **h–k**; microgametocyte, **l**. (**B**) *Plasmodium tribolonoti* from *Tribolonotus gracilis*, West Irian, Indonesia: meronts, **a–f**; macrogametocytes, **g–i**; microgametocyte, **j–l**. (Figures **d–g** and **k** modified from Telford, S. R., Jr., and Wellehan, J. F. X., Jr., *J. Parasitol.*, 91, 148, 2005, Figures 1–4 and 6, with permission.)

5.8 ± 0.5 × 4.8 ± 0.6 µm (5–6 × 4–6, N = 12), 27.5 ± 5.2 µm² (20–36), and 1.22 ± 0.11 (1.0–1.5). In contrast to gametocytes of *P. mackerrasae* from *E. stokesii*, in which vacuoles are seldom seen, the gametocytes in both *E. whitei* and *E. striolata* usually contained one to four vacuoles (Telford, 1979a).

Exoerythrocytic Merogony Three types of EE meronts are present in the life history of *P. mackerrasae* (Telford and Stein, 2000). Thrombocytes occasionally are infected in *E. stokesii* with extremely high parasitemias, which probably represent secondary EE meronts. Apparent phanerozoites in fixed tissues usually stain intensely (**Plate 41**, **c** and **d**) and measure 12–26 × 8–16 µm (17.2 ± 4.2 × 10.7 ± 2.7, N = 19), shaped from round at 13 × 13 µm to elongate at 26 × 8 µm, with most ovoid in shape. They occur most commonly in the heart (intertrabecular sites and pericardium) and commonly form clusters in blood vessel endothelium and connective tissue. They parasitize endothelium of pulmonary capillaries and occur in connective tissue of kidney, ovary, testis, striated muscle, colon, urinary bladder, and fat body. EE meronts in liver were in the endothelium of blood vessels or sinuses. Most found in the small intestine apparently occupied the lamina propria. None were present in the splenic pulp or connective tissue of the stomach and gall bladder. Extracellular meronts, frequently of great size and highly variable in shape, were found in circulating blood. Some were oval to elongate in shape, 23–53 × 12–35 µm. Minimum numbers of nuclei from five meronts were 55 in a meront 23 × 15 µm; 71 in one 16 × 12 µm; 132 in a meront 33 × 12 µm; 160 at 53 × 15 µm; and more than 300 in a meront 44 × 35 (Telford and Stein, 2000). There were many meronts of irregular form, usually large, sometimes occupying most of the microscopic field at ×1000 (**Plate 41**, **e** and **f**). Some were elongate and slender, others arachniform, with jagged cytoplasmic processes. The numbers of nuclei they contained certainly extended into the upper hundreds or low thousands. Some of the very large meronts apparently release merozoites into the bloodstream, for on several occasions blood smears showed a flood of free merozoites among the erythrocytes without presence of the large, irregular meronts.

Sporogony Unknown.

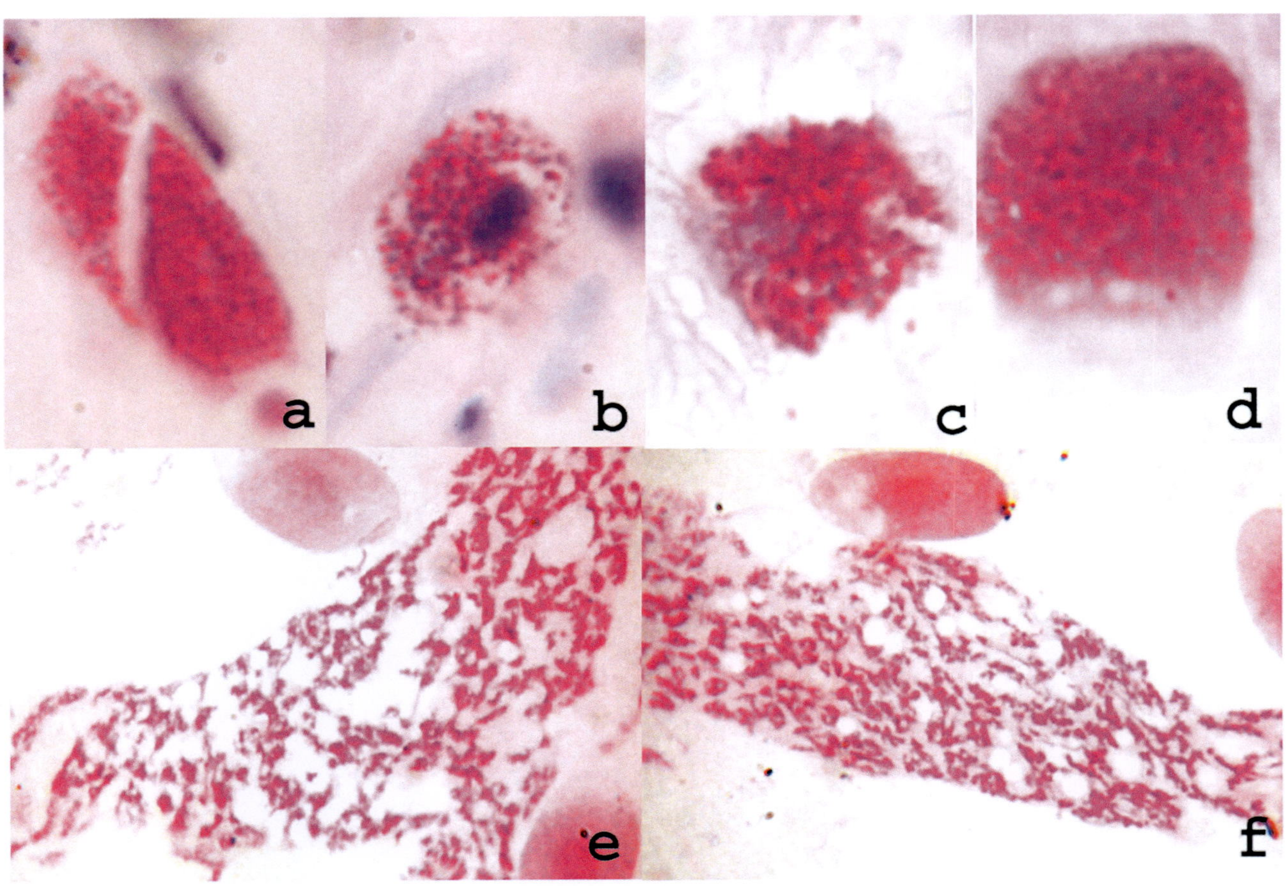

Plate 41 Exoerythrocytic meronts of *Plasmodium lionatum* of *Ptychozoon lionatum* (**a**, **b**) and *Plasmodium mackerrasi* (**c–f**) of *Egernia stokesii*: **a**, **b** in connective tissue of heart; **c**, **d** in connective tissue of kidney (**c**) and urinary bladder (**d**); **e** and **f** are macromeronts in circulating blood, perhaps dislodged from capillary endothelium.

Effects on Host Asexual stages are erythrocytic but never distort the host cell or nucleus or displace the nucleus. Infection by a single meront produces no hypertrophy of the host cell, but in multiply infected cells, erythrocyte length, width, and LW are enlarged, without affecting nuclear dimensions. Gametocytes never distort the host cell and rarely displace its nucleus or distort it. Erythrocytes and their nuclei, in intense infection of newly mature gametocytes, were no different in dimensions from uninfected cells except in LW, which was greater. Each infected cell had 1–5 (3.0 ± 1.3) additional parasites. In two other infections with larger gametocytes, one chronic with 1–3 (1.6 ± 0.8) parasites per cell and one active, intense infection with 0–4 (2.1 ± 0.9) additional parasites present per cell, erythrocyte lengths, widths, and LW were increased. Nucleus length and LW did not differ, only nucleus width was greater than that of normal erythrocytes. Relapses occurred among captive lizards from 1 to 20 months (8.7 months) postcapture, with parasitemias varying from 5% to 85%.

Remarks Comparisons of gametocyte dimensions from *E. stokesii* with those from *E. whitei* found differences in length, LW, and L/W. In comparison of the meront characters of length, width, LW, and nuclei between samples from *E. whitei* and *E. stokesii*, only the mean number of nuclei differed (Telford and Stein, 2000). Although fragments of EE meronts apparently torn loose from the endothelial lining of blood vessels have been seen occasionally in other saurian *Plasmodium* species, the huge EE meronts seen in circulating blood of *Egernia stokesii* infected by *Plasmodium mackerrasae* are unique. Their role in the life cycle of this species is unknown, but certainly they contribute to the maintenance of massive infections often seen in this host.

Plasmodium circularis Telford and Stein 2000 (**Plate 39**)

Diagnosis A *Plasmodium* (*Lacertamoeba*) species with immature meronts that form an unbroken ring encircling the host cell nucleus. Mature meronts are contracted into halteridial or dumbbell-shaped forms 7.5–26.0 × 3.0–6.0 μm, with LW 33.7–108 μm^2, that contain 19–52 nuclei. Black pigment granules are not prominent. Meront size relative to host cell nucleus size is 2.85, and to normal erythrocyte nuclei is 2.52. Gametocytes are rounded or oval, 7–12 × 3.6–11 μm, with LW 48–132 μm^2 and L/W 1.0–1.5. Pigment occurs as a few small, dispersed black granules. Gametocyte size relative to host cell nuclei is 2.63, and to normal erythrocyte nuclei is 1.79.

Type Host *Egernia stokesii* (Dumeril) (Sauria: Scincidae).

Type Locality Neuroodla, South Australia.

Other Hosts None known.

Other Localities South Australia: Hawker, The Four Mile, Quorn, Drakes Nob, Island Lagoon, and Chace Range.

Prevalence *P. circularis* was found in 7% of the *E. stokesii* at the type locality (Telford and Stein, 2000).

Morphological Variation After the third nuclear division, all meronts are elongate, and curve around the host cell nucleus. Before meronts begin segmentation, the erythrocyte nuclei are usually (84%) completely encircled by the immature meronts, which have apparently fused their elongated ends into an unbroken cytoplasmic ring. Rings formed by meronts are 32.3 ± 3.3 μm (26–38, N = 12) in circumference with their greatest width narrowing from 2.5–6 to 2–2.5 μm at its minimum. Ring-shaped meronts contain 23.3 ± 2.7 (21–28, N = 12) nuclei. At segmentation, meronts contract into elongate forms with broad ends, which most commonly are halteridial around host cell nuclei (48%). There are a few lentiform meronts polar in position; more elongate or dumbbell-shaped parasites are lateropolar or lateral to the erythrocyte nucleus. Occasional meronts are broken into two portions. Apparently, mature meronts are 15.6 ± 4.9 × 4.3 ± 0.7 μm (7.5–26.0 × 3.0–6.0, N = 25), with LW 66.2 ± 20.6 μm^2 (33.7–108) and 19–52 nuclei (26.2). Meront size relative to host cell nucleus is 2.85 ± 0.92 (1.5–4.6), and to normal erythrocyte nuclei 2.52 ± 0.78 (1.3–4.1). Gametocytes are 9.0 ± 1.3 × 7.3 ± 1.0 μm (7–12 × 3.6–11, N = 27), with LW 66.9 ± 18.3 μm^2 (48–132) and L/W 1.24 ± 0.15 (1.0–1.5). Gametocyte size relative to host cell nucleus size is 2.63 (1.7–4.7), and to normal erythrocyte nuclei is 1.79 (1.3–3.5). Gametocytes are sexually dimorphic in width, with macrogametocytes broader (7.7 ± 1.1, N = 16), and with greater LW (72.7 ± 21.0) than microgametocytes (6.8 ± 0.6, N = 11), LW 58.4 ± 8.6. The length and L/W ratio of gametocytes are similar, with macrogametocyte length 9.4 ± 1.5 and L/W 1.23 ± 0.17, in comparison to microgametocytes, 8.5 ± 0.7 and 1.26 ± 0.13. Gametocytes are usually polar but may lie lateral or lateropolar to the host cell nucleus. There are a few small, scattered black pigment granules.

Exoerythrocytic Merogony Unknown.

Sporogony Unknown.

Effects on Host Erythrocytes infected by meronts are enlarged in length, width, and LW. The length of host cell nuclei is greater, but width is similar to that of uninfected cells, and the LW is less. Host cells are distorted, with their nuclei displaced and often distorted. Gametocytes usually distort the host cell and displace its nucleus, often distorting the nucleus. Although the length and width of host erythrocytes are greater, LW is similar to that of normal cells. Host cell nuclei are smaller in length, width, and LW than those of normal cells.

Remarks *Plasmodium circularis* is distinguished from all other plasmodiids by the formation of immature meronts that completely encircle the host cell nucleus, with the apparent fusion of opposite tips of the cytoplasm into an unbroken ring. The gametocyte size relative to that of normal erythrocyte nuclei and the merozoite number are consistent with characteristics of the subgenus *Lacertamoeba* (Telford, 1988a), but the relative meront size slightly exceeds that typical of this subgenus. "Segmenters are evidently briefly present in circulating blood, and both maturing schizonts and gametocytes appear to be seasonal in their occurrence, present most commonly in the Australian spring from September to December" (Telford and Stein, 2000).

Plasmodium gracilis Telford and Wellehan 2005 (**Plate 40**)

Diagnosis A *Plasmodium* (*Carinamoeba*) species with meronts 3–6 × 3–5 µm, with LW 10.5–25.0 µm^2, that produce three to eight merozoites usually arranged as a fan or, rarely, in larger meronts, as a rosette. Meront size relative to the host cell nucleus averages 0.55, and to normal erythrocyte nuclei is 0.55. Pigment granules are dark golden. Gametocytes are 5.0–6.6 × 5.0–6.0 µm, with LW 25–40 µm^2 and L/W 1.0–1.2. Gametocyte size relative to host cell nucleus size averages 1.18, and to normal erythrocyte nuclei is 1.13. No vacuoles are present. Pigment is dispersed as blackish granules in macrogametocytes but usually forms one or two dark yellowish clumps in microgametocytes. Gametocytes are not sexually dimorphic in size or shape.

Type Host *Tribolonotus gracilis* De Rooij (Sauria: Scincidae).

Type Locality Indonesia, Irian Jaya presumably, no precise locality.

Other Hosts None known.

Other Localities None known.

Prevalence *P. gracilis* infected one of eight *T. gracilis* in mixed infection with *Plasmodium tribolonoti* (Telford and Wellehan, 2005).

Morphological Variation Meronts are 4.3 ± 0.7 × 3.5 ± 0.5 µm (3–6 × 3–5, n = 27), with LW 15.4 ± 4.0 µm^2 (10–25). Merozoites average 4.9 ± 1.4 (3–8), usually forming fans; rarely, nuclei of larger meronts are arranged as a rosette. Meront size relative to host cell nucleus is 0.55 (0.37–0.91), and to normal erythrocyte nuclei is 0.55 (0.37–0.89). Pigment appears yellow in mass, dark golden as individual granules. Gametocytes are 5.9 ± 0.4 × 5.5 ± 0.4 µm (5.0–6.6 × 5.0–6.0, n = 43), with LW 31.9 ± 4.2 µm^2 (25–40) and L/W 1.06 ± 0.07 (1.00–1.20). Gametocyte size relative to host cell nucleus size is 1.18 (0.69–1.58), and to normal erythrocyte nuclei is 1.13 (0.89–1.40). No vacuoles are present. Pigment forms as discrete blackish granules in macrogametocytes but usually as one or two dark yellowish clumps in microgametocytes. Gametocytes are not sexually dimorphic in size or shape.

Exoerythrocytic Merogony Unknown.

Sporogony Unknown.

Effects on Host Meronts parasitize erythrocytes, always polar to the nucleus of host cell. Cells are rarely distorted, but meronts may slightly distort one side of the nucleus and commonly displace it. They have no effect on dimensions of host erythrocytes or their nuclei. Gametocytes are always polar to the host cell nucleus and parasitize erythrocytes. Host cells are never distorted but commonly displace nuclei, often distorting them on one side by contact. Gametocytes have no effect on dimensions of host erythrocytes or their nuclei.

Remarks In the mixed infection with *P. tribolonoti, Plasmodium gracilis* was easily distinguished from the other species by its smaller meronts, merozoite numbers, and gametocyte size (Telford and Wellehan, 2005).

Plasmodium tribolonoti Telford and Wellehan 2005 (**Plate 40**)

Diagnosis A *Plasmodium* (*Lacertamoeba*) species characterized by rounded or oblong meronts 5–7 × 4–7 µm, with LW 22–42 µm^2, that produce 10–21 merozoites, irregularly arranged, in proerythrocytes. Meront size relative to host cell nucleus is 1.02, and to normal erythrocyte nuclei is 1.14. Pigment granules are dark golden. The mostly erythrocytic gametocytes are 6.5–9.0 × 5.5–7.5 µm, with LW 38–63 µm^2 and L/W 1.0–1.5. Gametocyte size relative to host cell nucleus size is 1.66, and to normal erythrocyte nuclei is 1.61. The dark greenish pigment is dispersed in macrogametocytes and usually appears as one or two dark greenish-yellow clumps in microgametocytes, occasionally dispersed around the margin.

Gametocytes are not sexually dimorphic in size or shape.

Type Host *Tribolonotus gracilis* De Rooij (Sauria: Scincidae).

Type Locality Indonesia, Irian Jaya presumably, no precise locality.

Other Hosts None known.

Other Localities None known.

Prevalence One of eight *Tribolonotus gracilis* was infected by *P. tribolonoti* (Telford and Wellehan, 2005).

Morphological Variation Mature or nearly mature meronts are 6.1 ± 0.7 × 5.3 ± 0.8 µm (5–7 × 4–7, n = 8), with LW 32.3 ± 6.8 µm^2 (22–42). Meronts produce 14.3 ± 3.6 (10–21) merozoites. Meront shape is rounded or oblong, with nuclei irregularly arranged. Their size relative to host cell nucleus is 1.02 (0.60–1.54), and to normal erythrocyte nuclei is 1.14 (0.78–1.50). Pigment granules are dark golden. Gametocytes are 7.2 ± 0.6 × 6.3 ± 0.5 µm (6.5–9.0 × 5.5–7.5, n = 41), with LW 45.5 ± 5.9 µm^2 (38–63) and L/W 1.15 ± 0.11 (1.00–1.50). Gametocyte size relative to host cell nucleus size is 1.66 (0.80–2.40), and to normal erythrocyte nuclei is 1.61 (1.37–2.23). Gametocytes are not sexually dimorphic in size or shape. In macrogametocytes, dark greenish pigment is dispersed but usually forms as one or two dark greenish-yellow clumps in microgametocytes or is occasionally dispersed around the margin. Vacuoles are rarely present.

Exoerythrocytic Merogony Unknown.

Sporogony Unknown.

Effects on Host Meronts parasitize proerythrocytes, only rarely distorting cells, without distortion of the nucleus but usually displacing it. Meronts are usually polar within the host cell, or lateral, but have no effect on dimensions of the host proerythrocytes or their nuclei. Gametocytes are always polar to the host cell nucleus. Most parasitize erythrocytes, rarely proerythrocytes (2%), and seldom distort the host cell, commonly displacing its nucleus, which is often distorted on one side by contact with the gametocyte. Gametocytes have no effect on dimensions of host erythrocytes or their nuclei.

Remarks Although the host species, *T. gracilis*, occurs in Papua New Guinea, the lizards examined probably originated in West Irian, where export of reptiles for the exotic pet trade is less restricted.

Plasmodium billbraya (Paperna and Landau) 1990 new comb.

Syn.: *Billbraya australis* Paperna and Landau 1990b

Diagnosis A *Plasmodium* (*Lacertamoeba*) species in which erythrocytic merogony is brief and intense, causing pronounced anemia, which then is supplanted by often intense gametocytemia that may infect erythrocytes by 6–11 young gametocytes, which mature slowly over several weeks and are replenished from exoerythrocytic merogony. Largest meronts observed are 8.0–11.2 × 5.6–8.8 µm, with estimated LW 45–98.6 µm^2, and contain six to eight nuclei. Mature gametocytes, variably shaped, are 5.6–19.2 × 4.0–11.2 µm, with estimated average LW 59–116 µm^2 and L/W 1.2–1.9. Gametocytes are sexually dimorphic in dimensions, with macrogametocytes greater in size than microgametocytes and more elongate in shape. Pigment forms as fine dark granules, dispersed in both meronts and gametocytes.

Type Host *Phyllodactylus marmoratus* (Gray) (Sauria: Gekkonidae).

Type Locality Cambraii, South Australia.

Other Hosts None known.

Other Localities Blowhole Creek, Fleurien Peninsula, South Australia.

Prevalence *Plasmodium billbraya* infected 1 of 4 *Phyllodactylus marmoratus* at Blowhole Creek in November 1981 and 1 of 22 (4.5%) collected from Cambraii in 1988 (Paperna and Landau, 1990b).

Morphological Variation Segmenting meronts were not observed by Paperna and Landau (1990b). The largest meronts reported were 8.0–11.2 × 5.6–8.8 µm and contained six to eight nuclei. The figures depict oval-to-elongate configurations. Estimated LW from the cited dimensions appears to be 45–98.6 µm^2. Gametocyte dimensions were provided for three samples over a 5-month period. A sample taken in February–March showed a macrogametocyte average of 13.7 × 7.8 µm (10.4–19.2 × 5.6–11.2, N = 29), indicating an average LW of 106.9 µm and L/W of 1.76 and microgametocytes of 8.5 × 7.0 µm (5.6–12.8 × 4.0–9.6, n = 17), with estimated LW 59.5 μm² and L/W 1.21. The sample from April had macrogametocytes 14.9 × 7.8 μm (12.0–18.4 × 7.2–11.2n N = 16), with estimated LW 116.2 μm² and L/W 1.91 and had microgametocytes 9.2 × 7.0 μm (5.6–15.2 × 4.0–8.8, N = 13), with estimated LW 64.4 μm² and L/W 1.31. Gametocytes are sexually dimorphic, with macrogametocytes larger and more elongate than microgametocytes. Pigment occurs as fine dark granules dispersed in both meronts and gametocytes.

Exoerythrocytic Merogony A single thrombocytic meront, 8.0 × 2.4 μm, with 11–12 nuclei was reported by Paperna and Landau (1990b). They described EE merogony occurring in "circulatory monocytes" in their definition of the genus *Billbraya* but provided no further information.

Sporogony Unknown.

Effects on Host Paperna and Landau (1990b) reported a prominent anemia resulting from hyperinfection of erythrocytes by asexual stages early during the course of infection. Proerythrocytes and erythrocytes are both infected by asexual stages and young gametocytes. Up to 95% of erythrocytes were infected in one infection and up to 12 young parasites occurred in multiply infected erythrocytes, although proerythrocytes rarely contained more than two parasites. Erythrocytes host to even massive multiple infections were only slightly hypertrophied, 18.2 × 9.4 μm in comparison to normal erythrocytes, 17.9 × 10.8 μm, and rarely showed distortion. Their nuclei usually retained their central position, rarely being displaced or distorted. Erythrocytes infected by one or two mature gametocytes were seldom enlarged, averaging 17.4 × 10.0 μm, nearly identical to normal erythrocytes. Multiple infections of either young macro- or microgametocytes, however, caused considerable hypertrophy, respectively 19.1 × 11.4 μm and 20.2 × 10.7 μm. Mature gametocytes occasionally displaced host cell nuclei laterally or to a polar position, apparently little affecting the shape of the nuclei.

Remarks The establishment of a new genus, *Billbraya*, for this plasmodiid species was criticized by Telford (1994) on the grounds that the elements of the generic diagnosis were applicable to other *Plasmodium* species as well (*P. mexicanum, P. saurocaudatum, P. giganteum, P. cnemidophori, P. balli*), with nothing applying exclusively to the Australian parasite. Here, it is considered to be a *Plasmodium* (*Lacertamoeba*) species (not *Paraplasmodium* as suggested earlier; Telford, 1994), and as such the specific name *australis* is preoccupied by *P. australis* Garnham 1966. The taxonomic designation proposed here eliminates both problems of generic and specific appellations and maintains the desire of Paperna and Landau to recognize R. S. (Bill) Bray. The species merits additional study, especially in the absence of data on mature meronts.

THE *PLASMODIUM* PARASITES OF SNAKES, SUBGENUS *OPHIDIELLA*

Plasmodium pythonis Fantham and Porter 1950

Diagnosis A *Plasmodium* (*Ophidiella*) species, parasitic in pythons, with large, oval meronts that fill the host erythrocyte and small, elongate gametocytes. Meronts are 12–21 × 11–16 μm, with estimated LW 145–337 μm², that contain up to 16 nuclei. Estimated meront size relative to normal erythrocyte nucleus size is 3.2–7.5. Nuclei tend to concentrate toward one end of the meront opposite from aggregated granules of light bronze pigment. Gametocytes are one-third or less the size of meronts, about 14 × 3–4 μm, with estimated LW 46–52 μm² and L/W 3.8–4.2. Gametocytes are approximately twice normal erythrocyte nucleus size. Macro- and microgametocytes are similar in dimensions and shape.

Type Host *Python sebae* (Gmelin) (Serpentes: Boidae).

Type Locality South Africa, no precise locality.

Other Hosts None known.

Other Localities None known.

Prevalence One of one *Python sebae* was infected by *Plasmodium pythonis* (Fantham and Porter, 1950).

Morphological Variation A large vacuole was present in the ovate trophozoites, which were 3.6–6.5 × 1.7–2.7 μm. Meronts were oval and 13.3–20.7 × 10.9–16.3 μm, which indicates an LW range of 145–377 μm², with size relative to that of normal erythrocyte nucleus size, 3.2–7.5. The number of nuclei varied, with a maximum of 16, and they were arranged in no particular form. The light bronze pigment was more or less clumped, sometimes at one end of the meront opposite from the nuclei, which tended to concentrate near one end of the meront. Pigment occasionally was located peripherally. Actual division of meronts was not observed. Gametocytes, with sexes similar in size and elongate shape, were approximately 14 × 3.3–3.7 μm, suggesting LW of 46–52 μm² and L/W 3.8–4.2. They were about twice the size of normal erythrocyte nuclei. Pigment granules were sometimes arranged in a couple of rows.

Exoerythrocytic Merogony Unknown.

Sporogony Unknown.

Effects on Host Meronts filled the host cell, enlarging it, displacing the erythrocyte nucleus and compressing it against the cell membrane. Gametocytes, smaller in size, apparently caused less hypertrophy of the host cell.

Remarks *Plasmodium pythonis* apparently is the only malarial parasite reported from pythons in Africa, Asia, and Australia. Another boid snake, however, *Corallus hortulana*, was reported by Pessôa et al. (1974a) as host to an undescribed *Plasmodium* species in Brazil.

Plasmodium melanoleuca Fantham and Porter 1950

Diagnosis A *Plasmodium* (*Ophidiella*) species, parasitic in African forest cobras, that is characterized by meronts 10.4–20.0 × 10.4–14.6 µm, with estimated LW 108–292 µm², that produce up to 16 merozoites. Estimated meront size relative to normal erythrocyte nucleus size is 1.5–4.1. Gametocytes, similar in size to the meronts, are 10.4–20.0 × 6.7–13.7 µm, with estimated LW 70–274 µm² and L/W 1.46–1.56. Gametocyte size relative to normal erythrocyte nucleus size is 1.0–3.8. Pigment forms as fine dark brown granules distributed in patches. There appears to be no sexual dimorphism in size or shape of gametocytes.

Type Host *Naja melanoleuca* Hallowell (Serpentes: Elapidae).

Type Locality Nyasaland (= Malawi), no precise locality.

Other Hosts None known.

Other Localities None known.

Prevalence *Plasmodium melanoleuca* parasitized both *Naja melanoleuca* examined by Fantham and Porter (1950).

Morphological Variation Trophozoites 2.6–5.0 × 1.5–2.6 µm contained a small vacuole. Meronts were 10.4–20.8 × 10.4–14.6 µm, with LW estimated at 108–292 µm². Up to 16 nuclei were present in meronts beginning merogony. Estimated meront size relative to normal erythrocyte nucleus size is 1.5–4.1. Gametocytes, described as large, were 10.4–20.0 × 6.7–13.7 µm, indicating LW values of 70–274 µm² and L/W ratios of 1.46–1.56. Estimated gametocyte size relative to normal erythrocyte nucleus size is approximately 1.0–3.8. There was no mention of sexual differences in gametocyte size or shape. Pigment was distributed in patches of fine brown granules in both meronts and gametocytes.

Exoerythrocytic Merogony Unknown.

Sporogony Unknown.

Effects on Host Erythrocytes host to either meronts or gametocytes were enlarged and their nuclei displaced, often severely.

Remarks *Plasmodium melanoleuca* is the only *Plasmodium* species yet described from elapid snakes. Because Neotropical elapids are semifossorial and secretive, presumably with limited exposure to possible arthropod vectors, it is unlikely that malarial parasites occur in them, but in contrast, in Africa, Asia, and Australia the Elapidae form a conspicuous faunal component, and *Plasmodium* infection might be expected.

Plasmodium bitis Fantham and Porter 1950

Diagnosis A *Plasmodium* (*Ophidiella*) species, parasitic in vipers, with meronts that considerably exceed the size of gametocytes. Meronts are 10.4–23.0 × 6.8–15.0 µm, with estimated LW 71–345 µm², and produce up to 24 nuclei, usually grouped in pairs. Meront size is approximately five times that of normal erythrocyte nucleus size. Bright yellow pigment granules, also often in pairs, are scattered among the nuclei. Gametocytes are 10.4–19.0 × 5.0–12.0 µm, with estimated LW of 84–187 µm² and L/W of 1.3–3.8. Estimated gametocyte size relative to normal erythrocyte nucleus size is approximately 1.1–2.5. Microgametocytes are elongate but are similar in size to the round or oval macrogametocytes. Yellow pigment granules tend to be scattered toward ends of microgametocytes and are more evenly dispersed in macrogametocytes.

Type Host *Bitis arietans* (Merrem) (Serpentes: Viperidae).

Type Locality Near King William's Town, Cape Province, South Africa.

Other Hosts Unknown.

Other Localities Unknown.

Prevalence *Plasmodium bitis* was found in a single, very large, *B. arietans*, 1 of 11 examined, but Fantham and Porter (1950) do not indicate clearly how many puff adders provided samples of blood.

Morphological Variation Amoeboid trophozoites were ovoid, about 3 × 2.2 μm. Meronts were 10.4 × 6.8–23.0 × 15.0 μm, with LW estimated at 71–345 μm², and produced up to 24 nuclei, "usually being grouped in twos, with bright yellow pigment granules, also often in twos, scattered among them" (Fantham and Porter, 1950). Completed merogony was not observed. Meront size was approximately five times that of normal erythrocyte nucleus size. Microgametocytes were elongate, "somewhat bean-shaped," with yellow pigment granules scattered at either end. Their dimensions were approximately 19 × 5 μm, with LW estimated at 102 μm², and L/W of 3.80, about 1.4 times normal erythrocyte nucleus size. The "round to oval" macrogametocytes were 10.4–15.6 × 8.1–12.0 μm, with estimated LW of 84–187 μm² and L/W of 1.30. Their estimated size relative to normal erythrocyte nucleus size is 1.1–2.5. Pigment granules were more evenly dispersed in macrogametocytes.

Exoerythrocytic Merogony Unknown.

Sporogony Unknown.

Effects on Host Host erythrocytes were enlarged by both meronts and gametocytes and were distorted, with displaced nuclei.

Remarks The report of a pigmented haemosporidian in *Bitis arietans* from Katanga (Huygelen and Mortelmans, 1958) probably represents another species because the gametocytes are halteridial in shape, differ in dimensions from *P. bitis*, and show no differences in size or shape between macro- and microgametocytes. *Plasmodium* parasites were reported from 13 of 164 (7.9%) Brazilian viperid snakes, *Bothrops moojeni*, by Pessôa et al. (1974a). *Plasmodium pessoai* in Costa Rica infects the bushmaster, *Lachesis muta*, as well as colubrid snakes (Ayala et al., 1978).

Plasmodium wenyoni Garnham 1965

Diagnosis A *Plasmodium* (*Ophidiella*) species with amoeboid younger asexual stages that produce variably shaped meronts with estimated dimensions 5.6–7.5 × 4.7–5.3 μm, with LW 26.3–39.8 μm², that produce 12–14, possibly more, merozoites. Meront size relative to host cell nucleus size is estimated at 1.10–1.31. Pigment forms as small dark dots but is inconspicuous. Estimated dimensions of gametocytes are 9.4–12.2 × 6.1–6.7 μm, with LW 63–78 μm² and L/W 1.40–2.64. Gametocyte size relative to host cell nucleus size is estimated to be 1.99–3.29. Macrogametocytes are similar in size to microgametocytes but are more elongate with irregular margins, in comparison to the ovoid microgametocytes, which have smooth margins. Pigment is dark brown and dispersed in prominent spherical granules.

Type Host *Thamnodynastes pallidus* (Linnaeus) (Serpentes: Colubridae).

Type Locality Brazil, no precise locality.

Other Hosts None known.

Other Localities None known.

Prevalence Unknown.

Morphological Variation Garnham (1965) described the younger asexual forms as amoeboid, with "tenuous and vacuolated" cytoplasm. Meronts are irregular in appearance, with estimated dimensions from two figured meronts that appear to be nearly mature, 5.6 × 4.7–7.5 × 5.3 μm, LW 26.3–39.8 μm², and with size relative to host cell nucleus size 1.10–1.31. Pigment apparently forms as a few small dark dots and is described as inconspicuous. Estimated dimensions of gametocytes (from figures) are 9.4 × 6.7 μm for one microgametocyte and 11.1–12.2 × 6.1–6.4 μm in two macrogametocytes, with LW calculated at 63–78 μm². The L/W ratio is lower for the microgametocyte, 1.40, than for the macrogametocytes, 1.91–2.64. Gametocyte size relative to host cell nucleus size is 1.99–3.29. Macrogametocytes are more elongate than microgametocytes, and their margins are irregular instead of smooth as in the ovoid microgametocyte, described as "subspherical" by Garnham. Pigment granules, spherical and dark brown, up to 24 in number, are dispersed in both sexes of gametocytes.

Exoerythrocytic Merogony Unknown.

Sporogony Unknown.

Effects on Host "Mature erythrocytes appear to be selected for invasion by this species, and they are little affected by the presence of the parasite" (Garnham, 1965). The figures show host erythrocytes apparently normal in size but with displaced, undistorted nuclei and apparently less affected by presence of meronts than by gametocytes.

Remarks The type and only slide was prepared by C. M. Wenyon from a *T. pallidus* that had died in the London Zoological Gardens in 1934. Wenyon had labeled the slide "Haemogregarine and Plasmodium," which immediately claimed Professor Garnham's attention when he received the slide three decades later. I viewed the slide in the late 1970s and, as is often the case with Wenyon's slides, the stain remained in excellent condition despite the passage of over 40 years.

Plasmodium tomodoni Pessôa and Fleury 1968 (**Plate 42**)

Diagnosis A *Plasmodium* (*Ophidiella*) species, parasitic in colubrid snakes, with large, rounded-to-elongate meronts that tend to curve around the erythrocyte nucleus and round-to-elongate, usually ovoid, gametocytes. Meronts are 10–21 × 7–10 μm, with LW 90–189 μm², and produce 27–66 merozoites. Meront size relative to host cell nucleus size averages 3.39, and to normal erythrocyte nucleus size is 2.73. Gametocytes are 10–20 × 8–13 μm, with LW 80–220 μm² and L/W 1.00–1.82. Gametocyte size relative to host cell nucleus size averages 3.03, and to normal erythrocyte nucleus size is 2.78. There is no sexual dimorphism in gametocyte size or shape. Pigment forms an irregular, prominent dark golden mass in mature meronts and is dispersed as small dark brown granules, often distributed along the periphery of gametocytes.

Type Host *Tomodon dorsatus* Duméril, Bibron and Duméril (Serpentes: Colubridae).

Type Locality Cananéia, São Paulo State, Brazil.

Other Hosts None known.

Other Localities None known.

Prevalence Not given in original description. Pessôa et al. (1974a) reported finding 2 of 179 (1.1%) *Tomodon dorsatus* parasitized by *P. tomodoni*.

Morphological Variation Pessôa and Fleury (1968) described mature meronts as 10–12 μm in size that could produce 30–40 merozoites. Mature macrogametocytes were 11–12 μm in diameter and microgametocytes were 9–10 μm. Pigment granules, dark brown and up to 26 in number, were described and figured as dispersed along the gametocyte margins. In a paratype slide of *P. tomodoni*, meronts are 13.8 ± 2.9 × 9.0 ± 1.0 μm (10–21 × 7–10, N = 25), LW 123.3 ± 27.2 μm² (90–189), with 41.8 ± 9.7 (27–66) merozoites. Meront size relative to host cell nucleus size is 3.39 ± 1.0 (2.00–6.11, N = 24), and to normal erythrocyte nucleus size is 2.73 ± 0.60 (2.00–4.19, N = 25). Meronts are variably formed but most often appear as rosettes or are elongate, curving around the host cell nucleus. Pigment, dispersed as small dark brown dots in younger meronts, usually forms a prominent irregular golden-brown mass, often central but variably positioned in mature meronts. Gametocytes are round to elongate, usually ovoid, 12.5 ± 2.8 × 9.5 ± 1.2 μm (10–20 × 8–13, N = 25), with LW 125.5 ± 39.2 μm² (80–220) and L/W 1.27 ± 0.22 (1.00–1.82). Gametocyte size relative to host cell nucleus size is 3.03 ± 1.00 (1.88–5.97), and to normal erythrocyte nucleus size is 2.78 ± 0.87 (1.77–4.88). There are no sexual differences in dimensions or shape of gametocytes. While large masses of dark golden pigment granules are commonly seen in mature gametocytes, pigment is usually dispersed as small dark brown granules and may be distributed along the periphery of the gametocyte.

Exoerythrocytic Merogony Unknown.

Sporogony Unknown.

Effects on Host Erythrocytes host to meronts are of normal size but have nuclei 18% smaller than uninfected erythrocytes, while cells host to gametocytes are hypertrophied by 13% over normal size, and their nuclei are not enlarged. Host erythrocytes to both stages are distorted, with displaced nuclei, and the latter are distorted to nearly the same degree, 40% and 48% by gametocytes and meronts, respectively.

Remarks Pessôa and Fleury (1968) apparently easily transmitted *P. tomodoni* by inoculation of infected blood to seven *T. dorsatus*, of which three became positive. Attempts to infect six other species of colubrid snakes, three *Thamnodynastes strigatus*, two *Xenodon merremi*, and one each of *Helicops modesta, Erythrolamprus aesculapii, Dryadophis bifossatus*, and *Liophis miliaris*, were unsuccessful, as was inoculation of infected blood into a lizard, *Ameiva a. ameiva*. These results suggest that *P. tomodoni* is highly host specific. Garnham and Duggan (1986) suggested that *P. tomodoni* might be a synonym of *Plasmodium wenyoni*, which it certainly is not given the much greater size of meronts and number of merozoites produced, as well as dimensional differences of the gametocytes.

Plasmodium pessoai Ayala, Moreno and Balaños 1978

Diagnosis A *Plasmodium* (*Ophidiella*) species, parasitic in colubrid and viperid snakes, with highly amoeboid trophozoites that often have pseudopodia that can extend across the diameter of the host erythrocyte. Mature meronts are shaped as bouquets or as flowers, with merozoites in more than a single layer within the meront, appearing as flower petals. Meronts are 7.0–9.8 μm in diameter,

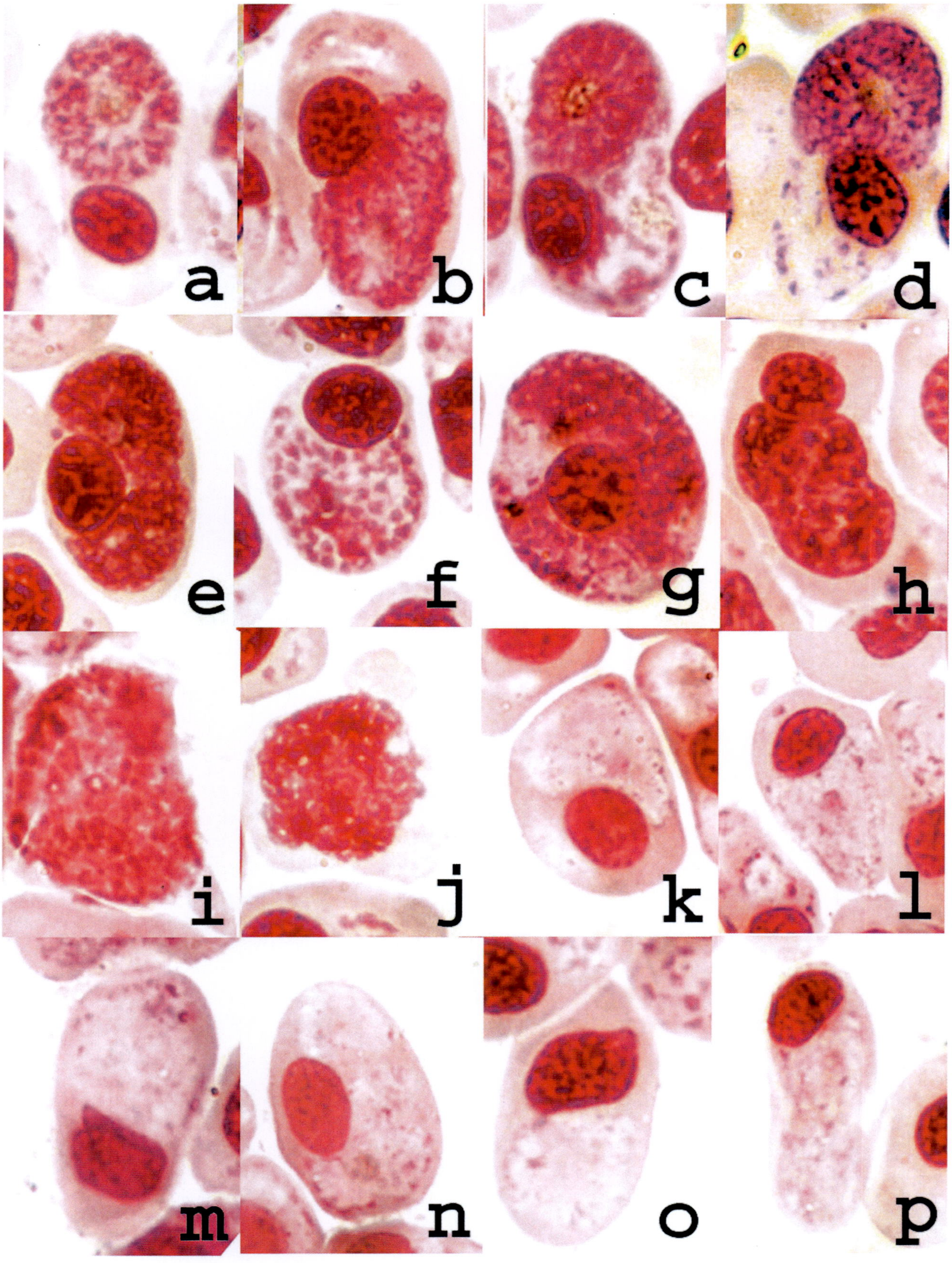

Plate 42 *Plasmodium tomodoni* from the colubrid snake *Tomodon dorsatus*, Brazil: meronts, **a–j**; macrogametocytes, **k–n**; microgametocytes, **o**, **p**.

with estimated LW 67–95 μm², and merozoites number 22–32. Estimated meront size relative to host cell nucleus size is 2.17, and to normal erythrocyte nucleus size is 2.04. Gametocytes are elongate, sausage-shaped, and average 10.4 × 4.6 μm, with LW range 27–75 μm² and L/W ratio 2.30. Estimated gametocyte size relative to host cell nucleus size is 1.93, and to normal erythrocyte nucleus size is 1.92. Sexual differences in gametocyte dimensions and shape are not apparent. Pigment forms a central or basal mass in meronts and is localized as small yellow-brown or golden granules either within a vacuole or dispersed within a limited portion of the gametocyte.

Type Host None stated in original description. Designated here as *Spilotes pullatus* (Linnaeus) (Serpentes: Colubridae).

Type Locality "La Selva" Tropical Studies Field Station near Puerto Viejo, Heredia Province, Costa Rica.

Other Hosts *Lachesis muta* (Linnaeus) (Serpentes: Viperidae).

Other Localities Bananito Sur, Limon Province, Costa Rica.

Prevalence *Plasmodium pessoai* parasitized a single *S. pullatus* examined and 1 of 70 (1.4%) *L. muta* (Ayala et al., 1978).

Morphological Variation As trophozoites enlarged following invasion of the host cell by 3 × 1.2 μm merozoites, they became amoeboid, often showing two to four long pseudopodia that could extend most of the diameter of the host cell. Mature meronts assumed configurations as a bouquet or a flower, circular or ovoid in shape, with merozoites arranged in two overlapping layers in the meront, like flower petals or "kernels of corn on cob as seen from one end" (Ayala et al., 1978). Segmenters averaged 8.2 μm (7.0–9.8, N = 10) in diameter, with the largest in the measured series 8.6 × 7.4 μm and 9.8 × 7.4 μm. Meront LW is estimated as 67–95 μm². Merozoite numbers could not be determined from the overlapping layers of merozoites, but seven ruptured segmenters contained 26.3 ± 4.1 (22–32) merozoites. Meront size relative to host cell nucleus size, estimated from figures, is 2.17, and to normal erythrocyte nucleus size is 2.04. The mass formed by golden-yellow pigment granules was located centrally or "at or near the apex" of the meront. Gametocytes, described as "small sausages," are elongate, with some showing "saw-toothed lateral borders." Gametocytes averaged 10.4 ± 1.9 × 4.6 ± 0.9 μm, with LW 48.1 ± 11.0 μm² (27–75, N = 32) and L/W 2.30. Gametocyte size relative to host cell nucleus size, estimated from figures, is 1.93, and to normal erythrocyte nucleus size is 1.92. No differences by sex in dimensions or shape are described. Pigment comprised 25–40 "small yellow-brown or golden granules usually gathered in a vacuole or in a single area of about 1/4 or 1/5 of the cytoplasm."

Exoerythrocytic Merogony Unknown.

Sporogony Unknown.

Effects on Host Host erythrocytes are barely enlarged, with little distortion except for "slight displacement of the nucleus by mature meronts or gametocytes."

Remarks The Annual Report of Gorgas Memorial Laboratory (GML), Panama, for 1964 reported the presence of *Plasmodium* infections in five species of colubrid snakes from the vicinity of Almirante on the Caribbean coast of Panama and adjacent to Costa Rica. Host species positive were *Clelia* sp. (one of one), *Drymarchon corais* (one of one), *Leptophis ahaetulla* (one of two), *Oxybelis aeneus* (three of seven), and *Pseustes poecilonotus* (one of one). As a staff member at GML, I did not see this material until Professor Garnham visited the laboratory in 1969 and strongly requested to see it. Gametocytes on a slide from *Oxybelis aeneus* resembled those of *P. pessoai*. Regrettably, the parasites have never been described despite the passage of 36 years.

ASIAN *HAEMOCYSTIDIUM* SPECIES

Haemocystidium simondi Castellani and Willey 1904 (Paperna and Landau) 1991

Diagnosis A *Haemocystidium* species characterized by round, oval, or lentiform, nonhalteridial gametocytes 8–14 × 6–12 μm, with LW 56–163 μm² and L/W 1.0–1.6. Pigment is comprised of black granules dispersed within the cytoplasm. Prominent cytoplasmic vacuoles are present in macrogametocytes.

Type Host *Hemidactylus leschenaulti* Duméril and Bibron (Sauria: Gekkonidae).

Type Locality Mamadu, near Vavuniya, Northern Province, Ceylon.

Other Hosts *Hemidactylus* sp. (= *H. brooki*, Garnham, 1966).

Other Localities Trincomalee, Ceylon (Dobell, 1910a); Hyderabad, India (Shortt, 1962).

Prevalence The only comment concerning prevalence is that of Dobell (1910a): "At Trincomalee (Sept.) nearly every individual [*H. leschenaulti*] was infected."

Morphological Variation In *H. brooki*, macrogametocytes are 12.5 × 9.8 μm (8–14 × 6–12, N = 13), with L/W 1.3 (1.0–1.6), and microgametocytes are 11.1 × 9.0 μm (8–14 × 7–10, N = 13), with L/W 1.1 (1.0–1.3) (Paperna and Landau, 1991). Estimated LW for macrogametocytes is 123 μm^2 (56–163), and microgametocytes are 100 μm^2 (56–140) (Telford, 1996a). Both sexes are round or oval; macrogametocytes may be lentiform (Paperna and Landau, 1991). Dimensions for microgametocytes of *H. simondi* in *Hemidactylus leschenaulti* were stated by Dobell (1910b) as "a length of 18 μ or more." Apparently using the same slides studied by Paperna and Landau (1991), Garnham (1966) reported microgametocytes of *H. simondi* in *H. brooki* to be spherical or ovoid and 10–12 μm in diameter, while macrogametocytes are slightly longer, 15 μm, and possess a large, oval vacuole.

Merogony Shortt (1962) found a single meront of *H. simondi* containing about 30 round merozoites in a Kupfer cell of the liver in *Hemidactylus* sp. (*H. brooki*).

Sporogony Unknown.

Effects on Host Paperna and Landau (1991) described only minor differences in average dimensions of erythrocytes infected by macrogametocytes, 15.6 × 12.7 μm, and microgametocytes, 17.4 × 12.9 μm, in comparison to that of normal erythrocytes, 16.4 × 13.9 μm, but their figures demonstrate pronounced distortion of host cells. Erythrocyte nuclei were displaced to polar position by the gametocytes but were neither distorted nor hypertrophied.

Remarks Mackerras (1961a) reported four Australian gecko species to be host to *Haemocystidium* species that appeared to most closely resemble *H. simondi* but admitted that other species could be present. Paperna and Landau (1991) distinguished all four as distinct species, none of them *H. simondi*. Because Dobell (1910b) interpreted small parasites in erythrocytes of *H. leschenaulti* as meronts with two to four merozoites, Wenyon (1915) synonymized *Haemocystidium* with *Plasmodium*. Later, Wenyon (1926) reinterpreted Dobell's figures as multiple infections of small gametocytes and placed *H. simondi* in *Haemoproteus*, synonymizing *Haemocystidium* with that genus (Telford, 1996a).

AUSTRALIAN *HAEMOCYSTIDIUM* SPECIES

Haemocystidium mackerrasae (Paperna and Landau) 1991

Diagnosis A *Haemocystidium* species with usually round gametocytes, nonhalteridial in position, 7–17 × 4–12 μm, with LW 29–202 μm^2 and L/W 1.0–2.8. Gametocyte dimensions are sexually dimorphic, with macrogametocytes larger and usually more elongate than microgametocytes, sometimes with irregular margins. Pigment granules are fine, sometimes coarse in macrogametocytes, and randomly dispersed.

Type Host *Heteronotia binoei* Gray (Sauria: Gekkonidae).

Type Locality Townsville, northern Queensland, Australia.

Other Hosts None known.

Other Localities Mornington Island, Gulf of Carpentaria, Queensland (Mackerras, 1961a).

Prevalence Three of 7 *H. binoei* infected from Mornington Island (Mackerras, 1961a); 1 of 29 (3.4%) from Townsville (Paperna and Landau, 1991).

Morphological Variation Macrogametocytes are 13.7 × 8.4 μm (9.6–16.8 × 5.6–12.0, N = 12), with L/W 1.8 (1.1–2.8). Estimated LW is 78 μm^2 (36–138) (Telford, 1996a). Dimensions of microgametocytes are 7.7 × 5.3 μm (7.2–11.2 × 4.0–7.2, N = 12), with L/W 1.6 (1.0–2.2) and estimated LW 41 μm^2 (29–81). Microgametocytes are usually oval and macrogametocytes variable in shape. In addition to greater dimensions, macrogametocytes sometimes have irregular margins. Pigment in both sexes forms as fine granules, dispersed randomly, with macrogametocytes showing a few coarser grains.

Merogony Unknown.

Sporogony Unknown.

Effects on Host Infected erythrocytes were little affected by *H. mackerrasae*, with only slight enlargement if any, except in the case of multiply infected cells, which could be hypertrophied in either dimension or distorted (Paperna and Landau, 1991). Erythrocyte nuclei usually remained central in the erythrocyte, even with multiple infections, seldom being displaced.

Remarks *Haemocystidium mackerrasae* (original spelling *mackerrasi*) is one of four *Haemocystidium* species that parasitize Australian geckoes and was confused with *H. simondi* by Mackerras (1961a).

Haemocystidium gehyrae (Paperna and Landau) 1991

Diagnosis A *Haemocystidium* species with usually round microgametocytes and more elongate macrogametocytes that sometimes assume a halteridial position around the erythrocyte nucleus. Gametocytes are 7–14 × 4–10 μm, with estimated LW 35–138 μm^2 and L/W ratios that average 1.33 in microgametocytes and 1.75 in macrogametocytes. Only shape differs sexually; dimensions are similar. Pigment granules are usually dispersed and fine in both sexes but sometimes are aggregated in microgametocytes and may form more coarsely in macrogametocytes.

Type Host *Gehyra australis* Gray (Sauria: Gekkonidae).

Type Locality Townsville, North Queensland, Australia.

Other Hosts None known.

Other Localities None known.

Prevalence One of 19 *G. australis* collected in 1986 was host to *H. gehyrae* (Paperna and Landau, 1991).

Morphological Variation Microgametocytes are 9.8 × 7.6 μm (7.2–12.0 × 4.8–10.4, N = 22), with estimated LW 74.5 μm^2 (35–125) and mean L/W 1.33 (N = 46) in both immature and mature microgametocytes. Macrogametocytes are 10.5 × 7.4 μm (9.0–14.4 × 4.0–9.6, N = 43), with estimated LW 77.7 μm^2 (36–138) and mean L/W 1.80. Unusually large macrogametocytes, attributed to *H. gehyrae*, can be 19.2–25.6 × 4.0–7.2, with L/W 3.4–4.8, which form as halteridia. Except for these exceptionally large macrogametocytes, dimensions of both sexes are similar, and they differ primarily in shape, microgametocytes being round or ovoid and macrogametocytes commonly elongate. Pigment granules are usually uniform in size and dispersed evenly but can sometimes be aggregated in males or coarser in macrogametocytes.

Merogony Unknown.

Sporogony Unknown.

Effects on Host Infected erythrocytes are hypertrophied with distorted and displaced nuclei. The effects are most severe in multiply infected cells with three or more parasites. Concurrent infection of cells with LEV caused "atrophy" (hypotrophy?) and additional distortion.

Remarks This species is known from a single infection that was followed for several months in the laboratory. The possibility that the very large macrogametocytes with halteridial configuration represent a separate species cannot be ruled out, but they might also indicate chronic-stage gametocytes.

Haemocystidium oedurae (Paperna and Landau) 1991

Diagnosis A *Haemocystidium* species characterized by predominantly elongate, halteridial gametocytes of both sexes that may assume this configuration while yet immature. A few gametocytes appear in rounded form. Gametocytes are 6–22 × 3–11 μm, with estimated LW 22–192 μm^2 in microgametocytes and 44–242 μm^2 in macrogametocytes, with L/W ratios similar, 1.2–5.5 in each sex. Macrogametocytes are larger on average than microgametocytes. Pigment in both gametocyte sexes is dispersed as fine and coarse granules, but in macrogametocytes can occasionally form aggregates. A variable number of vacuoles are present in both ends of the gametocytes.

Type Host *Oedura castelnaui* (Thominot) (Sauria: Gekkonidae).

Type Locality Townsville, North Queensland, Australia.

Other Hosts Not known.

Other Localities Fletcher View, 15 km east of Charter Towers (141 km east of Townsville), North Queensland.

Prevalence One of four *O. castelnaui* was infected by *H. oedurae* near Townsville (Paperna and Landau, 1991).

Morphological Variation Sexually undifferentiated gametocytes as small as 4.0–8.0 × 1.2–4.0 μm begin to form as halteridia. Mature microgametocytes are 8.0–20.0 × 2.8–9.6 μm, with estimated LW 22–192 μm^2 and L/W 3.10 (1.2–5.5, N = 20). Microgametocytes are predominantly halteridial (>70%), with a few, in polar position, rounded. Both fine and coarse pigment granules are dispersed in the cytoplasm. Mature macrogametocytes were larger than microgametocytes, averaging 18.0 × 6.8 μm (13.6–21.6 × 3.2–11.2, N = 43), with estimated LW 122 μm^2 (44–242) and L/W 2.65 (1.30–5.50), a similar range in shape but slightly less elongate, on average, than in microgametocytes.

Over 70% are halteridial, with the remainder rounded in a polar position in the host erythrocyte. Although pigment of macrogametocytes usually forms as dispersed fine granules, "larger macrogametocytes accumulated coarser pigment granules and exceptionally formed discrete aggregates" (Paperna and Landau, 1991). In both sexes, a variable number of vacuoles are present at both ends of the gametocytes.

Merogony Unknown.

Sporogony Unknown.

Effects on Host Host erythrocytes are enlarged and, with multiple infections, distorted. Nuclei in parasitized cells can retain their central position, be laterally displaced, or move into polar position.

Remarks Mackerras (1961a) reported a *Haemocystidium* sp. in 2 of 14 *Oedura tryoni* from Eidsvold, Mornington Island, Queensland, and figured a halteridial gametocyte but provided no dimensions for comparison with *H. oedurae*.

Haemocystidium underwoodsauri (Paperna and Landau) 1991

Diagnosis A *Haemocystidium* species characterized by round-to-oval microgametocytes and elongate halteridial macrogametocytes, which may, on completion of growth, nearly fill the host erythrocyte, completely encircling the host cell nucleus. Microgametocytes are 5.4–9.6 × 4.0–7.6 µm, with L/W ratio 1.0–1.8 and estimated LW 22–73 µm². Macrogametocytes are 14.4–19.2 × 3.2–9.6 µm, L/W 1.8–6.0, and estimated LW 46–184 µm². Pigment granules of variable sizes are dispersed in microgametocytes especially along margins, but occasionally in macrogametocytes may concentrate around small vacuoles. Large vacuoles are present in microgametocytes.

Type Host *Underwoodisaurus milii* (Bory de Saint-Vincent) (Sauria: Gekkonidae).

Type Locality Escarpment along the Murrey River at Manum, South Australia.

Other Hosts None known.

Other Localities None known.

Prevalence One of five *U. milii* was infected by *H. underwoodsauri* (Paperna and Landau, 1991).

Morphological Variation Microgametocytes are round or ovoid, 5.4–9.6 × 4.0–7.6 µm (N = 18), with L/W 1.0–1.8 and estimated LW 27–73 µm². In single occupancy of an erythrocyte, microgametocytes average slightly smaller, 7.3 × 6.3 µm (N = 9), and are rounder, with L/W 1.18, in comparison to those from double infections, which are 8.4 × 6.4 µm (N = 9), with L/W 1.34. Macrogametocytes are larger than microgametocytes, 16.2 × 7.4 µm (14.4–19.2 × 3.2–9.6, N = 13), and elongate, with L/W ratios 1.8–6.0. Estimated LW averages 120 µm² (46–184). Initially, macrogametocytes have halteridial shapes but with growth become bulky, nearly filling the host erythrocyte and encircling its nucleus or pushing it to one end of the host cell. A single macrogametocyte attained dimensions of 34.3 × 4.8 µm, and extracellular macrogametocytes were 24 × 8 µm. In both sexes, pigment granules are variably sized and dispersed but in microgametocytes tend to lie along the periphery, while in macrogametocytes some granules concentrate around a small vacuole. Microgametocytes usually have large vacuoles present in the cytoplasm.

Merogony Unknown.

Sporogony Unknown.

Effects on Host In single infections, gametocytes produced only slight hypertrophy of the host erythrocyte, with little distortion of the cell or its nucleus and only occasional displacement of the nucleus. Hypertrophy of host cells increased in double infections.

Remarks Paperna and Landau (1991) distinguished *H. underwoodsauri* from other halteridial species of *Haemocystidium* by the absence of large vacuoles in macrogametocytes and the nonhalteridial microgametocytes.

PALEARCTIC *HAEMOCYSTIDIUM* SPECIES

Haemocystidium phyllodactyli (Shortt) 1922

Diagnosis A *Haemocystidium* species with elongate gametocytes that in macrogametocytes, at least, may become halteridial in shape. Gametocytes are 13 × 6–15 × 5 µm, with estimated LW 79–83 µm² and L/W 2.30–2.94. The yellow or golden-brown pigment forms clusters and tends to aggregate about the two or more large vacuoles in the cytoplasm of both sexes. There are no sexual differences in dimensions.

Type Host *Phyllodactylus elisae* Werner (Sauria: Gekkonidae).

Type Locality Quritu, Persian Frontier (Iranian with India?).

Other Hosts None known.

Other Localities None known.

Prevalence *Haemocystidium phyllodactyli* is known from the type infection only (Shortt, 1922).

Morphological Variation Shortt (1922) provided only minimal dimensional data on *H. phyllodactyli*. Microgametocytes are "stout, sausage-shaped forms," 13.8 × 6.0 µm, with the largest forms, curling around the host cell nucleus, reaching 15.3 × 5.2 µm. Estimated LW values are 79–83 µm^2, using these measurements, with L/W ratios 2.30–2.94. There is no sexual difference in size or shape. Gametocytes of both sexes contain two or more large vacuoles, around which the yellow or golden-brown pigment granules tend to cluster in addition to aggregates of granules elsewhere within the cytoplasm.

Merogony Shortt (1922) described and figured a probable EE meront:

> In the wall of one of the air spaces of the lung an irregularly oblong body surrounded by a cyst-like wall was found. From one side of this cyst-wall a second similar but less densely-staining body projected outwards. Both bodies took a deep blue coloration with Giemsa stain and there was an indication of more darkly-staining areas in parts. Both bodies and the cystwall contained numerous and uniformly scattered grains of golden-brown pigment. The measurement of the cyst-wall in its greatest diameter was 11.3 microns and of the contained body 7 microns.

Sporogony Unknown.

Effects on Host There was little effect on host erythrocytes except slight nuclear displacement.

Remarks It is difficult to decide from Shortt's inadequate description whether gametocytes of both sexes can be described as "halteridial." Until the species is found again, it is probably best to consider microgametocytes as elongate and macrogametocytes halteridial. The presence of pigment in what appears to be EE meronts is also troubling.

Haemocystidium grahami Shortt 1922

Diagnosis A *Haemocystidium* species with round-to-ovoid, nonhalteridial gametocytes with estimated dimensions 19–23 × 14–15 µm, LW 275–353 µm^2, and L/W 1.27–1.49. Aggregates of dark pigment granules are scattered throughout the cytoplasm and form a border to small vacuoles when present.

Type Host *Agama nupta fusca* (= *Laudakia nupta*) (De Filippi) (Sauria: Agamidae).

Type Locality No precise locality stated: west and northwest Persia (Iran).

Other Hosts None known.

Other Localities None known.

Prevalence Not stated.

Morphological Variation The description of *H. grahami* by Shortt (1922) is nearly completely inadequate: "These mature forms measured in their greatest diameter 12 microns to 14.1 microns according to their shape. … Vacuoles and pigment were present as in the first-described species [*H. phyllodactyli*] but the vacuoles were neither so large nor so numerous." The illustrations show two mature gametocytes, one of each sex, in which aggregates of dark pigment granules are scattered throughout the cytoplasm, and small vacuoles appear to be bordered by pigment granules. Shortt provided no magnification or scale for the figures, but two of the figures of small parasites show apparently normal erythrocyte nuclei. Nuclei from uninfected erythrocytes of *Agama nupta fusca* from Pakistan average 6.6 ± 0.7 × 4.1 ± 0.4 µm (5–7 × 3.5–5, N = 10), with LW 26.6 ± 2.5 µm^2. Using these dimensions in comparison to the nuclei figured by Shortt, estimated size of the microgametocyte is 22.9 × 15.4 µm, LW 352.7 µm^2, and L/W 1.49, and that of the macrogametocyte is 18.7 × 14.7 µm, LW 274.9 µm^2, and L/W 1.27.

Merogony Unknown.

Sporogony Unknown.

Effects on Host Gametocytes of *H. grahami* cause apparent hypertrophy and distortion of host erythrocytes, displacement of nuclei laterally or to one pole of the cell, and some distortion by pressure on the nucleus.

Remarks Literature records of *H. grahami*, cited by Ovezmukhammedov (1987) from *Agama caucasica, A. lehmanni*, and *A. erythrogaster* from Turkmenistan and Uzbekistan, probably represent *Haemocystidium papernae* in view of the halteridial gametocytes figured by Ovezmukhammedov (Telford, 1996a).

Haemocystidium papernai Telford 1996 (**Plate 43**)

Diagnosis A *Haemocystidium* species with gametocytes 12–20 × 6–10 μm, LW 78–149 μm^2, and L/W 1.4–3.0, that usually assume a halteridial position within the host erythrocyte. There is no sexual dimorphism in gametocyte dimensions or in pattern of pigment dispersal. Merogony occurs in endothelium and connective tissue of lung, liver, spleen, and femoral muscles.

Type Host *Laudakia nupta* (De Filippi) (Sauria: Agamidae).

Type Locality Khadeji Falls, Karachi District, Sind Province, Pakistan.

Other Hosts *Trapelus agilis* (Agamidae).

Other Localities Rani Kot, Dadu District, Sind Province, Pakistan.

Prevalence Nineteen of 34 (55.9%) *L. nupta* and 1 of 5 *T. agilis* at the type locality; 1 of 2 *L. nupta* at Rani Kot.

Morphological Variation In active infection, gametocytes are 15.1 ± 1.4 × 6.8 ± 0.8 μm (12–20 × 6–8, N = 50), with LW 102.6 ± 13.1 μm^2 (78–140) and L/W 2.26 ± 0.35 (1.7–3.0). Dimensions are similar between sexes, with macrogametocytes 15.4 × 6.9 μm, LW 105 μm^2, and L/W 2.3, while microgametocytes are 14.9 × 6.7 μm, LW 100 μm^2, and L/W 2.2. Chronic-phase gametocytes are shorter, broader, and less elongate in shape, 14.4 ± 1.6 × 8.0 ± 0.9 μm (12–18 × 6–9.5, N = 25), with LW 114.8 ± 19.7 μm^2 (84–148.5) and L/W 1.82 ± 0.30 (1.4–2.5). Chronic-phase macrogametocytes are longer and broader than microgametocytes but similar in shape, 15.0 × 8.4 μm, LW 123 μm^2, and L/W 1.9 versus 13.7 × 7.8 μm, LW 106 μm^2, and L/W 1.8. Gametocytes in either active or chronic phase of infection are commonly halteridial in position within the host cell. Pigment granules, dispersed similarly in both sexes, number 24–36 in macrogametocytes and 17–40 in microgametocytes, with means of 29.4 and 28.9, respectively.

Merogony Meronts of *H. papernai* were common in connective tissue and the endothelium of blood vessels associated with the heart; in the pericardium, lungs, liver, spleen, kidneys; and between muscle masses of the femur (**Plate 43B, a–j**). None were present in hepatic parenchyma, splenic pulp, or bone marrow. The largest meronts were 17–20 × 12–15 μm, with 48–80 nuclei visible in single focal planes. Meronts were most commonly oval but in sections of femur showed attenuated ends (Telford, 1996a).

Sporogony Unknown.

Effects on Host Infected erythrocytes do not differ in length from normal cells but are wider, with greater LW, and distorted. Nuclei are displaced but similar in size and shape to nuclei of uninfected erythrocytes.

Remarks The course of infection, gametocyte growth, sex ratio of gametocytes, and seasonal prevalence were described by Telford (1996a).

Haemocystidium quettaensis Telford 1996 (**Plate 43**)

Diagnosis A *Haemocystidium* species with typically oval, never halteridial, gametocytes, 7.5–16 × 5–10 μm, with LW 45–150 μm^2 and L/W 1.0–2.6. Gametocytes are sexually dimorphic in dimensions, shape, and quantity of pigment, which tends to be distributed peripherally in clusters, in both sexes.

Type Host *Laudakia nupta* (De Filippi) (Sauria: Agamidae).

Type Locality Hazara Gunji village, Kalat District, Baluchistan Province, Pakistan.

Other Hosts None known.

Prevalence Two of two *Laudakia nupta* collected at the type locality were infected by *H. quettaensis*.

Morphological Variation Gametocytes are usually oval and never halteridial in position, 10.9 ± 1.8 × 7.9 ± 1.1 μm (7.5–16 × 5–10, N = 51), with LW 85.5 ± 19.4 μm^2 (45–150) and L/W 1.41 ± 0.34 (1.0–2.6). Macrogametocytes are 12.0 ± 1.1 × 8.0 ± 1.1 μm (9–16 × 5–10), with LW 96.2 ± 20.4 μm^2 (70–150) and L/W 1.53 ± 0.32 (1.1–2.6). Microgametocytes are 9.9 ± 1.2 × 7.7 ± 1.1 μm (7.5–13 × 5–9, N = 28), with LW 76.7 ± 13.4 μm^2 and L/W 1.32 ± 0.35 (1.0–2.4). Macrogametocytes have 39.8 pigment granules (25–70), in contrast to 30.7 (24–37) in microgametocytes.

Merogony Unknown.

Sporogony Unknown

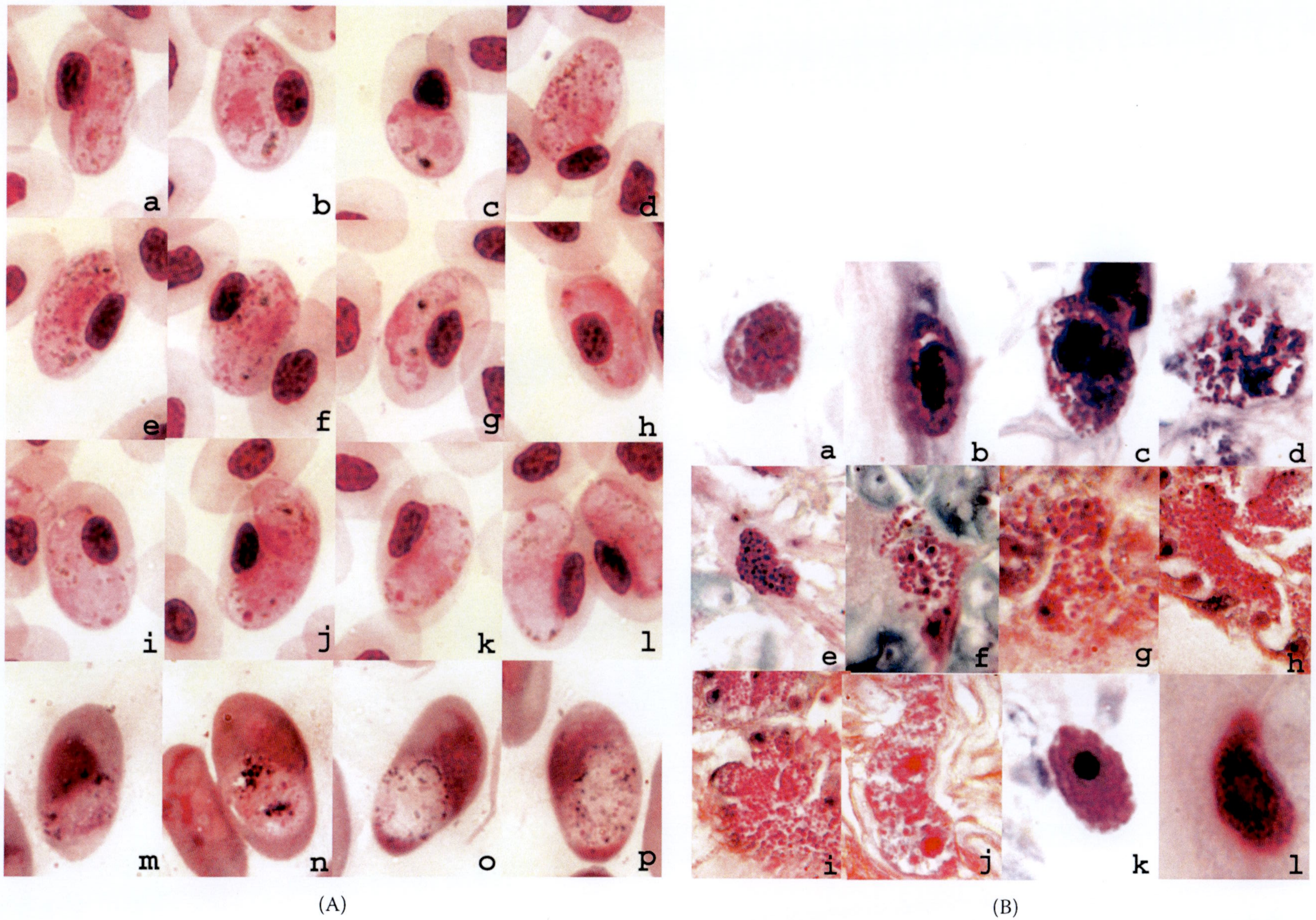

Plate 43 (**A**) *Haemocystidium papernae* (**a–l**) and *Haemocystidium quettaensis* (**m–p**) from *Stellio nupta*, Pakistan: macrogametocytes, **a–h**, **m**, **n**; microgametocytes, **i–l**, **o**, **p**. (**B**) Meronts of *Haemocystidium* spp.: from *H. papernae* heart (**a–d**), spleen (**e–i**), and lung (**j**); from *H. lygodactyli* heart (**k**, **l**). (Figure **j** modified from Telford, S. R., Jr., *Syst. Parasitol.*, 34, 197, 1996, Figure 63, with kind permission of Springer Science and Business Media.)

Effects on Host Erythrocyte length is greater in cells infected by macrogametocytes but not in those host to microgametocytes, in comparison to normal cells. Erythrocyte width and LW are increased by both gametocyte sexes, and cells are sometimes distorted by broadening. Erythrocyte nuclei dimensions are not altered, but nuclei are almost always displaced and sometimes distorted in shape.

Remarks Gametocyte size, shape, presence of sexual dimorphism in dimensions in active infection, and quantity of pigment and its distribution distinguish *H. quettaensis* from *H. papernai*. An additional character is the shape of immature gametocytes: elongate and often slender in *H. papernai*, oval in *H. quettaensis* (Telford, 1996a). At the time of description, their hosts were classified as distinct subspecies, *Agama nupta fusca* (*H. papernai*) and *Agama n. nupta* (*H. quettaensis*).

Haemocystidium tarentolae (Parrot) 1927 (Paperna and Landau) 1991

Diagnosis A *Haemocystidium* species with round to elongate, halteridial gametocytes that may, with age, broaden to nearly fill the host cell. Vacuoles are more common in younger gametocytes and may be replaced by a large, compressed cistern or folds indicating an empty cistern. Gametocytes are 8–18 × 4–12 μm, with L/W ratios 1.0–3.4, younger microgametocytes tending to be more elongate than macrogametocytes, with little difference in size. As macrogametocytes age, they enlarge and may become more elongate. Estimated LW is 32–146 μm^2, but halteridial macrogametocytes may range from 172 to 202 μm^2. Dark pigment granules vary from coarse to fine in size and are dispersed in the cytoplasm.

Type Host *Tarentola mauritanica deserti* (Sauria: Gekkonidae).

Type Locality El Kantara, Departement de Constantine, Algeria.

Other Hosts *Tarentola annularis* (Riding, 1930).

Other Localities Khartoum, Sudan (Riding, 1930); Banyuls-sur-Mer, France (Paperna and Landau, 1991).

Prevalence Parrot (1927) found *H. tarentolae* in two *T. mauritanica deserti* at the type locality. In Khartoum, 3 of 20 (15%) *T. annularis* were infected (Riding, 1930), and Paperna and Landau (1991) found 4 of 23 (17.4%) *T. mauritanica* infected in southwest France.

Morphological Variation Riding (1930) reported gametocytes from *T. annularis* to measure about 15 × 7 μm. His illustrations are consistent with those of Parrot (1927) and Paperna and Landau (1991). In the sample from southwest France, microgametocytes are described as variably shaped, round, and polar in position or halteridial. They measure 8.0–17.6 × 4.0–8.0 μm (N = 36), with L/W ratios 1.66 (1.0–3.4). Round or oval gametocytes are 8.0–16.8 × 6.4–12.0 μm (N = 29), with L/W 1.0–2.2. Estimated LW values for microgametocytes are 32–141 μm^2. In the younger macrogametocytes, 10.6 × 8.4 μm (8.0–12.8 × 6.4–9.6, N = 8), vacuoles are commonly present, but in those longer than 13 μm, only a few show vacuoles. Apparently, older macrogametocytes from subsequent blood films are larger, 14.6 × 8.4 μm (11.2–16.8 × 6.4–12.0, N = 9), than those found on earlier slides, 13.4 × 8.1 μm (9.6–15.2 × 6.4–9.6, N = 12). The older macrogametocytes are more elongate, with L/W 1.94 (1.7–2.7) than those observed earlier, less than 12 μm long, which average 1.70 (1.1–2.1). These more elongate macrogametocytes presumably are more commonly halteridial or bulky in form. Estimated LW values for round-to-elongate macrogametocytes are 61–146 μm^2 and for halteridial forms are 172–202 μm^2. Pigment is described as comprised of both coarse grains and fine, scattered granules.

Merogony Unknown.

Sporogony *Culicoides nubeculosus, Culex pipiens molestus*, and *Phlebotomus papatasi* could not be infected in the laboratory.

Effects on Host Infected erythrocytes are slightly hypertrophied and distorted, with minimal displacement of their nuclei, but as macrogametocytes age and increase to become bulky in form, host cells broaden, and nuclei are pushed to one side or end of the cell.

Remarks Levine (1984) renamed the *Haemocystidium* present in *T. annularis* as *Haemoproteus sudani*. This should be considered a synonym of *H. tarentolae*.

Haemocystidium edomensis (Paperna and Landau) 1991

Diagnosis A *Haemocystidium* species with halteridial gametocytes in both sexes, more commonly in macrogametocytes. Gametocytes are 9–21 × 3–11 μm, with L/W ratios 1.0–5.2 and estimated LW ranges 31–197 μm^2. Large vacuoles are more common in, and characteristic of, macrogametocytes and less often present in microgametocytes. Dark pigment granules may be coarse or fine, dispersed in microgametocytes and aggregated around the large vacuoles of macrogametocytes.

Type Host *Laudakia* (= *Agama*) *stellio* Hasselquist and Linnaeus (Sauria: Agamidae).

Type Locality Maaleh Edomin, 6 km east of Jerusalem, Cisjordan.

Other Hosts None known.

Other Localities None known.

Prevalence One of three *L. stellio* collected in March 1986 and one of five collected in November 1988 were infected by *H. edomensis* (Paperna and Landau, 1991).

Morphological Variation Microgametocytes are 13.0 × 6.2 μm (9.6–17.6 × 3.2–11.2, N = 26), with L/W 2.18 (1.1–5.2), 58% of which were elongated and halteridial. Estimated LW is 80.6 μm² (31–197). Some contain one to several small vacuoles. Dark pigment granules are scattered in several areas of the microgametocytes. Macrogametocytes are 15.0 × 6.0 μm (12.0–20.8 × 3.2–8.8, N = 38), with L/W 2.9 (2.0–4.4, N = 32), 85% of which are halteridial in shape. Estimated LW is 90 μm² (38–183). Macrogametocytes contain several large vacuoles, usually located in one or both ends of the cells, around which coarse brown pigment granules are concentrated in the larger macrogametocytes. In smaller macrogametocytes, pigment is more dispersed and with finer grains.

Merogony Unknown.

Sporogony *Haemocystidium edomensis* did not infect *Culicoides nubeculosus, Culex pipiens molestus*, and *Phlebotomus papatasi* in the laboratory.

Effects on Host Infected erythrocytes are slightly enlarged when host to single gametocytes and show greater hypertrophy when multiply infected. Nuclear displacement and distortion is uncommon, usually only in multiply infected cells.

Remarks Paperna and Landau (1991) listed only *H. kopki, H. phyllodactyli*, and *H. edomensis* as species with heavy concentrations of pigment associated with vacuoles in gametocytes. In *H. kopki* microgametocytes, the pigment lies within the vacuole (Telford, 1982d) rather than surrounding it.

Haemocystidium ptyodactyli (Paperna and Landau) 1991

Diagnosis A species of *Haemocystidium* with halteridial microgametocytes. Macrogametocytes are predominantly halteridial, but some assume rounded configurations. Dimensions are similar for microgametocytes, 16.8–22.4 × 2.4–10.0 μm, and macrogametocytes, 15.2–24.0 × 3.6–6.0 μm, but L/W ratios of 2.1–9.0 and estimated LW of 40–224 μm² in microgametocytes are greater than in macrogametocytes, 3.1–6.6 and 55–144 μm², respectively. The cytoplasm at the ends of microgametocytes is highly vacuolated, with coarse brown-to-black pigment granules either dispersed or aggregated in the vacuolated areas. Macrogametocytes are not heavily vacuolated, with one to four vacuoles only, and the coarse black pigment is aggregated at the distal ends of the cell.

Type Host *Ptyodactylus hasselquisti* (Donndorf) (Sauria: Gekkonidae).

Type Locality Roadside cave near Marg'e Nag'a, Central Jordan Valley, Cisjordan.

Other Hosts Unknown.

Other Localities Unreported.

Prevalence Prevalence is up to 60% "in the same habitat" (Paperna and Landau, 1991).

Morphological Variation The halteridial microgametocytes are 19.6 × 7.4 μm (16.8–22.4 × 2.4–10.0, N = 20), with L/W 4.20 (2.1–9.0). Estimated LW is 145 μm² (40–224). The cytoplasm is heavily vacuolated at the distal ends of the microgametocytes, and coarse granules of black pigment are either aggregated or dispersed in the vacuolated areas. Macrogametocytes are predominantly halteridial, 19.9 × 4.8 μm (15.2–24.0 × 3.6–6.0, N = 11), with L/W ratios 3.95 (3.1–6.6) and LW estimated at 96 μm² (55–144). Those of rounded shape (~15%) are 11.2–13.6 × 8.0–9.2 μm (N = 2), with L/W ratios of 1.40 and 1.47. Only one to four vacuoles are present in macrogametocytes, and pigment granules, coarse and black, form one or two aggregates at the distal ends.

Exoerythrocytic Merogony Dividing meronts occupied parasitophorous vacuoles in tissues of the spleen of a *P. hasselquisti* with an infection of trophozoites only (Paperna and Finkelman, 1999). Meronts were 16 × 18–18 × 20 μm and contained about 100 merozoites.

Sporogony Attempts to infect *Culicoides nubeculosus, Culex pipiens molestus, Phlebotomus papatasi*, and *P. dubosqi* in the laboratory were unsuccessful (Paperna and Landau, 1991).

Effects on Host Erythrocytes infected with *H. ptyodactyli* are little affected by presence of the parasite, except possibly for slight displacement of the nucleus.

Remarks The EE meront of *H. ptyodactyli* figured by Paperna and Finkelman (1999) differs mainly in appearance by occupying a parasitophorous vacuole from those of *H. kopki* and *H. papernae* reported by Telford (1996a) from connective tissue and endothelium of lung and femoral muscle. No vacuoles were visible in these species. In *H. lygodactyli* (Telford, 2005), EE meronts occurred in endothelium and connective tissue of the liver. Although most did not occupy vacuoles, a possible meront was seen within a vacuole in a hepatic parenchymal cell.

Haemocystidium kopki De Mello 1916 (Telford) 1982 (Plate 44)

Diagnosis A *Haemocystidium* species characterized by sexual dimorphism in the distribution of pigment, that of macrogametocytes in the form of dark granules dispersed throughout the cytoplasm, and in microgametocytes, aggregated in large golden masses within vacuoles in one or rarely two foci. Gametocyte dimensions vary by sex, infection phase, and host species, with mean length and width varying (10.1–21.9 × 8.4–12.8 μm), LW mean values 85–281 μm², and mean L/W ratios 1.21–1.72. Merogony occurs in endothelium and connective tissues of the lungs.

Type Host *Hemidactylus brooki* Gray (Sauria: Gekkonidae).

Type Locality Goa, India.

Other Hosts *Teratoscincus scincus, Teratoscincus microlepis* (Gekkonidae).

Other Localities 28 km west southwest of Nushki and Darzi Chah village, Chagai District, Baluchistan Province, Pakistan.

Prevalence *Haemocystidium kopki* was present in 6 of 17 *T. scincus* (35%) collected in 1976, in 1 of 3 examined in the early 1990s, and in 1 of 3 *T. microlepis* collected in 1976.

Morphological Variation Estimated dimensions of *H. kopki* in the type host (Telford, 1982d), based on figures of F. De Mello (1934), are 21.9 ± 2.1 × 12.8 ± 1.2 μm (17.9–25.4 × 10.3–15.1, N = 17), with LW 281.8 ± 40.0 μm² (209.4–365.8) and L/W 1.72 ± 0.25 (1.3–2.2). Gametocytes in an initial active infection of a *Teratoscincus scincus* juvenile are 13.7 ± 1.3 × 9.8 ± 1.0 μm (11–17 × 7–12, N = 85), with LW 134.3 ± 18.1 μm² (98–187) and L/W 1.40 ± 0.21 (1.1–2.1). In a relapse infection from days 31–46 of patency in the same juvenile gecko, gametocytes are smaller and more rounded, 10.1 ± 1.1 × 8.4 ± 0.8 μm (9–13 × 7–11, N = 53), LW 85.2 ± 14.6 μm² (60–132), and L/W 1.21 ± 0.15 (1.0–1.5). Chronic-phase gametocytes are 15.2 ± 2.7 × 10.5 ± 0.9 μm (11–23 × 8–12, N = 50), with LW 158.0 ± 26.4 μm² (110–230) and L/W 1.46 ± 0.34 (1.0–2.5). Gametocytes in a chronic infection of *Teratoscincus microlepis* are 16.7 ± 1.8 × 11.6 ± 1.3 μm (14–21 × 10–14, N = 24), LW 193.8 ± 29.9 μm² (150–280), and L/W 1.45 ± 0.23 (1.1–2.1). Gametocytes of chronic infections in *T. scincus* have greater width and consequently LW values than in the active infection, but are less wide and have lower LW than in the chronic infection of *T. microlepis* (Telford, 1982d). Except in the relapse infection in which there were no differences between macrogametocytes and microgametocytes, macrogametocytes in *T. scincus* are larger and more elongate in shape than microgametocytes in active and chronic infections. In *T. microlepis*, macrogametocytes have greater width than microgametocytes, but length and shape do not differ. Except in the relapse infection, size (LW) of macrogametocytes is greater than in microgametocytes in infections from both host species (Telford, 1982d).

Merogony Although meronts of *H. kopki* were not found in tissues of *T. scincus* and *T. microlepis* in the first material examined (Telford, 1982d), meronts were present in the endothelium and connective tissue of the lungs in two *T. scincus* with massive parasitemias examined later (Telford, 1996a). Dimensions were 12 × 9–27 × 6 μm, with meronts usually oval in shape but occasionally elongate. In three of the smaller meronts, 12–15 × 9 μm, which were less intensely stained than larger ones, 66–70 nuclei were counted in single focal planes. Several erythrocytes in peripheral blood from a chronic infection of *H. kopki* in *T. scincus* contained very large meronts, 17–19 × 12 μm, with more than 184 and more than 194 nuclei only, without visible cytoplasm or pigment (Telford, 1996a). Although F. De Mello (1934) reported meronts of *H. kopki* in the lungs of *Hemidactylus brooki* "in … either a monocyte or an endothelial cell … most frequently … in the pulmonary cells, sometimes in the intracellular spaces," this appeared doubtful earlier (Telford, 1982d). The presence of pulmonary meronts of *H. kopki* in *T. scincus* (Telford, 1996a) provides some credibility to his observation.

Sporogony Unknown. Oocysts did not develop in *Aedes aegypti* fed on a *T. scincus* with a massive infection of *H. kopki.*

Effects on Host In two active infections, host erythrocytes are distorted and 10%-16% larger in size than uninfected erythrocytes, but there is no difference in size or

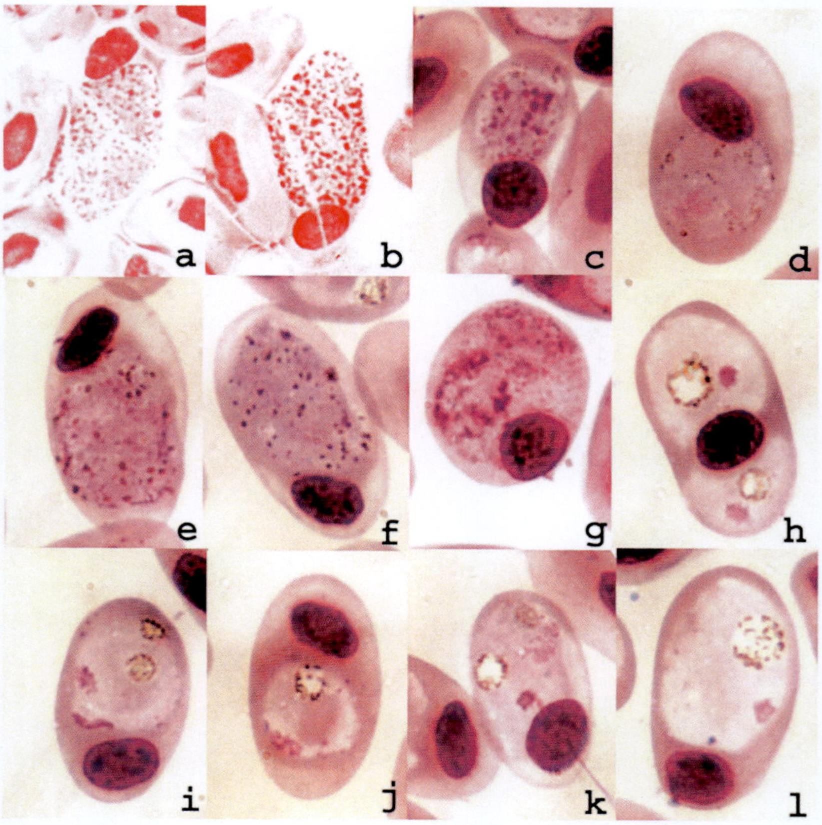

(A)

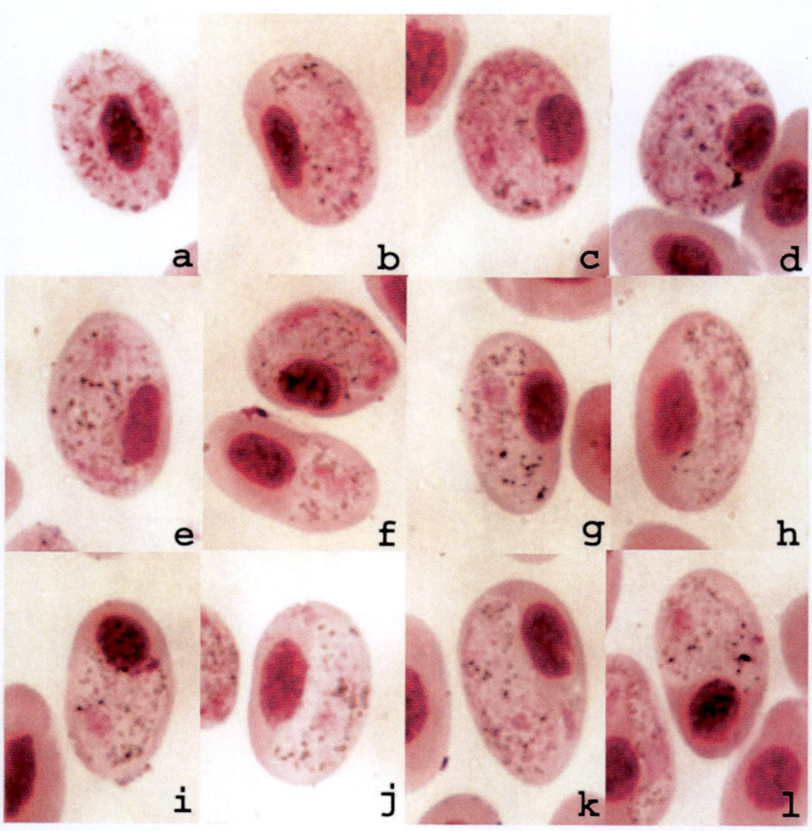

(B)

Plate 44 (**A**) *Haemocystidium kopki* from *Terato-scincus scincus*, Pakistan: erythrocytic meronts, **a**, **b**; macrogametocytes, **c–g**; microgametocytes, **h–l**. (Figures **a**, **b** modified from Telford, S. R., Jr., *Syst. Parasitol.,* 34, 197, 1996, Figures 87, 88, with kind permission of Springer Science and Business Media.) (**B**) *Haemocystidium lygodactyli* from *Lygodactylus capensis grotei*, Tanzania: macrogametocytes, **a–e**, **f** (top); microgametocytes, **f–l**.

shape of their nuclei, although all are displaced by parasite presence. Host erythrocytes in the relapse infection studied are 13% larger than normal cells and are all distorted; their nuclei are 9% greater in size, occasionally distorted and always displaced. In the chronic infections, host erythrocytes are enlarged by 19% and distorted; their nuclei are normal in size, sometimes distorted, and always displaced. Parasite-induced erythropoiesis was not present in active infections despite parasitemias exceeding 50%.

Remarks The aggregation of pigment into a large mass within a vacuole in microgametocytes only, in contrast to dispersed pigment in macrogametocytes, is unique to *H. kopki* and the principal reason for considering the parasites of *H. brooki* described by F. De Mello (1916) and the *Teratoscincus* spp. parasites to be conspecific, despite some differences in gametocyte size (Telford, 1982d). Growth of gametocytes and course of infection were reported also by Telford (1982d).

AFRICAN *HAEMOCYSTIDIUM* SPECIES

Haemocystidium lygodactyli Telford 2005 (**Plates 43 and 44**)

Diagnosis A *Haemocystidium* species, usually halteridial in position, with gametocytes 8–25 × 5–11 µm, LW 62–140 µm², and L/W ratio 1.1–3.9. Pigment is dispersed as black granules in both sexes. There is no significant sexual dimorphism in gametocyte dimensions. Elongate-to-oval meronts were found only in endothelium and connective tissue of lung.

Type Host *Lygodactylus capensis grotei* Sternfeld (Sauria: Gekkonidae).

Type Locality Northern slope of the Uluguru Mountains at edge of University Campus, Morogoro, Morogoro Region, Tanzania.

Other Hosts Unknown.

Other Localities Unknown.

Prevalence Five of 75 (6.7%) *Lygodactylus c. grotei* collected at the type locality were infected.

Morphological Variation Gametocytes are 16.3 ± 1.8 × 5.7 ± 1.1 µm (11–20 × 4–9.5, N = 50), with LW 93.0 ± 17.6 µm² (62–140), usually elongate, with L/W ratio 2.94 ± 0.58 (1.2–3.9). Pigment is dispersed as black granules and prominent. In active phase, there is no significant sexual dimorphism present in gametocyte dimensions: Microgametocytes are 16.5 ± 1.6 × 5.6 ± 1.1 µm (12.5–20 × 4–9, N = 25), LW 92.2 ± 20.5 µm² (62–140), L/W 3.04 ± 0.52 (1.7–3.9); macrogametocytes are 16.1 ± 2.0 × 5.9 ± 1.2 µm (11–19 × 5–9.5, N = 25), LW 93.9 ± 14.5 µm² (75–135), L/W 2.85 ± 0.63 (1.2–3.7). Chronic-phase gametocytes are 18.1 ± 3.2 × 8.7 ± 1.5 µm (8–25 × 5–11, N = 50), with LW 156.8 ± 41.3 µm² (80–250) and L/W 2.16 ± 0.53 (1.1–3.6). Mean gametocyte length and LW between sexes in chronic phase are similar, but microgametocytes are broader. Macrogametocytes are 17.8 ± 3.1 × 8.3 ± 1.8 µm (11–23 × 5–11, N = 25), with LW 148.9 ± 43.5 µm² (90–242) and L/W 2.24 ± 0.59 (1.1–3.6). Microgametocytes are 18.3 ± 3.3 × 9.0 ± 1.2 µm (11–25 × 7–11, N = 25), with LW 168.7 ± 34.1 µm² (99–250) and L/W 2.12 ± 0.38 (1.3–2.9). Macrogametocytes and microgametocytes in chronic phase are broader, have greater LW, and are less elongate than in active infection, but neither sex differs in gametocyte length by infection phase. Chronic-phase gametocytes are most commonly halteridial but rarely completely encircle the erythrocyte nucleus.

Merogony Meronts were found only in endothelium, connective tissue, and possibly a parenchymal cell of liver (**Plate 43B, k and l**). The shape was elongate to oval. Larger meronts, filled with nuclei, are 12.2 ± 1.8 × 6.9 ± 1.2 µm (10.0 × 5.0–16.0 × 9.0, N = 32), with LW 50–144 µm² (5.1 ± 23.1), and those apparently in cross section are 7.0 × 7.0–10.0 × 9.0 µm.

Sporogony Unknown.

Effects on Host In active phase, gametocytes did not alter length or width of infected erythrocytes or dimensions of their nuclei, but erythrocyte LW is greater than in uninfected cells. Host cells are often distorted by broadening (54%), with nuclei rarely displaced (4%), but nuclei are never grossly altered in shape. Gametocytes in chronic phase of infection increase the width and LW of host erythrocytes but do not alter nucleus size.

Remarks In two infections, gametocytes with more than one nucleus were present (Telford, 2005). Binucleate parasites were present at day 14 in one infection followed for 49 days and remained throughout the course of infection, with frequency of occurrence at 5% on day 14, 13% on day 21, and 34% on day 42. Gametocytes occasionally were seen with three nuclei present, and with normal gametocyte growth, binucleate parasites also increased in size, with most present on day 42 apparently mature gametocytes. Most were macrogametocytes, but binucleate microgametocytes were also present. Telford (2005) also presented data on the course of infection and sex ratio of gametocytes.

Haemocystidium opluri (Paperna and Landau) 1991

Diagnosis A *Haemocystidium* species characterized by microgametocytes more commonly oblong or oval than halteridial, and macrogametocytes that are halteridial to bulky in shape, filling much of the host erythrocyte. Gametocytes are 12–19 × 3–12 μm, with L/W ratios 1.1–5.8. Estimated LW is 41–230 μm². Macrogametocytes are less elongate and larger on average than microgametocytes. Fine pigment granules are distributed along the periphery of microgametocytes and are dispersed throughout the heavily vacuolated cytoplasm of macrogametocytes.

Type Host *Oplurus cuvieri* (Gray) (Sauria: Opluridae).

Type Locality Majunga, Madagascar.

Other Hosts *Oplurus quadrimaculatus*.

Other Localities Belo sur Tsiribihina and Fort Dauphin/Baie de Loukaio, Madagascar.

Prevalence Not stated.

Morphological Variation In *O. cuvieri* from Majunga, the usually oblong microgametocytes are 14.2 × 6.4 μm (12.8–18.4 × 3.2–11.2, N = 12), with L/W 2.7 (1.3–5.75) and estimated LW 90.9 μm² (41–206). A few oval microgametocytes are 10.4–13.6 × 9.6–11.2 μm, with L/W 1.1–1.2 and estimated LW 99.8–152.3 μm². Although not stated in the description, the figures indicate that some microgametocytes could be described as halteridial. Fine pigment granules are distributed along the periphery of the gametocytes. Macrogametocytes are halteridial to bulky, 16.6 × 8.5 μm (13.6–19.2 × 4.8–12.0, N = 15), with L/W 1.97 (1.3–3.5). Estimated LW is 141.1 μm² (65.3–230.4). The cytoplasm appears foamy and was highly vacuolated. Fine pigment granules are dispersed in the cytoplasm but do not form a border to the vacuoles. In *O. cuvieri* from Belo sur Tsiribhina, microgametocytes are 11.4 × 5.9 μm (9.6–14.4 × 4.0–7.2, N = 5), with L/W 2.0 (1.3–3.0) and estimated LW 67.3 μm² (38.4–103.7). There was "a higher concentration of coarse pigment granules" (Paperna and Landau, 1991). Macrogametocytes are oblong and halteridial-like, 13.3 × 7.1 μm (12.0–17.6 × 4.0–7.2, N = 8), with L/W 2.2 (1.5–4.4) and estimated LW 94.4 μm² (48–126.7). There were "heavy concentration of pigment granules" but no vacuoles in the cytoplasm. Gametocytes from *O. quadrimaculatus* were round or oval, rarely elongate, and not halteridial, smaller than in *O. cuvieri*. Microgametocytes are 8.2 × 5.9 μm (6.4–9.6 × 3.6–7.2, N = 6), with L/W 1.6 (1.0–2.6) and estimated LW 48.4 μm² (23.0–69.1). Pigment is comprised of dispersed fine and coarse granules, and the cytoplasm was "usually stain-resistant." Macrogametocytes, occasionally with irregular margins, are 10.8 × 7.4 μm (8.0–12.0 × 6.4–8.8, N = 21), with L/W 1.34 (1.0–2.3) and estimated LW 79.9 μm² (91.2–105.6). The cytoplasm was foamy in appearance, contained some vacuoles, and had "an extensive nuclear zone." Pigment is dispersed as fine granules.

Merogony Unknown.

Sporogony Unknown.

Effects on Host In the type infection, microgametocytes seldom affected host erythrocyte dimensions but could cause some lateral hypertrophy when surrounding the nucleus, displacing the latter when polar in position. Effects by macrogametocytes were similar, with larger macrogametocytes completely encircling the host cell nucleus or, when bulky in shape, displacing it. Nuclei otherwise appeared unaffected. In the infection from Belo sur Tsiribhina, host erythrocytes had smaller dimensions than did uninfected cells. Gametocytes from *O. quadrimaculatus* little affected host erythrocyte size except for occasional lateral expansion.

Remarks Size differences in gametocytes from *O. cuvieri* taken in different localities probably reflected difference in age of infection, as suggested by Paperna and Landau (1991), or perhaps in age of the gametocytes selected for measurement. The much smaller gametocytes from *O. quadrimaculatus* possibly represented host effect or infection age, but also might indicate a different *Haemocystidium* species present.

NEOTROPICAL *SAUROCYTOZOON* SPECIES

Saurocytozoon tupinambi Lainson and Shaw 1969 (**Plate 45**)

Diagnosis A *Saurocytozoon* species parasitic principally in lymphocytes, but occasionally in immature erythrocytes, with gametocytes 13–17 × 10–16 μm, with LW 90–206 μm² and L/W 1.1–1.9. Gametocyte cytoplasm in both sexes contains many dark "volutin" granules dispersed throughout, less conspicuous in microgametocytes than in macrogametocytes. Gametocytes are not sexually dimorphic in dimensions.

Type Host *Tupinambis nigropunctatus* Spix (Sauria: Teiidae) (syn. *T. teguixin* Linnaeus).

Type Locality Agua Preta, Utinga Forest, Pará State, Brazil.

Plate 45 (**A**) *Saurocytozoon tupinambi* from *Tupinambis teguixin*, Venezuela. Probable meronts: **a**, free fragment; **b–e**, in lymphocytes; macrogametocytes, **f–j**; microgametocytes, **k–l**. (Figures **a–c** and **e** from Telford, S. R., Jr., *Int. J. Parasitol.,* 8, 133, 1978, with permission, Elsevier.) (**B**) *Saurocytozoon mabuyi* from *Mabuya multifasciata*, Thailand. Probable meront, **a–c**; macrogametocytes, **d–h**; microgametocytes, **i–l**. (Figure **b** modified from Telford, S. R., Jr., *J. Parasitol.,* 69, 1141, 1983, Figure 2, with permission.)

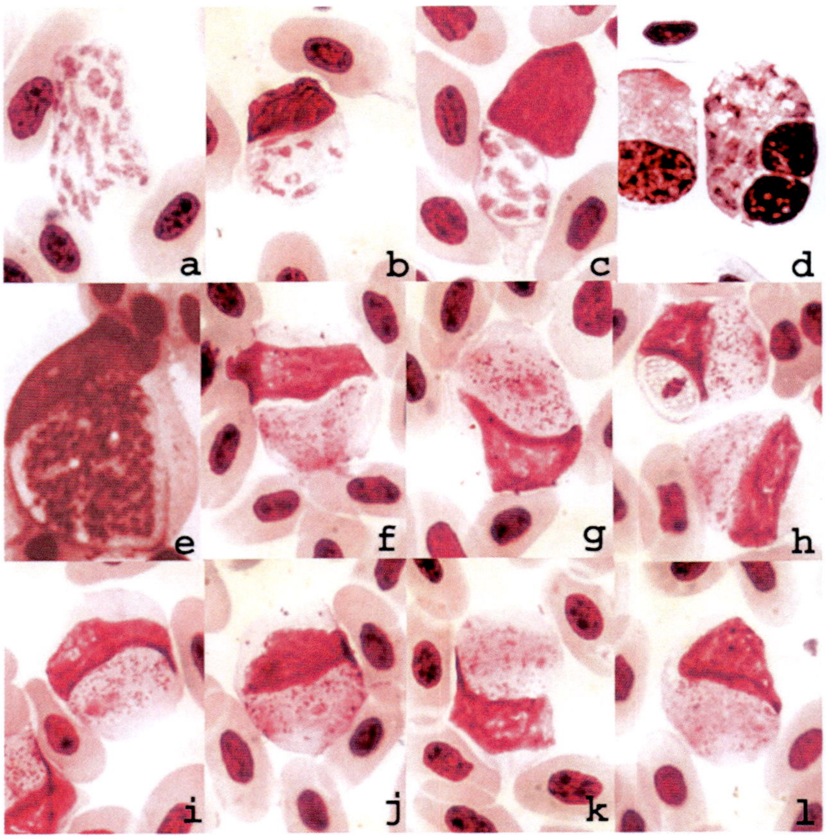

(A)

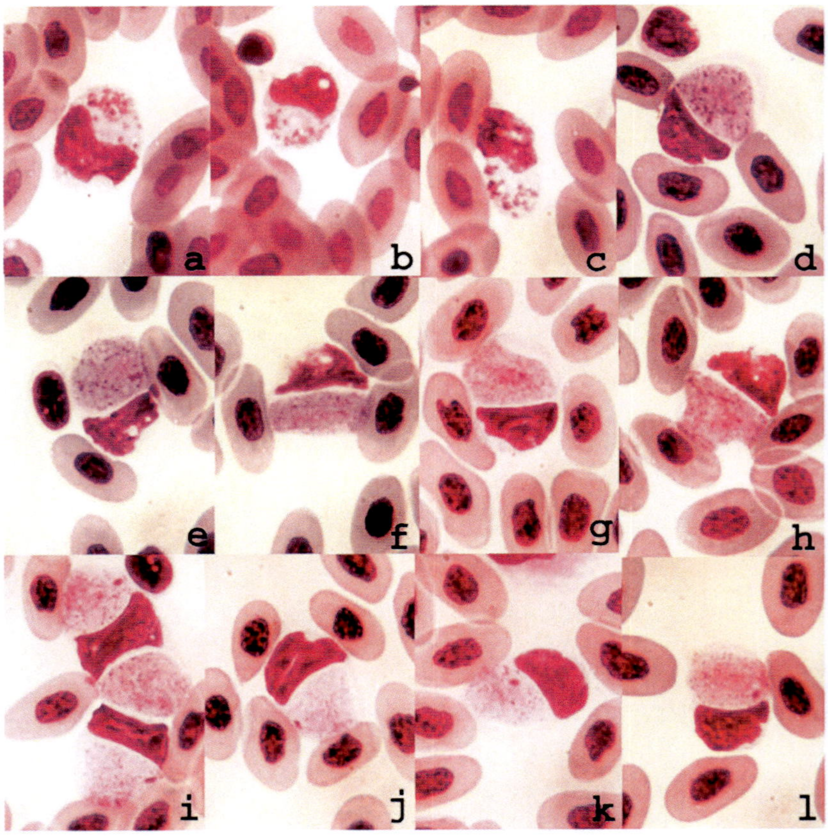

(B)

Other Hosts *Crocodilurus lacertinus* (Teiidae) (Lainson et al., 1974b).

Other Localities Bacarena, Pará State, Brazil (Lainson and Shaw, 1969a); Portuguesa and Cojedes states, Venezuela (Telford, 1978a).

Prevalence Five of 6 *T. nigropunctatus* in Brazil (Lainson and Shaw, 1969a) and 24 of 81 (29.6%) *T. teguixin* in Venezuela (Telford, 1978a) were infected by *S. tupinambi*.

Morphological Variation Lainson and Shaw (1969a) reported average dimensions of 15.6 × 13.2 µm for macrogametocytes and 15.6 × 12.6 µm for microgametocytes of *S. tupinambi* in Brazil. In an initial active infection of *S. tupinambi* in *T. teguixin* in Venezuela, mature gametocytes were 13.3 ± 1.3 × 8.7 ± 0.8 µm (10–15 × 7–10, N = 25), with LW 115.5 ± 15.3 µm^2 (90–150) and L/W 1.55 ± 0.21 (1.1–1.9). There is no difference between sexes in gametocyte dimensions: Microgametocytes were 13.5 ± 1.1 × 8.4 ± 0.8 µm (12–15 × 7–10, N = 12), with LW 113.5 ± 13.7 µm^2 (91–135) and L/W 1.62 ± 0.19 (1.0–1.2), and macrogametocytes were 13.2 ± 1.6 × 8.9 ± 0.8 µm (10–15 × 8–10, N = 13), with LW 117.3 ± 17.1 µm^2 (90–150) and L/W 1.48 ± 0.22 (1.1–1.8).

Merogony Lainson and Shaw (1969a) reported finding "a large piece of broken schizont ... in a spleen smear" from a host infected by *S. tupinambi*. Telford (1978a) followed an initial infection of *S. tupinambi* in a juvenile *T. teguixin* in Venezuela from the appearance of small uninucleate and binucleate parasites in lymphocytes to the appearance of gametocytes 40 days later. Larger meronts with 5–32 nuclei appeared in lymphocytes on day 8. On day 12, of 20 meronts found, 1 contained 102 nuclei, 7 had 18–32 nuclei, and the nucleus number in the remainder was 4–15. Most were in lymphocytes, the others in cells similar to them. On day 14, two larger meronts in lymphocytes had 44 and 52 nuclei and others 3–8. On day 29, 24- and 46-nucleate meronts were found, and two others contained 4 and 10 nuclei. On day 32, two meronts with 11 nuclei were found, as was a brood of "about 50 pyriform merozoites." The last meronts seen were on day 35 and had 18 and 14 nuclei. Fragments of larger meronts were found on some slides. While identity as *S. tupinambi* is indicated by presence of both meronts and gametocytes in the same type of host cells, sequential appearance of trophozoites, meronts, and gametocytes over 40 days, and correlation of lymphocyte density with parasitemia, the presence of an infection of *Plasmodium minasense tegui* in the same host prevents a definite identification of the lymphocytic meronts as *S. tupinambi* (Telford, 1978a).

Sporogony Landau et al. (1973) obtained sporogony of *S. tupinambi* in an experimental vector, the mosquito *Culex pipiens*. Ookinetes appeared 8–16 hours PF and were most abundant between 21 and 36 hours. Dimensions were 20–24 × 2–4.5 µm, with greatest width at the anterior end. Oocysts developed within the gut epithelium, expanding in size from 30–42 µm in diameter at 7 days to 62 µm over a period of 16 days, producing hundreds of long, thin sporozoites at completion of sporogony (Lainson et al., 1974b; Landau et al., 1973). Rupture of mature oocysts and invasion of the mosquito salivary glands was not observed (Landau et al., 1973), but degenerating oocysts were present.

Effects on Host Lymphocytes host to *S. tupinambi* became more rounded and enlarged as gametocytes grew, while their nuclei became elongated and narrow, situated closely along one side of the lymphocyte.

Remarks Although Landau et al. (1973) obtained mature oocysts containing sporozoites, these did not rupture, and no sporozoite invasion of the salivary glands occurred. This indicates that *Culex pipiens* is probably not a natural host to *S. tupinambi*. Completion of sporogony without oocyst rupture and salivary gland invasion has been reported for other saurian plasmodiid species when unsuitable hosts were involved (Petit et al., 1983; Klein, 1985).

A lymphocytic parasite of *Ameiva ameiva praesignis* in Estado Portuguesa, Venezuela, may be a *Saurocytozoon* species, perhaps *S. tupinambi*, which occurs in the same localities where infected *A. ameiva* were taken. Apparent gametocytes are somewhat smaller, 10.9 ± 1.9 × 6.6 ± 1.3 µm (9–16 × 5–10, N = 29), with LW 72.2 ± 19.7 µm^2 (45–140) and L/W 1.72 ± 0.43 (1.0–1.7). Each character overlaps in dimensions with *S. tupinambi* gametocytes. Prevalence in *A. ameiva* was low, 4 of 135 (3%) examined, and parasitemias light. These apparent gametocytes do not appear to be the same reported by Lainson et al. (2003), who found Brazilian *A. ameiva* infected with possibly three hematozoan species, which appeared to be species of *Lainsonia* or *Hemolivia*.

NEOTROPICAL/ORIENTAL *SAUROCYTOZOON* SPECIES

Saurocytozoon mabuyi Lainson, Landau, and Shaw 1974 (**Plate 45**)

Diagnosis A *Saurocytozoon* species that parasitizes scincid lizards of the genus *Mabuya* in both neotropical and oriental regions. Gametocytes in neotropical hosts average 11.0 × 8.0 µm and parasitize both lymphocytes and monocytes. Gametocytes in oriental hosts occur only in lymphocytes and are 7–16 × 6–11 µm, with LW 49–160 μm^2 and L/W 1.0–2.2. Gametocytes from both regions contain prominent azurophilic granules dispersed throughout the cytoplasm. There is no sexual dimorphism in gametocyte dimensions.

Type Host *Mabuya mabouya* (Lacépède) (Sauria: Scincidae).

Type Locality Ananindeua, Pará State, Brazil.

Other Hosts *Mabuya multifasciata* (Telford, 1983a).

Other Localities Southern Thailand and Singapore (Telford, 1983a).

Prevalence Lainson et al. (1974b) found *S. mabuyi* in 2 of 31 (6.5%) *Mabuya mabouya* in Brazil. Prevalence in *Mabuya multifasciata* from southern Thailand was 19 of 123 (15.4%) and 1 of 6 from Singapore (Telford, 1983a).

Morphological Variation Dimensions of Brazilian *S. mabuyi* averaged 11.0 × 8.0 µm, yet the range stated by Lainson et al. (1974b) was "9.0 × 9.0 µm to 12.5 × 9.0 µm," clearly an error that escaped the authors. Estimated LW and L/W values, based on the average, are 88 μm^2 and 1.38. In the total sample of mature *S. mabuyi* gametocytes from southeast Asia, gametocytes averaged 11.3 ± 1.5 × 8.7 ± 1.0 µm (7–16 × 6–11, N = 132), with LW 98.6 ± 18.1 μm^2 (49–160) and L/W 1.31 ± 0.25 (1.0–2.2). Macrogametocytes were 11.4 ± 1.6 × 8.7 ± 1.1 µm (7–16 × 6–11, N = 94), with LW 99.8 ± 19.6 μm^2 (49–160) and L/W 1.33 ± 0.26 (1.0–2.2), and microgametocytes were 11.0 ± 1.1 × 8.7 ± 0.9 µm (9–13 × 6–10, N = 38), with LW 95.8 ± 13.5 μm^2 (70–120) and L/W 1.27 ± 0.22 (1.0–2.0), indicating no sexual dimorphism in dimensions. As reported by Telford (1983a), on slides taken from a single host, gametocytes from cardiac blood were smaller and narrower than those from peripheral blood: 10.4 ± 2.0 × 8.1 ± 1.0 µm (N = 25), LW 84.9 ± 18.5 μm^2, L/W 1.31 ± 0.30 versus 11.0 ± 1.0 × 8.9 ± 1.0 µm (N = 25), LW 98.4 ± 12.0 μm^2, L/W 1.25 ± 0.20. Gametocytes from bone marrow were still smaller and rounder than those from cardiac blood, 8.1 ± 1.0 × 7.0 ± 1.0 µm (N = 25), LW 57.2 ± 14.5 μm^2, and L/W 1.18 ± 0.15. These last gametocytes showed no sexual differentiation and had no azurophilic granules in the cytoplasm. Gametocytes from cardiac blood also lacked azurophilic granules but were sexually differentiated by staining reaction.

Merogony Possible meronts with up to 24 nuclei were seen in lymphocytes from peripheral blood of one *M. multifasciata* (Telford, 1983a), but their identity as *S. mabuyi* has not been established. None were seen in smears or sections of liver, spleen, lung, intestine, heart, or bone marrow.

Sporogony Unknown.

Effects on Host Cells host to *S. mabuyi* in Brazilian skinks contain distorted nuclei, flattened against one side of the lymphocyte by the presence of the gametocyte (Lainson et al., 1974b). In southeast Asian skinks, smaller lymphocytes had nuclei compressed into "half- or quarter-moon shapes, and pressed firmly against" the cell margins (Telford, 1983a). Parasitized lymphocytes in the bone marrow were hypertrophied to sizes exceeding the largest normal lymphocytes. In larger lymphocytes, their nuclei were less compressed by gametocytes and more similar in size to the parasite.

Remarks Telford (1983a) found no quantitative or qualitative grounds to justify recognition of the parasite of *M. multifasciata* as a species distinct from *S. mabuyi*. The diversity of *Mabuya* species is far greater in southeast Asia and Africa than in the neotropics, and this suggests a relatively recent arrival of *Mabuya* in the Western Hemisphere, perhaps accompanied by *Saurocytozoon mabuyi*. A southeast Asian origin of the genus appears likely because the only hemosporidian parasites yet known from African skinks are *Plasmodium* species (Telford, 1983a).

THE *FALLISIA* SPECIES OF LIZARDS
NEOTROPICAL *FALLISIA* SPECIES

Fallisia effusa Lainson, Landau and Shaw 1974

Diagnosis A *Fallisia* species that typically parasitizes thrombocytes, only rarely lymphocytes. Meronts are ovoid or round, 10.0–14.0 µm in diameter, estimated LW 100–196 μm^2, and contain 50–150 merozoites. Gametocytes are an elongate oval, 8–10 × 5–6 µm, estimated LW 40–60 μm^2 and L/W 1.60–1.67. Macrogametocytes are larger than microgametocytes. During the merogonic phase of infection, nearly all thrombocytes may become infected, with many parasites in multiple infections, some of which are bizarre in appearance.

Type Host *Neusticurus bicarinatus* (Linnaeus) (Sauria: Teiidae).

Type Locality Riverine tropical rain forest at Anayindeua and Capanema, Pará State, Brazil.

Other Hosts None known.

Other Localities None known.

Prevalence *Fallisia effusa* was present in 26 of 32 (81.3%) *N. bicarinatus* examined (Lainson et al., 1974a).

Morphological Variation Meronts are round or oval, about 10.0–14.0 μm in diameter, suggesting an LW value of 100–196 μm², and produce 50–150 merozoites. Gametocytes are an elongate oval in shape. Macrogametocytes are larger than microgametocytes in average dimensions, respectively 10.0 × 6.0 μm and 8.0 × 5.0 μm. Estimated average LW values and L/W ratios, respectively, are 60.0 μm² and 40.0 μm², 1.67 and 1.60. There are one to three prominent vacuoles in the cytoplasm of both sexes, and small azurophilic granules are present. As parasitemia rises until nearly 100% of thrombocytes are infected, multiple infections appear containing any mixture possible of parasite stages, and a second type of meront appears that produces "a very small number of merozoites" (chronic-phase forms?) (Lainson et al., 1974a). Chronic infection is characterized by the presence of gametocytes only in thrombocytes and lymphocytes.

Exoerythrocytic Merogony Merogony within fixed tissue cells has not been described.

Sporogony Unknown.

Effects on Host In infection of thrombocytes, their limiting membrane thickens considerably, "often to such an extent that the infected cell takes on a cyst-like appearance ... particularly so in cells containing mature gametocytes" (Lainson et al., 1974a). Infected thrombocytes may become rounded and nearly twice the size of normal cells. Host cell nuclei may enlarge slightly and become distorted when pushed to the side or end of the infected thrombocyte. As infection progresses, "virtually 100% of the thrombocytes were parasitized with variable mixtures of trophozoites, schizonts and developing gametocytes" (Lainson et al., 1974a).

Remarks *Fallisia effusa* is the most productive species of the genus in terms of merozoite numbers produced in thrombocytes, with mature meronts containing up to 150 tiny merozoites. Although multiply infected cells are common in most *Fallisia* species, the numbers of parasites comprising multiple infections also seems to be greater in this species, with up to five gametocytes commonly seen in a single thrombocyte.

Fallisia modesta Lainson, Shaw and Landau 1974

Diagnosis A *Fallisia* species parasitic primarily in lymphocytes, rarely monocytes or thrombocytes. Meronts are round to oval, 5.0 × 5.0–9.0 × 7.5 μm, estimated LW 25–67.5 μm², that produce 12–24 merozoites. Gametocytes are round to oval, 6.0 × 6.0–8.5 × 5.5 μm, with estimated LW 36.0–46.8 μm² and L/W 1.00–1.55. Dimensions differ little between sexes.

Type Host *Tropidurus torquatus hispidus* (= *T. hispidus*) (Spix) (Sauria: Tropiduridae).

Type Locality Utinga, near Belém, Pará State, Brazil.

Other Hosts Not known.

Other Localities Not known.

Prevalence *Fallisia modesta* was described from a single specimen of *T. hispidus*.

Morphological Variation Meronts average 5.5 × 5.0 μm (5.0 × 5.0–9.0 × 7.5), with estimated LW 27.5 μm² (25.0–67.5), and contain usually 16 (12–24, N = 28) merozoites. Illustrations suggest that meront size is less than or equal to that of the infected lymphocyte nucleus. Microgametocytes average 7.0 × 5.5 μm (6.0 × 6.0–8.5 × 5.5) and macrogametocytes 7.5 × 5.5 μm (7.0 × 6.0–8.5 × 5.5). Estimated LW values and L/W ratios, respectively, are 38.5 μm² (36.0–46.8) and 1.27 (1.00–1.55) versus 41.3 (42.0–46.8) and 1.36 (1.17–1.55). Gametocyte dimensions are similar between sexes. Cytoplasmic vacuoles are absent in gametocytes. Azurophilic granules occur in both sexes and are especially prominent in microgametocytes.

Exoerythrocytic Merogony Unknown.

Sporogony Unknown.

Effects on Host Host cells are primarily lymphocytes; rarely, monocytes or thrombocytes are infected. The nuclei of infected lymphocytes are "strongly indented by the larger parasites, particularly the mature gametocytes and the cytoplasm becomes virtually non-existent" (Lainson et al., 1974a).

Remarks Scorza (1970a, 1970b, 1971c) described meronts and gametocytes from thrombocytes of *T. torquatus* (= *T. hispidus*) in Venezuela. Although meronts are 6–9 μm in diameter, they contain 24–26 merozoites, a larger number than in *F. modesta*, and gametocytes are larger, 9–10 × 7–8 μm. Because of the difference in gametocyte size, merozoite number, and host cell type, this is probably a different species from *F. modesta*. Scorza (1970a, 1970b) found infected thrombocytes in 4 of 54 *Plasmodium tropiduri* infections, and Telford (unpublished) found 2 of 10 infections of *P. tropiduri* contained gametocyte-infected thrombocytes in Venezuela. Whether the infections in Venezuela represent a species of *Fallisia* or are perhaps a thrombocytic cycle of *P. tropiduri*, as favored by Scorza and Telford, remains to be determined, as it can be, by genomic analysis.

Fallisia biporcati Telford 1998 (**Plate 46**)

Diagnosis A *Fallisia* parasite of lymphocytes and thrombocytes, meronts of which may exceed gametocytes in size. Meronts are 10.5–16 × 9–13 μm, with LW 94–208 μm^2, and produce 28–60 merozoites. Relative to host cell nuclei, meronts are 1.3–2.4, and to normal thrombocyte nuclei are 1.1–2.5. Gametocytes are 10–15 × 6–12 μm, with LW 82–150 μm^2 and L/W 1.0–2.2. Relative to host cell nucleus size, gametocytes are 0.7–2.2, and to uninfected thrombocyte nuclei are 1.0–1.8. There is no sexual dimorphism in gametocyte dimensions.

Type Host *Anolis b. biporcatus* (Wiegmann) (Sauria: Polychrotidae).

Type Locality Gaspar Sabanas on Rio Madroño, approximately 8 km north northwest of Chepo, Panama Province, Panama.

Other Hosts *Anolis lionotus*.

Other Localities El Aguacate, Panama Province, Panama.

Prevalence Overall prevalence of *F. biporcati* in the type host was 2 of 101 (2.0%) and 1 of 3 at the type locality; in *A. lionotus*, prevalence was 1 of 176 (1.1%) at El Aguacate.

Morphological Variation Meronts average 13.3 ± 2.0 × 11.5 ± 1.3 μm (10.5-16 × 9-13, N = 11), with LW 153.8 ± 36.7 μm^2 (94–208). Merozoite number averages 38.3 ± 9.1 (28–60). Relative to host cell nuclei, meronts average 1.83 ± 0.42 (1.3–2.4, N = 11), and to normal thrombocyte nuclei average 1.82 ± 0.43 (1.1–2.5). Gametocytes are 12.6 ± 1.3 × 9.0 ± 1.4 μm (10–15 × 6–12, N = 25), with LW 113.1 ± 19.2 μm^2 (82–150) and L/W 1.43 ± 0.31 (1.0–2.2). Relative to host cell nucleus size, gametocytes averaged 1.06 ± 0.32 (0.7–2.2, N = 24), and to uninfected thrombocyte nuclei averaged 1.34 ± 0.23 (1.0–1.8). Dimensions of gametocytes are not sexually dimorphic.

Exoerythrocytic Merogony Unknown.

Sporogony Unknown.

Effects on Host Dimensions of infected thrombocytes were enlarged by both meronts and gametocytes, but their nuclei were hypertrophied only by the latter. Both stages always distorted host cells and nuclei and displaced the latter.

Remarks The intrathrombocytic gametocytes of *F. biporcati* and two other species were reported as "exoerythrocytic gametocytes" of *Plasmodium* species by Telford (1970a) prior to the recognition of *Fallisia* as a distinct genus. Two interpretations of gametocyte presence in nonerythroid cells were suggested: "one or (probably) more malaria-like species which parasitize white blood cells exclusively or (2) … stages are part of the mechanism whereby latent infections can again give rise to patent parasitemia of the erythrocytes."

Fallisia poecilopi Telford 1998 (**Plate 46**)

Diagnosis A *Fallisia* species parasitic in thrombocytes and lymphocytes, with meronts 5.5–9 × 3–6 μm, LW 22–54 μm^2, oval to elongate in shape, filled with 20–51 merozoites in no particular arrangement. Meront size relative to host cell nucleus is 0.6–1.8, and to nuclei of uninfected thrombocytes is 0.6–1.4. Gametocytes are 7.5–14 × 6–11 μm, with LW 45–132 μm^2, and are round to elongate with L/W ratio varying from 1.0 to 1.6. Gametocyte size relative to host cell nucleus is 0.8–3.6, and to nuclei of uninfected thrombocytes is 1.0–2.8. There are no sexual differences in gametocyte dimensions or shape.

Type Host *Anolis poecilopus* (Cope) (Sauria: Polychrotidae).

Type Locality Panama, Panama Province, approximately 8 km north northwest of Chepo, Gaspar Sabanas on Rio Madrono (9°10′N, 79°06′W).

Other Hosts Not known.

Other Localities Panama, Colon Province, Santa Rita ridge along Rio Agua Clara.

Prevalence Overall, *F. poecilopi* parasitized 3 of 227 (1.3%) *Anolis poecilopus*, 2 of 47 (4.3%) from the type locality and 1 of 16 (6.3%) from the Santa Rita ridge.

Morphological Variation After the second nuclear division, meronts are 6.0–8.5 × 4.0–5.5 μm. Nuclei in meronts

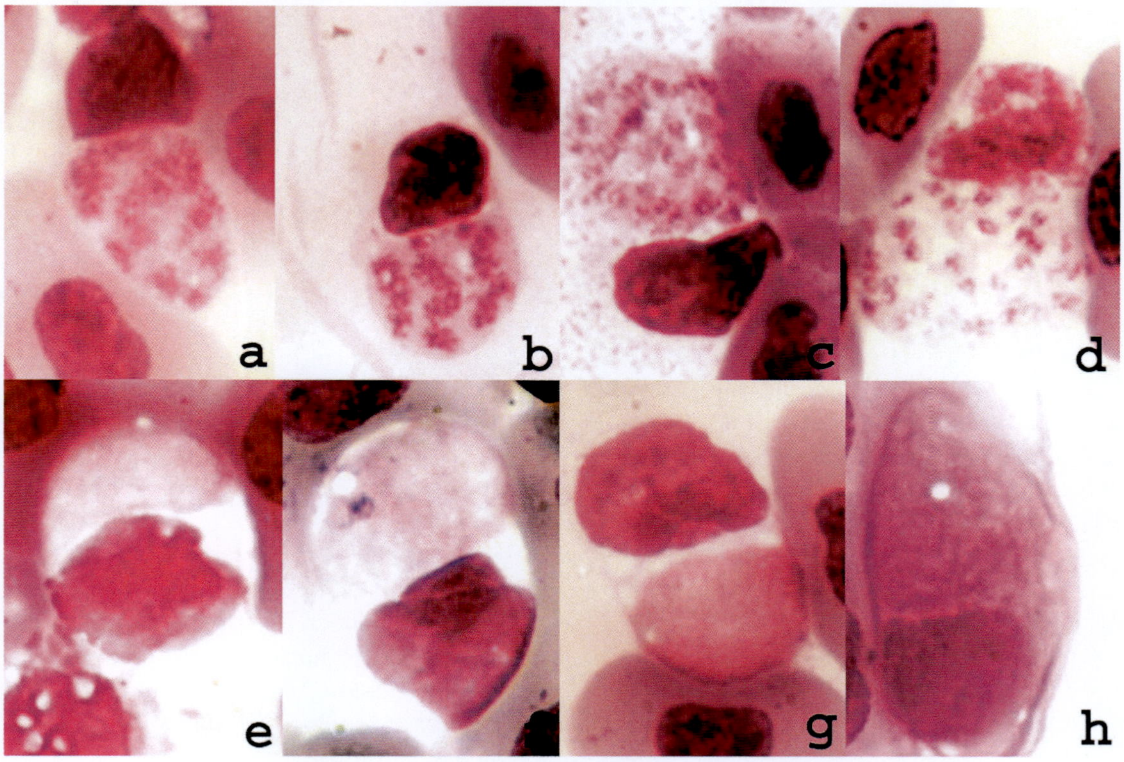

(A)

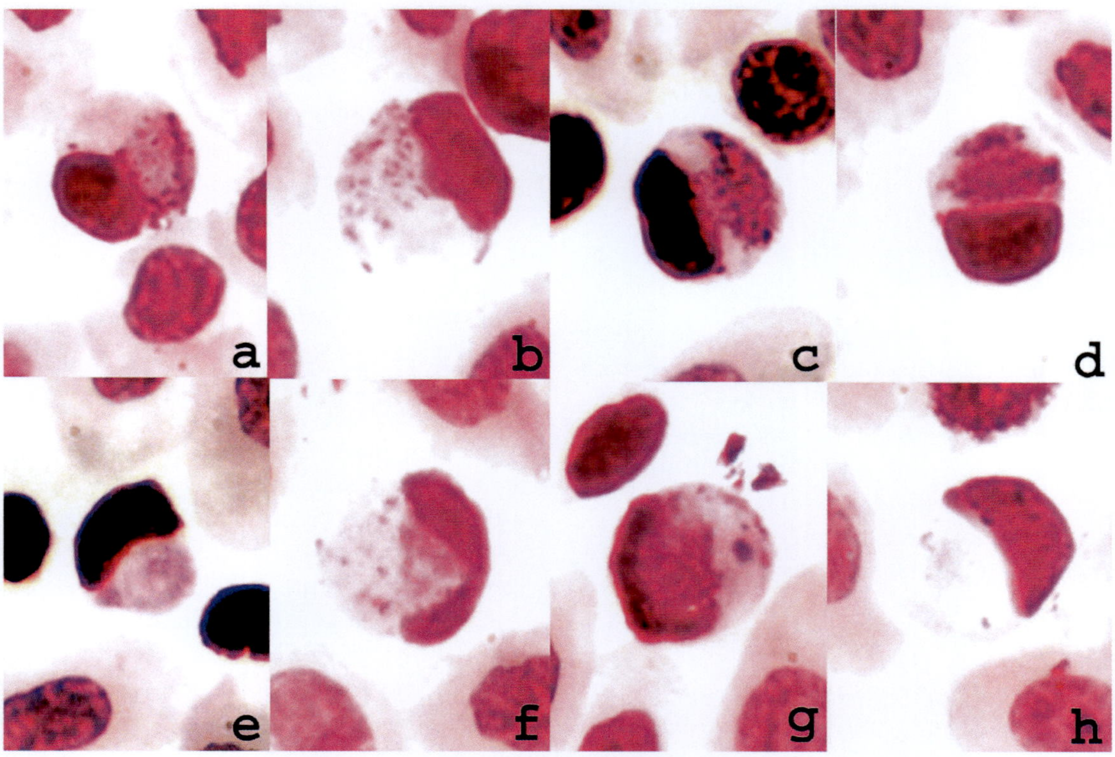

(B)

Plate 46 (**A**) *Fallisia biporcati* from *Anolis biporcatus*, Panama. Meronts, **a–d**; macrogametocytes, **e**, **f**; microgametocytes, **g**, **h**. (Figures **a** and **g** modified from Telford, S. R., Jr., *Syst. Parasitol.*, 40, 185, 1998, Figures 4 and 16, with kind permission of Springer Science and Business Media.) (**B**) *Fallisia poecilopi* from *Anolis poecilopus*, Panama. Meronts, **a–d**; macrogametocytes, **e**, **f**; microgametocytes, **g**, **h**. (Figure **d** modified from Telford, S. R., Jr., *Syst. Parasitol.,* 40, 185, 1998, Figure 33, with kind permission of Springer Science and Business Media.)

are tiny, densely packed, and intensely stained after four nuclear divisions and difficult to count. Segmentation was apparent at 7 × 5–6 μm, with 23–34 nuclei. Nearly mature or mature meronts are 5.5–9 × 3–6 μm (7.7 ± 0.9 × 4.7 ± 0.9, N = 17), with LW 22–54 μm^2 (36.5 ± 8.8), oval to elongate in shape, filled with nuclei or merozoites in no particular arrangement, with 20–51 (31.0 ± 10.7) nuclei. Meront size relative to host cell nucleus is 1.05 ± 0.30 (0.6–1.8, N = 17), and to nuclei of uninfected thrombocytes is 0.93 ± 0.22 (0.6–1.4, N = 17). Gametocytes with evidence of differential sexual staining are 7.5–14 × 6–11 μm (10.1 ± 1.6 × 8.0 ± 1.3, N = 32), with LW 45–132 μm^2 (82.0 ± 24.6), round to elongate with L/W ratio varying from 1.0 to 1.6 (1.27 ± 0.16). Gametocyte size relative to host cell nucleus is 1.38 ± 0.54 (0.8–3.6, N = 32) and to nuclei of uninfected thrombocytes is 1.73 ± 0.52 (1.0–2.8, N = 32). There are no significant sexual differences in gametocyte dimensions or shape.

Exoerythrocytic Merogony Unknown.

Sporogony Unknown.

Effects on Host Thrombocytes infected by meronts are larger. Thrombocyte nuclei are normal in size but are always altered in shape and displaced. Occasional rounded host cells may be lymphocytes. Thrombocytes infected by gametocytes and their nuclei are hypertrophied. Host cells and nuclei are always distorted and nuclei always displaced.

Fallisia dominicensis Telford 1998 (**Plate 47**)

Diagnosis A *Fallisia* parasite of thrombocytes, with meronts that approximate gametocytes in size. Meronts are 4–8 × 3–7 μm, with LW 12–56, and contain 8–22 merozoites. Meront size, relative to host cell nucleus, is 0.4–1.4, and to normal thrombocyte nuclei is 0.3–1.3. Gametocytes are also small, 5–9 × 4–7 μm, with LW 20–56 μm^2 and L/W 1.1–2.0. Relative to host cell nuclei, gametocyte size is 0.4–1.8, and to normal thrombocyte nuclei is 0.5–1.3. Gametocyte dimensions are not sexually dimorphic.

Type Host *Anolis c. cybotes* (Cope) (Sauria: Polychrotidae).

Type Locality Rio Seibo at Pedro Sanchez village, Seibo Province, Dominican Republic.

Other Hosts None known.

Other Localities None known.

Prevalence Overall, 3 of 83 (36%) *A. cybotes* were infected by *F. dominicensis* and 3 of 15 (20%) at the type locality.

Morphological Variation Meronts average 6.0 ± 1.1 × 4.8 ± 0.8 μm (4–8 × 3–7, N = 28), with LW 29.1 ± 9.5 μm^2 (12–56). Meronts produce 12.4 ± 3.5 (8–22) merozoites. Their size, relative to host cell nucleus, averages 0.69 ± 0.23 (0.4–1.4, N = 23), and to normal thrombocyte nuclei is 0.66 ± 0.21 (0.3–1.3). Gametocytes are 6.6 ± 1.2 × 5.0 ± 1.0 μm (5–9 × 4–7, N = 31), with LW 33.8 ± 12.0 μm^2 (20–56) and L/W 1.34 ± 0.19 (1.1–2.0). Relative to host cell nuclei, gametocyte size is 1.03 ± 0.37 (0.4–1.8, N = 22), and to normal thrombocyte nuclei is 0.76 ± 0.27 (0.5–1.3). There is no sexual dimorphism in gametocyte dimensions.

Exoerythrocytic Merogony Unknown.

Sporogony Unknown.

Effects on Host The size of thrombocytes and their nuclei are unaffected by meront presence, and gametocytes do not affect host cell size but cause shrinkage of the thrombocyte nucleus. Host cell nuclei are always displaced but seldom altered in shape by meronts; gametocytes always displace nuclei and commonly distort both host cells and nuclei.

Remarks *Fallisia dominicensis* possibly occurs on Jamaica, in *Anolis opalinus* (Telford, 1998a), but despite the thousands of anoles examined from both the Greater and Lesser Antilles, it is known with certainty only from the type host on Hispaniola.

Fallisia thecadactyli Telford 1998 (**Plate 47**)

Diagnosis A *Fallisia* species parasitic in thrombocytes and lymphocytes, with meronts 7–13 × 5–12 μm, LW 37–156 μm^2, oval, oblong, or triangular in shape, often flattened on the side adjacent to the host cell nucleus, filled with 28–61 nuclei or merozoites in no particular arrangement. Meront size relative to host thrombocyte nucleus is 0.8–3.6, and to nuclei of uninfected thrombocytes is 0.9–3.9. Gametocytes are round, oval, triangular, or elongate, 7–15 × 5–11 μm, with LW 40–154 μm^2 and L/W ratio 1.1–2.2. Gametocyte size relative to host cell nucleus is 0.7–2.3, and to nuclei of uninfected thrombocytes is 1.0–3.8. There are no sexual differences in gametocyte dimensions or shape.

Type Host *Thecadactylus rapicaudus* (Houttuyn) (Sauria: Gekkonidae).

Type Locality Panama, Colon Province, Quebrada Bonita.

Other Hosts None known.

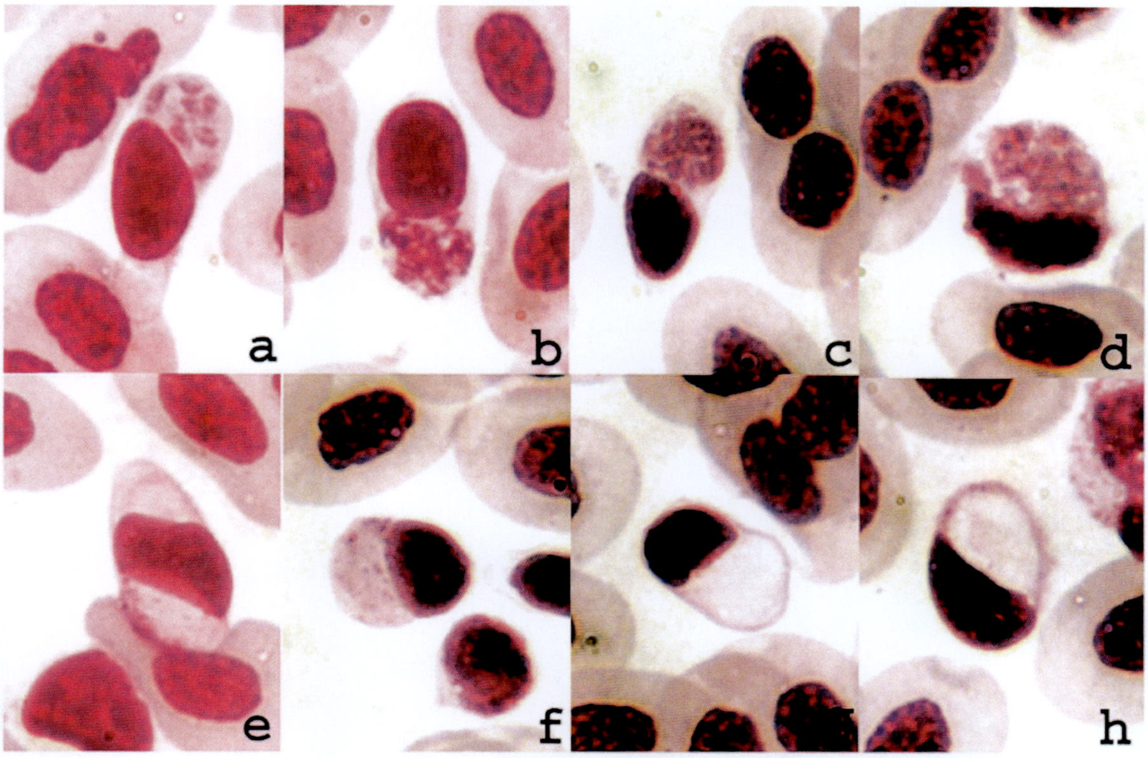

(A)

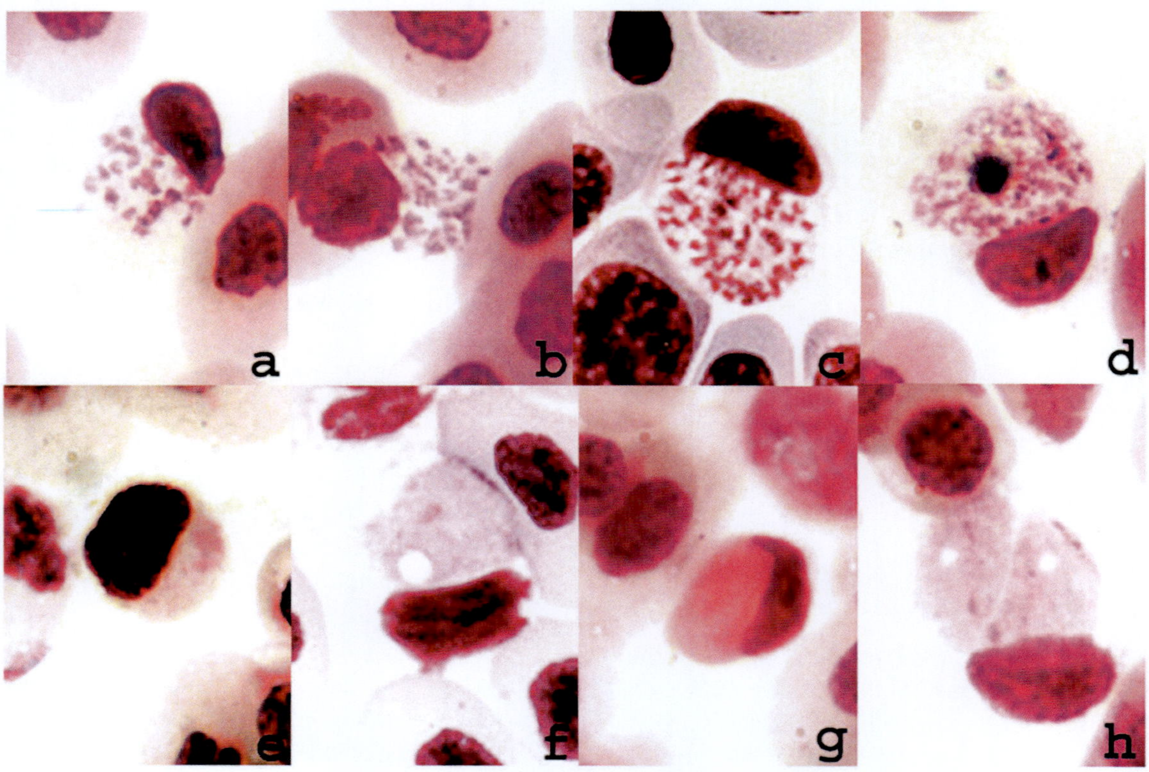

(B)

Plate 47 (**A**) *Fallisia dominicensis* from *Anolis cybotes*, Dominican Republic. Meronts, **a–d**; macrogametocytes, **e**, **f**; microgametocytes, **g**, **h**. (**B**) *Fallisia thecadactyli* from *Thecadactylus rapicaudus*, Panama. Meronts, **a–d**; macrogametocytes, **e**, **f**; microgametocytes, **g**, **h**. (Figures **a**, **c**, and **h** modified from Telford, S. R., Jr., *Syst. Parasitol.,* 40, 185, 1998, Figs. 42, 45, and 48, with kind permission of Springer Science and Business Media.)

Other Localities Venezuela, Estado Cojedes, Municipio Manrique, Tierra Caliente.

Prevalence Overall, 2 of 25 (8%) of Panamanian *T. rapicaudus* were infected by *F. thecadactyli*, 2 of 14 (14.3%) from the vicinity of the type locality; 1 of 22 from Venezuela (4.5%) were infected, 1 of 4 from Tierra Caliente.

Morphological Variation Immature meronts with 10–18 nuclei are 7–13 × 3–7 µm. Nearly mature or mature meronts are oval, oblong, or triangular in shape, 10.3 ± 2.2 × 8.0 ± 2.2 µm, (7–13 × 5–12, N = 13), with LW 86.6 ± 38.6 μm^2 (37–156), often flattened on the side adjacent to the host cell nucleus, filled with nuclei or merozoites in no particular arrangement, and average 40.2 ± 12.6 (26–61). Meront size relative to host thrombocyte nucleus is 1.84 ± 0.88 (0.8–3.6, N = 10), and to nuclei of uninfected thrombocytes is 2.14 ± 0.97 (0.9–3.9, N = 13). Gametocytes with differential sexual staining are 10.4 ± 2.1 × 7.0 ± 1.5 µm (7–15 × 5–11, N = 29), with LW 74.8 ± 27.5 μm^2 (40–154); round, oval, triangular, or elongate, with L/W 1.51 ± 0.3 (1.1–2.2). Gametocyte size relative to host cell nucleus is 1.36 ± 0.47 (0.7–2.3, N = 29), and to nuclei of uninfected thrombocytes is 1.85 ± 0.67 (1.0–3.8, N = 29). There are no sexual differences in gametocyte dimensions or shape.

Exoerythrocytic Merogony Unknown.

Sporogony Unknown.

Effects on Host Thrombocytes infected by meronts are enlarged more than twice the uninfected cell size, and distorted, but the host nuclei, always distorted and displaced, are normal in size. Thrombocytes infected by gametocytes are hypertrophied to more than double the size of uninfected cells, and their nuclei are larger than the nuclei of uninfected cells. Host cells and nuclei are always distorted, and nuclei are always displaced. Host cells are thrombocytes and lymphocytes.

Fallisia simplex Lainson, Shaw, and Landau 1975 (**Plate 48**)

Diagnosis A *Fallisia* parasite of thrombocytes, geographically variable in dimensions. The meronts may exceed gametocytes in size. Meronts from Brazilian hosts are 6.5 × 6.5–10.0 × 6.5 µm, with estimated average LW 50.7 μm^2, and contain 8–28 merozoites. Gametocytes are 6.5 × 5.2–10.4 × 7.8 µm, with estimated average LW 33.8–81.1 μm^2 and L/W 1.3. Meronts from Guyana populations of the same host are 7–15 × 5–10 µm with LW 42–130 μm^2 and produce 22–62 merozoites. Gametocytes are 6–9 × 4–6 µm, with LW 28–54 μm^2 and L/W 1.3. There is no sexual dimorphism in gametocyte dimensions.

Type Host *Plica umbra* (Linnaeus) (Sauria: Iguanidae).

Type Locality Belém, Pará State, Brazil.

Other Hosts None known.

Other Localities Vicinity of Georgetown, Guyana (Telford, 1973).

Prevalence Twenty of 235 (8.5%) of Brazilian *P. umbra* were infected by *F. simplex* (Lainson et al., 1975) and 1 of 7 from Guyana (Telford, 1973).

Morphological Variation In Brazilian *P. umbra* (Lainson et al., 1975), meronts averaged 7.8 × 6.5 µm (6.5 × 6.5–10.0 × 6.5), with LW, estimated from the average, 50.7 μm^2. Meronts produce an average of 18 merozoites (8–28). Gametocytes are 6.5–10.4 × 5.2–7.8 µm. Macrogametocytes average 8.2 × 6.5 µm (6.5 × 5.2- 10.4 × 7.8), with LW 53.3 μm^2 (33.8–81.1) and L/W 1.30–1.33). Microgametocytes average 7.9 × 5.6 µm (6.5 × 5.2–9.1 × 7.2), with LW 44.2 μm^2 (33.8–65.5) and L/W 1.41 (1.26–1.30). In *P. umbra* from Guyana, meronts are 10.6 ± 2.2 × 7.2 ± 1.5 µm (7–15 × 5–10, N = 25), with LW 76.4 ± 22.0 μm^2 (42–130), and contain 40.9 ± 10.8 (22–62) merozoites. Gametocytes are 6.9 ± 0.8 × 5.2 ± 0.7 µm (6–9 × 4–6), N = 25), with LW 36.0 ± 6.5 μm^2 (28–54) and L/W 1.34 ± 0.26 (1.0–2.0). Gametocyte dimensions do not differ by sex.

Exoerythrocytic Merogony Unknown.

Sporogony Unknown.

Effects on Host Lainson et al. (1975) described distortion of the infected thrombocytes from the elongate shape of normal cells to oval or almost spherical shapes, often hypertrophied. Thrombocyte nuclei were "indented, displaced or squashed to the periphery." Thrombocytes in the Guyana host were similarly affected by parasite presence.

Remarks The Guyana infection was reported as a *Plasmodium tropiduri* infection of thrombocytes (Telford, 1973). Meronts of this strain are larger and produce more merozoites than do the Brazilian infections, and gametocytes are smaller. Statistical comparison cannot be done, however, without more precise information on the Brazilian parasites, so it is probably best to consider the two strains conspecific in the absence of genome data on both.

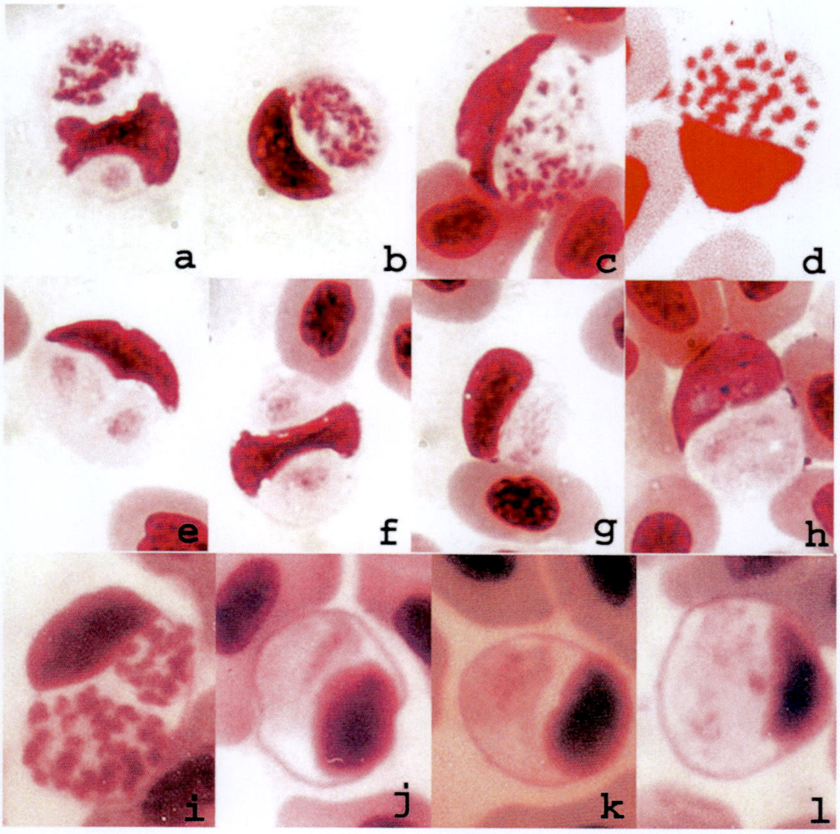

(A)

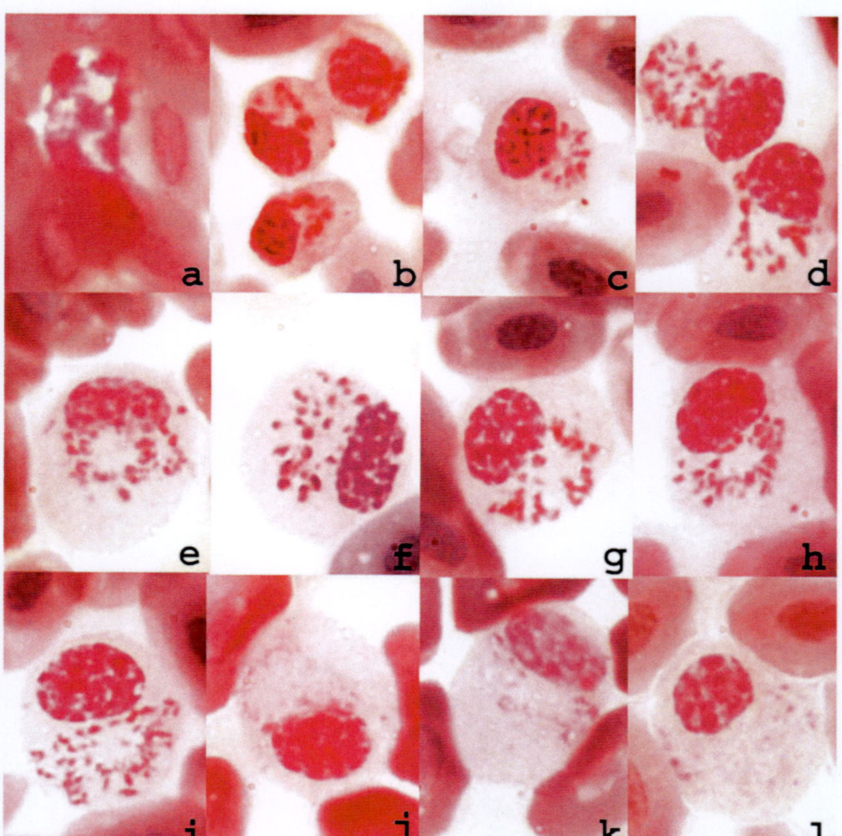

(B)

Plate 48 (**A**) *Fallisia simplex* from *Plica umbra*, Guyana (**a–h**), and *Fallisia siamense* from *Draco maculata*, Thailand (**i–l**). Meronts, **a–d**, **i**; macrogametocytes, **e**, **f**, **j**, **k**; microgametocytes, **g**, **h**, **l**. (Figure **a** from Telford, S. R., Jr., *Int. J. Parasitol.,* 3, 829, 1973, with permission, Elsevier.) (**B**) *Fallisia kantoensis* sp. nov. from *Eumeces latiscutatus*, Honshu, Japan. Exoerythrocytic meront free from host cell, **a**; meronts, **b–i**; macrogametocytes, **j**, **k**; microgametocyte, **l**.

ASIAN *FALLISIA* SPECIES

Fallisia siamense (Telford) 1986 (**Plate 48**)

Diagnosis A *Fallisia* species parasitic in thrombocytes that produces 18–64 merozoites in meronts usually larger than gametocytes, 8–15 × 4–10 μm, with LW 40–144 μm². Gametocytes are 7–12 × 4–9 μm, with LW 35–96 μm² and L/W 1.0–3.5. There is no sexual dimorphism in gametocyte dimensions.

Type Host *Draco maculatus* Gray (Sauria: Agamidae).

Type Locality Ramintra, near Bangkok, Thailand.

Other Hosts None known.

Other Localities None known.

Prevalence In 1 of 14 (7.1%) *D. maculatus* examined between 1976 and 1980.

Morphological Variation Mature meronts average 11.1 ± 1.2 × 7.8 ± 1.2 μm (8–15 × 4–10, N = 36), with LW 87.7 ± 24.0 μm² (40–144). Merozoites average in number 40.0 ± 12.1 (18–64). Gametocytes average 9.1 ± 1.5 × 6.1 ± 1.1μm (7–12 × 4–9, N = 50), with LW 55.4 ± 13.2 μm² (35–96) and L/W 1.56 ± 0.45 (1.0–3.3). Gametocytes do not differ in vdimensions by sex.

Exoerythrocytic Merogony Unknown.

Sporogony Unknown.

Effects on Host Thrombocytes host to meronts are hypertrophied to twice or more normal size and rounded, with nuclei severely distorted, usually compressed and flattened against the cell membrane. Gametocytes cause less hypertrophy, but occasionally stretch thrombocytes laterally when elongated in shape. Occasionally, gametocytes appear to have pushed through the thrombocyte membrane without rupturing it. Thrombocyte nuclei are sometimes flattened against the side of the host cell but often are nearly normal in shape.

Remarks *Fallisia siamense* was originally described as a *Plasmodium* (*Fallisia*) species, but later Telford (1988a) recognized *Fallisia* as a plasmodiid genus following the experimental transmission of an avian *Fallisia* species by Gabaldon et al. (1985) that demonstrated absence of an erythrocytic cycle.

Fallisia kantoensis sp. nov. (**Plate 48**)

Diagnosis A *Fallisia* species parasitic in polychromatophilic erythroblasts. Meronts are 7–15 × 6–9 μm, with LW 49–126 μm², and produce 12–38 merozoites. Gametocytes are 10–15 × 7–10 μm, with LW 84–150 μm² and L/W 1.25–1.88, and are larger than meronts. There is no apparent sexual dimorphism in gametocyte dimensions.

Type Host *Eumeces latiscutatus* (Hallowell) (Sauria: Scincidae).

Type Locality Hanno, Saitama Prefecture, Honshu, Japan.

Other Hosts None known.

Other Localities None known.

Prevalence One of 18 (5.6%) *E. latiscutatus* from the type locality was infected by *F. kantoensis*.

Morphological Variation Meronts of *F. kantoensis* are variably shaped, usually ovoid, 11.1 ± 2.5 × 7.5 ± 1.0 μm (7–15 × 6–9, N = 13), with LW 84.7 ± 27.0 μm² (49–126), and produce 27.0 ± 6.1 (12–38, N = 15) merozoites. Gametocytes, which are larger than meronts and much less common, are 13.3 ± 1.5 × 8.8 ± 0.9 μm (10–15 × 7–10, N = 10), with LW 117.2 ± 20.5 μm² (84–130) and L/W 1.53 ± 0.19 (1.25–1.88). In the small sample of gametocytes, there is no apparent sexual dimorphism in dimensions.

Exoerythrocytic Merogony A single EE meront free from its host cell on a slide of cardiac blood was 17 × 8 μm. It contained several clumps of six to eight round nuclei.

Sporogony Unknown.

Effects on Host Both meronts and gametocytes distorted and slightly enlarged host cells and distorted and displaced their nuclei. Meronts appeared to occupy positions lateral to the nucleus, and gametocytes were either lateral or lateropolar.

Remarks Host cells appear to correspond to the polychromatophile erythroblasts of Maximow and Bloom (1953) or polychromasic "stem cell type" normoblasts of Pienaar (1962, p. 39), on the basis of nuclear structure ("checkerboard" appearance), staining reaction of the cytoplasm, and shape and size of cell and nucleus. Hapantotype blood film is deposited in the USNPC, Beltsville, Maryland, no. 100328.

AUSTRALIAN *FALLISIA* SPECIES

Fallisia copemani Paperna and Landau 1990

Diagnosis A *Fallisia* species parasitic primarily in lymphocytes, with meronts that may exceed gametocytes in size, 7.2–11.2 × 4.0–8.0 μm, with estimated LW 61.6 μm². Meronts produce 12–32 merozoites, the number of which affects meront dimensions. Gametocytes are 6.0–11.2 × 3.2–7.2 μm, with estimated LW 44.0 μm² and L/W 1.57. Microgametocytes are slightly larger than macrogametocytes.

Type Host *Carlia rhomboidalis* (Peters) (Sauria: Scincidae).

Type Locality Daintree Forest, northern Queensland, Australia.

Other Hosts None known.

Other Localities None known.

Prevalence One of nine *C. rhomboidalis* was infected by *F. copemani*.

Morphological Variation Overall, meronts are 8.8 × 7.0 μm (N = 21), with average LW estimated at 61.6 μm², and contain 12–32 merozoites. Meronts with 12–17 nuclei are 8.6 × 6.4 μm (N = 10), with 17–24 nuclei are 8.9 × 7.8 μm (N = 4), and with 24–32 nuclei are 9.0–7.4 μm (N = 7). Gametocytes in singly infected lymphocytes are 6.0–11.2 × 3.2–7.2 μm and average 8.3 × 5.3 μm (N = 36), with estimated LW 44.0 μm² and L/W 1.57. Macrogametocytes average 7.7 × 5.4 μm (6.0–11.2 × 3.2–7.2, N = 20), with estimated LW 41.6 μm² and L/W 1.43. Microgametocytes average 9.1 × 5.25 μm (6.4–9.6 × 3.6–6.4, N = 16), with estimated LW 47.8 μm² and L/W 1.73, apparently slightly larger and more elongated than macrogametocytes (no statistical evaluation). Average dimensions in multiply infected host cells are 7.9 × 4.7 μm (4.8–11.2 × 2.8–5.6, N = 23), LW 37.1 μm², and L/W 1.68 for macrogametocytes and 8.4 × 4.2 μm (6.4–9.6 × 3.6–5.6, N = 4), LW 35.3 μm², and L/W 2.00 for microgametocytes. In both sexes, gametocytes in multiple infections are slightly smaller and more elongated than in singly infected cells.

Exoerythrocytic Merogony Unknown.

Sporogony Unknown.

Effects on Host The heavily infected host reached maximum parasitemias of 84% and 64% of lymphocytes, with lows of 5% and 8% during an observation period of 71 days. The numbers of circulating lymphocytes were apparently depleted at times of high parasitemia. Parasitized lymphocytes were nearly three times the size of normal cells when meronts were present and twice normal size with gametocyte infection. Lymphocyte nuclei often remained normal in size; they could be compressed but not indented. A few parasites occupied thrombocytes, but lymphocytes were the dominant type of cell host to the parasites.

Remarks The apparent *Fallisia* species reported by Thompson and Hart (1946) in about 30 of 60 *Carlia fusca* (= *Leiolopisma fuscum*) from Goodenough Island in the D'Entrecastaux Island Group near New Guinea had meronts that produced at segmentation 15–20 merozoites and apparent gametocytes 8.0 × 4.0 μm, similar to that of *F. copemani*. The lesser number of merozoites probably indicates specific difference from the parasite of *Carlia rhomboidalis*.

PROGARNIA SPECIES

Progarnia archosauriae Lainson 1995

Diagnosis A plasmodiid parasite of Neotropical crocodilians that undergoes merogony in leukocytes, thrombocytes, and erythroid cells and gametogony in nonerythroid cells. Meronts produce 8–22 merozoites. Mature gametocytes are spherical or subspherical and are 6–7 μm in diameter.

Type Host *Caiman c. crocodilus* (Linnaeus) (Crocodilia: Alligatoridae).

Type Locality Marajó Island, Pará State, Brazil.

Other Hosts None known.

Other Localities None known.

Prevalence Seven of 19 (36.8%) young *Caiman crocodilus* were infected by *P. archosauriae* (Lainson, 1995).

Morphological Variation Dimensional data for meronts were not given in the species description, and the only reference to gametocyte size is, "Mature gametocytes are spherical to sub-spherical, and measure 6–7 μm in diameter" with "no appreciable size difference between" sexes (Lainson, 1995). Meronts produce 8–22 merozoites in lymphocytes and monocytes.

Cell Types Parasitized Asexual stages parasitize most commonly lymphocytes and monocytes, less commonly thrombocytes, and proerythrocytes of the erythroid series. Gametocytes were reported from lymphocytes, monocytes, and thrombocytes, and on a single occasion, a basophil.

Exoerythrocytic Merogony Unknown.

Sporogony Unknown.

Effects on Host Gametocytes may occupy the entire host cell and "markedly indent the nucleus" (Lainson, 1995), particularly in thrombocytes, but otherwise appear to have little effect on the host *Caiman*.

Remarks Genome data are badly needed for *Progarnia archosauriae* (along with DNA analysis of *Fallisia* and *Plasmodium* (*Garnia*) species) to verify Lainson's (1995) speculations regarding its systematic and evolutionary significance. The parasite certainly merits generic distinction, but where within the still-emerging classification of haemosporinids is very much open to question.

HAEMOPROTEUS SPECIES OF SNAKES

Haemoproteus mesnili (Bouet) 1909 Telford 2007 (**Plate 49**)

Diagnosis A *Haemoproteus* species with usually elongate gametocytes, 11–29 × 5–10 μm, LW 72–247 μm^2, and L/W 1.1–4.4. Sexual dimorphism in gametocyte dimensions are pronounced in chronic infection, where macrogametocytes are longer, with greater LW and less rounded shape than microgametocytes. Both sexes are heavily pigmented with black granules, 31–62 dispersed throughout in macrogametocytes and 20–46 often clumped or concentrated near ends of microgametocytes. Gametocytes in active infection are commonly halteridial in position within erythrocytes, rarely so in chronic infection. Megalomeronts form in cardiac muscle, and other meront generations are found only in the spleen and its connective tissue. Developing meronts form as square-to-rectangular cytomeres, becoming ovoid when mature, containing hundreds of small rounded merozoites.

Type Host *Naja n. nigricollis* Reinhardt (Serpentes: Elapidae).

Type Locality Odienne, Ivory Coast.

Other Hosts *Naja haje, Sepedon haemachatus.*

Other Localities Morogoro, Morogoro Region, Tanzania (Telford, 2007); Gaoua, Upper Volta (Bouet, 1909a); Bamako, Mali (Leger and Leger, 1914); Accra, Ghana (Macfie, 1919); and Deleib Hill and Nasser on River Sobat, Sudan (Wenyon, 1909a).

Prevalence Seven of eight (88%) *N. nigricollis* collected at Morogoro, Tanzania (Telford, 2007).

Morphological Variation In active infection, macrogametocytes 18.3 ± 4.0 × 7.6 ± 1.4 μm (11–29 × 5–10, N = 27), with LW 137.5 ± 36.2 μm^2 (84–247) and L/W 2.51 ± 0.70 (1.1–4.2). Microgametocytes are 17.0 ± 2.2 × 7.0 ± 1.4 μm (12–23 × 5–10, N = 24), with LW 117.6 ± 30.2 μm^2 (72–184) and L/W 2.53 ± 0.71 (1.2–4.4). In chronic infection, macrogametocytes are 18.9 ± 2.1 × 9.2 ± 1.3 μm (12–21 × 7–12, N = 25), with LW 167.1 ± 28.4 μm^2 (120–220) and L/W 2.00 ± 0.36 (1.2–2.6). Microgametocytes are 15.0 ± 2.6 × 8.8 ± 1.5 μm (10–20 × 6–12, N = 25), with LW 132.2 ± 32.8 μm^2 (93–240) and L/W 1.76 ± 0.45 (1.0–2.7).

Merogony Megalomeronts are present in cardiac muscle, elongate, 55–65 × 22–35 μm, the largest with 100 large merozoites in a single focal plane. A very large meront, 383 × 107 μm, apparently ruptured from a site in cardiac muscle, found within the heart cavity. Subsequent generations of trophozoites and meronts occur only in the spleen and its connective tissue. Uninucleate parasites are within parasitophorous vacuoles in cells of the spleen, 7.5–10 × 3–8 μm, with nuclei 3–5 × 2.5–3. The smallest meronts are oval or round, 8 × 8–17 × 11 μm, with 12–16 square-to-rectangular, deeply basophilic cytomeres. With 17–32 cytomeres, meronts are 12 × 11–36 × 11 μm. Meronts 20 × 16–26 × 22 μm contained 51–57 cytomeres. Maturing meronts are smaller, 9–25 × 7.5–15 μm, containing mostly small, rounded nuclei with a few larger, irregularly shaped chromatin masses, none with the relatively straight margins of cytomeres. In single focal planes, 40–50 nuclei are present. The ovoid mature meronts are 12 × 10–15 × 12 μm and contain hundreds of rounded merozoites.

Sporogony Unknown.

Remarks Dimensions of *H. mesnili*, where given by Bouet (1909a), Wenyon (1909a), Leger and Leger (1914), and Macfie (1919), were consistent with the samples from Tanzanian *N. nigricollis* (Telford, 2007). Wenyon (1909a) described *H. najae* from the Sudan but later (1926) recognized that *Plasmodium mesnili* of Bouet (1909a) had priority.

Haemoproteus balli Telford 2007 (**Plate 49**)

Diagnosis A *Haemoproteus* species with rounded or ovoid gametocytes, never halteridial, 8–14 × 6–10 μm, with LW 48–112 μm^2 and L/W 1.0–2.0. Pigment granules

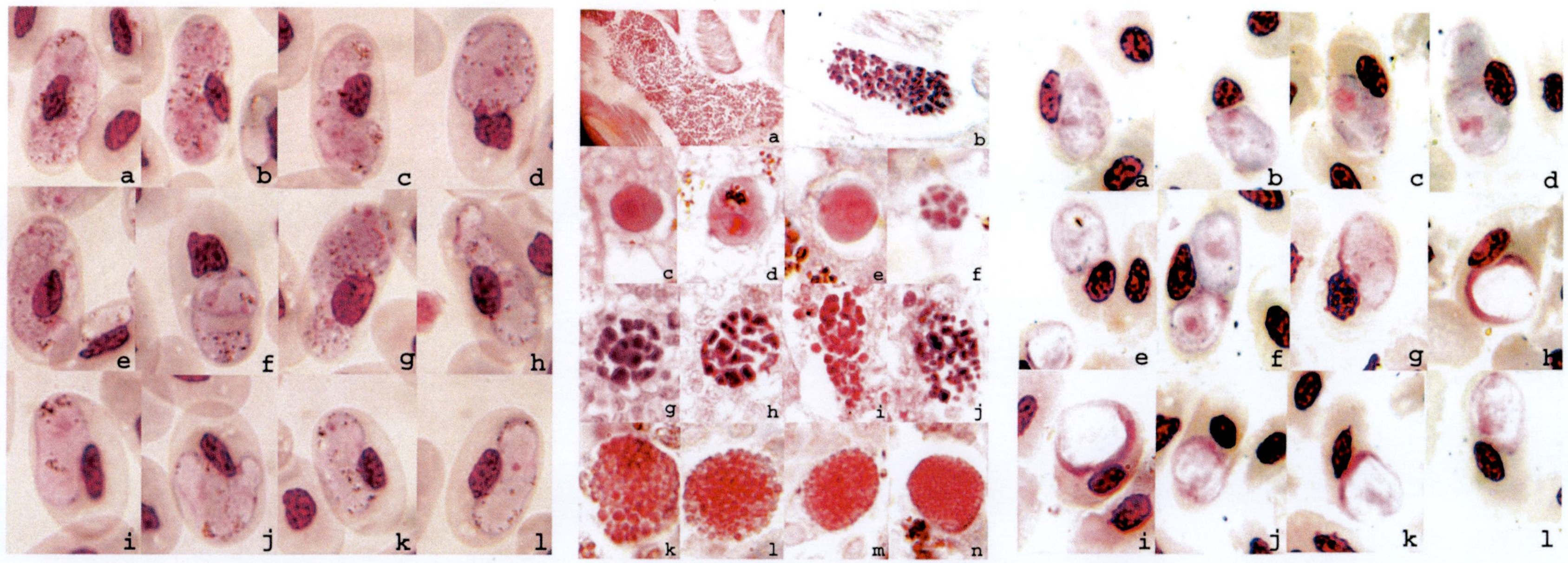

(A) (B) (C)

Plate 49 (**A**) *Haemoproteus mesnili* from *Naja nigricollis*, Tanzania. Macrogametocytes, **a**, **g**; microgametocytes, **h–l**. (Figures **d**, **g–j**, and **l** modified from Telford, S. R., Jr., *J. Parasitol.,* 93, 673, 2007, Figures 4, 7, 8, 11, 12, 14, with permission.) (**B**) Merogony of *Haemoproteus mesnili* in heart (**a**, **b**) and spleen (**c–n**) of *Naja nigricollis*: **a**, **b**, possible primary meronts; **c–e**, uninucleate parasites; **f–j**, meronts containing cytomeres; **k–n**, mature meronts. (Figures **a–c**, **f**, **g–i**, **l**, **m** modified from Telford, S. R., Jr., *J. Parasitol.,* 93, 673, 2007, Figures 21, 23, 26, 31, 32, 36, 39, 40, 42, 43, with permission.) (**C**) *Haemoproteus balli* from *Naja haje*, Kenya: **a–g**, macrogametocytes; **h–j**, microgametocytes. (Figures **c**, **e**, and **f** from Telford, S. R., Jr., *J. Parasitol.,* 93, 673, 2007, Figures 48, 49, and 52, with permission.)

dispersed in both sexes and few in number, three to ten only. Microgametocytes have an elongated peripheral band of concentrated chromatin, occasionally fragmented in four to six poorly defined masses, which apparently represents or contains the nucleus.

Type Host *Naja h. haje* (Linnaeus) (Serpentes: Elapidae).

Type Locality Near Lake Baringo, Marigat, Kenya.

Other Hosts None known.

Other Localities None known.

Prevalence One of one *N. haje* collected at the type locality (Ball, 1967a).

Morphological Variation In active infection, macrogametocytes are 11.4 ± 1.6 × 7.4 ± 0.6 μm (8–14 × 6–9, N = 25), with LW 84.7 ± 15.9 μm^2 (48–112) and L/W 1.55 ± 0.18. Microgametocytes are 10.2 ± 1.2 × 8.0 ± 0.9 μm (8–14 × 6–10, N = 25), with LW 81.7 ± 14.6 μm^2 (51–112) and L/W 1.29 ± 0.19 (1.0–1.9).

Merogony Unknown.

Sporogony Unknown.

Remarks Ball (1967) reported this species as *H. mesnili*. However, the gametocytes differ strikingly in dimensions, those of *H. balli* being shorter and more rounded than in *H. mesnili*. Pigmentation is far heavier, with nearly ten times the number of granules in *H. mesnili* than in *H. balli*. *Haemoproteus mesnili* is commonly halteridial in position, in contrast to *H. balli,* which is never halteridial. Gametocyte nuclei in both sexes of *H. mesnili* are discrete, small bodies; the nucleus of *H. balli* microgametocytes is contained within or forms as an elongated, peripheral band of chromatin. Erythrocytes host to *H. mesnili* are hypertrophied; those containing *H. balli* gametocytes are similar in dimensions to normal cells.

HAEMOPROTEUS SPECIES OF NORTH AMERICAN TURTLES

Haemoproteus degiustii sp. nov.

Synonym: *Haemoproteus "metchnikovi"* (Simond) 1901 Hewitt 1940 (**Plate 50**)

Diagnosis A *Haemoproteus* species with gametocytes 8–18 × 4.5–10 μm, LW 56–162 μm^2, and L/W 1.0–2.0. Dark golden pigment is in individual granules or small clumps, not dispersed over the entire gametocyte. Megalomeronts present in the spleen form pseudocytomeres that produce large numbers of very small merozoites. Host erythrocytes are not hypertrophied by the presence of gametocytes, which are seldom halteridial in shape.

Type Host *Chrysemys picta marginata* Agassiz (Testudines: Emydidae).

Type Locality Pontiac Lake Recreation Area, Pontiac, Oakland County, Michigan.

Other Hosts *Pseudemys rubriventris, Pseudemys concinna suwanniensis, Trachemys scripta elegans, Graptemys geographica, G. barbouri, Emydoidea blandingi, Apalone s. spinifera, Apalone spinifera emoryi, Apalone ferox*, and *Chelydra serpentina.*

Other Localities Plymouth County, Massachusetts, and Cape May County, New Jersey (Haskell and Telford, unpublished); Pope or Massac County, Illinois (Marquardt, 1966); Burleson and Brazos counties, Texas (Wang and Hopkins, 1965); Wisconsin, Minnesota, Louisiana (DeGiusti, 1965); Decatur County, Georgia, and Gilchrist County, Florida (Telford, unpublished).

Prevalence Illinois, 1 of 14 (7.1%) *T. scripta elegans* (Marquardt, 1966); Texas, 4 of 19 (21%) *T. scripta elegans*, 1 of 1 *Apalone s. emoryi*, 1 of 6 *Chelydra serpentina* (Wang and Hopkins, 1965); Massachusetts, 8 of 82 (9.8%) *P. rubriventris* (Haskell and Telford, unpublished); New Jersey, 6 of 23 (26.1 %) *P. rubriventris* (Haskell and Telford, unpublished); *P. concinna suwanniensis* Florida, 1 of 26 (3.8%) (Telford); Maryland (?), 1 of 6 *T. scripta elegans* obtained from a seafood market in Maryland were reported by Hewitt (1940) to be infected by *H. "metchnikovi."*

Morphological Variation Hewitt (1940) described macrogametocytes as having about 20 pigment granules and averaging 9.5 × 5.8 μm (N = 25); microgametocytes were 9.0 × 5.3 μm (N = 25). Wang and Hopkins (1965) reported dimensions of *H. degiustii* gametocytes as microgametocytes 8–15 × 4.5–5.3 μm, with 12–18 light yellow pigment granules, and macrogametocytes 9.5–15 × 4.5–6 μm, with 9–16 pigment granules. Parasitemias of their samples appeared to be active at 2–6% in the three host species and comprised of both mature and developing gametocytes. Chronic-phase gametocytes in *P. rubriventris* from Massachusetts averaged 11.1 ± 2.4 × 8.6 ± 0.9 μm (8–18 × 7–10, N = 25), with LW 96.5 ± 26.5 μm^2 (56–162) and L/W 1.30 ± 0.25 (1.0–2.0). Macrogametocytes were 10.0 ± 2.6 × 8.3 ± 0.6 μm (8–18 × 7–9, N = 13), with LW 83.9 ± 26.1 μm^2 (56–90) and L/W 1.21 ± 0.25 (1.0–2.0). Microgametocyte

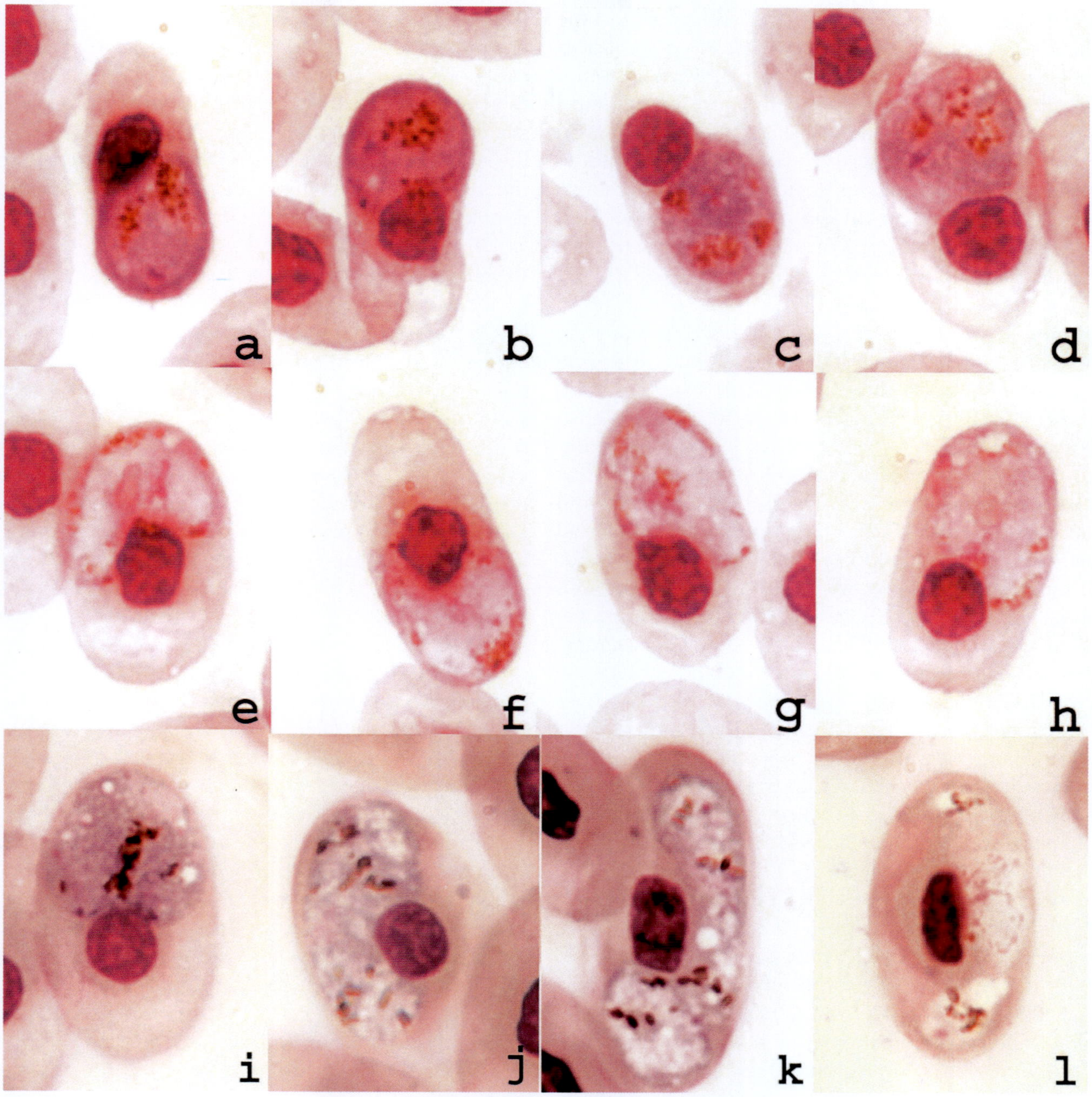

Plate 50 *Haemoproteus degiustii* sp. nov. from *Pseudemys rubriventris*, Massachusetts (**a–h**); *Graptemys barbouri* (**i**), and *Apalone ferox* (**j–l**), Georgia; **a–d**, **i–k**, macrogametocytes; **e–h**, **l**, microgametocytes.

dimensions were 11.4 ± 1.2 × 8.2 ± 1.0 µm (10–15 × 7–10, N = 12), with LW 101.7 ± 17.5 µm² (77–124) and L/W 1.30 ± 0.18 (1.1–1.7). Pigment appeared as dark golden granules or small clumps, usually clustered rather than dispersed. Granules averaged 22.9 and varied from 18 to 31. Gametocytes were similar in appearance in a single infection of *Pseudemys concinna suwanniensis* in North Florida.

Merogony Megalomeronts were present only in the spleen "packed with extremely small merozoites" (DeGiusti, 1965), their production requiring 3–4 weeks (DeGiusti and Dobrzechowski, 1974). Dimensions of mature meronts were stated by Garnham (1966) as "about 40–50 µm in diameter," which appears to be the only published data on meront dimensions. Four weeks after infection, young gametocytes were visible in experimental infections of *Chrysemys picta* (DeGiusti and Dobrzechowski, 1974). During development of the meronts, multinucleate syncytia, termed *pseudocytomeres* (Sterling and DeGiusti, 1972) form, and merozoites are produced by budding from the pseudocytomeres. Sterling and DeGiusti characterized the nucleated islands of *H. "metchnikovi"* meronts as pseudoctyomeres, similar in their formation to those of *Haemoproteus columbae* as described by Bradbury and Galucci (1971), in contrast to cytomeres as defined by Wenyon (1926), for which multinucleated masses segment into 15 or more unpigmented masses (cytomeres) with a single nucleus each, which then divides, forming multinucleate bodies that fill the cytoplasm of the host cell, each cytomere within its own cyst wall or membrane.

Sporogony *Haemoproteus "metchnikovi"* is transmitted by the tabanid fly *Chrysops callidus* to *Chrysemys picta marginata* in Michigan (DeGiusti, Sterling, and Dobrzechowski, 1973; Sterling and DeGiusti, 1974; DeGiusti and Dobrzechowski, 1974). Oocysts, about 20-μm diameter at maturity and containing 100–200 sporozoites, form beneath the basement membrane of the midgut epithelium and with growth protrude into the hemocoel (Sterling and DeGiusti, 1974). Transmission to *C. picta* occurs from mid-June through late September, with over 50% of the *Chrysops callidus* population infected (DeGiusti and Dobrzechowski, 1974). In mid-June to mid-July, about 20–25% of *C. callidus* have sporozoite infections in the salivary glands. Infections of *C. callidus* salivary glands could be massive, with over 1000 sporozoites present, sometimes filling the lumen of the gland (DeGiusti et al., 1973). New populations of gametocytes appear in *C. picta* each month from April to October, with peak gametocytemias in April and May (DeGiusti, 1965). Sporozoite dimensions are 9–12 × 1–1.4 μm in maximum width (Sterling and DeGiusti, 1974) and are crescent shaped with "blunt anterior and tapered posterior end" and a centrally located nucleus. Following transmission, young gametocytes appear in peripheral blood in 30–32 days, with mature gametocytes present in about 3 months (DeGiusti et al., 1973). Apart from *C. picta*, only *Trachemys s. elegans* could be infected experimentally by sporozoites of *H. "metchnikovi"* (DeGiusti and Dobrzechowski, 1974). Only low parasitemias appeared in this last host, while *Chelydra serpentina* and *Apalone spinifera* were refractory to infection.

Effects on Host Host erythrocytes in *H. "metchnikovi"* samples from Massachusetts and New Jersey were little affected by the presence of chronic-phase gametocytes, with no significant hypertrophy of erythrocytes or their nuclei. The host cell was rarely distorted, but sometimes the erythrocyte nuclei were distorted and usually were displaced but not hypertrophied by the parasite. About 35% of parasitized erythrocytes showed some degree of dehemoglobinization. In experimental infections (DeGiusti and Dobrzechowski, 1974), the peak parasitemias appeared in 3–4 weeks following patency with massive infections of erythrocytes that could contain up to 10 young gametocytes per host cell. In week 3 of patency, lymphocytes in the peripheral blood gradually declined in numbers, then sharply rose in abundance in parallel with a rapid decline in parasitemia. As a result of this cell-mediated immunity, the parasitemia stabilized by 8 weeks of infection and remained constant over 9 months of observation.

Remarks The very conservative taxonomic philosophy of Wenyon, who was reluctant to recognize new species, has influenced many subsequent students of parasitic protistans to an absurd degree for three-quarters of a century. Because Wenyon (1926) thought all turtle hemoproteids were probably *Haemoproteus "metchnikovi,"* Hewitt (1940) used this name for the *Haemoproteus* he found in *Trachemys scripta elegans*. Without discussion of differences noted by the authors in admittedly scanty descriptions, Garnham (1966) considered species parasitizing Australian chelids, Indian trionychids, North American emydids and trionychids, African testudinids, and Middle American kinosternids all to represent *H. "metchnikovi,"* characterizing it as a species with a "low degree of host specificity." Garnham did, however, admit the possibility that studies on their life cycles might alter this view. Mackerras (1961a) did not accept the view of Wenyon and recognized *H. chelodinae* as the species present in Australian chelid turtles, pending comparison with Indian material. The practice of lumping all reported *Haemoproteus* parasites of turtle species from Asia, Africa, Australia, and North America into a single taxon, *H. "metchnikovi,"* was commented on by DeGiusti and Dobrzechowski (1974) and most recently by Lainson and Naiff (1998). Given the evidence of host specificity reported by DeGiusti and Dobrzechowski (1974), who could not infect chelydrid and trionychid hosts experimentally with sporozoites that readily infected the natural emydid host *C. picta* and less readily another emydid, *Trachemys s. elegans*, it is time to stop applying the name *"metchnikovi"* to the *Haemoproteus* parasites of North American turtles as well as other species described, no matter how poorly, from Africa and Australia. Accordingly, the name *Haemoproteus "metchnikovi"* (Simond) 1901 is here restricted to the parasite described from the trionychid species *Chitra indica* at Agra, India. It is appropriate to recognize a new taxon, *Haemoproteus degiustii* sp. nov. for the hemoproteid species that occurs in North American emydid turtles, previously termed *H. "metchnikovi."* Genomic comparisons could clarify the relationships of the hemoproteids reported from trionychids and chelydrids of Texas (Wang and Hopkins, 1965) more easily than laborious determinations of their vectors followed by experimental infection studies.

HAEMOPROTEUS SPECIES OF NEOTROPICAL TURTLES

Haemoproteus geochelonis Lainson and Naiff 1998

Diagnosis A *Haemoproteus* species with round or broadly reniform gametocytes 4.4–8.1 × 3.7–7.4 µm, LW estimated at 31.8–35.1 µm² and L/W 1.0–1.5. One vacuole, rarely two, is present in gametocytes, and there are 8–16 bacilliform pigment granules.

Type Host *Geochelone denticulata* (Linnaeus) (Testudines: Testudinidae).

Type Locality Near Monte Dourado, Pará State, Brazil.

Other Hosts None known.

Other Localities None known.

Prevalence One of one *G. denticulata* examined was infected by *H. geochelonis* (Lainson and Naiff, 1998).

Morphological Variation Macrogametocytes average 6.6 × 5.4 µm (4.4–8.1 × 3.7–7.4, N = 33), with L/W 1.2 (1.0–1.5) and with LW estimated from the average, 35.1 µm². Microgametocytes average 6.0 × 5.3 µm (5.2–8.0 × 4.4–6.0, N = 37), with L/W 1.1 (1.0–1.5) and estimated LW at 31.8 µm². Usually one, rarely two, vacuoles occurs in the cytoplasm, and 8–16 bacilliform pigment granules are present.

Merogony Unknown.

Sporogony Unknown.

Effects on Host Host erythrocytes are neither enlarged nor distorted, and their nuclei are unaltered in shape or position.

Haemoproteus peltocephali Lainson and Naiff 1998

Diagnosis A *Haemoproteus* species with gametocytes 6.7–12.6 × 6.0–11.1 µm, L/W 1.0–2.0, and estimated LW 68.4–74.7 µm². Young gametocytes are amoboid in shape, with spiky pseudopodia present in microgametocytes. There are 12 to more than 30 dark greenish-black bacilliform pigment granules in gametocytes, with vacuoles variably present.

Type Host *Peltocephalus dumerilianus* (Schweigger) (Testudines: Pelomedusidae).

Type Locality Rio Negro near Barcelos, northern Amazonas State, Brazil.

Other Hosts None known.

Other Localities None known.

Prevalence Four of eight *P. dumerilianus* examined were infected by *H. peltocephali* (Lainson and Naiff, 1998).

Morphological Variation Young gametocytes are amoeboid in shape, with spiky pseudopodia present in microgametocytes. Macrogametocytes are 9.7 × 7.7 µm (7.4–12.6 × 6.2–11.1), with L/W 1.2 (1.0–1.4) and LW estimated from the average 74.7 µm². Microgametocytes are 9.0 × 7.6 µm (6.7–12.5 × 6.0–9.0), with L/W 1.2 (1.0–2.0) and estimated LW 68.4 µm². Pigment granules are bacilliform in shape, dark greenish-black, and number from 12 to more than 30, averaging 21. Vacuoles are variably present in gametocyte cytoplasm.

Merogony One small meront with eight merozoites was found in a monocyte on a smear from the spleen.

Sporogony Unknown.

Effects on Host Infected erythrocytes are not enlarged or distorted, and their nuclei appear unaffected by parasite presence.

HAEMOPROTEUS SPECIES FROM AFRICAN TURTLES

Haemoproteus testudinis (Laveran) 1905

Diagnosis A *Haemoproteus* species with ovoid or reniform gametocytes 10–12 µm in length, or elongate, halteridial gametocytes 20 × 7–8 µm. There is no alteration in shape of host erythrocyte nuclei.

Type Host *Testudo pardalis* (Testudines: Testudinidae).

Type Locality South Africa, no precise locality.

Other Hosts None known.

Other Localities None known.

Prevalence Not stated.

Morphological Variation No data other than that given in the diagnosis are available for this species.

Merogony Unknown.

Sporogony Unknown.

Effects on Host Host erythrocyte nuclei are not altered in shape.

Remarks This species has not been reported since its inadequate description by Laveran (1905).

Haemoproteus cajali Pittaluga 1912

Diagnosis A *Haemoproteus* species with round-to-elongate gametocytes, some of which form as halteridia. Gametocyte dimensions are unknown. Pigment forms as coarse yellow-brown grains. Host erythrocytes are little changed from uninfected cells.

Type Host *Clemmys africana*. (This host species is not listed as a synonym for any currently recognized species in Reptile-database.org.)

Type Locality Spanish Guinea (Equatorial Guinea).

Other Hosts None known.

Other Localities None known.

Prevalence Unknown.

Morphological Variation No morphological data are available for this species.

Merogony Unknown.

Sporogony Unknown.

Effects on Host Parasitized erythrocytes are little affected.

Remarks This species has not been reported since its inadequate description (G. Pittaluga, 1912).

Haemoproteus roumei (Bouet) 1909

Diagnosis A *Haemoproteus* species with macrogametocytes that may be spherical, 12.6 μm in diameter, or elongate and halteridial, 12.6–16.2 × 10.8 μm. Spherical microgametocytes are 9 μm in diameter and elongate forms are 14.4 × 3.6 μm. Pigment forms as compact masses and as grains dispersed in islands within the cytoplasm of microgametocytes.

Type Host *Kinixys belliana* Gray (Testudines: Testudinidae).

Type Locality Tombougou, Ivory Coast.

Other Hosts None known.

Other Localities Guinea and Porto-Novo, Dahomey (Benin) (Joyeux, 1913).

Prevalence In Guinea, Joyeux (1913) found 19 of 34 *K. belliana* infected with *H. roumei*.

Morphological Variation The only morphological data available are those stated in the definition.

Merogony Unknown.

Sporogony Unknown.

Effects on Host Host cells may be increased in size, with displaced nuclei, but cell shape is not distorted.

Remarks This species has not been reported since Joyeux reported it in 1913.

Haemoproteus baluzuci Travassos Santos Diaz 1953

Diagnosis A *Haemoproteus* species with halteridial gametocytes in both sexes. Macrogametocytes are elongate and larger than microgametocytes, 19.1–24.8 × 6.6–7.6 μm. Microgametocytes are more often ovoid or rounded, 11.9–18.8 × 5.9–8.9 μm, and contain fewer pigment granules, rarely exceeding 15 in number, in comparison to 12–30 in macrogametocytes. In both sexes, pigment granules are dispersed throughout the cytoplasm but tend to aggregate toward the poles in microgametocytes.

Type Host *Kinixys belliana zuluensis* Hewitt (Testudines: Testudinidae).

Type Locality Vicinity of Alto Limpopo, Mozambique.

Other Hosts None known.

Other Localities None known.

Prevalence Unknown.

Morphological Variation Macrogametocytes are predominantly halteridial in shape, with numerous vacuoles of varying size in the cytoplasm. Dimensions are 19.1–24.8 × 6.6–7.5 μm, estimated LW 126–186 μm^2 and L/W 2.9–3.3. The 12–30 large pigment granules are scattered throughout the cytoplasm. Microgametocytes are commonly rounded or ovoid, with elongate forms tending toward halteridial configuration. Their size is smaller than the macrogametocytes, 11.9–18.8 × 5.9–8.9 μm, with estimated LW 70–167 μm^2 and L/W 2.0–2.1. Pigment is less abundant than in macrogametocytes, rarely exceeding 15 granules, and is disseminated but tends to aggregate at the poles. Few if any vacuoles are present in microgametocytes.

Merogony Unknown.

Sporogony Unknown.

Effects on Host Infected erythrocytes are similar in length to uninfected cells but are hypertrophied in their width.

Remarks This species has not been reported since its description in 1953.

HAEMOPROTEUS SPECIES FROM ASIAN TURTLES

Haemoproteus "metchnikovi" (Simond) 1901

Diagnosis A *Haemoproteus* species with round or ovoid gametocytes 6–10 μm in diameter, occasionally forming blunt prolongations. Gametocytes are not halteridial. Pigment forms as dispersed fine grains in macrogametocytes and coarse granules irregularly distributed in microgametocytes.

Type Host *Chitra indica* (Gray) (Testudines: Trionychidae).

Type Locality Agra, India.

Other Hosts None known.

Other Localities None known.

Prevalence All 20 *C. indica* examined were infected by *H. "metchnikovi"* (Simond, 1901).

Morphological Variation The description is inadequate and does not provide any useful information for analysis.

Merogony Unknown.

Sporogony Unknown.

Effects on Host Host cells are little altered by gametocytes, and their nuclei are not displaced.

Remarks There are no additional reports of this species.

Haemoproteus caucasica Krasil'nikov 1965

Diagnosis A *Haemoproteus* species with ovoid or lentiform-to-elongate, halteridial gametocytes, with estimated size 10–18 × 5.5–6 μm, LW 55–99 μm^2, and L/W 1.6–3.5. Fine pigment granules are dispersed in the cytoplasm of macrogametocytes, but in microgametocytes coarser granules appear singly or in groups of two or more. Several distinct, variably shaped vacuoles are present in macrogametocytes.

Type Host *Testudo graeca* (Testudines: Testudinidae).

Type Locality Southeastern Republic of Georgia.

Other Hosts None known.

Other Localities None known.

Prevalence Unclear.

Morphological Variation Although no scale bar or magnification was given for the figures, it is possible to estimate the size of the figured gametocytes by using the dimensions of the "*Plasmodium smirnovi*" trophozoite in the figures, said to be 2–3 μm in size. Two macrogametocytes are estimated to be 10 × 6 μm, with LW 60 μm^2 and L/W 1.67, and 18 × 5.5 μm, with LW 99 μm^2 and L/W ratio 3.27. Two microgametocytes can be estimated at 10 × 5.5 μm, LW 55 μm^2, and L/W 1.82, and 16 × 5.5 μm, LW 88 μm^2, and L/W 2.91.

Merogony Unknown.

Sporogony Unknown.

Effects on Host Not clear.

Remarks As stated elsewhere (Telford, 1994), the illustrations of "*Plasmodium smirnovi*" are not convincing in view of its presence in mixed infection with *Haemoproteus caucasica*. Until a better description and illustrations are available that prove the identification as a *Plasmodium* species is correct, *Plasmodium smirnovi* Krasil'nikov 1965 should be considered a synonym of *H. caucasica*.

HAEMOPROTEUS SPECIES FROM AUSTRALIAN TURTLES

Haemoproteus chelodinae (Johnston and Cleland) 1909 Mackerras 1961

Diagnosis A *Haemoproteus* species with round-to-reniform gametocytes 7–12 × 7–9 μm, with estimated LW 50–99 μm² and L/W 1.0–1.7. Macrogametocytes are larger than microgametocytes and contain one to three prominent round vacuoles that are not apparent in microgametocytes. Pigment is dispersed as individual granules or rods or aggregated into irregular masses or clumps.

Type Host *Chelodina longicollis* (Shaw) (Testudines: Chelidae).

Type Locality Near Sydney, Australia.

Other Hosts *Chelodina oblonga, Emydura macquarii, Emydura latisternum, Emydura krefftii,* and *Elseya dentata* (Mackerras, 1961a).

Other Localities Brisbane, Mt. Nebo, and Eidsvold in Queensland, Perth in Western Australia.

Prevalence One of two *C. longicollis* collected near Sydney was infected by *H. chelodinae* (T. H. Johnston and Cleland, 1909).

Morphological Variation Dimensions in *C. longicollis* were reported as 4 × 3 (immature) to 12 × 7 μm, oval or rounded to kidney shaped, by T. H. Johnston and Cleland (1909), and in *C. oblonga*, average size was 11 × 8 μm (T. H. Johnston and Cleland, 1912). Mackerras (1961a) described macrogametocytes as round, 11 × 9 μm, containing one to three "sharply defined vacuoles", and microgametocytes as smaller, about 8 × 7 μm, without vacuoles. Pigment in macrogametocytes was formed as "rods or granules usually scattered through the cytoplasm but sometimes aggregated into irregular masses." In microgametocytes, "pigment was usually in clumps." Illustrations by Mackerras show an ovoid macrogametocyte with two large round vacuoles and a round microgametocyte. Their respective dimensions, calculated from the scale bar, were 9.7 × 7.1 μm and 7.7 × 7.1 μm. Dimensions, estimated from the sparse published data, are 7–12 × 7–9 μm, LW 50–99 μm², and L/W 1.0–1.7, with macrogametocytes larger than microgametocytes.

Merogony Unknown.

Sporogony Unknown.

Effects on Host "Only slight enlargement of the infected cells occurred" (Mackerras, 1961a). Figures show erythrocyte nuclei of normal size and position within the infected cells.

Remarks Both T. H. Johnston and Cleland (1909) and Mackerras (1961a) placed turtle hemoproteids within *Haemocystidium*, considered by Shortt (1922) and Wenyon (1926) to be a synonym of *Haemoproteus*. *Haemoproteus* **is clearly distinguished from** ***Haemocystidium*** **by the structure of its meronts, which have nuclei arranged in cytomeres. In** ***Haemocystidium*****, meronts are similar to those of** ***Plasmodium*** **EE stages, without any indication of cytomere formation.**

2
THE HEMOGREGARINES

Hemogregarines (Apicomplexa: Adeleiorina) are the most common, widely distributed, and speciose of reptilian hemoparasites and infect each of the orders of living reptiles, including the Rhynchocephalia (the tuatara). Three families of hemogregarines are presently recognized, the Hepatozoidae, Haemogregarinidae, and Karyolysidae, each distinguished by very different developmental patterns in their invertebrate hosts in which sporogony occurs. Perhaps 400 species have been described, mostly under the fallacious basis of "presence in a different host indicates specific identity" (Telford, 1984b). Regrettably, almost all species described before the late 1960s were known from the erythrocytic stages alone or, in the case of a few, from circulating cells and divisional forms in fixed tissues of various organs. Ball (1967) attempted to discourage their description just from the stages in blood cells alone, and this has contributed to a reduction in the number of species descriptions since that time, except when accompanied by data on sporogony in a competent invertebrate host. Indeed, most hemogregarines cannot be reliably placed within one of the four genera that occur in reptiles without this knowledge because the usual stages found in blood cells are the gamonts, many of which are very similar in appearance. When used in combination with a description of sporogonic stages, gamont morphology can be helpful, but its use alone is not acceptable. The species accounts below include only those with enough data to ensure their generic identity.

Most hemogregarine species were described as *Haemogregarina* before it was realized, except by a very few students of hemogregarines, that the sporogonic pattern is essential for both generic and familial identification. These species included several hemococcidian species that are only distantly, if at all, related to hemogregarines. The hemococcidia use invertebrates only for transmission, with sporogony occurring in the vertebrate host alone.

Haemogregarina

Haemogregarina stepanowi Danilewskyi 1885 was the first hemogregarine described from reptiles and is the type species of the Haemogregarinidae Neveu-Lemaire 1901. In the reptilian species of this family, gamonts ingested by leeches fed on an infected reptile undergo gametogenesis and intracellular sporogony within intestinal cells of the leech. Four aflagellate microgametes form during microgametogenesis. The zygote formed from fertilization of the macrogamete by one of the microgametes develops into an oocyst that lacks sporocysts, the eight sporozoites arising from a single germinal center, that is, the oocyst is "monosporoblastic" (Siddall and Desser, 1991). The sporozoites infect leech tissues outside the intestine, sites described by Siddall and Desser (1991) as "anastomosing lacunae of the circulatory system," producing primary meronts that may contain up to 250 merozoites. These meronts release merozoites, the infective stage for the reptile, that concentrate near the tip of the proboscis and await the next blood meal of the leech when the merozoites will enter the reptile's circulatory system. The merozoites form preerythrocytic meronts within fixed cells of lung, liver, and spleen and produce about 18 merozoites (Siddall and Desser, 1991). The merozoites then enter erythrocytes and become premeronts that divide into eight erythrocytic merozoites. On rupture of the host cell, these merozoites may invade erythrocytes and form more erythrocytic meronts or gamonts.

Siddall (1995) recognized 19 species of *Haemogregarina*, all in turtle hosts, and two other genera that parasitize fish, *Cyrillia* and *Desseria*. He suggested that all other *Haemogregarina s.l.* species described from "snakes, crocodilians, lizards, and birds" (and, presumably, from the tuatara) should be considered as *Hepatozoon* species. In view of the paucity of life cycle studies among reptilian hemogregarine species, it is likely that *Haemogregarina* will be found to occur in reptiles other than turtles that have contact with leeches. An example, possibly, is *Haemogregarina floridana* from aquatic snakes in Florida (Telford et al., 2001), placed in that genus by the presence of erythrocytic meronts in circulating blood that apparently are sequestered in the lung when nearing maturity.

Hepatozoon

The genus *Hepatozoon* was established for intraleukocytic parasites of rats when the life cycle in both rodents and mites (*Laelaps echidninus*) was demonstrated by Miller (1908). This followed reports of leukocytic and erythrocytic parasites in dogs and various rodents within the previous 3 years by several workers (Wenyon, 1926). In the mite, the presence of large oocysts containing many sporocysts within which sporozoites formed was the distinctive generic character. Infected mites eaten by rats released sporozoites that entered the rodent's circulatory system and formed meronts within cells of the liver, probably only in endothelium rather than hepatic parenchymal cells. Their merozoites, after several cycles within liver cells, invaded leukocytes or erythrocytes, and gamonts, which were the infective stage for mites, developed.

The presence of *Hepatozoon* in reptilian hosts was not demonstrated until Hoare (1932) described the sporogony of *H. pettiti* (Thiroux, 1910, 1913) in tsetse flies fed on infected Nile crocodiles. One other species of *Hepatozoon*, *H. caimani*, has been proven to parasitize crocodilians (Lainson, Paperna, and Naiff, 2003), although there are records of other hemogregarines in crocodiles for which sporogony is unknown. Shortly thereafter, Robin (1936a, 1936b) demonstrated the sporogonic pattern of *Hepatozoon mesnili* from a saurian host, *Gecko verticillatus* (= *G. gecko*). To this date, sporogonic data have become available for only eight species of *Hepatozoon* that parasitize lizards. It was not until 1967 that a life cycle proving *Hepatozoon* identity for a parasite of snakes was demonstrated: *Hepatozoon rarefaciens* (Sambon and Seligmann, 1907) in Mexican snakes (Ball, Chao, and Telford, 1967). Since 1967, a sporogonic pattern characteristic of *Hepatozoon* has been reported for 23 other hemogregarine species. Following the revision of *Haemogregarina* by Siddall (1995), Smith (1996) transferred all of the remaining *Haemogregarina s.l.* species to *Hepatozoon*, in accordance with Siddall's recommendation to that effect.

Most life cycles of reptilian *Hepatozoon* species have been studied usually using mosquitoes from laboratory colonies of various *Culex* and *Aedes* species and, rarely, *Anopheles* species (see species accounts). There are apparently only eight vector species that have been found naturally infected by *Hepatozoon* oocysts or sporocysts of reptilian origin: the tsetse fly *Glossina palpalis* (Chatton and Roubaud, 1913; Macfie, 1916); the ticks *Amblyomma dissimile* (Ball et al., 1969; Telford, unpublished) and *Hyalomma cf. aegyptium* (Paperna et al., 2002); the mites *Ophionyssus* sp. (Ramanandan Shanavas and Ramachandran, 1990) and *Hirstiella* sp. (J. E. Lewis and Wagner, 1964); the reduviid bugs *Triatoma arthurneivae* (Rocha e Silva, 1975) and *Triatoma rubrovaria* (Osimani, 1942; Talice, 1929); and the phlebotomine *Lutzomyia vexator occidentis* (Ayala, 1970a), its *Hepatozoon* species infective for both snakes and lizards. Adler and Theodor (1925) thought *Hepatozoon* sporogonic stages they found in *Phlebotomus* were of gecko origin. Sporogony occurred in leeches, *Haementeria lutzi*, fed experimentally on Brazilian snakes (Pessôa and Cavalheiro, 1969a, 1969b), but no infections resulted in experimental snakes, which casts doubt on the capacity of leeches to vector *Hepatozoon* species (Smith, 1996). Garnham (1954) found sporogonic stages of a hemogregarine he named *Hepatozoon argentis* in the soft tick *Argas brumpti*, but identification of the vertebrate host was not proven, although *Agama mossambica* was thought to be the host. In addition to mosquitoes used successfully as experimental hosts for sporogony are mites *Ophionyssus scincorum* (Allison and Desser, 1981).

The life history pattern of *Hepatozoon* species in reptiles may be described as follows, bearing in mind that deviations from the pattern are known for some species (Smith, 1996), in particular for those species that require two vertebrate hosts in the life cycle: In the three species thus far demonstrated (*H. domerguei, H. sipedon, H. sirtalis*), transmission to the second, final vertebrate host depends on its predation on the first vertebrate host, itself infected by ingestion of the infective invertebrate host. Laboratory studies have clearly demonstrated that many *Hepatozoon* species, particularly of snakes, are able to infect the vertebrate final host directly from the invertebrate in which sporogony has occurred. How this infection takes place in nature is still an open question in my opinion, given the great diversity in species and their dietary habits, and the varying ecology of the final vertebrate hosts. The weight of opinion would require ingestion either from predation on infected first vertebrate hosts or by direct ingestion of the infected invertebrates. The mere presence of monozoic or dizoic cysts of *Hepatozoon* in the tissues of a vertebrate *is not*, per se, evidence that the role of these cysts is to transmit the parasite to a final vertebrate host. These cysts are commonly seen in liver and lungs of snakes that do not form part of the food chains of other snake species. Their role, in fact, may be to provide continuing infection of the snake by becoming meronts when the initial rounds of merogony in snake tissues following infection have ended in a purely erythrocytic parasitemia of mature gamonts, gradually diminishing in intensity. Their function might also be a defense against immunity, if there is such a response directed at the erythrocytic infection. Since pathology even in massive infections usually appears mild, this is not very likely. Salivary transmission of *Hepatozoon matruhensis* to the colubrid snake *Psammophis schokari* was apparently accomplished by bites of *Culex pipiens* (Ebraheem et al., 2006) and *C. neavei* (Rashdan and El-Sebaii, 2006), with patency of 28–35 and 31–35 days, respectively, following feeding by infected mosquitoes. Detailed description of sporozoite location within mosquitoes is needed, however. In recent studies of sporogony in mosquitoes of 15 *Hepatozoon* species of Florida snakes, oocysts or sporocysts of three species (*H. pictiventris, H. sauritus, H. polytopis*) were easily found in proboscides removed from mosquitoes prior to dissection (Telford et al., 2001, 2004, 2005b). In another study (Telford et al., 2008), an oocyst with mature sporocysts of *Hepatozoon horridus* was found within the salivary glands of a mosquito fed on the snake host. These findings at least suggest that salivary transmission is possible during feeding by mosquito vectors in nature. Sporogony in proximity to or within the proboscis would not have to occur in a high percentage of such extremely abundant vectors to maintain a high prevalence of *Hepatozoon* species in natural snake populations.

The typical life history of *Hepatozoon* species studied with the use of experimental mosquito vectors begins quickly after ingestion of erythrocytic gamonts in the blood meal. Macro- and microgamonts leave their host cells and penetrate the intestinal wall, entering the hemocoel, where pairing in szygy occurs; gametogenesis, probably within fat body cells of the hemocoel, produces two to four biflagellated microgametes, one of which fertilizes the macrogamete; and a zygote forms. The zygote becomes an expanding oocyst containing multiple germinal centers (polysporocystic) that form sporocysts, which in turn produce sporozoites that range in number from 4 to 50 or more, varying by *Hepatozoon* species. Oocysts may form in the hemocoel of the abdomen and thorax or within the head, and while some species show no preferred site, developing in all three locations, others may be restricted to a particular portion of the vector body. When infection of the vertebrate host is accomplished, by forced ingestion in experimental situations, sporozoites leave their sporocysts, apparently penetrate the intestinal wall, and enter the circulatory system of the vertebrate. Sporozoites reach the liver or lungs in most *Hepatozoon* species, where they form meronts in hepatic parenchymal cells, Kupfer cells, and endothelial cells of capillaries or sinus walls. The first generation of meronts, macromeronts, forms larger and fewer macromerozoites than does the second meront generation, their progeny, which produce more and smaller micromerozoites in usually larger micromeronts. It is not clear whether either type of meront has the capacity to produce subsequent generations of its same type, but given the profusion of meronts of both types often seen in reptile tissues, this would appear to be possible. Ultimately, micromerozoites are released from their meronts and enter erythrocytes of the circulating blood.

A somewhat different mode of gametogony was described by Robin (1936a, 1936b) in mosquitoes fed on geckoes infected with *Hepatozoon mesnili*. Gametogony began in the lumen of the gut, and four nuclei resulted from division in the microgametocyte. Without formation of a flagellated microgamete accompanied by cytoplasmic division, one nucleus entered the cytoplasm of the macrogametocyte before fusion of the two gametocytes. The zygote formed from fertilization then penetrated the gut wall into the hemocoel, where the common pattern of oocyst formation and sporogony ensued. Ball and Oda (1971) suggested that this different mode of sporogony could be adequate justification for taxonomic division of the genus *Hepatozoon*. Miller (1908) described a similar mode of gametogony in mites infected by the rodent hemogregarine *Hepatozoon perniciosus* (= *H. muris*), a different pattern from that found in other mammalian *Hepatozoon* species (Smith, 1996). Another variation in the life cycle pattern of *Hepatozoon* species was described by Allison and Desser (1981) for *Hepatozoon lygosomarum*, which undergoes sporogony in the mite *Ophionyssus scincorum*. Although szygy was not apparent in the intestinal lumen of the mites 2–3 days following feeding, and gametogenesis was not observed, uninucleate oocysts were found in the walls of the intestinal ceca, where the polysporocystic oocysts characteristic of *Hepatozoon* developed. Another *Hepatozoon* species, *H. kisrae*, also utilizes an acarinid, *Hyalomma cf. aegyptium*, for sporogony, but the pattern follows that observed for *Hepatozoon* species of snakes, with oocysts forming within the tick hemocoel, but on the gut surface (Paperna et al., 2002). In yet another variation of the gametogonic process, uniflagellated microgametes are formed during microgametogenesis within the mosquito hemocoel in *Hepatozoon gracilis, H. aegypti*, and *H. mehlhorni* (Bashtar et al., 1984c, 1987, 1991). Production of uniflagellated microgametes was described by Smith and Desser (1997b) as "a retention of the ancestral condition." It is clearly premature to divide the genus *Hepatozoon* into additional genera based on these sorts of variation in gametogony and sites of zygote formation and sporogony without supporting evidence from genomic analysis, hopefully of many more species than the few thus far examined for either gametogonic or genomic characters.

Morphology

Intrinsic Stages Within erythrocytes, *Hepatozoon* gamonts are often called "banana-shaped." Unlike gamonts of

Haemogregarina species, gamonts of most *Hepatozoon* species are not recurved on themselves, forming two limbs of similar length that extend for most of the host cell length. Both ends are usually blunt, one or both tapering slightly, and one end slightly wider than the other. I and others have considered the broader end as "anterior" and the narrower end as "posterior." In some species, the posterior end has a pronounced taper, and in these there may be a recurved "tail," which can sometimes reach the gamont midbody. Cytoplasm of the gamont may be uniform or show a profusion of rounded vacuoles. In some species, azurophilic granules are scattered within the cytoplasm. There are hemogregarine species, supposedly *Hepatozoon*, with elongate narrow gamonts, apparently somewhat rigid, that stretch the host erythrocyte in its length dimension and narrow its width. The length, maximum width, size calculated as length times maximum width (LW), and the length/width ratio (L/W) are often useful taxonomic characters when used in conjunction with sporogonic characters.

Gamonts free of host cells are elongate and slender with a nucleus slightly anterior to midbody and usually have a somewhat tapered posterior end. Depending on the species, apparently mature gamonts may be as short as 10 µm in length and can exceed 20 µm, but most are 14–18 µm. Gamont nuclei are usually compact bodies at midbody or anterior in the second quarter of the gamont, sometimes entering or occupying the anterior first quarter. They are often squarish in shape, covering the width of the gamont, or somewhat triangular, described as "saddle shaped," extending or not the full width of the gamont. The nuclei in some species may be elongate and narrow, extending from the first or second quarter past midbody into the third quarter of the gamont. Young gamonts, often termed *trophozoites* in the literature, are elongate but shorter and narrower than the mature forms, with prominent nuclei that often cover most of the parasite body. Nuclei in these stages usually show some irregularity in shape, with spaces visible between masses or bands of chromatin, and the cytoplasm is often vacuolated. Irregularly formed nuclei are also common in gamonts of mature size, and occasionally nuclei are fragmented into small pieces or into two halves, but apparent meronts within erythrocytes are known only for the crocodilian parasite *Hepatozoon pettiti* (Hoare, 1932). Host cells are often distorted and enlarged, with nuclei displaced or elongated and narrowed, but seldom can be described as lysed.

The hypertrophy of host erythrocytes may be dramatic, as in *Hepatozoon rarefaciens* and *H. fusifex*, which represent the extremes of flattening or shape distortion of the host cell, respectively. Dehemoglobinization of the erythrocyte may be characteristic of some species and reaches its maximum in a compact, contracted mass adjacent to the gamont and surrounded by very thin, transparent cytoplasm, as in *Hepatozoon sipedon* and *H. sirtalis.* Gamont sex is seldom apparent, but some species have been described as showing different staining intensity, some gamonts darker and others paler, which is probably indicative of sexual dimorphism. As in the other three genera of hemogregarines, encapsulated gamonts are often seen in circulating erythrocytes. The function of encapsulation is unproven but could be to improve the survival of gamonts during digestion of a blood meal by a competent vector or perhaps as a defense against immune response by the vertebrate host.

Extrinsic Stages Oocysts are usually spherical or ovoid and appear as a thin wall surrounding the sporocysts. A residual body is sometimes apparent. They can attain a diameter exceeding 300 µm, depending on the number of sporocysts characteristic of the *Hepatozoon* species. Mature oocysts are fragile and often ruptured during dissection of the infected invertebrate host. The sporocysts are also spherical or ovoid to somewhat elongate in shape, again surrounded by a cyst wall. As with oocysts, their diameter depends on the number of sporozoites that form within, both sporocyst size and sporozoite number often characteristic of a *Hepatozoon* species. The smallest sporozoite numbers are those of *H. sipedon* and *H. sirtalis,* which produce 8 and 4–8, respectively, within sporocysts that can number more than 300 in a single, large oocyst. Sporozoites are usually elongate and narrow, with one tapered end and a prominent nucleus near midbody, usually arranged in no particular manner within the sporocyst, although some appear to form in rows. A residual body is typical in sporocysts and may serve as a focus in arrangement of the sporozoites.

Karyolysus

Two genera of hemogregarines comprise the Karyolysidae, *Karyolysus* Labbé 1894 and *Hemolivia* Petit, Landau, Baccam and Lainson 1990. *Karyolysus* is a saurian parasite, reported so far only from lizards of the genera *Lacerta* and *Podarcis* (Lacertidae). *Karyolysus* species are transmitted by an ingested mite vector, *Sauronyssus saurarum,* in which transovarian transmission occurs in the infected female mite through its eggs to the larvae that represent the next generation. Gamonts in the lizard erythrocytes, when ingested by a mite, quickly leave their host cells (Svahn, 1975b) and form pairs within the first day postingestion (PI). The macrogametocyte becomes rounded as a macrogamete within 48 hours postingestion. Microgametocytes in contact with the macrogametocyte divide once to produce two biflagellated microgametes, one of which fertilizes the macrogamete. This zygote forms an oocyst in either epithelial cells of the mite stomach (*K. lacertae*) or the podosomal region of the hemocoel (*K. lacazei*), as described by Reichenow (1921) and confirmed by Svahn (1975b). The latter author found that oocysts of *K. latus* usually developed in the stomach epithelium but could be found in the anterior portion of the hemocoel as well. About 16 motile sporokinetes are formed within the oocyst by a single germinal center. Released from the oocyst, these sporokinetes enter the eggs of the mite and apparently encyst, becoming sporocysts in which 16–32 sporozoites form. The eggs hatch into mite larvae, then molt into the nymphal stage about 24 hours after hatching. When nymphs take their first blood meal about a week later, the sporocysts contain mature sporozoites, and the infection is transmitted when the infected nymphs are eaten by the lizard host. Meronts form in the capillary endothelium of the liver, lungs, heart, and spleen of the lizard and undergo several cycles of merogony before infected erythrocytes appear in the lizard's blood about 30 days later. The first generations of meronts are the primary meronts and produce short and stout macromerozoites, which then become micromeronts in endothelial cells and form more slender micromerozoites that enter erythrocytes and become gamonts.

The use of the term *sporokinete* by Reichenow (1921), Svahn (1975b), and others was criticized by Barta and Desser (1989) and Siddall and Desser (1991), taking the strict position that the stage formed by the oocysts, the product of the sexual process, "are by definition sporozoites and the subsequent asexual stages are meronts" (Desser, 1993), despite their formation in the invertebrate host rather than the vertebrate. Although Svahn's (1975) photographs may suggest otherwise, "typical sporocystic walls around sporocysts of *Karyolysus* species in the nymphal mite hosts" were not demonstrated.

Karyolysus gamonts often produce severe distortion and sometimes apparent lysis of the nucleus in host erythrocytes, thus the generic name. But, not all infected red cells show this distortion, and taxonomic assignment of hemogregarines from saurian genera other than *Lacerta* or *Podarcis* from Europe should be based on identification from sporogonic stages or, eventually, by DNA comparisons. Some authors have applied *Karyolysus* to hemogregarines of snakes, which almost certainly are *Hepatozoon*, not *Karyolysyus*.

Hemolivia

Hemolivia stellata, described from the neotropical anuran *Bufo marinus*, the marine toad, in Brazil (Petit, Landau, Baccam, and Lainson, 1990), is a hemogregarine that undergoes sporogony in a tick, *Amblyomma rotondatum*, in which the zygote forms a star-shaped oocyst that produces sporokinetes in intestinal cells, which then enter other intestinal cells and become sporocysts that give rise to sporozoites. This pattern contrasts with *Karyolysus*, for which sporokinetes are produced in the adult mite and are transovarially transmitted through the mite egg to the next generation of mites. Petit et al. (1990) suggested that transmission to a new anuran host might utilize any of the known modes of transmission for hemogregarines: ingestion of sporocyst-infected ticks, an infected paratenic host infected by contamination from the external environment or directly from sporocyst-contaminated surroundings, even predation of a vertebrate host in which cysts containing cystozoites are present in its tissues. Experimental toads were infected by ingestion of ticks containing sporocysts or toad liver containing cystozoites. Gamonts appeared in erythrocytes in 40–45 days postingestion and also accumulated in pigmented reticuloendothelial cells. Intraerythrocytic meronts containing 2–24 merozoites were also observed in the capillaries of liver, lungs, and spleen, more rarely in the heart and circulating blood, and unpigmented cells of the reticuloendothelial system. The last were smaller and contained 2–14 merozoites. Encysted meronts were found in erythrocytes in the liver and spleen, and cysts containing a single cystozoite, rarely two, were found both within erythrocytes in these tissues and in reticuloendothelial cells with or without pigment.

The host range of *Hemolivia* was extended into reptiles by Smallridge and Paperna (1997) in their description of *Hemolivia mariae* from the Australian scincid lizard *Tiliqua rugosa*. The hemogregarine is transmitted by the tick *Amblyomma limbatum*, which commonly infests the lizards. Oocysts formed, following ingestion of blood meals containing *H. mariae* gamonts, in epithelial cells of the tick gut. Oocysts are stellate in form, with three to five arms filled with crystalline bodies around which the sporokinetes form. Sporokinetes disperse from oocysts and enter intestinal cells as well as cells of other organs in the tick, occupying parasitophorous vacuoles "with a wavy wall or a narrow space bound by a single lamella" (Smallridge and Paperna, 1997). These form sporocysts with a hard wall, which produce 8–20 sporozoites. Although yet undescribed, sporozoites become thick-walled meronts in the tissues of the lizard, and their merozoite progeny infect erythrocytes of the circulating blood, forming gamonts within opaque cyst walls and young meronts (figures of Smallridge and Bull, 2001b) that apparently reach maturity within blood vessels of the lizard's organs.

Landau and Paperna (1997) reassigned *Hepatozoon mauritanicum* (Brumpt 1938) Michel (1973), a tick-transmitted hemogregarine of the tortoise *Testudo graeca*, to *Hemolivia* after reexamining the original histological sections of infected *Hyalomma aegyptium*. The irregularly shaped oocysts were not stellate in form, but oocysts containing sporokinetes and sporocysts containing sporozoites were found in tick intestinal cells. This reappraisal of sporogony, as well as the gamonts, meronts, and cysts containing single cystozoites within erythrocytes, was adequate evidence of identity as a *Hemolivia* species.

Although acarine-transmitted *Hepatozoon* species have been described in saurian hosts (i.e., *H. kisrae,* Paperna et al., 2002; and *H. octosporei,* Ramanandan Shanavas and Ramachandran, 1990), these form oocysts within the hemocoel of the tick or mite host. Hemogregarine species such as *Hepatozoon lygosomarum* (Allison and Desser, 1981), which forms oocysts and sporocysts within the walls of the intestinal ceca of mites, should be closely studied to rule out the production of sporokinetes by oocysts before sporocysts appear in the cells of the gut wall. *Hemolivia* species will probably be found elsewhere in the world given their presence in the herpetofaunas of Brazil, Mediteranean Europe, and Australia.

Ultrastructure

In reptiles, the first ultrastructural study of a hemogregarine was that of Stehbens and Johnston (1967) on an unidentified parasite of an Australian gecko. The ultrastructure of gamonts, gametogenesis, and sporogonic development from oocyst to sporozoite of *Haemogregarina balli* were examined by Siddall and Desser (1990). Vivier et al. (1972) investigated the ultrastructure of some aspects of sporogony by *Hepatozoon domerguei.* Bashtar et al. (1984b, 1984c) described the fine structure of gamonts, merozoites, gametogony, oocysts, sporocysts, and sporozoites of *Hepatozoon aegypti* from an Egyptian snake. The ultrastructure of *Hepatozoon mocassini* gamonts was described by Nadler and Miller (1985) and its gametogenesis and sporogony by Lowichik et al. (1993). Gametogony and sporogonic development of *Hepatozoon sipedon* was described by Smith and Desser (1997b). Sporogony in these studies took place in mosquito invertebrate hosts. For a tick-borne *Hepatozoon,* Paperna et al. (2002) found similarity in development of *Hepatozoon kisrae* to that found in *Hepatozoon* species in mosquito hosts, including the formation of crystalloid material within pockets of the endoplasmic reticulum in sporocysts, that then was divided among the sporozoites as in the mosquito-borne species. This procedure differed in the sporogony of *Hemolivia mariae* (Smallridge and Paperna, 2000a, 2000b), also tick borne, in which the crystalloid material develops in the apparent absence, or perhaps early disappearance, of an extensive endoplasmic reticulum and is distributed among sporokinetes before sporozoite formation. In contrast to *Hepatozoon* and *Hemolivia,* this crystalloid material is formed during merogony of *Haemogregarina* species in the leech host, not during sporogony, and merozoites rather than sporozoites are the infective stages for their vertebrate hosts (Desser et al., 1995). A thick wall, largely impervious to stains, surrounding gamonts also distinguishes gamonts of *Hemolivia* and *Karyolysus* from those of reptilian *Hepatozoon* species (Boulard et al., 2001); this encapsulation wall differs between *Hemolivia* and *Karyolysus* by the presence of conspicuous sutures in the former genus and the apparent absence of sutures in the latter (Beyer and Sidorenko, 1984). As described by Desser (1993), the ultrastructure of hemogregarine gamonts for the few species examined is quite similar among all: His example of the anuran parasite *Haemogregarina magna* (Paterson et al., 1988) described the pellicle of the gamont as

> composed of an outer plasmalemma and an inner complex of two closely apposed membranes. The components of the apical complex … include two preconoidal rings, a conoid, and an elaborate polar ring complex … [that] consists of a stout, electron-lucent polar ring and approximately 78 posteriorly directed, radially arranged "tine-like" structures. These widen anteriorly and fuse as they merge into the polar ring.

Some variation in the appearance of these structures in other species studied, as cited, does occur, but the general plan of organization is very similar for all and provides little taxonomic distinction. For details of fine structure, the original literature should be consulted.

Species Accounts

APICOMPLEXA, Adeleorina

Hepatozoidae

NORTH AMERICAN *HEPATOZOON* SPECIES

Hepatozoon rarefaciens (Sambon and Seligmann, 1907) Ball, Chao, and Telford, 1967 (**Plate 51**)

Diagnosis A *Hepatozoon* species characterized by gamonts 11–18 × 2.5–9 µm and spherical oocysts 58–200 µm that contain up to 64 spherical sporocysts 21–59 µm in diameter, producing 13–42 sporozoites. Merogony is primarily in liver cells and hepatic sinuses. Macromeronts average 24 × 17 µm and produce up to 14 merozoites, and micromeronts contain 30–50 merozoites. Sporogony occurs in the hemocoel of the vector, primarily in the abdomen. Infected erythrocytes may be markedly hypertrophied and distorted, reaching dimensions two or three times greater in each axis than uninfected erythrocytes.

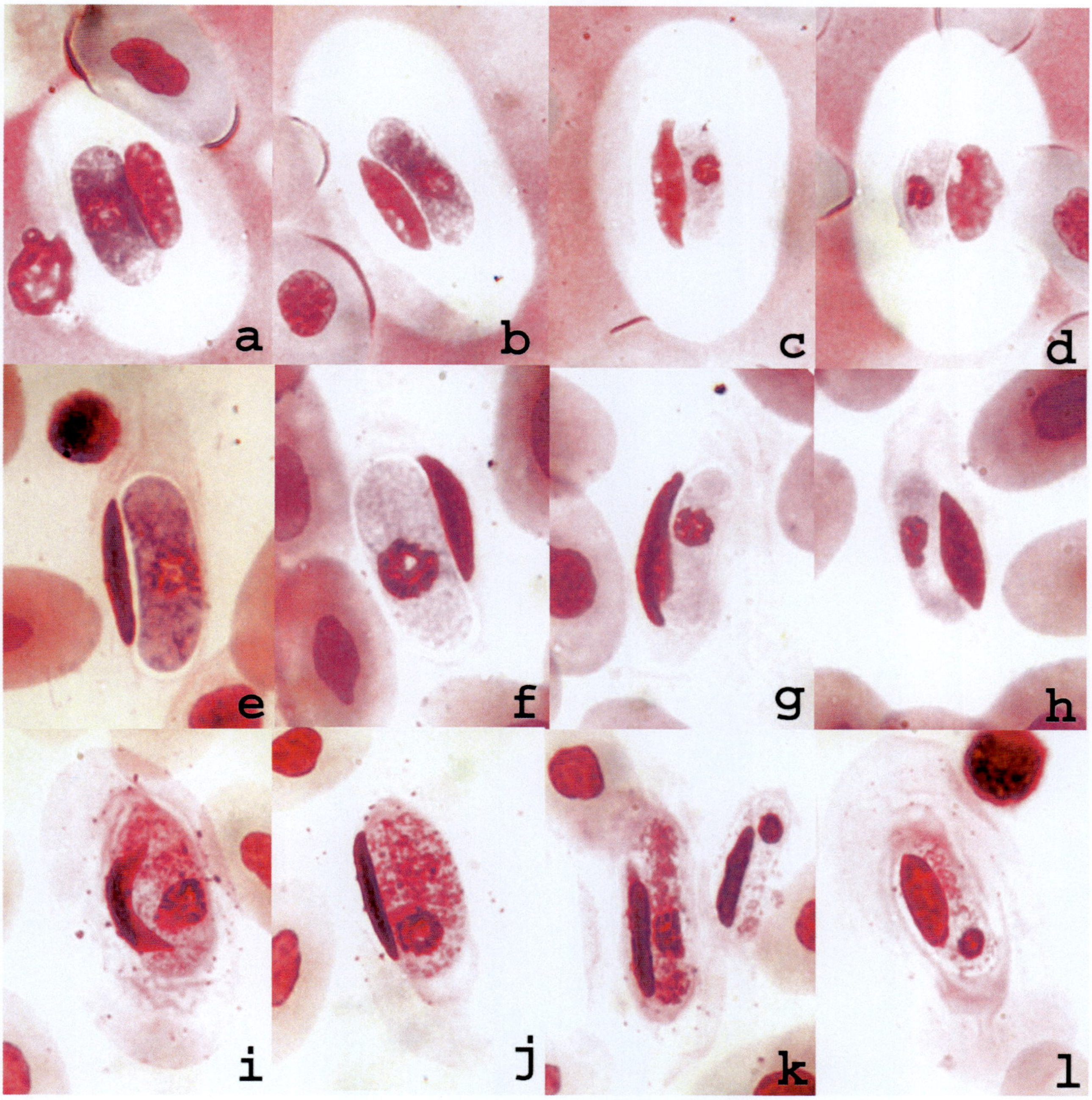

Plate 51 *Hepatozoon rarefaciens* from *Drymarchon corais* spp: **a–d**, *D. corais couperi*, Florida; **e–h**, *D. corais erebennus*, Texas; **i–l**, *D. corais rubidus*, Mexico.

Type Host *Drymarchon corais couperi* Holbrook (Serpentes: Colubridae).

Other Hosts *Drymarchon corais rubidus, D. corais erebennus.*

Type Locality Not stated. Here restricted to southeastern United States.

Other Localities Alachua County, Florida, and Texas (Telford); Estados Colima and Nayarit, Mexico (Ball et al., 1967).

Prevalence Four of 8 *D. corais couperi* from northern Florida (Telford); 19 of 20 *D. corais rubidus* from vicinity of Colima and Tepic, western Mexico (Ball et al., 1967).

Morphological Variation Dimensions of *H. rarefaciens* gamonts in the type infection in *D. corais couperi* (termed *meronts* by Sambon, 1909c) were 13–17 × 4–6 μm. Sambon (1909) described the presence of two types of gamonts with different staining reactions, one staining dark with a less-obvious nucleus and the other pale with a deeply staining, more compact nucleus. The same difference in staining reaction was present in *H. rarefaciens* from *D. corais couperi* in northern Florida. In the latter material, gamonts averaged 14.5 ± 1.0 × 4.9 ± 0.8 μm (13–16 × 3.5–6, N = 25), with LW 71.1 ± 14.6 μm² (45.5–93) and L/W 3.01 ± 0.36 (2.5–3.8). The deeply staining gamonts were 15.3 ± 0.3 × 5.4 ± 0.6 μm (15–16 × 4–6, N = 13), with LW 82.4 ± 8.2 μm² (60–93) and L/W 2.88 ± 0.34 (2.5–3.8), while those that had pale cytoplasm averaged 13.5 ± 0.5 × 4.3 ± 0.5 μm (13–14.5 × 3.5–5, N = 12), with LW 58.8 ± 8.9 μm² (45.5–72.5) and L/W 3.16 ± 0.33 (2.7–3.7). In *D. corais rubidus* infections, on which the redescription of *H. rarefaciens* is based (Ball et al., 1967), mature gamonts averaged 15.4 × 5.5 μm (11 × 3–22 × 10, N = 40) indicating an average LW of 84.7 μm². Ball et al. (1967) did not comment on the different staining reactions present in gamonts, but their figures show gamonts with both light and dark cytoplasm. Gamont nuclei were occasionally terminal but usually were central in position. All gamonts figured by Sambon (1909c) show central nuclei, while nuclei in the sample from a Florida host were about equal in the numbers of gamonts showing nuclei at midpoint or in the second quarter of the gamont.

Sporogony Sporogony was obtained experimentally in three species of mosquitoes, *Culex tarsalis, Aedes sierrensis*, and *Anopheles albimanus*, all from laboratory colonies of California origin. Uninucleate oocysts were present in the mosquito hemocoel in 72–78 hours postfeeding (PF), and oocysts contained sporocysts in 10–14 days. Mature oocysts were present in 16–29 days, with motile sporozoites demonstrable at 16 days. Average oocyst size in *C. tarsalis* was 129 × 129 μm (58 × 58 – 200 × 200), and there were on average 20 sporocysts (3–64) contained within oocysts. Sporocyst size varied from 21 × 21 to 59 × 59 μm, with an average of 39 × 39 μm, and had an LW of 1521 μm². Sporocysts produced 29 (13–42) sporozoites. Oocysts obtained from an experimental infection of *H. rarefaciens* in *Boa constrictor* infected from *D. corais rubidus* (Ball et al., 1967) were similar to those derived directly from the natural host: Oocyst size was 120 × 120 μm (89 × 89–155 × 155); they contained 17 sporocysts (4–30); sporocysts averaged 42 × 42 μm (28 × 28 to 56 × 56) and produced 26 (15–34) sporozoites.

Merogony Meronts form primarily in hepatic cells, with cysts containing them present in hepatic sinuses. Up to 14 macromerozoites 13 × 3 μm were observed in macromeronts, which averaged 24 × 17 μm. Micromeronts, 21 × 10 μm on average, produced 30–40 micromerozoites, 9 × 1.5 μm in average size. Other sites where meronts occurred were in endothelial cells or capillaries of the lung, spleen, pancreas, and heart.

Effects on Host *Hepatozoon rarefaciens* was described from a parasitemia of 60% in a *D. corais couperi* (Sambon, 1909c). Young gamonts had little effect on host cells, but mature gamonts, described as encapsulated, occupied erythrocytes "enormously enlarged, completely dehemoglobinized, and thinned out like wafers" (Sambon, 1909c), which might attain a size of 32–42 by 12–18 μm, nearly three times normal erythrocyte dimensions of 16 × 11 μm. Host cell nuclei were hypertrophied and were "closely adherent" to the gamont. Sambon (1909c) mentioned the presence of gamonts in leukocytes, perhaps a consequence of the massive parasitemia. Encapsulated gamonts were reported also by Ball et al. (1967). Hypertrophy of *D. corais rubidus* erythrocytes varied from slight to great, with shape considerably distorted. Hypertrophied erythrocytes host to gamonts averaged 34 × 15 μm but could reach 48 μm in length, in comparison to normal uninfected erythrocyte dimensions of 17 × 10 μm. In *D. corais couperi* from north Florida, hypertrophied erythrocytes infected by gamonts varied from 30 to 37 × 18 to 24 μm, in comparison to normal erythrocytes, which were 19.5–22 × 10–13 μm. Erythrocyte nuclei from infected cells were 11–20 × 3–6 μm in comparison with 6.5–9 × 4.5–6 μm in normal cells.

Remarks The infection of a juvenile *Boa constrictor* with *H. rarefaciens* derived from *D. corais rubidus* by Ball et al. (1967) was the first experimental infection of a snake host from one family (Boidae) by a *Hepatozoon* species obtained by a capable insect vector from the natural host that belonged to another snake family (Colubridae). Parasite morphology and developmental pattern in both mosquito and unnatural snake hosts remained the same during a second vector transmission of *H. rarefaciens* from the infected boa to another boa. The prepatent period was 47 days following first ingestion of infected mosquitoes, and 20 days after the last feeding of the first transmission, and was comparable in the second transmission. Multiple feedings of infected mosquitoes were done to increase the chances of successful infection of the snakes. This study was also the first demonstration of sporogony for a reptilian *Hepatozoon* species of the Western Hemisphere.

In a second study on experimental transmission of *H. rarefaciens* to an unnatural host species, Chao and Ball (1969) transmitted *H. rarefaciens* to a different colubrid snake species, *Pituophis c. catenifer* from California. A subsequent feeding by *C. tarsalis* on the infected *P. catenifer* resulted in sporogonic stages identical to those seen when mosquitoes were fed on the natural colubrid host, *D. corais rubidus*, or the unnatural experimental host, *B. constrictor.*

In each of the experimental hosts, the greatly hypertrophied host cells were not seen: Gamonts were similar to those present in *D. corais rubidus* infections, which showed more modest effects on host erythrocytes. Identity of the gamonts as *H. rarefaciens* was confirmed by the presence of morphologically identical sporogonic stages of mosquitoes fed on the natural host and both of the experimental unnatural host species. This may represent an effect of briefly observed, active infection in contrast to long-established, chronic infection of the snake hosts.

Hepatozoon rarefaciens is the only *Hepatozoon* species for which sporogony has been achieved in vitro (Ball and Chao, 1973). Using a *Culex pipiens* cell line, sporogony from erythrocytic gamonts was completed in 20 days at 23–24°C, comparable to 16–19 days in the mosquito hemocoele (Ball et al., 1967).

Hepatozoon fusifex Ball, Chao, and Telford, 1969 (**Plate 52**)

Diagnosis A *Hepatozoon* species characterized by the presence of gamonts 9.2–17.3 × 2.3–8.1 µm in erythrocytes of normal size, hypertrophied gametocytes containing gamonts 15–23 × 4.6–11.5 µm, and gamonts 15–18.4 × 4.6–9.2 µm that occupy greatly elongated, spindle-shaped erythrocytes that are leukocytozoid in appearance. Merogony occurs primarily in capillary endothelium of the lung but in heavy infections, in the liver, spleen, kidney, brain and heart as well. Macromeronts average 22 × 18 µm and produce up to 18 merozoites. Micromeronts are 29 × 19 µm and may form 55–60 micromerozoites. Sporogony occurs naturally in the hemocoel of the ixodid tick *Amblyomma dissimile* and experimentally in *Culex tarsalis* and *Aedes togoi*. Nearly spherical oocysts 178 × 165–282 × 234 µm contain 146–250 elliptical sporocysts that produce 15–35 sporozoites.

Type Host *Boa constrictor imperator* Daudin (Serpentes: Boidae).

Other Hosts None known.

Type Locality Vicinity of Colima, Estado Colima, Mexico.

Other Localities 19.8 km north of Culiacan, Estado Sinaloa, and Tepic, Estado Nayarit, Mexico.

Prevalence *H. fusifex* parasitized 55 of 57 (96.5%) *Boa constrictor* collected near Colima, 2 of 2 in Sinaloa, and the only boa collected at Tepic.

Morphological Variation In normally shaped erythrocytes, *H. fusifex* gamonts averaged 14.6 × 4.9 µm (9.2–17.3 × 2.3–8.1, N = 50). Gamonts in greatly hypertrophied erythrocytes were 18.7 × 8.7 µm (15–23 × 46–115, N = 50), and in fusiform erythrocytes were 16.4 × 6.3 µm (15–18.4 × 4.6–9.2, N = 50) (Ball et al., 1969). Nuclei usually occupied the second quarter of the gamont in all three forms.

Sporogony In *Culex tarsalis,* oocysts were nearly spherical (L/W 1.06) to slightly ovoid (L/W 1.21), 235 × 205 µm (170 × 165–282 × 234), and contained an average of 202 (146–250) sporocysts. Sporocysts were elliptical, averaging 40 × 26 µm (35 × 24–45 × 28), with L/W ratios 1.46–1.61. Sporocysts produced an average of 27 (15–35) sporozoites.

Merogony Meronts primarily occupied capillary endothelium of the lung, as well as the liver, heart, brain, and kidney, and in the last organ were seen in glomeruli as well. Macromeronts averaged 22 × 18 µm and produced up to 18 macromerozoites, 12.5 × 2.7 µm. Micromeronts were 29 × 19 µm and formed 55–60 micromerozoites 11 × 14 µm on average.

Effects on Host The major effect of *H. fusifex* on its hosts appears to be that of gross distortion and hypertrophy of host erythrocytes. Except for double infections, host erythrocytes appeared little affected by the presence of gamonts with "normal" *Hepatozoon* appearance. Uninfected erythrocytes averaged 18.4 × 11.1 µm (16.1–20.7 × 9.2–15). Parasitized erythrocytes that were hypertrophied but not fusiform were 38.9 × 20 µm (27.6–49.5 × 15–28.8). In extreme cases, these erythrocytes showed their outer margins thrown into several wave-like folds. The spindle-shaped host cells measured 51.6 × 10.7 µm (34.5–69 × 8.1–13.8), and in these cells the gamonts were intermediate in size, 16.4 × 6.3 µm, in comparison to 18.7 × 8.7 µm in the greatly hypertrophied erythrocytes and 14.6 × 4.9 µm in erythrocytes of normal appearance.

Remarks The presence of three very distinctive morphological forms of *Hepatozoon* in the boa constrictors from western Mexico raised the question of whether a single species, *H. fuxifex*, was present, or if these were the result of the presence of multiple species (Ball et al., 1969). Only one type of sporocyst could be obtained in either of the mosquito host species or in the tick *Amblyomma dissimile*, regardless of the morphological forms present in the blood of the donor snake. *Culex tarsalis*, in fact, could not be infected when the boa showed all three types of infected cells. When fed on boas showing only the small gamont forms present, elliptical sporocysts resulted, and these small erythrocytic forms were the only type observed in experimentally infected boas for as long as 26 months following infection. *Aedes togoi*, however, did become infected after feeding on a boa with all three types of infected cells present and showed only elliptical sporocysts of *H. fusifex*. Naturally infected *Amblyomma dissimile* were removed from a boa with the three types of infected cells present, and those ticks infected experimentally showed only elliptical sporocysts present. In later studies, Dr. Ball adopted the very conservative position of referring to *H. fusifex* as *Hepatozoon* sp., despite the presence of only elliptical sporocysts in the infected mosquitoes that were used for transmission.

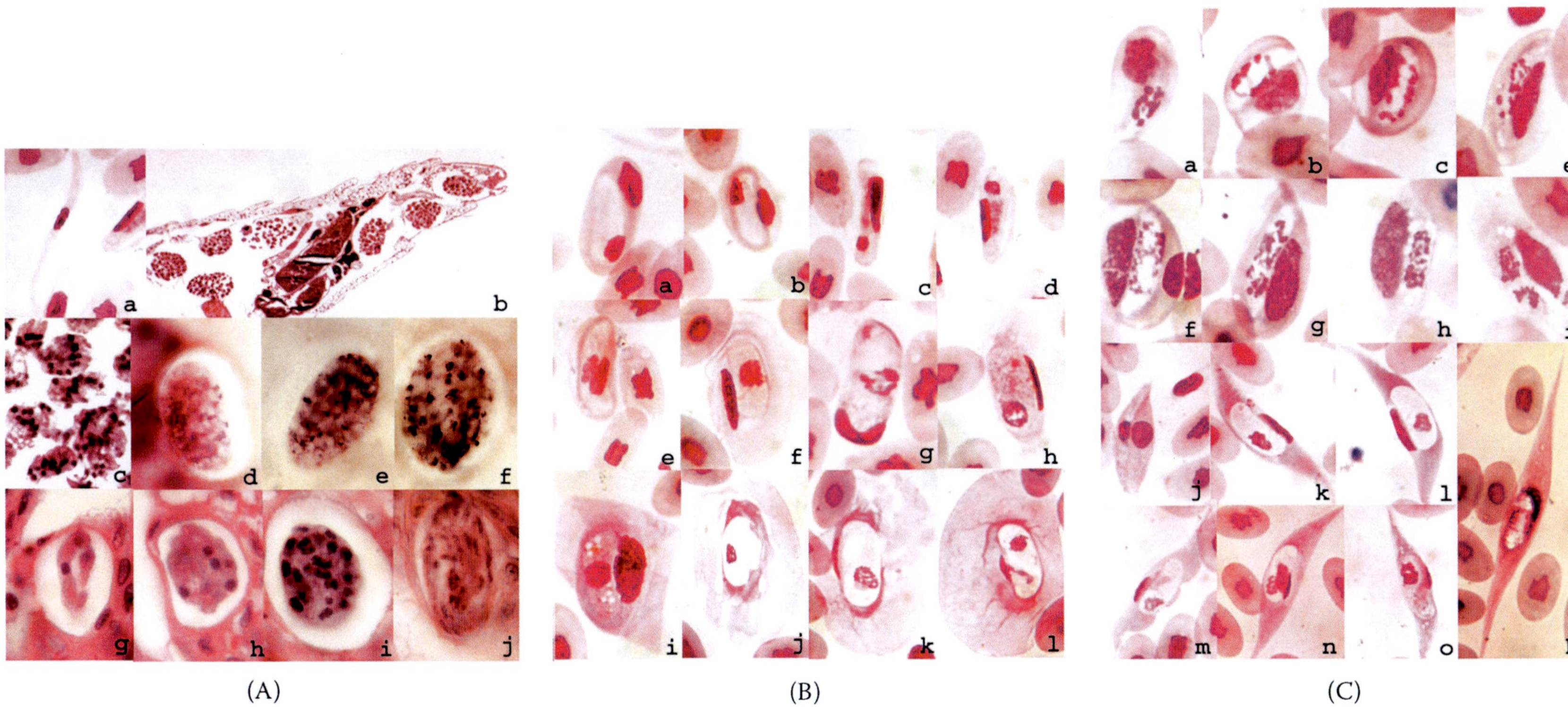

Plate 52 *Hepatozoon fusifex* from *Boa constrictor*, Mexico. (**A**) **a**, free gamont; **b**, mature oocysts; and **c**, sporocysts in hemocoel of *Aedes togoi*; **d–f**, mature sporocysts within natural vector, *Amblyomma dissimile*, Mexico; **g, h**, macromeronts, and **i, j**, micromeronts in lung of *B. constrictor*. (**B**) a–e, gamonts in slightly altered erythrocytes, and **f–l**, in greatly hypertrophied host cells. (**C**) **a–j**, young gamonts in immature host cells that mature into erythrocytes of fusiform shape, **k–p**.

With the perspective of time and experience with a large number of *Hepatozoon* species, I believe that a single species, *Hepatozoon fusifex*, was present in the original material obtained in Colima in 1963. This species may have produced varying effects on host cells perhaps because of the erythrocytic stage of maturation. Duration of infection of the host snake is apparently not an explanation because the greatly hypertophied host cells were present in a massively infected juvenile boa. A series of intermediate stages from normal erythrocytes infected with small gamonts to either of the grossly distorted, hypertrophied host cells was mentioned by Ball et al. (1969) as supporting the presence of *H. fusifex* only in the Colima boas. In one infection obtained in 1966, immature erythrocyte precursor cells as young as erythroblasts and basophilic proerythrocytes could be found infected with young gamonts, and a series of these immature host cells strongly suggested that the fusiform erythrocytes resulted when very young cells of the erythrocyte series became infected. The likelihood of this infection of immature host cells would be increased as parasitemia increased, placing stress on the erythropoietic system. It is appropriate to mention that experimental infections of *H. rarefaciens* in the same laboratory also resulted in host cells that were not hypertrophied. With the genomic methods available now, this problem has become amenable to solution.

As with *H. rarefaciens*, *H. fusifex* was transmitted to unnatural vertebrate hosts by feeding them infected *C. tarsalis*. Booden, Chao, and Ball (1970) transmitted *H. fusifex* (as *Hepatozoon* sp.) to the lizard *Anolis carolinensis*, and Oda, Chao, and Ball (1971) infected *Pituophis c. catenifer* with the same parasite. Serial transfer of *H. fusifex* to additional *P. catenifer* was accomplished during a period of 1.5 years. Another saurian species, *Sceloporus occidentalis*, was experimentally infected both directly from *Boa constrictor* and indirectly from *P. catenifer*. Prepatent periods were from 25 to 45 days in these experiments. Duration of parasitemia in these unnatural hosts were 1 month in *S. occidentalis*, 3–4 months in *A. carolinensis*, and at least 1.5 year in *P. catenifer*, and the parasite remained capable of infecting mosquitoes, in each case producing the elliptical sporocysts characteristic of *H. fusifex*.

The capacity of this parasite of boa constrictors from western Mexico to infect not only a natural vector, the tick *Amblyomma dissimile*, but also *Culex tarsalis* of California origin and *Aedes togoi* from Japan and, by ingestion of these vectors, a colubrid snake (*P. catenifer*) and phrynosomatid lizard (*S. occidentalis*) from California, and a polychrotid lizard (*A. carolinensis*) from the southeastern United States testifies to the versatile capacity of some *Hepatozoon* species to infect arthropods and vertebrates of greatly different phyletic relationships.

Hepatozoon sipedon Smith, Desser and Martin, 1994 (**Plate 53**)

Diagnosis A *Hepatozoon* species characterized by elongate, thin gamonts 19.0 × 3.7 µm, never recurved, spherical oocysts 262.6 µm containing up to 816 spherical sporocysts, 18.2 µm, that produce eight sporozoites. Sporogony occurs within fat bodies of the mosquito hemocoele. Ingestion of infected vectors by frogs produces dizoic cysts in their livers within 7 days, which are the infective stage for the final snake hosts. Merogony occurs in liver, lung, and kidneys of snakes fed infected frogs. Macromeronts average 49.0 × 40.3 µm by 15 days postinfection and produce hundreds of merozoites. These produce micromeronts 55.1 × 44.6 µm in liver, lung, kidneys, and heart of the snake host. Micromeronts also produce hundreds of merozoites, which enter erythrocytes between 30 and 40 days postinfection.

Type Host *Nerodia s. sipedon* (Linnaeus) (Serpentes: Colubridae).

Other Hosts None known.

Type Locality Queen's University Biological Station, approximately 63 km north of Kingston, Ontario, Canada.

Other Localities None reported.

Prevalence *Hepatozoon sipedon* infected 18 of 26 (69.2%) *N. sipedon* at the type locality.

Morphological Variation Smith et al. (1994b) provided only mean values for dimensions of gamonts, meronts, and sporogonic characters. Gamonts were 19.0 ± 0.9 × 3.7 ± 0.5 µm (N = 20), indicating an LW value of 70.3. A topotypic infection sent to the author by G. F. Bennett in 1990 had the following characteristics: Gamonts averaged 20.0 ± 0.7 µm × 4.2 ± 0.3 µm (19–22 × 3.5–5.0, N = 24), with LW 83.5 ± 7.6 µm^2 (70–105) and L/W 4.83 ± 0.38 (4.2–5.7).

Sporogony Oocysts averaged 262.6 µm (N = 20) and contained up to 816 (mean 591, N = 5) sporocysts 18.2 µm (N = 40). Eight sporozoites only were found in sporocysts and averaged 20.9 × 3.4 µm (N = 40). Sporogony occurred in fat bodies of the hemocoel of two species of experimental vectors, *Culex pipiens* and *C. territans*, both of which were thought to be natural vectors (Smith et al., 1994b). Mature oocysts were present in both *Culex* species at 28 days PF but were not found in eight *Aedes aegypti* examined after 28 days.

Merogony Macromeronts averaged 49.0 ± 4.0 × 40.3 ± 5.5 µm (N = 20) by 15 days postinfection, and micromeronts were 55.1 ± 3.1 × 44.6 ± 4.0 µm (N = 20) by 30 days

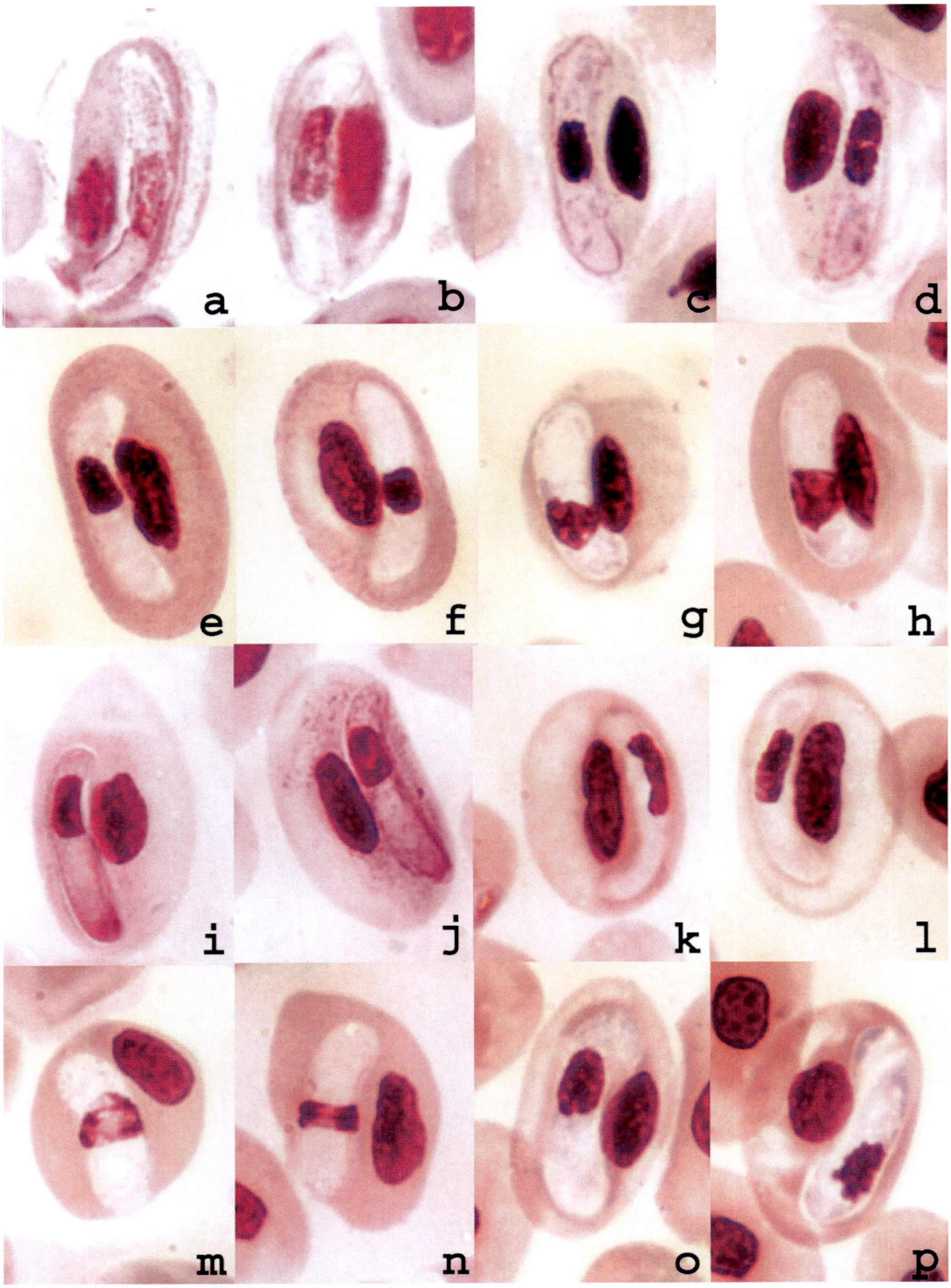

Plate 53 Hepatozoon species from North American snakes: **a, b**, *H. sipedon* of *Nerodia sipedon*, Canada. All from Florida: **c, d**, *H. sirtalis* of *Thamnophis sirtalis*; **e, f**, *H. moccasini* of *Agkistrodon piscivorus*; **g, h**, *H. pictiventris* of *Nerodia fasciata pictiventris*; **i, j**, *H. fasciatae* of *N. fasciata pictiventris*; **k, l**, *H. seminatrici* of *Seminatrix pygaea*; **m, n**, *H. sistruri* of *Sistrurus miliaris barbouri*; **o, p**, *H. horridus* of *Crotalus horridus*.

postinfection. Both types of meronts produced hundreds of merozoites, 20 × 2.9 μm in the case of macromeronts and 15.2 × 2.4 μm for micromeronts. Dizoic cysts in frogs were 22.0 ± 1.1 × 18.4 ± 1.4 μm (N = 20).

Effects on Hosts *Nerodia sipedon* infected naturally showed maximum parasitemias not exceeding 0.2%. Five juvenile *N. sipedon* experimentally infected by ingestion of tissues from frogs that had ingested infected mosquitoes died within 60 days following the appearance of erythrocytic parasites. Their parasitemias at death exceeded 90%. An experimentally infected *Thamnophis sirtalis*, however, developed a parasitemia that did not exceed 5%. In the heavily infected snakes, one to four gamonts were present in erythrocytes, and rarely, neutrophils and lymphocytes became infected. Mortality was apparently restricted to juvenile *N. sipedon* inasmuch as adult snakes longer than 25 cm and similarly experimentally infected had parasitemias of less than 90% and did not show adverse effects by *H. sipedon* infection (Smith et al., 1994b).

Remarks Although direct transmission of infection to snakes by ingestion of infected vectors has been reported for other *Hepatozoon* species, *H. sipedon* could not be transmitted directly. Infection could only be accomplished by feeding tissues containing dizoic cysts from frogs that had eaten infected mosquitoes. Neither tissue nor erythrocytic stages appeared in snakes directly fed infected mosquitoes. The ingestion of liver from an *N. sipedon* that contained meronts as well as erythrocytic parasites did not result in infection of the recipient snake. No sporogonic development was observed in leeches, *Placobdella ornata*, syntopic with *N. sipedon*, when fed on infected snakes (Smith et al., 1994b).

Hepatozoon sirtalis Telford, Wozniak, and Butler, 2001 (**Plate 53**)

Diagnosis A *Hepatozoon* species characterized by elongate, thin gamonts 17–22 × 3–6 μm not recurved, with LW 51–120 μm² and L/W 3.3–6.7. Nucleus position is usually at midbody or entirely within second quarter of gamont. Nearly spherical oocysts 248–416 × 248–396 μm contain several hundred nearly spherical sporocysts 12–18 × 12–18 μm, with LW 156–315 μm² and L/W 1.0–1.4, that contain four to eight sporozoites. Erythrocytic cytoplasm is thin and colorless around the margin and contracted around the gamont.

Type Host *Thamnophis s. sirtalis* Linnaeus (Serpentes: Colubridae).

Other Hosts *Thamnophis sauritus sackenii, Coluber constrictor priapus, Coluber constrictor helvigularis,* and *Nerodia fasciata pictiventris.*

Type Locality Gainesville, Alachua County, Florida.

Other Localities Liberty County, Florida.

Prevalence *H. sirtalis* was found in 7 of 25 (28%) *T. sirtalis* in north Florida, 1 of 20 (5%) *T. sauritus sackenii* statewide, 1 of 26 (4%) *Coluber constrictor priapus* in Alachua County, and 1 of 3 *C. constrictor helvigularis* in west Florida, 1 of 4 *Nerodia f. pictiventris* in Marion County, and 1 of 9 *Nerodia f. pictiventris* in Alachua County.

Morphological Variation *H. sirtalis* gamonts in the type host were 20.0 ± 0.8 × 4.1 ± 0.5 μm (17–22 × 3–6, N = 75), with LW 82.1 ± 11.4 μm² (51–120) and L/W 4.9 ± 0.6 (3.3–6.7). In accidental host *Nerodia fasciata pictiventris*, gamonts were 19.7 ± 0.6 × 3.9 ± 0.4 μm (18–21 × 3–5, N = 25), with LW 77.5 ± 6.9 μm² (60–90) and L/W 5.0 ± 0.6 (4.2–6.7).

Sporogony Sporogony is primarily within the abdominal hemocoel of the experimental vector *Aedes aegypti*, rarely in the thorax or head. Fresh oocysts were 304 ± 97.0 × 297.3 ± 85.5 μm (218–416 × 248–396, N = 3); stained oocysts were 196.0 ± 52.0 × 189.0 ± 50.9 μm (139–267 × 139–257, N = 5). One flattened oocyst contained 341 sporocysts. Fresh sporocysts were 14.2 ± 1.5 × 13.7 ± 1.3 μm (12–18 × 12–18, N = 25), with LW 197.1 ± 40.4 μm² (156–315) and L/W 1.04 ± 0.04 (1.0–1.4); stained sporocysts were 14.7 ± 2.0 × 13.2 ± 1.9 μm (12–21 × 10–18, N = 36), with LW 196.1 ± 55.0 μm² (130–357) and L/W 1.12 ± 0.11 (1.0–1.5). Sporocysts contained 5.7 ± 1.5 (4–8, N = 36) sporozoites.

Merogony Unknown.

Effects on Host Infected erythrocytes were hypertrophied in length with greater LW but width was similar to uninfected cells. Nuclei of infected cells were greater in length and less in width than in uninfected erythrocytes. Infected cells were always distorted, with thin and colorless margins extending to the cytoplasm contracted around the gamont in active infections.

Remarks *H. sirtalis* was transmitted inadvertently to a *Nerodia fasciata pictiventris*, negative for gamonts during 28 months of captivity, by ingestion of a frog, *Rana sphenocephala,* also negative for *Hepatozoon* gamonts, approximately 2 months before erythrocytic infection in the snake was detected. Gamonts were identical to those of the type infection of *H. sirtalis* in *Thamnophis sirtalis* taken at the same site 3 years earlier. The infection, followed for 9 months, never achieved a parasitemia of 1%, and tissues taken at necropsy were negative for merogonic stages.

Hepatozoon sirtalis and *H. sipedon* are unique among *Hepatozoon* species from snakes for which sporogony is known. Both form large, spherical oocysts that contain hundreds of small, spherical sporocysts with very low numbers of sporozoites, eight only in *H. sipedon* and four to eight in *H. sirtalis*. Both the small size of sporocysts and low numbers of sporozoites differ greatly from those described for *Hepatozoon* species that can infect snakes directly by ingestion of the infected vector. *Hepatozoon sipedon* certainly and *H. sirtalis* apparently are transmitted to the final vertebrate host by ingestion of an infected intermediate vertebrate host, itself infected by eating the vector. This difference in mode of transmission, indirect (*H. sipedon, H. sirtalis*) versus direct (all other *Hepatozoon* species of snakes with known sporogony), is probably of taxonomic significance. Gamont morphology also is very similar for *H. sipedon* and *H. sirtalis*, both producing elongate, slender gamonts that cause a condensation of erythrocyte cytoplasm around the gamont and a thinness along the cell margin (Smith et al., 1994b; Telford et al., 2001).

Hepatozoon mocassini (Laveran) 1902, Nadler and Miller 1984 (**Plate 53**)

Diagnosis A *Hepatozoon* species with slender, nonrecurved gamonts, 14–18 × 4–6 μm, oocysts 98–116 μm that contain 32–320 spherical-to-elongate, usually ovoid sporocysts 32–62 × 18–32 μm. Sporocysts produce 7–42 sporozoites. Merogony occurs in hepatic parenchymal cells. Macromeronts average 15.3 × 12.1 μm and contain up to 4 macromerozoites and micromeronts 12.1 × 10.4 μm that produce approximately 30 micromerozoites. Sporogony occurs in the hemocoel of abdomen and thorax and in the head of experimental mosquito hosts.

Type Host *Agkistrodon piscivorus* Lacépède (Serpentes: Viperidae).

Other Hosts None known.

Type Locality Gainesville, Alachua County, Florida.

Other Localities In Louisiana: Manchac Swamp, Tangipahoa Parish (Nadler and Miller, 1984), Hebert Center in Plaquemines Parish and Sarny Swamp in St. Charles Parish (Lowichik and Yaeger, 1987). In Florida: Payne's Prairie, Gainesville, Alachua County, and Osceola National Forest, Columbia County (Telford and Moler, unpublished).

Prevalence At Hebert Center, Plaquemines Parish, Louisiana, 95 of 103 (92.2%) *A. piscivorus leucostoma* were infected by *H. mocassini* (Lowichik and Yaeger, 1987). In Florida, Langmann (1899) found hemogregarines, later described as *Haemogregarina mocassini* by Laveran (1902), in all 26 *A. piscivorus* examined from Gainesville, Alachua County. Ten of 13 (76.9%) *A. piscivorus* examined from 2003 to 2005 from the northern edge of Payne's Prairie, Gainesville, were infected by *H. mocassini* (Telford and Moler, unpublished).

Morphological Variation In Louisiana, Nadler and Miller (1984) reported gamont dimensions as 15.8 × 4.0 μm (14–17 × 4, N = 75), with calculated LW 63.2 μm^2 (56–68) and L/W 3.95 (3.50–4.25). Gamont nuclei were described as "eccentric," 4.5 × 3.0 μm (N = 10), with LW estimated as 13.5. Nucleus position in the illustrated gamonts is midbody in most, second quarter in the remainder. A sample from Alachua County, Florida, showed gamonts that are 16.8 ± 0.6 × 4.8 ± 0.4 μm (16–18 × 4–5.5, N = 25), LW 80.6 ± 7.7 μm^2 (66–99) and L/W ratio 3.53 ± 0.36 (2.91–4.25). Nuclei averaged 4.8 ± 0.4 × 3.2 ± 0.4 μm (4.0–5.5 × 3.0–4.0, N = 25), with LW value 15.6 ± 2.2 μm^2 (12.0–22.5). Gamonts from Columbia County are 15.6 ± 0.6 × 5.4 ± 0.4 μm (14.5–16.5 × 4.5–6.0, N = 25), LW 83.6 ± 7.4 μm^2 (69.7–96.0) and L/W 2.92 ± 0.22 (2.58–3.44), with average nucleus size 4.6 ± 0.5 × 4.6 ± 0.5 μm (4.0–5.5 × 4.0–5.5, N = 25) and LW 20.9 ± 2.9 μm^2 (16.0–27.5). Nuclei were usually situated at midbody of the gamont (48%) or in the second quarter (32%) in those from Columbia County but were more commonly in the second quarter (64%) or extended into the first (32%) in the Alachua County sample. No gamonts were recurved, and this was not reported in the infections from Louisiana, but the figures of Laveran (1902) show gamonts of *H. mocassini* with one recurved end.

Sporogony Nadler and Miller (1984) described oocysts as 144 μm (98–196) in diameter and containing 61–320 sporocysts. Sporocysts were spherical to ovoid, 31.0 × 28.5 μm (37.5 × 45.0–22.5 × 22.5, N = 70). They produced 7–18 sporozoites, 18.8 × 4.0 μm (N = 10). The oocysts developed in the hemocoel of the abdomen usually, sometimes in the thorax. Lowichik et al. (1993) described gamonts as penetrating the gut wall and entering abdominal fat bodies, where szygy occurred by 24 hours PF. At 2 days PF, microgamonts formed four microgametes. Macrogamonts had also differentiated at 2 days, and zygotes were observed at 3 days. By day 12 PF, sporocysts contained developing sporozoites, and mature oocysts were present on day 17 PF. Oocysts were 93.3 ± 16.6 μm (N = 5) in diameter. Only intact oocysts were found in the abdomen of infected *Aedes aegypti*, the experimental host.

Aedes aegypti fed on *H. mocassini* from Columbia County, Florida (Telford and Moler, unpublished), survived poorly, most dying in the second week PF, with only 2 of 300 living and infected at 18 days. Some free sporocysts were found in the head and thorax, but many were present, along with oocysts, in the abdominal hemocoel. Oocysts were nearly spherical to ovoid, 216.3 ± 54.3 × 196.6 ± 57.1 μm (115–180 × 80–262.5, N = 8), and contained 32–257 sporocysts. Sporocysts were spherical to almost elongate, 40.1 ± 9.8 × 23.9 ±

3.4 µm (22.5–62.5 × 18–38, N = 90), with LW 964.7 ± 290.4 µm² (450–1686) and L/W ratio 1.70 ± 0.42 (1.00–2.63). Sporocysts contained 30.3 ± 5.4 (20–42, N = 45) sporozoites.

Merogony Nadler and Miller (1984) found meronts of *H. mocassini* in liver parenchymal cells of *A. piscivorus*. Macromeronts averaged 15.3 × 12.1 µm (N = 10) and contained up to four macromerozoites, 14.0 × 3.5 µm (N = 15). Micromeronts were 12.1 × 10.4 µm (N = 10) and produced approximately 30 micromerozoites 8.0 × 1.5 µm (N = 10).

Effects on Host Infected erythrocytes in *A. piscivorus* from Louisiana were significantly hypertrophied in length, width, and volume. Distortion of host cells and displacement of their nuclei appear to be minimal in the figures of Nadler and Miller (1984). In infections from Florida snakes, the erythrocytes of the Columbia County sample were hypertrophied by 11%, their nuclei were normal in size, but all erythrocytes were distorted to some degree. Infected erythrocytes in *A. piscivorus* from Alachua County were considerably more hypertrophied, 38%, and their nuclei were about 9% larger in size than uninfected cell nuclei. All parasitized erythrocytes were distorted. There was no indication of dehemoglobinization of infected erythrocytes in either infection from Florida snakes.

Remarks Nadler and Miller (1984) fed mosquitoes infected with *H. mocassini* to an uninfected *A. piscivorus leucostoma*, and gamonts appeared in erythrocytes 43 days later. Infection of a different host species, two of the colubrid snake *Nerodia r. rhombifera*, captive born and free of congenital infection, were each fed six infected mosquitoes (Lowichik et al., 1993). After 5 weeks, one snake showed immature gamonts in erythrocytes, and at 8 weeks both snakes had patent infections, with parasitemias of 0.2% and 0.3%.

Although the gamonts of *H. mocassini* are similar in appearance, the Florida samples have greater average widths (4.8 and 5.4 µm) than those from Louisiana snakes (4.0 µm), with this reflected in the lesser LW value and greater L/W ratio estimated for the latter sample. Dimensions of sporocysts, however, indicate that sporocyst length of *H. mocassini* from Florida is greater and width somewhat less, 40.1 × 23.9 µm versus 31.0 × 28.5 µm, resulting in more elongate sporocysts with a higher L/W ratio, 1.70, in comparison to 1.09 in sporocysts from Louisiana infections. The sporocyst LW is similar, 965 µm² versus 884 µm², respectively. The major difference is in sporozoite numbers produced by sporocysts, with the maximum of 18 in Louisiana sporocysts less than the minimum of 20 in those from Florida, and in significantly higher average numbers, 30.3 in the latter sample than in the Louisiana sporocysts, 18.8. Only genomic comparison can determine whether there is geographic variation in sporozoite numbers or the presence of two otherwise similar *Hepatozoon* species.

Hepatozoon pictiventris Telford, Wozniak and Butler 2001 (**Plate 53**)

Diagnosis A *Hepatozoon* species characterized by elongate, nonrecurved gamonts, 11–16 × 4–6 µm, with attenuated ends and nuclei 3–6 × 3–5 µm, usually present in the second quarter of the gamont. In the experimental mosquito hosts, sporogony takes place primarily in the thoracic hemocoel, the head, and proboscis, seldom in the abdomen. Oocysts are nearly spherical, 55–184 × 46–155 µm, and contain 2–28 sporocysts 18–60 × 18–48 µm, also nearly spherical, that produce 14–78 sporozoites.

Type Host *Nerodia fasciata pictiventris* (Cope) (Serpentes: Colubridae).

Type Locality Palm Beach Gardens Estates, Palm Beach County, Florida.

Other Hosts None known.

Other Localities Palm City, Martin County, Florida.

Prevalence Nine of 11 (82%) *N. fasciata pictiventris* from the vicinity of the type locality, and 10 of 16 (63%) overall from southeast Florida were infected by *H. pictiventris*.

Morphological Variation Gamonts are elongate and not recurved, 13.7 ± 1.0 × 4.7 ± 0.4 µm (11–16 × 4–6 N = 75), LW 64.8 ± 7.3 µm² (46–83) and L/W 2.9 ± 0.4 (2.0–3.8). Free gamonts are 19.6 ± 1.3 × 2.4 ± 0.5 µm (18–22 × 2–30, N = 10), with strongly attenuated ends. Nuclei are 4.3 ± 0.8 × 4.0 ± 0.4 µm (3–6 × 3–5, N = 75), with LW 17.5 ± 3.6 µm² (12–30), usually (75%) situated in the second quarter of the gamont, commonly (25%) extending into the third quarter, with a few (11%) of those in the third quarter at the middle of the gamont.

Sporogony Sporogony occurred in the experimental vector *Aedes aeqypti* predominantly within the hemocoele of the thorax, in the head, and in the proboscis, rarely in the abdomen. Stained oocysts from the abdomen are nearly spherical, 90.7 ± 26.3 × 81.0 ± 26.4 µm (65–146 × 55–143, N = 12), and contain 4–28 (9.8 ± 7.5) spherical sporocysts, 31.3 ± 5.6 × 27.6 ± 4.9 µm (23–40 × 20–38, N = 19), LW 887.6 ± 298.7 µm² (500–1520) and L/W 1.14 ± 0.13 (1.0–2.9), with 32.6 ± 10.7 (15–45, N = 12) sporozoites contained within. Oocysts from the thorax are nearly spherical, 109.3 ± 30.2 × 94.0 ± 23.5 µm (58–184 × 50–155, N = 53), and contain 3–27 (12.1 ± 6.1, N = 53) nearly round sporocysts, 35.5 ±

7.3 × 31.2 ± 5.7 μm (24–60 × 19–48, N = 105), LW 1138.3 ± 444.6 μm² (475–2880) and L/W 1.14 ± 0.14 (1.0–1.8), which contain 33.4 ± 13.3 (14–78, N = 91) sporozoites. Stained oocysts from the head are nearly spherical, 109.4 ± 32.2 × 95.2 ± 29.3 μm (55–175 × 46–151, N = 35), and contain 7.4 ± 3.6 (2–15, N = 16) nearly spherical sporocysts, 33.9 ± 7.2 × 29.4 ± 5.5 μm (18–52 × 18–42, N = 89), LW 1026.0 ± 365.2 μm² (324–1976) and L/W 1.16 ± 0.16 (1.0–1.6), which contain 26.6 ± 9.9 (14–55, N = 38) sporozoites. Fresh oocysts from the head are also nearly spherical, 128.7 ± 24.2 × 112.0 ± 22.8 μm (70–175 × 52–151, N = 21), and contain slightly more ovoid sporocysts 36.5 ± 5.5 × 31.0 ± 4.8 μm (26–48 × 24–42, N = 51), LW 1145.0 ± 299.1 μm² (653–1764) and L/W 1.19 ± 0.18 (1.0–1.6). An oocyst present in the proboscis contained four sporocysts. Stained sporocysts from the proboscis are 37.9 ± 6.2 × 35.7 ± 5.7 μm (20–47, N = 23), LW 1381.8 ± 395.7 μm² (400–2021) and L/W 1.06 ± 0.9 (1.0–1.3), and contained 29.2 ± 7.9 (15–49, N = 19) sporozoites. Free sporozoites are 23.2 ± 3.3 × 4.2 ± 0.9 μm (N = 10).

Merogony A single macromeront 17 × 11 μm, within a cyst 25 × 18 μm, that contained six macromerozoites was found in sections of liver and lung from three infections. No micromeronts were seen.

Effects on Host The cytoplasm of infected erythrocytes is often thin and colorless (46%) in comparison to normal cells but is sometimes normal in appearance. Infected cells are commonly distorted (39%). Erythrocyte length and LW are greater in infected cells, but their width and nucleus size are similar to uninfected erythrocytes.

Remarks The following remarks were made:

> Five of 31 mosquitoes from which the proboscis was dissected separately from the head contained 1–21 sporocysts in the proboscis. In each case oocysts or sporocysts were present in the head dissection. No proboscides were infected in 15 mosquitoes dissected < 22 days postfeeding, 5 of 16 were positive at 22–24 days. ... The presence of many mature sporocysts and an oocyst within the proboscides of infected mosquitoes is the first evidence placing the infective stages of an *Hepatozoon* species into proximity for possible transmission by bite of a non-acarine vector directly to a new host. (Telford et al., 2001)

Hepatozoon fasciatae Telford, Wozniak and Butler 2001 (Plate 53)

Diagnosis A *Hepatozoon* species characterized by elongate gamonts, not recurved, 16–18 × 3–5 μm, with nuclei 3–7 × 2–4 μm, situated in the second quarter of the gamont, often extending into the first quarter. Sporogony occurs primarily within the abdomen of experimental mosquito hosts. Oocysts are 72–277 × 60–257 μm and contain 30–158 usually ovoid sporocysts 15–50 × 14–30 μm that produce 12–38 sporozoites.

Type Host *Nerodia fasciata pictiventris* (Cope) (Serpentes: Colubridae).

Type Locality Florida, Alachua County, Gainesville, vicinity of Payne's Prairie.

Other Hosts *Nerodia f. fasciata, Nerodia floridana.*

Other Localities Palm City, Martin County, and Alachua, Alachua County, Florida; South Carolina, no precise locality.

Prevalence Six of nine *N. fasciata pictiventris* from Alachua County and four of four *N. fasciata pictiventris* and one of two *N. floridana* from Marion County were infected by *H. fasciatae.*

Morphological Variation Gamonts are elongate and not recurved, 16.5 ± 0.6 × 3.7 ± 0.4 μm (16–18 × 3–5 N = 25), LW 61.7 ± 6.3 μm² (48–71) and L/W 4.5 ± 0.5 (3.6–5.7). Gamont nuclei are 5.1 ± 0.8 × 2.9 ± 0.3 μm (3–7 × 2–4, N = 25), with LW 14.4 ± 2.7 μm² (10–21), always within the second quarter of the gamont, and commonly (72%) extending into the first quarter, seldom (8%) into the third quarter of the gamont. Free gamonts are 18.8 ± 2.0 × 3.5 ± 0.6 μm (16–22 × 2.5–4.5, N = 10). Gamont dimensions from an infection in *N. fasciata fasciata* from South Carolina are nearly identical with those from the type locality in Alachua County, Florida: 16.8 ± 0.8 × 3.8 ± 0.4 μm (15–18 × 3–4.5, N = 25), LW 63.2 ± 6.5 μm² (48–76.5) and L/W 4.5 ± 0.6 (3.75–5.83), with nuclei 5.8 ± 0.9 × 2.3 ± 0.4 μm (4.5–7.5 × 2.0–3.0), LW 13.1 ± 2.7 μm² (9.0–18.0).

Sporogony Sporogony in the experimental host *Aedes aeqypti* is primarily within the hemocoel of the abdomen, less commonly in the thorax, and rarely in the head. The nearly spherical stained oocysts are 187.8 ± 37.3 × 177.0 ± 34.1 μm (108–277 × 99–257, N = 25), with the L/W ratio 1.06 ± 0.07 (1.0–1.3), and contain 84.9 ± 40.6 (30–158, N = 11) sporocysts. Stained sporocysts from the abdomen and thorax are ovoid, 27.5 ± 6.3 × 22.1 ± 4.0 μm (15–45 × 14–30, N = 30), LW 622.7 ± 221.3 μm² (210–1215) and L/W 1.25 ± 0.23 (1.0–1.9), with 21.1 ± 6.3 (12–38, N = 25) sporozoites 19.5 ± 1.7 × 2.9 ± 0.3 μm (N = 10) contained within. Fresh oocysts from the abdomen are nearly spherical, 172.9 ± 43.5 × 167.3 ± 43.8 μm (72–250 × 60–233, N = 28), and contain rounded or usually more elongate sporocysts, 30.4 ±

8.5 × 22.9 ± 2.7 µm (22–50 × 17–28, N = 30), LW 703.0 ± 234.4 µm^2 (394–1187) and L/W 1.33 ± 0.35 (1.0–2.1).

Merogony A single oval macromeront, 20 × 13 µm, from the lung was enclosed in a cyst 20 × 19 µm and contained eight macromerozoites. Micromeronts, small, oval or pear-shaped, contained within cysts 17–32 × 14–26 µm^2, are 20.3 ± 3.0 × 14.6 ± 2.3 µm (15–25 × 10–20, N = 17) and contain over 30 micromerozoites 5.6 ± 0.5 × 1.3 ± 0.5 µm (5–6 × 1.0–1.5, N = 7).

Effects on Host The length and LW of infected erythrocytes are greater than uninfected cells, but width and nucleus dimensions do not differ. The erythrocyte cytoplasm is often thin and colorless (36%) in comparison to normal cells. Infected cells are rarely distorted (4%), and there is no difference in dimensions of host cell nuclei.

Remarks Another species of *Hepatozoon* described from *Nerodia fasciata*, *Haemogregarina bradfordi* Sambon, 1909, apparently has longer gamonts (18–20 µm, Sambon, 1909c) and a smaller gamont nucleus, 4 × 3 µm. It differs by the contraction of host cell cytoplasm around the gamont and the erythrocyte nucleus, similar to that seen in erythrocytes parasitized by *Hepatozoon sirtalis*, which does not occur in cells parasitized by *H. fasciatae*. The gamont dimensions of *H. bradfordi* are consistent with those of *H. sirtalis*, and Telford et al. (2001) suggested that Sambon may have described a cross infection by *H. sirtalis* into *N. fasciata*. Gamonts apparently of *H. sirtalis*, showing the cytoplasm contraction, have been seen in an *N. fasciata pictiventris* in Marion County, Florida (Telford, unpublished).

Hepatozoon seminatrici Telford, Wozniak and Butler 2001 (Plate 53)

Diagnosis A *Hepatozoon* species characterized by elongate, nonrecurved gamonts 14–20 × 3–5 µm with nuclei 4–7 × 2–5 µm usually occupying the second quarter of the gamont and commonly extending into the first quarter. Sporogony takes place primarily within the hemocoel of the abdomen in experimental mosquito hosts. Oocysts are nearly spherical, 119–252 × 243 µm, and contain 88–215 ovoid-to-elongate sporocysts 22–56 × 10–34 µm that produce 8–54 sporozoites. Merogony occurs primarily in the liver and lung.

Type Host *Seminatrix p. pygaea* (Cope) (Serpentes: Colubridae).

Type Locality Vicinity of Payne's Prairie, Gainesville, Alachua County, Florida.

Other Hosts None known.

Other Localities None known.

Prevalence One of four *S. pygaea* was infected by *H. seminatrici*.

Morphological Variation Gamonts are elongate, not recurved, 16.2 ± 1.1 × 4.0 ± 0.5 µm (14–20 × 3–5 N = 50), LW 63.9 ± 8.7 µm^2 (45–85) and L/W 4.1 ± 0.5 (3.2–5.2). Free gamonts are 14–18 × 2–4 µm (15.5 ± 1.6 × 2.5 ± 0.9, N = 5). Gamont nuclei are 5.3 ± 0.6 × 2.8 ± 0.6 µm (4–7 × 2–5, N = 50), with LW 14.8 ± 3.6 µm^2 (10–30), always situated in the second quarter of the gamont, commonly (54%) extending into the first quarter, seldom (10%) into the third quarter of the gamont.

Sporogony Sporogony in the experimental host *Aedes aegypti* is primarily within the hemocoel of the abdomen, less commonly in the thorax, and rarely in the head. Stained oocysts are nearly spherical, 163.4 ± 29.5 × 155.7 ± 29.2 µm (119–228 × 119–218, N = 18), and contain ovoid-to-elongate sporocysts 41.9 ± 8.4 × 29.2 ± 2.4 µm (30–56 × 23–34, N = 25), LW 1220.5 ± 266.1 µm^2 (864–1848) and L/W 1.45 ± 0.33 (1.0–2.3), with 27.7 ± 10.1 (8–54, N = 55) sporozoites, 20.3 ± 2.1 × 3.4 ± 0.3 µm (N = 10), contained within. Fresh oocysts, nearly spherical, are 209.4 ± 22.8 × 202.8 ± 22.3 µm (177–253 × 173–243, N = 17) and contain 178.4 ± 39.6 (88–215, N = 8) sporocysts. Fresh sporocysts are 32.1 ± 6.1 × 20.5 ± 4.3 µm (22–49 × 10–28, N = 21), LW 668.5 ± 219.5 µm^2 (300–1097) and L/W 1.62 ± 0.43 (1.1–3.0).

Merogony Macromeronts only were observed in the liver and lung. They were contained within cysts 22 ± 3.0 × 16.4 ± 3.3 µm (16–32 × 9–21, N = 36), with LW 376.3 ± 106.7 µm^2 (184–608) and L/W 1.42 ± 0.27 (1.0–2.3), and were similar in dimensions in both tissues. Meronts are 18.8 ± 2.7 × 11.7 ± 2.6 µm (13–24 × 8–18, N = 36), with LW 224.2 ± 75.0 µm^2 (120–432) and L/W 1.65 ± 0.28 (1.2–2.3). Meronts produce 3.8 ± 1.0 (3–8, N = 145) macromerozoites, with no difference between liver (4.0 ± 1.1, N = 77) and lung (3.6 ± 1.0, 68). Merozoite dimensions are 18.6 ± 1.5 × 3.9 ± 0.2 µm (N = 10). One macromeront was found in cardiac muscle, 29 × 28 µm, with 11 nuclei, contained within a cyst 39 × 38 µm. Cysts containing an apparently fully formed cystozoite were more common (11%) in liver than in lung (5%); the respective frequencies of other zoite numbers were similar in 100 cysts each in liver and lung: 2, 36% versus 39%; 3, 27% versus 37%; 4, 15% in each tissue; 5, 7% versus 2%; 6, 4% versus 2%.

Effects on Host The cytoplasm of infected erythrocytes is often thin and colorless (40%) in comparison to normal cells, and infected cells are usually distorted (80%). Erythrocyte length, width, LW, and nucleus length are greater than in uninfected erythrocytes; nucleus length and LW are less in infected cells than in uninfected, and nucleus width is no different.

Remarks The host of *H. seminatrici* is a small, completely aquatic snake that probably only rarely is exposed to possible dipteran vectors and most improbably to acarinid species. If infection is by direct inoculation from a hematophagous vector, then leeches are the most likely candidates. Because the diet of *S. pygaea* includes small anurans, infection could be acquired by ingestion of an infected intermediate host, as with *Hepatozoon sipedon* and *H. sirtalis*.

Hepatozoon sistruri Telford, Butler and Telford 2002 (**Plate 53**)

Diagnosis A *Hepatozoon* species characterized by broadly elongate, nonrecurved gamonts 12.6–15.8 × 4.7–6.3 µm and nuclei 2.6–5.3 × 2.5–5.3 µm, situated primarily in the second quarter and at midbody of the gamont. Sporogony occurs within the head and hemocoel of the thorax and abdomen in experimental mosquito hosts. Oocysts are spherical to usually ovoid, 92–245 × 82–240 µm, and contain 12–42 sporocysts that are spherical to usually ovoid, 25–50 × 20–50 µm, that produce 19–70 sporozoites.

Type Host *Sistrurus miliarius barbouri* Gloyd. (Serpentes: Viperidae).

Type Locality Palm Beach Gardens, Palm Beach County, Florida.

Other Hosts None known.

Other Localities None known.

Prevalence One of three *S. miliarius barbouri* examined from Palm Beach County was infected by *H. sistruri*.

Morphological Variation Gamonts are broadly elongate and not recurved, 14.1 ± 0.8 × 5.6 ± 0.5 µm (12.6–15.8 × 4.7–6.3, N = 25), LW 79.4 ± 9.6 µm^2 (62–100), and L/W 2.5 ± 0.2 (2.2–2.8). Free gamonts are 19.5–24 × 3–3.5 µm. Gamont nuclei are narrow, 3.7 ± 0.6 × 4.2 ± 0.8 µm (2.6–5.3 × 2.6–5.3, N = 25), with LW 15.6 ± 3.0 µm^2 (9.7–22.2), always situated in the second quarter of the gamont, rarely extending into the first quarter (4%), but commonly (52%) at midbody of the gamont.

Sporogony Sporogony in the experimental host *Aedes aegypti* occurs within the head and hemocoel of the thorax and abdomen. Oocysts are spherical to usually ovoid, 163.6 ± 45.1 × 154.7 ± 46.9 µm (92–245 × 82–240, N = 9), and contain 12–42 (27.4 ± 13.9, N = 5) spherical to usually ovoid sporocysts, 39.7 ± 6.2 × 33.5 ± 5.8 µm (25–50 × 20–50, N = 62), LW 1348.9 ± 374.9 µm^2 (500–2500) and L/W 1.20 ± 0.19 (1.0–1.8), that produce 45.7 ± 17.1 (19–70, N = 12) sporozoites.

> The type infection of *H. sistruri* had a relatively high parasitemia, 5%, with double infections of erythrocytes common. Mortality of *Aedes aegypti* fed upon the infection was extremely high, especially within the first 24 hr, when over 50% of 129 died. ... At dissection, most of the oocysts in the abdomen were associated with or within fat bodies and contained only large sporoblasts. Many oocysts with formed sporocysts had not completed sporozoite formation. A second feeding had similar results on initial mortality, with all 47 alive at removal from the feeding cage dead within 1 week. (Telford et al., 2002)

Merogony Undescribed.

Effects on Host The cytoplasm of infected erythrocytes is normal in appearance, but infected cells are always distorted. Infected erythrocytes are similar in length to uninfected cells but have greater width and LW, with their nuclei larger than those of uninfected cells.

Remarks Pigmy rattlesnakes collected in Volusia County in central Florida and Alachua-Marion counties in north Florida are infected by a *Hepatozoon* with gamonts that are very similar to *H. sistruri*. Except in nucleus size, all gamont characters have slight but significant differences. Without comparison of sporogonic characters, these populations cannot be identified either as distinct to or conspecific with *H. sistruri*.

Hepatozoon horridus Telford, Moler and Butler 2008 (**Plate 53**)

Diagnosis A *Hepatozoon* species characterized by gamonts with vacuolated cytoplasm, 13.0–17.0 × 4.0–6.0 µm, with broadly rounded ends. Gamont nuclei are 4.0–6.0 × 3.0–5.0 µm, located in the second quarter of the gamont, rarely extending into the first or third quarters.

Type Host *Crotalus horridus* Linnaeus (Serpentes: Viperidae).

Other Hosts None known.

Type Locality State Road 250, 3.6 km west of Taylor, Baker County, Florida.

Other Localities None reported.

Prevalence One of eight *C. horridus* from southern Georgia and northern Florida was infected by *H. horridus*.

Morphological Variation Gamonts of medium width with broadly rounded ends, not recurved, 15.7 ± 0.9 × 5.1 ± 0.6 μm (13.0–17.0 × 4.0–6.0, N = 25), with LW 80.2 ± 11.9 μm^2 (63–102) and L/W 3.10 ± 0.33 (2.6–4.0). Nuclei 5.0 ± 0.6 × 3.7 ± 0.6 μm (4.0–6.0 × 3.0–5.0), with LW 18.2 ± 3.7 μm^2 (13.5–25.0), always present in second quarter of gamont, seldom extending into first quarter (8%) or third quarter (12%). Gamont cytoplasm contains many small, rounded vacuoles.

Sporogony Sporogony in the experimental vector *Aedes aegypti* occurs within the head, but more commonly in the hemocoel of the thorax and abdomen. Oocysts are spherical, usually, to ovoid, 106.0 ± 40.5 × 97.9 ± 36.8 μm (40–212 × 32–192, n = 52) and L/W 1.09 ± 0.10 (1.0–1.5), containing 20.4 ± 13.7 (4–61, n = 31) sporocysts. Sporocysts are spherical to elongate, 24.0 ± 4.2 × 20.4 ± 2.4 μm (14–45 × 11–25, n = 66), LW 491.9 ± 116.4 μm^2 (196–900), and L/W 1.18 ± 0.22 (1.0–2.3), containing 13.0 ± 3.8 (8–21, n = 36) sporozoites. In one of five mosquitoes dissected, one oocyst was found within the salivary gland

Merogony Unknown.

Effects on Host Gamonts: The erythrocyte cytoplasm rarely appears dehemoglobinized (4%), and infected cells are often distorted (48%). Infected erythrocytes and their nuclei are similar in dimensions to uninfected cells.

Remarks The oocyst found within the salivary gland is a unique observation that suggests salivarian transmission for this *Hepatozoon* species is possible.

Hepatozoon sauritus Telford, Wozniak and Butler 2001 (Plate 54)

Diagnosis A *Hepatozoon* species with slender and elongate gamonts, nonrecurved, 13–19 × 2.5–5.5 μm, their nuclei are 3–7.5 × 1.5–5.5 μm, almost always situated in the second quarter of the gamont, commonly extending into either the first or third quarter. The anterior end of gamonts is slightly narrower and broadly pointed in contrast to the more distinctly rounded posterior end. Sporogony occurs in the head and hemocoel of the thorax and abdomen, rarely in the proboscis of experimental mosquito hosts. Oocysts are spherical to usually ovoid, 65–283 × 60–275 μm, and contain 5–222 ovoid sporocysts, 18–53 × 15–40 μm, that produce 8–52 sporozoites, the average number of which varies among host species from 17.0 to 56.0 μm. The pulmonary meronts contain 10–21 macromerozoites and 22–57 micromerozoites.

Type Host *Thamnophis sauritus sackenii* (Kennicott) (Serpentes: Colubridae).

Type Locality Palm Beach Gardens Estates, Palm Beach County, Florida.

Other Hosts *Coluber constrictor priapus, C. constrictor helvigularis, Elaphe obsoleta quadrivittata, E. obsoleta rossalleni, E. guttata guttata, Thamnophis s. sirtalis, Diadophis p. punctatus* (Telford et al., 2004). *Crotalus horridus* (Telford et al., 2008).

Other Localities Within Florida, *C. constrictor* ssp. in Alachua, Levy, Marion, Liberty, and Palm Beach counties; in *E. obsoleta*, Alachua and Marion counties; in *E. guttata*, Alachua and Columbia counties; in *T. sirtalis*, Alachua and Lake counties; in *D. punctatus*, Alachua County; morphologically identical gamonts indicate the presence of *H. sauritus* in *T. sauritus* of Lee County.

Prevalence Overall prevalence in known host species in Florida, 35%: 2 of 9 in *T. sauritus;* 54% in *C. constrictor*, 47% in *E. obsoleta*, 15% in *E. guttata*, 20% in *T. sirtalis*, and 7% in north Florida *D. punctatus*. Prevalence by county: 2 of 9 in *T. sauritus*, Palm Beach; in *C. constrictor priapus* 2 of 5, Palm Beach, 12 of 26 (46%), Alachua, 2 of 5, Marion; in 1 of 1, Levy; in 2 of 4 *C. constrictor helvigularis*, Liberty; in *E. obsoleta quadrivittata* 7 of 9, Alachua, and 3 of 3, Marion; in *E. guttata* 1 of 3, Alachua, and 1 of 1, Columbia;

In *T. sirtalis,* 9 of 22 (41%) in Alachua and 1 of 1 in Lake; in *D. punctatus,* 1 of 17 (6%) in Alachua (Telford, unpublished).

Morphological Variation Gamont dimensions by host are 15.8 ± 0.9 × 3.7 ± 0.5 μm (14–18 × 3–5.5, N = 50), LW 57.9 ± 9.5 μm^2 (45–93), L/W 4.40 ± 0.70 (3.1–5.8), with nuclei 5.3 ± 0.5 × 2.9 ± 0.4 μm (4–6.5 × 2–4), LW 14.9 ± 2.5 μm^2 (10–20) in *T. sauritus*; 15.7 ± 0.8 × 4.0 ± 0.5 μm (14–18 × 3–5, N = 170), LW 62.7 ± 8.1 μm^2 (45–85), L/W 4.00 ± 0.54 (2.8–5.8), with nuclei 4.8 ± 0.7 × 2.8 ± 0.4 μm (3.5–7.5 × 2–4), LW 13.5 ± 2.4 μm^2 (8–21) in *C. constrictor*; 16.9 ± 0.9 × 4.0 ± 0.5 μm (14–19 × 2.5–5, N = 96), LW 67.0 ± 10.3 μm^2 (42.5–87.5), L/W 4.32 ± 0.58 (3.3–6.8), with nuclei 5.4 ± 0.9 × 2.8 ± 0.5 μm (3–7.5 × 1.5–5), LW 15.1 ± 3.7 μm^2 (9–35) in *E. obsoleta*; 16.2 ± 1.0 × 3.9 ± 0.4 μm (14–18 × 3–5, N = 50), LW 63.2 ± .8 μm^2 (44.6–76.5), L/W 4.17 ± 0.56 (3.0–5.7), with nuclei 5.0 ± 0.8 ×

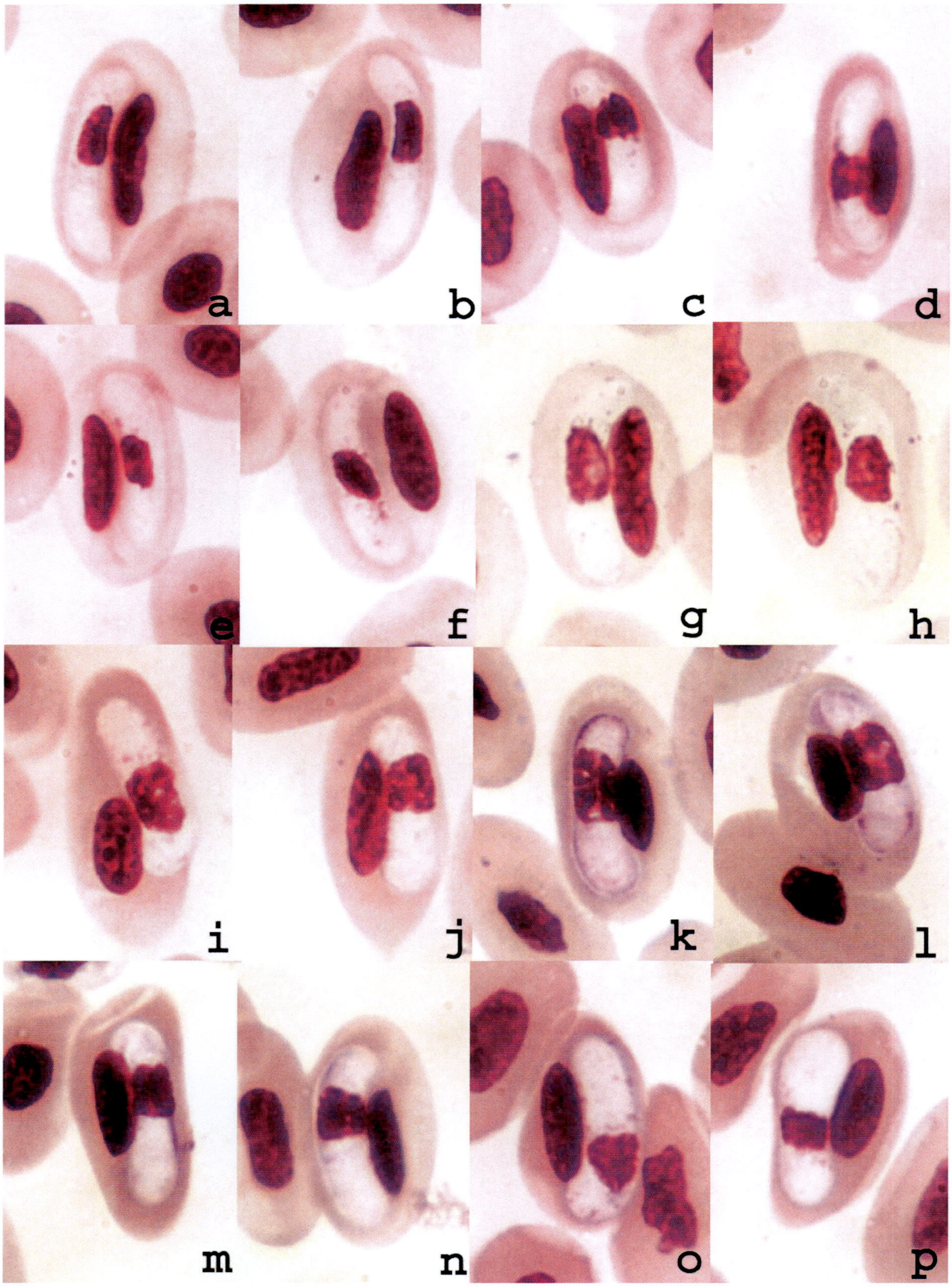

Plate 54 Hepatozoon species from Florida snakes: **a, b**, *H. sauritus* of *Coluber constrictor priapus*; **c, d**, *H. polytopis* of *C. constrictor priapus*; **e, f**, *H. confusus* of *C. constrictor priapus*; **g, h**, *H. priapus* of *C. constrictor priapus*; **i, j**, *H. mansoni* of *Masticophis flagellum*; **k, l**, *H. guttata* of *Elaphe guttata*; **m, n**, *H. punctatus* of *Diadophis punctatus*; **o, p**, *H. punctatus* of *Rhadinaea flavilata*.

2.8 ± 0.3 µm (4–7.5 × 2–3.5), LW 14.2 ± 2.3 µm² (9–19.3) in *E. guttata*; 16.7 ± 0.8 × 4.0 ± 0.3 µm (15–19 × 3–5, N = 75), LW 66.3 ± 6.3 µm² (51–81), L/W 4.24 ± 0.39 (3.2–5.7), with nuclei 5.4 ± 0.9 × 2.5 ± 0.4 µm (3–7.5 × 2–3.5), LW 13.3 ± 2.7 µm² (7.5–24.5) in *T. sirtalis*; 17.0 ± 0.9 × 3.8 ± 0.6 µm (16–19 × 3–5, N = 25), LW 65.0 ± 11.5 µm² (48–90), L/W 4.53 ± 0.65 (3.6–5.7), with nuclei 5.4 ± 0.8 × 2.4 ± 0.4 µm (4.5–7 × 2–3.5), LW 12.8 ± 1.9 µm² (9–17.5) in *D. punctatus*. Gamont nuclei are nearly always in the second quarter of the gamont (97%) and commonly extend into either the first (38%) or third quarter (23%), very rarely extending from the first to the third quarter. The anterior end of the gamonts typically is slightly tapered to a broad point, in contrast to the posterior end, which is broadly rounded.

Sporogony Sporogony takes place in the head and hemocoel of the thorax and abdomen in the experimental mosquito host *Aedes aegypti*, and in 1 of 33 infected mosquitoes, mature sporocysts were found in the proboscis. Overall, the spherical to usually ovoid oocysts are 65–283 × 60–275 µm (N = 59), with L/W ratio 1.04–1.13, and contain 5–222 (N = 37) primarily ovoid sporocysts. Among the four host species for which sporogony was obtained, oocysts averaged larger from *E. obsoleta*, 196 ± 62 × 180 ± 72 µm (N = 3), containing 89.8 ± 80.3 (19–144, N = 4) sporocysts, and smallest from *T. sauritus*, 116 ± 24 × 111.4 ± 23.5 µm (N = 15), containing 34.5 ± 11.5 (15–52, N = 14) sporocysts. In *C. constrictor*, oocysts were 146 ± 53 × 139.4 ± 53.2 µm (N = 11) with 57.0 ± 38.3 (5–90, N = 4) sporocysts, and from *T. sirtalis* 176 ± 59 × 170 ± 60.5 µm (N = 30), containing 63.3 ± 57.8 (8–222, N = 15) sporocysts. Oocyst shape was nearly spherical to usually ovoid, with L/W ratios averaging in the same host sequence, 1.13, 1.04, 1.06, and 1.05, ranging between 1.0 and 1.3 from all host species. Sporocyst dimensions were smaller from *T. sauritus,* which was studied from fixed material, 28.6 ± 8.1 × 20.8 ± 3.3 µm (18–52 × 15–29, N = 90), than from the other three host species, studied unfixed. Sporocyst dimensions were very similar in *C. constrictor* (38.4 ± 5.9 × 26.0 ± 5.0 µm, 22–53 × 17–40 µm, N = 65), *E. obsoleta* (35.1 ± 6.0 × 28.0 ± 4.7 µm, 27–45 × 17–38 µm, N = 30), and *T. sirtalis* (39.0 ± 9.0 × 24.0 ± 2.2 µm, 25–53 × 20–28 µm, N = 50). In the same host sequence, the L/W ratios of sporocysts were 1.41 (1.0–2.9), 1.52 (1.0–2.3), 1.29 (1.0–2.4), and 1.64 (1.0–2.4). Fewer sporozoites were produced by *H. sauritus* from *T. sauritus*, 17.0 ± 7.1 (8–40, N = 90) than from *T. sirtalis*, 36.0 ± 9.6 (19–52, N = 27), but numbers from *T. sauritus* were more similar to those from *C. constrictor*, 25.0 ± 5.8 (16–38, N = 63) and *E. obsoleta*, 24.4 ± 6.6 (16–36, N = 17).

Merogony Merogony occurred in the lung only of the type host, *T. sauritus*. Macromeronts 18.0 ± 2.2 × 13.6 ± 1.8 µm (15–22 × 10–17, N = 20) were enclosed within cysts 20.7 ± 3.2 × 15.9 ± 1.8 µm (17–28 × 12–19) and contained 16.3 ± 3.0 (10–21) macromerozoites 9.2 ± 0.9 × 2.1 ± 0.2 µm (N = 7). Micromeronts were 27.0 ± 3.4 × 15.6 ± 2.1 µm (21–35 × 13–20, N = 21), contained within cysts 30.8 ± 3.4 × 18.9 ± 3.1 µm (25–36 × 14–25), and produced 32.1 ± 9.3 (22–57) micromerozoites 7.1 ± 1.2 × 1.5 µm (N = 7).

Effects on Host Infected erythrocytes in all hosts are longer than uninfected cells, reflected in a greater LW in most infections, but the width of erythrocytes containing gamonts is usually similar to normal cells. Nuclei of infected cells are longer in most infections but are similar in width to uninfected cells, which is commonly reflected in greater LW values for infected cell nuclei. Infected erythrocytes are usually distorted (93%), with dehemoglobinization variable, often absent but sometimes severe in individual infections.

Remarks The comparison of samples from the type infection in *T. sauritus* of southern Florida with infections of *C. constrictor, E. obsoleta*, and *T. sirtalis* of northern Florida found a single haplotype ("B") present in a nucleotide sequence of 530-bp alignment of the 18S ribosomal RNA (rRNA) gene (Telford et al., 2004). This supported the conclusion based on gamont and sporogonic characters that a single species of *Hepatozoon* infects multiple species of colubrid snakes in nature, ranging from the northern to the southern part of the state. Most of the other 13 species of *Hepatozoon* described from Florida snakes on the basis of sporogonic characters and gamont morphology are restricted, as presently known, to single-host species, the exceptions being *H. polytopis* and (rarely) *H. sirtalis*. Telford et al. (2004) reported variation in both gamont and sporogonic characters within and among species host to *H. sauritus*. Sporozoite numbers produced by sporocysts from sylvatic and suburban *T. sirtalis* differed significantly, 36.0 ± 9.6 (N = 27) and 24.7 ± 5.7 (N = 83), respectively, although sporocyst dimensions were similar. This may reflect geographic variation rather than habitat influences on sporogony in view of the about 125 km distance between sample areas.

Hepatozoon polytopis Telford, Butler and Telford 2005 (**Plate 54**)

Diagnosis A *Hepatozoon* species characterized by short and broad, nonrecurved gamonts, 10–15 × 3–6 µm, with nuclei 2.5–6 × 2.5–4.5 µm, that are situated in the second quarter of the gamont and often extend into the first quarter. Young gamonts are often multinucleate. Oocysts are spherical to ovoid, 62–240 × 57–190 µm, and contain 3–103 spherical-to-ovoid sporocysts, 28–73 × 25–58 µm, that produce 22–64 sporozoites. Sporogony takes place within the head and hemocoel of the thorax and abdomen.

Type Host *Coluber constrictor priapus* Dunn and Wood (Serpentes: Colubridae).

Type Locality Jupiter Farms, Palm Beach County, Florida.

Other Hosts *Thamnophis sauritus sackenii.*

Other Localities None known.

Prevalence *Hepatozoon polytopis* infected 3 of 5 *C. constrictor*, in the vicinity of the type locality, and 10 of 13 (77%) *T. sauritus*.

Morphological Variation In the type host, gamonts are usually short and broad, not recurved, 12.8 ± 1.0 × 4.6 ± 0.7 μm (10.0–15.0 × 3.5–6.0, N = 50), with LW 58.5 ± 8.8 μm^2 (42–84) and L/W 2.84 ± 0.49 (1.8–3.7). Gamont nuclei are 4.5 ± 0.6 × 3.4 ± 0.5 μm (3.0–6.0 × 2.5–4.5), with LW 15.1 ± 2.8 μm^2 (10.0–24.0), always present in the second quarter of the gamont, usually extending into the first quarter (45%) and seldom into the third quarter (11%). Young gamonts occasionally are multinucleate, but the characteristic broadened shape of mature parasites was observed only in uninucleate gamonts. In *T. sauritus*, gamonts are 13.0 ± 0.9 × 4.3 ± 0.4 μm (11.0–15.0 × 3.0–5.0, N = 100), with LW 55.4 ± 6.7 μm^2 (37.5–70.0) and L/W 3.07 ± 0.39 (2.3–4.2). Gamont nuclei in this host are 4.1 ± 0.6 × 3.7–0.5 μm (2.5–5.5 × 2.5–4.5), LW 15.1 ± 3.4 μm^2 (6.3–22.5). Gamont length, width, LW, and nucleus LW were similar between the samples from *C. constrictor* and *T. sauritus*, but gamont shape was thinner in the latter host, as indicated by the slightly greater L/W ratio. Nucleus length in the sample from *C. constrictor* was slightly greater and nucleus width slightly less than in gamonts from *T. sauritus*, but nucleus size (LW) was similar. Position of the nucleus in the first quarter of the gamont was 45% in the type host and less frequent (25%) in *T. sauritus*

Sporogony Sporogony in the experimental host *Aedes aegypti* occurs within the head and hemocoel of the thorax and abdomen. Sporocysts were present in three of the five proboscides examined, numbering 5, 5, and 4. The spherical-to-ovoid oocysts are 122.1 ± 38.3 × 104.9 ± 31.4 μm (62–240 × 57–190, N = 34) and L/W 1.17 ± 0.19 (1.0–1.9) and contain 31.3 ± 21.4 (3–103, N = 39) sporocysts. The sporocysts, also spherical to ovoid in shape, are 38.0 ± 7.1 × 33.9 ± 6.0 μm (28–73 × 25–58, N = 79), LW 1325.1 ± 520.3 μm^2 (756–4168) and L/W 1.12 ± 0.11 (1.0–1.4), with 42.9 ± 12.5 (22–64, N = 30) sporozoites contained within. Occasional sporocysts are twice the dimensions of typical sporocysts.

Merogony Unknown.

Effects on Host The cytoplasm of infected erythrocytes is often thin, appearing partially dehemoglobinized (86%) in both host species, with infected cells usually distorted (92%). Infected erythrocytes are longer than uninfected cells, with greater LW, but are similar in width and nuclear dimensions.

Remarks As reported by Telford et al. (2005), gamonts morphologically similar to *H. polytopis* infected a *Diadophis p. punctatus* from the type locality. Sporogony was not obtained, but:

> A nucleotide sequence (530 bp) alignment of the 18S rRNA gene was generated for hemogregarines present in the three snake species for which samples were taken (*C. constrictor priapus, T. sauritus sackenii, D. punctatus*). Both the single sample from *C. constrictor priapus* and 5 from *T. sauritus sackenii* demonstrated the same haplotype, designated "C." The haplotype of the hemogregarine found in *D. punctatus* was unique, and was designated "I," clearly a distinct species from that present in the other two hosts despite the similarities in gamont morphology.

Hepatozoon confusus Telford, Butler and Moler 2005 (**Plate 54**)

Diagnosis A *Hepatozoon* species characterized by slender, nonrecurved gamonts with both ends rounded, 14–17 × 3.5–5 μm, and nuclei 2.5–4.5 × 4–6 μm that always occupy the second quarter of the gamont and commonly extend into the first or third quarters as well. Oocysts are spherical to ovoid, 52–278 × 50–278 μm, and contain 7–111 spherical-to-ovoid sporocysts 21–38 × 20–33 μm that produce 12–32 sporozoites.

Type Host *Coluber constrictor priapus* Dunn and Wood (Serpentes: Colubridae).

Type Locality North side of Paynes Prairie, Alachua County, Florida.

Other Hosts *Coluber constrictor helvigularis.*

Other Localities Liberty, Levy, Marion, and Palm Beach counties, Florida.

Prevalence Twelve of 43 (28%) *Coluber constrictor* overall were infected by *H. confusus*, 5 of 26 (19%) in Alachua, 2 of 4 in Liberty, 2 of 5 in Marion, 1 of 1 in Levy, and 2 of 5 in Palm Beach counties.

Morphological Variation The gamonts are slender with both ends rounded and not recurved, 15.6 ± 0.7 × 4.1 ± 0.4 μm (14.0–17.0 × 3.5–5.0, n = 25), with LW 64.3 ± 7.3 μm² (52–80) and L/W 3.8 ± 0.39 (2.8–4.4). Gamont nuclei are 5.0 ± 0.5 × 2.7 ± 0.3 μm (2.5–4.4 × 4.0–6.0), with LW 13.5 ± 1.7 μm² (11.0–16.5), always present in the second quarter of the gamont, commonly extending into the first quarter (28%) and into the third quarter (24%).

Sporogony Sporogony in the experimental vector *Aedes aegypti* takes place within the head and hemocoel of the thorax and abdomen. The spherical-to-ovoid oocysts are 115.5 ± 51.1 × 108.9 ± 48.5 μm (52–278 × 50–278, n = 47) and L/W 1.06 ± 0.06 (1.0–1.2) and contain 25.0 ± 24.6 (7–111, n = 30) sporocysts. The sporocysts are spherical to ovoid, 27.6 ± 3.2 × 25.2 ± 2.7 μm (21–38 × 20–33, n = 78), LW 701.3 ± 145.5 μm² (420–1125) and L/W 1.09 ± 0.10 (1.0–1.4), and produce 20.2 ± 5.1 (12–32, n = 30) sporozoites.

Merogony Unknown.

Effects on Host The cytoplasm of infected erythrocytes is sometimes thin, appearing partially dehemoglobinized (28%), with infected cells usually distorted (92%). Infected erythrocytes are longer than uninfected cells but similar in width and LW (64%); erythrocyte nuclei are longer but narrower with similar LW.

Hepatozoon priapus Telford, Butler and Moler 2005 (Plate 54)

Diagnosis A *Hepatozoon* species characterized by robust gamonts with broadly rounded ends, nonrecurved, 17–20 × 3.5–6 μm, with nuclei 5–7 × 2.5–4 μm, situated in the second quarter of the gamont and often extending into the third quarter. The spherical-to-ovoid oocysts are 55–123 × 47–115 μm and contain 6–31 sporocysts, also spherical to ovoid, 19–50 × 16–38 μm, that contain 5–18 sporozoites.

Type Host *Coluber constrictor priapus* Dunn and Wood (Serpentes: Colubridae).

Type Locality Gainesville, Alachua County, Florida.

Other Hosts *Coluber constrictor helvigularis.*

Other Localities Liberty and Palm Beach counties, Florida.

Prevalence Seven of 43 (16%) *Coluber constrictor* overall were infected by *H. priapus*, 4 of 26 (15%) in Alachua, 2 of 5 in Palm Beach, and 1 of 4 in Liberty counties.

Morphological Variation The gamonts are robust with broadly rounded ends, not recurved, 18.0 ± 0.9 × 4.2 ± 0.6 μm (17.0–20.0 × 3.5–6.0, n = 25), with LW 76.4 ± 10.9 μm² (59–105) and L/W 4.31 ± 0.53 (2.9–5.4). Gamont nuclei are 6.0 ± 0.5 × 3.0 ± 0.3 μm (5.0–7.0 × 2.5–4.0), with LW 17.9 ± 1.9 μm² (13.7–21.0). Nuclei are always present in the second quarter of the gamont, seldom extending into the first quarter (8%) but often into the third quarter (20%).

Sporogony Sporogony in the experimental vector *Aedes aegypti* occurs within the head and the hemocoele of the thorax and abdomen. Oocysts are spherical to ovoid, 92.5 ± 32.5 × 86.0 ± 32.0 μm (55–123 × 47–115, n = 5) and L/W 1.08 ± 0.10 (1.0–1.3), and contain 14.0 ± 11.8 (6–31, n = 4) sporocysts. The spherical-to-ovoid sporocysts are 26.3 ± 6.6 × 23.3 ± 5.5 μm (19–50 × 16–38, n = 45), LW 641.2 ± 321.5 μm² (320–1500) and L/W 1.13 ± 0.19 (1.0–2.2), and produce 12.6 ± 3.4 (5–18, n = 29) sporozoites.

Merogony Unknown.

Effects on Host Cytoplasm of the infected erythrocytes is always thin, appearing dehemoglobinized, with infected cells always distorted. The infected erythrocytes are much longer than uninfected cells and wider, with greater LW; the erythrocyte nuclei are longer than uninfected cells but similar in width and LW.

Remarks *Hepatozoon priapus* produces fewer sporozoites on average per sporocyst than do any of the other *Hepatozoon* species described from Florida except *H. sirtalis.*

Hepatozoon mansoni (Sambon and Seligmann) 1907 Telford and Telford 2002 (Plate 54)

Diagnosis A *Hepatozoon* species characterized by broadly elongate, nonrecurved gamonts 13–16 × 4–6 μm, with nuclei 3–6 × 3–6 μm usually situated in the second quarter of the gamont, often extending into the first quarter. In experimental mosquito hosts, sporogony occurs in the head and thorax predominantly, rarely in the abdomen. The spherical-to-ovoid oocysts are 79–198 × 69–178 μm and contain 7–64 spherical-to-ovoid sporocysts 19–48 × 18–44 μm that produce 12–32 sporozoites.

Type Host *Masticophis f. flagellum* Shaw (Serpentes: Colubridae).

Type Locality Ocala National Forest, Marion County, Florida, about 4.5 km south of County Road 316 on Forest Road 67.

Other Hosts Experimental: *Tantilla relicta neilli.*

Other Localities Bradford County, Florida. Roudabush and Coatney (1937) reported the species but without locality data.

Prevalence Infections of *H. mansoni* were present in the single *M. flagellum* examined from Bradford County and in three of seven from Marion County, Florida.

Morphological Variation Gamonts are broadly elongate and not recurved, 14.7 ± 0.6 × 4.7 ± 0.4 μm (13–16 × 4–6 N = 25), LW 69.5 ± 7.6 μm^2 (56–88) and L/W 3.1 ± 0.2 (2.6–3.6). Gamont nuclei are 4.8 ± 0.6 × 3.9 ± 0.5 μm (3–6 × 3–5, N = 25), with LW 18.5 ± 2.7 μm^2 (12–23), usually (75%) situated in the second quarter of the gamont, commonly extending into the first quarter (21%), rarely (4%) at midbody of the gamont. In an experimental host, *Tantilla relicta neilli,* gamonts were 14.8 ± 0.5 × 4.7 ± 0.4 μm (14–16 × 4–6 N = 25), LW 70.0 ± 7.6 μm^2 (56–90) and L/W 3.1 ± 0.3 (2.5–3.8). Nuclei of the gamonts were 5.1 ± 0.5 × 4.1 ± 0.3 μm (4–6 × 3–5, N = 25), with LW 20.9 ± 3.1 μm^2 (15–30), mostly (92%) situated in the second quarter of the gamont, seldom extending into the first quarter. Dimensions of the gamonts were similar to those in the type host for all characters with the exception of nucleus LW, which was greater in *T. relicta.*

Sporogony Sporogony in the experimental vector *Aedes aegypti* occurs within the head and hemocoele of the thorax, rarely within the abdomen. The spherical-to-ovoid oocysts are 144.2 ± 28.7 × 126.1 ± 25.0 μm (79–198 × 69–178, N = 30) and L/W 1.15 (1.0–1.6) and contain 7–64 (29.2 ± 16.6, N = 15) sporocysts. Sporocysts are spherical to ovoid, 33.1 ± 6.3 × 29.8 ± 5.7 μm (19–48 × 18–44, N = 58), LW 1017.8 ± 384.4 μm^2 (342–2112) and L/W 1.11 ± 0.10 (1.0–1.6), and produce 20.2 ± 4.9 (12–32, N = 28) sporozoites, 12.1 ± 2.1 × 2.5 ± 0.4 μm (N = 8).

Merogony Tissues have not been examined. In a road-killed host, cysts 17–20 × 14–15 μm, which contained two or three zoites, 15–19 × 3–4 μm, with nuclei 3 × 3 μm, were found in hepatic blood, perhaps broken free from a crushed organ.

Effects on Host The cytoplasm of infected erythrocytes is commonly thin, appearing partially dehemoglobinized (64%) or contracted into a central mass, with infected cells always distorted. Infected erythrocytes were longer and more slender than uninfected cells but did not differ in LW. Gamont nuclei were smaller in LW than those of uninfected cells. In the experimental infection of *T. relicta neilli,* cytoplasm of the erythrocyte seldom (8%) appeared partially dehemoglobinized, and no cells showed contraction of the cytoplasm as in *M. flagellum.* Infected cells were commonly distorted (44%). Dimensions of infected erythrocytes were greater in length, width, and LW, but their nuclei did not differ in LW from those of uninfected cells.

Remarks One *Tantilla relicta neilli* Telford, negative on initial examination and subsequently, was inoculated per os with oocysts and sporocysts from infected mosquitoes. Erythrocytes showed patent infection by gamonts at 62 days PI. Serial sections of organs were negative for meronts.

Hepatozoon guttata Telford, Butler and Telford 2002 (Plate 54)

Diagnosis A *Hepatozoon* species characterized by broadly elongate, nonrecurved gamonts 10–17 × 3.5–6.0 μm and nuclei 3.5–5.0 × 3.0–4.5 μm, situated in the second quarter of the gamont and commonly extending into the first quarter. In experimental mosquito hosts, oocysts are spherical to usually ovoid, 45–155 × 40–152.5 μm, and contain 2–32 spherical-to-ovoid sporocysts 20–55 × 17.5–47.5 μm that produce 14–89 sporozoites.

Type Host *Elaphe g. guttata* (Linne) (Serpentes: Colubridae).

Type Locality Jupiter, Palm Beach County, Florida.

Other Hosts None known.

Other Localities In Florida: Vero Beach, Indian River County; Palm Beach Gardens Estates and Jupiter Farms, Palm Beach County.

Prevalence *Hepatozoon guttata* infected 11 of 15 (73.3%) *E. guttata* examined from southeastern Florida.

Morphological Variation Gamonts are broadly elongate and not recurved, 14.6 ± 0.8 × 4.6 ± 0.4 μm (13–17 × 3.5–6 N = 25), LW 67.2 ± 7.3 μm^2 (49–77), and L/W 3.2 ± 0.3 (2.8–4.3). Free gamonts are 19.5–24 × 3–3.5 μm. Gamont nuclei are 4.6 ± 0.4 × 3.9 ± 0.4 μm (3.5–5 × 3–4.5, N = 25), with LW 17.9 ± 2.6 μm^2 (10–23), always situated in the second quarter of the gamont, commonly extending into the first quarter (44%), rarely (4%) at midbody of the gamont. Gamonts from a chronic infection were shorter, 13.4 ± 1.6 × 4.6 ± 0.5 μm (10–16 × 4–6, N = 25), LW 62.7 ± 11.7 μm^2 (40–82.5), with a lower L/W ratio of 2.91 ± 0.35 (2.08–3.38), but did not differ in other characters from the active type infection.

Sporogony Sporogony in the experimental host *Aedes aegypti* is within the head and hemocoel of the thorax, only rarely in the abdominal hemocoel. The spherical to usually ovoid oocysts are 87.2 ± 22.3 × 75.9 ± 19.9 μm (45–155 × 40–152.5, N = 78) and L/W 1.16 (1.0–1.7) and contain 2–32 (7.1 ± 4.9, N = 78) spherical-to-ovoid sporocysts. Sporocysts are 34.8 ± 7.3 × 31.0 ± 6.2 μm (20–55 × 17.5–47.5, N = 50), with LW 1115.5 ± 462.8 μm^2 (350–2612) and L/W 1.12 ± 0.12 (1.0–1.8), and produce 45.7 ± 15.6 (14–89, N = 75) sporozoites, 8.0 ± 1.0 × 1.8 ± 0.3 μm (N = 5).

Merogony Dizoic cysts were present in the liver and lung of a chronically infected host. A single macromeront, 20 × 12 μm within a cyst 28 × 12 μm in the lung, contained six merozoites, 9.5 × 2.5 μm, as well as several uninucleate parasites 10.5–11 × 3 μm, which were possibly newly invaded macromerozoites.

Effects on Host The cytoplasm of infected erythrocytes is rarely thin (8%), appearing partially dehemoglobinized, and infected cells are usually distorted (68%). Infected erythrocytes are longer than uninfected cells, with greater LW, but are similar in width; their nuclei are smaller in length and width than those of uninfected cells. Infected erythrocytes in a chronic infection did not differ from normal cells in dimensions, but their cytoplasm commonly (50%) appeared darker than in uninfected erythrocytes.

Hepatozoon punctatus Telford, Wozniak and Butler 2001 (**Plate 54**)

Diagnosis A *Hepatozoon* species characterized by nonrecurved gamonts 12–16 × 4–6 μm with nuclei 3–7 × 3–6 μm usually situated in the second quarter of the gamont. Oocysts are nearly spherical, 168–195 × 165–185 μm and contain 25–64 nearly spherical sporocysts, 20–32 × 16–31 μm, that produce 15–33 sporozoites. Merogony occurs in the liver. Macromeronts contain 3–10 macromerozoites, and micromeronts produce 14–40 micromerozoites. Sporogony occurs in the abdominal and thoracic hemocoel of experimental mosquito hosts.

Type Host *Diadophis p. punctatus* (Linnaeus) (Serpentes: Colubridae).

Type Locality Palm Beach Gardens Estates, Palm Beach County, Florida.

Other Hosts *Rhadinaea flavilata*.

Other Localities Jupiter Farms, Palm Beach County, Florida.

Prevalence One of three *D. punctatus* examined from the vicinity of the type locality and three of six collected at Jupiter Farms were infected by *H. punctatus*. One of one *R. flavilata* collected at Jupiter Farms.

Morphological Variation Gamonts are broadly elongate, not recurved, 13.4 ± 1.0 × 5.1 ± 0.5 μm (12–16 × 4–6 N = 25), LW 68.0 ± 9.7 μm^2 (54–90), and L/W 2.7 ± 0.3 (2.2–3.6). Free gamonts are 18.0 ± 2.4 × 4.1 ± 0.6 μm (14–20 × 3–5, N = 7). Gamont nuclei are 4.5 ± 0.9 × 4.3 ± 0.7 μm (3–7 × 3–6, N = 25), with LW 19.3 ± 4.3 μm^2 (14–30), usually (80%) situated in the second quarter of the gamont.

Sporogony Sporogony in the experimental vector *Aedes aeqypti* occurs within the hemocoele of abdomen and thorax. Oocysts are nearly spherical, 180.7 ± 9.8 × 175.0 ± 8.9 μm (168–195 × 165–185, N = 6), and contain 25–64 (40.7 ± 15.2) sporocysts. Sporocysts are nearly spherical, 25.1 ± 3.0 × 22.7 ± 3.0 μm (20–32 × 16–31, N = 57), LW 575.2 ± 135.9 μm^2 (320–992), and L/W 1.11 ± 0.13 (1.0–1.5), and produce 20.1 ± 4.7 (15–33, N = 22) sporozoites, 12.1 ± 2.1 × 2.5 ± 0.4 μm (N = 8).

Merogony Merogony takes place in the liver. Macromeronts in a natural infection are 18.3 ± 5.2 × 12.5 ± 3.6 μm (12–27 × 8–16), LW 230.3 ± 82.7 μm^2 (108–320), and L/W 1.60 ± 0.7 (1.0–2.5), and contain 7.5 ± 2.6 (3–10) macromerozoites. In experimental infection, macromeronts are 19.4 ± 4.5 × 11.7 ± 1.9 μm (14–29 × 9–17), LW 226.0 ± 74.6 μm^2 (154–493), and L/W 1.71 ± 0.51 (1.0–2.3), and contain 3.6 ± 1.2 (3–8, N = 19) macromerozoites, 11.9 ± 1.4 × 2.5 ± 0.5 μm (N = 5). In 33 cysts, 42.4% contained one (21.1%) or two cystozoites, 33.3% had three, 21.2% had four, and only one (3%) contained eight macromerozoites. In a natural infection, micromeronts are 28.3 ± 3.9 × 18.1 ± 2.4 μm (20–33 × 15–22), LW 519.9 ± 129.8 μm^2 (300–726), and L/W 1.57 ± 0.1 (1.3–1.8), and contain 25.2 ± 8.0 (14–40) micromerozoites, 6.5 ± 1.5 × 1.7 ± 0.4 μm (N = 5).

Effects The cytoplasm of infected erythrocytes is similar in appearance to normal cells, but infected cells are always distorted. The dimensions of infected erythrocytes and their nuclei do not differ from uninfected cells.

Remarks Oocysts obtained in mosquitoes fed on the type infection of *H. punctatus* were inoculated by pipette into the esophagus of a *D. punctatus* collected in Gainesville, Florida, that had remained negative for over 5 months. Erythrocytes did not show patent infection before death of the snake at 110 days PI, but apparently mature macromeronts and young micromeronts were found in histological sections of its liver (Telford et al., 2001).

Hepatozoon sauromali Lewis and Wagner 1964 (**Plate 55**)

Diagnosis A *Hepatozoon* species characterized by broad gamonts 10–21 × 3.3–7.7 μm, with one end occasionally recurved, and nuclei, thin and band-like to broad and saddle-shaped, 2.2–7.7 μm in diameter. Nuclei are situated at mid-body of the gamont and commonly extend into the second quarter. Sporogony occurs in mites, *Hirstiella* sp., in which oocysts contain many sporocysts about 12 μm in diameter that produce 10–12 sporozoites. Dimorphic meronts develop in the liver.

Type Host *Sauromalus hispidus* Stejneger (Sauria: Iguanidae).

Type Locality Angel de la Guarda Island, Baja California Sur, Mexico.

Other Hosts *Sauromalus varius, S. australis*, and *S. obesus*.

Other Localities Granite Island, Las Arrestras, San Esteban Island, and San Lorenzo Island in the Gulf of California, Baja California, Mexico (J. E. Lewis and Wagner, 1964). In Southern California, Palm Springs and the San Jacinto Mountains, Riverside County (Telford, 1970d), and vicinity of Barstow, San Bernardino County (Mitchell, 1965).

Prevalence Six of 7 *S. obesus* collected in the San Jacinto Mountains were infected by *H. sauromali* at elevations of 245 and 915 m (Telford, 1970d), and 14 of 14 (100%) adult *S. obesus* from San Bernardino County were infected (Mitchell, 1965).

Morphological Variation In their sparse description, J. E. Lewis and Wagner (1964) described gamonts as "Oval, elongated, one end sometimes recurved. ... Gametocytes from 10 by 6 to 18 × 5 μ." Telford (1966) reported gamonts of three slightly different sizes: 17.3 × 6.4 μm (15.4–18.7 × 5.5–7.7, N = 20), LW 110.7 μm^2, and L/W 2.70, with nuclei 5.0 μm (3.3–6.6) in diameter; 15.2 × 6.1 μm (13.2–17.6 × 5.5–7.7, N = 20), LW 92.7 μm^2, and L/W 2.49, nuclei 5.2 μm (3.3–7.7); and 17.0 × 4.0 μm (15.4–20.9 × 3.3–5.5, N = 20), LW 68.0 μm^2, and L/W 4.25, with nuclei 3.8 μm (2.2–5.5) in diameter. The three forms differed somewhat in staining reaction, but all are consistent in dimensions with the type description. Nuclei could be narrow and band-like, extending the full width of the gamont, or broad and saddle-like. Nuclei sometimes extended from midbody well into the second quarter of the gamont. Recurving of the gamont was not recorded in the gamonts from *S. obesus*.

Sporogony J. E. Lewis and Wagner (1964) provided very few details of sporogony in the mite *Hirstiella* sp., except that "Mature cyst contains many sporocysts. Sporocyst averages 12 μ, contains 10–12 sporozoites."

Merogony According to Lewis and Wagner (1964), meronts develop "only in liver. Mature meront rounded or elliptical with about 24 nuclei. ...Mean size of cyst 26 × 30 μ." Mitchell (1965) found macromeronts of *H. sauromali* in *S. obesus* that contained 2–6 merozoites, 10 μm in length, and micromeronts with 24 merozoites. Telford (1966) reported hepatic meronts in *S. obesus* as "similar to those described by Lewis and Wagner (1964)." Both authors described ovoid micromeronts with micromerozoites arranged along one side of the meront, often fan-like. Apparent macromeronts in *S. obesus* were 11–20 × 7–16 μm and contained seven to usually ten macromerozoites.

Effects on Host Infected erythrocytes are hypertrophied, with nuclei usually displaced laterally. In *S. obesus*, the size of infected erythrocytes averages 26% greater than uninfected cells. Infected cells are always distorted.

Remarks J. E. Lewis and Wagner (1964) crushed infected mites and attempted to infect lizards of four genera by intraperitoneal, intramuscular, and oral routes. The lizards used were species of *Uta, Sceloporus,* and *Phrynosoma* (Phrynosomatidae) and the anguid *Gerrhonotus*. Intraperitoneal and intramuscular routes were unsuccessful. Only *Sceloporus* species became infected, 30 days PF, with 4 of 23 lizards fed mites patent for small, intraerythrocytic gamonts.

NEOTROPICAL *HEPATOZOON* SPECIES

Hepatozoon tupinambis (Laveran and Salimbeni) 1909 (**Plate 55**)

Diagnosis A *Hepatozoon* species characterized by gamonts 14.5–19.0 × 5.0–8.0 μm, not recurved, and enclosed within a usually clear capsule. Oocysts are ovoid, approximately 128 × 105 μm, and contain many ovoid sporocysts about 18 × 17–25 × 18 μm that produce 7–30 sporozoites. Merogony occurs in the liver and lung of the saurian host and sporogony in the hemocoel of mosquitos. Parasitized erythrocytes and their nuclei are greatly hypertrophied, with nuclei often divided into two or more portions.

Type Host *Tupinambis teguixin* (Linnaeus) (Sauria: Teiidae).

Type Locality Ilha do Governador, Guanabara State, Brazil.

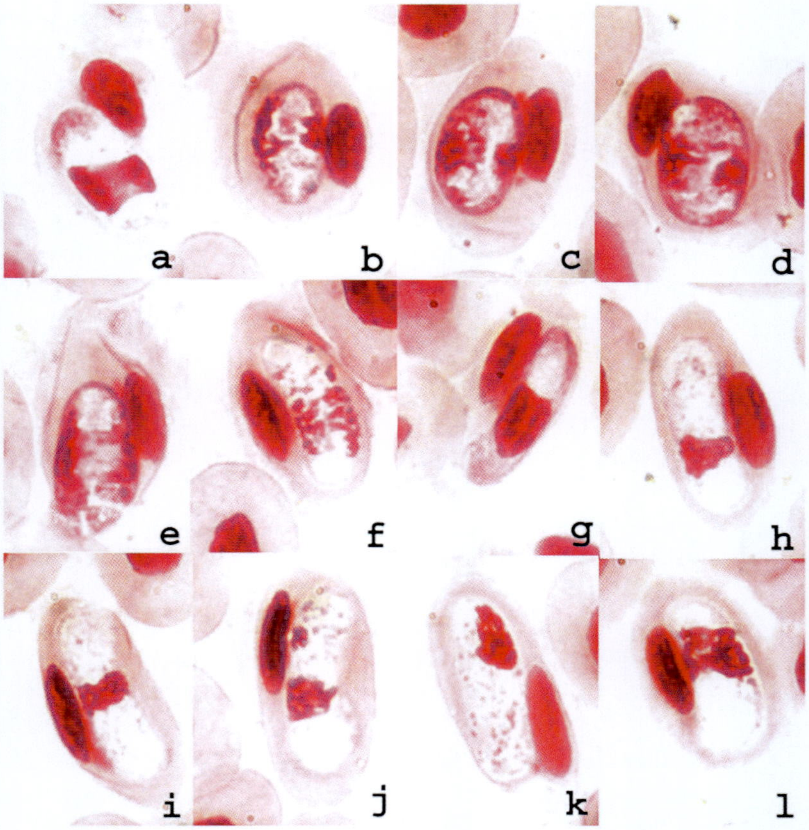

(A)

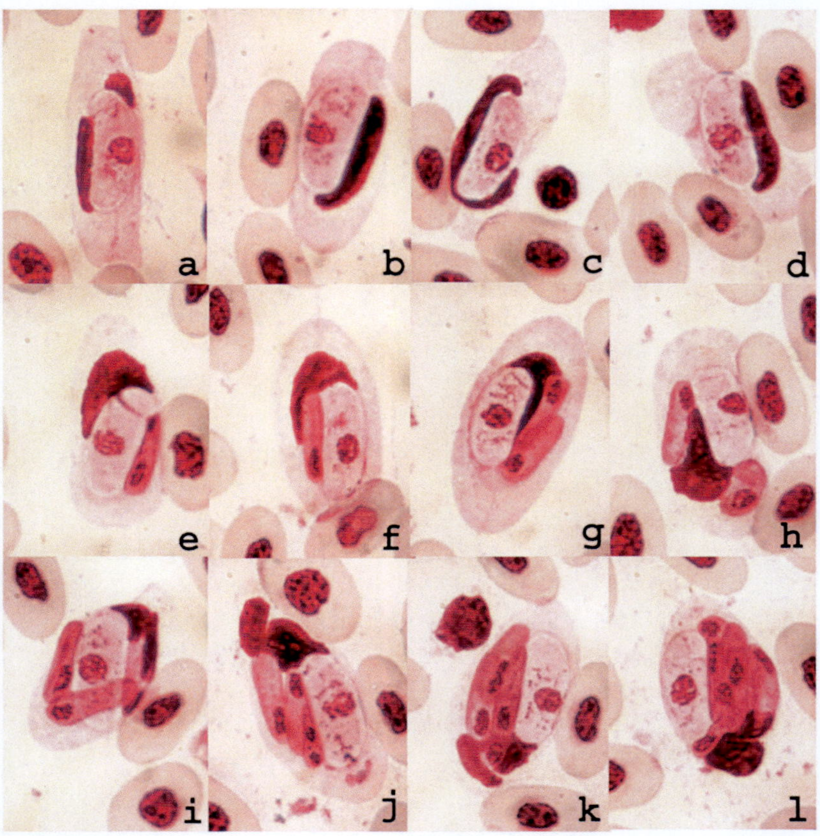

(B)

Plate 55 (**A**) *Hepatozoon sauromali* from *Sauromalus obesus,* California. Young gamonts, **a–e**; mature gamonts, **f–l**. (**B**) *Hepatozoon tupinambis* from *Tupinambis teguixin,* Venezuela. Mature single gamonts, **a–d**; mature gamonts with associated accessory gamonts, **e–l**.

Other Hosts None known.

Other Localities In Brazil, Ilha Solteira, São Paulo State, and Fortaleza, Ceara State (Pessôa et al., 1974b). In Venezuela, Cojedes State, Municipio Cojedes, and Portuguesa State, Municipio Piritu (Telford). Also in Colombia, no precise locality (Telford).

Prevalence *Hepatozoon tupinambis* infected three of ten (30%) *T. teguixin* from Ilha Solteira and three of seven from Fortaleza (Pessôa et al., 1974b). In Venezuela, prevalence was 12 of 26 (46.2%) at Finca la Coromoto, Municipio Cojedes, and 6 of 21 (28.6%) at San Jorge, Municipio Piritu (Telford).

Morphological Variation The gamonts of *H. tupinambis* were described by Laveran and Salimbeni (1909) as cylindrical, with rounded ends, and dimensions 14–16 × 5–6 μm. Laveran (1909) commented, "Au dernier stade de développement, une des extrémites plus effilee que l'autre se replie sur le corps," which indicates that the posterior end is recurved against the body of the gamont. This recurved portion was not mentioned by Pessôa et al. (1974b) and was not seen in any of the hundreds of gamonts I observed in Venezuelan tegu lizards. Pessôa et al. did not add to the description of *H. tupinambis*, but did provide additional description of its characteristic effects on host cells (as described in the Effects on Host section). In *T. teguixin* from Venezuela, gamonts are 16.0 ± 1.0 × 6.3 ± 0.7 μm (14.5–19.0 × 5.0–8.0, N = 25), with LW 101.7 ± 15.0 μm^2 (75–133) and L/W ratio 2.50 ± 0.26 (2.00–3.00). Both ends of the gamonts are broadly rounded with no indication of a recurved posterior portion. The nucleus is 3.9 ± 0.9 × 4.1 ± 0.9 μm (3.0–7.5 × 3.0–6.0), LW 16.1 ± 6.9 μm^2 (9.0–45.0), and is usually situated at midbody (56%) or within the second quarter of the gamont (40%), rarely extending into the first quarter. Gamonts are contained within a clear, rarely opaque, capsule within the host erythrocyte.

In the Venezuelan infections, the gamont is usually accompanied by smaller "accessory" gamonts within the erythrocyte, 12.3 ± 0.6 × 3.6 ± 0.7 μm (11–13 × 2.5–5.0, N = 10), that have a prominent nucleus, 4.6 ± 1.3 × 2.1 ± 0.4 μm (3.0–8.0 × 1.5–3.0), located at one end, which is presumably the first quarter of the gamont. The accessory gamonts number one to seven, in addition to the large gamont, with their frequency in a sample of 101 parasitized erythrocytes as follows: 1 in 37, 2 in 31, 3 in 16, 4 in 10, 5 in 4, 6 in 2, and 7 in 1. The cytoplasm of the accessory gamonts stains deep red with Giemsa stain, in contrast to that of the large gamont, in which only the nucleus is stained red. In one of these gamonts, of a total of five in one parasitized cell, there were four distinct "micronuclei" that stained dark blue. Neither Laveran and Salimbeni (1909) nor Pessôa et al. (1974b) reported the presence of these accessory gamonts.

Sporogony Pessôa et al. (1974b) obtained sporogony of *H. tupinambis* in the hemocoel of the mosquito *Culex fatigans* but provided an inadequate description of the sporogonic stages. Oocysts appear to be ovoid, with a calculated size (from a figure) of about 128 × 105 μm. The single illustrated oocyst contained many ovoid sporocysts with estimated size of 18 × 17–25 × 18 μm, that produce 7–30 sporozoites. Attempts to infect leeches (*Haementeria gracilis*), ticks (*Amblyomma agammum*), or triatomid bugs (*Rhodnius prolixus* and *Panstrongylus megistus*) were unsuccessful.

Merogony Pessôa et al. (1974b) provided little information on the meronts that were found in liver and lung, and their photographs are poor. In a subsequent article, Pessôa et al. (1974c) gave a clear photograph of a dizoic cyst in the liver of an experimental snake host.

Effects on Host Perhaps the most obvious characteristic of *H. tupinambis* is its effect on infected erythrocytes. These are always distorted and greatly hypertrophied to about three times (LW average 431.2 ± 60.7 μm^2, N = 25) the size of uninfected erythrocytes (LW 144.4 ± 14.9 μm^2, N = 10). The cells have a thin, dehemoglobinized cytoplasm extending outward from the parasite and nucleus that stains light pink with Giemsa. Nuclei of infected erythrocytes are usually about twice the normal length (16.1 ± 3.0 μm versus 6.5 ± 0.9 μm) but are nearly the same width, 3.5 ± 0.6 μm versus 3.8 ± 0.5 μm, respectively, and are often oddly distorted, even lysed, and are commonly broken into two or more fragments. Both previous descriptions (Laveran and Salimbeni, 1909; Pessôa et al., 1974b) reported similar effects on host erythrocytes.

Remarks Pessôa et al. (1974b), using *Culex fatigans* infected 24 days earlier with *H. tupinambis* from a *T. teguixin*, inoculated laboratory-born rattlesnakes *Crotalus durissus terrificus* per os with triturated mosquitoes. Both juvenile snakes had patent infections 42 days postinfection. Initially, infected erythrocytes showed some hypertrophy and distortion, but with time the effect on host cells diminished. Necropsy of one experimental snake at 100 days revealed some merogonic cysts in the liver.

The presence of the accessory gamonts in erythrocytes with a normal gamont of *H. tupinambis* is unique. Initially thought to represent an erythrocytic merogony, it appears more likely that the smaller, slender, deeply stained parasites with nearly terminal nuclei represent microgamonts that have entered erythrocytes already host to a gamont of normal size and appearance. This identification as microgamonts is supported by the presence of four distinct micronuclei seen in one microgamont. Perhaps this accumulation of microgamonts, if that is their identity, may be related to increasing the probability of szygy with the normal gamont following the blood meal of a suitable vector.

Hepatozoon carinacauda Pessôa and Cavalheiro 1969

Diagnosis A *Hepatozoon* species with gamonts 23–25 × 5–6 μm and oocysts 200–250 μm in diameter that contain 140–150 sporocysts about 24–28 × 20–26 μm, which produce 10–40 sporozoites. Micromeronts may exceed 70 μm in diameter and contain 150–160 micromerozoites. Sporogony can occur in the leech *Haementeria lutzi*.

Type Host *Helicops carinicaudus* (Wied) (Serpentes: Colubridae).

Type Locality Votoporanga, São Paulo State, Brazil.

Other Hosts None known.

Other Localities None known.

Prevalence Pessôa et al. (1974a) reported two of two *Helicops carinicaudus* were infected by *Hepatozoon carinacauda*.

Morphological Variation The description of *H. carinacauda* by Pessôa and Cavalheiro (1969c) is inadequate. Gamonts measure 23–25 × 5–6 μm, which indicates an LW of 115–150 μm^2 and L/W ratio of 4.17–4.60. Nuclei are rounded, 3–5 μm in diameter, and are located in the second quarter of the gamont. Enlarged, broad gamonts were found in the blood of the snake host a month after initial examination. These were 29–30 × 12–13 μm, with estimated LW values of 348–390 μm^2 and L/W 2.31–2.42. Dimensions of these larger gamonts, calculated from the figures, show gamonts 25–26 × 10–13 μm and LW values considerably smaller, 250–338 μm^2, with L/W ratios 2.00–2.50. The gamont nuclei also lie within the second quarter and are similar in dimensions, 2.8–4.1 × 5.9–8.3 μm, to those of the smaller gamonts shown, 3.5–4.1 × 4.5–6.2 μm. With so few gamonts of either type available for comparisons, the calculations from the figures are nevertheless reasonably consistent with the stated dimensions, and the larger gamonts, occupying erythrocytes with pycnotic nuclei, could be senile or at least of greater age than the smaller gamonts.

Sporogony Sporulated oocysts were observed in the coelom of the leech *Haementeria lutzi* fed 36–40 days earlier on infected *Helicops carinacaudus.* Oocysts were 200–250 μm in diameter and contained about 140–150 sporocysts. Sporocyst dimensions were not stated by Pessôa and Cavalheiro, but calculations of length and width from illustrations suggest a size of 24.1–27.5 × 20.3–25.9 μm, with L/W ratios 1.06–1.25, indicating subspherical to ovoid shapes. Apparently, 10–40 sporozoites were produced by sporocysts.

Merogony Pessôa and Cavalheiro (1969c) do not describe meronts containing macromerozoites of *Hepatozoon carinacauda*, but meronts up to 70 μm in diameter contain 150–160 micromerozoites. Merogony appears to occur most commonly in the lung, but meronts were found also in the liver and intestine.

Effects on Host Erythrocytes host to gamonts are hypertrophied greatly by the larger gamonts, with nuclei displaced laterally or to one end of the cell. Nuclei were described as pycnotic.

Remarks The descriptions of gamonts, meronts, and sporogonic stages, and the illustrations provided by Pessôa and Cavalheiro (1969c) are barely adequate to justify recognition of the species *H. carinacauda.* It is included here because it is clearly a *Hepatozoon* species that can undergo sporogony in a leech, which could indicate the vector group responsible for transmission in nature, given that *Helicops carinicaudus* is an aquatic snake.

Hepatozoon plimmeri (Sambon) 1909

Diagnosis A *Hepatozoon* species with distinctly encapsulated gamonts 10–15 × 2.5–4.5 μm, the posterior ends recurved for about 4 μm. Nearly spherical oocysts 120–150 μm in diameter contain 20–150 broadly ovoid-to-spherical sporocysts of 25–30 μm that produce 12–50 sporozoites. Sporogony occurs in the hemocoel of culicine mosquitoes. Micromeronts develop in tissues of the lung and liver.

Type Host *Bothrops jararaca* (Wied) (syn. *Lachesis lanceolatus* Lacépède) (Serpentes: Viperidae).

Type Locality Brazil, no precise locality.

Other Hosts *Bothrops moojeni* (Pessôa et al., 1971).

Other Localities São Paulo State, Brazil.

Prevalence In *B. jararaca*, 3 of 13 (23.1%) from São Paulo State (Pessôa, 1967) were infected by *H. plimmeri*. Overall prevalence in this host was reported as 31 of 131 (23.7%) and in *B. moojeni* as 13 of 164 (7.9%) by Pessôa et al. (1974a).

Morphological Variation The gamonts of *Haemogregarina plimmeri* were described by Sambon (1909) as 14–15 × 4–4.5 μm, with a recurved posterior end of about 4 μm, enclosed within "thick, sausage-shaped capsules." Gamont nuclei were "more or less median" in position, 4 × 3.5–4 μm. Sambon's figure shows nuclei mostly occupying the second quarter of the gamont, entering the first and the third quarters as well. Pessôa (1967) reported gamonts of *H. plimmeri* from *B. jararaca* to be 10–14 × 2.5–3.0 μm, with a slender capsule and showing a recurved posterior end. Gamont nuclei in Pessôa's figures are second quarter, extending into the first and perhaps the third, as shown by Sambon. No dimensions were given of gamonts specifically from *B. moojeni* (Pessôa et al., 1971a), but the figures show gamonts with dimensions calculated from the scale bar ranging from 7.1 to 16.8 × 3.6 to 6.4 μm. Disregarding the two smaller gamonts in Figure 3, which appear to be immature, the two larger gamonts figured by Pessôa et al. (1971a) are 11.8–16.8 × 5.4–6.4 μm. These are more in agreement with those reported from *B. jararaca*.

Sporogony Sporogony was obtained in the mosquito *Culex dolosus* (Pessôa et al., 1971a). Early oocysts appeared to be inserted into the gut wall. Later, oocysts were found in the hemocoel. Oocysts were nearly spherical, 120–150 μm in diameter, and contained 120–150 ovoid-to-spherical sporocysts 25–30 μm in diameter, which produced 12–50 sporozoites, 15–16 × 3.5–4.0 μm in size.

Merogony Meronts formed in tissues of the lung and liver in *B. jararaca* (Pessôa, 1967). Only "macrocistos" (= micromeronts) containing micromerozoites were observed, with little description provided.

Effects on Host Sambon (1909) reported that nuclei of infected erythrocytes were displaced. Most of the figures of Pessôa (1967) and Pessôa et al. (1971) show little apparent hypertrophy of the host cells, but lateral displacement of their nuclei usually resulted from the gamont assuming a central position in the cell, closely associated with the erythrocyte nucleus.

Remarks Pessôa (1967) followed Hoge (1965) in identifying *Lachesis lanceolatus* of Lacépède (1789) and Sambon (1909) as *Bothrops jararaca* rather than as *Bothrops atrox* as other authors have done. It is not absolutely certain that the hemogregarines of *B. jararaca* (Pessôa, 1967) and *B. moojeni* (Pessôa et al., 1971) are conspecific, and this identity needs to be confirmed by DNA analysis. It is impossible to resolve the identity question from the characteristically poor descriptions and illustrations of Pessôa, but here he is given the benefit of the doubt. De Biasi et al. (1972) reported congenital transmission of hemogregarines from *Bothrops moojeni*, which presumably are *Hepatozoon plimmeri*.

Hepatozoon strigatus (Pessôa) 1967

Diagnosis A *Hepatozoon* species characterized by thick gamonts 14–18 × 6–8 μm with nuclei that occupy the second quarter and extend into the first quarter of the gamont. Oocysts are 120–150 μm in diameter and contain 20–50 spherical sporocysts 30–40 μm in diameter that produce 30–40 sporozoites. Micromeronts develop in tissues of the liver and lung. Host erythrocytes are hypertrophied with pale, thin cytoplasm toward the periphery and elongated, laterally displaced nuclei. Sporogony occurs in culicine mosquitoes.

Type Host *Thamnodynastes strigatus* (Günther) (Serpentes: Colubridae).

Type Locality São Paulo State, Brazil.

Other Hosts None known.

Other Localities None known.

Prevalence One of nine *T. strigatus* from São Paulo State was infected by *Haemogregarina strigatus* (Pessôa, 1967). Pessôa et al. (1974a) reported overall prevalence as 45 of 192 (23.4%).

Morphological Variation Gamonts of *H. strigatus* were 14–18 × 6–8 μm, with calculated nucleus dimensions 5.3–6.2 × 3.8–4.3 μm. Although not mentioned in the original description by Pessôa (1967), one of the figures of *H. strigatus* in Pessôa et al. (1974a) shows a prominent recurved posterior portion of a gamont. Nuclei occupy most of the second and enter into the first quarter of the gamont.

Sporogony Oocysts were 120–150 μm in diameter (Pessôa et al., 1970) and contained 20–50 spherical sporocysts. Sporocysts measured 30–40 μm in diameter and produced 30–40 sporozoites 10–12 μm in length. Sporogony was obtained in two *Culex* species, *C. fatigans* and *C. dolosus*, with oocysts developing in the hemocoel.

Merogony Micromeronts were found in tissues of the liver and lung, with the hepatic meronts larger than those in the lung, up to 50 μm in diameter, and contained hundreds of nuclei and micromerozoites.

Effects on Host Infected erythrocytes were enlarged, rounded or irregular in shape, with nuclei elongated and laterally displaced. The peripheral cytoplasm was thin and pale in comparison to the cytoplasm immediately surrounding the gamont and erythrocyte nucleus.

Remarks Infection could not be transmitted by bite of infected mosquitoes or by injection of a large number of oocysts and sporozoites into snakes. However, *H. strigatus* appeared in the blood of snakes about 20 days after feeding them infected mosquitoes.

Hepatozoon miliaris (Pessôa) 1968

Pessôa (1968) described erythrocytic gamonts from *Liophis miliaris* collected in São Paulo State, Brazil, as *Haemogregarina miliaris*. Gamonts were 13–14 × 3.0–3.5 μm, with a nucleus 5–6 μm in length, situated anteriorly and occupying most of the first and second quarters of the gamont. Using the scale provided, the two gamonts illustrated are 14.5 × 4.5 μm and 12.7 × 4.1 μm, with calculated LW 65.3 μm^2 and 52.1 μm^2, and L/W ratios 3.22 and 3.10, respectively. Their nuclei had dimensions of 5.5 × 2.7 μm and 5.5 × 3.2 μm, with calculated LW 14.9 μm^2 and 17.6 μm^2, respectively. The dimensions calculated from the figures are consistent with the stated dimensions. The host erythrocytes appear to be normal in size with only slight nuclear displacement. Later, Pessôa and Cavalheiro (1969b) described sporogony of a hemogregarine he identified as *H. miliaris* from *Liophis miliaris* collected in the city of São Paulo. It is not clear that these were the hosts used for the description of *H. miliaris*, but the figures show distinctly different hemogregarine gamonts. Dimensions calculated from the figures indicate that the two gamonts were 20.7 × 5.6 μm and 18.5 × 5.9 μm, with LW values 115.9 μm^2 and 109.2 μm^2 and L/W ratios 3.36 and 3.14, respectively. Nuclei were narrow bands occupying the posterior portion of the second quarter to midbody of the gamonts, with lengths of 2.6 μm and 3.7 μm and widths the same as those of the gamonts. Their respective LW values were 14.6 μm^2 and 21.8 μm^2. Clearly, the gamonts figured in the 1968 and 1969 articles belonged to different species in my opinion. Pessôa and Cavalheiro (1969b) recognized the differences in the gamonts but attributed them to differences in infection phase. While this is possible, perhaps, with gamont size, it is very unlikely that gamont nucleus morphology and position would differ so greatly within the same *Hepatozoon* species. Accordingly, while Pessôa and Cavalheiro (1969b) demonstrated sporogony of a *Hepatozoon* species, it is not at all certain that the sporogonic development is that of *H. miliaris*.

Sporogony of one of the two *Hepatozoon* species from *Liophis miliaris* occurred in the leech *Haementeria lutzi*. Gamonts apparently in szygy were present in the "estomago" (intestinal ceca?) of the leeches dissected 24 hours PF. Microgametes were found 5 days PF, and between 5 and 10 days probable zygotes were present, still within the "stomach." After that, oocysts were present in the coelom and blood sinuses. At 25 days to 1 month, small-to-very large oocysts 200 or more μm in diameter (?) were observed. These contained ovoid-to-spherical sporocysts about 26 μm in diameter. Sporozoite numbers within sporocysts varied greatly, 7–9 to about 40.

"*Haemogregarina pallida*" Pessôa, Sacchetta, and Cavalheiro 1971

Haemogregarina pallida was described by Pessôa et al. (1971b) as the first species of the genus *Haemogregarina* to be found in snakes, in *Thamnodynastes pallidus*. However, the results obtained by feeding *Culex fatigans* and *C. dolosus* on the infected snake produced only multisporocystic oocysts typical of *Hepatozoon* species. The illustrations of infected erythrocytes show rather clearly the presence of erythrocytic meronts, but by itself, this is inadequate evidence of generic identity. Oocysts of *Hepatozoon* that were described were 90–110 μm in diameter and contained 40–50 sporocysts 10–20 μm in diameter that produced 30–50 sporozoites. Oocysts were found on the gut wall of the mosquitoes, and some sporozoites were free in the hemocoel. Eight leeches, *Haementeria lutzi*, fed on the infected snake were examined up to 50 days PF but remained negative. Pessôa et al. did not distinguish clearly between gamonts and erythrocytic premeronts or meronts. Despite providing minimal data on sporogonic stages, the remainder of the description is too confusing and inadequate to recognize the *Hepatozoon* species present. Pessôa et al. (1974a) reported the examination of 30 *Thamnodynastes pallidus*, 8 infected by *Hepatozoon* and 1, the type infection, by "*Haemogregarina pallida*." The identity of the parasite with erythrocytic meronts remains unknown, but the evidence suggests the presence of a mixed infection in the host of "*Haemogregarina pallida*." The species is discussed here only because of the unique finding of erythrocytic meronts of a hemogregarine in a snake host.

Hepatozoon caimani (Carini) 1909, Lainson, Paperna and Naiff 2003

Diagnosis A species of *Hepatozoon* with elongate, slender gamonts 16.2 × 2.5–25 × 4 μm, recurved within a capsule at dimensions of 10 × 3.8–13.8 × 3.8 μm, with nuclei at midbody that may extend into the first quarter of the gamont. In experimental mosquito hosts, sporogony occurs on the midgut wall beneath the elastic membrane of the hemocoel surface. Oocysts are nearly spherical, up to 260 μm in diameter, and contain an estimated 80–100 spherical sporocysts 20–30 μm in diameter that produce 12–24 sporozoites. Meronts of a single type develop in the lamina propria of the small intestine, are 13 × 9.6–20.7 × 18 μm, and produce 6–10 crescent-shaped merozoites. Monozoic and dizoic cysts occur primarily in the liver but other organs as well in naturally infected caimans, and monozoic, dizoic, and hexazoic cysts appear in anurans following ingestion of infected mosquitoes. The caiman final vertebrate host can be infected either directly by ingestion of infected mosquitoes or indirectly from cyst-infected vertebrates.

Type Host *Caiman latirostris* Daudin (Crocodylia: Alligatoridae).

Type Locality Rio de Janeiro State, Brazil.

Other Hosts *Caiman c. crocodilus, Caiman c. yacare* (Lainson et al, 2003).

Other Localities Bragança, Para State and Mato Grosso State, Brazil (Lainson et al., 2003).

Prevalence Lainson et al. (2003) reported a prevalence of *H. caimani* as 46 of 60 (77%) in young *Caiman c. crocodilus* of Bragança, Para State, Brazil.

Morphological Variation Gamonts of *H. caimani* (Lainson et al., 2003) are encapsulated and recurved, 20.7 × 3.0 μm (16.2 × 2.5–25.0 × 4.0, N = 25) when extracellular and 12.2 × 4.3 μm (10.0 × 3.8–13.8 × 3.8, N = 50) when within the capsule. Estimated LW and L/W of free gamonts are 62.1 μm^2 and 6.90, respectively. The capsules respond variably to stain. Nucleus dimensions were not stated, but figures show broad-to-narrow nuclei that extend from midbody well into the second quarter of the gamont and some into the first quarter. Some nuclei may appear as a "widely dispersed reticulum."

Sporogony *Culex dolosus* (Pessôa et al., 1972) and *Culex fatigans* (Lainson et al., 2003) have been shown to support the complete sporogony of *H. caimani*. In *C. fatigans*, mature oocysts were present in large numbers at 21 days PF and were located on the hemocoel side of the mosquito midgut, enclosed by the "elevated elastic membrane of the midgut surface" (Lainson et al., 2003). Oocysts are nearly spherical, up to 260 μm in diameter, and contain "an estimated 80–100 spherical sporocysts," 20–30 μm in diameter. Sporocysts contain an estimated 12–24 crescentic sporozoites, 19–22 × 4–5 μm (N = 25), and bud off from a "conspicuous residual body."

Merogony In eight naturally infected caimans, no meronts were found in liver, lungs, spleen, or kidney. Monozoic and dizoic cysts were present in these tissues, with dimensions of 14 × 10 μm (12.5 × 6.25–21.5 × 21, N = 25), and they contained zoites that were 12.5 × 3.7 μm (10.0 × 1.25–15.0 × 4.0). Infected mosquitoes were fed to the anurans *Leptodactylus fusca* and *Rana catesbeiana,* in which monozoic, dizoic, and hexazoic cysts were present in abundance in the liver 28 days PI in *L. fusca* and 23 days PI in *R. catesbeiana*. Some cysts were found in the lungs and spleen, and the cells host to cysts "appeared to be ... reticuloendothelial cells." Cysts in the anurans were "indistinguishable" from those found in naturally infected caimans and were ovoid to spherical in shape, 15.0 × 10.0 μm (14.5 × 12.0–21.0 × 20.0, N = 25) in their dimensions. One or two, rarely four to six, zoites were present in the cysts. Erythrocytic parasites did not appear in the anuran hosts.

Infected mosquitoes fed to caimans produced patent gamont infections in peripheral blood at 82 days PI. Monozoic and dizoic cysts were common in their livers, rare in the other organs, but no meronts were found. Caimans fed infected anurans and examined at 13–14 days PI had abundant developing and mature meronts present in the lamina propria of the small intestine only, but there was no patent infection of erythrocytes. Meronts in tissue smears were 15.8 × 13.0 μm (13.0 × 9.6–20.7 × 18.0, N = 25) and in tissue sections were 16.5 × 12.2 μm (14.0 × 14.0–22.2 × 11.8, N = 25). Meronts produced 6–10 crescentic merozoites about 11.2 × 2.0 μm (9.6 × 2.2–16.0 × 2.2) and perhaps more in some meronts. Meronts apparently were monomorphic. At 52 days PI, erythrocytes were negative for gamonts, but at 79 days gamonts were present.

Effects on Host Erythrocytes are seldom enlarged by presence of a single gamont, but multiply infected cells are hypertrophied and distorted. The nuclei of infected erythrocytes are displaced laterally, but appear in figures to be neither enlarged nor distorted.

Remarks Caimans may become infected directly from ingestion of infected hematophagous arthropods or by feeding on other vertebrates (including smaller caimans probably) that have ingested infected arthropods and become hosts to monozoic, dizoic, or hexazoic cysts in their viscera. Lainson et al. (2003) believed that the latter mode of transmission may be the predominant mode for *H. caimani*. Both modes of infection produce patent erythrocytic infections by gamonts, and this appears to be unique among *Hepatozoon* species thus far studied. Other distinctive traits found in *H. caimani* are the restriction of merogony to the lamina propria of the small intestine and apparent formation of a single type of meront. The presence of macro- and micromeronts is typical in other reptilian *Hepatozoon* species where merogony is known.

ASIAN *HEPATOZOON* SPECIES

Hepatozoon mesnili Robin, 1936 (**Plate 56**)

Diagnosis A *Hepatozoon* species characterized by falciform gamonts 15–17 × 4–7 µm with the anterior third of the gamont recurved on the median third, in which the nucleus is situated. Gamonts may be extended within host cells, with dimensions 26–33 × 4.5–7.5 µm. Gamont nuclei show prominent spaces between blocks of chromatin, giving an appearance of being vacuolated. Oocysts are spherical, 150–450 µm, and contain a highly variable number of ovoid sporocysts that average 33 × 18.5 µm and produce 12–16 sporozoites. Meronts occur in the intestinal wall, liver, and especially the lung, apparently forming only seldom more than six merozoites. Sporogony occurs in the abdomen, thorax, and head of capable mosquito hosts.

Type Host *Gekko verticillatus* Laurenti (= *Gekko gecko* Linnaeus) (Sauria: Gekkonidae).

Other Hosts None known.

Type Locality Vicinity of Saigon, Vietnam.

Other Localities Southern Thailand (Telford).

Prevalence Twenty-three of 50 (46%) in *G. gecko* from the type locality (Robin, 1936a) and 6 of 16 (37.5%) in southern Thailand.

Morphological Variation In the Vietnamese type material, gamonts occurred in two forms within erythrocytes: falciform, in which the anterior third of the gamont were recurved, usually on the middle and posterior thirds, and oval. The falciform gamonts sometimes did not show a bent portion. Gamont length, without including the recurved portion, was 15–16 × 6–7 µm anteriorly, 4–5 µm width posteriorly, and gamonts free from their host cells were elongate, 25–30 × 3–4 µm (Robin, 1936a). Nuclei, often vacuolated, were 6–7 µm and occupied the width of the gamont in the middle third, with the anterior portion narrower than the posterior end. Chromatin was arranged in transverse bands or in masses along the periphery of the nucleus, and nuclei appeared to be vacuolated. Nuclei were located in the middle third of the elongate, free gamonts. Oval gamonts with a nearly central nucleus were 14 × 6 µm and were always found in infected geckoes in variable proportions to the falciform type. Robin (1936a) thought the oval gamonts might be degenerating forms.

Mature gamonts in the Thai infections were not falciform but were extended within the erythrocyte, stretching it longitudinally and narrowing its width. One end commonly showed a slight bend, particularly in smaller gamonts. Dimensions of gamonts in an active infection were 29.6 ± 2.3 × 6.1 ± 0.9 µm (26–33 × 4.5–7.5, N = 23), with LW 181.4 ± 29.3 µm^2 (121.5–240) and L/W 4.20 ± 0.80 (3.60–6.40). Nuclei occupied the second third of the gamont, extending into the posterior third commonly, and occasionally into the broader anterior third of the gamont. Nucleus dimensions were 15.7 ± 1.9 × 6.0 ± 0.8 µm (12–20 × 4.5–7.5), with LW 94.7 ± 16.5 µm^2 (58.5–123.7). Nuclei occupied the entire width of the gamont. All nuclei seen had large, irregular areas free of chromatin (the vacuolation of Robin, 1936a) and frequently had strip-like bands of chromatin blocks along both sides of the nucleus. Gamonts occasionally were seen as shorter, broader forms with oval, rounded, or squarish nuclei that stained less intensely than in elongate gamonts, with fewer pale areas within the nuclei. Dimensions of two such gamonts were 21 × 11 µm and 27 × 8.5 µm.

Sporogony Oocysts are spherical, 170–450 µm in diameter, average about 250 µm, and contain sporocysts, the number of which varies according to oocyst size. Sporocysts, 33 × 18.5 µm, produce 12–16 sporozoites between days 10 and 15 PF. Fifteen days is minimum sporulation time; development in *Culex fatigans* requires approximately 3 weeks and in *Stegomyia albopicta* about 30 days.

Merogony Although Robin (1936a) described meronts, all measured 24–28 × 17 µm and produced 4–6 merozoites. His figures suggest that these were macromeronts and macromerozoites, and he may have missed the micromeronts, although his comment "Il est exceptionnel de trouver des figures de merogonie á plus de 6 schizozoites" suggests that he may have seen micromeronts without realizing their significance. The lung is apparently the primary site for merogony of *H. mesnili*.

Effects on Host Parasitized erythrocytes show hypertrophy in both length and width and displacement of the nucleus, but the nucleus is not distorted (Robin, 1936a). In Thai infections, erythrocytes host to gamonts averaged

Plate 56 (**A**) *Hepatozoon mesnili* from *Gekko gecko,* Thailand. Young gamonts, **a, b**; mature gamonts, **c–l**. (**B**) *Hepatozoon ayorgbor* from *Python regius,* Ghana. Probable immature gamonts, **a, b**; mature gamonts, **c–g**; free gamont, **h**.

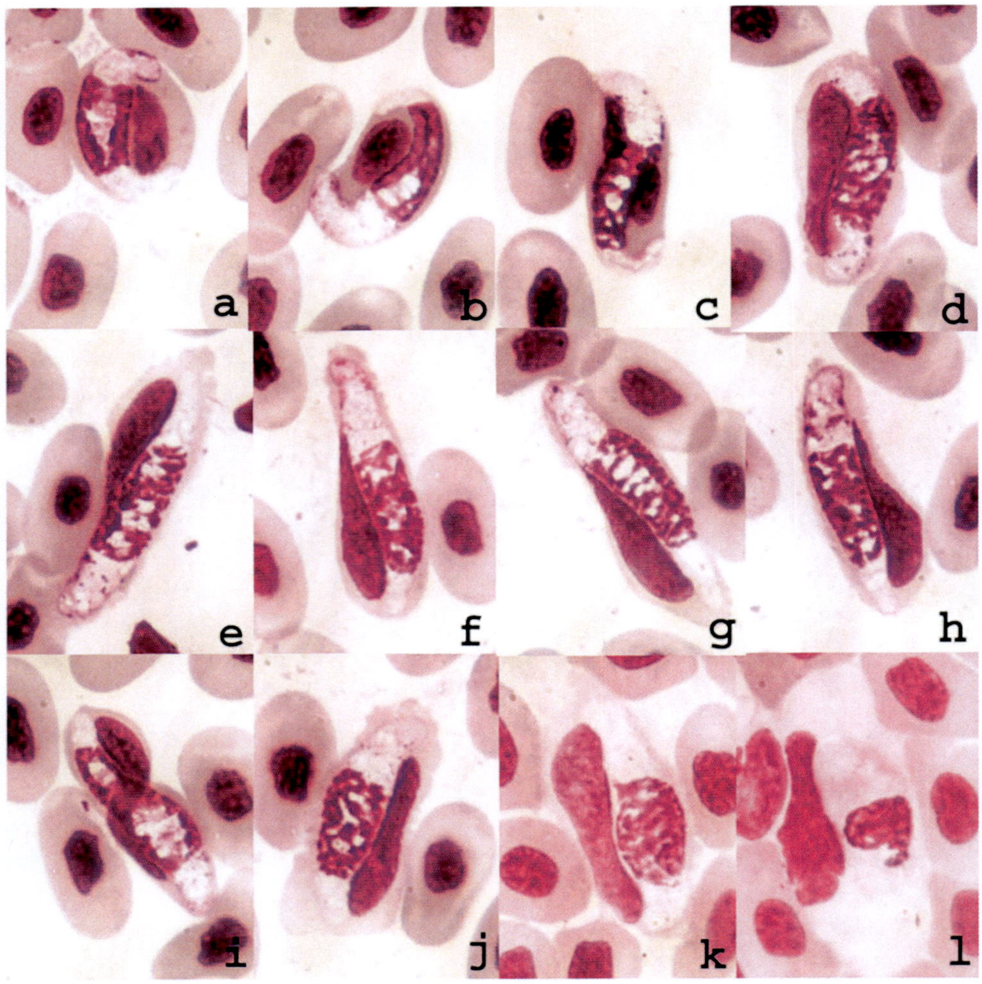

(A)

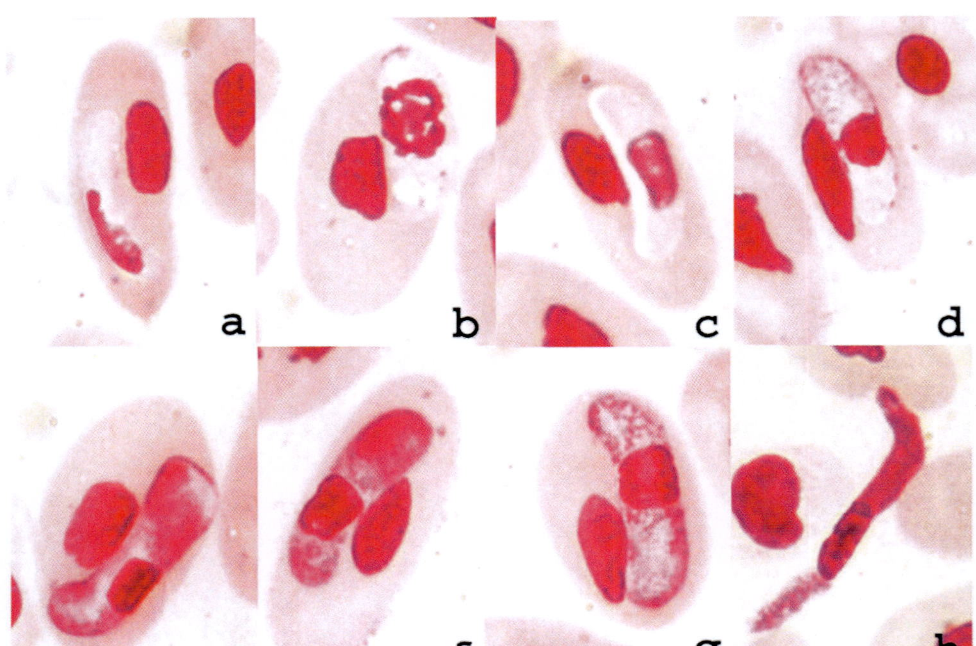

(B)

1.7 times larger than the size of uninfected cells, and nuclei of infected cells were 2.7 times larger on average. Erythrocytes host to the broader gamonts usually had paler cytoplasm, and their margins appeared to be undergoing lysis. All host cells were distorted, elongated, and narrowed, in the case of extended gametocytes, and their nuclei were always distorted, usually elongated and enlarged, displaced laterally by the gamonts.

Remarks Robin (1936a, 1936b) described the fusion of gamonts in pairs, production of microgametes, fertilization of the macrogamete, then formation of a zygote, all occurring within the lumen of the mosquito intestine. The zygote then passes through the intestinal wall into the hemocoel. Attempts to infect geckoes by bite were unsuccessful, but 22 days after infection per os, all five experimental geckoes showed patent infection (Robin, 1936b).

Gametogenesis and zygote formation by *H. mesnili* within the mosquito intestinal lumen as described by Robin (1936a, 1936b) is a different pattern from that found in *H. rarefaciens*, in which gametogenesis and zygote formation occur in the hemocoel (Ball and Oda, 1971), a difference of possible systematic significance.

The only apparent difference in gamont morphology between the Vietnamese and Thai infections is that of falciform gametocytes within erythrocytes, considered characteristic by Robin (1936a, 1936b), but which were rarely seen in the Thai material, except in smaller (immature?) gamonts. Robin's elongate, free gamonts were nearly identical in dimensions to the erythrocytic mature gamonts of Thai infections. It is possible that smears were not made immediately from blood obtained from geckoes by Robin. The presumably delicate, hypertrophied host cells could well have ruptured before or during the smearing process, resulting in the presence of free elongate gamonts on the blood films. The appearance of chromatin in gamont nuclei appears to be very similar, with prominent open spaces between chromatin blocks and their common alignment along the margins of the nuclei or in transverse bands. This is distinctive for *H. mesnili* among described species of *Hepatozoon*.

Hepatozoon octosporei Shanavas and Ramachandran 1990

Diagnosis A *Hepatozoon* species with encapsulated and recurved gamonts 17.3–19.5 × 3.8–4.9 µm, with nuclei 6.0–7.5 × 3.8–4.5 µm, situated at midbody and extending into the second quarter of the gamont. Sporogony occurs in the hemocoele of *Ophionyssus* sp. mites, forming spherical-to-ovoid mature oocysts 60–78 µm that contain eight elliptical-to-ovoid sporocysts 23.5–27.0 × 18.0–24.0 µm that produce 8–25 sporozoites. Merogony occurs in hepatic parenchymal cells. Macromeronts form two to six macromerozoites, and micromeronts produce 20–28 micromerozoites.

Type Host *Mabuya carinata* (Schneider) (Sauria: Scincidae).

Type Locality Piravom, Ernakulam District, Kerala State, India.

Other Hosts None known.

Other Localities None known.

Prevalence Shanavas and Ramachandran (1990) report the presence of *H. octosporei* in 28 *M. carinata* at the type locality, with no mention of the total number examined from there, and its absence from 62 skinks from other localities in Kerala State.

Morphological Variation Mature gamonts are encapsulated, with a recurved narrow end (posterior) and a slight curving of the broadened, rounded end. Dimensions from a "minimum" of 20 specimens apparently were taken for all mean values. Gamonts are 18.6 × 4.3 µm (17.3–19.5 × 3.8–4.9), with LW and L/W estimated at 80.0 µm^2 and 4.33, respectively. Gamonts newly emerged from host erythrocytes are 16.5–19.5 × 2.3–3.0 µm. Gamont nuclei are 6.9 × 4.2 µm (6.0–7.5 × 3.8–4.5) and are situated at midbody, extending into the second quarter of the gamont.

Sporogony Sporogony takes place in the hemocoel of *Ophionyssus* sp. mites, which occur on the skink hosts in nature. Zygotes appear 3–4 days PF with young oocysts present on day 5 PF, and sporoblast division is evident by day 8 PF. Sporozoites differentiate by day 13 or 14 PF, with mature oocysts present on day 14. Oocysts are spherical or ovoid, 63.8 µm (60–78) in diameter, and contain eight elliptical-to-ovoid sporocysts 32.0 × 21.5 µm (23.5–27.0 × 18.0–24.0) that produce 8–24 sporozoites, "a majority with 10–20," accompanied by a residual body. The sausage-shaped sporozoites are 21.0 × 4.6 µm (19.5–22.5 × 4.1–4.9).

Merogony Dimorphic meronts form in hepatic parenchymal cells. Macromeronts, spherical or ovoid, are 16.4 × 14.1 µm (15–18 × 11.3–18.0) and contain two to six sausage-shaped macromoerozoites, 15.5 × 3.7 µm (13.5–17.3 × 3.0–5.3). Micromeronts are also spherical to ovoid in shape and slightly larger, 20.8 × 18.4 µm (17.3–24.0 × 16.5–21.8), and produce 20–28 spindle-shaped micromerozoites, 9.0 × 1.7 µm (8.3–9.8 × 1.5–1.9).

Effects on Host Parasitemia varied from 0.4% to 16%, presumably in natural infections. Erythrocytes containing mature gametocytes are hypertrophied by about 59% in average LW. Dehemoglobinization is apparent along one side of host cells, and their nuclei are pycnotic, elongated, more slender, and "adherent to the capsule of the parasite" (Shanavas and Ramachandran, 1990). In experimental infections, four of seven skinks that ingested 20 or more infected mites died 2–5 days PI and the remainder at 25–28 days PI, prior to patency. Five skinks that ingested 12 or fewer infected mites survived the infection. Two to 3 days prior to death, the skinks stopped feeding and "became restless." Hepatic tissue was destroyed, and liver sinusoids were "dilated and congested with blood. Mononuclear leukocytes aggregated at the foci of infection. Extensive liver damage and/or parasite toxemia could be the cause of death" (Shanavas and Ramachandran, 1990).

Remarks Experimentally infected *M. carinata* showed patent blood infections 30–34 days PI, with one infected mite adequate to produce infection. Early meronts were seen by 5 days PI, and "fully formed macromerozoites" on day 9 PI, and were predominant on days 4 (inconsistent with day 9) and 17 PI. On day 21 PI, micromeronts with 10–18 nuclei were present, and on day 30 micromeronts contained 20–28 micromerozoites, "just after the appearance of the blood forms." Meronts of both types were present in the liver up to 42 days following onset of blood parasitemia.

This study provides valuable and rarely reported information on the course of laboratory-induced infection by the natural vector in the natural vertebrate host of a *Hepatozoon* species.

AUSTRALASIAN *HEPATOZOON* SPECIES

Hepatozoon breinlii Mackerras, 1961

Diagnosis A *Hepatozoon* species characterized by slender gamonts 14–16 × 1.5–2.5 µm, with a large, elongate nucleus situated near the midpoint of the gamont. Merogony is in the liver. Macromeronts are 22 × 19 µm and contain up to 6 merozoites, and micromeronts are of similar size and produce over 30 merozoites. Oocysts of varying size develop in the mosquito *Culex fatigans* and contain 4 to 50 or more ovoid-to-spherical sporocysts about 25 µm in diameter that produce 10–20 sporozoites.

Type Host *Varanus tristis orientalis* Fry (Sauria: Varanidae).

Other Hosts *Varanus v. varius, Varanus gouldii.*

Type Locality Innisfail, Queensland, Australia.

Other Localities Eidsvold, Queensland, Australia.

Prevalence *H. breinlii* parasitized four of four *V. tristis orientalis* taken at the type locality.

Morphological Variation Gamonts measured 14–16 × 1.5–2.5 µm within erythrocytes, and when free of host cell were 16 × 2 µm (14.5–18 × 1.5–2.5, N = 15).

Sporogony *H. breinlii* underwent sporogony in the hemocoele of *Culex fatigans* (Mackerras, 1962). Oocysts varied considerably in size and produced 4 to over 50 sporocysts, with smaller oocysts maturing as early as 14 days postinfection. Sporocysts were ovoid or spherical, about 25 µm in diameter, and contained 10–20 sporozoites 18–20 × 4.5 µm in size. Occasionally, two oocysts were observed sharing a single cyst wall.

Merogony Meronts were present in the liver of *V. tristis* killed on days 29 and 70 postinfection. Macromeronts (termed "Y-meronts" by Mackerras, 1962) were 22 × 19 µm and contained up to six macromerozoites. Micromeronts ("X-type meronts") were of similar size to macromeronts but produced over 30 micromeronts.

Effects on Host The slender gamonts of *H. breinlii* caused only slight enlargement of host erythrocytes: Normal cells were 14–17 × 9–11.5 µm, those containing gamonts were 16–19 × 8.5–10 µm. Erythrocyte nuclei were sometimes slightly flattened and displaced by gamont presence.

Remarks Mackerras (1962) described the early stages of sporogony in great detail, from the pairing of gamonts in the hemocoel and among fat bodies on the second day postfeeding, to maturation of sporocysts. Experimental infections in *V. tristis orientalis* became patent on days 27, 28, and 32, following ingestion of infected mosquitoes.

Hepatozoon lygosomarum (Doré 1919) Allison and Desser 1981

Diagnosis A *Hepatozoon* species characterized by gamonts 13.6–16.8 × 4.2–6.3 µm, with a slightly recurved end, and nearly spherical oocysts that contain 30–50 spherical sporocysts, 15.0–16.3 µm in diameter, that produce four to six sporozoites. Sporogony occurs in the wall of intestinal ceca of the mite *Ophionyssus scincorum*. Dimorphic meronts develop in the hepatic parenchymal cells and endothelial cells of the spleen. Macromeronts 18–25 µm in diameter contain three to six macromerozoites, and micromeronts 21–28 µm produce 16–32 micromerozoites.

Erythrocytes host to gamonts are markedly hypertrophied, with laterally displaced nuclei.

Type Host *Oligosoma nigriplantare* (Peters) (syn. *Lygosoma moco* of Doré, 1919, and Laird, 1951) (Sauria: Scincidae).

Type Locality Makara District, Wellington, New Zealand.

Other Hosts None known.

Other Localities Birdlings Flat, Banks Peninsula, and Lake Alexandrina district, South Canterbury, New Zealand (Allison and Desser, 1981).

Prevalence Five of nine *O. nigriplantare* from Makara River, Wellington (Laird, 1951), 3 of 60 (5.0%) from Birdlings Flat, and 1 (?) of 8 from the Lake Alexandrina district (Allison and Desser, 1981) were infected by *H. lygosomarum*.

Morphological Variation Mature gamonts were 15.4 × 5.6 μm (13.6–16.8 × 4.2–6.3, N = 20). Estimated LW is 86.2 μm^2 and L/W ratio is 2.75. Although no dimensions were provided by Allison and Desser (1981), nuclei of gamonts appear to be large, their width nearly or completely that of the gamont body, their length 40–45% of gamont length. Nuclei are located in the middle third of the gamont, perhaps extending into the first quarter of the gamont. The posterior end of the gamont is slightly recurved against the stout gamont body.

Sporogony Sporogony was observed in the mite *Ophionyssus scincorum*. The site of szygy and gamogony are not stated, but at least szygy can be inferred to occur within the gut lumen from the comment "3 days after feeding still showed many free gametocytes in the gut contents. … On days 2 and 3 after feeding, gametocytes were seen in apposition but gametogenesis was not observed" (Allison and Desser, 1981). Uninucleate oocysts appeared in the intestinal cecal walls on days 4 and 5 PI, and these tripled in size from days 5 to 10. Sporulated sporocysts were present on days 9–10 (usually 10) PI. Nearly spherical oocysts, no dimensions given, contained 30–50 spherical sporocysts 15.6 μm (15–16.3) in diameter, with estimated LW about 243 μm^2, which usually produced four sporozoites, with six sometimes present. Sporozoite dimensions were 17.5 × 4.5 μm (16–18 × 4–5, N = 5).

Merogony Meronts in the hepatic parenchyma and splenic endothelium were dimorphic. The slightly smaller macromeronts were 20.7 μm (18–25, N = 9) in diameter and contained three to six macromerozoites 17.8 × 4.9 μm (16.8–18.9 × 4.7–5.3, N = 4). The more commonly seen micromeronts were 23.3 μm (21–28.4, N = 9) in diameter and contained 16–32 micromerozoites 10.9 × 2.7 μm (10.5–12.6 × 2.1–3.3, N = 5). Nuclei of macromerozoites "were larger and more distinct" than those of micromerozoites.

Effects on Host Host erythrocytes were slightly to distinctly hypertrophied, to about 10% larger, as shown by the figures, with their nuclei displaced laterally. Allison and Desser (1981) commented that a heavily infected female skink, kept for 6 weeks, maintained the same parasitemia, gave birth to four healthy young, and "exhibited no overt signs of illness."

Remarks Attempts at transmission to uninfected lizards by ingestion of infected mites were unsuccessful. Distinctive features in sporogony of this *Hepatozoon* species are the possible formation of the zygote within the intestinal lumen, sporogony within the wall of the intestinal ceca of the mite, and formation of usually four sporozoites within the sporocysts, a very small number in comparison to most *Hepatozoon* species for which sporogony has been obtained.

Hepatozoon tuatarae (Laird) 1950

Diagnosis An apparent *Hepatozoon* species with gamonts 12.5–16.7 × 3.3–4.4 μm, surrounded by a reniform capsule. The tapered posterior end of the gamont is recurved to form a short tail. The nucleus is irregular in form, vacuolated, and extends across the width of the gamont, slightly behind midgamont, and is about one-third of the gamont length. Parasitized erythrocytes are neither enlarged nor distorted, but their nuclei may be displaced.

Type Host *Sphenodon punctatus* (Gray) (Rhynchocephalia).

Type Locality Stephen Island, Cook Straits, New Zealand.

Other Hosts None known.

Other Localities Middle Trio Island, Cook Straits.

Prevalence Six of 31 (19.4%) tuataras from Stephen Island and 1 of 4 from Middle Trio Island were infected by *H. tuatarae* (Laird, 1950).

Morphological Variation Laird (1950) found gamonts to average 14.7 × 3.9 μm (12.5–16.7 × 3.3–4.4, N = 6). Estimated from these averages, LW is 57.3 μm^2 and L/W 3.77. Gamonts are recurved to form a short and tapered tail.

Their nuclei are large, extending over about one-third of gamont length and the entire width of the gamont, to the flexure point of the gamont. Nuclei are vacuolated and irregular in shape. Gamonts are contained within a distinct reniform capsule.

Merogony Unknown. Laird (1950) illustrated two small parasites that appeared to be binucleate, but these may have been young gamonts.

Sporogony Unknown. Laird (1950) commented that the tick *Aponomma sphenodonti* commonly infests tuataras.

Effects on Host Erythrocytes parasitized by *H. tuatarae* are normal in size and appearance, but their nuclei may be slightly displaced to one side. Laird reported that parasitemias in infected tuataras were very low, not exceeding 1 parasite in 40,000 erythrocytes, and thought that the presence of gamonts apparently phagocytosed by mononuclear leukocytes and large lymphocytes in each infection indicated the presence of chronic infections.

Remarks Despite the lack of sporogonic evidence that the tuatara hemogregarine actually is a *Hepatozoon* species, it is considered thus in accordance with Smith (1996). It is included in this book because of the unique position of its host species within the Class Reptilia. Desser (1978) provided a color plate illustrating a gamont of *H. tuatarae* (M. A. Peirce, personal communication, not seen by me).

AFRICAN *HEPATOZOON* SPECIES

Hepatozoon ayorgbor Sloboda, Kamler, Bulantová, Votýtpka and Modrý 2007 (**Plate 56**)

Diagnosis A *Hepatozoon* species characterized by broadly elongate gamonts 11–13 × 2–3.5 µm, with nuclei located in the second quarter of the gamont, extending slightly into the first quarter. Free gamonts are 20–25 × 1.5–2.5 µm, with nuclei slightly posterior to midbody. Sporogony occurs within the hemocoel of thorax and abdomen in experimental mosquito hosts. Ovoid oocysts contain up to 1013 spherical-to-ovoid sporocysts that produce 14–38 sporozoites. Infected erythrocytes may be slightly hypertrophied.

Type Host *Python regius* Shaw (Serpentes: Boidae).

Type Locality Ghana.

Other Hosts Experimental hosts are *Lamprophis fuliginosus* and *Boa constrictor*.

Other Localities None known.

Prevalence Sloboda et al. (2007) found 43 of 55 (78%) *P. regius* infected.

Morphological Variation The broadly elongate gamonts average 12.2 × 2.9 µm (11–13 × 2–3.5, N = 30), with estimated LW 35.4 µm^2 and L/W 4.21, and are not recurved. Gamont nuclei may be broad, extending most of the width of the gamont, or thinner, elongated, and lateral, 5.2 × 1.6 µm (4–6.5 × 1.5–2), with estimated LW 8.3 µm^2. Nuclei are situated in the second quarter of the nucleus and may extend into the first quarter.

Sporogony Oocysts, within abdominal and thoracic hemocoel of the experimental mosquito vector *Culex quinquefasciatus* are ovoid in shape, 251.5 × 247.7 µm (58–638 × 58–638, N = 165), with average L/W 1.01, and contain 228.5 (33–1013, N = 17) sporocysts. Sporocysts are spherical to ovoid, 35.5 × 23.7 µm (20–53 × 18–42, N = 30), with average L/W ratio 1.50. Sporozoite number averages 21.3 (14–38, N = 30).

Merogony A single micromeront found in the liver of an experimental *P. regius* measured 18 × 11 µm and contained 26 merozoites (Sloboda et al., 2007, Figure 12), while the meront in Sloboda et al.'s Figure 16 is a micromeront, and Figures 14 and 17 are possibly macromeronts in the liver and lung of the experimental host *L. fuliginosus*.

Effects on Host The gamonts of *H. ayorgbor* have little effect on dimensions of host cells except for a slight narrowing in erythrocyte width and lengthening of the nucleus of infected cells. The nucleus is usually somewhat displaced laterally by gamont presence.

Remarks Although Sloboda et al. dissected 100 ticks, *Aponomma latum*, removed from 50 of the series of 55 *P. regius*, with infestation levels of 7.6 and 6.2 for male and female snakes, respectively, no evidence of sporogony was detected. Six lizards, *L.* (*Lepidodactylus?*) *lugubris* were inoculated with sporocysts per os or intraperitoneally, but none became infected with *H. ayorgbor*.

Hepatozoon pettiti (Thiroux) 1910, Hoare 1932 (**Plate 57**)

Diagnosis A *Hepatozoon* species with elongated, thin gamonts 18–25 × 2.5–5.0 µm, encapsulated and recurved, with both limbs usually nearly equal in length. Capsule dimensions are 9–12 × 4–6 µm. The gamont nucleus is small, 3.0–5.5 × 1.5–3.0 µm, and anterior in position. Oocysts are irregularly rounded, 258–300 × about 237 µm, and contain over 100 spherical sporocysts 38–53 µm in

Plate 57 Hemogregarines of crocodilians. (**A**) *Hepatozoon pettiti* from *Crocodylus niloticus,* Zimbabwe. Immature gamonts, **a, b**; mature gamonts, **c–j**; encapsulated gamont, **k**; free gamont, **l**. (**B**) *Haemogregarina crocodilinorum* from *Alligator mississippiensis,* Florida. Immature parasites, **a, b**; premeronts, **c, d**; erythrocytic meronts, **e–i**; mature gamonts, **j–l**.

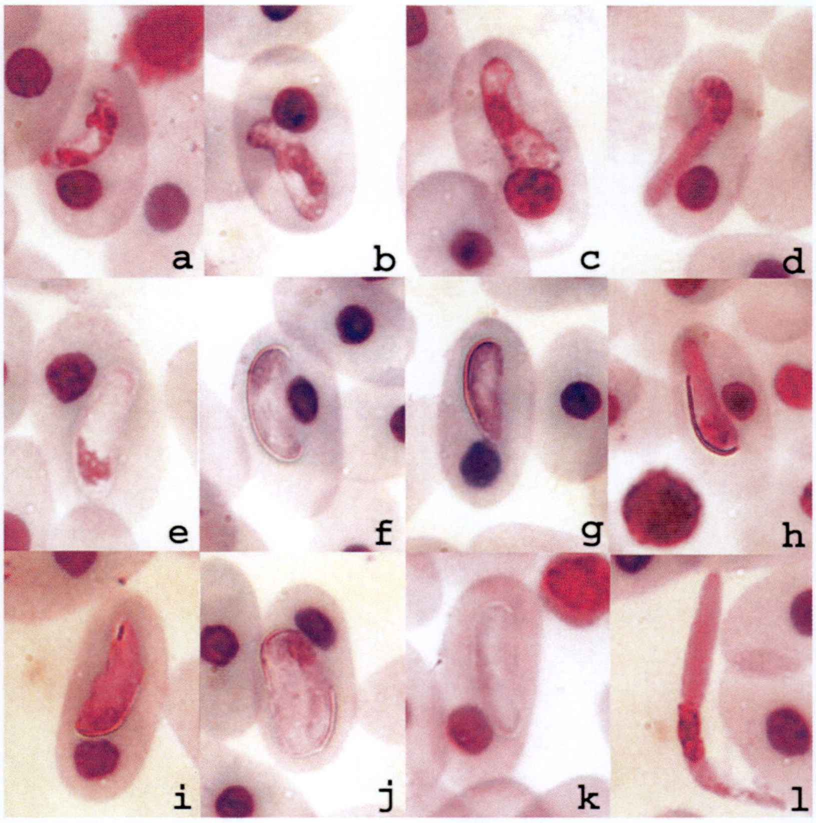

(A)

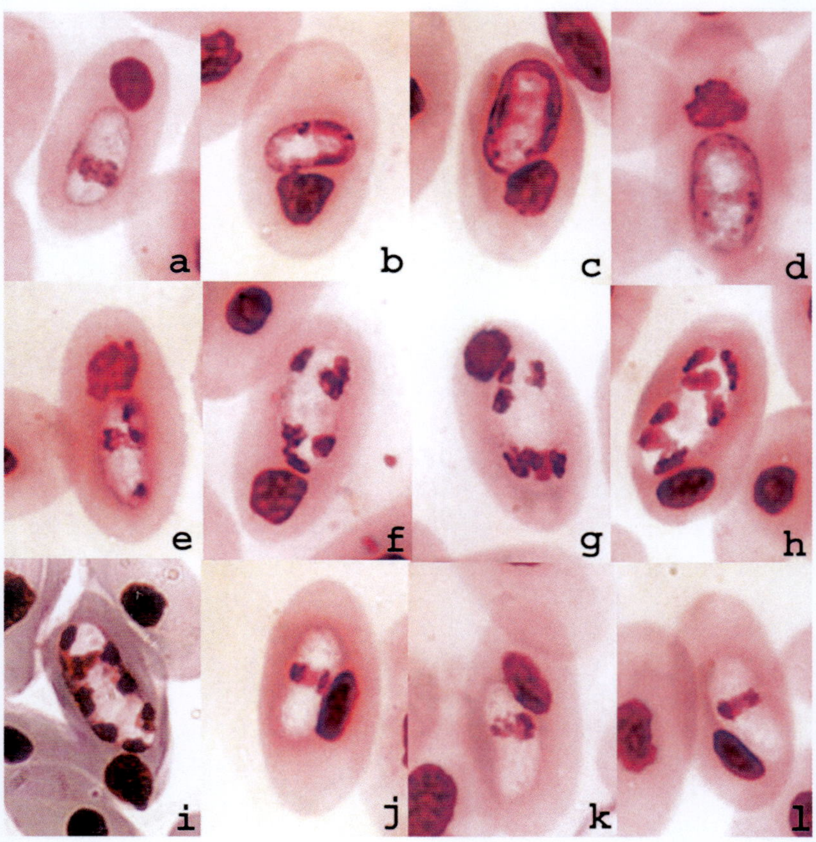

(B)

diameter that contain 25–28 or more sporozoites. Merogony occurs in erythrocytes and in Kupfer cells of the liver. Sporogony takes place in the hemocoel of the tsetse fly *Glossina palpalis*.

Type Host *Crocodylus niloticus* Laurenti (Crocodylia: Crocodylidae).

Type Locality Sénégal.

Other Hosts None known.

Other Localities Mali (Leger and Leger, 1914); Sudan (Wenyon, 1909a); Lake Victoria, Uganda (Hoare, 1932); Save River Valley, Mozambique (Pienaar, 1962); Gonore Zhou, Zimbabwe (Telford).

Prevalence Precise data are not available, but Hoare (1932) commented that all of the crocodiles he examined ("a fair number … in Lake Victoria") were infected by *H. pettiti*.

Morphological Variation Hoare (1932) reported that gamonts were "slender, encapsulated sausage-shaped forms with an elongated nucleus, measuring about 11.5 µ in length. … Some of the parasites are fairly broad and doubled up. … The limbs of the doubled up parasite apparently fuse together." The dimension of 11.5 µm represents that of the capsule. In a series of *H. pettiti* gamonts examined from Zimbabwe, opaque capsules were 9.5–12.0 × 4.0–5.5 µm, but in those capsules that were clear, every gamont with nondispersed nuclear material showed two limbs of the gamont body, nearly equal in size, with the slightly shorter limb not containing the nucleus. The delineation of the limbs was often difficult to discern, and these were not measured. Gamont length, the sum of both limbs, is 20.5 ± 1.8 µm (16–22, N = 25), and width of the primary limb containing the nucleus, and measured near the nucleus, is 3.4 ± 0.5 µm (2.5–5.0). Gamont LW is 69.9 ± 11.0 μm^2 (52.5–96.0), and the L/W ratio is 6.94 ± 1.09 (3.80–8.40). A single free gamont measured 23.5 × 3.0 µm, comparable to the dimensions of the encapsulated gamonts. Gamont nuclei are always in the first quarter of the gamont, often terminal, and are 4.1 ± 0.6 × 2.1 ± 0.4 µm (3.0–5.5 × 1.5–3.0, N = 25), with LW 8.9 ± 2.0 μm^2 (6.0–13.5). Smaller, slender gamonts often show dispersed nuclear material, which apparently aggregates at one end as maturity approaches.

Sporogony Oocysts form in the hemocoel of the tsetse fly *Glossina palpalis* (Hoare, 1932), apparently completing sporogony by 20 days PF. The oocysts are irregularly rounded, 258–300 × about 237 µm, and contain over 100 sporocysts. Sporocysts are spherical, 38–53 µm in diameter. Hoare (1932) did not count sporozoites, but his figures indicate that the number produced exceeds 25–28. Sporozoites measure "about 18 µ in length" and are "banana-shaped." Similar sporogonic stages of a *Hepatozoon* species were reported from *G. palpalis* in West Africa by Chatton and Roubaud (1913) and Macfie (1916); these species are probably *H. pettiti*.

Merogony Hoare (1932) did not distinguish between macro- and micromeronts. His Figures 49–51 appear to be macromeronts containing 11–13 merozoites, and the micromeronts are probably "a rounded body containing over thirty nuclei arranged on the periphery." Figures 44–47 of Hoare (1932) indicate that meronts form in erythrocytes as well, and he believed that on rupture the "meronts" enter Kupfer cells that line hepatic sinusoids, where merogony continues. None were seen in the liver parenchymal cells.

Effects on Host Parasitized erythrocytes are about 20% larger than uninfected cells, but there is little effect on size and shape of host cell nuclei other than displacement to one of the ends of the cell (80%) or laterally.

Remarks The erythrocytic meronts figured by Hoare (1932) need further investigation. Because of the presence of polysporocystic oocysts, there is no question that the generic assignment to *Hepatozoon* is correct as that genus is presently distinguished.

Hepatozoon domerguei Landau, Chabaud, Michel, and Brygoo 1970

Diagnosis A *Hepatozoon* species with short, narrow, and recurved, encapsulated gamonts. Gamonts average 14 × 3 µm in dimensions. Nuclei are situated in the second quarter of the gamont. In experimental mosquito hosts, oocysts of two types develop, a smaller, early maturing oocyst less than 100 µm in diameter that contains 4–20 sporocysts and larger oocysts that mature later, are up to 260 µm in diameter, and produce many more sporocysts. Sporozoites number 8–40. The smaller oocysts produce infections comprised of dimorphic meronts and erythrocytic gamonts; the larger oocysts form cysts in the tissues of vertebrate intermediate hosts, containing zoites that divide by endodyogeny, which on ingestion by a second vertebrate host produce infections with dimorphic meronts and erythrocytic gamonts. Macromeronts appear early, primarily in the liver, and contain about 30 macromerozoites, followed by micromeronts in the lung and other tissues, that produce over 100 micromerozoites.

Type Host *Madagascarophis colubrina* (Schlegel) (Serpentes: Colubridae).

Type Locality Diego Suarez, Madagascar.

Other Hosts Experimental: *Lioheterodon modestus* (Colubridae), *Python sebae* (Boidae), *Oplurus sebae* (Sauria: Opluridae), and (early meronts and cysts only) *Podarcis muralis* and *Lacerta sicula* (Sauria: Lacertidae).

Other Localities Unknown.

Prevalence Unknown.

Morphological Variation Gamonts are enclosed within a capsule permeable to stain and average 14 × 3 µm, with one end recurved (Landau et al., 1972). Gamont LW and L/W, calculated from the mean dimensions, are 42 µm^2 and 4.67, respectively. Although variable in form, the nucleus of an illustrated mature, encapsulated gamont is situated in the second quarter of the gamont and is about 3.8 × 3.3 in size. A figure of a free gamont has dimensions of approximately 22.3 × 3.3 µm, with the nucleus 6.7 × 3.3 µm, clearly situated in the second quarter. Landau et al. (1972) reported the mean length of free gamonts as 19 µm and commented that the morphology of gamonts in the blood appeared to be identical in all of the hosts studied.

Sporogony Landau et al. (1972) obtained sporogony of *H. domerguei* in two mosquito species, *Culex pipiens fatigans* and *Anopheles stephensi*. Morphology of the sporogonic states appeared to be identical in the two mosquito hosts except for size of the mature oocysts, which was greater in *C. pipiens*. Free gamonts were present in the mosquito hemocoel by 6 hours PF. Pairing of gamonts sometimes occurred free in the hemocoel of the mosquito, sometimes within a fat body. Szygy could be observed from 12 hours to day 4 PF, with oocysts first apparent from day 2 PF. Microgametocytes produced two biflagellated microgametes with flagella 25 µm in length. Oocysts formed from the zygote developed throughout the body cavity of the mosquito host but most often were in the abdominal hemocoel. Two groups of oocysts can develop, distinguished by maturation time, oocyst size, and number of sporocysts produced. Some mature as early as 10 days PF (Landau et al., 1972) at a smaller size, 40–80 µm in diameter, and contain fewer sporocysts, 4–20. Other oocysts develop more slowly, maturing in 20–25 days at a larger size, up to 260 µm in diameter, and contain many more sporozoites (number not reported). The two populations of oocysts appear to have different roles. Although no dimensions of sporocysts were reported by Landau et al. (1972), the two size groups of oocysts were said to produce sporocysts and sporozoites that are morphologically identical. Experimental infections, established by the early-maturing, smaller oocysts produced a typical infection with primary macromeronts and secondary micromeronts in the tissues, followed by gamonts in the erythrocytes. The later-maturing, larger oocysts, in mixed populations with the oocysts of earlier maturation in the mosquitoes, resulted in experimental infections that were comprised of tissue meronts and erythrocytic gamonts, but in addition had small cysts in the tissues containing zoites. Sporozoites number 8–40 per sporocyst, and sporozoite length averaged 19 µm (Landau et al., 1972), Sporozoite arrangement varied by sporocyst shape, round or elongate, but sporocysts of both shapes could be found in the same oocyst.

Merogony In experimental Madagascan lizards *Oplurus sebai*, macromeronts ("merontes primaries") form predominantly in the liver, in reticuloendothelial or parenchymal cells, beginning nuclear division on day 7 PI, with some mature meronts present on day 8. These contain about 30 relatively large macromerozoites and remain present in the liver at least until day 16 PI. In an unnatural saurian host, the European lacertid *Podarcis muralis*, meronts were less commonly seen; no mature meronts were found between days 12 and 23 PI and by day 56 PI were no longer present, although cysts were numerous. The cysts were ovoid, 20 × 11–25 × 18 µm in tissue smears and 21–23 × 12 µm in sections. Zoites within cysts divided by endodyogeny, numbered usually 1–6, sometimes 12 in *O. sebae* and the snake hosts, and 1–4, very rarely more, in the lacertid species. Micromeronts ("merontes secondaires") usually occupied endothelial cells of the vascular wall of all tissues but especially of the lung. Their presence coincided with the appearance of gamonts in the erythrocytes. At maturity, micromeronts were ovoid in shape, averaged 24 × 17 µm in size, and contained more than 100 micromerozoites that were destined to invade erythrocytes.

Ingestion of infected mosquitoes by the snakes *Madagascarophis colubrina* and *Lioheterodon modesta* and the lizard *Oplurus sebae* of Madagascar produced typical *Hepatozoon* infections that included macro- and micromeronts in the tissues and erythrocytic gamonts, as well as cysts, in each species. *Python sebae* from Central Africa were infected by the hepatic cysts from *O. sebae* and showed normal infections of *H. domerguei* and cysts as well. Infections produced in *P. sebae* by infected mosquitoes fed on *M. colubrina* also had meronts and gamonts, but in some snakes no cysts were found, and in others cysts were present. Cysts from tissues of *P. sebae* fed to the European *P. muralis* produced cysts in the lizards but no meronts or gamonts. The cysts from this host, however, produced infections in *P. sebae* that showed all four stages (macro- and micromeronts, gamonts, and cysts). Cysts only were also found in *Lacerta sicula* infected from *P. muralis* that showed both meronts and cysts, but *Lacerta viridis* was refractory to infection by *H. domerguei* from mosquitoes.

Gamonts found in both *Oplurus sebae* and *Python sebae* were infective for mosquitoes and subsequently for *O. sebae*. Figures 1 and 7 of Landau et al. (1972) diagram the extensive experimental results employing both sporozoite and cystic-induced infections by *H. domerguei*.

Effects on Host The effects on hosts by *H. domerguei* have not been described, but the figures indicated a rounding of an erythrocyte infected by a mature gamont and displacement of its nucleus.

Remarks *Hepatozoon domerguei* is the first *Hepatozoon* species described with two demonstrated modes of infection, directly from ingestion of infected mosquitoes and indirectly by ingestion of tissues from a vertebrate intermediate host containing cysts enclosing infective zoites. In addition, two groups of different size oocysts, maturing at different times and producing different numbers of sporocysts claimed to be identical in their morphology, apparently have progeny with different functions: The smaller, early maturing oocysts produce infections characterized by dimorphic meronts and erythrocytic gamonts; the larger, later-maturing oocysts produce cysts containing infective zoites in the host tissues, which when ingested by a second vertebrate host can produce typical *Hepatozoon* infections showing dimorphic meronts and erythrocytic gamonts. The many experimental infections apparently confirm conspecificity of the two types of oocysts. The morphological data provided in the series of studies is nearly nonexistent, and it is difficult to characterize *H. domerguei* except by its performance in experimental infections. It is possible that the type host, *Madagascarophis colubrina*, is host to more than one *Hepatozoon* species because Brygoo (1963a) figured apparently mature gamonts, some with recurved portions, that had dimensions (estimated from the scale bar) of 16.0 × 6.5 μm, 17.8 × 5.5 μm, and 18.5 × 5.5 μm, as well as two smaller gamonts, 12.9 × 4.9 μm and 15.1 × 6.2 μm, more consistent with the mean dimensions of 14 × 3 μm provided by Landau et al. (1972). The two largest of Brygoo's gamonts have calculated LW values of 101.8 μm^2 and 97.9 μm^2, over twice the mean LW, 42 μm^2, calculated for *H. domerguei*. The discrepant LW values support the possibility of mixed infections and consequent confusion of their sporogonic stages.

Hepatozoon gracilis (Wenyon) 1909, Bashtar, Abdel-Ghaffar, and Shazly 1987

Diagnosis A *Hepatozoon* species characterized by elongate, very thin gamonts, 18–22.5 × 0.9–1.4 μm, and oocysts that produce 8–50 apparently ovoid sporocysts approximately 22 × 19 μm, which contain 8–24 sporozoites. Merogony occurs in hepatic parenchymal cells, where macromeronts produce 3–16 macromerozoites, and micromeronts contain 25–50 micromerozoites. Sporogony occurs in the hemocoel of experimental mosquito hosts. Infected erythrocytes are hypertrophied and elongated, with their nuclei displaced usually laterally or to one end of the cell.

Type Host *Mabuya quinquetaeniata* (Lichtenstein) (Sauria: Scincidae).

Type Locality Wau, Bahr El Ghazal Province, Sudan.

Other Hosts None known.

Other Localities Abu-Rawash and Kirdasah areas, Egypt.

Prevalence *Hepatozoon gracilis* infected 168 of 420 (40%) of Egyptian *M. quinquetaeniata* (Bashtar et al., 1987).

Morphological Variation Wenyon (1909a) described *Haemogregarina gracilis* as a narrow and elongated hemogregarine, 16 or 17 × about 1.5 μm, enclosed within a transparent cyst. The nuclei (in figures) were elongate and about one-third of the gamont body in length. Gamonts as described from Egyptian skinks by Bashtar et al. (1987) were similar in proportions, 21.3 ± 1.5 × 1.14 ± 0.20 μm (18–22.5 × 0.9–1.4). Estimated LW is 32.4 μm^2 (16–32), and L/W is 18.7 (16.1–20.0). Gamont nuclei averaged 10.2 ± 0.5 × 1.13 ± 1.19 μm, with estimated LW 11.5 μm^2, and appear to occupy the second quarter of the gamont, extending into the first quarter.

Sporogony Gamogony began with szygy and differentiation into macro- and microgamonts on day 2 PI within the hemocoel of the experimental mosquito host *Culex pipiens molestus*. On day 3, differentiation was completed, and microgamonts produced four microgametes. Fertilization of macrogametes occurred on day 4, and zygote growth into an oocyst took place from days 5 to 8 PI. On days 8 and 9, oocysts ruptured, releasing sporoblasts into the hemocoel, with nuclear division occurring during days 10–12. Oocysts, for which no dimensions were provided, had on average 28 ± 14.2 (8–50) sporoblasts that by day 16 were fully formed sporocysts, which contained on average 17.0 ± 8.4 (8–24) sporozoites. Dimensions of sporocysts were not stated, but illustrations indicate a size of approximately 22 × 19 μm, with estimated LW 418 μm^2 and L/W ratio 1.16. Sporozoite dimensions averaged 10.3 ± 0.6 × 2.08 ± 0.13 μm.

Merogony Wenyon (1909a) reported the presence of meronts in the liver that were of two types, one that produced

8–16 macromerozoites and larger meronts that contained an "enormous number" of micromerozoites. Bashtar et al. (1987) reported that "micromeronts" produced on average 11.0 ± 4.2 (3–16) macromerozoites 13.5 ± 0.5 × 1.02 ± 0.13 μm in size, while "macromeronts," in size up to 22.7 ± 1.2 × 18.3 ± 0.5 μm, contained 40.0 ± 7.8 (25–50) micromerozoites, 16.7 ± 0.9 × 1.13 ± 1.21 μm. Merogony apparently occurs exclusively in the liver, "mainly" in hepatic parenchymal cells, and meronts are enclosed within a parasitophorous vacuole.

Effects on Host The elongate, thin gamonts cause slight hypertrophy and elongation of host erythrocytes, displacing the host cell nuclei laterally or, sometimes, to one end of the cell.

Remarks Experimental infection of uninfected skinks was accomplished both by intraperitoneal inoculation of sporozoites or per os. The oral route produced patent infection in 35 days, and the parenteral route in "4 weeks" (Bashtar et al., 1987).

PALEARCTIC *HEPATOZOON* SPECIES

Hepatozoon aegypti Bashtar, Boulos and Mehlhorn 1984

Diagnosis A *Hepatozoon* species characterized by gamonts 13–16 × 2.5–3.0 μm, and spherical oocysts that contain 15–75 ovoid sporocysts approximately 26–42 × 31–37 μm that produce 8–46 sporozoites. Merogony occurs only in the lung, primarily in capillary endothelium. Macromeronts contain 8–15 or fewer macromerozoites, and micromeronts produce 25–40 micromerozoites. Sporogony occurs in the hemocoel of experimental mosquito hosts. Infected erythrocytes are slightly elongated, and their nuclei are usually displaced laterally or to one end of the cell.

Type Host *Spalaerosophis diadema* (Schlegel) (Serpentes: Colubridae).

Type Locality Moudureit El-Tahrir, Egypt (by designation).

Other Hosts None known.

Other Localities Al-Haddin, Egypt.

Prevalence *Hepatozoon aegypti* infected 22 of 105 (20.9%) *S. diadema* collected in the two stated localities.

Morphological Variation Gamonts and younger stages are enclosed usually within a parasitophorous vacuole in the erythrocyte, which sometimes cannot be discerned. Mature gamonts are 13–16 × 2.5–3.0 μm, with estimated LW 32–48 μm^2 and L/W ratio around 5.2. The nucleus apparently lies within the second quarter of the gamont and extends into the first quarter. Dimensions of the nucleus are 4.5 × 2.5–3.0 μm, with estimated LW 10–15 μm^2.

Sporogony Both szygy and gamogony occurred within the hemocoel of the experimental mosquito host *Culex pipiens molestus*. Szygy was observed on day 2 PI, and differentiation by sex began then. Microgamonts produced four microgametes, and fertilization was observed on days 3 and 4. The zygote developed into an oocyst in 5–8 days PI, and nuclei appeared in finger-like infoldings between days 8–12 PI. Sporocysts formed on days 13–15, and sporozoites were visible on day 16, often arranged in two rows each side of a residual body. Oocyst dimensions were not stated in the description, but a spherical oocyst containing sporulated sporocysts in the figures measured about 130 μm in diameter. Oocysts produced an average of 52 (15–75) ovoid sporocysts. Dimensions of sporocysts in the figures are approximately 26–42 × 21–37 μm, with LW estimated at 546–1554 μm^2 and L/W ratios 1.14–1.24. Sporocysts contained 26 (8–46) sporozoites, 9–12 × ~2 μm in size.

Merogony Merogony occurs exclusively in the lung, primarily within capillary endothelial cells. Apparent macromeronts, sometimes spherical to ovoid in shape, are about 17 × 6.5 μm and produce 8–15 or fewer macromerozoites. Micromeronts are larger, about 27 × 21 μm, and form 25–40 micromerozoites. Merozoites are similar in size in both types of meront, about 9 × 2 μm, with nuclei 4 × 1.5 μm. Meronts are contained within parasitophorous vacuoles.

Effects on Host Gamonts have little effect on host erythrocytes except for a slight elongation of the cell and displacement of the nucleus laterally or to one end of the host cell. Bashtar et al. (1984a) believed that *H. aegypti* had no pathological effect on its snake host despite the development of massive parasitemias exceeding 70%.

Remarks Experimental infections of *H. aegypti* in uninfected snakes (presumably *S. diadema*) were accomplished both per os and by intraperitoneal inoculation. Patency appeared after 6 weeks following oral infection and after 5 weeks when the intraperitoneal route was used. Both infection routes provided infections capable of transmission to mosquitoes. Attempts to infect *Aedes aegypti* and unidentified mites or "snake ticks" were not successful, and lizard species could not be infected.

Hepatozoon mehlhorni Bashtar, Abdel-Ghaffar and Shazly 1991

Diagnosis A *Hepatozoon* species with gamonts 17.2 × 5.4 μm and ovoid oocysts 135 × 120 μm that produce 11–60 ovoid sporocysts that contain 5–12 sporozoites. Sporogony occurs in the hemocoel of experimental mosquito hosts. Meronts occur in the pulmonary endothelium, cells of the spleen, and hepatic parenchymal cells and are dimorphic. Macromeronts, 18.2 × 13.5 μm, produce 2–14 macromerozoites, and the larger micromeronts, 30.2 × 22.6 μm, contain 16–40 micromerozoites. Infected erythrocytes are significantly enlarged, with laterally displaced nuclei.

Type Host *Echis carinatus* (Schneider) (Serpentes: Viperidae).

Type Locality Siwah Oasis, Egypt (by designation).

Other Hosts None known.

Other Localities Bahariah Oasis, Egypt.

Prevalence *Hepatozoon mehlhorni* parasitized 48 of 75 (64%) *E. carinatus* collected from the two known localities.

Morphological Variation Erythrocytic gamonts averaged 17.2 ± 1.6 × 5.4 ± 0.5 μm, with estimated LW 92.9 μm^2 and L/W ratio 3.19. Nuclei occupied the second quarter of the gamont, sometimes the middle third, and could extend into the first quarter. Nucleus dimensions averaged 6.3 ± 0.6 × 5.4 ± 0.5 μm, with estimated LW 34.0 μm^2. The parasitophorous vacuole containing the gamont sometimes could not be discerned.

Sporogony Szygy occurred in the hemocoel of the experimental mosquito host *Culex pipiens* on day 2 PI, with gamogony apparently completed on day 3. Microgamonts produced two to four microgametes, and fertilization of macrogametes was observed on day 4. Between days 5 and 8, zygotes developed into ovoid oocysts, which reached an average size of 135 ± 2.6 × 120 ± 1.8 μm. Sporoblasts developed between days 8 and 10 PI and became ovoid sporocysts by day 14. On average, oocysts produced 35 (11–60) sporocysts, 16.8 ± 1.3 × 13.1 ± 0.6 μm, with estimated LW 220 μm^2 and L/W 1.28. Sporocysts contained eight (5–12) sporozoites 12.6 ± 1.2 × 4.1 ± 0.3 μm.

Merogony Meronts were found in pulmonary endothelium, the spleen, and hepatic parenchymal cells. Two types of meront were present, macromeronts 18.2 ± 0.6 × 13.5 ± 0.5 μm that contained 8 (2–14) macromerozoites 15.1 ± 0.12 × 6.2 ± 0.8 μm and micromeronts 30.2 ± 1.7 × 22.6 ± 1.2 μm that produced 28 (16–40) micromerozoites 17.2 ± 0.7 × 5.0 ± 0.2 μm in size.

Effects on Host Host erythrocytes were enlarged by about 40%, and their nuclei were displaced laterally. Parasitemias reached 800 parasites per 10^3 erythrocytes in some vipers, and in one the parasitemia was 96%.

Remarks Experimental infections were established in clean vipers by intraperitoneal inoculation of a homogenate of whole infected mosquitoes, 20–25 days PF on infected snakes. Patent infections appeared 4–6 weeks later.

A *Hepatozoon* species described without sporogonic stages from Pakistani *Echis carinatus* (Mohiuddin et al., 1967), *H. echisi*, has shorter and thinner gamonts, 14.0 ± 1.30 × 2.8 ± 0.6 μm^2 (N = 100), which probably indicates a different species from *H. mehlhorni*. The merogonic pattern of *H. echisi* is also different, with merogony apparently restricted to the pulmonary endothelium. Meronts apparently are not dimorphic as in *H. mehlhorni* and produced usually 22–25, up to 38, merozoites 6 × 1.5 μm in size. The merozoite size is less than one-half the size of either macro- or micromerozoites in the Egyptian saw-scaled vipers.

Hepatozoon seurati (Laveran and Pettit) 1911 (Foley and Catanei) 1925

Diagnosis A *Hepatozoon* species with encapsulated and recurved gamonts 12–16 × 3–5 μm, ovoid oocysts 152 × 126 μm that contain 35–85 sporocysts that produce three to eight sporozoites and monomorphic meronts that form up to ten merozoites. Gamont nuclei occupy the second quarter and extend into the first quarter of the gamont. In some infections, hypertrophied host erythrocytes are severely dehemoglobinized except for a central contraction of the cytoplasm around the gamont and closely associated host cell nucleus.

Type Host *Cerastes cornutus* Linnaeus (Serpentes: Viperidae).

Type Locality Vicinity of Chellala and Bou Saada, Laghouat, Algeria.

Other Hosts *Cerastes cerastes* (Abdel-Ghaffar, Bashtar, and Shazly, 1991).

Other Localities Egypt (Abdel-Ghaffar et al., 1991).

Prevalence *Hepatozoon seurati* infected 37 of 154 (24%) *C. cornutus* from the vicinity of Laghouat, Algeria (Foley and Catanei, 1925).

Morphological Variation In *C. cornutus,* gamonts were enclosed within a clear capsule when mature, which was not easily visible with younger gamonts. Dimensions of the recurved gamonts were 12–16 × 3–5 µm and of free gamonts calculated from figures of Foley and Catanei (1925) as 14–15 × 2.5–4.5 µm. Gamont nuclei dimensions, from figures, are about 4–6 × 2.5–5 µm and are situated in the second quarter, extending into the first quarter of the gamont. Recurved portions of the gamonts appear to be about one-third to one-half of the gamont length.

Sporogony Sporogony of *H. seurati* from *Cerastes cerastes* took place in the myxocoel of the mosquito *Culex pipiens molestus* (Abdel-Ghaffar et al., 1991). Gamonts free from host cells penetrated the midgut wall on day 1 PF, and szygy was evident on day 2. Gametogony occurred on day 3 PI, with microgamonts producing two to four microgametocytes with a single flagellum about 20.4 µm in length. On day 4, fertilization of macrogametes produced zygotes that became ovoid oocysts 152 × 126 µm. Sporoblasts appeared by day 10 PI, and 60 (35–85) sporocysts 20 ± 0.8 × 16 ± 0.6 µm containing sporozoites were present between days 14 and 18 PI. Sporocysts produced five (three to eight) sporozoites 14 × 3.2 µm on average.

Merogony Foley and Catanei (1925) reported the formation of only one type of meront in *C. cornutus,* in liver, lung and spleen, with dimensions that reached 30 × 16 µm, containing up to ten large merozoites accompanied by a residual body. Cysts, containing what were apparently two to four zoites, were also present in the tissues of the infected snakes.

Effects on Host Parasitized erythrocytes usually appeared normal in size but with their nuclei sometimes displaced laterally or to one pole of the cell. In about one-third of the infections, Foley and Catanei (1925) described considerably hypertrophied host erythrocytes that were nearly completely dehemoglobinized except for a central contraction of the cytoplasm surrounding the closely associated gamont and erythrocyte nucleus. Heavy infections were often associated with a pronounced eosinophilia.

Remarks *Hepatozoon seurati* is distinguished from most *Hepatozoon* species by the presence of sporocysts that produce eight or fewer sporozoites, apparently monomorphic meronts, and in some infections, enlarged host erythrocytes that are severely dehemoglobinized except for a central portion of normal cytoplasm contracted around the host cell nucleus adjacent to the gamont.

Abdal-Ghaffar et al. (1991) collected several potential arthropod vector species from *C. cerastes* and their burrows and fed them in the laboratory on infected vipers. Mites (*Haemolaelaps aegyptius* and *H. centrocarpus*) examined from 4 to 18 days PF did not show parasites or their developmental stages. Ixodid and argasiid ticks, *Hyalomma impeltetum* and *Argas* sp., respectively; sand flies (*Phlebotomus* sp.); and the mosquito *Aedes aegypti* showed no indication of gamogony or sporogony 4–8 days PF. Only *Culex pipiens molestus* became infected and produced sporozoites. The authors commented that "the area of viper distribution had an epidemic populations of … *Culex pipiens molestus.*"

Hepatozoon najae (Laveran) 1902

Diagnosis A *Hepatozoon* species with gamonts 14–18 × 3.5–6.0 µm. Oocysts develop in the hemocoel of mosquito vectors and form 20–50 sporocysts that produce 14–37 sporozoites. Merogony occurs in liver parenchyma, spleen cells, and endothelial cells of capillaries, including those of the lung, where macromeronts and micromeronts form. Erythrocytes host to gamonts are longer and slightly wider and show a 30–80% hypertrophy over uninfected cells.

Type Host *Naja tripudians* (= *Naja naja*).

Type Locality Vicinity of Pondicherry, Madras, India.

Other Hosts *Naja nigricollis* (Bashtar and Abdel-Ghaffar, 1987); *Naja haje* (Ball, 1967a).

Other Localities Aswan, Egypt (Bashtar and Abdel-Ghaffar, 1987); near Lake Baringo, Marigat, Kenya (Ball, 1967a); Morogoro, Tanzania (Telford).

Prevalence In Tanzania, four of eight *N. nigricollis* were infected by *Hepatozoon najae.*

Morphological Variation Laveran (1902) described gamonts from *Naja tripudians* as 14 × 3 µm when intracellular, with one end rounded and the other pointed. Gamonts free of erythrocytes were 21–22 × 3 µm. In *Naja nigricollis* of Egypt, Bashtar and Abdel-Ghaffar (1987) gave dimensions of 17.7 × 3.1 µm, with a nucleus 8.1 × 2.7 µm that "often appeared as a roughly transverse band." The cytoplasm contained a few vacuoles. In Tanzanian *N. nigricollis,* gamonts were 16.8 ± 0.8 × 4.4 ± 0.6 µm (15–18 × 3.5–6.0, N = 25), with LW 73.4 ± 11.1 µm^2 (56–108) and L/W 3.93 ± 0.5 (3.00–4.86). Nuclei averaged 5.3 ± 0.7 × 3.5 ± 0.6 µm (4.0–7.5 × 2.5–4.5, N = 25), with LW 17.8 ± 3.6 µm^2 (12.5–24.8). Nuclei usually occupied the second quarter of the gamont and overlapped into the first quarter (84%), seldom being confined to the first or second quarter only. In a different host, *Naja haje* of Kenya, gamonts were slightly shorter and wider than in *N. nigricollis*, averaging 15.0 ± 0.4 × 5.2 ± 0.3 µm (14–16 × 5.0–6.0, N = 25), with a similar LW value, 77.9 ± 4.8 µm² (70–90), and a lower L/W ratio, 2.9 ± 0.20 (2.50–3.20). Gamont nuclei were similar in length but narrower in width, 5.6 ± 0.5 × 2.2 ± 0.5 µm (5.0–7.0 × 2.0–4.0, N = 25), LW 12.5 µm² (10–24). Nuclear position was similar to *N. nigricollis* gamonts, with 80% occupying the second quarter and overlapping the first, seldom confined to first or second quarters.

Sporogony Gamogony and sporogony took place in the hemocoel of *Culex pipiens molestus* (Bashtar and Abdel-Ghaffar, 1987), with fully formed sporozoites present in 20–25 days following ingestion of infected blood. Although not stated, the figures of Bashtar and Abdel-Ghaffar suggest that the type of host cell within the hemocoel used for gamogony and sporogony was probably a fat body cell. On days 2 and 3 postingestion, parasites "associated in pairs within a parasitophorous vacuole in a host cell" and differentiated into "micro- and macrogamonts." Microgamonts were 9 × 14 µm; their nucleus divided into three or four daughter nuclei that became the same number of microgametes, pear shaped and 4 × 2 µm, on day 4. The macrogamont rounded up and increased in size to 30 µm in diameter. Fertilization occurred on day 4 or 5, forming the zygote, which then became an oocyst. Between days 10 and 15, 20 to 50 sporoblasts, on average 39, formed, becoming sporocysts. Sporozoites, 10 × 2.2 µm, developed in sporocysts between days 20 and 25, averaging 28 per sporocyst, with a range of 14–37.

Merogony Bashtar and Abdel-Ghaffar (1987) found meronts of *H. najae* in "liver parenchyma, spleen cells, endothelial cells of blood capillaries, and within some capillaries in the lung." Macromeronts 20 × 9 µm produced 6–20 (average 12) macromerozoites, and micromeronts 35 × 20 µm contained 25–50 (average 42) micromerozoites. Without stating the meront type, merozoites were 12 × 2.5 µm, with a nucleus 5 × 2 µm.

Effects on Host Gamonts caused host erythrocytes to become more elongated and enlarged from average dimensions of 17 × 9.2 µm to 21.7 × 13 µm, indicating an 80% hypertrophy in gametocyte size (LW). Their nuclei were always displaced laterally within the erythrocyte and in figures show a more rounded shape than in uninfected cells. Infected cells in both *N. nigricollis* of Tanzania and *N. haja* of Kenya were longer and slightly broader than normal erythrocytes, with size hypertrophied by 57% and 36%, respectively

Remarks The *Hepatozoon* gamonts observed by Wenyon (1909a) in *Naja nigricollis* and *N. haje* in the Sudan may have been the same as described in those species above. Whether the name *Hepatozoon najae* that was originally given to the parasite of *Naja naja* in India is appropriate for the hemogregarine present in two other cobra species in eastern Africa from the Mediterranean region down to central Africa cannot be established without genomic comparison or the sporogony of each parasite population in the three cobra species.

Hepatozoon kisrae Paperna, Kremer-Mecabell and Finkelman 2002

Diagnosis A *Hepatozoon* species with stout, banana-shaped gamonts 12–15 × 2–5.5 µm and a centrally located nucleus 4–5.5 µm in width. Macromeronts with 4–16 macromerozoites and micromeronts that produce 16–32 micromerozoites develop primarily in pulmonary endothelium and sometimes in the liver. Cystozoic cysts containing two to four, rarely eight, zoites occupy macrophages within the liver. Sporogony occurs in the ixodid tick *Hyalomma cf. aegyptium*. Mature oocysts are spherical or nearly so, 200–230 × 230 µm, and contain over 100 ovoid-to-ellipsoidal sporocysts 34 × 24–61 × 23 µm that produce 16–35 sporozoites.

Type Host *Laudakia stellio* (Linnaeus) (= *Agama stellio*) (Sauria: Agamidae).

Type Locality Kisra, southeast Samaria, Palestine.

Other Hosts None known.

Other Localities Lifte Valley, Jerusalem, and Duyuk, Israel (Desser and Yekutiel, 1986/87).

Prevalence Nine of 10 *L. stellio* were infected at the type locality (Paperna et al., 2002).

Morphological Variation Apparently, mature gamonts were encapsulated by a "hard cyst" and were stout and banana shaped, 12–15 × 2–5.5 µm (N = 9). Estimated LW is 24–83 µm², and L/W is 2.73–6.00). The centrally located nucleus was 4–5.5 µm in width, "multilobed or homogeneously packed" (Paperna et al., 2002). The gamont cytoplasm could be "homogeneously pale, or had a variable quantity of vacuoles and granules."

Sporogony Sporogony occurs in the ixodid tick *Hyalomma cf. aegyptium*. Seven of ten *L. stellio* from the type locality were infested with ticks, of which six lizards had

Hepatozoon-infected ticks. Oocysts of *H. kisrae* were 67–74 × 47–54 µm before sporulation and occurred on the hemocoel surface of the tick gut in engorged nymphs. Aggregates of sporocysts 43–50 × 24–27 µm were formed by sporoblasts. Over 100 sporocysts were present in mature oocysts 200–230 × 230 µm that were spherical to subspherical in shape. Sporocysts were ovoid to ellipsoidal, 34 × 24–61 × 23 µm, and contained 16–35 sporozoites. Mature oocysts containing sporocysts were found in unengorged adult male and female ticks and in incompletely engorged nymphs. Uninfected *L. stellio* that were fed sporocyst-infected ticks, five from naturally infected and one from an experimentally infected tick, showed patent erythrocytic infections 32–40 days PI in five lizards, and in one lizard after 90 days. Gamonts matured by days 37 in two hosts, by day 40 in two, and by day 60 in one lizard. Two geckoes, *Ptyodactylus hasselquisti*, fed simultaneously with the agamid lizards on tick viscera containing sporocysts did not show infection by day 40 PF. Engorged tick larvae were not found infected either when still engorged or as hungry nymphs up to 40 days following feeding.

Merogony Merogony occurred in the lungs, with meronts 29–39 × 15–22 µm developing within endothelial cells. Occasional meronts were seen within hepatic endothelial cells. Macromeronts produced 4–16 macromerozoites 11–16 × 3–4 µm, and micromeronts produced 16–32 micromerozoites, 8–11 × 1.5–2 µm. Cystozoic cysts were found apparently in macrophages within "melanomacrophage centers" of the liver. Dizoic cysts were 19–24 × 12–17 µm (N = 7), and those containing four zoites were 20–28 × 10–15 µm (N = 7). Occasional octozoic cysts were found. Zoites apparently formed by endodyogeny were 12–15 × 2–5 µm (N = 12). Exceptionally, zoites could reach 22 × 3.8 µm and had tapering, deeply staining, anterior ends. A blue-staining inclusion possibly represented a crystalline body in some dizoites.

Effects on Host Not described.

Remarks Transovarian transmission did not occur in *Hyalomma cf. aegyptium*. A second species of *Hepatozoon* with distinctive gamonts that parasitized *L. stellio* in sites away from Kisra was reported by Paperna et al. (2002) to develop in mosquitoes, *Culex pipiens*, but has not been described. Although morphologically indistinguishable from *Hyalomma aegyptium* that feed predominantly on tortoises, none of the ticks reared from those obtained from *L. stellio* would feed on tortoises at any life history stage. *Hyalomma aegyptium* is the proven vector for *Hemolivia mauritanica* of tortoises (Michel, 1973; Landau and Paperna, 1997; Široký et al., 2007).

Haemogregarinidae

NORTH AMERICAN *HAEMOGREGARINA* SPECIES

Haemogregarina crocodilinorum Börner 1901, Khan, Forrester, Goodwin and Ross, 1980 (Plate 57)

Diagnosis A *Haemogregarina* species with lentiform gamonts 8.0–11.0 × 4.0–6.0 µm, with one extremity recurved, and erythrocytic meronts 11.0–20.0 × 7.0–9.0 µm that produce 6–12 merozoites. Sporogony is in the leech *Placobdella multilineata*, in which unisporocystic oocysts produce six to eight sporozoites.

Type Host *Alligator mississippiensis* (Daudin) (Crocodilia: Alligatoridae).

Type Locality Southeastern United States.

Other Hosts None known.

Other Localities Known from Florida, Arkansas, North Carolina, and South Carolina (Khan et al., 1980).

Prevalence In Florida, 64 of 67 (53.1%); Arkansas, 6 of 21 (28.6%); North Carolina, 6 of 7; and South Carolina, 1 of 4 alligators were infected by *H. crocodilinorum* (Khan et al., 1980).

Morphological Variation Gametocytes with recurved, presumably posterior portions, are 10.0 × 5.0 µm (8.0–11.0 × 4.0–6.0, N = 40). Gamonts emerged from erythrocytes are 17.2 × 2.3 µm (14.0–21.0 × 2.0–3.0, N = 40). Erythrocytic meronts are 15.6 × 8.1 µm (11.0–20.0 × 7.0–9.0) and contain 6–12, usually 8, nuclei.

Merogony Nonerythrocytic merogony has not been described.

Sporogony Free gamonts, sometimes in szygy, were present in gut contents of engorged leeches collected 7 days earlier. Two types were seen, one lightly and the other more intensely stained. The lightly stained became microgametocytes and formed up to four nonflagellated microgametes. Oocysts were found within epithelial cells of the leech intestinal wall in two of four leeches dissected on days 112 and 139. One impression smear showed an oocyst that contained six nuclei, apparently destined to become sporozoites. Nine intracellular unisporocystic oocysts, none of which appeared mature, containing up to eight nuclei were found in sections (Khan et al., 1980).

Effects on Host Infected erythrocytes are enlarged in width, but their length is not affected by gamont presence. Host cell nuclei remain normal in size but are always displaced into a polar position.

Remarks Khan et al. (1980) attempted transmission of *H. crocodilinorum* to two yearling alligators by feeding them each 14 leeches removed from wild-caught alligators 2 weeks earlier, following digestion of blood meals in the leeches. Neither became positive over a 6-week period.

Despite the clear statement by Khan et al. (1980) that unisporocystic oocysts are produced by this species, Smith (1996) considered *H. crocodilinorum* to be a *Hepatozoon* species rather than *Haemogregarina*. The presence of unisporocystic oocysts within a leech host accompanied by erythrocytic merogony is adequate evidence of generic identity as *Haemogregarina* for *H. crocodilinorum*.

Haemogregarina balli Paterson and Desser 1976

Diagnosis A *Haemogregarina* species with lentiform gamonts 8.0–14.5 × 3.0–6.5 μm and erythrocytic meronts 14.5–21 × 5.0–10.0 μm that produce six or eight merozoites. Microgamonts have a club-shaped nucleus and a short, recurved tail that is not present in macrogamonts, which have a round, centrally positioned nucleus. Tissue meronts are predominantly hepatic and contain 13–25 nuclei. Oocysts located on the microvillar surface of epithelial cells in the gastric ceca of the leech host produce nonsporocystic sporozoites that enter endothelial cells of anastomosing lacunae in the circulatory system of the leech, where about 250 merozoites are formed. Free merozoites enter tissues of the proboscis and are the infective stage, transmitted by bite, to the turtle host.

Type Host *Chelydra s. serpentina* (Linnaeus) (Testudines: Chelydridae).

Type Locality Lake Sasajewan, Algonquin Park, Ontario, Canada.

Other Hosts *Chrysemys picta marginata* and *Clemmys insculpta* (Siddall and Desser, 1992, 2001).

Other Localities Wolf Hunt Pond, Algonquin Pond, Ontario.

Prevalence All 37 *C. serpentina* examined were infected by *H. balli* (Desser, 1973) and 6 of 11 (55%) of *C. picta marginata* (Siddall and Desser, 2001).

Morphological Variation Gamonts average 12.6 × 5.3 μm (8.0–14.5 × 3.0–6.5, N = 48) and are lentiform in shape. Estimated from the averages, LW is 66.8 μm^2, and L/W is 2.4. Microgamonts stain pale blue with red inclusions and have club-shaped nuclei and a short, recurved tail. Macrogamonts have no recurved tail, a central, spherical nucleus, and a flocculent, basophilic cytoplasm. Erythrocytic meronts, rarely seen in peripheral blood, are 18.6 × 7.2 μm (14.5–21 × 5.0–10.0, N = 46) with six or eight merozoites arranged around a central residual mass. Immature meronts appear as "serpentine" forms with two broad arms that fuse together before nuclear division begins. Merozoites are 11.7 × 2.5 μm (9.0–13.0 × 2.0–2.5, N = 20), with nuclei 5.9 × 2.5 μm (5.0–6.5 × 2.0–2.5, N = 20).

Merogony Tissue meronts are predominantly in liver cells but occur also in the lung and spleen. In section, meronts are oval or circular, with the largest averaging 15.4 μm (9.0–23.5, N = 87) in diameter. Meronts produce 18 (13–25, N = 11) merozoites.

Sporogony Sporogony occurs in the leeches *Placobdella parasitica* and *P. ornata*. Micro- and macrogametocytes associate on the microvillar surface of epithelial cells in the intestinal ceca of the leech, where gamete formation occurs. One of four nonmotile gametes formed by a microgametocyte fertilizes a macrogamete, and the remaining three remain on the surface of the developing oocyst as compact, residual nuclei. Oocysts develop on the microvillar surface, and produce eight unisporocystic sporozoites, which form around a germinal center. Almost mature oocysts with eight nuclei but undifferentiated cytoplasm are 13.0 × 9 μm (10.5–14.5 × 8.0–10.5, N = 5). A fully differentiated oocyst in a smear from leech intestine was 22.0 × 14.5 μm (Paterson and Desser, 1976). Sporozoites are 26.5 × 6.0 μm (N = 4), with a nucleus centrally located, 5.0 × 4.0 μm. Liberated sporozoites are found in leech tissues. Sporozoites give rise to meronts, most commonly in the anastomosing lacunae of the leech circulatory system (Siddall and Desser, 1991). Meronts contain at maturity about 250 merozoites. Free merozoites occur in the tissues of the anterior two-thirds of the proboscis, most concentrated near the tip, and are the infective stage for the turtle host.

Effects on Host Erythrocytes infected by gametocytes of *H. balli* are neither hypertrophied nor distorted but may be slightly elongated. Cells containing meronts appear to be slightly enlarged (Desser, 1973, figures). The nuclei of erythrocytes host to either stage are displaced laterally or toward one pole of the erythrocyte but are not greatly distorted by parasite presence.

Remarks *Haemogregarina balli* is the only *Haemogregarina* species for which unequivocal transmission to the reptilian host has been demonstrated. Reichenow (1910) claimed to have transmitted *H. stepanowi*, but as Siddall and Desser (1991) pointed out, the experimental host was wild caught and the prepatent period extremely short. The prepatent period of *H. balli* in a laboratory-reared turtle required 12 weeks. Siddall and Desser (2001) transmitted *H. balli*, designated thus on the basis of blood-stage similarities, by bite of infected leeches, *Placobdella ornata*, from naturally infected *Chrysemys picta marginata* (Emydidae) to *Chelydra serpentina*, which had been hatched and raised in the laboratory.

Haemogregarina sp. (**Plate 58**)

Diagnosis A species of *Haemogregarina* with slender, recurved gamonts 29–35 × 3–4.5 µm, of which the anterior limb comprises 48–54% of the total length. The small nucleus, 5–7.5 × 2–5 µm, is located at approximately midbody of the gamont. Erythrocytic meronts with small, compact nuclei are 13–17 × 4.5–9 µm and contain three to eight nuclei. Erythrocytic meronts with larger, squarish or rectangular nuclei become rounded and then undergo many nuclear divisions to produce very large, usually ovoid-to-rounded meronts that contain up to 150 nuclei or more within the thinly stretched host erythrocyte membrane. Merogony directly from sporozoites occurs in tissues of the *Placobdella* spp. vectors, with merozoites from these meronts presumably infective for the turtle host. Erythrocytes infected by gamonts commonly appear elongated and more slender than uninfected cells but are hypertrophied only by about 10%.

Type Host *Macrochelys temminckii* (Troost) (Testudines: Chelydridae).

Type Locality Spring Creek, Decatur County, Georgia.

Other Hosts *Graptemys barbouri* (Emydidae).

Other Localities Santa Fe, Econfina, Shoal, Yellow, and Escambia rivers of northern Florida.

Prevalence The prevalence of *Haemogregarina* sp. in 115 *M. temminckii* was 100% in every locality sampled between 2001 and 2007, with sample size by locality: Spring Creek, 78; Santa Fe, 15; Shoal River, 2; Econfina River, 9; Yellow River, 6; and Escambia River, 5. Two of 11 *G. barbouri* from Spring Creek were also infected.

Morphological Variation Gamonts have a slender appearance and are recurved into two limbs within the erythrocyte. Gamont dimensions are 32.4 ± 1.3 × 3.5 ± 0.4 µm (29.5–35 × 3–4.5, N = 25), with the anterior limb comprising 51.5% (48.5–53.7) of the total length. Gamont LW averages 114.9 ± 13.7 µm^2 (93–150.8), and L/W is 9.25 ± 1.0 (7.4–11.2). Nuclei are situated at gamont midbody and are 6.2 ± 0.7 × 2.8 ± 0.6 µm (5–7.5 × 2–5, N = 25), with LW 17.4 ± 5.4 µm^2 (10–35). There is no obvious sexually distinctive staining reaction, but those gamonts with heavy presence of azurophilic granules may be macrogamonts, in contrast to possible microgamonts that lack the suffusion of granules.

Merogony Meronts produced by sporozoites are present in tissues of the anterior third of the leech body, possibly corresponding to the "anastomosing lacunae" of the leech circulatory system, as described by Siddall and Desser (1991). Preerythrocytic meronts within the turtle host have not been found. The most abundant stage present in circulating blood is usually the premeront, which is distinguished by the presence of a prominent, central nucleus. Two types of erythrocytic meronts occur in the circulating blood. One type corresponds to those described for other *Haemogregarina* species: elongate and slightly curved, with 3–8 (4.7 ± 1.5, N = 14) small, compact nuclei (**Plate 58g,h**) that are a flattened ovoid in shape and dimensions of 14.9 ± 1.2 × 6.8 ± 1.0 µm (13–17 × 4.5–9, N = 14). The second type is similar, initially, in dimensions to the first type but contains, when first distinguishable as meronts, four larger, roughly square-to-rectangular nuclei (**Plate 58f**) that, as division occurs, form a rounded meront that then becomes elongate with nuclei arranged linearly in one or two rows when 6–12 nuclei have been formed. Division continues and eventually produces very large meronts with approximately 150 nuclei contained within the very thin and stretched erythrocyte membrane. These large meronts reach their peak in abundance in late summer and fall but remain visible in circulating blood year-round despite declining abundance of trophozoites and prematuration meronts in spring when gamonts usually comprise most of the erythrocytic parasite population. The large meronts were present in 102 of 115 infections in the type host, *M. temminckii*, and in both of the infected *Graptemys barbouri* of 11 examined at the type locality.

Sporogony Oocysts and sporocysts were not found in eight sectioned leeches, *Placobdella* spp., although two contained meronts in the leech tissues.

Effects on Host Gamont-infected erythrocytes appear often to cause an elongation and thinning of host erythrocytes, but hypertrophy in their length and width averages only 5% in each dimension, and their size (LW) is enlarged by only about 10%. Nuclei in erythrocytes host to gamonts are displaced into usually polar or lateropolar positions and average slightly smaller in size, about 5% less, than in uninfected cells. Cells infected by type 1 meronts are slightly

Plate 58 *Haemogregarina* sp. from *Macrochelys temminckii*, Georgia. (**A**) Primary meront in leech tissues, **a**; trophozoites, **b, c**; premeront, **d**; primary erythrocytic meronts, **e–h**; gamonts, **i–l**. (**B**) Secondary erythrocytic meronts, **a–h**.

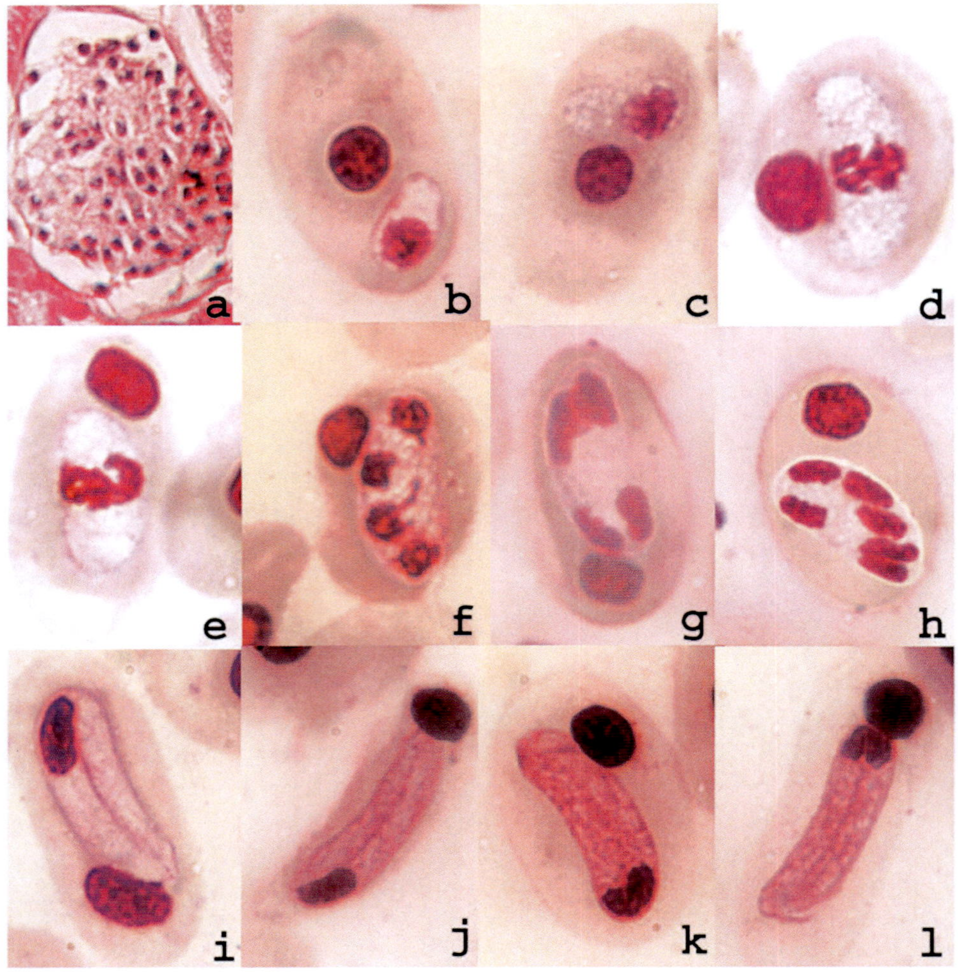

(A)

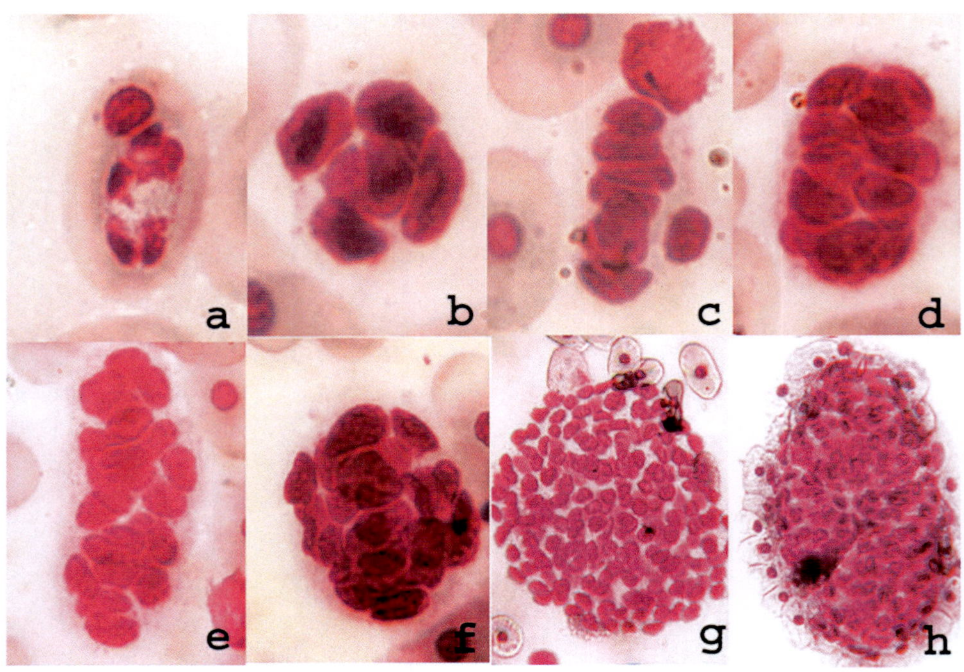

(B)

rounded, less elongated than those host to gamonts, with their nuclei similarly displaced.

Remarks This *Haemogregarina* species is distinguished from all other described species by the production of very large meronts in the circulating blood, and is under description (Telford et al., submitted).

PALEARCTIC *HAEMOGREGARINA* SPECIES

Haemogregarina stepanowi Danilewsky 1885 (**Plate 59**)

Diagnosis A *Haemogregarina* species with gamonts 29–37 × 3.0–5.5 µm, strongly recurved into two nearly equal limbs within the host erythrocyte. Erythrocytic meronts are 8–17 × 5.5–11 µm in size and produce four to eight merozoites while in circulating blood. Primary meronts in fixed tissue cells have not been described. Erythrocytic meronts reportedly are sequestered in the bone marrow of the turtle host, where 13–24 macromerozoites form within the erythrocyte. Macromerozoites give rise to micromeronts in erythrocytes that form either macro- or microgamonts, the infective stages for the leech vector *Placobdella catenigera*. Sporogony in the leech produces eight nonsporocystic sporozoites that enter the leech hemocoele and produce infection when inoculated into a turtle during feeding by the leech.

Type Host *Emys orbicularis* (Linnaeus) (Testudines: Emydidae).

Type Locality Europe, no specific locality.

Other Hosts None reported (see Remarks).

Other Localities Portugal (França, 1910, cited by Wenyon, 1926); Zelezin, Bulgaria.

Prevalence Unknown.

Morphological Variation This description was prepared from an active infection in *E. orbicularis* collected in Bulgaria. Trophozoites usually occupied a polar position in erythrocytes, little affecting the host cell nucleus. As size increased and trophozoites became premeronts or gamonts, the nucleus was usually forced into a polar or occasionally lateral position in the erythrocytes. Trophozoites were slightly curved with a large nucleus at one end and a prominently vacuolated cytoplasm. In premeronts, the vacuoles disappeared, and the nucleus became located centrally, often spanning the width of the parasite at midbody. Premeronts averaged 10.8 ± 1.0 × 5.1 ± 0.8 µm (9.5–13 × 4.5–6.5, N = 16), with nuclei 4.6 ± 1.5 × 3.8 ± 0.8 µm (3–7 × 2–5). Erythrocytic meronts in circulating blood contained two to eight nuclei, none of which had any evidence of cytoplasmic division into merozoites. Meronts containing two nuclei were 8–16 × 6–7 µm and with four to six nuclei were 12–16 × 6–7 µm. Two meronts that had seven and eight nuclei measured 14.5 × 5.5 µm and 15 × 11 µm, respectively. Gamonts within erythrocytes were recurved into two limbs, with the anterior limb comprising 54% of gamont length on average, varying 50–59%. Total length of gamonts averaged 32.0 ± 1.9 µm (29.0–36.5, N = 25), with maximum width 4.2 ± 0.7 µm (3.0–5.5), LW 135.6 ± 27.9 µm² (88.5–195.2), and L/W ratio 7.8 ± 1.2 (6.0–11.2). The gamont nucleus was located at or near the bend at midgamont, usually lying slightly more in the anterior limb. Nucleus size averaged 5.5 ± 0.8 × 3.5 ± 0.5 µm (4–7 × 2.5–4.5, N = 25), with LW 19.1 ± 3.3 µm² (12.0–24.8). Gamonts were commonly enclosed within an opaque capsule 17.6 ± 1.1 × 6.5 ± 1.2 µm (17–19 × 5.5–8.0, N = 6).

Merogony Sporozoites inoculated by the leech vector enter erythrocytes and become U-shaped vermicular trophozoites (Reichenow, 1910), then with growth the vermicular arms fuse and meronts are formed. Erythrocytes containing these meronts sequestrate in the bone marrow of the turtle. The meronts become macromeronts and produce 13–24 macromerozoites, which enter into erythrocytes, form micromeronts, and divide into six micromerozoites. Micromerozoites also enter erythrocytes and differentiate into either macrogamonts with small nuclei or microgamonts with large nuclei. The two gamont types are the infective stage for the leech host.

Sporogony Within the leech *Placobdella catenigera*, gametogony takes place in the leech intestine, where macro- and microgametocytes round up, fuse in szygy, and are enclosed within a membrane (Reichenow, 1910). A large, oval body or macrogamete is formed by the macrogametocyte, and the microgametocyte shrinks into a small, oval body. Within the small body, the nucleus divides, forming four microgametes, one of which fertilizes the macrogamete to produce a zygote. The zygote becomes a nonsporocystic oocyst in which eight sporozoites develop; then, rupture of the oocyst releases them into the intestinal tract of the leech from which the sporozoites enter the hemocoele, reaching the dorsal blood vessel of the circulatory system. When the leech feeds, sporozoites are injected through the proboscis into the turtle.

Effects on Host Pathology within infected turtle hosts has not been described. Parasitized erythrocytes appear to be only slightly longer and often wider than uninfected cells.

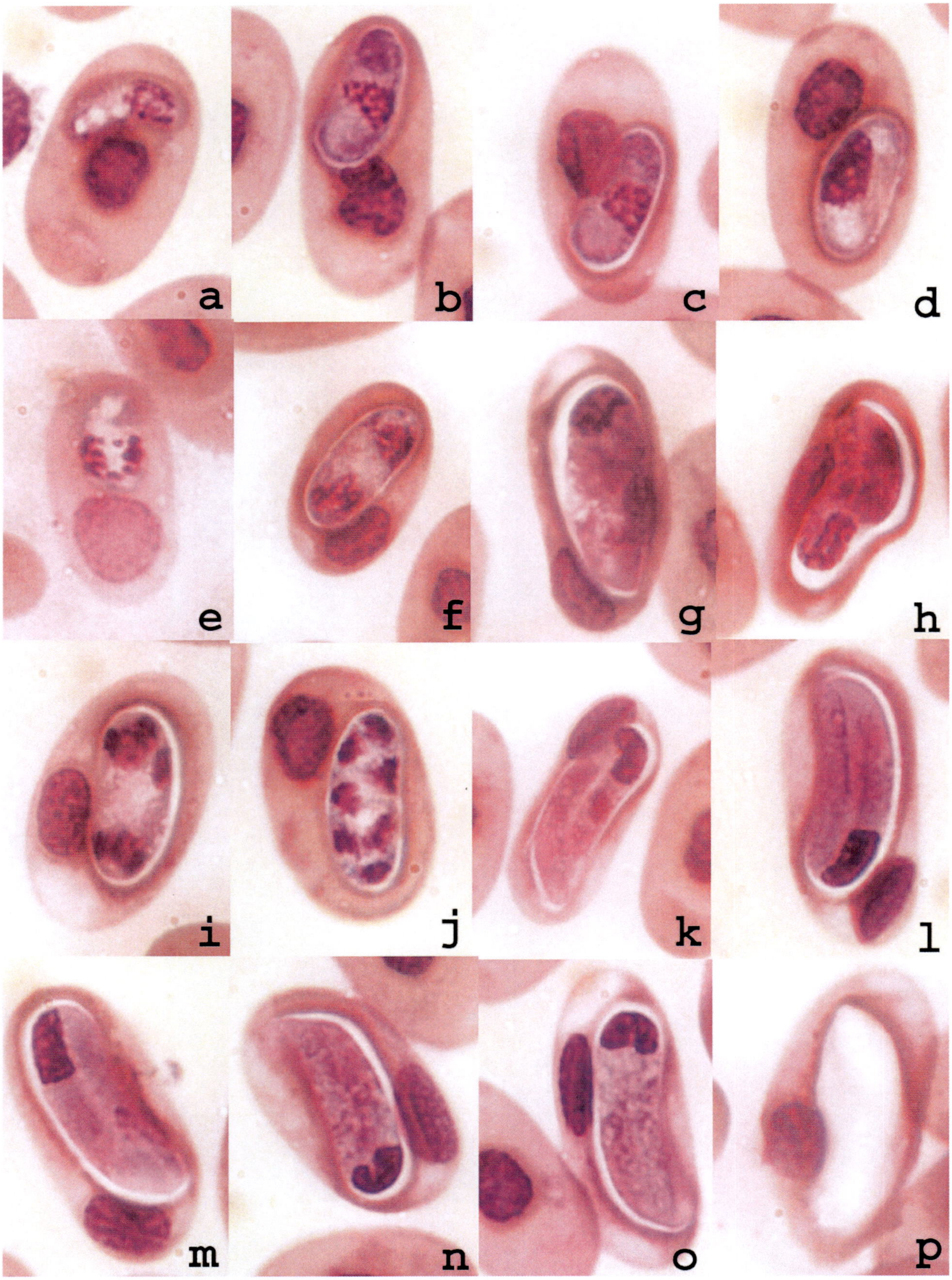

Plate 59 *Haemogregarina stepanowi* from *Emys orbicularis,* Bulgaria. Trophozoite, **a**; premeronts, **c, d**; erythrocytic meronts, **b, e–j**; gamonts, **k–o**; encapsulated gamont, **p**.

Remarks Despite his exhaustive description of the vertebrate phase of the life cycle, Reichenow (1910) did not describe the primary merogony in various fixed cells of the lung, liver, and spleen that occurs in *Haemogregarina balli* (Paterson and Desser, 1976; Desser, 1993), and he did not find the postsporogonic meronts in the anastomosing lacunae of the leech circulatory system, which produce up to 250 merozoites that are the infective stage for the turtle when fed on by the infected leech (Siddall and Desser, 1991). It is perhaps more likely that these two events were overlooked by Reichenow rather than the existence of two such different life cycle patterns among *Haemogregarina* species. However, neither Koidzumi (1910) nor Hahn (1909) described merogony as occurring in any cells other than erythrocytes in the *Haemogregarina* species from Japanese and North American turtles, respectively, that they studied. A short duration of primary merogony in fixed cells is a likely explanation. The application of the specific name *H. stepanowi* to turtle parasites of different continents (Hahn, 1909, and others) is not justified without evidence of conspecificity from genome analysis.

ASIAN *HAEMOGREGARINA* SPECIES

Haemogregarina choudhuryi Ray and Bhattacharjee 1984

Diagnosis A *Haemogregarina* species with lentiform gamonts, not recurved, 8.5 × 2.0–3.5 μm, and LW 15.5–20.6 μm². Erythrocytic meronts are 10.0 × 5.5 μm, with LW 38.5 μm², and produce six merozoites. Nonerythrocytic merogony occurs in capillary endothelium of the lung, where oval meronts 16.5 × 10.0 μm contain 26–35 merozoites. Sporogony in leeches occurs on the endothelial surface of intestinal ceca, with nonsporocystic oocysts forming eight to ten sporozoites.

Type Host *Lissemys p. punctata* (Bonnaterre) (Testudines: Trionychidae).

Type Locality Balitha village, Bankura District, West Bengal, India.

Other Hosts None known.

Other Localities None known.

Prevalence Three of five *L. punctata* were infected by *H. choudhuryi* at the type locality.

Morphological Variation Lentiform gamonts average 8.0 × 2.5 μm, with LW 18.7 μm² (N = 10). Microgamonts are 8.5 × 2.0 μm, with LW 15.5 μm² (N = 10), and macrogamonts, more elongated and slightly curved with blunt ends, are 8.5 × 3.5 μm, with LW 20.6 μm² (N = 10). Erythrocytic meronts are 10.0 × 5.5 μm, with LW 38.5 μm², and produce six merozoites arranged around a central residual mass. Merozoites are 5.7 × 1.0 μm with a central oval nucleus.

Merogony Nonerythrocytic merogony occurs only in the endothelium of the lung capillaries. At maturity, meronts are oval in shape, 16.5 × 10.0 μm, and contain 20–35 merozoites, 4.0 × 1.2 μm, that are formed around a central residual mass.

Sporogony Sometime following a blood meal by leeches, *Helobdella nociva*, micro- and macrogamonts are located on the microvillar surface of epithelial cells in the intestinal ceca. After becoming gametocytes, microgametocytes produce four nonmotile gametes, which fuse with macrogametes on the endothelial surface of the gastric ceca. Mature oocysts are 15.5 × 12.5 μm and contain eight to ten sporozoites that average 10.5 × 2.5 μm. R. Ray and Bhattacharjee (1984) stated, "Sporozoites liberated from the oocyst are found in the tissues of the leech."

Effects on Host Parasitized erythrocytes are hypertrophied, with the nucleus displaced toward a polar position, or are even "totally ejected in case of double infection" (R. Ray and Bhattacharjee, 1984).

Remarks While Siddall (1995) considered *H. choudhuryi* to be *Haemogregarina billeti* Simond 1901, it is impossible to identify this species with other *Haemogregarina* species described from the same region, and in some cases, the same host species, without genomic comparison of material from those localities where they were described. These species were *H. gangetica* Misra (1976) (redesignation of *H. simondi* Misra et al., 1974), *H. laverani* Simond (1901), *H. malabarica* De Mello (1932), *H. mesnili* Simond (1901), *H. vittatae* Robertson (1908), and *H. xavieri* De Mello (1932). Because the description, although short, of *H. choudhuryi* is relatively complete, with data provided for gamonts, primary and secondary meronts, sites of development in the turtle host, and sporogony in the leech vector, it is more desirable, in my opinion, to recognize *H. choudhuryi* as a valid species until appropriate comparisons can be made with the other described forms from the vast Indian subcontinent and vicinity.

AFRICAN *HAEMOGREGARINA* SPECIES

Haemogregarina pelusiensi Pienaar 1962

Diagnosis A *Haemogregarina* species with gamonts, recurved, 12.5–24 × 3.4–9 μm, enclosed within a translucent hard capsule, lentiform, 9–17 × 5–9 μm. The cytoplasm of the gamont is vacuolated, particularly anteriorly. Erythrocytic meronts are rare and 15–17 × 12–13 μm when two or three nuclei are present; mature meronts are undescribed. Sporogony occurs beneath the intestinal epithelium of the leech vector.

Type Host *Pelusios s. sinuatus* Smith (Testudines: Pelomedusidae).

Type Locality Rio do Save, Mozambique.

Other Hosts None known.

Other Localities Near Pietersburg, North Transvaal, South Africa (Paperna, 1989).

Prevalence Pienaar (1962) examined a single infected *P. sinuatus*, and Paperna (1989) found three of these turtles positive for *H. pelusiensi*. Neither author mentioned specimens found negative.

Morphological Variation Pienaar (1962) described gamonts ("sporonts") as 12.5–15.5 × 3.4 μm but provided no dimensions of meronts. Paperna (1989) reported gamonts with somewhat greater dimensions, 13–24 × 5–9 μm, "stout, with 1 end tapering and tightly bent and enlarged in a bean-shaped, 9–17 × 5–9 μm translucent hard capsule" (Paperna, 1989). Both authors reported vacuolated ("foamy" of Pienaar) cytoplasm in gamonts. Dividing meronts with two or three nuclei within erythrocytes are 15–17 × 12–13 μm (Paperna, 1989).

Merogony Merogony within fixed cells of host tissues has not been described.

Sporogony In the leech *Placobdella multistrigata*, Paperna (1989) found macro- and microgametocytes associated and zygotes attached to the intestinal epithelium. Zygotes, undivided oocysts, and oocysts in early division were present in the connective tissue beneath the intestinal epithelium. Paperna reported the occurrence of advanced oocysts in the leech connective tissue, with over 40 nuclei and 66 × 22 μm in size, and one oocyst 61 × 44 μm that contained over 55 sporozoites. These were either in connective tissue in proximity to the intestine "or in a more distant location" (Paperna, 1989). Sporozoites were 19–26 × 3–7 μm in size.

Effects on Host According to Pienaar (1962), erythrocytes parasitized by *H. pelusiensi* were considerably hypertrophied, and their nuclei were displaced laterally or to a pole of the cell. Infected erythrocytes were rounder and often distorted (Paperna, 1989).

Remarks When compared with the merogony of *Haemogregarina balli* in *Placobdella ornata* in Canada (Siddall and Desser, 1991), it is apparent that Paperna (1989) did not find mature oocysts containing sporocysts but rather described meronts, a product of sporozoites, as oocysts in extraintestinal locations, and the "sporozoites" were actually merozoites.

Karyolysidae

KARYOLYSUS SPECIES OF PALEARCTIC LIZARDS

Karyolysus lacazei (Labbé) 1894, Svahn 1975 (**Plate 60**)

Diagnosis A *Karyolysus* species with slender, recurved gamonts, without capsule, that bend somewhat around the erythrocyte nucleus. Gamonts are 17.7–27.7 × 2.7–4.6 μm, with estimated LW 79.2 μm^2 and L/W 6.1. The gamont nucleus is elongate, 3.8 × 8.1 × 1.9–3.8 μm, with estimated LW 15.7 μm^2. Nuclei are sometimes well defined but can be diffuse in appearance. One type of meront only, 11.5–30.8 × 7.7–18.5 μm, occurs in tissues of several organs and produces 16–32 merozoites. Oocysts, which mature in the podosomal region of the hemocoel of the mite vector, average 28 × 24 μm in size and produce 16 sporokinetes. Sporocysts formed in the mite egg average 24.0 × 14.9 μm and contain 16 sporozoites.

Type Host *Lacerta a. agilis* Linnaeus (Sauria: Lacertidae).

Type Locality France.

Other Hosts *Lacerta vivipara, L. viridis, Podarcis muralis.*

Other Localities Skillingaryd, Småland, Sweden; Tejn and Melsted, Bornholm, west of Vordingberg, Zealand, Swanninge, Funen, and Anholt, Denmark (Svahn, 1974).

Prevalence Overall, Svahn (1974) found infections of *K. lacazei* in seven localities in Denmark and one in Sweden. Among 11 Danish localities, *K. lacazei* was present in *L. agilis* in 6, with 38 of 55 (69.1%) lizards infected. Three lizards only harbored mixed infections. Three of eight *L. vivipara* from Melsted, Bornholm, Denmark, were host to *K. lacazei.*

Morphological Variation Gamonts from Scandinavian *L. agilis* are 22.0 μm (17.7–27.7, N = 200) × 3.6 μm (2.7–4.6, N = 75), with LW and L/W estimated from averages 79.2 μm^2

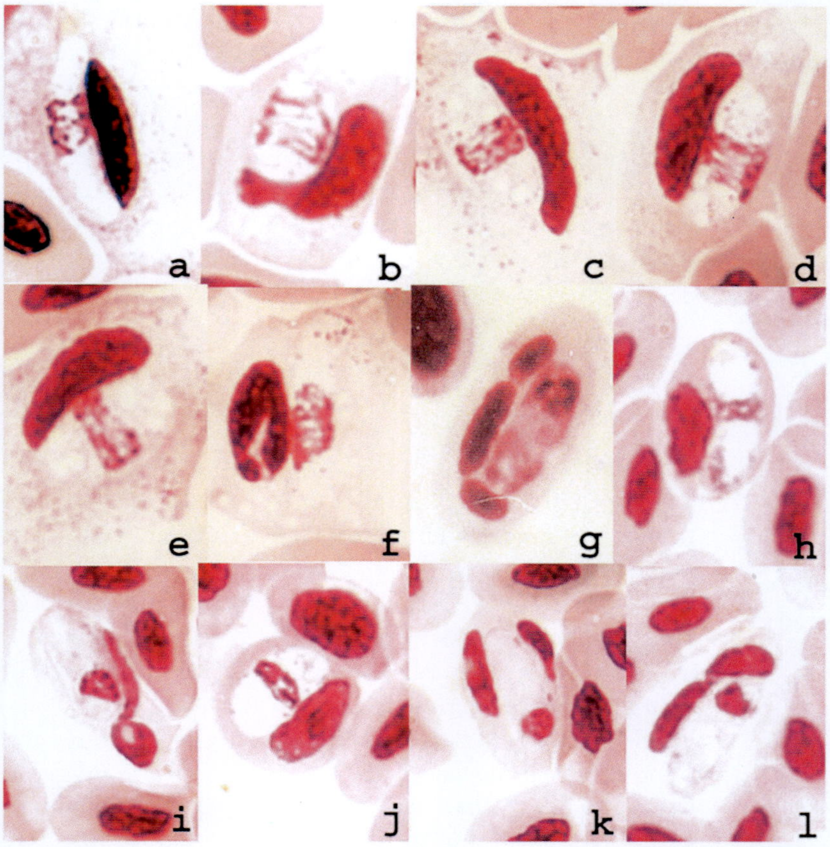

(A)

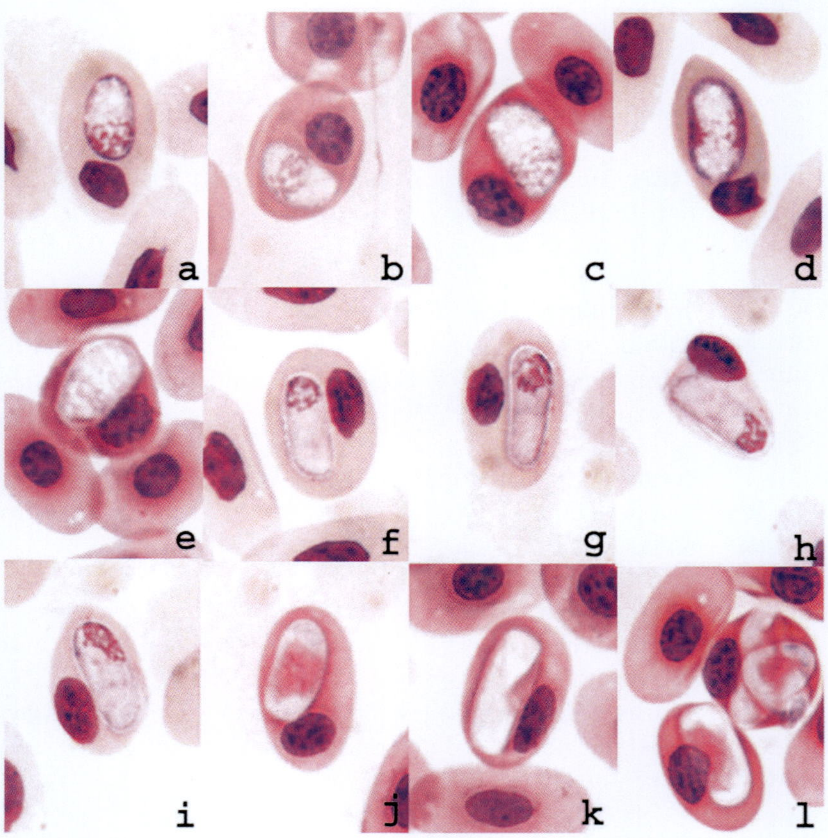

(B)

Plate 60 (**A**) *Karyolysus* spp. of European lacertid lizards. *Karyolysus latus* gamonts from *Lacerta agilis*, Russia, **a–e**, and *Karyolysus lacazei* gamonts from *Podarcis muralis*, Switzerland, **f–l**. (**B**) *Hemolivia mauritanicum* from *Testudo marginata,* Greece. Immature parasites, **a, b**; probable premeronts, **c–e**; mature gamonts, **g–i**; encapsulated gamonts, **j–l**.

and 6.1, respectively. Svahn (1975b) described gamonts as "slender, wormlike shape, and bent." Presumably "bent" indicates recurved, an interpretation supported by Reichenow's (1920) figures of *K. lacazei*, which depict recurved gamonts. As in Reichenow's figures, the nucleus is elongate, "sometimes well defined, sometimes more diffuse," variably positioned, and measures 5.4 × 2.9 μm (3.8–8.1 × 1.9–3.8, N = 25). In comparisons with gamont dimensions given by Labbé (1894) from French lizards (25–28 μm) and Reichenow (1921) from Spanish *Podarcis muralis* (25–32 μm), Svahn (1975b) thought the gamonts from Scandinavian *L. agilis*, 18–28 μm, agreed well.

Merogony Svahn (1975b) found only one type of meront present in sections of liver, lung, heart, spleen, ovary, and on one occasion, intestine. Meronts were ovoid, measured 19.9 × 12.1 μm (11.5–30.8 × 7.1–18.5, N = 20), and produced 26–32 slender, elongate merozoites, 8–12 μm. Labbé (1894) also reported the presence of only one type of meront, in bone marrow and spleen, 15–20 μm in length, containing 15–20 merozoites.

Sporogony Sporogony of *K. lacazei* from Scandinavian lizards occurred in the mite *Sauronyssus saurarum*. Ingested by adult mites, gamonts left erythrocytes immediately after ingestion of the blood meal. Free macrogametocytes averaged 19.1 × 2.8 μm, microgametocytes were 14.2 × 2.6 μm. Fertilization of macrogametes by microgametes and zygote formation occurred within 48 hours PF, with zygotes measuring 9.2–11.5 μm, rounding up as a sphere 17 μm in diameter. Depending on temperature, development could be arrested at this point for as long as 8 days. With higher temperatures, at 21–30°C, oocysts reached a maximum size of 21 × 15–38 × 38 μm, averaging 28 × 24 μm. Development of zygote to oocyst occurred within the mite hemocoel in the podosomal regions. Within oocysts, 16 nuclei formed that became elongated vermicular, motile sporokinetes, 35 × 8–42 × 8.5 μm. These matured after a second blood meal by the mite, invaded eggs within the mite, and encysted as sporocysts, 21.5–26.2 × 11.5–18.5 μm, averaging 24.0 × 14.9 μm. Division within sporocysts produced 16 sporozoites, 20.0 × 5.4–24.6 × 4.2 μm. Sporocysts were mature within nymphal mites about 10 days following oviposition of eggs by female mites, during which period hatching occurred and the larvae molted within 1 day to nymphs. Lizards infected by ingesting only a few experimentally infected nymphs became patent for *K. lacazei* 30–34 days later (Svahn, 1975b).

Effects on Host Parasitemia by *K. lacazei* is often "very heavy," but Svahn (1974, 1975b) did not state parasitemias as percentages of erythrocytes infected. Host cells are hypertrophied and their cytoplasm apparently severely dehemoglobinized. Erythrocyte nuclei are displaced, swollen, and less intensely stained or sometimes compressed and darkly stained.

Remarks Because sporogony of *Karyolysus* is essentially similar, where known, for all species of the genus, only points of difference are described in the several accounts of *Karyolysus* species.

Karyolysus lacertae (Danilewsky) 1886 Reichenow 1913 Syn. *Karyolysus lacertarum* Labbé 1894

Diagnosis A *Karyolysus* species with a short, recurved posterior part of the gamont and an anteriorly placed nucleus, contained within a capsule somewhat resistant to penetration by stains. Gamonts are 12.3–14.6 × 3.9–4.6 μm, with estimated LW 58.5 μm^2 and L/W 3.2. The gamont nucleus averages 3.7 × 3.0 μm in size, with estimated LW 11.1 μm^2. Two types of meronts occur within tissues of organs, an ovoid meront 13.1–17.7 × 8.5–13.1 μm that produces 8–30 merozoites with a large dense nucleus, and round or ovoid meronts 11.5–20.0 × 9.2–13.8 μm that contain 32 to more than 100 merozoites with round dense nuclei. Oocysts form within epithelial cells of the mite stomach.

Type Host *Lacerta a. agilis* Linnaeus (Sauria: Lacertidae).

Type Locality Kharkov, Russia.

Other Hosts *Podarcis muralis, Lacerta viridis, L. ocellata, L. vivipara*, many *L. saxicola* ssp., *L. strigata, L. taurica, L. sicula,* and *L. truncata*.

Other Localities France (Labbé 1894); vicinity of Madrid, Spain (Reichenow, 1920, 1921); Tejn, Bornhelm, Denmark (Svahn, 1974, 1975b); and many Russian localities (Ovezmukhammedov, 1987).

Prevalence Two of 19 (10.5%) *L. agilis* from Tejn, Bornholm, Denmark, were infected by *K. lacertae*, both in mixed infection with *K. lacazei* (Svahn, 1974, 1975b).

Morphological Variation Gamonts of *K. lacertae* in *L. agilis* from Denmark (Svahn, 1975b) average 13.6 × 4.3 μm (12.3–14.6 × 3.9–4.6, N = 53), with LW and L/W, estimated from averages, 58.5 μm^{2}and 3.2, respectively. Nuclei, located in the anterior end of the gamont, average 3.7 × 3.0 μm, with estimated LW 11.1 μm^2.

Merogony Two kinds of meronts occur in tissues of the liver, lung, heart, spleen, and ovary. Svahn (1975b) reported the presence of meronts in smears taken from the tip of the lizard tail, which indicates presence of meronts in striated muscle or the capillaries associated with it as well. Meronts that form large merozoites were considered by Reichenow (1913, 1921) to be "asexual," the merozoites forming new meronts in the tissues. "Sexual" meronts formed numerous smaller merozoites that entered erythrocytes and became gamonts. Svahn (1975b) found asexual meronts in lung and spleen. Ovoid in shape, they averaged 15.7 × 11.4 µm (13.1–17.7 × 8.5–13.1, N = 15) and contained 8–30 merozoites about 10 × 3 µm. Sexual meronts occurred in all tissues listed; they were round or ovoid, 14.8 × 11.5 µm (11.5–20.0 × 9.2–13.8m N = 20), and contained about 32 merozoites, 9 µm in length. Two apparently unusually larger meronts in liver smears measured 26.9 × 17.7 µm and 29.2 × 19.2 µm.

Sporogony When ingested by the mite *Sauronyssus saurarum*, the encapsulated gamonts of *K. lacertae* take longer to leave the host erythrocyte than do gamonts of *K. lacazei* and *K. latus* (Svahn, 1975b). Free macrogametocytes of *K. lacertae* are 15.0 × 3.3 µm (11.5–17.7 × 2.7–4.2, N = 25), and microgametocytes are 15.3 × 2.0 µm (13.1–16.9 × 1.5–2.7, N = 15). Sporogony of *K. lacertae* appears similar to that of *K. lacazei*, described above, except that the zygotes enter epithelial cells of the mite stomach, where oocysts form, in contrast to *K. lacazei,* which develops in the hemocoel. Oocysts and sporokinetes of the two species are similar in dimensions, and their sporokinetes could not be distinguished in mixed infections. Sporocysts of *K. lacertae,* however, are larger than those of *K. lacazei*, 30.9 × 18.9 µm (28.5–33.1 × 16.2–22.3, N = 12), and sporozoite numbers were more variable, 16–32. Sporogony required 5–7 days after the parent mite took her second blood meal. Up to 100 sporocysts of *K. lacertae* were found in one nymphal mite, in contrast to the usual few sporocysts present in the mites (Svahn, 1975b).

Effects on Host Experimental infections of *K. lacertae* in laboratory lizards were often very heavy (Svahn, 1975b). Although hosts showed no disease symptoms, leukocytosis and anemia resulted from infection. Gamonts caused hypertrophy of infected erythrocytes and apparently some dehemoglobinization (Svahn, 1975b), with nuclei swollen or sometimes compressed and strongly displaced.

Remarks Reichenow (1913, 1920, 1921) elucidated the life cycle of *Karyolysus lacertae* in *Sauronyssus saurarum* and *Podarcis* (*Lacerta*) *muralis*, with great emphasis on the cytology of life history stages in the mite vector. The studies by Svahn (1974, 1975b) verified Reichenow's conclusions in all respects and provided additional information on the distribution and ecology of *K. lacertae* and three other species of *Karyolysus* in Scandinavian lizards.

Karyolysus latus Svahn 1975 (**Plate 60**)

Diagnosis A *Karyolysus* species without capsule, usually with a broad, oval shape, sometimes lentiform, with recurved posterior end visible in older gamonts. Gamonts are 9.6–17.7 × 3.5–10.0 µm, with estimated LW 76.7–107.3 µm^2 and L/W 2.1–2.2. Volutin granules are often present at one or both ends of gamont. Gamont nuclei are usually central but occasionally terminal, diffuse in macrogamonts and more compact in microgamonts. Meronts occur primarily in lung or in liver. One type of meront is present, ovoid, 13.1–20.8 × 8.5–12.3 µm, containing 8–24 nuclei.

Type Host *Lacerta a. agilis* Linnaeus (Sauria: Lacertidae).

Type Locality Snogeholm, Scania, Sweden.

Other Hosts *Lacerta vivipara.*

Other Localities Reported from Maglehem, Degeberga, Hallamölla, and Brösarp, all in Scania, Sweden, and from Gilleleje, Denmark (Svahn, 1974).

Prevalence *Karyolysus latus* was found in 24 of 74 (32.0%) *L. agilis* from five Swedish localities; in a single *L. vivipara* from Degeberga, Sweden; and in 1 of 1 *L. agilis* from Denmark (Svahn, 1974).

Morphological Variation Macrogametocytes, which are strongly basophilic in staining, are 14.9 × 7.2 µm (13.1–17.7 × 5.0–10.0, N = 475), with LW and L/W estimated from averages 107.3 µm^2 and 2.1, respectively. Nuclei are large, usually central but occasionally terminal, 6.0 × 4.3 µm (3.1–5.4 × 4.6–7.7, N = 25) and estimated LW 25.8 µm^2, diffuse in structure. Microgamonts are smaller, 13.0 × 5.9 µm (9.6–15.4 × 3.5–7.7, N = 377), with estimated LW and L/W, respectively, 76.7 µm^2 and 2.2, and stain faintly blue. Their nuclei are smaller and more compact than those of macrogamonts and average 3.6 × 4.5 µm (3.1–4.6 × 3.8–6.2, N = 20), with estimated LW 16.2 µm^2.

Merogony Meronts, of one type only, occur in the lung and liver, most commonly in the former site. Their dimensions vary, 13.1–20.8 × 8.5–12.3 µm, and they contain 8–24 nuclei. Pulmonary merozoites are about 9 µm in length.

Sporogony Lacking a capsule around them, gamonts are liberated from host cells immediately after a blood meal by a mite. Macrogametocytes of *K. latus* are 21.8 × 6.0 µm

(16.1–25.4 × 3.8–7.7, N = 20), and microgametocytes are 17.8 × 4.1 µm (12.3–23.8 × 3.5–5.4, N = 10). Sporogony follows the same course as for *K. lacazei* and *K. lacertae*. Zygotes are 13.8–14.6 µm, forming oocysts usually in epithelial cells of the stomach as in *K. lacertae* but also in the podosoma and anterior opistosoma of the coelom, as in *K. lacazei*. Oocysts of *K. latus* average 36 × 24 µm (25 × 15–52 × 35, N = 15) and contain 16 nuclei that, when mature sporokinetes, measured 50 × 8–64 × 8.5 µm. Sporokinetes contain abundant volutin granules, which give them a reddish hue. Sporocysts formed when sporokinetes invade mite eggs are 28.4 × 20.3 µm (23.1–32.3 × 15.4–28.5, N = 12) and produce 32 sporozoites 19.2 × 7.7–24.0 × 5.4 µm, 7 days after the second feeding by the maternal mite.

Effects on Host Erythrocytes infected by gamonts of *K. latus* are hypertrophied and dehemoglobinized to some degree. Host cell nuclei are always displaced, usually elongated, compressed, and more intensely stained than nuclei from uninfected cells. Gamont presence sometimes divides erythrocyte nuclei into two portions.

Remarks There is technically a question concerning the validity of the name *latus* inasmuch as Svahn used the name in 1974, prior to the formal description that appeared in 1975. A stable taxonomy would not be well served to consider *Karyolysus latus* to be a *nomen nudum*, and Svahn's intention is followed here.

Karyolysus minor Svahn 1975

Diagnosis A *Karyolysus* species characterized by small, oval or pear-shaped gamonts, without capsule. The posterior end may be recurved, and the nucleus is band-shaped. Gamonts are 4.6–7.7 × 3.1–5.4 µm, with estimated LW 23.4 µm^2 and L/W 1.5. Gamonts can occur in leukocytes as well as in erythrocytes.

Type Host *Lacerta a. agilis* Linnaeus (Sauria: Lacertidae).

Type Locality Degeberga, Scania, Sweden.

Other Hosts None known.

Other Localities Mols, Jutland, and Svanninge, Funen, Denmark.

Prevalence *Karyolysus minor* infected one of five *L. agilis* at the type locality, one of one at Swanninge, and one of six at Mols (Svahn, 1974).

Morphological Variation Gamonts average 6.0 × 3.9 µm (4.6–7.7 × 3.1–5.4, N = 76), with LW and L/W estimated from the mean 23.4 µm^2 and 1.54, respectively. The cytoplasm stains faintly red with Giemsa stain. The nucleus is diffuse, sometimes band-shaped, sometimes appearing just as associated granules.

Merogony Unknown.

Sporogony Unknown.

Effects on Host Erythrocytes infected by the small gamonts of *K. minor* are normal in size, shape, and staining reaction. Erythrocyte nuclei appear normal and are not displaced. Leukocytes can be infected; Svahn (1975b, Figure 20) depicted a probable lymphocyte containing a gamont.

Remarks The name *Karyolysus minor*, as *K. latus* above, was published by Svahn in 1974, but the formal description did not appear until 1975.

HEMOLIVIA SPECIES

Hemolivia mauritanicum (Sergent and Sergent) 1904 Landau and Paperna 1997 (**Plate 60**)

Diagnosis A *Hemolivia* species of tortoises that forms strongly recurved gamonts 9–12 × 5–7 µm in cysts that average 12.2 × 6.2 µm within erythrocytes. Meronts, 13–17 × 7–11 µm, form within reticuloendothelial cells of the liver, lungs, spleen, bone marrow, and muscles, and in erythrocytes within these organs. Cysts containing one to six zoites occur in the same organs of the host as do meronts. Sporogony takes place in stomach epithelial cells of the ixodid tick *Hyalomma aegyptium,* where irregularly shaped oocysts produce sporokinetes that give rise to sporocysts containing on average 16 sporozoites.

Type Host *Testudo graeca* Linnaeus (Testudines: Testudinidae).

Type Locality North Africa.

Other Hosts *Testudo marginata* (Široký et al., 2004, 2005).

Other Localities Bulgaria, Greece, and Turkey (Široký et al., 2005).

Prevalence Široký et al. (2005) found *H. mauritanica* in 2 of 14 (14%) *T. graeca* in Bulgaria and in 24 of 26 (92%) in Turkey, as well as in 38 of 47 (81%) of *T. marginata* in Greece.

Morphological Variation The mature gamont, free from the erythrocyte, averages 18.4 μm in length (Michel, 1973), with 5.9 μm maximum width calculated from Figure 3 of Michel. Within the host cell, it is recurved on itself in a cyst on average 12.2 × 6.2 μm. The nucleus occupies the entire gamont width and is about equal in length to its width. The two gamont limbs are about equal in length. The anterior end of the free gamont is broadly pointed, and the posterior end forms a more narrow point. Azurophilic granules and numerous small, rounded vacuoles occur in the cytoplasm of the gamont. Siroký et al. (2007) reported mature gamonts to be 9–12 × 5–7 μm (N = 22), "elongated oval to cylindrical, sometimes bean-shaped," encased within a stain-resistant capsule. Gamont nuclei are in a polar position.

Sporogony Sporogony takes place in the ixodid tick *Hyalomma aegyptium*, as reported by Brumpt (1938). Michel (1973) described sporogony beginning with free gamonts present in the stomach contents at 7 days PF, rarely seen associated in pairs. Histological sections showed many gamonts within epithelial cells of the stomach, notably in the apical portion of the cell. At 12 days, gamonts were visibly paired; female gametocytes appeared rounded and closely associated with more elongate male gametocytes. The latter produced two microgametes that then divided to form four gametes. Both gametogenesis and fertilization occurred within the epithelial cells. Young oocysts had formed by 12 days PF. At 26 days PF, oocysts had grown to 80 μm in diameter, and sporogony began with sporoblast formation. At maturation (Michel, 1973), oocysts produced elongate mature sporocysts approximately 26 × 10 μm (from Michel's figure) that contained an average of 16 sporozoites. The sporozoites had dimensions of 16 × 2 μm, with a large crystalloid body at one of the poles. Sporocyst sizes in *H. aegyptium*, collected from *T. graeca* and *T. marginatum*, were similar from each *Testudo* species, elongately ellipsoidal in shape, 31.5 × 16.4 μm (28–35 × 12.5–21), and contained "numerous banana-shaped sporozoites" (Siroký et al., 2005). Oocyst shape was not described. At maturity, oocysts appeared to rupture, releasing sporozoites into the epithelial cells. Landau and Paperna (1997) reexamined and reinterpreted Brumpt's material from *H. aegyptium*, and although no details were provided, their figures appear to support the production of sporokinetes rather than sporocysts by the oocysts. One figure, however, apparently shows sporokinetes within the oocyst and a sporocyst either adjacent to the oocyst or within it. Oocyst shape was described as irregular rather than stellate, which is characteristic of *Hemolivia stellata* of neotropical toads and *Hemolivia mariae* of Australian skinks.

Merogony Michel (1973) described meronts found in a *T. graeca* experimentally infected by ingestion of 31 positive *H. aegyptium*. Gamonts appeared 30 days PF, and tissues taken 4 months later showed many meronts within reticuloendothelial cells of the liver, lungs, spleen, bone marrow, and muscles. Mature meronts averaged 16.8 × 12.2 μm in size and contained 12–16 merozoites arranged in two groups at each pole of the residual cytoplasmic mass. Michel considered these to be comparable in their morphology to the "merontes secondaires" of *Hepatozoon domerguei*, described by Landau et al. (1972), that produce gamonts (i.e., micromeronts). Michel also found cysts in liver and lungs that averaged 15.3 × 10.7 μm and usually contained two cystozoites, rarely four or six. Mature meronts within reticuloendothelial cells and erythrocytes were reported by Siroký et al. (2007), mostly in the spleen, liver, lungs, and kidneys. Mature meronts were 14.2 × 9.3 μm (13–17 × 7–11, N = 17) and contained 7–12 elongate merozoites, mostly arranged in parallel within the long axis of meronts. Cysts were found in the same organs as meronts and contained usually two, rarely one, elongate cystozoites. Cysts are 14.8 × 7.9 μm (12–18 × 6–10) when dizoic and 13 × 6–8 μm when monozoic.

Effects on Host Pathological effects by *H. mauritanica* on its host were described as mild by Siroký et al. (2007) in the tortoise with the greatest parasitemia, a gametocytemia of 21.3%. There was "mild focal to multifocal lower nephron necrosis and hepatocellular hydropic degeneration."

Remarks New material of the tick stages should be examined to confirm the formation of sporokinetes by oocysts that then invade new epithelial cells and produce sporocysts. The available description provides few details.

Hemolivia mariae Smallridge and Paperna 1997 (**Plate 61**)

Diagnosis A *Hemolivia* species with encapsulated gamonts 18 × 5 μm with estimated LW 90 μm² and L/W 3.6. Stellate oocysts 142–201 μm with three to five arms develop in a parasitophorous vacuole within gut epithelial cells of the tick vector. Sporokinetes that develop within oocysts enter other gut cells and into extraintestinal locations, where they become hard-walled oval sporocysts, 29–34 × 14–18 μm, giving rise to 8–20 sporozoites.

Type Host *Tiliqua rugosa* Gray (Sauria: Scincidae).

Type Locality Mount Mary, South Australia.

Other Hosts *Egernia stokesi* (J. Stein, unpublished).

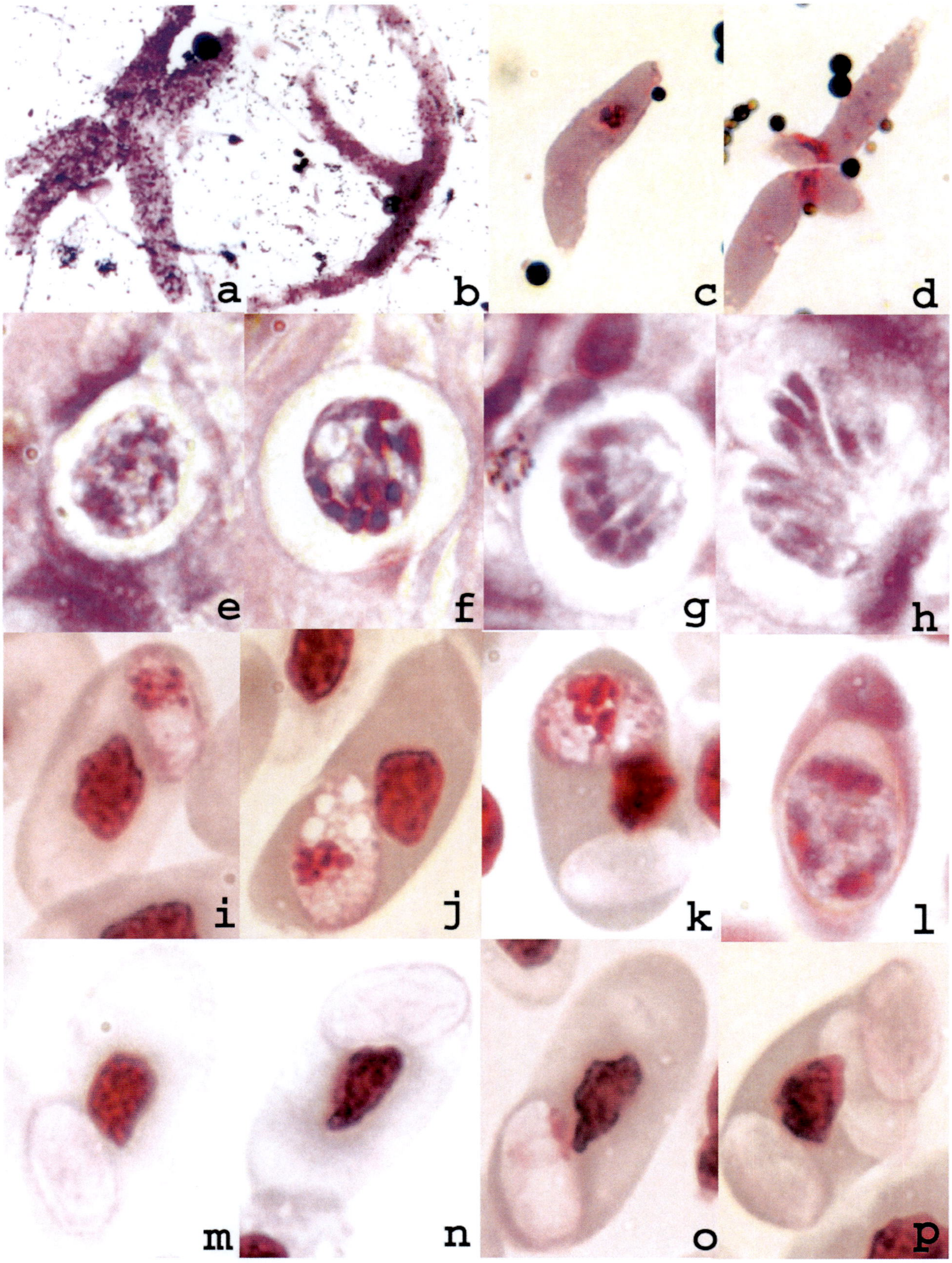

Plate 61 *Hemolivia mariae* from *Egernia stokesiia,* Australia. Oocysts in tick vector, **a, b**; free sporokinetes from tick, **c, d**; hepatic meronts in lizard host, **e–h**; trophozoite, **i**; premeronts, **j, k**, with encapsulated gamont also in **k**; erythrocytic meront, **l**; encapsulated gamonts, **m–p**

Other Localities In *Egernia stokesi*, Chace Range, Quom, Neuroodla, The Four Mile, Drakes Nob, and Hawker, all in South Australia (J. Stein, unpublished).

Prevalence In a population of *T. rugosa* near the type locality, *Hemolivia mariae* infected 66 of 567 (11.5%); prevalence did not fluctuate during 5 years of sampling (Smallridge and Bull, 2001b). There was no difference between sexes, but *H. mariae* was more common in adults (12.8%) than in subadults and juveniles (1.6%), and prevalence among adults declined with increasing body length. Although adults were more heavily infested with ticks than younger lizards, prevalence did not differ between lizards infested and those not infested by ticks at capture. Tick infestation did not differ between sexes of lizards. In a series of 39 *E. stokesi* that were hemoparasite positive, 23 were infected by *H. mariae* (Telford).

Morphological Variation Smallridge and Paperna (1997) reported only the mean size of gamonts, 18 × 5 μm, in the type host, *Tiliqua rugosa*. Estimated LW and L/W from this average were 90 μm^2 and 3.6, respectively. In *Egernia stokesi*, the more commonly seen encapsulated gamonts, with rounded ends and presumably recurved, are nearly opaque, 9.0 ± 0.3 × 5.0 ± 0.3 μm (8.5–9.5 × 4.5–6.0, N = 25) (Telford). The thick capsule surrrounding the erythrocytic gamonts is 30 nm thick (Smallridge and Paperna, 2000b), and the gamont nucleus is at one end. Gamonts are contained within parasitophorous vacuoles.

Merogony Although Smallridge and Bull (2001a) illustrated immature meronts in erythrocytes of *Tiliqua rugosa*, there apparently has been no description of either erythrocytic meronts or those within other tissues of the host. Both types of meronts were found by Juergen Stein in *Egernia stokesii* and are illustrated here along with stages from ticks but are not described, pending his eventual publication.

Sporogony Gamonts, ingested usually by the tick *Amblyomma limbatum* or by *Aponomma hydrosauri*, leave the capsule within tick intestinal cells. Szygy apparently occurs while still encapsulated, within the intestinal cells, and within the parasitophorous vacuole (Smallridge and Paperna, 2000b). Both the capsule wall and that of the vacuole disappear as development continues. Stellate oocysts with three to five arms develop from the zygote within intestinal cells. Large numbers of sporokinetes form around crystalloid material within the oocyst, and when mature are released and scatter throughout the tick intestine, entering cells of the same type as those that host oocysts (Smallridge and Paperna, 2000a). A parasitophorus vacuole forms around sporokinetes, and a thick sporocyst wall forms. Sporozoites develop from the sporokinete and densely pack the sporocyst around a residuum (Smallridge and Paperna, 2000a). When ticks containing mature sporocysts engorge on lizards and they are ingested by the lizards, infection results and is patent in 6.9 (4–9) weeks (Smallridge and Bull, 1999).

Effects on Host Erythrocytes infected by gamonts of *H. mariae* appear normal in size and structure. Infection intensity appears to be at a low level, 0.01–2.50%, both between years and throughout a year (Smallridge and Bull, 2001b), and infections did not seem to affect their host's body condition.

Remarks Infections of *H. mariae* in *Egernia stokesi* appear to be more common, of much greater intensity, and of longer duration than those in the type host, *Tiliqua rugosa* (J. Stein, personal communication).

3

THE HEMOCOCCIDIA OF LIZARDS

Apicomplexa: Eimeriorina

The Hemococcidia, family Lankesterellidae Nöller 1920, represented in reptiles by three genera, *Lankesterella* Labbé 1899, *Schellackia* Reichenow 1919, and *Lainsonia* Landau 1973, are true coccidians, related to the intestinal parasites of the Eimeriidae. The course of gametogony and sporogony within cells of the intestinal wall in lizards differs little from the intestinal coccidians except in the absence of sporocyst formation in the lankesterellids. Instead of transmission by direct contact with sporocysts in a fecally contaminated external environment, sporozoites enter circulating blood cells (leukocytic, erythrocytic, or both, depending on the species) and then are ingested by acarine or dipteran vectors. Transmission is accomplished at least for saurian hosts by predation on the infected invertebrate. There is no development or multiplication within the vector; it plays a mechanical role only in the hemococcidian life history. Infections by hemococcidia can be transmitted experimentally by ingestion of the infected arthropod or by intraperitoneal injection of infected blood (Bonorris and Ball, 1955; Landau et al., 1974; Lainson et al., 1976; Bristovetsky and Paperna, 1990; Paperna and Finkelman, 1996). Jordan and Friend (1971), Klein et al. (1988a), and Bristovetzky and Paperna (1990) transmitted *Schellackia* infections by feeding infected mites, ticks, mosquitoes, or sand flies to lizards. Bristovetsky and Paperna (1990) and Finkelman and Paperna (1998) fed infected blood and liver to uninfected lizards and accomplished transmission to the recipients.

Evidence that sporozoites acquired from infected blood meals can accumulate in epithelial cells of the mosquito or mite gut was provided by Reichenow (1919), Landau (1973), Klein et al. (1988a), and Paperna (1993). The life cycle in the vertebrate host can be summarized as follows: Sporozoites of *Schellackia* species ingested or inoculated enter the intestinal epithelial cells and form meronts; the resulting merozoites become gamonts that produce macro- and microgametes; fertilization results in a zygote, usually in the lamina propria of the intestinal wall; and oocysts then form and produce eight sporozoites only, without formation of sporocysts. When released from the oocysts, the sporozoites enter white or red blood cells and await ingestion by an arthropod for transmission. Variation in intestinal site and blood cell type is a species characteristic and is described in the species accounts.

At present there are nine or possibly ten valid species of *Schellackia* described from lizards of the families Polychrotidae, Phrynosomatidae, Agamidae, Gekkonidae, Opluridae, and Lacertidae, and undescribed species have been seen in the Teiidae, Scincidae, and Chamaeleonidae. Host specificity appears to be very high, just as among intestinal coccidians, and genome comparisons of the probably many *Schellackia* species found in saurian faunas of such areas as the American Southwest and Mexico would be the most efficient method of species differentiation. Sporozoite morphology and dimensions lack sufficient, consistent variation to be very useful in taxonomy of the hemococcidians.

Two species of *Lainsonia* are known, *L. iguanae* from *Iguana iguana* (Landau, 1973) and *L. legeri* in the tegu *Tupinambis teguixin* (Landau et al., 1974). *Lainsonia* differs from *Schellackia* by the development of sporozoites through asexual and sexual reproduction within reticuloendothelial cells of the liver, spleen, lungs, kidney, and capillaries of the brain. Sporozoites either enter white or red blood cells or accumulate in reticuloendothelial cells, entering an infective diapause (Lainson et al., 1976). As in *Schellackia*, eight sporozoites are produced by oocysts. *Lainsonia iguanae* sporozoites ingested by mosquitoes accumulated in cells of the mosquito stomach wall (Landau, 1973). *Lainsonia legeri* was transmitted by intraperitoneal inoculation and ingestion of infected blood (Landau et al., 1974).

Lankesterella differs from *Schellackia* and *Lainsonia* by producing 32 or more sporozoites within oocysts and is primarily a parasite of anurans. Two poorly described species have been reported in lizards: *L. millani* (Alvarez Calvo, 1975) in *Lacerta lepida nevadensis* of Spain and *L. baznosanui* in *Lacerta vivipara* of Romania (Chiriac and Steopoe, 1977). The latter species supposedly develops within capillaries and venules of the ovaries, suprarenal gland, and corpus luteum, rather strange sites for a hemococcidian. Insufficient information is available for further comment on these parasites.

The several studies on ultrastructure of *Schellackia* species have found few differences in structure of meronts, gamonts, and sporozoites, none of them major, from *Lankesterella* and intestinal coccidians. Ostrovska and Paperna (1987a) examined the fine structure of *Schellackia cf. agamae* gamonts and merogony of the same species (1987b). Although merozoites were formed by exogenesis, there was an indication that some merozoite primordium, budding into subpellicular inclusions, might be developing by endogenesis. Klein et al. (1988a) found nothing distinctive in the fine structure of *Schellackia golvani*, in comparison to other *Schellackia*, *Lankesterella*, *Eimeria*, *Toxoplasma*, and *Haemogregarina* species. Paperna (1993) examined *Culex molestus* and *Aedes aegypti* 5–7 days following feeding on *Laudakia stellio* with chronic *Schellackia cf. agamae* infections. Sporozoites were found in parasitophorous vacuoles within intestinal epithelial cells without modification of basic structures seen in sporozoites from blood and liver. There was some slight damage to host cells by sporozoites. *Schellackia ptyodactyli* fine structure was examined by Paperna and Finkelman (1996); again, the similarity of merogonic stages, microgametes, and sporozoites to those of *Eimeria* species was confirmed. Macrogamonts diverged somewhat in the appearance of their wall-forming bodies from that of *Schellackia cf. agamae* and the pattern typical of eimerians.

Species Accounts

SCHELLACKIA SPECIES OF NORTH AMERICAN LIZARDS

Schellackia occidentalis Bonorris and Ball 1955 (**Plate 62**)

Diagnosis A *Schellackia* species parasitic in erythroid and leukocytic cells of North American lizards of the genera *Sceloporus* and *Uta*. Merogony and gamogony occur in epithelial cells of the small intestine, and oocysts develop in the lamina propria. A refractile body is usually present in sporozoites.

Type Host *Sceloporus occidentalis becki* Van Denburgh (Sauria: Phrynosomatidae).

Type Locality Santa Cruz Island, California.

Other Hosts *Sceloporus occidentalis biseriatus, Uta stansburiana hesperis* (Bonorris and Ball, 1955); *Sceloporus undulatus* (Jordan and Friend, 1971).

Other Localities Los Angeles County, California (Bonorris and Ball, 1955); Fargo, Georgia (Jordan and Friend, 1971); Alachua (Klein et al., 1988a), Putnam, and Hillsborough counties, Florida (Telford).

Prevalence *Schellackia occidentalis* infected two of eight *Sceloporus occidentalis becki* and two of three *Uta stansburiana hesperis* from the type locality (Bonorris and Ball, 1955). Annual prevalence in 5761 *S. undulatus* examined from 1956 to 1970 varied 7–28%, average 21% at Fargo, Georgia (Jordan and Friend, 1971), and in Florida 4 of 270 (1.5%) *S. undulatus* were infected (Telford).

Sporozoites Sporozoites in erythrocytes of *S. occidentalis becki* or *U. stansburiana hesperis* averaged 7.8 × 3.6 μm (5.6–9.6 × 2.8–5.6) and did not contain a reserve vacuole (Bonorris and Ball, 1955). None were seen in leukocytes. In *S. undulatus* from north Florida, a single refractile body is variably present, even within the same infection (Telford). Erythrocytic sporozoites were 6.9–8.7 × 4.1–7.0 μm, leukocytic sporozoites were 7.7–10.9 × 4.8–7.7 μm, and free sporozoites were 9.0–10.8 × 3.8–5.1 μm in newly infected *S. undulatus* (Klein et al., 1988a). Shape varied from elongate to oval in leukocytes and spherical, lentiform, or teardrop-shaped in erythrocytes. Bonorris and Ball (1955) reported *Schellackia occidentalis* only from erythrocytes. In *S. undulatus*, erythrocytes were host to 81.5% of sporozoites and leukocytes to 15.2%, and 3.3% were free in 26 films of circulating blood (Klein et al., 1988a). Leukocyte types were not identified.

Merogony Meronts (as "schizonts") were present in epithelium of the small intestine, between the gut lumen and nuclei of epithelial cells (Bonorris and Ball, 1955). Mature meronts were spherical, 11.2 μm in diameter, and contained 13–20 merozoites.

Sporogony Bonorris and Ball (1955) found immature oocysts only in the lamina propria of the intestine. They were approximately 10 μm in diameter and had a thin refractile wall.

Transmission *Schellackia occidentalis* was transmitted by intraperitoneal inoculation of infected blood, with

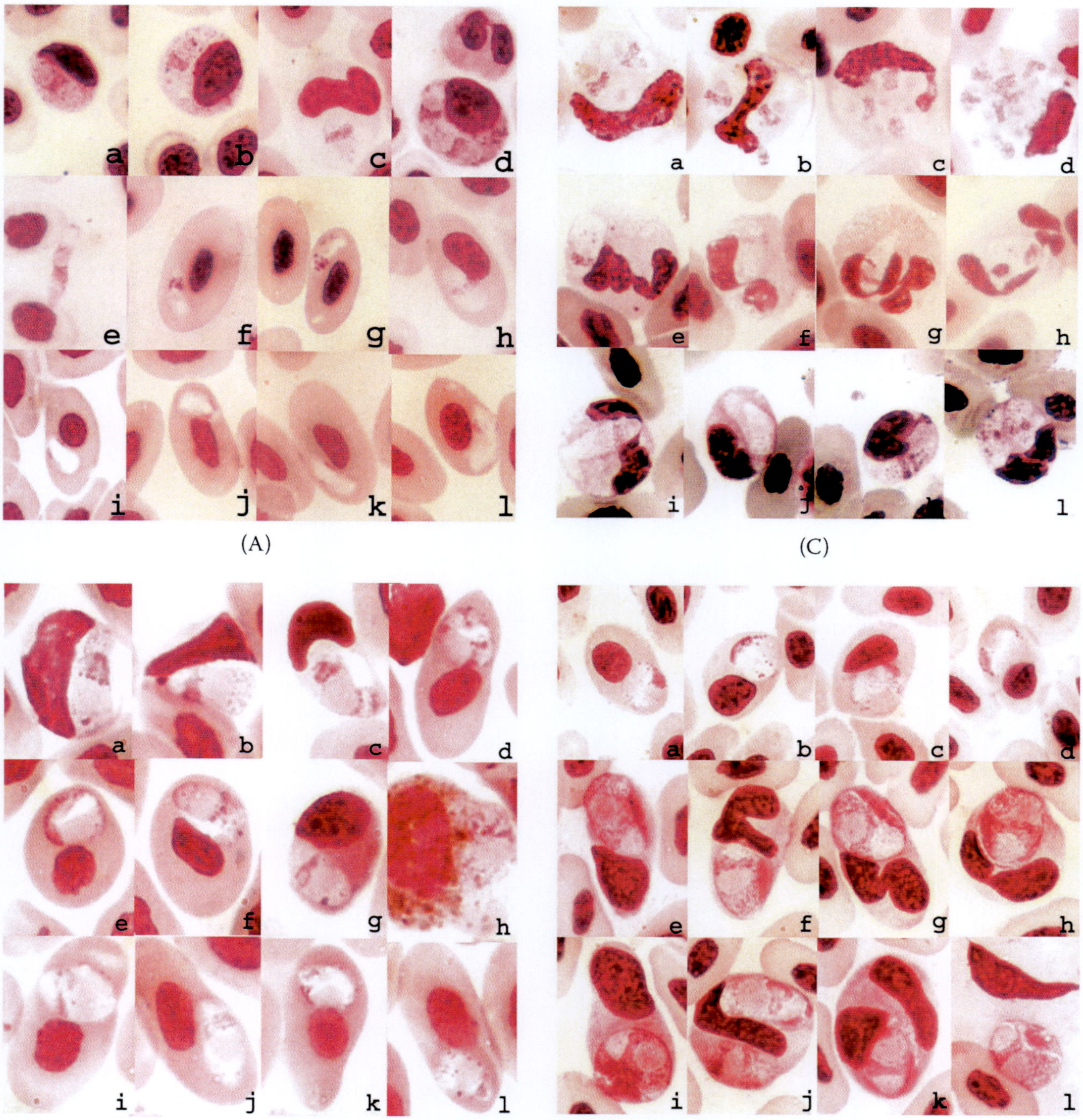

Plate 62 Hemococcidian spp. of New World lizards. (**A**) *Schellackia occidentalis* sporozoites in leukocytes, **a–d**; a free sporozoite, **e**; and in erythrocytes, **f–h**, of *Sceloporus occidentalis,* California, and in erythrocytes of *Sceloporus undulatus*, Florida, **i–l**. (**B**) Sporozoites of *Schellackia landaue* in *Polychrus marmoratus,* Venezuela, **a–f**, and in *Polychrus gutturosus*, Panama, **g–l**. Host cells are leukocytes, **a–d**, **g–h**, and erythrocytes, **e**, **f**, and **i–l**. (**C**) *Schellackia cf. golvani* sporozoites in leukocytes of *Anolis cybotes,* Hispaniola, **a–d**; *Anolis carolinensis*, Florida, **e–h**; and in *Anolis sagrei*, Florida, **i–l**. (**D**) Sporozoites of *Lainsonia iguanae* in erythrocytes of *Iguana iguana,* **a–d**, and in leukocytes of *Lainsonia legeri,* **e–l**, in *Tupinambis teguixin*, both from Venezuela.

three of four experimental hosts positive after 79–97 days (Bonorris and Ball, 1955). When mites *Geckobiella texana* from infected lizards were fed to lizards free from either mite infestations or *Schellackia* infection, patent infection was observed in the lizards 30–45 days later. Jordan and Friend (1971) placed laboratory-hatched *S. undulatus* in jars with lizards infected by *Schellackia* and infested with *G. texana*, and three of four became positive in 40–47 days. Mites engorged on a positive lizard were fed to a laboratory-hatched lizard, and it became positive 28 days later. Klein et al. (1988a) infected *S. undulatus* by ingestion of four different arthropods that had previously fed on infected lizards: *Culex erraticus* fed 7 days previously and force-fed to three lizards produced no infections, but when fed to lizards 1 day following their blood meal, infected nine of nine; *Aedes aegypti* 6 hours postfeeding infected three of four lizards; *Lutzomyia vexator* fed 1–7 days earlier infected eight of eight *S. undulatus*; and *Geckobiella texana* removed from infected lizards produced infections in two of three lizards to which they were fed. At 18–24°C, infections required a 37-day prepatent period, but at 32°C only needed 7–17 days. *Schellackia occidentalis* fed via arthropod to *Anolis carolinensis* did not produce infection in the experimental lizards, but this species was readily infected experimentally by its own parasite species, *Schellackia cf. golvani*.

Effects on Host The usually low-level parasitemias by *Schellackia occidentalis* in its lizard hosts have no apparent effect on the host other than occasional displacement of nuclei within host cells. Both Bonorris and Ball (1955) and Klein et al. (1988a) reported free sporozoites on slides of circulating blood, which suggests that some infected cells are hemolyzed by the parasite, but the generally low parasitemia minimizes the effect of blood cell loss.

Remarks Presence or absence of a single reserve vacuole or refractile body is probably not a good taxonomic character to distinguish species because its presence is variable within the same infection. The constant presence of two vacuoles in sporozoites of an apparent *Schellackia* species parasitizing *Uma* species in southern California (Telford, 1966), however, is distinctive as a taxonomic character.

SCHELLACKIA AND *LAINSONIA* SPECIES OF NEOTROPICAL/NORTH AMERICAN LIZARDS

Schellackia landaue Lainson, Shaw, and Ward 1976 (**Plate 62**)

Diagnosis A *Schellackia* species parasitic in erythroid and leukocytic cells of neotropical lizards of the genus *Polychrus*. Merogony and gametogony occur in intestinal epithelial cells, and oocysts develop in the lamina propria. During intestinal development, asexual multiplication occurs also within lymphocytes and monocytes in the spleen and liver, apparently by endodyogeny. Sporozoites may contain one or two refractile bodies and can accumulate in reticuloendothelial cells of liver, lung, and spleen.

Type Host *Polychrus marmoratus* (Linneus) (Sauria: Polychrotidae).

Other Hosts *Polychrus gutturosus*.

Type Locality Capanéma, Pará State, Brazil.

Other Localities Araure, Portuguesa State, Venezuela; El Aguacate, Panama Province, Panama (Telford).

Prevalence *Schellackia landaue* parasitized 17 of 148 (11.5%) *P. marmoratus* in Brazil (Lainson et al., 1976), 1 of 6 *P. marmoratus* in Venezuela (Telford, 1980), and 27 of 66 (40.4%) *P. gutturosus* in Panama (Telford, 1977).

Sporozoites Sporozoites within or newly emerged from oocysts are 13.0 × 3.0 μm, with "one end broader and more acutely pointed than the other" (Lainson et al., 1976). The nucleus is band-like and usually situated centrally. One or two (70%) refractile bodies of unequal size may be variably placed in the sporozoite either anterior and posterior to the nucleus, or together at one end. Within monocytes or lymphocytes, sporozoites are less elongated, becoming "stumpy bean or sausage-shape," 9.5 × 5.5 μm. Most sporozoites are erythrocytic (up to 84%) and assume an oval or spherical shape, 6.5 × 5.5 μm, with clearly visible refractile bodies, polar or lateropolar in position within erythrocytes. Accumulated sporozoites within reticuloendothelial cells of viscera, most commonly those containing pigment, are 9.5 × 5.5 μm, and resemble sporozoites in circulating leukocytes. Sporozoites within intestinal cells of infected mosquitoes are 13.0 × 3.0 μm and contain only one refractile body.

Erythrocytes (mostly), lymphocytes, and monocytes in peripheral blood, as well as reticuloendothelial cells in the lung, spleen, and liver are parasitized by *S. landaue*.

Merogony Two distinct types of meronts form within the apex of epithelial cells of the small intestine. Micromeronts are 10.0 × 10.0–25.0 × 15.0 μm and produce 10–50 micromerozoites 4.5 × 1.5 μm. Macromeronts are 12.0 × 10.0–30.0 × 25.0 μm and produce from 6 to 10 macromerozoites in the smallest meronts to as many as 100 in the largest, 11.0 × 2.5 μm. In extraintestinal sites, within lymphocytes and monocytes, apparent endodyogeny forms pairs of organisms 5.0 × 2.5 μm "clearly not sporozoites," distinguished by absence of refractile bodies and with a different staining reaction. Within the spleen, some cells contained 8, 10, and 16 of these organisms, and apparent division into multiple pairs within single cells was also seen in the liver (Lainson et al., 1976).

Gametogony Gametocyte formation occurs in the same site as merogony, the apical portion, above the nucleus, of epithelial cells in the small intestine. Microgametocytes, usually oval in shape, are 11.0 µm in diameter to 27.0 × 10.0 µm and produce about 20 microgametocytes to approximately 150, depending on size. Free microgametes are 10.0–11.0 µm and have two flagella, about 10 µm in length. Macrogametocytes may reach 15.0 × 12.5 µm and contain a prominent nucleus 3.8–5.0 µm in diameter. As they grow, macrogametocytes migrate toward the basal region of the host cell. Fertilization occurs within the epithelial cell.

Sporogony Zygotes enter the lamina propria beneath the epithelial cells and a cell wall forms around the oocysts, which average 14.3 × 13.3 µm (13.7 × 12.3–15.0 × 15.0 , N = 50), and the eight sporozoites are formed, associated with a residual body.

Transmission Lainson et al. (1976) accomplished transmission to uninfected *P. marmoratus* by feeding each lizard 20 mosquitoes, *Culex pipiens fatigans*, fed 14 days earlier on a heavily infected lizard. Sporozoites had accumulated within epithelial cells of the mosquito midgut. At 23 days postingestion (PI), merogonic and gametogonic stages were present in lizard intestinal epithelium, and at 30 days, endodyogeny and some mature oocysts were found. By 45 days, sporozoites appeared in erythrocytic and leukocytic cells of the peripheral blood, and oocysts and free sporozoites were abundant in subepithelial tissues. Infected mosquitoes fed to one each of two other lizard species, *Tropidurus torquatus* (Tropiduridae) and *Ameiva ameiva* (Teiidae), did not show infection by 45 days postingestion. No natural vector of *S. landaue* is known.

Effects on Host The only apparent effect on infected hosts is some degree of nuclear displacement within the host cells. Intestinal epithelial cells, however, are destroyed by the parasite (Lainson et al., 1976).

Remarks Sporozoites present in erythrocytes of *P. gutturosus* in Panama predominantly appear oval or spherical in shape, as in the type host, *P. marmoratus*. This similarity of form plus congenerity of the hosts are the basis for identification of the Panamanian *Schellackia* species as *S. landaue*.

Schellackia golvani Rogier and Landau 1975 (**Plate 62**)

Diagnosis A *Schellackia* species parasitic in nonerythroid blood cells of lizards in the genus *Anolis*. Merogony and microgametogony occur in the epithelial cells of the lower small intestine, just below the brush border. Macrogametocytes develop in the lamina propria, where oocysts form. A single refractile body is present in sporozoites.

Type Host *Anolis marmoratus* Duméril and Bibron (Sauria: Polychrotidae).

Type Locality Island of Guadeloupe, Caribbean.

Other Hosts *Anolis* species on Hispaniola (*cybotes, armouri, ricordii, coelestinus, distichus, chlorocyanus*), Jamaica (*lineatopus, opalinus, grahami, sagrei*), and Grand Cayman (*conspersus*), in Florida (*sagrei, carolinensis*) and Georgia (*carolinensis*).

Other Localities Hispaniola, Jamaica, Grand Cayman Islands (Telford), Georgia (Jordan and Friend, 1971), and Florida (Telford, 1978c).

Prevalence *S. golvani* was reported from 18 of 52 (34.6%) *A. marmoratus* (Rogier and Landau, 1975). Other Caribbean *Anolis* species found infected (Telford) are as follows: *cybotes*, 10 of 97 (10.3%); *coelestinus*, 1 of 6; *ricordii*, 1 of 1; *distichus*, 2 of 27 (10.3%); *armouri*, 1 of 11 (9.1%); *chlorocyanus*, 2 of 6; *lineatopus*, 8 of 78 (10.3%); *opalinus*, 1 of 52 (1.9%); *grahami*, 9 of 39 (23.1%); *sagrei*, 1 of 36 (2.8%); *conspersus*, 3 of 42 (7.1%). In Florida, infected were 2 *Anolis sagrei* (Hillsborough County) of 300 *A. sagrei* statewide and 31 of 1539 (2.0%) *A. carolinensis* in Alachua, Marion, Gilchrist, Polk, Sarasota, Charlotte, Lee, Collier, Monroe, and Dade counties (Telford). At Fargo, Georgia, Jordan and Friend (1971) found 4 of 6000 (0.07%) of *A. carolinensis* infected between 1956 and 1970, 6 of 312 (1.9%) on Sapelo Island, 81 of 509 (15.9%) on Cumberland Island, and 15 of 50 (30.0%) on Seahorse Island off the Florida coast.

Sporozoites Rogier and Landau (1975) described sporozoites of *S. golvani* in the type host as 8.25 × 3.28 µm (7.65–10.71 × 3.06–3.81), with a pale blue refractile body ("crystalloide"). Sporozoite dimensions from *Anolis carolinensis* of Alachua County, Florida, were 8.8–11.5 × 4.6–7.7 µm (Klein et al., 1988a). In *A. marmoratus*, *S. golvani* parasitized polymorphonuclear leukocytes primarily, but also infected lymphocytes and monocytes. In *A. carolinensis* and the various Caribbean species, the characteristic host cells for sporozoites are polymorphonuclear leukocytes (neutrophils).

Merogony Meronts form beneath the brush border of epithelial cells in the lower small intestine. Ovoid in form, they attain a size of 6.42 × 3.15 µm (4.59–7.65 × 3.06–4.59) and produce 15–20 merozoites. These disseminate in the lumen of the small intestine before reentering epithelial cells. Those that become microgametocytes remain in the same site as the meronts (i.e., just below the brush border). Merozoites that become macrogametocytes migrate to the basal membrane and develop within the lamina propria (Rogier and Landau, 1975).

Sporogony Oocysts form within the lamina propria, reaching a diameter of 7.25 μm in smears and 5.7 μm in section and produce eight sporozoites, in association with a residual body (Rogier and Landau, 1975).

Transmission Natural vectors of *S. golvani* are unknown. Jordan and Friend (1971) found no *Geckobiella texana* mites on the 6000 *A. carolinensis* they examined between 1956 and 1970. Klein et al. (1988a) fed mosquitoes, *Culex erraticus*, and sand flies, *Lutzomyia vexator*, on *A. carolinensis* infected with putative *S. golvani*, and after 2–29 days postfeeding of the mosquitoes and 5–6 days postfeeding for the sand flies, fed them to 12 and 3 *A. carolinensis*, respectively, negative for *Schellackia* infections. Mosquitoes kept at 18–24°C produced infections in five of seven lizards 21–25 days later (in four) and 81 days in one after ingesting infected mosquitoes. In three of five lizards fed *C. erraticus* maintained at 32°C, patent infections appeared 10–12 days postingestion. Three of three anoles fed *L. vexator* kept at 18–24°C for 5–6 days were positive after 21–23 days. Although mosquitoes and sand flies fed 1–6 days earlier on an infected anole and maintained at either 18–24°C or 32°C (five mosquitoes) were fed to nine *Sceloporus undulatus*, no infections were transmitted.

Effects on Host There is no discernible effect on hosts by *Schellackia golvani* infection.

Remarks In addition to the infections found in *Anolis* species of both the Caribbean and Florida-Georgia, five Panamanian *Anolis* species and one each from Honduras and Guatemala (Telford, 1977) were infected by *Schellackia* sporozoites in polymorphonuclear leukocytes that are very similar to *S. golvani*. Although genome analysis may eventually define taxonomic diversity in these hemococcidia, no harm is done by considering them to be *S. golvani sensu lato*.

Lainsonia iguanae Landau 1973 (**Plate 62**)

Diagnosis A *Lainsonia* species with sporozoites present in both erythrocytes and leukocytes of circulating blood in *Iguana iguana*. Merogony, gametogony, and sporogony occur primarily in reticuloendothelial cells of the liver, especially in those containing pigment, and less commonly in the spleen. Two refractile bodies are found in sporozoites, one on each side of the nucleus.

Type Host *Iguana iguana* (Linnaeus) (Sauria: Iguanidae).

Type Locality Exu, Pernambuco State, Brazil.

Other Hosts None known.

Other Localities La Mère and Cayenne, French Guiana, and Bélem, Pará State, Brazil (Landau, 1973); Araure, Portuguesa State, Venezuela (Telford); and Panama Province, Panama (Telford, 1977).

Prevalence Landau (1973) reported three iguanas from the type locality and single specimens from the two localities in French Guiana and at Bélem, Brazil positive for *L. iguanae*. Overall, 2 of 29 (6.9%) in Panama (Telford, 1977) and 2 of 69 (6.9%) in Venezuela, both from Araure, Estado Portuguesa (Telford), were positive.

Sporozoites Landau (1973) described sporozoites as globular, ovals with rounded ends, 11 × 7 μm. Most of the sporozoite cytoplasm is occupied by the two similar size refractile bodies (crystalloides) between which the triangular nucleus is inserted.

Cells Host to Sporozoites Both erythrocytes and leukocytes, especially monocytes, in circulating blood and reticuloendothelial cells in the viscera are parasitized by sporozoites.

Merogony Meronts, in reticuloendothelial cells of the liver, were 12 μm in mean diameter. Landau (1973) depicted a meront with at least 24 nuclei present.

Gametogony Microgametocytes, usually round but sometimes oval, measured about 15.3 μm in diameter, and macrogametocytes were round, 17–20 μm in diameter. Both develop in reticuloendothelial cells of the liver.

Sporogony As in *Schellackia*, the oocysts (dimensions not given) produce eight sporozoites but develop in reticuloendothelial cells of the liver and spleen rather than in intestinal epithelium.

Transmission Mosquitoes, *Aedes aegypti*, fed on an iguana with *L. iguanae* showed sporozoites present in cells of the stomach, with similar morphology to those present in macrophages of the iguana host.

Effects on Host No pathology has been described for *L. iguanae*.

Lainsonia legeri Landau, Lainson, Boulard, and Shaw 1974 (**Plate 62**)

Diagnosis A *Lainsonia* species with sporozoites in circulating blood confined to monocytes and lymphocytes. Merogony, gametogony, and sporogony occur primarily in reticuloendothelial cells of the lungs, kidney, and capillaries of the brain. Leukocytic sporozoites in circulating blood contain a single prominent refractile body, but sporozoites within macrophages in all of the organs may contain one or two, rarely three, refractile bodies.

Type Host *Tupinambis nigropunctatus* Spix (syn. *T. teguixin* Linnaeus) (Sauria: Teiidae).

Type Locality Bélem, Pará State, Brazil.

Other Hosts None known.

Other Localities Portuguesa and Cojedes states, Venezuela (Telford).

Prevalence In Brazil, not stated by Landau et al. (1974). In Venezuela, 21 of 81 (25.9%) *T. teguixin* collected in Estados Portuguesa and Cojedes were infected by *L. legeri* (Telford).

Sporozoite Dimensions In leukocytes of circulating blood, sporozoites average 9.5 × 4.0 μm, and in macrophages of various organs they average 9.5 × 3.8 μm (9.2–10.8 × 3.1–4.5) (Landau et al., 1974). The leukocytic sporozoites contained a single refractile body, 3 μm in diameter, in the center of the sporozoite, bordering the transverse nucleus. In macrophages of the viscera, there are one, two, or rarely three refractile bodies, the largest averaging 3 μm, and the others much smaller, if present. Both ends of sporozoites are rounded.

Cells Host to Sporozoites Monocytes and lymphocytes in circulating blood and macrophages in visceral sites are hosts.

Merogony In sections of lung, meronts average 14.5 × 7.8 μm and, on impression smears, contained about 25 merozoites.

Gametogony On impression smears of kidney and lung, macrogametocytes measure 14–17 μm in diameter, but in smears of brain obtained by compression between two slides, macrogametocytes extending along capillaries could attain a length of 31 μm and breadth of 11 μm. Microgametocytes in vascular endothelium of the lung averaged 16.2 × 9.5 μm (12.5–18.5 × 7.5–12.5, N = 11). About 150 microgametes are formed, 4.5 × 0.7 μm on average, with flagella 7.5 μm.

Sporogony Mature oocysts 12.5 μm (10.8–13.9) in diameter are distributed in the liver, spleen, lung, and kidney. Most are enclosed by a capsule 1.0–1.5 μm in thickness. Oocysts contain eight sporozoites and a residual body.

Transmission Landau et al. (1974) transmitted *L. legeri* to uninfected young *T. nigropunctatus* by intraperitoneal injection and by esophageal ingestion of infected blood. With ingestion of blood, meronts and immature oocysts were found in the viscera at necropsy after 22 days, immature oocysts were circulating in the blood at 28 days, and sporozoites were in the blood cells at 36 days postingestion. At 49 days postingestion, mature oocysts were found in the viscera at necropsy, and sporozoites were present in both blood and tissues. At 48 days in the lizard infected by intraperitoneal inoculation of blood, sporozoites were present in the circulating blood.

Effects on Host Overt pathology has not been described in infections of *L. legeri*.

SCHELLACKIA SPECIES OF ASIAN LIZARDS

Schellackia orientalis Telford 1993 (**Plate 63**)

Diagnosis A *Schellackia* species present in both erythroid and leukocytic cells of east and southeast Asian lizards of the genus *Takydromus* (Lacertidae). Merogony and gamogony occur in epithelium of the small intestine, and oocysts form in the lamina propria. Sporozoites contain a single refractile body.

Type Host *Takydromus tachydromoides* (Schlegel) (Sauria: Lacertidae).

Type Locality Hanno, Saitama Prefecture, Honshu, Japan.

Other Hosts *Takydromus sexlineatus.*

Other Localities Kawaguchi, Tsurukawa, Hiyoshi, and Tokyo, Honshu, Japan (Telford, 1997b); Ramintra, near Bangkok, Thailand (Telford, 1993).

Prevalence At the type locality, overall prevalence from 1965 to 1967 of *S. orientalis* was 1.8% among 1275 *T. tachydromoides.* Prevalence in adult lizards was 2.4% (N = 752) and 1.0% (N = 523) in juveniles. Prevalence varied 0.4–12.5%

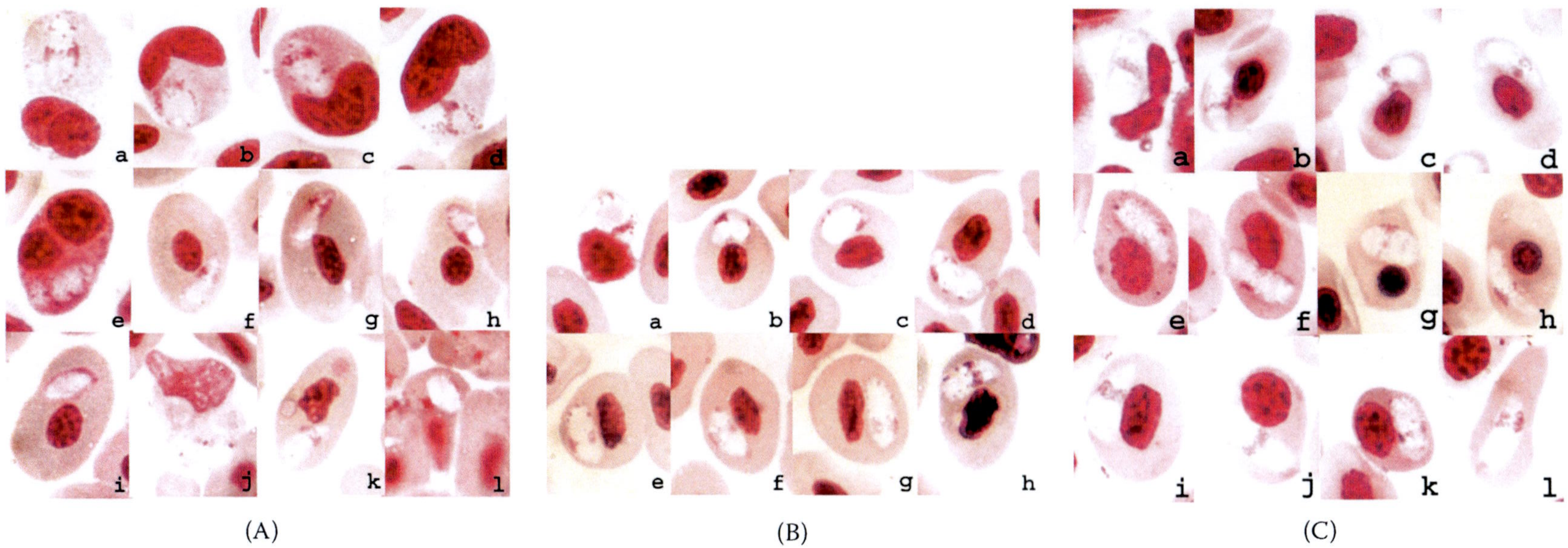

Plate 63 *Schellackia* spp. of Old World lizards. (**A**) Sporozoites of *Schellackia orientalis* in *Takydromus tachydromoides,* Honshu, Japan, **a–i**, and in *Takydromus sexlineatus*, Thailand, **j–l**. Host cells **a-e**, **j** are leukocytes; others are erythrocytes. (**B**) Sporozoites of *Schellackia cf. bolivari* in *Podarcis muralis* of Switzerland. Host cell in **a** is a thrombocyte; others are erythrocytes. (**C**) *Schellackia calotesi* sporozoites in *Calotes mystaceus,* Burma, **a–h**, and *Calotes versicolor*, Thailand, **i–l**. Host cell **a** is a leukocyte; others are erythrocytes.

among five other localities in central Honshu (Telford, 1997b). In Thailand, 2 of 20 (10%) *T. sexlineatus* were infected in December 1976 and 3 of 20 (15%) in April 1980.

Sporozoites In *T. tachydromoides,* erythrocytic sporozoites are 6.9 ± 0.7 × 3.3 ± 0.8 μm (5–8 × 2–5, N = 54) and are never rounded in shape. Leukocytic sporozoites are wider, 6.9 ± 0.9 × 4.4 ± 0.6 μm (5–9 × 3–6, N = 31), and broadly pyriform, rarely oval or rounded. In *T. sexlineatus*, leukocytic sporozoites are longer and broader than erythrocytic sporozoites, 7.2 ± 1.1 × 4.3 ± 0.5 μm (5–9 × 3–5.5, N = 42) versus 6.0 ± 0.8 × 3.6 ± 0.5 μm (4.5–7.5 × 3–4.5, N = 21), respectively. A single refractile body, slightly bluish when well stained, is present at approximately the midpoint of sporozoites. Sporozoite nuclei are highly variable in form, from narrow bands to broad areas of chromatin along one or both margins at the pointed end of the sporozoite. Sporozoites parasitize most types of nonerythroid cells, including lymphocytes, macrophages, monocytes, thrombocytes, azurophils, polymorphonuclear leukocytes (neutrophils), and apparent stem cells. Macrophages (48%) and monocytes (33%) are most commonly parasitized, and four to five sporozoites can be seen in macrophages. Immature erythroid cells (erythroblasts, proerythrocytes) are sometimes infected, but most sporozoites following the initial stages of infection occur singly in erythrocytes.

Merogony Merogony occurs profusely in the epithelium of the lower small intestine, uncommonly in the duodenum and upper and middle intestine. Meronts average 10.4 ± 1.9 × 8.7 ± 1.7 μm (8–13 × 7–12, N = 12) and produce 14.9 ± 3.5 (10–20, N = 15) merozoites, approximately 5 × 1 μm. Groups of meronts are common. Microgamonts 6–15 × 4.5–9 μm, elongate and sometimes lobulate, nearly fill the parasitophorous vacuole. Macrogamonts, which were not clearly distinguished from zygotes, were 7.5–9 × 6.5–7 μm and occupied the center of large parasitophorus vacuoles.

Sporogony Oocysts occur in the lamina propria and are 9.6 ± 1.4 × 7.7 ± 1.1 μm (8–11 × 6–9, N = 11). The eight sporozoites formed within oocysts are 5.5–6 × 1 μm and are accompanied by a residual body.

Transmission The only ectoparasite observed associated with *T. tachydromoides* during a 3-year field and laboratory study (Telford, 1997b) was the tick *Ixodes nipponensis*. Sporozoites were not seen in smears of wild-caught ticks, and there was no correlation between evidence of tick infestation and presence of *S. orientalis* in lizards (Telford, 1997b). Intraperitoneal inoculation of infected blood produced infections 17–42 days later in juvenile lizards.

Effects on Host Parasitized host cells differ little from uninfected cells except for occasional displacement of erythrocyte nuclei.

Remarks *Schellackia orientalis* (two host spp.) is a component of the *Takydromus* host-parasite complex in eastern and southeastern Asia that includes *Plasmodium sasai* (three hosts), *Trypanosoma takydromi* (two hosts), and *Eimeria takydromi* (three hosts), which reflects the common ancestry of *T. tachydromoides* (Japanese main islands), *T. smaragdinus* (Ryukyu Islands), and *T. sexlineatus* (southeast Asia) (Telford, 1993, 1997b).

Schellackia calotesi (Syn. *Gordonella calotesi* Ray and Sarkar 1969) Levine 1980, Finkelman and Paperna 1998 (**Plate 63**)

Diagnosis A *Schellackia* species parasitic in erythrocytes and monocytes of southern and southeastern Asian agamid lizards of the genus *Calotes.* Merogony and gametogony occur in the mucosal epithelial layer of the anterior small intestine, and oocysts develop in both epithelial cells and the lamina propria. Sporocysts contain up to two refractile bodies.

Type Host *Calotes mystaceus* Duméril and Bibron (Sauria: Agamidae).

Type Locality Chiang Mai, Thailand.

Other Hosts *Calotes versicolor.*

Other Localities India (H. N. Ray and Sarkar, 1969); Kon Kaen, Thailand (Finkelman and Paperna, 1998); Sabu Daung Village, Hlegu Township, and Rangoon (Telford), Burma.

Prevalence In Thailand, *S. calotesi* was found in 1 of 3 *C. versicolor* from Kon Kaen and 2 of 34 (5.9%) *C. mystaceus* from Chiang Mai (Finkelman and Paperna, 1998). In Burma, at Sabu Daung Village, 1 of 6 *C. versicolor* and 10 of 41 *C. mystaceus* (24.4%) were infected, and in Rangoon, 1 of 124 (0.8%) *C. versicolor* and 6 of 69 (8.7%) *C. mystaceus* were host to *S. calotesi* (Telford).

Sporozoites Finkelman and Paperna (1998) described sporozoites of *S. calotesi* as 8.4–10.5 × 2.6–5.0 μm, with a "somewhat anterior" nucleus that preceded two refractile bodies. Sporozoites in *C. mystaceus* of Rangoon were similar in size to those from Thai hosts, but the number of refractile bodies varied from apparently none to two, with one of the latter number sometimes on each side of the nucleus. Position of the nucleus was in the anterior third of the sporozoite and was often indicated by nothing more than a couple of chromatin granules on each side of the sporozoite, occasionally as a loose band of chromatin granules across the width of the sporozoite. In the material described from Thailand (Finkelman and Paperna, 1998), sporozoites occupied erythrocytes in the circulating blood and were often found either within or next to free hepatic melanomacrophages. In a heavy infection of Burmese *C. mystaceus*, sporozoites were commonly seen in apparent monocytes as well as in erythrocytes, the usual host cell type.

Merogony Meronts develop and gamonts form in the "mucosal epithelial layer" of the upper small intestine (Finkelman and Paperna, 1998). Meronts produced up to ten "emerging merozoites." Microgamonts were 11.2–14.0 × 6.2–8.0 µm, and macrogamonts 7.0–9.8 × 8.4–9.0 µm, and were contained within expanded parasitophorus vacuoles, as were the meronts.

Sporogony Mature oocysts, which were not contained within vacuoles, were found in both the epithelial layer and in the lamina propria, but no sporulating forms were seen in the Thai hosts (Finkelman and Paperna, 1998). Smears prepared from the upper small intestine of a massive infection in a Burmese *C. mystaceus* had some octonucleate oocysts present (Telford).

Transmission Finkelman and Paperna (1998) fed blood and liver from an infected *C. versicolor* to two apparently uninfected specimens of each host species. Infection developed in only one *C. mystaceus*.

Effects on Host Unknown.

Remarks H. N. Ray and Sarkar (1969) described a "new haemosporidian" from *Calotes versicolor* in India, *Gordonella calotesi,* in abstract only at the Third International Congress of Protozoology. Levine (1980) considered *Gordonella* to be a synonym of *Schellackia* and recognized the species as *S. calotesi*. *Calotes versicolor* has a considerable geographic range, from Pakistan throughout southern India into Thailand and Burma, and it is possible that the same *Schellackia* species occurs throughout the range. Until genomic analysis settles the question, it is probably best to use the name *S. calotesi* but with the type host and locality based on the information formally published by Finkelman and Paperna (1998) and accompanied by a detailed description.

SCHELLACKIA SPECIES OF AFRICAN LIZARDS

Schellackia brygooi Landau 1973

Diagnosis A *Schellackia* species parasitic in erythroid and leukocytic cells of peripheral blood and in reticuloendothelial cells of organs in Madagascan lizards of the genus *Oplurus*. Merogony and microgametogony occur in the apical portion of epithelial cells of the small intestine. Macrogametocytes form oocysts in the reticuloendothelial cells of the lamina propria and sometimes at the base of epithelial cells. A single refractile body is present in sporozoites.

Type Host *Oplurus sebae* (Duméril and Bibron) (Sauria: Opluridae).

Type Locality Majunga, Madagascar.

Other Hosts *Oplurus cyclurus.*

Other Localities Tuléar and Ihotry, Madagascar.

Prevalence Unknown.

Sporozoites Examined unfixed, sporozoites in erythrocytes average 13 µm and are elongate and slender. In stained preparations, some erythrocytic sporozoites are 10.7 × 2.3 µm and are not surrounded by a membrane. Most sporozoites are more globular, 8 × 3.8–9.2 × 3 µm, with one end more pointed than the other. At the more rounded end is a round refractile body occupying the full width of the sporozoite. The nucleus, formed by small grains of chromatin arranged as a transverse band, is located at the middle of the sporozoite. Sporozoites within monocytes are more globular and average 10 × 4 µm in size.

Merogony Meronts develop in the apical portion of intestinal epithelial cells and, as nuclear division proceeds, occupy almost all of the space between the cell nuclei and their apex. At maturity, meronts contain 20–40 elongate and thin merozoites.

Sporogony Oocysts form following fertilization in the lamina propria of the small intestine and sometimes within the epithelial cells. There is no residual body. Their diameter in section is 8.0 µm, and eight sporozoites are produced.

Transmission In the mosquito *Culex p. pipiens,* sporozoites, apparently unchanged in morphology from those occupying erythrocytes, accumulate in epithelial cells of the mosquito gut without altering the cells. *Culex pipiens fatigans*, fed on an infected *Oplurus cyclurus*, were ingested a month later by juvenile *O. sebae* that were apparently negative. After 29 days, identical sporozoites invaded erythrocytes. Merogony and sporogony were observed in intestinal cells, and sporozoites accumulated in reticuloendothelial cells of other organs, but divisional stages were not found outside the intestinal wall.

Effects on Host No pathology in the saurian host was observed.

Remarks Landau (1973) suggested *Chamaeleo brevicornis* as a probable host to *S. brygooi*, but this is unlikely in view of the high specificity of other *Schellackia* species for their vertebrate hosts.

SCHELLACKIA SPECIES OF LIZARDS IN THE MEDITERRANEAN REGION

Schellackia bolivari Reichenow 1919 (**Plate 63**)

Diagnosis A *Schellackia* species parasitic in European lacertids of the genera *Acanthodactylus* and *Psammodromus*. Sporozoites occupy either erythrocytes or lymphocytes, depending on host species, and contain two refractile bodies. Meronts produce 10–16 merozoites in epithelial cells of the middle small intestine, where microgametocytes also develop and produce microgametes that migrate to fertilize macrogametes in the lamina propria, where oocysts mature. Transmission is by ingestion of infected mites, *Liponyssus saurarum*.

Type Host *Acanthodactylus erythrurus* (Schinz) (= *A. vulgaris*) (Sauria: Lacertidae).

Type Locality Madrid, Spain.

Other Hosts *Psammodromus hispanicus*.

Other Localities None known.

Prevalence Unknown.

Sporozoites The elongate but short, 5.2-µm, sporozoites occupy erythrocytes in *A. erythrurus* and lymphocytes of *P. hispanicus*. Conspecificity was proven by cross-infection experiments (Reichenow, 1919). Two refractile bodies are present in sporozoites.

Merogony Meronts develop in the apical portion of epithelial cells in the middle small intestine of the saurian host and produce 10–16 merozoites.

Sporogony Merozoites destined to become microgametocytes remain in the epithelial cells, while those that will form macrogametocytes migrate to the lamina propria. Following microgametogony, the biflagellated microgametes also migrate to the lamina propria, where they fertilize macrogametes. Eight sporozoites, associated with a residual body, are produced and enter erythrocytes or lymphocytes, depending on the host species.

Transmission The mite *Liponyssus saurarum* acquires sporozoites of *S. bolivari* while feeding on infected lizards. These accumulate in cells of the stomach epithelium in the mite and are transmitted to lizards by ingestion of the mite (Reichenow, 1919).

Effects on Host The small size of sporozoites produces no significant alteration of host cells except in the case of multiple infections.

Remarks *Schellackia bolivari* is unique in the fact that different blood cell types are parasitized in the two known saurian hosts, *Acanthodactylus erythrurus* and *Psammodromus hispanicus*, both of which belong to the Lacertidae. Genomic analysis should determine if the *Schellackia* species that parasitize *Lacerta* and *Podarcis* species in Europe are conspecific with *S. bolivari*.

Schellackia ptyodactyli Paperna and Finkelman 1996

Diagnosis A *Schellackia* species with short sporozoites in erythrocytes, thrombocytes, granulocytes, and lymphocytes of the circulating blood of the gecko *Ptyodactylus hasselquistii*. Merogony occurs in the apical ends of epithelial cells in the upper small intestine, and gametogony in the basal layer of the epithelium. Oocysts form in the epithelium but are more common in the lamina propria, where they may occur in unpigmented macrophages. Two large refractile bodies are visible in sporozoites.

Type Host *Ptyodactylus hasselquistii* (Donndorf) (Sauria: Gekkonidae).

Type Locality Lower Jordan Valley, 20 km north of Jericho, Cis-Jordan.

Other Hosts Unknown.

Other Localities Unknown.

Prevalence Along the rift escarpment of the lower Jordan Valley, prevalence of *S. ptyodactyli* varied among host age groups and by season, from 5% to 58% at one locality and 13% to 38% in another location.

Sporozoites Sporozoites 4.0–4.6 × 1.0–1.3 μm in the lamina propria were present either within cells, including macrophages, or were extracellular. In circulating blood, sporozoites parasitized erythrocytes and at least three types of nonerythroid cells, thrombocytes, granulocytes, and lymphocytes and were observed in the liver, lungs, and spleen, mostly within blood cells or free. In experimental infections, nonerythroid cells were commonly infected early, but later in the infection erythrocytes were the principal host cells to sporozoites.

Merogony Merogony occurred in epithelial cells of the upper small intestine, at the apical end of the cells. Multiply infected cells were common in a heavy infection, containing up to 13 zoites. Mature meronts were rounded, 7–8 × 6–7 μm, and produced up to 36 merozoites distributed around a large residual body.

Sporogony Microgamonts were 9–12 × 7–10 μm and occupied the basal layer of epithelium; macrogamonts measured 11–12 × 7–10 μm. Both zygotes and young oocysts also occurred in the epithelium, but nonsporulated oocysts were more common in the lamina propria. Apparent host cell residues in the lamina propria contained inclusions within which the oocysts were found, and these residues were also present in macrophages. Oocysts were smaller than macrogamonts, 7–10 μm in diameter, perhaps as an artifact of fixation.

Transmission Experimental infections were accomplished by intraperitoneal inoculation of infected blood. Young merogonic stages were evident by 8 days PI, and meronts, merozoites, and young macrogamonts were seen by day 11. In a gecko examined 10 days PI, mature macrogamonts were present, and sporozoites were found in intestinal and extraintestinal tissues. One gecko showed infected blood cells 21–34 days PI. *Gekkobia* sp. mites and trombiculid mite larvae infested geckoes in their natural habitats, but neither contained sporozoites.

Effects on Host Parasitemia in both natural and experimental infections was low, from less than 0.1% to 5% of blood cells infected, and persisted at a low level, less than 2%, for over 1 year. The very small sporozoites do not appear to alter host erythrocytes but may push thrombocyte nuclei toward one end of the cell, as illustrated by Paperna and Finkelman (1996).

Remarks As Paperna and Finkelman (1996) pointed out, *S. ptyodactyli* is the only *Schellackia* species yet described from gekkonid lizards.

Schellackia cf. agamae Bristovetzky and Paperna 1990

Diagnosis A *Schellackia* species parasitic in agamid lizards of the eastern Mediterranean region. Sporozoites occupy erythrocytic and nonerythroid cells and contain two refractile bodies. Merogony and gametogony occur in the mucosal epithelium of the duodenum. Meronts produce 8–32, rarely 4, merozoites. Zygotes develop into oocysts in the epithelial cells near the basal membrane, and oocysts mature in the lamina propria. Sporozoites more commonly occupy nonerythroid cells earlier than erythrocytes, and may parasitize hepatocytes and macrophages of the liver, then aggregate in melanin-containing macrophages as erythrocytes are invaded. Sporozoites increase in size with age, independent of host cell type.

Type Host *Laudakia stellio* (Linnaeus) (= *Agama stellio*) (Sauria: Agamidae).

Type Locality Western outskirts of Jerusalem, Israel.

Other Hosts None known.

Other Localities Probably Egypt (Omran et al., 1981).

Prevalence Unknown.

Sporozoites Bristovetsky and Paperna (1990) have shown that sporozoites vary in size by site and by age of infection. Between 14 and 20 days PI, sporozoites in the lamina propria of the intestine were 7.3 ± 1.8 × 1.8 ± 0.6 μm (4.2–9.8 × 1.4–2.8, N = 15), in the liver, 8.5 ± 1.3 × 1.9 ± 0.7 μm (6.3–11.2 × 1.4–2.8, N = 15), and in blood cells (nonerythroid?) were 5.2 ± 1.1 × 1.3 ± 0.4 μm (3.3–7.5 × 1.1–3.8, N = 15). From days 25–50, hepatic sporozoites were 9.3 ± 1.7 × 2.3 ± 1.0 μm (7.0–12.6 × 1.4–4.2, N = 15) and in erythrocytes (presumably) were 11.4 ± 1.4 × 1.4 ± 0.1 μm (8.8–13.3 × 1.3–1.5, N = 15). In presumably chronic infection phase between 150 and 230 days PI, sporozoites were 10.7 ± 2.2 × 1.4 μm (7.0–14.0 × 1.4, N = 15) in the liver, and in erythrocytes were 10.9 ± 1.9 × 1.4 μm (7.0–13.2 × 1.4, N = 15). Sporozoites were observed in macrophages and hepatocytes within the tissues and in polymorphonuclear leukocytes and erythrocytes in the blood. Although not stated by Bristovetsky and Paperna, their illustrations show two prominent refractile bodies present in sporozoites.

Merogony In the mucosal epithelium of the duodenum, merogony began prior to 5 days PI, continued somewhat when gametogony began in days 11–14 PI, and finished when oocysts and sporozoites were present between 14 and 78 days PI. Mature meronts averaged smaller at 5 days PI, 8.8 ± 3.3 × 8.6 ± 2.3 µm, than at 7 days PI, 14.9 ± 8.4 × 11.5 ± 6.5 µm. Usually, 8–32 merozoites were formed; rarely 4 were present.

Sporogony Microgametocytes at 11 and 14 days PI, when dividing had 4–40 nuclei present, and when mature contained biflagellate microgametes. Macrogametocytes developed at the same time as microgametocytes and were sometimes seen in the same host cell. Mature microgametocytes were 14.0 ± 4.3 × 12.9 ± 4.5 µm, and macrogametocytes were smaller, 9.4 ± 3.8 × 6.7 ± 2.0 µm. Zygotes, present at 11–18 days PI, were found in the epithelial cells near the basal membrane and averaged 11.9 ± 1.7 × 11.5 ± 1.6 µm in size. Zygotes developed into oocysts in the epithelial cells between 14 and 18 days PI, but oocysts matured in the lamina propria at an average size of 14.2 ± 5.1 × 12.3 ± 2.8 µm and produced eight sporozoites accompanied by a residuum that contained a refractile body.

Transmission Bristovetzky and Paperna (1990) transmitted *Schellackia cf. agamae* to uninfected *L. stellio* by four routes: ingestion of liver or blood from lizards infected at least 25 days; subcutaneous inoculation of infected blood; ingestion of four or five *Aedes aegypti* or *Culex molestus* 4 and 7 days following their feeding on infected lizards; and ingestion of 20–30 larval ticks, *Hyalomma aegyptium,* 23 days after they had fed on infected lizards. All methods accomplished infection, as determined from necropsy or blood examination at varying intervals following inoculation or ingestion. Dermanyssid mites (unidentified) removed from naturally infected, wild-caught lizards were found to contain sporozoites. Infections in experimental lizards appeared in erythrocytes at 25 days PI and later, and persisted at their initial level of parasitemia as long as 370 days. Unsuccessful attempts were made to infect the lizards *Podarcis sicula* (Lacertidae) and *Mabuya vittata* (Scincidae) and a snake, *Coluber jugularis* (Colubridae). Reinfection during the chronic phase of infection in *L. stellio* from 152 to 330 days PI was unsuccessful, but in one lizard in which the infection had disappeared at 260 days PI, a heavy infection by meronts and gametocytes was present in intestinal tissues, without sporozoites present.

Effects on Host No pathology resulting from infection of *L. stellio* by this *Schellackia* species was described by Bristovetzky and Paperna (1990).

Remarks Laveran and Pettit (1909) described a *Schellackia* species as *Haemogregarina agamae* from *Agama colonorum* (= *A. agama*) in Senegal. The erythrocytic parasites were present in two size groups, 8–9 × 2–3 µm and 10–13 × 3 µm. Presence of a refractile body in the apparent sporozoites was not mentioned. Later, in the same host species, Rogier (1977) reported what was apparently the same *Schellackia* species, *S. agamae*, and described merogony and sporogony from experimental infections but gave no details of sporozoites. Given the difficulty in identification with certainty of a *Schellackia* species from morphological characters alone, Bristovetzky and Paperna (1990) were wise to designate the *Schellackia* in *L. stellio* as *Schellackia cf. agamae.* This is supported by the geographical distance between Central and West Africa where *A. agama* occurs and the Israel-Egypt area inhabited by *L. stellio*, as well as the very high level of host specificity seen in almost all *Schellackia* species. A comparison by genome analysis of the two parasites is probably necessary to evaluate their (unlikely) conspecificity.

4

KINETOPLASTIDA, TRYPANOSOMATIDAE

Trypanosoma

Approximately 80 species of *Trypanosoma* are described from reptilian hosts, represented by 11 in chelonians, 48 in lizards, 20 in snakes, and 2 in crocodilians. Some of these, perhaps, are synonyms. There are many reports of trypanosomes simply as *Trypanosoma* sp. (Brygoo, 1966b), some of which are probably valid species. Inadequate descriptions containing few morphometric data were provided for many of them. Although there is great variability in form, with adequate morphological data a species can usually be distinguished from others, especially in the case of saurian trypanosomes. Life cycles and natural vectors are poorly known, and only *Trypanosoma platydactyli* has been examined biochemically.

Taxonomic descriptions of trypanosomes should provide certain measurement data (Telford, 1995b): maximum length, excluding narrow cytoplasmic projections and the free flagellum; maximum width; the distance from anterior end to middle of the nucleus; distance from the kinetoplast to the midpoint of the nucleus and to the anterior and posterior ends; nucleus length and width. Although of little taxonomic use, length of the free flagellum should be stated. In a comparison of coefficients of variation for these characters among 14 species of trypanosomes from four saurian families (Telford, 1995a), body length and width were the least variable, and the greatest variability was found in the free flagellum, the distance from the kinetoplast to the posterior, and kinetoplast to the nucleus. There was a similar degree of variability of individual taxonomic characters among the 14 *Trypanosoma* species compared. Two useful characters for comparison, derived from measurements, are the average locations of the nucleus and kinetoplast, expressed as a percentage of the body length. The nuclear index (length/width) and body shape index (length/maximum width), expressed as ratios, are also helpful.

The trypanosome life history stage present in circulating blood of reptiles is the trypomastigote. The forms present in culture or in the gut of vectors may be promastigotes, amastigotes, spheromastigotes, epimastigotes, or metacyclic trypanosomes, depending on the species of trypanosome. The trypanosomes of most aquatic reptiles apparently develop within leeches. The trypomastigotes ingested with a blood meal transform into epimastigotes, which divide by binary fission and become infective metacyclic trypanosomes. *Trypanosoma grayi* of African crocodiles utilizes tsetse flies as vectors and also follows this pattern, but transmission results from ingestion of the tsetse fly vector rather than by its bite. A more complex developmental sequence occurs in other species that are transmitted by dipteran hosts. In phlebotomine sand flies, the ingested trypomastigotes first become promastigotes or amastigotes, develop into spheromastigotes, and finally become epimastigotes, the apparent infective stage. This pattern was reported also in leeches fed on the Indian turtle *Lissemys punctata* (R. Ray, 1987), but the epimastigotes transform into metacyclic trypanosomes.

Within leeches fed on infected aquatic turtles and snakes, trypomastigotes transform into epimastigotes in the leech crop or stomach. Division by binary fission may result in daughter cells unequal in size. The resulting epimastigotes become transitional forms with the kinetoplast located near the nucleus (Woo, 1969). Metacyclic trypomastigotes are the infective stage. This is the pattern of development found in the first reptilian trypanosome life cycle studied, by Robertson (1908, 1909). Transformation into epimastigotes of *Trypanosoma vittatae* occurred immediately in the crop following feeding by the leeches *Glossiphonia* sp. and *Limnatis granulosa* on the turtle *Emyda vittata* of Ceylon.

The epimastigotes then became slender trypanosomes of variable size with a narrow membrane and a short flagellum and with the kinetoplast located very close and anterior to the nucleus (Robertson, 1908). Woo (1969) fed the leeches *Placobdella parasitica* and *P. rugosa* on the North American turtle *Chrysemys picta marginata* infected by *Trypanosoma chrysemydis*. Trypomastigotes became epimastigotes in 4 days postingestion (PI) at 21–23°C. Transitional stages were seen from 8 to 13 days, and by 22 days some metacyclic trypanosomes appeared, which became abundant by day 34. When leeches were held at a higher temperature (31°C), development was more rapid: Epimastigotes were seen by day 2, transitional forms on day 4, and metacyclic trypomastigotes appeared by days 14–16 PI. When denied a second blood meal, leeches lost their infections between days 28 and 35 and could not infect turtles when fed on day 36. Infected leeches fed a second time, kept at 22–24°C for about 35 days, were starved for an additional 14 days. These retained metacyclic trypanosome infections and when fed once more produced an infection in a turtle. At 3 days following the last blood meal, epimastigotes, transitional forms, and metacyclic trypanosomes were present in the leeches, which indicated a cyclic development in the leech host. In addition to bite by leeches, infection was possible by subcutaneous inoculation of metacyclic trypanosomes from the crops and gastric ceca of the leeches or from in vitro cultures. Infection was not possible from the ingestion of infected leeches by turtles. Woo could not infect mosquitoes with *Trypanosoma chrysemydis*.

Brumpt (1914) fed the leeches *Placobdella braziliensis* and *P. catenigera* on the aquatic snake *Helicops modestus* infected by *Trypanosoma brazili*. Epimastigotes were present after 4 days and metacyclic trypomastigotes by 19 days. Brumpt suggested that infection occurred by ingestion of the leech because development occurred in the leech stomach without invasion of the proboscis sheath. Pessôa and Fleury (1969) found only promastigote and epimastigote stages of *Trypanosoma hogei* from the snake *Rachidelus brazili* in the leech *Haementeria lutzi*. In some trypanosome parasites of snakes, division of trypomastigotes by binary fission in the bloodstream of the host can occur: *Trypanosoma butantanense* in the snake *Xenodon merremi* (Arantes and da Fonseca, 1931) and *Trypanosoma hydrae* of *Nerodia fasciata* (Ayala, Atkinson, and Vakalis, 1983). It was possible to infect snakes with *T. hydrae* by intraperitoneal inoculation of infected blood, with infections patent in 1 week.

The tsetse fly *Glossina fuscipes* transmits *Trypanosoma grayi* of African crocodiles in Uganda (Hoare, 1929, 1931a, 1931b). Lloyd and Johnson (1924) found *T. grayi* in *Glossina palpalis* and *G. tachinoides* in Nigeria, but never in *G. morsitans*, a vector of mammalian trypanosomiasis. Because crocodiles were rare or absent in many sites in dry season when flies became infected, the source of infection was thought to be *Varanus exanthematicus*. These monitor lizards, infected by trypanosomes, basked in trees near resting sites of the flies. *Trypanosoma grayi* infections in flies were thought by Lloyd et al. (1924) to possibly be comprised of both *Trypanosoma kochi* from crocodiles and *T. varani* from monitors, with *T. grayi* perhaps representing a composite of both species rather than a distinct species. Minter-Goedbloed et al. (1993) found that *T. varani* easily developed experimentally in a sand fly host, *Phlebotomus dubosqi*, but not in the tsetse flies *Glossina morsitans* and *G. palpalis*. *Trypanosoma grayi*-like trypanosomes could infect tsetse flies but not sand flies. Molyneux (1977) reported finding *T. grayi* in *G. tachinoides* and *G. palpalis* in Upper Volta. Although the trypanosomes develop in tsetse flies in a similar manner to the leech-transmitted *Trypanosoma* species, the mode of transmission differs. When trypomastigotes are ingested by tsetse flies, they transform into epimastigotes in the midgut and begin to divide. As numbers increase, epimastigotes enter the hindgut, where they become metacyclic trypomastigotes (Hoare, 1972). For 2–3 days after ingestion, the blood meal in the midgut that contains the epimastigotes is surrounded by a peritrophic membrane. At 4–5 days postingestion, many that have entered the hindgut reach the extraperitrophic space and attach to the epithelium (Hoare, 1931a). All have escaped from within the peritrophic membrane and have migrated anteriorly within the extraperitrophic space within 6–8 days, reinvading the midgut and anterior hindgut. Division of epimastigotes continues along the midgut from days 9 to 34, with transformation to metacyclic trypanosomes, which then migrate posteriorly to complete development in the ileum of the hindgut, with infection then restricted to the ileum. Infected tsetse flies enter the open mouths of basking crocodiles and defecate while feeding or are crushed when crocodiles close their jaws. Metacyclic trypanosomes reach the bloodstream of crocodiles through mucous membranes and become trypomastigotes.

Although an anuran parasite, *Trypanosoma rotatorium*, can develop in mosquitoes (Desser et al., 1973), transmission of reptilian trypanosomes by mosquitoes has not been demonstrated. *Trypanosoma salamantae*, from the snake *Epicrates cenchris crassus*, began development in *Culex dolosus* within 24 hours postingestion, but did not reach the infective stage (de Biasi et al., 1975). Infection per os by cultural forms of *Trypanosoma phylodriasi*, isolated from *Philodryas nattereri*, of two laboratory-born *Bothrops alternatus*, with a prepatent time of 5 days was also reported by de Biasi et al. (1975). *Culex pipiens fatigans* were fed by Brygoo (1963b) on a chameleon infected by *Trypanosoma therezieni*. Dividing epimastigotes were found in the gut of the mosquitoes in 18–24 hours but disappeared between 48 and 60 hours. The inoculation of chameleons with engorged mosquitoes within 6 hours after a blood meal produced infection, but inoculations after 24 hours were not successful. Attempts by Woo (1969) and Christensen and Telford (1972) to infect mosquitoes with trypanosomes from turtles and geckoes, respectively, were unsuccessful.

Only phlebotomine sand flies appear to transmit trypanosomes of terrestrial reptiles. In all species thus far studied, the developmental pattern of saurian trypanosomes is similar. After ingestion, trypomastigotes become amastigotes, then spheromastigotes, resulting ultimately in epimastigotes, the infective stage. Within sand flies, the parasite apparently multiplies in all stages, but division of trypomastigotes within lizard hosts is rare, if it occurs at all. The apparent absence of division by trypomastigotes in the blood of infected lizard hosts possibly explains the typically low parasitemias in infected lizards. Shortt and Swaminath (1931) in Assam studied the development of *Trypanosoma phlebotomi* from *Hemidactylus frenatus* in the sand fly *Sergentomyia babu shortti*. Within 3–5 hours postingestion, trypanosomes were ovoid or nearly spherical. In the midgut, amastigotes divided within hyaline cysts that each contained 40–60 amastigotes by 24 hours. Some flagellated forms appeared in 35–40 hours, and by 72 hours, most were flagellated. By 113 hours, motile epimastigotes were present in the midgut and hindgut. Division within secondary cysts in the midgut continued to produce epimastigotes, which entered the hindgut. Geckoes became patent in 17–25 days after ingestion of infected flies. Shortt and Swaminath (1928) had referred to this particular trypanosome earlier as *T. hemidactyli* and the sand fly as *Phlebotomus minutus*. Epimastigotes of *Trypanosoma thecadactyli* were present primarily in the hindgut of *Lutzomyia trinidadensis* in 3–14 days postingestion, with some persisting through day 7 in the midgut (Christensen and Telford, 1972). When *Sergentomyia bedfordi* were fed on *Mabuya striata* in Ethiopia (Ashford et al., 1973), amastigotes of *Trypanosoma boueti* infected the midgut and hindgut at 48 hours, became spheromastigotes by 72 hours, and continued division in the midgut. Epimastigotes then appeared soon in the midgut and hindgut. When the remainder of the blood meal was voided at 4 to 5 days, few flagellates remained in the midgut, but the hindgut was tightly packed with epimastigotes following evacuation of the blood meal remnants.

Development of *Trypanosoma platydactyli* in the esophagus and midgut occurred in *Sergentomyia minuta* fed on infected *Tarentola mauritanica* (Adler and Theodor, 1935), and infection of a gecko may have resulted when it was fed an infected sand fly. *Trypanosoma scelopori* and *Trypanosoma gerrhonoti* developed in the sand fly *Lutzomyia vexator occidentis*, with epimastigotes present usually in the cardia and occasionally in both the stomach and esophagus (Ayala and McKay, 1971). The presence of developed infections in the esophagus, midgut, and cardia (*T. scelopori, T. gerrhonoti, T. platydactyli*) as well as in the hindgut (*T. phlebotomi, T. thecadactyli, T. boueti*) of vectors probably indicates differing modes of transmission of saurian *Trypanosoma* species that are independent of zoogeography and taxonomy of the vectors (Old World *Sergentomyia* and New World *Lutzomyia*). Although attempts to transmit reptilian trypanosomes by nonvectorial means have rarely been successful, Brygoo (1963b) successfully transmitted a saurian trypanosome by the inoculation of infected blood. *Trypanosoma therezieni* was transferred by inoculation from *Chamaeleo brevicornis* to four other species of chameleons. Chameleons were also infected by feeding them liver from a *Chamaeleo lateralis* previously infected with *T. therezieni* by inoculation.

With the exception of *Trypanosoma therezieni* (Brygoo, 1963b), there is little evidence of pathogenicity by reptilian trypanosomes. Although the natural host of *T. therezieni, Chamaeleo brevicornis*, appears to be unaffected by the infection, infected blood inoculated into three other *Chamaeleo* species resulted in fatal infections. Two large species, *C. parsoni* and *C. verrucosus*, weighing between 100 and 300 g, died in 25–35 days postinoculation. A small species, *Chaemaeleo lateralis*, 15 g or less in weight, died within 11–18 days. After 6 days, epimastigotes appeared in the blood of *C. lateralis*, followed by their rapid proliferation to high parasitemia until death, when Brygoo described the blood as resembling "a pure culture of trypanosomes accompanied by some blood cells" (Telford, 1984b). No divisional stages were seen in other tissues, but emboli comprised of trypanosomes were present in pulmonary capillaries. The infections in *C. lateralis* showed a decrease in erythrocytes from 8–9 × 10^5 before infection to 2–3 × 10^5 per mm^3 at death, 13–14 days later. Only trypomastigotes were present in the natural infections of *C. brevicornis*.

Most species of reptilian trypanosomes are known from a single host: 77 saurian species of 28 genera are reported hosts of 48 species of *Trypanosoma*, an average of 1.6 hosts per species of *Trypanosoma;* 14 species of trypanosomes are known from 19 chelonian host species of 14 genera (1.4 hosts per species); and the 22 species of *Trypanosoma* from snakes are reported from 24 host species and 22 genera of hosts, 1.1 hosts per trypanosome species. Conclusions based on the examination of blood slides certainly provide less than an accurate estimate of trypanosome occurrence and prevalence in naturally infected hosts, and most reports are based on the finding of trypanosomes on thin blood smears. Low parasitemias are typical of reptilian hosts infected by trypanosomes. Blood cultures or microscopic examination of freshly drawn, liquid blood under a coverslip would provide more accurate data. Very few studies have explored host specificity experimentally. *Trypanosoma chrysemydis* was transmitted to the emydid turtles *Graptemys geographica* and *Chrysemys picta,* and to the chelydrid *Chelydra serpentina,* by inoculation of crop and cecal contents from infected leeches, but two other emydids, *Eymdoidea blandingi* and *Clemmys guttata,* were refractory to infection (Woo, 1969). The two species resistant to infection are less aquatic than the others and probably rarely have contact with leeches. A snake (*Mimophis mahafalensis*), a gecko (*Phelsuma lineatum*), a gerrhosaurid lizard (*Zonosaurus* sp.), a turtle (*Pyxis arachnoides*), and a frog (*Rana mascariensis*) could not be infected by inoculation of *T. therezienii* from *Chamaeleo brevicornis*, but other chameleons (*C. oustaleti, C. lateralis, C. verrucosus*, and *C. parsoni*) were susceptible (Brygoo, 1963b). Grewal (1955) similarly failed to infect five lizard species (*Agama* sp., *Acanthodactylus* sp., *Anguis fragilis, Scincus scincus*, and *Lacerta viridis*) by inoculation from cultures of *Trypanosoma garnhami* but had similar negative results from the same cultures with the natural host *Hemidactylus brookii.*

Species Accounts

TRYPANOSOMA SPECIES OF NEOTROPICAL LIZARDS

Trypanosoma plicae Lainson, Shaw and Landau 1975

Diagnosis A large, monomorphic *Trypanosoma* species with an elongate nucleus and an elongate, broad shape. Dimensions are 49–64 × 10–21 µm. The positions of the kinetoplast and nucleus relative to body length average 94.1% and 66.9%, respectively; the nuclear index is 6.0, and the shape index is 2.5.

Type Host *Plica umbra* (Linnaeus) (Sauria: Iguanidae).

Other Hosts None known.

Type Locality Belém, Pará State, Brazil.

Other Localities None reported.

Prevalence *Trypanosoma plicae* was found in 27 of 235 (11.5%) *P. umbra* examined by Lainson et al. (1975).

Morphological Variation Trypomastigotes are elongate and broad, with a pointed anterior and bluntly rounded posterior end. Dimensions average 40.5 × 15.9 µm (49–64 × 10–21, N = 15), with the average shape index 2.5. The kinetoplast is situated 38.1 µm (25–47) from the anterior end, 9.3 µm (6–12, N = 10) behind the nucleus, and 2.4 µm (0.5–5) from the posterior end. The nucleus is "long, thin and diffuse," 14.4 × 2.4 µm (12–17 × 2–3, N = 10), and situated 27.1 µm (19–35, N = 10) from the anterior end. The nuclear index averages 6.0. Relative positions of the kinetoplast and nucleus to the body length average 94.1% and 66.9%, respectively. The undulating membrane is prominent and ends in a free flagellum that averages 16.4 µm (12–28).

Invertebrate Host Unknown. No development was observed in the mosquito *Culex pipiens fatigans* (Lainson et al., 1975).

Trypanosoma superciliosae Walliker 1965

Diagnosis A monomorphic, elongate, slender, and very large *Trypanosoma* species with total length 87–124 µm and average body length estimated at 96.2 µm. Body width is 9–19 µm. Estimated positions of the kinetoplast and nucleus relative to body length are 64.9% and 60.3%, respectively. The nucleus is elongate, 9–24 µm, with nuclear index 6.1, and shape index 6.8.

Type Host *Uranoscodon superciliosa* (Linnaeus) (Sauria: Iguanidae).

Other Hosts None known.

Type Locality Near Codajaz, Amazonas, Brazil.

Other Localities None known.

Prevalence Unknown.

Morphological Variation Trypomastigotes are monomorphic, elongate, slender, and very large, with total length including the free flagellum 112.3 µm (87–124, N = 20). Body length, estimated from subtracting mean length of the free flagellum from the mean total length, is 96.2 µm. Dimensions stated by Walliker are width of body at center of nucleus, 14.2 µm (9–19); posterior edge of body to posterior edge of kinetoplast, 33.8 µm (26–43); kinetoplast diameter, 0.9 µm (0.5–1.0); anterior edge of kinetoplast to posterior edge of nucleus, 2.35 µm (0–7); nucleus length, 16.8 µm (9–24); maximum nucleus breadth, 2.75 µm (2–4); anterior edge of nucleus to anterior end of body, 49.6 µm (37–66); free flagellum, 16.1 µm (5–25). The nuclear index is 6.1. The relative positions of kinetoplast and nucleus to body length are estimated at 64.9% and 60.3%, respectively, and the shape index is about 6.8. The undulating membrane is prominent, and shows about six waves.

Invertebrate Host Unknown.

Remarks Telford (1996c) referred to this species as both *T. superciliosae* and *T. uranoscodoni*. This was an editing error, and there is no *Trypanosoma uranoscodoni*. There are many other editing errors in this article. *Trypanosoma superciliosae* is the largest trypanosome described from saurian hosts.

Trypanosoma anolisi Telford 1996 (**Plate 64**)

Diagnosis A monomorphic *Trypanosoma* species with a small compact nucleus, elongate to rounded in shape, 19–42 × 14–28 μm. The relative positions to body length of the kinetoplast and nucleus average among host species are 51–82% and 54–78%, respectively. The nucleus averages 2.1–3.4 × 1.6–2.3 μm in size, with the average nuclear and shape indices among host species varying as 1.1–1.9 and 1.3–2.5, respectively. The dimensions vary by host species, location, and among infections within the same host species.

Type Host *Anolis limifrons* Cope (Sauria: Polychrotidae).

Other Hosts *Anolis frenatus, A. biporcatus, A. tropidogaster, A. lemurinus.*

Type Locality El Aguacate, Panama Province, Panama.

Other Localities Frijolito Creek and Quebrada Juan Grande near Gamboa, Canal Zone, Panama, and Belize.

Prevalence At the type locality, 1 of 29 (3%) *A. limifrons,* 2 of 78 (3%) *A. biporcatus*, and 1 of 18 (6%) *A. tropidogaster* were infected by *T. anolisi.* At Frijolito Creek, 3 of 74 (4%) *A. limifrons*, 4 of 7 (57%) *A. frenatus*, and 2 of 12 (17%) *A. frenatus* at Quebrada Juan Grande were infected, as were 2 of 4 *A. lemurinus* from Belize.

Morphological Variation In the type infection, trypomastigotes are monomorphic, elongate to rounded in shape, with an anterior cytoplasmic projection sometimes present. Dimensions average 24.6 ± 1.4 × 17.9 ± 2.1 μm (22.0–27.0 × 15.0–21.0, N = 17), with shape index 1.39 ± 0.17 (1.19–1.62). The kinetoplast is prominent, round or ovoid, 1–1.5 μm in diameter, and is situated 20.2 ± 3.2 μm (14.0–24.0) from the anterior end, 5.2 ± 3.0 μm (2.0–13.0) from the posterior end of the body, and 4.1 ± 1.5 μm (2.0–6.5) posterior to the nucleus. The compact nucleus is rounded to roughly oval, often tapering to a point at one end, 2.1 ± 0.4 × 1.9 ± 0.4 μm (1.5–3.0 × 1.5–3.0), with nuclear index 1.14 ± 0.18 (1.0–1.5), situated 19.3 ± 2.0 μm (15.0–22.0) from the anterior end of the body. Relative to body length, the kinetoplast and nucleus are situated at 82.1% and 78.5%, respectively. The undulating membrane is weakly developed, often lying across the body, but commonly lies along the margin. The free flagellum averages 12.3 ± 2.1 μm (10.0–14.0). The cytoplasm stains moderately, but the nucleus is lightly stained and difficult to discern.

Among the six host species of *T. anolisi* (Telford, 1996c), the range of average dimensions is 20.5–33.4 × 16.5–19.1 μm (overall, 19–42 × 14–28 μm). The relative positions of the kinetoplast and nucleus to body length vary 51–82% and 54–78%, respectively; the nucleus length is 2.1–3.4 μm, the nuclear index is 1.1–1.9, and the shape index is 1.3–2.5.

Invertebrate Host Unknown.

Remarks In similar vein to his remarks about *Trypanosoma serveti*, Telford (1996c) commented:

> The populations of trypanosomes with small compact nuclei that parasitize *Anolis* species in Middle America probably represent more than one species, and it is frustrating to be unable to establish this upon the basis of stable and consistent morphometric characters, without the use of sophisticated biochemical procedures.

Trypanosoma serveti Peláez and Streber 1955 (**Plate 64**)

Diagnosis A monomorphic *Trypanosoma* species with an elongate, narrow nucleus and slender to broad in shape, 20–45 × 8–29 μm in dimensions. The relative positions to body length of the kinetoplast and nucleus average among host species 45–93% and 26–82%, respectively. The nucleus averages 8.0–13.3 × 2.9–3.9 μm in size, with the average nuclear and shape indices among host species varying 1.3–6.0 and 1.0–4.4, respectively. The dimensions vary by host species, geography, and among infections within the same host species.

Type Host *Sceloporus teapensis* Günther (Sauria: Phrynosomatidae).

Other Hosts *Sceloporus variabilis, Anolis tropidonotus, Corytophanes hernandezi, Anolis limifrons, A. frenatus, A. poecilopus,* and *A. fuscoratus.*

Type Locality San Andres Tuxtla, Veracruz State, Mexico.

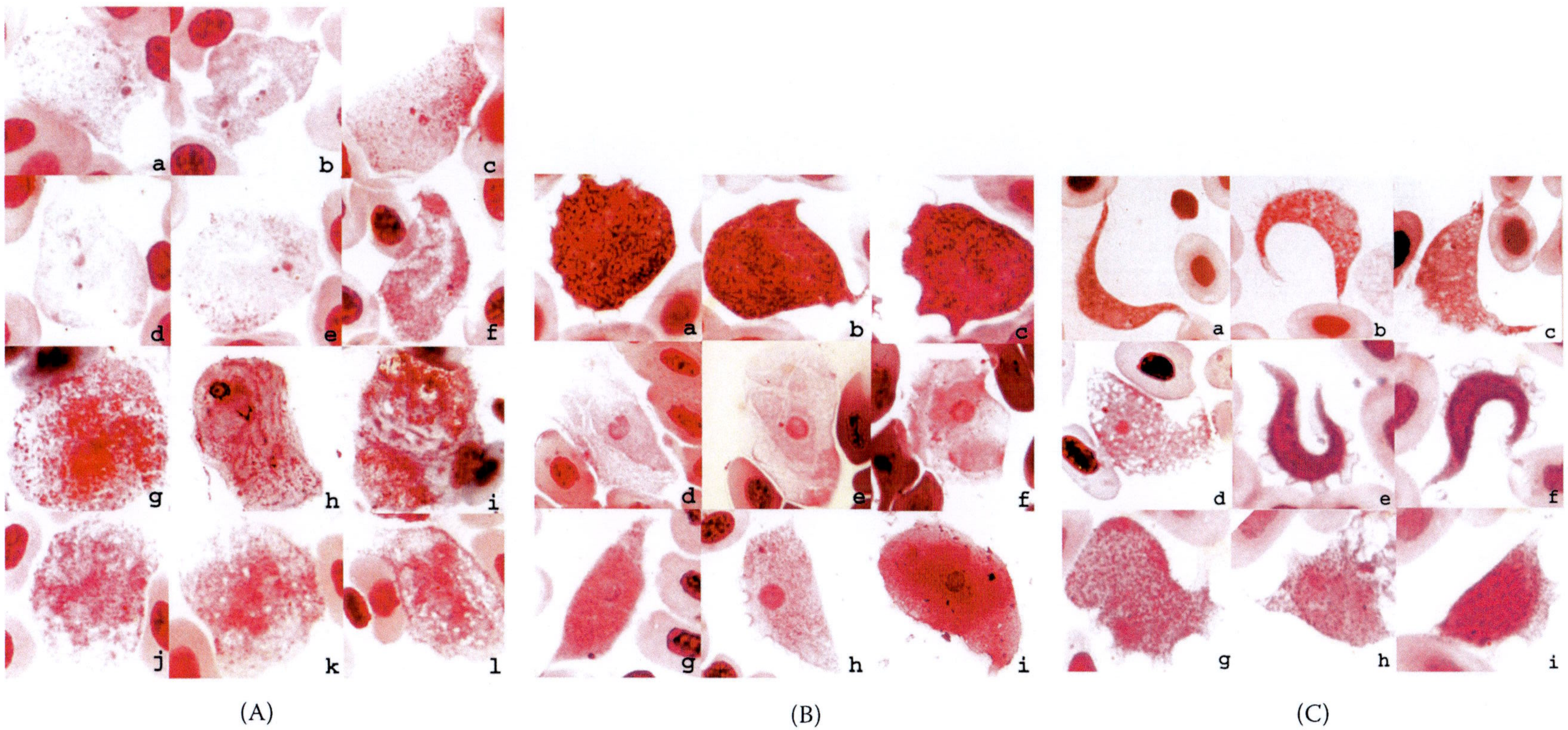

Plate 64 *Trypanosoma* spp. of New World lizards. (**A**) *Trypanosoma anolisi* from *Anolis frenatus,* **a**, **b**, *and A. limifrons,* Panama, **c**; *Trypanosoma serveti* from *Anolis frenatus,* Panama, **d–f**; *Trypanosoma fairchildi* from *Anolis capito,* Panama, **g–i**; *Trypanosoma plicaplicae* from *Plica plica,* Guyana, **j–l**. (**B**) *Trypanosoma thecadactyli* from *Thecadactylus rapicaudus,* Panama, **a–c**; *Trypanosoma gonatodi* from *Gonatodes albogularis,* Panama, **d–f**; *Trypanosoma torrealbai* from *Gonatodes taniae,* **g**, **h**, and *Phyllodactylus ventralis,* Venezuela, **i**. (**C**) *Trypanosoma scelopori* from *Sceloporus occidentalis,* California, **a–c**; *Trypanosoma poinsetti* from *Sceloporus poinsetti,* Texas, **d–f**; *Trypanosoma serveti* from *Corytophanes hernandezi,* Mexico, **g–i**.

Other Localities Panama: Barro Colorado Island (Guerrero et al., 1977); Frijolito Creek and Quebrada Juan Grande, near Gamboa (Telford, 1996c); Peruvian Amazonas (Guerrero and Ayala, 1977).

Prevalence At the type locality, prevalence was 8 of 43 (18.8%) in *S. teapensis* (Peláez and Streber, 1955); 7 of 37 (19%) in *S. variabilis*, 28 of 281 (10%) in *A. tropidonotus*, and 1 of 11 (9%) *C. hernandezi* (Lowichik et al., 1988). In Panamanian lizards, *T. serveti* was present in 3 of 12 *A. frenatus* (25%) at Quebrada Juan Grande, in 1 of 7 (14%) at Frijolito Creek, and in 4 of 26 (15%) overall; 1 of 60 (2%) *A. poecilopus* at Frijolito Creek and 1 of 227 (0.4%) overall; at Barro Colorado, in 2 of 296 (0.7%) *A. limifrons*, in 1 of 26 (4%) *A. frenatus* in 1974, and in 3 of 144 (2%) *A. limifrons* in 1975 (Guerrero et al., 1977). One of 16 (6%) *A. fuscoauratus* in Peruvian Amazonas was parasitized by a *T. serveti*-like trypanosome (Guerrero & Ayala, 1977).

Morphological Variation As redescribed by Telford (1996c) from the precise figures of Peláez and Streber (1955), *T. serveti* trypomastigotes are monomorphic but variably shaped, elongate and slender to broad in appearance, with short anterior cytoplasmic projection. Dimensions are 35.2 ± 3.3 × 16.8 ± 1.7 μm (27.9–38.6 × 14.3–22.1, N = 15). The kinetoplast is small, situated 27.1 ± 4.1 μm (14.3–31.1) from the anterior end, and 11.0 ± 3.7 μm (4.3–16.4) from the posterior end of the body. The nucleus is elongated, rarely oval, and often somewhat diffuse between ends, 9.6 ± 1.2 × 2.9 ± 0.7 μm (7.5–11.4 × 1.8–4.3), with the nuclear index 3.47 ± 0.96 (1.8–6.0). The nucleus is situated 6.4 ± 0.8 μm (4.9–7.9) anteriorly to the kinetoplast and 22.3 ± 3.6 μm (16.4–28.4) from the anterior end of the body. Positions of the kinetoplast and nucleus relative to the body length are 80.0% and 63.4%, respectively. The posterior end of the body is rounded. The undulating membrane is well-developed, with 5–12 folds usually visible. The free flagellum averages 8.1 ± 1.9 μm (5.0–11.4). The cytoplasm appears granular and stains moderately to deeply.

Among the seven host species of *T. serveti* (Telford, 1996c), average dimensions range 23.5–37.2 × 13.6–24.3 μm (overall 20–45 × 8–29 μm). The relative positions of the kinetoplast and nucleus to body length vary 45–93% and 26–82%, respectively, the nucleus length is 4–20 μm, the nuclear index is 1.3–6.0, and the shape index is 1.0–4.4.

Invertebrate Host Unknown.

Remarks *Trypanosoma serveti* is a highly variable species when evaluated by the standard morphometric characters used in taxonomic descriptions. Even at the type locality where four lizard species serve as host within a rather limited area, significant differences were present in seven of nine characters compared among samples from *Anolis tropidonotus*, *Sceloporus variabilis*, and *Corytophanes hernandezi*. All nine characters differed when compared with the type host at that locality, *S. teapensis*, although another source of variability was introduced by utilizing both measurements from figures in the type description and those obtained directly from blood films. Telford (1996c) concluded that:

> When the "best" taxonomic characters (i.e., of least variability) differ as greatly as do those between samples from the same host at the same locality (*A. tropidonotus* in Mexico, *A. frenatus* in Panama), it is probably wise to attribute the many significant differences among samples from the six host species to the effects of host and geography, rather than to taxonomic difference at the species level among the trypanosomes. Isoenzyme analysis might well provide a different interpretation.

Accordingly, all trypanosomes with elongate and narrow nuclei in hosts of the genera *Anolis, Sceloporus*, and *Corytophanes* from Mexico to Panama were considered to be *T. serveti* as well as the trypanosomes reported from *A. fuscoauratus* of Peru.

Trypanosoma fairchildi Telford 1996 (**Plate 64**)

Diagnosis A monomorphic *Trypanosoma* species with a broad, tongue shape and both ends rounded, 36.0–59.0 × 23.0–35.0 μm, with a shape index of 1.59. The kinetoplast and nucleus are situated at 75.6% and 64.5% relative to body length, respectively. The nucleus is variably shaped, with the nuclear index 1.71.

Type Host *Anolis capito* Peters (Sauria: Polychrotidae).

Other Hosts None known.

Type Locality Frijoles River, 4.8 km north of Gamboa, Canal Zone, Panama.

Other Localities None known.

Prevalence *Trypanosoma fairchildi* infected one of two *A. capito* examined from the Frijoles River and one of nine from all localities (11%).

Morphological Variation Trypomastigotes are monomorphic, broad and tongue-shaped in appearance, with both anterior and posterior ends broadly rounded, and without cytoplasmic projections. Dimensions average 46.9 ± 9.5 × 29.5 ± 3.9 μm (36.0–59.0 × 23.0–35.0, N = 10). The shape index is 1.59 ± 0.21 (1.33–1.93). The round or oval kinetoplast is prominent, up to 1.5 μm in diameter, and situated 34.5 ± 9.3 μm (25.0–50.0) from the anterior end, 14.3 ± 0.7 μm (7.5–30.0) from the posterior end of the body, and 9.0 ± 4.7 μm (3.0–18.5) posterior to the nucleus. The nucleus is oval to triangular and compact, often tapering to a point at one end, 2.4 ± 0.5 × 1.4 ± 0.3 μm (1.5–3.0 × 1.0–2.0), with the nuclear index 1.71 ± 0.33 (1.2–2.0). The nucleus is situated 30.7 ± 8.4 μm (19.0–42.0) from the anterior end of the body. Relative to body length, the kinetoplast and nucleus are situated at 75.6% and 64.5%, respectively. The undulating membrane is poorly developed, difficult to discern, usually lying across the body, rarely along the margin. The free flagellum was visible in one specimen and 5 μm in length. The cytoplasm stains deeply, as does the nucleus, which is difficult to discern.

Invertebrate Host Unknown.

Trypanosoma plicaplicae Telford 1996 (**Plate 64**)

Diagnosis A monomorphic, broadly elongate *Trypanosoma* species 26.0–45.0 × 17.0–24.0 μm, with a shape index of 1.56. The positions of the kinetoplast and nucleus relative to body length are 75.8% and 56.2%, respectively. The nucleus is ovoid, tapering to a point at one end, with the nuclear index 2.32.

Type Host *Plica plica* (Linnaeus) (Sauria: Iguanidae).

Other Hosts None known.

Type Locality Vicinity of Georgetown, Guyana.

Other Localities None known.

Prevalence *Trypanosoma plicaplicae* infected one of ten *P. plica* examined.

Morphological Variation Trypomastigotes are monomorphic, broad to broadly elongate in shape, with no anterior cytoplasmic projections. The posterior end of the body is usually rounded, sometimes elongated with a broad end. Dimensions average 30.9 ± 5.4 × 19.9 ± 2.1 μm (26.0–45.0 × 17.0–24.0, N = 11); the shape index is 1.56 ± 0.25 (1.18–1.89). The kinetoplast is prominent, up to 1.5 μm in diameter, situated 23.3 ± 3.1 μm (19.0–31.0) from the anterior end, 10.3 ± 3.7 μm (6.0–17.0) from the posterior end of body, and 8.4 ± 2.9 μm (5.5–16.0) posterior to the nucleus. The nucleus is ovoid and compact, tapering to a point at one end, 7.2 ± 1.8 × 3.2 ± 0.8 μm (5.0–11.0 × 2.0–4.5), with the nuclear index 2.32 ± 0.6 (1.5–3.5). The nucleus is situated 17.5 ± 4.9 μm (11.0–29.0) from the anterior end of the body. Relative positions of the kinetoplast and nucleus to body length average 75.8% and 56.2%, respectively. Undulating membrane is poorly developed with three to five folds usually visible. The free flagellum averages 8.8 ± 4.1 μm (4.0–14.0). The cytoplasm appears granular and stains lightly.

Invertebrate Host Unknown.

Trypanosoma thecadactylii Christensen and Telford, 1972 (**Plate 64**)

Diagnosis A monomorphic *Trypanosoma* species 19–24 μm in body length and 6.5–12.5 μm in width, ovoid or slightly triangular, occasionally elongate in shape, with a prominent posterior cytoplasmic projection up to 60% of body length and occasionally a short anterior projection. Kinetoplast is situated at 90.4% and nucleus at 71.7% relative to anterior end. The nucleus averages 2.9 × 2.6 μm, with a deep-staining, eccentric chromatin mass resembling a half-moon, and surrounded by a clear, perinuclear area; nuclear index is 1.2.

Type Host *Thecadactylus rapicaudus* (Houttyn) (Sauria: Gekkonidae).

Type Locality Madden Forest near Madden Dam, Canal Zone, Republic of Panama.

Other Hosts None known.

Other Localities Panama: Fort Sherman and Quebrada Bonita Cave, Canal Zone; Gaspar Sabana, Chepo; Sasardi, San Blas Territory.

Prevalence Nine of ten (90%) in Madden Forest and type locality; six of seven in Quebrada Bonita Cave; single records elsewhere.

Morphological Variation *Trypanosoma thecadactyli* varies in body length (BL), excluding cytoplasmic projections, 19–24 μm (21.9, N = 20), and maximum width (MW) at level of nucleus, 6.5–12.5 μm (9.1). The posterior cytoplasmic projection is 2–15 μm (9.3), extending about 10–60% of body length, and the anterior projection, when present, is a maximum of 4.5 μm (1.7). The free flagellum extends 9–20 μm (12.4). The kinetoplast is situated 14.5–20.5 μm (19.3) from the anterior end (kinetoplast anterior, KA), 3.5–5.0 μm posterior to center of the nucleus (kinetoplast to nucleus center, KN), and 1–4 μm (2.0) from the posterior end (kinetoplast to posterior end, KP). The nucleus is situated 11–16.5 μm (14.4) from the anterior end (nucleus anterior, NA). Positions of the kinetoplast (K%) and nucleus (N%) relative to anterior end are 90.4% and 71.7% of body length, respectively. The nucleus averages 2.9 × 2.6 μm and is round, with about half of the nuclear chromatin deeply stained, resembling a half-moon, and is surrounded by a clear perinuclear area. Cytoplasm of the trypanosome stains deeply with Giemsa stain.

Invertebrate Host Laboratory-reared sand flies, *Lutzomyia trinidadensis*, were fed on *Thecadactylus rapicaudus* infected with *T. thecadactyli* and were examined 1–14 days postfeeding. Between 3 and 14 days, 53% of the sand flies were infected, with flagellates fairly evenly distributed throughout the alimentary tract posterior to the cardia, and were found in the pylorus, hindgut, and rectal ampulla, occasionally persisting in the midgut through day 7. Two-thirds of the infections were restricted to the hindgut after digestion of the blood meal. Most of the flagellates were epimastigotes, but a few in most infections were trypomastigotes. Massive concentrations of flagellates often distorted the pylorus and other areas of hindgut. The digestive tracts in 18 of 88 (20%) wild-caught *L. trinidadensis* collected from four Panamanian localities contained flagellates. Hindgut infections were present in all positive sand flies, four had midgut infections, and two each had infected cardia and malpighian tubules. The rectal ampulla and portions of the hindgut commonly were enlarged because of massive infections.

Remarks *Lutzomyia trinidadensis* and *Thecadactylus rapicaudus* have a close association in nature, both favoring tree buttresses as microhabitat in Panama, where *L. trinidadensis* is the dominant sand fly species present in buttresses. Geographic range of the sand fly and the gecko coincide well, from southern Mexico into Brazil. The host gecko feeds readily on *L. trinidadensis* in the laboratory, and the characteristic posterior station infections in the sand flies suggest ingestion as the likely mode of transmission in nature (Christensen and Telford, 1972). *Culex pipiens quinquefasciatus* fed on infected geckoes did not become infected, and the gecko *Gonatodes albogularis* could not be infected with *T. thecadactyli* either by forced ingestion or intraperitoneal infection of flagellates from sand flies.

Trypanosoma gonatodi Telford 1979 (**Plate 64**)

Diagnosis A broad and elongate monomorphic *Trypanosoma* species shaped like a leaf or tongue, with an indistinct undulating membrane and dimensions 34–47 × 17–28 μm. The large kinetoplast is situated at 67.8%, well behind the nucleus at 48.1% from the anterior end. The nucleus is round and averages 5.5 × 5.7 μm, with a nuclear index of 0.98.

Type Host *Gonatodes albogularis* (Duméril and Bibron) (Sauria: Gekkonidae).

Other Hosts None known.

Type Locality Sasardi about 5 km west of Mulatupo, San Blas Territory, Panama.

Other Localities None reported.

Prevalence *Trypanosoma gonatodi* parasitized 13 of 46 (28.3%) *G. albogularis* from the type locality.

Morphological Variation The monomorphic trypomastigotes are broad and elongate, leaf-like or tongue-shaped, with an undulating membrane that is seldom distinct. Dimensions are 39.0 ± 3.7 × 22.5 ± 3.2 μm (34–47 × 17–28, N = 20), with the free flagellum, when visible (in 25%), averaging 9.0 ± 3.2 μm. The nucleus is nearly round, 5.5 ± 0.8 × 5.7 ± 0.5 μm (4–7 × 5–7), with a nuclear index of 0.98, and is situated 18.9 ± 3.9 μm (12.0–26.0) from the anterior end. The kinetoplast is situated in the posterior third of the body, 26.4 ± 6.6 μm (10–38) from the anterior end and 13.1 ± 4.9 μm (2.5–21.0) from the posterior.

Invertebrate Host Unknown.

Trypanosoma torrealbai Telford 1995 (**Plate 64**)

Diagnosis A monomorphic *Trypanosoma* species with dimensions 27.0–38.0 × 13.0–23.0 μm, broadly elongate in shape, with a prominent undulating membrane. The inconspicuous kinetoplast is situated, relative to the body length, at 83.9–88.0%, and the nucleus is at 45.7–47.0%. The nucleus is ovoid, often with a cleared space surrounding it, with a nuclear index of 0.84–0.85.

Type Host *Gonatodes taniae* Roze (Sauria: Gekkonidae).

Other Hosts *Phyllodactylus ventralis*.

Type Locality Rancho Grande, Aragua State, Venezuela.

Other Localities Fundo Vega Honda, Municipio Guanare, Portuguesa State, Venezuela.

Prevalence *Trypanosoma torrealbai* parasitized 2 of 3 *G. taniae* at the type locality, 1 of 15 (6.7%) *P. ventralis* at Fundo Vega Honda, and 1 of 86 (1.2%) overall in Portuguesa, Falcon, and Guarico states.

Morphological Variation Trypomastigotes are monomorphic, broadly elongate in shape. In *G. taniae*, dimensions are 30.7 ± 2.7 × 16.4 ± 1.7 μm (27.0–35.0 × 13.0–19.0, N = 20), with the kinetoplast situated at 26.9 ± 2.3 μm (23.0–31.0) from the anterior end, 7.0 ± 2.2 μm (1.5–11.0) from the posterior, behind the nucleus, which is located 14.5 ± 2.6 μm (9.0–19.5) from the anterior margin. The nucleus is 5.4 ± 0.5 × 6.5 ± 0.5 μm (5.0–6.0 × 5.0–7.0), with nuclear index 0.84 ± 0.11 (0.7–1.0). In *P. ventralis*, dimensions are 32.7 ± 3.2 × 18.7 ± 2.7 μm (28.0–38.0 × 13.0–23.0, N = 20); the kinetoplast is at 27.4 ± 3.2 μm (23.0–33.0) from the anterior end, and 8.1 ± 1.9 μm (6.0–13.0) from the posterior margin. The nucleus, often with a clear area surrounding it, is situated at 14.9 ± 2.1 μm (12.0–18.5) from the anterior end and measures 4.9 ± 0.5 × 5.8 ± 0.6 μm (4.0–6.0 × 5.0–7.0), nuclear index 0.85 ± 0.09 (0.7–1.0). Relative positions of the nucleus and kinetoplast to body length are 47.0% and 88.0%, respectively, in *G. taniae* and 45.7% and 83.9%, respectively, in *P. ventralis*. The undulating membrane is prominent, with the average lengths of the free flagellum 7.6–10.4 μm in the two hosts. A short posterior projection of the cytoplasm is commonly seen.

Invertebrate Host Unknown.

TRYPANOSOMA SPECIES OF NORTH AMERICAN LIZARDS

Trypanosoma scelopori Ayala, 1970 (Plate 64)

Diagnosis A monomorphic *Trypanosoma* species 34.5–47.0 μm in body length and 3.0–14.0 μm in width, elongate, with a distinct anterior cytoplasmic projection, and with the posterior end of the body usually forming a slender point. Kinetoplast is situated at 68.2% and the nucleus at 60.1% relative to the anterior end of the body. The nucleus, spherical to oval and compact, averages 4.0 × 3.4 μm, with nuclear index 1.20. The kinetoplast is prominent and often surrounded by a vacuole.

Type Host *Sceloporus o. occidentalis* Baird and Girard (Sauria: Phrynosomatidae).

Type Locality University of California Hopland Field Station, Hopland, Mendocino County, California.

Other Hosts *Crotaphytus collaris* (Crotaphytidae).

Other Localities Skagg Springs, Sonoma County, California; Newberry Springs, 24 km east of Barstow, San Bernardino County, California (*C. collaris*).

Prevalence *T. scelopori* was found in 3 of 827 (0.4%) *S. occidentalis* from Mendocino County (Ayala, 1973); 6 of 46 (13.0%) *S. occidentalis* at Skagg Springs (Ball and Chao, unpublished, vide Ayala, 1970c); 1 of 80 (1.3%) at the type locality (Telford).

Morphological Variation Ayala's (1970c) description of *T. scelopori* provided these dimensions: BL 45.1 μm (40–49, N = 60); MW 8.1 μm (3.2–11.5); posterior end to nucleus 21.5 μm (17–26); NA 23.6 μm (20–26); nucleus diameter 3.3 μm (2.5–4.2); free flagellum 3.0 μm (1.5–6.0). As redescribed by Telford (1996c) from topotypic material, dimensions were 40.8 ± 2.8 × 7.8 ± 2.4 μm (34.5–47.0 × 3.0–14.0, N = 58); NA 27.8 ± 2.1 μm (25–36); nucleus spherical to oval, 4.0 ± 0.5 × 3.4 ± 0.5 μm (3.0–6.0 × 2.5–5.0); KA 27.8 ± 2.1 μm (22.5–32.0), 3.3 ± 0.7 μm (2.0–7.5) posterior to nucleus; and free flagellum 3.5 ± 1.7 μm (1.0–8.0). Dimensions of *T. scelopori* from *C. collaris* were BL 37.8 μm (34–43, N = 20); MW 6.2 μm (3–9); NA 23.6 μm (19–27); KN 2.4 μm (1–4); KA 25.6 μm (23–30); KP 12.2 μm (11–14); nucleus 4.3 × 3.4 μm (3–5 × 2–5); free flagellum 4.1 μm (1–11).

Invertebrate Host Ayala (1970d) did not clearly distinguish the extrinsic development of *T. scelopori* from that of *T. gerrhonoti* in the sand fly *Lutzomyia vexator occidentis*. By 18 hours postfeeding, epimastigotes, pyriform flagellates, and spheromastigotes, many of the last two forms undergoing binary fission, and forms similar to bloodstream trypomastigotes were present within the blood meal in the stomach. By 64 hours, digestion was nearly completed, with forms similar to those found at 18 hours present, as was a new form, a long, thin epimastigote. A number of short trypomastigotes were also seen. Occasional flagellates were present in the cardia. After 64 hours postfeeding, flagellates in abundance were found attached to the anterior stomach or posterior cardia. After 6 or more days, the cardia was filled with thin epimastigotes. In 43 of 57 infected sand flies dissected, infections were limited to the cardia, with esophageal flagellates seen on two occasions. In 6 of 14 sand flies, thousands of epimastigotes filled the entire midgut, stomach, and cardia. Attempts to transmit *T. scelopori* by inoculation of flagellates from the cardia were unsuccessful.

Remarks Ayala did not describe the position of the kinetoplast except that it was "located immediately behind and nearly touching nucleus" (Ayala, 1970c). He also described the nucleus as round but showed an oval nucleus in some of his figures. Some of the discrepancy between the measurement sets of trypomastigotes is due to the apparent inclusion by Ayala of the anterior cytoplasmic projection in body length and its exclusion from body length by Telford. In both host species, two somewhat distinct forms of *T. scelopori* trypomastigotes occur: a broader form, more lightly stained, and a narrow form, intensely stained by Giemsa stain (Ayala, 1970c; Telford, 1996c). There were, however, no differences in the most important taxonomic characters: body length and position of nucleus and kinetoplast relative to the anterior end of the trypanosome (Telford, 1996c). Although Ayala (1970d) was unable to transmit *T. scelopori* by inoculation, epimastigote forms were seen in the stomach of one *L. vexator occidentis* 10 days following its feeding on an infected lizard (Ayala, 1970c). The anterior station of *T. scelopori* in infected sand flies suggests that transmission by bite may be the normal route.

Trypanosoma poinsetti Telford 1996 (**Plate 64**)

Diagnosis A monomorphic, elongate, and relatively slender *Trypanosoma* species with short anterior cytoplasmic projections, 31.0–42.0 × 8.0–13.0 μm. Positions of the kinetoplast and nucleus relative to the body length are 70.6% and 61.9%, respectively. The nucleus is round to ovoid with a nuclear index of 1.49.

Type Host *Sceloporus poinsetti* Baird and Girard (Sauria: Phrynosomatidae).

Other Hosts None known.

Type Locality Enchanted Rock, Llano County, Texas.

Other Localities None known.

Prevalence Five of 17 (29%) *S. poinsetti* collected at the type locality were infected with *T. poinsetti*.

Morphological Variation Trypomastigotes are monomorphic, elongate, and relatively slender, with average shape index 3.39 (2.46–4.67), and with short anterior cytoplasmic projections. The body length averages 35.7 ± 3.0 × 10.7 ± 1.3 μm (31.0–42.0 × 8.0–13.0, N = 25). The kinetoplast is prominent, approximately 1.5 μm in diameter, and situated 25.2 ± 2.0 μm (21.5–29.5) from the anterior end, and 10.6 ± 1.9 μm (6.0–14.0) from the posterior end of the body. The nucleus is nearly spherical to ovoid, compact, 4.3 ± 0.7 × 2.9 ± 0.3 μm (3.0–5.5 × 2.5–3.5), with nuclear index 1.49 ± 0.3 (1.1–2.2). It is situated 3.1 ± 0.6 μm (2.5–4.5) anteriorly to the kinetoplast and 21.9 ± 1.8 μm (19.0–26.0) from the anterior end of the body. Positions of the kinetoplast and nucleus relative to body length average 70.6% and 61.9%, respectively. The posterior end of the body is broadly pointed. The free flagellum averages 3.5 ± 0.9 μm (2.0–5.0). The undulating membrane is well-developed, with seven to ten folds usually visible. The cytoplasm appears granular and stains moderately; the nucleus stains lightly.

Invertebrate Host Unknown.

Trypanosoma urosauri Telford 1996

Diagnosis A monomorphic, elongate, and slender *Trypanosoma* species 29.0–38.0 × 5.0–9.0 μm. Positions of the kinetoplast and nucleus relative to the body length are 67.1% and 61.9%, respectively. The nucleus is round to ovoid with a nuclear index of 1.27.

Type Host *Urosaurus graciosus* Baird and Girard (Sauria: Phrynosomatidae).

Other Hosts None known.

Type Locality Enchanted Rock, Llano County, Texas.

Other Localities None known.

Prevalence One of 34 *U. graciosus* (3%) from the type locality was infected by *T. urosauri*.

Morphological Variation Trypomastigotes are monomorphic, elongate, and slender, with average shape index 5.58, and a broadly pointed anterior end. Dimensions average 34.3 ± 3.3 × 6.3 ± 1.4 μm (29.0–38.0 × 5.0–9.0, N = 6). The kinetoplast is prominent, approximately 1.5 μm in diameter, and situated 23.0 μm (17.5–29.0) from the anterior end, 13.7 ± 1.8 μm (11.0–16.0) from the posterior end of body. The nucleus is spherical to oval, compact, 2.4 ± 0.7 × 1.9 ± 0.5 μm (2.0–3.5 × 1.5–2.5), with nuclear index 1.27 ± 0.2 (1.0–1.4), situated 3.3 ± 1.4 μm (1.5–5.5) anteriorly to the kinetoplast and 22.7 ± 2.3 μm (19.0–26.0) from the anterior end of the body. Positions of the kinetoplast and nucleus relative to body length average 67.1% and 61.9%, respectively. The posterior end of the body is broadly pointed. Undulating membrane is well-developed with four or five folds usually visible. The free flagellum is 7.8 ± 3.6 μm (5.0–14.0). The granular cytoplasm stains moderately and the nucleus lightly.

Invertebrate Host Unknown.

Remarks A mathematical error in calculating kinetoplast distance from the anterior end was present in the description of *T. urosauri* (Telford, 1996c), resulting in an anterior position relative to the nucleus, which has been corrected here.

Trypanosoma gerrhonoti Ayala and McKay, 1971

Diagnosis A monomorphic *Trypanosoma* species 50–70 µm in body length by 4–16 µm in width, tapering gradually to both ends. The nucleus is round or elliptical, situated 30–41 µm from the anterior end and 16–25 µm from the posterior end of the body, with kinetoplast 1–2 µm behind the nucleus. Faint pellicular striations occasionally are visible on the surface. The nucleus is situated at 59% and kinetoplast at 61% relative to the anterior end of body. The free flagellum is short, averaging 3.1 µm.

Type Host *Gerrhonotus m. multicarinatus* Blainville (Sauria: Anguidae).

Type Locality University of California Hopland Field Station, Hopland, Mendocino County, California

Other Hosts None known.

Other Localities Twin Sisters Peak, Solano County (Ayala and McKay, 1971), and Mts. Diablo and San Leandro, Contra Costa County (McKay, 1936, vide Ayala, 1970c), California.

Prevalence Two of 51 (3.9%) of *G. multicarinatus* in Solano County (McKay, 1936, vide Ayala, 1970c) and 2 of 29 (6.9%) at Hopland, Mendocino County (1), and Twin Sisters Peak, Solano County (1), California (Ayala and McKay, 1971).

Morphological Variation Trypanosomes averaged 58.0 × 6.8 µm (50–70 × 4–16, N = 55) as described by McKay (1936, vide Ayala, 1970c) and 58.8 × 9.3 µm (51–66 × 6–14, N = 100) in Ayala's description (1970a, 1971). The nucleus, which averaged 3.5 µm, was situated 34.6 (30–41) µm from the anterior end and 20.4 (16–25) µm from the posterior end, with the kinetoplast 1–2 µm posterior to the nucleus. The free flagellum was short, 3.1 µm (0.5–5.5). Four to 12 folds were visible in the undulating membrane.

Invertebrate Host As mentioned, Ayala (1970d) did not distinguish the extrinsic development of *T. gerrhonoti* from that of *T. scelopori* in the sand fly *Lutzomyia vexator occidentis*, stating that development "of trypanosomes from both lizard species was so similar that it can be described at the same time." Both trypanosome species evidently adopt an anterior station in infected sand flies.

Remarks Hatchling *G. multicarinatus* were inoculated with flagellates from the cardia of *L. vexator occidentis* fed on an infected *G. multicarinatus* 11–12 days earlier, and infections of *T. gerrhonoti* appeared in two of five lizards within 3 weeks postinoculation.

TRYPANOSOMA SPECIES OF AFRICAN LIZARDS

Trypanosoma cnemaspi Telford 1995 (**Plate 65**)

Diagnosis A monomorphic *Trypanosoma* species, 19.0–32.0 × 9.0–17.0 µm, usually broad and leaf-like, occasionally rounded or ovoid, with a rounded posterior end of the body and rarely anterior cytoplasmic projections. The kinetoplast is prominent, situated at 87.3–90.9% and the nucleus is 57.0–58.9% from the anterior end. The elongated and narrow nucleus averages 10.0 × 2.6 µm, and the nuclear index is 2.3–7.0.

Type Host *Cnemaspis barbouri* Perret (Sauria: Gekkonidae).

Other Hosts *Hemidactylus platycephalus.*

Type Locality Kimboza Forest, 1 km north of the Ruvu River below Kibungo Village, south side of the Uluguru Mountains, Morogoro Region, Tanzania.

Other Localities None known.

Prevalence *Trypanosoma cnemaspi* was found in 8 of 23 (35%) *C. barbouri* and in 9 of 30 (30%) *H. platycephalus.*

Morphological Variation The monomorphic trypomastigotes are broad and leaf-like in appearance but are occasionally rounded or ovoid, with intensely stained cytoplasm. The posterior end of the body is typically rounded, and there are rarely any anterior cytoplasmic projections. The dimensions average 23.4 ± 2.2 × 12.2 ± 1.9 µm (19.0–32.0 × 9.0–17.0, N = 65). The free flagellum averages 9.6 ± 1.9 µm (5.0–15.0). The prominent kinetoplast is approximately 1.5 µm in diameter and is situated 20.9 ± 1.8 (17.0–26.0) µm from the anterior end and 2.6 ± 1.9 (0.5–11) µm from the posterior end of body. The nucleus is elongated and narrow, 10.0 ± 1.4 × 2.6 ± 0.6 (6.0–14.0 × 2.0–4.0) µm, with the nuclear index 3.9 ± 0.9 (2.3–7.0). It is situated 8.8 ± 1.7 (1.8–11.5) µm anteriorly to the kinetoplast and 13.6 ± 2.1 (8.5–20.0) µm from the anterior end of the body. The axostylar body, if present, is not distinct.

Invertebrate Host Unknown.

Trypanosoma uluguruensis Telford 1995 (**Plate 65**)

Diagnosis A monomorphic, usually broad and leaf-like *Trypanosoma* species, sometimes rounded or ovoid, with a rounded posterior end and dimensions 25.0–49.0 × 10.0–24.0 µm. The prominent kinetoplast is situated at 86.1–87.8% and the nucleus 56.6–60.2% from

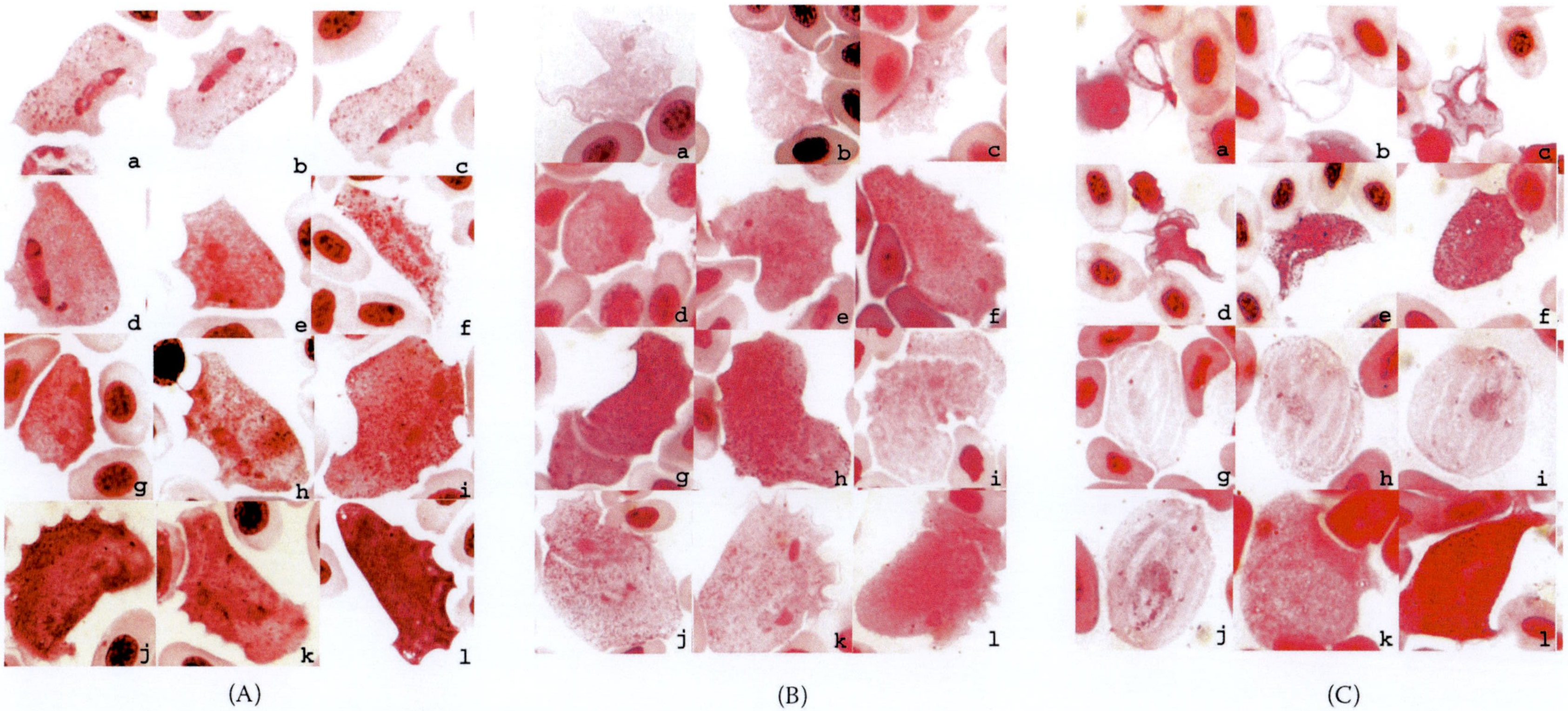

Plate 65 *Trypanosoma* spp. of African lizards. (**A**) *Trypanosoma cnemaspi* from *Cnemaspis barbouri,* **a–d**, and *Hemidactylus platycephalus*, **e**, **f**; *Trypanosoma uluguruensis* from *Hemidactylus platycephalus*, **g–l**, all Tanzanian. (**B**) *Trypanosoma kimbozae* from *Hemidactylus platycephalus*, **a–d**; *Trypanosoma lygodactyli* from *Lygodactylus luteopicturatus*, **e–h**; *Trypanosoma cordyli* from *Cordylus tropidosternum,* **i–l**, all Tanzanian. (Figure **a** modified from Telford, S. R., Jr., *Syst. Parasitol.*, 31, 37, 1995, Figure 31, and Figure **k** modified from Telford, S. R., Jr., *Syst. Parasitol.,* 31, 61, 1995, Figure 7, with kind permission of Springer Science and Business Media.) (**C**) *Trypanosoma martini* from *Mabuya quinquetaeniata,* Kenya, **a–f**; *Trypanosoma boueti* from *Mabuya maculilabris,* **g**, and *Mabuya striata,* **h–j**, Congo; *Trypanosoma striatae* sp. nov. from *Mabuya striata,* Tanzania, **k**, **l**.

the anterior end. The nucleus is elongated and narrow and averages 15.3 × 3.1 µm, and the nuclear index is 2.3–9.0.

Type Host *Hemidactylus platycephalus* Peters (Sauria: Gekkonidae).

Other Hosts *Lygodactylus luteopicturatus, Lygodactylus capensis grotei, Lygodactylus williamsi.*

Type Locality North slope of the Uluguru Mountains at Morogoro Botanical Garden, Morogoro, Morogoro Region, Tanzania.

Other Localities Morogoro and University Campus and Mindu Mountain near Morogoro, Kimboza Forest, on the south side of the Uluguru Mountains in Morogoro Region, and Bahari Beach, 21 km north of Dar-es-Salaam, Tanzania.

Prevalence Prevalence in the type host was 7 of 8 at the type locality, 5 of 8 on Mindu Mountain, 4 of 19 (21%) in Morogoro, 2 of 6 on the University Campus, Morogoro; 11 of 30 (37%) in Kimboza Forest; and 2 of 32 (6%) at Bahari Beach; prevalence in *L. luteopicturatus* was 3 of 24 (12%) in Morogoro, 13 of 45 (29%) on the northern slope of the Uluguru Mountains at the edge of University Campus, 1 of 21 (5%) at Bahari Beach, and 1 of 1 collected in Kimboza Forest; prevalence in *L. capensis grotei* was 10 of 40 (25%) on the northern slope of the Uluguru Mountains at the edge of the University Campus and 2 of 35 (6%) at Morogoro; prevalence in *L. williamsi* was 1 of 15 (7%) in the Kimboza Forest.

Morphological Variation Trypomastigotes are monomorphic, broad, and leaf-like in appearance, occasionally rounded or ovoid, with intensely stained cytoplasm. Dimensions average 34.9 ± 4.4 × 17.2 ± 2.9 µm (25.0–49.0 × 10.0–24.0, N = 115). Anterior cytoplasmic projections are rarely present. The free flagellum averages 9.4 ± 3.1 µm (3.0–22.0). The kinetoplast is prominent, approximately 1.5 µm in diameter, situated 30.2 ± 4.5 µm (17.5–45.5) from the anterior end and 4.9 ± 3.2 µm (0.1–14.0) from the posterior end of the body. The nucleus is elongated and narrow, 15.3 ± 2.6 × 3.1 ± 0.6 µm (7.0–24.0 × 2.0–5.0), with the nuclear index 5.0 ± 1.2 (2.3–9.0), and situated 12.0 ± 2.6 µm (6.5–17.5) anteriorly to the kinetoplast and 20.3 ± 3.7 µm (11.0–30.5) from the anterior end of the body. Often, only the roughly triangular ends of the nucleus are well stained, separated by a poorly or unstained central portion, and closely associated with or overlying an apparent axostylar body. The posterior end of the body is typically rounded.

Invertebrate Host Unknown.

Trypanosoma kimbozae Telford 1995 (**Plate 65**)

Diagnosis A monomorphic *Trypanosoma* species that is broadly elongate with a prominent anterior projection usually present. Dimensions are 25.0–49.0 × 10.0–16.0 µm, and the cytoplasm does not stain intensely. The kinetoplast is not prominent and is situated at 72.5%, with the nucleus at 69.0% from the anterior end of the body. The nucleus is ovoid and averages 3.3 × 2.3 µm, and the nuclear index is 1.0–2.0.

Type Host *Hemidactylus platycephalus* Peters (Sauria: Gekkonidae).

Other Hosts None known.

Type Locality Kimboza Forest, 1 km north of the Ruvu River below Kibungo Village, south side of Uluguru Mountains, Morogoro Region, Tanzania.

Other Localities None known.

Prevalence Prevalence of *T. kimbozae* was 6 of 40 (15%).

Morphological Variation Trypomastigotes are monomorphic, broadly elongate, with the anterior end usually attenuated into a prominent projection and the posterior end broadly pointed; the cytoplasm does not stain intensely. The dimensions average 40.0 ± 6.4 × 12.6 ± 1.6 µm (25.0–49.0 × 10.0–16.0, N = 24). The free flagellum averages 6.3 ± 1.2 µm (4.0–9.0). The kinetoplast is not prominent, approximately 1.5 µm in diameter, and situated 29.1 ± 5.4 µm (19.0–38.0) from the anterior end and 11.2 ± 2.1 µm (6.0–14.0) from the posterior end of the body. The nucleus is ovoid, 3.3 ± 0.4 × 2.3 ± 0.4 µm (2.5–4.0 × 1.5–3.0), with the nuclear index 1.5 ± 0.2 (1.0–2.0), and situated 1.7 ± 0.6 µm (0–2.5) anteriorly to the kinetoplast and 27.8 ± 5.5 µm (17.0–37.0) from the anterior end of body. The axostylar body is not present.

Invertebrate Host Unknown.

Trypanosoma lygodactyli Telford 1995 (**Plate 65**)

Diagnosis A monomorphic, broad, and leaf-like *Trypanosoma* species, 25.0–39.0 × 11.0–22.0 µm, with a typically rounded posterior end, rarely with anterior cytoplasmic projections. The prominent kinetoplast is situated at 83.1–88.7% and the nucleus at 60.4–61.5% from the anterior end. The spherical-to-oval nucleus averages 2.6 × 1.8 µm, with a nuclear index of 0.8–2.0.

Type Host *Lygodactylus luteopicturatus* Pasteur (Sauria: Gekkonidae).

Other Hosts *Lygodactylus capensis grotei, Hemidactylus platycephalus.*

Type Locality Morogoro, Morogoro Region, Tanzania, northern slope of the Uluguru Mountains at University Campus.

Other Localities Kimboza Forest, south side of the Uluguru Mountains, Morogoro Region, Tanzania.

Prevalence *Trypanosoma lygodactyli* parasitized 8 of 121 *L. luteopicturatus* (7%), 3 of 75 *L. capensis grotei* (4%), and 2 of 123 *H. platycephalus* (2%) overall. At the type locality, prevalence in the type host was 6 of 45 (13%), and was 2 of 24 (8%) in nearby Morogoro; 2 of 35 (6%) in *L. capensis grotei* at Morogoro; 1 of 40 (3%) in *H. platycephalus* at the type locality; and 2 of 30 (7%) in the Kimboza Forest.

Morphological Variation Trypomastigotes are monomorphic and usually broad and leaf-like in appearance, occasionally rounded or ovoid, with intensely stained cytoplasm, and rarely with anterior cytoplasmic projections. Dimensions average 31.5 ± 3.3 × 17.1 ± 2.6 μm (25.0–39.0 × 11.0–22.0, N = 38). The free flagellum is 10.0 ± 3.1 μm (5.5–18.0). The prominent kinetoplast is approximately 1.5 μm in diameter and situated 27.9 ± 3.0 μm (21.0–37.0) from the anterior end, 4.2 ± 2.5 μm (0.5–11) from the posterior end of body. The compact nucleus is spherical to oval, 2.6 ± 0.9 × 1.8 ± 0.5 μm (1.0–5.0 × 1.0–2.5), with a nuclear index of 1.5 ± 0.3 (0.8–2.0), and is situated 10.8 ± 4.2 μm (4.0–20.5) anteriorly to the kinetoplast and 20.7 ± 5.0 μm (10.0–34.0) from the anterior end of the body. It is closely associated with the axostylar body when the latter is visible. The posterior end of the body typically is rounded.

Invertebrate Host Unknown.

Trypanosoma petteri Brygoo 1966

Diagnosis A *Trypanosoma* species described as polymorphic, with a slender form 43 × 8 μm, with an ovoid nucleus situated at 37% and kinetoplast at 74% and sharply pointed ends. A broad, leaf-like form 50 × 10 μm, with a more elongate nucleus and a kinetoplast located virtually on the posterior margin possibly is not conspecific.

Type Host *Phelsuma madagascariensis* Gray (Sauria: Gekkonidae)

Other Hosts None known.

Type Locality Ankara, Madagascar.

Other Localities None reported.

Prevalence Unknown.

Morphological Variation Brygoo (1966b) provided only mean dimensions from a series of 10 specimens. If conspecific, the trypomastigotes are clearly dimorphic, with some slender and elongated, with pointed ends, and larger, leaf-like forms with one or both ends rounded. Apparent intermediate forms were considered to be a third distinct type. The smaller, more slender trypomastigotes of "la forme classique" averaged 43 × 8 μm, with the greater dimension of the nucleus 4.5 μm, and situated at 16 μm or 37% relative to body length. The kinetoplast, surrounded by a clear area, is located 16 μm behind the nucleus, at about 74%, and 15 μm anterior to the posterior edge of the body. Both ends of the trypomastigotes are sharply pointed, and the undulating membrane shows nine or ten distinct, narrow waves. The second, leaf-like form is larger, 50 × 10 μm, and the nuclei appear to be elongate and considerably larger than in the narrow, typical form. The kinetoplast is situated very close to the rounded posterior margin in the specimens illustrated, no further than 0.8–3.0 μm from it. The undulating membrane is distinct but forms perhaps one-half as many, and larger, waves than in the slender form. The anterior end can be broadly pointed. In the possible intermediate form, the extremities are not as finely pointed as in the typical form. The position of the kinetoplast and size and position of the nucleus are more similar to the slender form than to the broad type.

Invertebrate Host Unknown.

Remarks The differing positions of the kinetoplast strongly suggest the presence of two species of *Trypanosoma* in the original material, the slender form *T. petteri* and at least a second species, the broad leaf-like form. The two least variable, and therefore most useful, taxonomic characters in gekkonid trypanosomes are the relative positions of the kinetoplast and the nucleus to the body length (Telford, 1995a).

Trypanosoma garnhami Grewal 1955

Diagnosis A monomorphic *Trypanosoma* species, broadly elongate and leaf-like in shape, with dimensions 18.2–38.9 × 12.1–19.7 μm, a prominent free flagellum, and a variably distinct axostyle. Relative to body length, the kinetoplast is situated at 64.9% and the nucleus at 53.4%. The weakly staining nucleus is variable in shape, from spherical to elongate.

Type Host *Hemidactylus brookii angulatus* Gray (Sauria: Gekkonidae).

Other Hosts None known.

Type Locality Nairobi (?), Kenya.

Other Localities None reported.

Prevalence Unknown.

Morphological Variation The monomorphic trypomastigotes are broad and leaf-like, 25.3 ± 6.4 × 15.8 ± 2.6 µm (18.2–38.9 × 12.1–19.7, N = 10). The undulating membrane is variably distinct, and the free flagellum averages 14.7 ± 3.5 µm (9.5–20.6) in length. The kinetoplast is small and located 16.3 ± 5.9 µm (8.5–29.5) posterior to the anterior margin, and 8.9 ± 4.8 µm (0.6–16.1) from the posterior margin. The nucleus lies on average 5.8 ± 5.1 µm (1.7–13.0) from the kinetoplast, 14.8 ± 2.2 µm (12.8–17.7) from the anterior margin of the body. Position of the kinetoplast and nucleus relative to the body length are 64.9% and 53.4%, respectively. The nucleus is variable in shape, spherical, oval or elongate, and is located at the anterior end of the axostyle when the latter is visible. Nucleus has an average length of 4.3 ± 2.3 µm (1.8–8.2), stains weakly, and sometimes appears as two side-by-side bodies when division is imminent.

Invertebrate Host The natural invertebrate host is unknown. Grewal (1955) found that the triatomid bug *Rhodnius prolixus* and the soft tick *Ornithodoros parkeri* could be infected experimentally with *T. garnhami*. In the arthropods, most of the forms present 4 days PI were leptomonads, and a few amastigotes were seen. At 7 days PI, the numbers of flagellates had increased, still mostly leptomonads commonly seen dividing, and amastigotes were more common, along with some apparent transitional crithidia. On day 17, leptomonads were 8.3 × 2.7–11.0 × 1.7 µm, with flagella 16.2–22.3 µm in length. Amastigotes were round or oval, 6.0 × 6.0 µm or 6.0 × 4.5–6.5 × 5.5 µm in size, some showing a free flagellum. Metacyclic trypanosomes appeared in the gut, at a size of 7.8 × 2.2–10.6 × 2.2 µm. Four days postinoculation of infected feces into the European lizard *Lacerta viridis*, flagellates were observed in a culture of heart blood. The inoculation of culture forms into several species of gekkonid, agamid, anguid, lacertid, and scincid lizards did not produce infections.

Remarks It is unlikely that either hemipterans or soft ticks play a vectoral role in nature for this trypanosome.

Trypanosoma zonuri Telford 1995

Diagnosis A monomorphic, delicate, and fragile-appearing *Trypanosoma* species, usually elongate, 22.5–35.5 × 7.0–11.0 µm, rarely with cytoplasmic projections. The closely associated kinetoplast and nucleus are situated on average, relative to body length, 67.8% and 67.5%, respectively. The nucleus is round or ovoid, with nuclear index 1.3.

Type Host *Cordylus cordylus sensu lato* (Sauria: Cordylidae).

Other Hosts None known.

Type Locality Approximately 21 km west of Ngorongoro Crater, Arusha Region, Tanzania.

Other Localities None reported.

Prevalence Two of seven (29%) *C. cordylus* were infected by *T. zonuri* (Telford, 1995c).

Morphological Variation The monomorphic trypomastigotes, delicate and fragile in appearance with clear, unstained cytoplasm, are usually elongate, 29.6 ± 3.0 × 8.9 ± 1.0 µm (22.5–35.5 × 7.0–11.0, N = 25), rarely with cytoplasmic projections. Free flagellum is 10.3 ± 1.7 µm (8–16). The kinetoplast is prominent, approximately 1 µm in diameter, situated 20.1 ± 2.6 µm (15.5–25.0) from the anterior end and 9.5 ± 1.3 µm (7–12) from the posterior end of body. The nucleus is rounded, usually, or ovoid, 1.5 ± 0.3 × 1.1 ± 0.2 µm (1.0–2.0 × 1.0–1.5), with a nuclear index of 1.3 ± 0.3 (1.0–2.0). It is adjacent to and often in contact with the kinetoplast and is situated 20.1 ± 2.9 µm (14.0–24.5) from the anterior end. Positions of the kinetoplast and nucleus, respectively, relative to body length, are 58–74% (67.8 ± 4.1) and 52–74% (67.5 ± 5.2).

Invertebrate Host Unknown.

Remarks Specific identity of the saxicolous host is uncertain. It is possibly the same species recorded as *Cordylus* sp. from Arusha and southern Kenya described by Spawls et al. (2002) but is much darker in coloration, almost black. Only a single morphological type of trypanosome was observed in each infection, despite the extended periods of observation of 11 and 12 months.

Trypanosoma cordyli Telford 1995 (Plate 65)

Diagnosis A monomorphic, broad, and leaf-like *Trypanosoma* species 32.0–62.0 × 19.0–31.0 µm, intensely staining, with no cytoplasmic projections. Relative positions of the kinetoplast and nucleus to body length are 63.4% and 57.7%, respectively. The nucleus is usually ovoid, with a nuclear index 1.6.

Type Host *Cordylus tropidosternum* (Cope) (Sauria: Cordylidae).

Other Hosts None known.

Type Locality 3.8 km west of Rondo Forestry Station, Rondo Forest, Lindi District, Lindi Region, Tanzania.

Other Localities None reported.

Prevalence The only *C. tropidosternum* collected at the type locality, 2 August 1984, was infected by *T. cordyli*. Five from Tanga Region and one from Morogoro Region were negative for trypanosomes.

Morphological Variation Trypomastigotes are monomorphic, broad, and leaf-like in appearance, with intensely stained cytoplasm. Their shape is often rounded or ovoid, 45.3 ± 7.4 × 24.6 ± 3.3 µm (32.0–62.0 × 19.0–31.0, N = 20), without cytoplasmic projections. The free flagellum averages 6.2 ± 1.9 µm (4–10). The kinetoplast is prominent, approximately 1.5 µm in diameter, and situated 28.5 ± 4.1 µm (22.0–35.0) from the anterior end and 16.8 ± 4.7 µm (9–27) from the posterior end of the body. The nucleus is usually ovoid, 4.3 ± 0.5 × 2.8 ± 0.5 µm (3.0–5.0 × 2.0–3.5), with a nuclear index 1.6 ± 0.3 (1.1–2.0). It is usually slightly anterior to the kinetoplast, 25.9 ± 3.8 µm (20.0–32.5) from the anterior end. The positions of the kinetoplast and nucleus, relative to body length, are 63.4 ± 5.8% (53–73%) and 57.7 ± 7.3% (47–75%), respectively.

Invertebrate Host Unknown.

Remarks The host was positive for *T. cordyli* at initial examination on 8 August 1984 and remained positive until 25 February 1988, although parasites were often absent on slides taken at irregular intervals during this period. Telford (1995c) reported that:

> Individual *T. cordyli* commonly appeared to be binucleate, with nuclei well separated … up to 15 or more µm apart, while in others the nuclei were connected by a band of similarly reddish but lighter staining material about 4 µm wide. These were at first thought to be dividing trypomastigotes, but only a single kinetoplast was evident in all, closely associated with one of the nuclei. No explanation is obvious, unless in this species the nucleus divides before the kinetoplast.

The host is a semiarboreal species that occurs on tree trunks from ground level to heights exceeding 10 m, where it occupies tree holes or crevices.

Trypanosoma domerguei Brygoo 1965

Diagnosis A monomorphic, elongate, and broad, leaf-like *Trypanosoma* species 19.6–29.2 × 14.3–21.6 µm, with the kinetoplast situated, relative to body length, at about 62.2% and the nucleus at 30.9%. The nucleus is ovoid to elongate, 5.9–14.6 µm in length.

Type Host *Oplurus sebae* Duméril and Bibron (Sauria: Opluridae).

Other Hosts None known.

Type Locality Ampijoroa, Madagascar.

Other Localities None reported.

Prevalence *Trypanosoma domerguei* parasitized 5 of 14 (35.7%) *O. sebae* overall (Brygoo, 1965b).

Morphological Variation The monomorphic trypomastigotes are elongate and broad, leaf-like, often circular in shape, 23.3 × 17.4 µm (19.6–29.2 × 14.3–21.6, N = 10). The kinetoplast, surrounded by a clear area, is situated posterior to the nucleus, on average 3.2 µm (1.3–6.3, N = 9) from its posterior edge, and 5.6 µm (2.3–10) from the posterior end of the body. The distance of the kinetoplast from the anterior end, estimated from averages, is about 14.5 µm or 62.2% relative to the body length. The nucleus is ovoid to elongate, 6.8 µm (5.9–14.6) in the greater dimension, with a distance between its anterior border and the anterior edge of the body of 7.2 µm (5.3–11.3) or 30.9% of the body length. The free flagellum averages 6.7 µm (4.9–11, N = 8) in length.

Invertebrate Host Unknown.

Remarks Brygoo (1965b) remarked that the dimensions may not be exact because of the difficulty in interpreting the anterior and posterior borders of the trypomastigotes.

Trypanosoma therezieni Brygoo 1963

Diagnosis A monomorphic, elongate, and slender *Trypanosoma* species 28–33 × 4.5–6.5 µm, with a shape index of 5.8. The relative positions of the kinetoplast and nucleus to body length are estimated at 62.5% and 34.3%, respectively. The nucleus shape is ovoid to quadrangular.

Type Host *Chamaeleo brevicornis* Günther (Sauria: Chamaeleonidae).

Other Hosts None known.

Type Locality Ampanaherana, Canton Fandrandava, Subprefecture Fianarantsoa, Madagascar.

Other Localities Perinet, Madagascar.

Prevalence At the type locality, 2 of 8 *C. brevicornis* were infected and at Perinet, 6 of 57 (10.5%).

Morphological Variation The monomorphic trypomastigotes are elongate and slender, with a finely pointed anterior end and a posterior that is less sharp to blunt. Dimensions are 32 × 5.5 µm (28–33 × 4.5–6.5), with a shape index 5.8. The kinetoplast is situated an estimated 20 µm (17–22) from the anterior end, 7 µm (5–11) from the posterior end, and 9 µm (7–10) behind the nucleus. The nucleus is located 11 µm (10–12) from the anterior end. It is described as oval or quadrangular, without stated dimensions, and as calculated from the figures, appears to be 2.1–3.6 × 1.4–1.8 µm in size. Positions of the kinetoplast and nucleus relative to body length are approximately 62.5% and 34.3%, respectively. The undulating membrane is prominent and forms four to six waves, terminating in a long free flagellum of 17 µm (15–22).

Invertebrate Host Unknown.

Remarks As stated elsewhere (Telford, 1995b), Brygoo (1963b) has reported the only successful transmission of a saurian trypanosome by inoculation of infected blood. *Trypanosoma therezieni* from infected *C. brevicornis* was inoculated into *Chamaeleo parsoni, C. verrucosus*, and *C. lateralis*, and chameleons were also infected by ingestion of liver from a *C. lateralis* previously infected by blood inoculation. *Trypanosoma therezieni* is apparently nonpathogenic to the natural host, *C. brevicornis*, but infections were always fatal in the species inoculated with infected blood. *Chamaeleo parsoni* and *C. verrucosus* that weighed 100–300 g died 25–35 days PI. *Chamaeleo lateralis*, much smaller and usually less than 15 g, died within 11–18 days PI. After 6 days, epimastigotes appeared in the blood of *C. lateralis,* and the parasitemia increased until death, when the blood resembled "a pure culture of trypanosomes accompanied by some blood cells" (Telford, 1984b).

Trypanosoma martini Bouet 1909, Syn. *T. mabuiae* (Wenyon 1909) (**Plate 65**)

Diagnosis A dimorphic *Trypanosoma* species represented by a broader, longer form with a prominent undulating membrane that often appears with a posterior portion spirally twisted on itself, often forming a circular configuration, and a delicate, smaller and straight, very narrow form with a poorly developed undulating membrane. The larger form is 19–36 × 4–10 µm, with shape index 5.49. The nucleus is ovoid and small, with nuclear index 2.18. Relative positions of the kinetoplast and nucleus to the body length are 78% and 75%, respectively. The delicate smaller form is 17–24.5 × 2–3 µm, shape index is 9.54, and has a round-to-ovoid nucleus with nuclear index 1.46. The kinetoplast and nucleus are posterior, with positions relative to the body length 0.88% and 0.83%, respectively.

Type Host *Mabuya maculilabris* (Gray) (Sauria: Scincidae).

Other Hosts *Mabuya perroteti* (Bouet, 1909b), *M. quinquetaeniata* (Wenyon, 1909a), *M. reddoni* (Franca, 1911).

Type Locality Ivory Coast, West Africa (vide Wenyon, 1926).

Other Localities Sudan: Wau, Bahr-el-Ghazal Province (Wenyon, 1909a), in *M. quinquetaeniata.* In *M. raddoni* from Guinea Bissau (as Portuguese Guinea; Franca, 1911); Kenya: near Lake Victoria (Garnham, 1950), in *M. maculilabris*; Pole, West Pokot District (M. Mutinga, collector), in *M. quinquetaeniata.*

Prevalence Unknown.

Morphological Variation Bouet (1909b) reported dimensions of *T. martini*, as calculated from his characters, as 53 × 7–8 µm, kinetoplast to posterior end as 16 µm and to nucleus 0 µm, nucleus to anterior end as 35 µm, nucleus diameter was 2 µm, and free flagellum was 20 µm. Wenyon (1909a) described a dimorphic flagellate as *Trypanosoma mabuiae*, with broad and slender forms, the former 30–40 × 8.5 with a prominent undulating membrane, but did not distinguish the nucleus and kinetoplast. The slender form was 20–25 × 2–2.5 µm, with his figures showing a long free flagellum arising from a posterior kinetoplast, slightly behind the nucleus, and a weakly developed undulating membrane. Garnham (1950) found only a single form, with a well-developed undulating membrane, total length 52 µm. The calculated body length is 37 µm, comprised by nucleus 5 µm to posterior tip, a 4-µm nucleus, and nucleus to anterior end 28 µm. Body width was 7.5 µm, and it was 15 µm to the free flagellum. Garnham stressed the spiral turning of the posterior body on itself, as described by Bouet (1909b), and thought that the Kenyan flagellate could represent a small race of *T. martini*.

A dimorphic trypanosome present in *Mabuya quinquetaeniata* of West Pokot District, Kenya has a larger and broader form nearly identical to the figures of Garnham (1952), with a prominent and floppy undulating membrane and a body usually turned in a spiral manner on itself, often forming a circular configuration. Body dimensions are 31.8 ± 3.8 × 6.1 ± 1.3 µm (19–36 × 4–10, N = 24). The small kinetoplast is situated 6.8 ± 1.8 µm (3.0–9.5) from the posterior end, and 24.9 ± 3.8 µm (13–31) from the sharp anterior tip. It is usually positioned immediately behind the nucleus, 1.0 ± 0.5 µm (0–1.5), and occasionally appears to be actually on the midpoint or end of the nucleus. The small nucleus is 2.6 ± 0.7 × 1.3 ± 0.3 µm (1.5–4.0 × 1.0–1.5), usually ovoid in shape, and is situated 23.8 ± 3.7 µm (12.5–30.0) behind the anterior end. The nuclear index averages 2.18 and the body shape index 5.49. Positions of the kinetoplast and nucleus relative to body length average 78% and 75%, respectively. The cytoplasm stains lightly. The free flagellum, usually visible, averages 10.5 ± 2.8 µm (5–17, N = 21). A much less common, smaller, and thin form has a less-well-developed and sometimes-difficult-to-discern undulating membrane. In a small series (N = 4) of intact specimens, body length averages 21.1 µm and maximum width 2.3 µm (17–24.5 × 2–3). The kinetoplast distance from the anterior end is 18.5 µm, and from the posterior, pointed tip, is 2.5 µm, 0.9 µm behind the nucleus. The nucleus is 1.6 × 1.1 µm and is located 17.5 µm from the anterior end. The free flagellum, when visible, averages 11.0 µm. Positions of the kinetoplast and nucleus relative to the body length are 88% and 83%, respectively. This slender form appears to be delicate, and most found were damaged to some degree.

Invertebrate Host Unknown.

Remarks The sample from Kenyan *M. quinquetaeniata* is remarkably similar to the trypomastigotes figured by Garnham (1952) and the smaller, thin forms shown by Wenyon (1909a). Whether the East African trypanosome is conspecific with the apparently larger *T. martini* of West Africa cannot be determined except by genomic or isoenzyme analysis. The size difference is possibly indicative only of measurements from a single, large trypanosome, and study of a series could well show similar dimensions. If different species are present, then *Trypanosoma mabuiae* Wenyon would be the appropriate designation for the East Africa flagellate.

Trypanosoma boueti Martin 1907 (**Plate 65**)

Diagnosis A *Trypanosoma* species characterized by a broad elongate body with two to four refractile longitudinal ridges on its surface, 18–33 × 11–23 µm in two host species, with a shape index averaging 1.35–1.47. Relative positions of the kinetoplast and nucleus to the body length average 65–67% and 49–52%, respectively. The nucleus is ovoid in shape, with a nuclear index of 1.40–1.46.

Type Host *Mabuya raddoni* (Gray) (Sauria: Scincidae).

Other Hosts *Mabuya maculilabris* (Schwetz, 1931 as *T. martini*); *M. striata* (Ashford et al., 1973).

Type Locality French Guinea (Guinea).

Other Localities Senegal (Leger and Leger, 1914); Congo: Kisangani (= Stanleyville of Schwetz, 1931), Kinshasa (Telford); Ethiopia: Sabeta, 25 km southwest of Addis Ababa, and Kutaber, 400 km north of Addis Ababa (Ashford et al., 1973); Tanzania: Morogoro (Telford).

Prevalence Congo, in *M. maculilabris*: at Kisangani, 6 of 73 (8.2%) (Schwetz, 1931) and at Kinshasa 4 of 26 (15.4%) (Telford). In Morogoro, 9 of 153 *M. striata* were positive for trypanosomes (Telford), but two species were present, and the discard of very lightly infected slides before species could be distinguished prevents any statement of prevalence. In Ethiopia, about 25% of more than 100 *M. striata* examined were infected (Ashford et al., 1973).

Morphological Variation Ashford et al. (1973) described trypomastigots in *M. striata* as averaging 28 × 16 µm, with a large, circular nucleus 4.5 µm in diameter. Calculated dimensions from their nine figures of *T. boueti* provided averages of 27.5 × 16.5 µm (20–33 × 13–22), with kinetoplast and nucleus locations relative to body length 67.6% and 55.5%, respectively. Two to four refractile longitudinal ridges or striations along the body were visible. The size, positions of the kinetoplast and nucleus, and presence of the longitudinal striations are consistent with samples from Kinshasa and Morogoro. Dimensions of the series from *M. maculilabris* of Kinshasa are 24.1 ± 2.1 × 16.7 ± 2.7 µm (18.5–27 × 11.5–23, N = 20). The kinetoplast is located 15.8 ± 2.2 µm (10.5–20) from the anterior end, 9.2 ± 2.0 µm

(5.5–15) from the posterior, and 3.9 ± 0.8 µm (1.5–5.0) behind the nucleus. The nucleus is ovoid, 6.0 ± 0.7 × 4.2 ± 0.6 µm (5–7 × 3.5–5), and is situated 11.9 ± 1.9 (9–16) from the anterior end. Relative positions of the kinetoplast and nucleus to the body length are 65% and 49%, respectively. The nuclear index is 1.46, and the body shape index is 1.47. The free flagellum was visible in only 10% of trypomastigotes and averaged 3.5 µm, and the undulating membrane was difficult to discern. In *M. striata* from Morogoro, dimensions are 23.8 ± 2.3 × 17.7 ± 1.7 µm (19–28 × 14.5–21, N = 20). The kinetoplast is situated at 16.2 ± 3.0 µm (10–22) from the anterior edge, 7.8 ± 1.6 µm (4.5–10) from the posterior end, and 3.8 ± 1.4 µm (1.5–9) behind the nucleus. The ovoid nucleus is 5.6 ± 0.6 × 4.0 ± 0.4 µm (4–6.5 × 3–5) and lies 12.4 ± 2.5 µm (8.5–18) from the anterior end. The positions of the kinetoplast and nucleus relative to the body length are 67% and 52%, respectively. The nuclear index is 1.40, and the body shape index is 1.35. The undulating membrane is not distinct, and the free flagellum, visible in 20% of trypomastigotes, averages 7.3 µm in length.

Invertebrate Host Ashford et al. (1973) fed sand flies, *Sergentomyia bedfordi*, on *Mabuya striata* infected by *T. boueti* and readily obtained infections in them. Small clumps of 2–17 amastigotes, 3.5 µm in diameter, were present in midgut contents by 48 hours PF. By 60 hours PF, amastigote aggregations measured 50 µm in diameter, comprised of "hundreds of parasites." Spheromastigotes appeared about 72 hours PF and shortly thereafter became epimastigotes as the aggregates broke apart, with division continuing. When the gut contents were voided 4–5 days PF, most of the flagellates left the midgut and formed "an immobile palisade" around the rectal glands of the hind gut. Dissections of 388 wild-caught *S. bedfordi* found natural flagellate infections in 42 (11%). Some flies contained only amastigotes or spheromastigotes in the midgut; others had epimastigotes in the midgut and hindgut. The epimastigotes, 9–25 µm in length, resembled those produced in laboratory infections, as did the amastigotes and spheromastigotes found. Field observations showed close association between *S. bedfordi* and *M. striata*, and sand flies were observed feeding on the skinks "under culverts and in rock cracks." The posterior station infections of *S. bedfordi* indicate natural transmission to skinks when sand flies are ingested.

Remarks The trypanosomes figured by Schwetz (1931) from *M. maculilabris* are not *T. martini* as he identified them, but are similar to *T. boueti* in the larger nucleus size, distance between kinetoplast and nucleus (they are adjacent in *T. martini*), development of the undulating membrane, trypomastigote shape, and calculated relative positions of the kinetoplast (71–83%) and nucleus (62–63%). Schwetz did not describe the presence of the refractile ridges seen in *T. boueti* from Ethiopia, Congo, and Tanzania.

Trypanosoma striatae sp. nov. (**Plate 65**)

Diagnosis An elongate, broad, monomorphic *Trypanosoma* species, 25–34 × 14–25 µm, shape index 1.70, without longitudinal striations on the body. The nucleus is long and narrow, 9–15 × 3–4.5 µm, with nuclear index 3.50. The kinetoplast is located well behind the nucleus; their positions relative to body length are 84.0% and 51.5%, respectively.

Type Host *Mabuya striata* (Peters) (Sauria: Scincidae).

Other Hosts *Mabuya maculilabris.*

Type Locality Morogoro, Morogoro Region, Tanzania.

Other Localities Amani, Eastern Usambara Mountains, Tanga Province, Tanzania.

Prevalence Undetermined.

Morphological Variation Monomorphic trypomastigotes from *M. striata* are elongate and broad, 30.4 ± 2.7 × 18.1 ± 2.9 µm (25–34 × 14.5–25, N = 12), without longitudinal or transverse striations on the body. The kinetoplast is located near the posterior end, 5.5 ± 4.0 µm (0–13) from it, and 25.5 ± 3.0 µm (19–30) from the anterior end, well behind the nucleus, 10.6 ± 2.4 µm (5.5–14). The nucleus is elongate and narrow, 12.1 ± 1.9 × 3.5 ± 0.5 µm (9–15 × 3–4.5), at 15.7 ± 3.7 µm (11.5–23.5) from the anterior end. Relative positions of the kinetoplast and nucleus to the body length are 84.0% (68.2–96.7) and 51.5% (42.4–78.3), respectively. The nuclear index is 3.50 (2.25–4.67), and the shape index is 1.70 (1.25–2.28). The undulating membrane can be difficult to discern and terminates in a free flagellum, 9.9 ± 1.5 µm (8–12, N = 7), visible in about one-half of the trypomastigotes. The cytoplasm stains deeply and contains many red and black apparent granules.

Invertebrate Host Unknown.

Remarks *Trypanosoma striatae* is the only trypanosome in African scincid hosts with an elongated, narrow nucleus. It differs from *T. boueti* in nucleus size and shape, absence of striations on the body, greater shape and nuclear ratios, a slightly greater length, and a more posteriorly situated kinetoplast. The broad form of *T. martini* is similar in body length but narrower in width, with a much greater shape index, a far smaller and more posterior nucleus adjacent to the kinetoplast, and a more conspicuous, better developed undulating membrane. The inadequately described *Trypanosoma mochli* has a more obvious undulating membrane and a small, ovoid nucleus situated almost adjacent to the kinetoplast at the posterior end of the body. One or two possible longitudinal myonemes are present in the body of *T. mochli* (Van den Berghe et al., 1964), and it is more slender, apparently resembling *T. martini* in appearance rather than *T. boueti* and *T. striatae*. Hapantotype blood film is deposited in the U.S. National Parasite Collection (USNPC), Beltsville, Maryland, no. 100338.

Trypanosoma betschi Brygoo 1966

Diagnosis A monomorphic *Trypanosoma* species highly variable in shape, rounded to rectangular, 20.2–31.8 × 14.0–25.6 µm. The body shape ratio averages 1.35. The nucleus appears as a clear area in the cytoplasm, 17.8–16.3 × 3.1–5.4 µm, with average nuclear ratio 2.74. Relative to body length, the kinetoplast and nucleus are positioned at 72.4% and 48.4%, respectively, on average.

Type Host *Gerrhosaurus madagascariensis* (Gray) (Sauria: Gerrosauridae)

Other Hosts None known.

Type Locality Ankara, Madagascar.

Other Localities None known.

Prevalence Two of two *G. madagascariensis* from the type locality were infected by *T. betschi*, but 94 from Mahabo on the east coast of Madagascar were uninfected (Brygoo, 1966a).

Morphological Variation Because of the variability of body shape in *T. betschi*, Brygoo (1966a) provided only the "plus grande dimension" (i.e., the body length as construed here), which averaged 26 µm, varying from 20 to 30 µm. Ten clear figures included with the description were measured to provide at least estimates of the taxonomic characters. Body length is very close to that stated by Brygoo, 25.7 µm (20.2–31.8). Maximum width is 19.3 µm (14.0–25.6). The kinetoplast is quite distinct and variable in position relative to the nucleus, 18.4 µm (12.4–26.4) from the anterior edge, 12.8 µm (4.3–21.7) from the posterior, and 6.8 µm (1.6–12.4) from the center of the nucleus, and is commonly in contact with the clear area comprised by the nucleus. The kinetoplast, when separated from the nuclear area, is surrounded by a narrow clear zone. The nucleus appears as a clear space, round or oval to elliptical, 11.0 × 4.2 µm (7.8–16.3 × 3.1–5.4), with its center point 12.4 µm (8.5–17.1) from the anterior edge of the body. The nuclear ratio is 2.74 (1.57–4.00). Body shape is quite variable, although still monomophic, from almost circular to rectangular, with rounded margins to nearly straight lines. The shape index is low, 1.35 (1.13–1.64). Positions of the kinetoplast and nucleus relative to body length average 72.4% (53.3–100) and 48.4% (36.7–68.8), respectively. The undulating membrane is usually distinct, with one to six waves apparent, or none, and terminates in a free flagellum 11.6 µm (3.1–22.5).

Invertebrate Host Unknown.

Remarks This is the only *Trypanosoma* species recorded from gerrhosaurid hosts, which justifies the calculation of dimensional estimates from Brygoo's (1966a) illustrations. It is probable that the actual nuclear chromatin is somewhat smaller in area, but it was not distinguished by a deeper-staining reaction from the cytoplasm; indeed, it stained less heavily than the surrounding cytoplasm of the body.

TRYPANOSOMA SPECIES OF MEDITERRANEAN AND MIDDLE EASTERN LIZARDS

Trypanosoma platydactyli Catouillard 1909

Diagnosis A *Trypanosoma* species, delicate in appearance, with dimensions 26.0–38.5 × 8.0–13.0 µm. The nucleus and kinetoplast are positioned similarly relative to the body length, 49.9% and 51.0%, respectively. The small nucleus is ovoid with nuclear index 1.47. Both ends of the body are pointed, and the cytoplasm stains weakly.

Type Host *Tarentola mauritanica* (Linnaeus) (Sauria: Gekkonidae).

Other Hosts *Cyrtopodion kotchyi.*

Type Locality Tunis, Tunisia.

Other Localities Gafsa, Tunisia (Chatton and Blanc, 1914a); Malta (Adler and Theodor, 1935); Southern Italy (Maroli et al., 1983) and Catania (Adler and Theodor, 1931); Spain (Wood, 1935); Banyuls-sur-Mer, Pyrénées Orientales, France (Rioux et al., 1979).

Prevalence In Tunis, 10 of 33 (30.3%) *Tarentola mauritanica* examined by Chatton and Blanc (1914a) were infected by *Trypanosoma platydactyli*.

Morphological Variation The monomorphic, slender trypomastigotes, delicate in appearance, have a thin cytoplasm that does not stain heavily. Both ends are sharply pointed. Catouillard (1909) described *T. platydactyli* from Tunis as having a length including the flagellum of 36–56 μm. Telford (1995a) suggested that the body length alone, without the free flagellum, which measured 12–18.5 μm, probably varied from 24 to 38 μm. Width, including the undulating membrane, was 8–14 μm, and the nucleus was 1.4 × 2.0 μm. The illustration showed close association of the kinetoplast and nucleus. Dimensions of a series of *T. platydactyli* from Banyuls-sur-Mer, France, are consistent with the type description: body dimensions of 31.9 ± 3.1 × 9.8 ± 1.2 μm (26.0–38.5 × 8.0–13.0, N = 17). The kinetoplast and nucleus are about the same distance posterior from the anterior end, 16.3 ± 2.7 μm (10.0–20.5) and 16.0 ± 2.9 μm (9.0–20.5), respectively. Their positions relative to the body length, respectively, average 51.0% and 49.9%. The kinetoplast is 15.6 ± 2.6 μm (12.0–20.5) from the posterior edge and 0.3–1.0 μm from the nucleus. The nucleus is ovoid and small, 1.5 ± 0.4 × 1.1 ± 0.3 μm (1.0–2.5 × 0.7–1.5), with a nuclear index that averages 1.47. The free flagellum, when distinct, averages 9.5 μm (5.0–15.0).

Invertebrate Host Adler and Theodor (1935) found development of *T. platydactyli* to occur in the esophagus and midgut of the phlebotomine sand fly *Sergentomyia minuta* fed on infected *Tarentola mauritanica*. A close association between this vector, the trypanosome, and the saurian hosts in southern Italy, *T. mauritanica* and *C. kotchyi*, was described by Maroli et al. (1987).

Trypanosoma turcici Marinkelle and Al-Mahdawi 1980

Diagnosis A monomorphic *Trypanosoma* species characterized by an extremely slender form, 28.5–34.6 × 2.0–2.9 μm. The nucleus is ovoid, with a nuclear index of 0.7–0.8. The kinetoplast is minute and situated relative to the body length at 67.3%; the nucleus is at 56.2%. The free flagellum is prominent, the undulating membrane narrow, and a series of small vacuoles extend from the kinetoplast to the posterior end of the body.

Type Host *Hemidactylus turcicus* (Linnaeus) (Sauria: Gekkonidae).

Other Hosts None known.

Type Locality Al-Madaein, Baghdad Province, Iraq.

Other Localities Baghdad City and Yusufiya, Baghdad Province, and Swera, Wasit Province, Iraq.

Prevalence "Nearly 10%" of 200 *H. turcicus* were infected by *T. turcici* (Marinkelle and Al-Mahdawi, 1980).

Morphological Variation The monomorphic trypomastigotes are slender with attenuated ends, 32.2 ± 1.2 × 2.4 ± 0.2 μm (28.5–34.6 × 2.0–2.9, N = 200), and a prominent free flagellum 18.0 ± 1.4 μm (13.7–21.0). The kinetoplast is minute and punctiform, located slightly posterior to the nucleus, 0.4–0.7 μm from the nucleus center. Estimated position of the kinetoplast from the anterior end is 21.8 μm, and from the posterior end is 12.2 ± 0.6 μm (10.5–13.9). The ovoid nucleus is 0.9–1.2 μm in length, with the nuclear index 0.7–0.8, and is situated 18.1 ± 1.2 μm (15.0–21.5) from the anterior end and 13.1 ± 0.7 μm (11.0–15.0) from the posterior end. Positions of the kinetoplast and nucleus relative to body length are 67.3% and 56.2%, respectively. The ratio of body length to width is 13.4. The undulating membrane is narrow, forming two to four waves. Nine to 18 small vacuoles arranged in a straight row extend from the posterior end of the body to the kinetoplast, and occasionally a few vacuoles occur anterior to the nucleus.

Invertebrate Host Unknown.

Remarks The very high ratio of body length to body width (L/W) is the highest reported among described *Trypanosoma* species from saurian hosts (Telford, 1995a).

TRYPANOSOMA SPECIES OF ASIAN LIZARDS

Trypanosoma gekkonis Telford, 1995

Diagnosis A monomorphic *Trypanosoma* species, usually elongate, pointed at each end, the anterior sometimes forming a cytoplasmic projection and the posterior usually broader but still extended. Body length 42.0–59.0 μm and maximum width 10.5–14.0 μm. The kinetoplast is prominent, situated 26–48 μm from the anterior end and 8.5–26.5 μm from the posterior end. The nucleus is broadly elongate to rounded, 5.0–7.0 × 2.0–5.5 μm, situated 23.0–46.5 μm from the anterior end, and 2.0–6.5 μm anteriorly to the kinetoplast. The nuclear index is 1.1–2.5. Relative to the anterior end, the kinetoplast is situated at 71.6% and the nucleus at 65.9%. The free flagellum is 5–14 μm.

Type Host *Gekko gecko* Linnaeus (Sauria: Gekkonidae).

Type Locality Southern Thailand, no precise locality.

Other Hosts None known.

Other Localities None reported.

Prevalence Two of ten (20.0%) subadult *G. gecko* were infected; five adults were negative.

Morphological Variation Body length averaged 50.3 ± 5.7 μm (42.0–59.0, N = 6), and maximum width was 12.3 ± 1.4 μm (10.5–14.0). The kinetoplast is 36.2 ± 9.7 μm (26–48) from the anterior end and 14.4 ± 6.8 μm (8.5–26.5) from the posterior end of body. The nucleus is 6.1 ± 1.0 × 3.9 ± 1.2 μm (5.0–7.0 × 2.0–5.5), with nuclear index 1.7 ± 0.6 (1.1–2.5), is 33.3 ± 8.9 μm (23.0–46.5) from the anterior end of body and 5.0 ± 1.8 μm (2.0–6.5) anteriorly to the kinetoplast. The free flagellum is 9.5 ± 3.7 μm (5–14).

Invertebrate Host Unknown.

Remarks The host series was obtained from a dealer in Thailand.

Trypanosoma ryukyuense Miyata 1977

Diagnosis A polymorphic *Trypanosoma* species characterized by the presence of prominent spiral ridges around the body, more distinct in the slender forms. Slender forms are 31.1–45.0 × 8.1–14.2 μm. The kinetoplast is conspicuous and situated at 52.8%, and the round-to-elliptical nucleus, 1.7–3.6 μm in length, is at 49.3% relative to the anterior body end. The broad forms are 25.8–40.0 × 16.4–30.0 μm, and their nuclei are 2.8–3.6 × 2.2–3.1 μm.

Type Host *Eublepharis kuroiwae kuroiwae* (Namie) (Sauria: Gekkonidae).

Other Hosts None known.

Type Locality Yona, Kunigami-son, Okinawa, Ryukyu Islands, Japan.

Other Localities Benoki, Okinawa (Miyata et al., 1987).

Prevalence All four *E. kuroiwae* examined by Miyata (1977) were host to *T. ryukyuense.* Miyata et al. (1987) found all 5 *E. kuroiwae* collected at Benoki, Okinawa, infected, as were 15 of 16 (93.8%) from the type locality, Yona.

Morphological Variation Trypomastigotes are polymorphic with body shape varying from elongate and slender, with several prominent spiral ridges around the body, to broad, leaf-like forms in which the spiral ridges are not conspicuous. Dimensions of the slender form are 37.2 ± 4.5 × 10.6 ± 1.5 μm (31.1–45.0 × 8.1–14.2, N = 15), and its ends are pointed. The free flagellum averages 11.9 ± 3.3 μm (8.1–22.2). The conspicuous kinetoplast is situated at 19.7 ± 3.0 μm (15.3–25.6) from the anterior end and 17.6 ± 2.5 μm (13.8–20.8) from the posterior end. The kinetoplast is located 4.4 ± 1.3 μm (2.1–6.4) behind the nucleus. The nucleus is round or elliptical, 2.8 ± 0.6 μm (1.7–3.6) in length, surrounded by a clear space, and is situated 18.4 ± 3.0 μm (14.2–23.9) from the anterior end of the body. Relative positions of the kinetoplast and nucleus from the anterior end are 52.8% and 49.3%, respectively. The broad trypomastigotes are 33.3 ± 3.6 × 23.3 ± 3.7 μm (25.8–40.0 × 16.4–30.0, N = 15), with nuclei 3.2 ± 0.2 × 2.9 ± 0.3 μm (2.8–3.6 × 2.2–3.1), and average nuclear index 1.1. The free flagellum averages 8.9 μm (2.8–15.3). The kinetoplast is situated 9.4 μm (3.4–18.9) from the nucleus.

Invertebrate Host Unknown.

Remarks Miyata (1977) reported the presence of trypomastigotes intermediate in shape between the slender and broad forms, as well as the rare presence of smaller trypanosomes that lacked spiral ridges, which he considered to be younger stages. The spiral ridges appear to undulate in a manner similar to the undulating membrane as the trypomastigote moves.

Trypanosoma pertenue brooki De Mello and Lobo 1938

Diagnosis A slender, monomorphic *Trypanosoma* species, delicate in appearance, 15–42 × 2–6 μm. The kinetoplast is minute, situated at 64.0% relative to the body length, and the ovoid nucleus is at 54.5%. The free flagellum is variably visible.

Type Host *Hemidactylus brookii* Gray (Sauria: Gekkonidae).

Other Hosts None known.

Type Locality Socorro Village, Bardes, India.

Other Localities None known.

Prevalence Unknown.

Morphological Variation Trypomastigotes are monomorphic and slender, with pointed ends. Dimensions of the series examined by I. F. De Mello and Lobo (1938) are 29.7 ± 6.1 × 5.0 ± 9.3 μm (15–42 × 2–6, N = 31). The kinetoplast is minute, situated 19.0 ± 4.3 μm (10–27) from the anterior and 10.9 ± 3.1 μm (5–16) from the posterior end and 1.5 ± 0.6 μm (0.5–3.0) behind the nucleus. The nucleus is ovoid, 1.3 ± 0.4 μm (1.0–2.0) in diameter, and situated

16.2 ± 4.2 μm (7–25) from the anterior end. Relative positions of the kinetoplast and nucleus to the body length are 64.0% and 54.5%, respectively. When visible, the free flagellum averages 10.0 ± 4.5 μm (5–16).

Invertebrate Host Unknown.

Remarks The authors of this species considered it to be sufficiently similar to *Trypanosoma pertenue*, inadequately described by Robertson (1908) from *Hemidactylus triedrus* of Ceylon to justify its designation as a "variety" of *T. pertenue*.

Trypanosoma takydromi Telford 1982 (**Plate 66**)

Diagnosis A monomorphic *Trypanosoma* species, elongate and broad, sometimes rounded and leaf-like in shape with dimensions 25–48 × 12–33 μm and shape index 1.60–1.80. Average positions of the kinetoplast and nucleus relative to body length in the two host species vary at 68.2–80.4% and 43.8–68.2%, respectively. The nucleus is elongate and narrow, with nuclear indices 3.20–3.80.

Type Host *Takydromus smaragdinus* Boulenger (Sauria: Lacertidae).

Other Hosts *Takydromus sexlineatus.*

Type Locality Amami Island, Ryukyus, Japan.

Other Localities Vicinity of Bangkok, Thailand.

Prevalence One of 81 (1.2%) *T. smaragdinus* from Amami Island and 3 of 40 (7.5%) *T. sexlineatus* from Thailand were infected by *Trypanosoma takydromi* (Telford, 1982b).

Morphological Variation Trypomastigotes are elongate and broad to rounded, leaf-like. Dimensions in the type host are 35.9 ± 3.4 × 22.9 ± 2.7 μm (31–43 × 19–33, N = 33), with average shape index 1.60. The kinetoplast is situated 28.8 ± 2.9 μm (22–35) from the anterior end and 8.2 ± 2.8 μm (1–15) from the posterior end, and is 9.8 ± 3.5 μm (0.5–16) posterior to the nucleus. The nucleus is 12.9 ± 3.0 × 3.5 ± 0.6 μm (6.2–19 × 2.5–4.5) and is located 19.0 ± 3.9 μm (11.5–28.5) from the anterior end. Average positions of the kinetoplast and nucleus relative to the body length are 80.4% and 52.8%, respectively. The nuclear index averages 3.80. The undulating membrane is usually prominent and shows four to six undulations. The free flagellum is seldom visible (39%) and is short, 4.2 μm on average (2–6). In *Takydromus sexlineatus*, dimensions are 35.9 ± 7.8 × 20.2 ± 3.7 μm (25–48 × 12–26, N = 13), with an average shape index 1.80. The kinetoplast is situated 24.2 ± 5.0 μm (18–33) from the anterior end, 11.4 ± 4.6 μm (4–18) from the posterior end, and 8.7 ± 2.8 μm (5–13) behind the nucleus.

The nucleus averages 11.9 ± 2.1 × 3.8 ± 0.5 μm (9–16 × 3–4.5) and lies 15.4 ± 4.7 μm (9–26) from the anterior end. Relative to the body length, the average positions of the kinetoplast and nucleus are 68.2% and 43.8%, respectively. The average nuclear index is 3.20. The free flagellum, visible in 54% of the trypomastigotes measured, averages 3.2 μm (2.5–4.3).

Invertebrate Host Unknown.

Remarks *Trypanosoma takydromi* is another component of the host parasite complex that may have parasitized a common ancestor of the three *Takydromus* species that occupy Honshu, the Ryukyu Islands, and mainland southeast Asia during the Pliocene prior to geological activity that gave rise to the Ryukyu archipelago (Telford, 1982a, 1997b). It has not yet been reported from *T. tachydromoides* in the main Japanese islands.

Trypanosoma scincorum Telford, 1982 (**Plate 66**)

Diagnosis A monomorphic, flat and leaf-like, broadly elongate *Trypanosoma* species, usually with broadly rounded posterior end narrower than anterior, rarely with an acuminate posterior cytoplasmic projection. Three to five waves are formed by undulating membrane when visible. Body length, exclusive of any cytoplasmic projection, is 17–34 μm, with maximum width 12–20 μm; the nucleus is elongate, 5–13 × 2–5 μm, with nuclear index 1.2–4.0, situated in middle third of the body, anterior to the kinetoplast, which is 8–21 μm from the anterior and 2.5–18.0 μm from the posterior end. Relative to the anterior end, the kinetoplast is situated at 64.9%, and the nucleus center is at 44.4%, on average. The free flagellum, when visible, is usually 6–8 μm but may extend up to 20 μm.

Type Host *Mabuya multifasciata* (Kuhl) (Sauria: Scincidae).

Type Locality Gyobu Reservoir, 9 km north of Taikkyi, Taikkyi Township, Rangoon Division, Burma.

Other Hosts *Mabuya macularia, Sphenomorphus maculatus, Sphenomorphus sabanus.*

Other Localities Hlegu Township, Rangoon Division, Burma; southern Thailand; Sepilok Lait, Sabah, Malaysia.

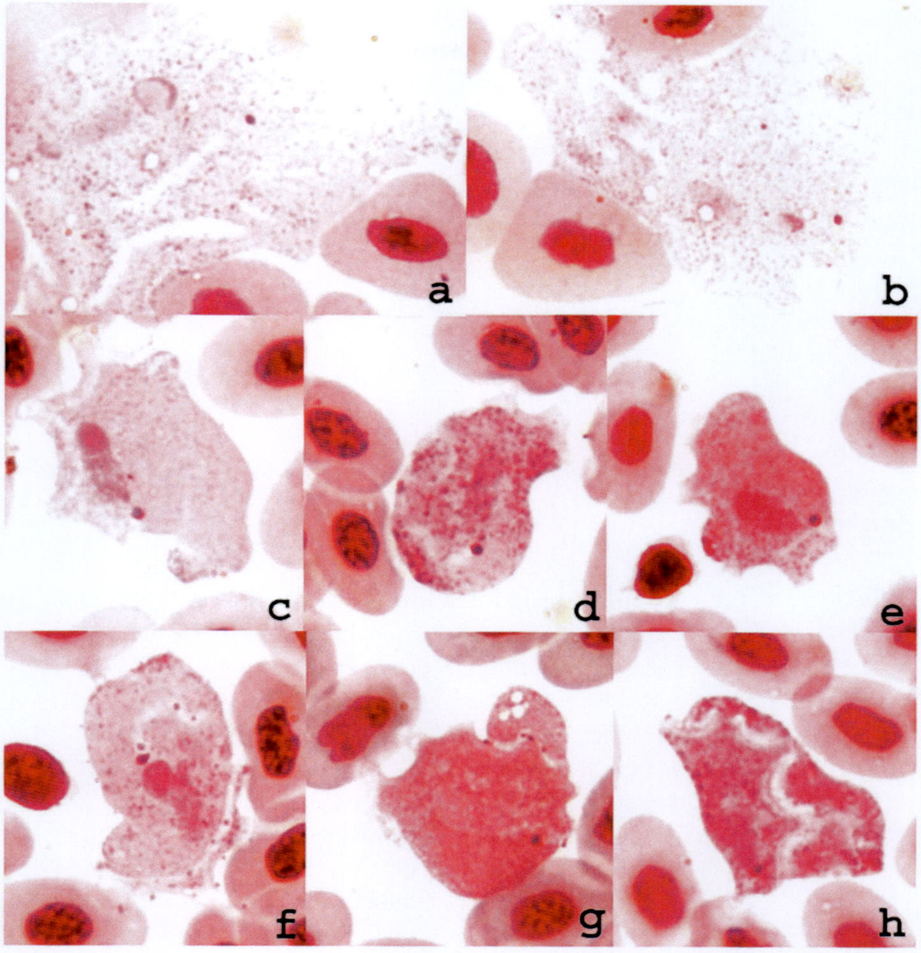

(A)

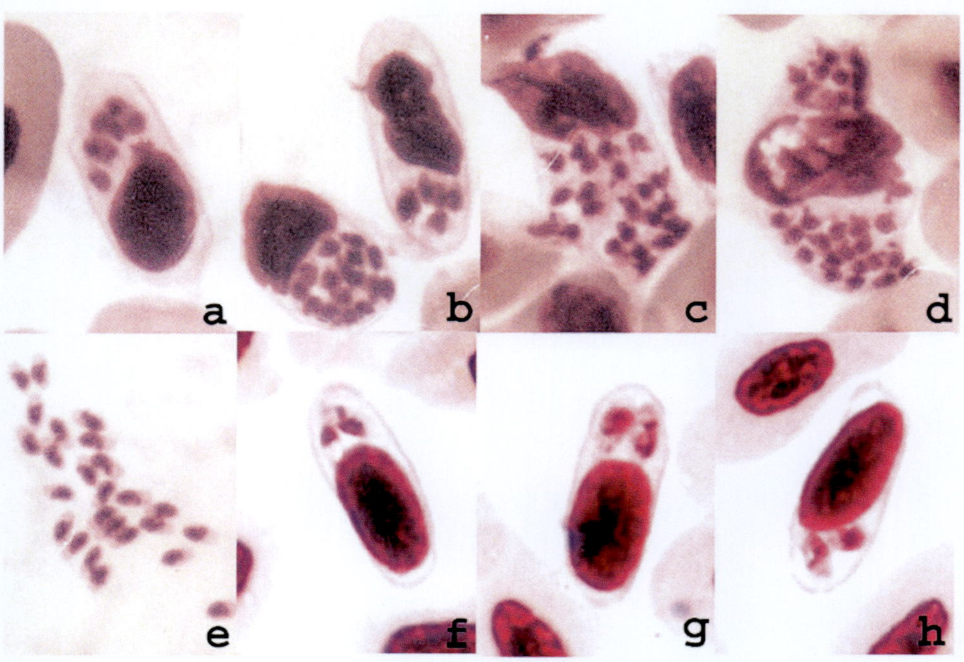

(B)

Plate 66 *Trypanosoma* and *Sauroleishmania* spp. from Asian lizards. (**A**) *Trypanosoma takydromi* from *Takydromus smaragdinus*, Ryukyu Islands, Japan, **a**, **b**; *Trypanosoma scincorum* from *Mabuya multifasciata*, Thailand, **c–h**. (**B**) *Sauroleishmania* spp. amastigotes in thrombocytes of Pakistani lizards, *Teratoscincus scincus*, **a–d**, free amastigotes from ruptured cell, **e**, and *Trapelus agilis*, **f–h**. (Figures **c–e** from Telford, S. R., Jr., *Parasitology*, 79, 317–324, 1979, Figures a, c, and e, with permission, Cambridge University Press.)

Prevalence In *M. multifasciata*, Burma, 3 of 27 (11.1%); Thailand, 21 of 123 (17.1%); in *M. macularia*, Burma 1 of 1; Thailand, 1 of 36 (2.8%); in *S. maculatus*, Burma 1 of 7 (14.3%); in *S. sabanus*, Malaysia 1 of 7 (14.3%).

Morphological Variation An analysis of variation in ten characters of possible taxonomic importance in *T. scincorum* and three *Trypanosoma* species of Tanzanian geckoes (Telford, 1996c) found an absence of host effect on morphology of *T. scincorum* for eight characters. Only two characters, maximum body width and nucleus width, varied significantly among samples from three host species (*M. multifasciata, M. macularia, S. maculatus*). Comparisons were made by host species, geographical area, or sample composition (single or composite samples). The characters used were the six least variable directly obtained from measurements of at least 20 trypanosomes: body length (BL), maximum width (MW), kinetoplast anterior (KA), nucleus anterior (NA), nucleus length (NL), and nucleus width (NW). Four characters were derived from these: shape index (BL/NL, SI), K%, N%, and nucleus index (NL/NW, NI). In comparisons among host species, irrespective of origin, BL (in µm) averaged 23.1 in *M. multifasciata*, 24.1 in *M. macularia*, and 21.2 in *S. maculatus*. In the same host sequence, averages for other characters were MW (in µm), 15.4, 16.7, 14.8; KA (in µm), 14.9, 15.8, 15.4; NA (in µm), 9,9, 10.5, 10.0; NL (in µm), 7.8, 8.4, 6.6; NW (in µm), 3.6, 3.7, 3.0; K%, 64.4, 65.9, 72.7; N%, 42.9, 43.9, 47.2; SI, 1.52, 1.46, 1.43; NI, 2.24, 2.32, 2.29. Reference should be made to the article cited (Telford, 1996c) for the results of other comparisons of *T. scincorum*, that is, composite or single samples; same host, different areas; or different hosts, single or total samples, same geographic areas. In *T. scincorum*, the characters BL and MW, BL and NA were not found to be correlated, but BL and KA (0.621), KA and NA (0.839), and K% and N% (0.784) were correlated. Other characters showed correlations below 0.500 or negative correlations (Telford, 1996c).

Invertebrate Host Unknown.

Remarks *T. scincorum* will probably be found in other scincid lizards of southeast Asia given its known presence in Burma, Thailand, and northern Borneo in four species of two scincid genera.

TRYPANOSOMA SPECIES OF AUSTRALIAN LIZARDS

Trypanosoma phylluri Mackerras 1961

Diagnosis A monomorphic, leaf-like *Trypanosoma* species, delicate in appearance, with broad anterior and pointed posterior ends, 36–48 × 7–15 µm. The kinetoplast is closely associated with the small ovoid nucleus situated 52–56% relative to the body length.

Type Host *Phyllurus platurus* (Shaw) (Sauria: Gekkonidae).

Other Hosts None known.

Type Locality French's Forest, near Sydney, Australia.

Other Localities Mosman and Narrabeen, vicinity of Sydney (Mackerras, 1961a).

Prevalence *Trypanosoma phylluri* infected 11 of 68 (15.9%) of *P. platurus* collected north of Sydney (Mackerras, 1961a).

Morphological Variation The monomorphic, leaf-like, weakly staining trypomastigotes are delicate in appearance, broad at the anterior end and pointed at the posterior end. Dimensions are 36–48 × 7–15 µm, usually 14 µm, with the kinetoplast closely associated with the nucleus, 0–1 µm distant from it and 14–21 µm from the posterior end of the body. The ovoid nucleus is small, 2 × 1 µm, situated 20–25 µm from the anterior end. Relative position of both nucleus and kinetoplast to the body length is estimated to be 52–56%. The free flagellum is short, 2–6 µm in length.

Invertebrate Host Unknown.

TRYPANOSOMA SPECIES OF NEOTROPICAL SNAKES

Trypanosoma brazili Brumpt 1914 Brygoo 1965

Diagnosis A monomorphic, elongate, and slender *Trypanosoma* species with pointed ends, 40.3–78.5 × 5.7–9.4 µm. The average body shape index is 8.04. The kinetoplast has a flattened ovoidal shape and lies perpendicular to the long axis of the trypomastigote. The nucleus is ovoid, 3.2–7.1 µm in length. Positions of the kinetoplast and nucleus relative to body length average 73.2% and 44.8%, respectively.

Type Host *Helicops modestus* Günther (Serpentes: Colubridae).

Other Hosts None known.

Type Locality Brazil, no precise locality.

Other Localities None reported.

Prevalence Unknown.

Morphological Variation Trypomastigotes are monomorphic, elongate, and slender, with pointed ends, 62.7 × 7.8 µm (40.3–78.5 × 5.7–9.4, N = 9), with an average body shape index 8.04. The kinetoplast is a flattened ovoid in shape, with its long axis perpendicular to the long axis of the trypomastigotes, and it occupies a clear area in the cytoplasm. It is situated 45.9 µm (29.2–56.9) from the anterior end of the body, 12.1 µm (7.2–17.4) from the posterior, and 17.8 µm (10.7–24.9) behind the nucleus. The ovoid nucleus is bordered by a clear area and is 4.9 µm (3.2–7.1) in its long axis. It is located 28.1 µm (18.5–32) from the anterior end. Positions of the kinetoplast and nucleus relative to body length average 73.2% and 44.8%, respectively. The undulating membrane is well-developed and shows on average five or six waves, terminating in a free flagellum 10.7 µm (4.6–17.4) that is commonly not visible.

Invertebrate Host Brumpt (1914) obtained development of *T. brazili* in the leech *Placobdella brasiliensis* over a period of 19 days. The developmental stages on days 2, 4, 11, and 19 have been illustrated by Brygoo (1965a).

Remarks *Trypanosoma brazili* was poorly described by Brumpt (1914) and its taxonomy confused by Arantes and Da Fonseca (1931), who recognized two species from *Xenodon* (= *Ophis*) *merremi*, *T. butantanense* and *T. merremi*, then further muddled by Lavier (1943), who referred to *T. brazili* as *T. vitali*. Brygoo (1965a) provisionally recommended recognition of two species, *T. brazili* from *Helicops modestus* and *T. butantanensis* from *Xenodon merremi*, with *T. merremi* considered to be a synonym of *T. butantanense*. Brygoo pointed out that one could consider *T. brazili* to be a *nomen nudum* because the only precision in the description by Brumpt was "ressemble beaucoup au *Trypanosoma leptodactyli* adulte." Brygoo redescribed *T. brazili* from Brumpt's original slides.

Trypanosoma hogei Pessôa 1968

Diagnosis A monomorphic, elongate, and slender *Trypanosoma* species with pointed ends, 36.5–46.5 × 7–9 µm, with body shape index 5.2. The nucleus is 4.5–5 × 2–2.5 µm; nuclear index is 2.0–2.3. The positions of the kinetoplast and nucleus relative to body length are 68–73% and 46–48%, respectively.

Type Host *Rachidelus brazili* Boulenger (Serpentes: Colubridae).

Other Hosts None known.

Type Locality Quintana, São Paulo, Brazil.

Other Localities None known.

Prevalence Overall prevalence was 6 in 21 (28%) *R. brazili* reported by Pessôa et al. (1974a).

Morphological Variation Trypomastigotes are monomorphic, elongate, and slender, with pointed ends. When fixed, the body is rarely extended, usually taking a circular configuration. Total length, including the free flagellum of 5–8 µm, is 41.5–54.5 µm. Body dimensions are 36.5–46.5 × 7–9 µm, with the kinetoplast located 26.5–31.5 µm from the anterior end, 10–15 µm from the posterior, and 4.5–5 µm behind the nucleus. The nucleus is 4.5–5 × 2–2.5 µm and lies 17.5–21.5 µm from the anterior end of the body. The body shape index is 5.2, and the nuclear index is 2.0–2.3. The positions of the kinetoplast and nucleus relative to body length are 68–73% and 46–48%, respectively. The undulating membrane is well-developed and prominent.

Invertebrate Host Pessôa and Fleury (1969) reported that *T. hogei* developed readily in leeches, *Haementeria* sp., fed on *R. brazili*.

Remarks There is remarkable agreement in body length obtained by subtracting the minimum and maximum free flagellum lengths from the minimum and maximum total lengths, and totaling the distance of kinetoplast from the posterior and from the nucleus length plus nucleus distance from the anterior end. The minimum and maximum values thus calculated are 36.5 and 46.5 for each method, respectively, which adds confidence to either method of estimation.

Trypanosoma constrictor Pessôa and Fleury 1969

Diagnosis A monomorphic, elongate, and moderately broad *Trypanosoma* species with pointed ends, 54.5–82.0 × 9.0–15.0 µm, body shape index 5.5–6.4. The nucleus is ovoid, 5.5–7.1 × 3.2–5.0 µm, with nuclear index 1.4–1.7. Relative to body length, the positions of the kinetoplast and nucleus are 69–73% and 52–57%, respectively.

Type Host *Boa constrictor amarali* (Stull) (Serpentes: Boidae).

Other Hosts None known.

Type Locality Mato Grosso State, Brazil, no precise locality.

Other Localities None known.

Prevalence One of ten *B. constrictor* examined was infected by trypanosomes, but the origin of the nine negative snakes was not mentioned by Pessôa and Fleury (1969). Pessôa et al. (1974a) later reported overall prevalence of *T. constrictor* as 4 of 29 (14%) examined at Instituto Butantan.

Morphological Variation The monomorphic trypomastigotes are elongate and moderately broad in appearance, with pointed ends, 54.5–82.0 × 9.0–15.0 µm, with body shape index 5.5–6.4. The kinetoplast is situated 39.8–59.6 µm from the anterior end, 20.0–30.0 µm from the posterior, and 4.3–5.5 µm behind the nucleus. The ovoid nucleus is 5.5–7.1 × 3.2–5.0 µm, with nuclear index 1.4–1.7, and is located 30.0–47.0 µm from the anterior end of the body. The relative positions of kinetoplast and nucleus to body length are 69–73% and 52–57%, respectively. The undulating membrane is well-developed and can show up to nine waves. It terminates in a short free flagellum, 1.5–2.0 µm when visible.

Invertebrate Host Unknown. Pessôa and Fleury (1969) did not obtain development of *T. constrictor* in leeches.

Remarks The attempted transmission of *T. constrictor* by inoculation of infected blood into two *Helicops modestus* was unsuccessful.

Trypanosoma salamantae Pessôa and Fleury 1969

Diagnosis A moderately broad, elongate, monomorphic *Trypanosoma* species with pointed ends, 60.0–73.0 × 11.0–15.0 µm, and body shape index 4.9–5.5. The nucleus is ovoid, 4.0–6.0 × 2.5–3.5 µm, with nuclear index 1.60–1.71. The kinetoplast and nucleus positions relative to body length are 62–63% and 45–48%, respectively.

Type Host *Epicrates cenchria crassus* (Cope) (Serpentes: Boidae).

Other Hosts None known.

Type Locality Cajurú, São Paulo State, Brazil.

Other Localities City of São Paulo, Brazil.

Prevalence Two of 16 (12.5%) *E. cenchria* examined were host to *T. salamantae*; overall prevalence was later stated as 1 of 37 *E. cenchria* (Pessôa et al., 1974a).

Morphological Variation Trypomastigotes are monomorphic, elongate, and moderately broad, with pointed ends. Dimensions are 60.0–73.0 × 11.0–15.0 µm, with the body shape index 4.9–5.5. The kinetoplast is situated 37.0–46.0 µm from the anterior end, 23.0–27.0 µm from the posterior, and 3.0–4.0 µm behind the nucleus. The nucleus is ovoid, 4.0–6.0 × 2.5–3.5 µm, with nuclear index 1.60–1.71, and is located 27.0–35.0 µm from the anterior end. Relative positions of the kinetoplast and nucleus to body length are 62–63% and 45–48%, respectively. The undulating membrane is well-developed, forming seven to nine waves, and terminating in a long free flagellum, 12.0–17.0 µm.

Invertebrate Host Pessôa and Fleury (1969) were unable to infect leeches with *T. salamantae*.

Remarks The attempted infections of two *E. cenchris* and one *Boa constrictor amarali* by *T. salamantae*-infected blood were unsuccessful.

Trypanosoma erythrolampri Wenyon 1909

Diagnosis A slender *Trypanosoma* species 28 × 4.2 µm with an undulating membrane lacking pronounced waves and having a rod-shaped kinetoplast. Kinetoplast and nucleus are situated relative to body length at 85% and 68%, respectively. The nucleus is small, 2.7 µm in diameter, and the body shape index is 5.9.

Type Host *Erythrolamprus aesculapi* Linnaeus (Serpentes: Colubridae).

Other Hosts None known.

Type Locality South America, no precise locality.

Other Localities None reported.

Prevalence Unknown.

Morphological Variation Wenyon (1909b) described trypanosomes of two types, wide and narrow forms. In the wide form, the dimensions are 28 × 4.2 µm, with the free flagellum 5.6 µm. The nucleus is 2.7 µm in diameter, 18.9 µm posterior to the anterior body margin and 14.7 µm from the posterior end. The rod-shaped kinetoplast is situated 23.8 µm from the anterior and 9.8 µm from the posterior end, 4.9 µm behind the nucleus. The locations of kinetoplast and nucleus relative to body length, respectively, are 85% and 68%, and the body shape index is 6.7. The undulating membrane lacks prominent waves. The narrow form is similar in body length, 23.1 µm, but is only

2.8 μm in width, a shape index of 8.3. The kinetoplast is 13.3 μm from the anterior end and located relative to body length 57.5%. It is located adjacent or anterior to the nucleus. Wenyon (1926) referred to the slender form as crithidial in structure and suggested this was a result of postmortem changes.

Invertebrate Host Unknown.

Remarks The broader form is considered here to be typical of *T. erythrolampri*, given the possibility of postmortem changes in some trypanosomes that affected the kinetoplast and nucleus positions. The host had died in the London Zoological Gardens.

TRYPANOSOMA SPECIES OF NORTH AMERICAN SNAKES

Trypanosoma hydrae Ayala, Atkinson and Vakalis 1983

Diagnosis An elongate and thin *Trypanosoma* species, 31.5–41.2 × 1.4–3.8 μm, with finely pointed ends and a shape index of 15.7. Positions of the kinetoplast and nucleus relative to body length average 74.4% and 51.5%, respectively. The nucleus is small, 2.0–3.0 × 1.2–1.5 μm, elliptical in shape, with a nuclear index 1.6–2.0.

Type Host *Nerodia fasciata confluens* Blanchard (Serpentes: Colubridae).

Other Hosts Experimental only: *Thamnophis p. proximus* and *Elaphe obsoleta lindheimeri*.

Type Locality Manchac Swamp, 5 km northeast of Laplace, Saint John the Baptist Parish, Louisiana.

Other Localities None known.

Morphological Variation Trypomastigotes are elongate and thin, 36.0 ± 2.5 × 2.3 ± 0.4 μm (31.5–41.2 × 1.4–3.8, N = 37), with both ends sharply pointed. The kinetoplast is situated 26.8 μm (24.4–28.6) from the anterior end, 9.2 ± 1.2 μm (7.1–12.6) from the posterior, and 7.9 ± 2.2 μm (3.4–16.3) behind the nucleus. The nucleus is elliptical, 2.0–3.0 × 1.2–1.5 μm, and is located 18.5 ± 2.9 μm (13.5–23.8) from the anterior end. Positions of the kinetoplast and nucleus relative to body length are 74.4% and 51.4%, respectively. The shape index is 15.7, and the nuclear index is 1.6–2.0. The undulating membrane is moderately developed and terminates in a short free flagellum that averages 3.9 ± 1.1 μm (1.3–5.8).

Invertebrate Host Unknown.

Remarks The intraperitoneal inoculation of infected blood into two other species of colubrid snakes, *Thamnophis proximus* and *Elaphe obsoleta*, and a juvenile *Nerodia fasciata* produced patent infections 1 week later, and parasitemias reached a level almost comparable to that in the natural host and persisted for at least 2 months in each snake. Attempts to infect four of the viperid snake *Agkistrodon piscivorus leucostoma*, four turtles *Chrysemys scripta elegans*, and three lizards *Anolis carolinensis* were unsuccessful.

Trypanosoma yaegeri Ayala, Atkinson and Vakalis 1983

Diagnosis A large, elongate, and slender *Trypanosoma* species, 45.9–80.3 × 3.3–6.3 μm, with sharply pointed ends and a shape index of 16.0. The kinetoplast and nucleus are situated, relative to body length, at 61.2% and 50.5%, respectively. The nucleus is round to elliptical, 2.5–6.0 × 2.0–4.4 μm, with the nuclear index about 1.3. Fixed specimens are usually coiled in a circle with overlapping extremities.

Type Host *Agkistrodon piscivorus leucostoma* Troost (Serpentes: Viperidae).

Other Hosts None known.

Type Locality Sarpy Swamp, about 2 km from Norco, Saint Charles Parish, Louisiana.

Other Localities None reported.

Prevalence Unknown.

Morphological Variation The large trypomastigotes are elongate and slender with finely pointed ends, 70.3 ± 9.1 × 4.4 ± 0.8 μm (45.9–80.3 × 3.3–6.3, N = 20). The small, rectangular kinetoplast is located 43.0 μm (38.3–45.6) from the anterior end, 27.3 ± 6.8 μm (7.1–34.7) from the posterior, and 9.6 ± 3.9 μm (4.8–22.9) behind the nucleus, usually lying adjacent to the body margin. The nucleus is round to "irregularly elliptical," 2.5–6.0 × 2.0–4.4 μm, surrounded by a cleared space and occupying most of the body width. It is situated 35.5 ± 4.2 μm (26.7–41.8) from the anterior end. Positions of the kinetoplast and nucleus relative to the body length are 61.2% and 50.5%, respectively. The shape index is 16.0, and the nuclear index is about 1.3. The undulating membrane is well developed, showing 7–12 waves, and ends in a short free flagellum, 7.7 ± 1.6 μm (4.9–8.9). When fixed, the trypomastigotes are usually coiled in a circular shape with the extremities overlapping.

Invertebrate Host Unknown.

Remarks This trypanosome was originally reported by Marquardt and Yaeger (1967) but was not described (Ayala et al., 1983).

Trypanosoma thamnophis Fantham and Porter 1953

Diagnosis A monomorphic *Trypanosoma* species 53–66 × 3.3–4.4 μm, with blunt posterior and pointed anterior ends. Positions of the kinetoplast and nucleus relative to body length are 91.6% and 51.1%, respectively; the body shape index is about 14.4 and the nuclear index 1.60. The ovoid nucleus is 4–5 × 2.5–3.1 μm.

Type Host *Thamnophis sirtalis* (Linnaeus) (Serpentes: Colubridae).

Other Hosts None known.

Type Locality Lantier, Quebec Province, Canada.

Other Localities None reported.

Prevalence Unknown.

Morphological Variation Fantham and Porter (1953) reported only length and width, nucleus dimensions, and free flagellum length for *T. thamnophis.* Ayala et al. (1983) calculated positions of kinetoplast and nucleus and the dimensional data from the two illustrations in the type description, and these estimates are utilized here as averages, with the original measurements within parentheses. The trypanosome is monomorphic, elongate, and slender with blunt posterior and pointed anterior ends and a well-developed undulating membrane, 56.0 × 3.9 μm (53.3–66 × 3.3–4.4) in body length, with the shape index about 14.4 (15.0–16.2). The kinetoplast is situated 51.3 μm from the anterior end, 4.8 μm from the posterior, and 22.7 μm behind the nucleus. The nucleus is ovoid, 4–5 × 2.5–3.1 μm, 28.6 μm from the anterior end. The positions of the kinetoplast and nucleus relative to body length are 91.6% and 51.1%, respectively. The nuclear index is about 1.60. The free flagellum is 17.4 μm but was 24.4–25.5μm in the original description.

Invertebrate Host Unknown.

Remarks Fantham and Porter (1953) mentioned the presence of a second smaller trypanosome in *Thamnophis sirtalis* from Montreal West, 16.8–21.1 × 2–3.3 μm. Their figures suggest a young trypomastigote that had not yet attained the configuration and dimensions of mature trypomastigotes. Ayala et al. (1983) did not comment on this form.

Trypanosoma floridana sp. nov. (**Plate 67**)

Diagnosis A monomorphic, usually elongate *Trypanosoma* species 38.0–49.0 × 4.0–10.0 μm, with a broadly pointed posterior end, and body shape index of 6.29, on average. Positions of the kinetoplast and nucleus, relative to body length, average 87.0% and 54.7%, respectively. The usually ovoid nucleus is 4.5–8.0 × 3.0–6.0 μm, with the average nuclear index 1.40.

Type Host *Nerodia floridana* (Goff) (Serpentes: Colubridae).

Other Hosts *Nerodia fasciata pictiventris.*

Type Locality Payne's Prairie, Alachua County, Florida.

Other Localities None reported.

Prevalence *Trypanosoma floridana* infected two of three *N. floridana* and two of four *N. fasciata pictiventris* at the type locality.

Morphological Variation Trypomastigotes are monomorphic and usually elongate, 43.4 ± 2.9 × 6.9 ± 1.6 μm (38.0–49.0 × 4.0–10.0, N = 24), without cytoplasmic projections and with a broadly pointed posterior end. The body shape index averages 6.29. The kinetoplast is prominent, ovoid, approximately 1 × 1.5 μm, and situated 37.7 ± 2.9 μm (33.5–44.0) from the anterior end, 5.8 ± 1.2 μm (4–8) from the posterior, and 14.0 μm behind the nucleus. The nucleus is usually ovoid, with a deeply stained center and often with irregular borders, 6.1 ± 0.9 × 4.4 ± 0.8 μm (4.5–8.0 × 3.0–6.0), with L/W index 1.4 ± 0.2 (1.0–1.8), and situated 23.7 ± 2.6 μm (19.0–29.0) from the anterior end. Position of the kinetoplast and nucleus, respectively, relative to body length, are 82–91% (87.0 ± 0.02), and 45–62% (54.7 ± 0.04). The free flagellum is 9.1 ± 1.6 μm (7–12).

Invertebrate Host Unknown.

Remarks In body length, *T. floridana* is intermediate to *T. thamnophis* and *T. hydrae*, far smaller than *T. yaegeri,* and maximum width is greater in *T. floridana* than in the other species. Position of the nucleus is approximately central in all four species, but the kinetoplast is posterior and near the end of the body in *T. floridana* and

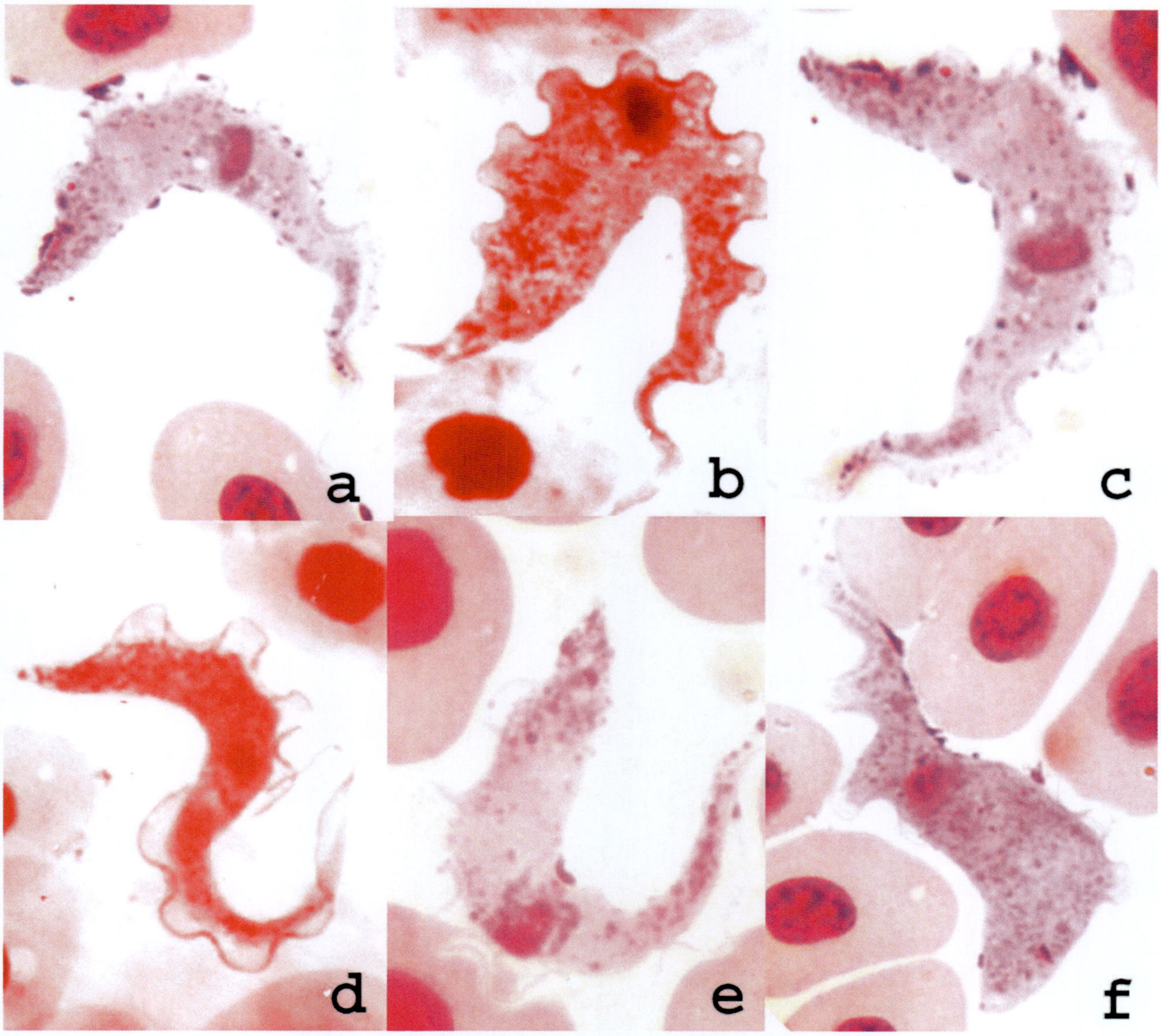

Plate 67 *Trypanosoma floridana* sp. nov. from *Nerodia floridana,* **a–c**, **e**, **f**, and *Nerodia fasciata pictiventris* at the type locality, Paynes Prairie, Gainesville, Florida.

T. thamnophis, but in *T. yaegeri* and *T. hydrae* it is much closer to the nucleus. Nuclei are larger and more ovoid in *T. floridana* and *T. yaegeri*, much smaller and elliptical in the other two species. The shape of the posterior end also differs, attenuated and pointed in *T. yaegeri* and *T. hydrae,* bluntly rounded in *T. thamnophis,* and broadly pointed in *T. floridana*.

Parasitemias in *N. fasciata pictiventris* were much lower than in *N. floridana*. Five trypanosomes from one of the *N. fasciata pictiventris* were measured, with these averages for characters: BL, 46.0 ± 2.4 μm; MW, 6.6 ± 1.7 μm; BL/MW, 7.3 ± 1.9; KA, 38.9 ± 1.7 μm; KP, 8.1 ± 0.7 μm; KN, 14.4 ± 0.9 μm; NA, 24.5 ± 2.1 μm; NL, 5.6 ± 0.5 μm; NW, 4.0 ± 0.4 μm; NL/NW, 1.4 ± 0.1; free flagellum, 7.0 ± 1.8 μm. Positions of the kinetoplast and nucleus, respectively, are 84.6% and 53.3%. Hapantotype blood films are deposited in the U.S. National Parasite Collection (USNPC), Beltsville, Maryland, nos. 92848–92849.

TRYPANOSOMA SPECIES OF AFRICAN SNAKES

Trypanosoma clozeli Bouet 1909

Diagnosis A large and broad, monomorphic *Trypanosoma* species, 82–122 × 15–25 μm, with the body shape index 4.9–5.5. Relative positions of the kinetoplast and nucleus to the body length are 64% and 57–61%, respectively, in the two known host species. The nucleus is 4–5 μm in diameter. The undulating membrane forms a spiral around the body.

Type Host *Natrix anoscopus* (= *Tropidonotus ferox* Günther, vide Brygoo, 1966b) (Serpentes: Colubridae).

Other Hosts *Grayia smithi.*

Type Locality Ivory Coast, West Africa.

Other Localities None known.

Prevalence Unknown.

Morphological Variation In the type host, the monomorphic trypomastigotes are large and broad, 82–122 × 15–25 µm, with the kinetoplast immediately behind the nucleus at 2 µm, 57–77 µm from the anterior end, and 25–45 µm from the posterior. The nucleus is 5 µm in diameter and located 50–70 µm from the anterior end. The undulating membrane forms a spiral around the body and terminates in a free flagellum 10 µm in length. The body shape index appears to be 4.9–5.5. The positions of the kinetoplast and nucleus relative to body length are 63.9% and 57–61%, respectively. In *Grayia smithi,* trypomastigotes are about 107 µm in length, with the kinetoplast 37.8 µm from the posterior end, 68.7 µm from the anterior, and 3 µm behind the nucleus. The nucleus is 4–5 µm in diameter and is located 61.2 µm from the anterior end. Relative positions of kinetoplast and nucleus appear to be 64% and 57%, respectively.

Invertebrate Host Unknown.

Remarks The minimal data given in the type description hardly justify the inclusion of *T. clozeli* in the accounts of species, except for the fact that it appears to be one of the largest reptilian trypanosomes described in the literature.

Trypanosoma voltariae Macfie 1919

Diagnosis A monomorphic *Trypanosoma* species 42 × 4 µm on average, with shape index 10.5. The nucleus averages 3 × 2 µm, with nuclear index 1.5. Positions of the kinetoplast and nucleus, relative to body length, are 59.5% and 50.0%, respectively. In fixed specimens, the posterior portion of the body is bent over at a right angle, which produces a somewhat circular configuration.

Type Host *Naja nigricollis* Reinhardt (Serpentes: Elapidae).

Other Hosts None known.

Type Locality Ghana, West Africa.

Other Localities None known.

Prevalence Unknown.

Morphological Variation The monomorphic trypomastigotes average 50 µm, of which the free flagellum comprises 8 µm. Body size averages, then, 42 × 4 µm, which indicates a shape index of 10.5. The kinetoplast is 25 µm from the anterior end, 17 µm from the posterior, and 4 µm behind the nucleus. The nucleus is 3 × 2 µm and is located 21 µm from the anterior. The nuclear index is 1.50. Positions of the kinetoplast and nucleus relative to body length are 59.5% and 50.0%, respectively. Trypomastigotes are elongate and slender, with the pointed posterior portion "bent at a right angle to the rest of the body," forming a somewhat circular configuration in fixed specimens. The anterior portion "frayed out into a number of striae which blended with the undulating membrane" (Macfie, 1919). The undulating membrane is well developed.

Invertebrate Host Unknown.

Remarks Wenyon (1909a) described a trypanosome from *Naja nigricollis* collected on the River Sobat in Sudan with a total length of 50 µm, "the body is looped in a characteristic spiral manner," as *Trypanosoma najae.* The slender form, with the circular configuration shown in Wenyon's figures, has the same total length as *T. voltariae,* 50 µm, and similarly shows a well-developed undulating membrane and prominent free flagellum. It is very likely the same species described from *N. nigricollis* of West Africa, *T. voltariae,* but is known only from living flagellates. Wenyon was unable to locate any trypanosomes on stained slides. If the trypanosomes of *Naja nigricollis* from Sudan and from West Africa prove to be conspecific, then *T. najae* would have priority over the better-described *T. voltariae.*

Trypanosoma haranti Brygoo 1965

Diagnosis A monomorphic, elongate *Trypanosoma* species with pointed anterior and rounded posterior ends, 20.7–21.1 × 4.9–8.8 µm. The kinetoplast is immediately behind the tiny nucleus, 1.2–2.0 µm in diameter. Positions of the kinetoplast and nucleus relative to body length average 62.8% and 59.1%, respectively. The body shape index averages 4.35.

The posterior one-third of the body appears clear in contrast to the anterior portion of the body.

Type Host *Liopholidophis lateralis* Duméril and Bibron (Serpentes: Colubridae).

Other Hosts None known.

Type Locality Perinet, Madagascar.

Other Localities None known.

Prevalence Unknown.

Morphological Variation Trypomastigotes are monomorphic and elongate, with a pointed anterior end and a rounded posterior. Dimensions are 27.4 × 6.3 μm (20.7–31.1 × 4.9–8.8, N = 10), with the kinetoplast 17.2 μm (11.8–20.6) from the anterior, 8.8 μm (8.0–10.9) from the posterior end, and 0.6 μm (0.2–1.6) behind the nucleus. The nucleus is ovoid to somewhat irregular in shape and small, 1.7 μm (1.2–2.0) in diameter, and is located 16.6 μm (11.6–19.4) from the anterior end. The positions of the kinetoplast and nucleus relative to the body length are 62.8% (57.0–66.2) and 59.1% (56.0–62.4), respectively. The posterior one-third of the body appears as a clear zone, contrasting with the homogeneous and more deeply stained anterior portion. The undulating membrane appears simple in development but can form two to four waves, terminating in a clearly visible free flagellum averaging 7.0 μm (5.1–10.3). The body shape index averages 4.35.

Invertebrate Host Unknown.

Remarks The host genus is endemic to Madagascar, and *T. baranti* is the only trypanosome described from Madagascan snakes.

TRYPANOSOMA SPECIES OF ASIAN SNAKES

Trypanosoma primeti Mathis and Leger 1909

Diagnosis A *Trypanosoma* species that occurs in large and small forms. The larger form is 82.5 × 14 μm, with nucleus 4.0 μm. Positions of kinetoplast and nucleus relative to body length are 61% and 57%, respectively, and the body shape index is 5.9. The small form is 44 × 7 μm with a 2-μm nucleus. Kinetoplast and nucleus positions relative to body length are 59% and 55%, respectively, and the shape index is 6.3.

Type Host *Xenochropus piscator* (Schneider) (Serpentes: Colubridae).

Other Hosts None known.

Type Locality Ran-nuoc, Vietnam.

Other Localities Hoa-khe, Vietnam.

Prevalence Unknown.

Morphological Variation Mathis and Leger (1909) described both large and small forms of *T. primeti.* In the large form, body length and width, respectively, are 82.5 μm and 14 μm, with a free flagellum of 23 μm. The nucleus is 4 μm in diameter and is situated at 47 μm from the anterior margin, 31 μm from the posterior end. The kinetoplast is 50 μm from the anterior, 3 μm behind the nucleus, and 28 μm from the posterior end. Locations of the kinetoplast and nucleus relative to body length are 61% and 57%, respectively. The body shape index is 5.9. The small form is 44 × 7 μm, with a free flagellum 13 μm in length. The nucleus is 2.5 μm in diameter and lies 24 μm behind the anterior end, 17.5 μm from the posterior. The kinetoplast is at 26 μm from the anterior margin, 15.5 μm from the posterior, and 2 μm behind the nucleus. Relative positions to body length are 59% and 55%, respectively for kinetoplast and nucleus. The body shape index is 6.3.

Invertebrate Host Unknown.

Remarks The comparative locations of kinetoplast and nucleus and the shape indices are probably sufficiently similar to support the conclusion of Mathis and Leger that a single species of *Trypanosoma* parasitizes *X. piscator* in the Vietnamese localities where their material originated.

TRYPANOSOMA SPECIES OF NORTH AMERICAN TURTLES

Trypanosoma chrysemydis Roudabush and Coatney 1937

Diagnosis A monomorphic, elongate, and slender *Trypanosoma* species with pointed ends, 46.8–50.0 × 3.2–4.1 μm, and shape index 13.1 on average. The nucleus is spherical to ovoid, with estimated nuclear index 1.0–1.9. Relative positions of the kinetoplast and nucleus to body length are 64.7% and 39.1%, respectively.

Type Host *Chrysemys belli marginata* (Agassiz) (Testudines: Emydidae) (= *Chrysemys picta marginata*).

Other Hosts *Chelydra serpentina* (Roudabush and Coatney, 1937), *Graptemys geographica* (Woo, 1969). Experimental: *Apalone spinifera, A. ferox, A. mutica* (Jefferson, 1965).

Type Locality Ames, Iowa.

Other Localities Peru, Nebraska (Roudabush and Coatney, 1937); Ontario, Canada (Woo, 1969).

Prevalence Unknown.

Morphological Variation The monomorphic trypomastigotes are elongate and slender with pointed ends, 48.6 × 3.7 μm (46.8–50.0 × 3.7–4.1, N = 4, Roudabush and Coatney, 1937). The kinetoplast is located 44.0 μm from the anterior end, 4.6 μm from the posterior, and 12.4 μm behind the nucleus. The nucleus, dimensions not stated in the type description, is 19.0 μm from the anterior end. The nuclear index, estimated from figures, is 1.0–1.9. Positions of the kinetoplast and nucleus relative to body length average 64.7% and 39.1%, respectively. The body shape index averages 13.1. The undulating membrane is well-developed and forms three to six waves, ending in a free flagellum of 13.1 μm on average.

Invertebrate Host *Trypanosoma chrysemydis* develops in the leeches *Placobdella multilineata, P. ornata, P. rugosa,* and *P. parasitica* (Jefferson, 1965; Woo, 1969; Siddall and Desser, 1992). *Placobdella parasitica* and *P. rugosa* fed on infected *Chrysemys picta* and maintained at 22–24°C had metacyclic stages in the crop in 22 days PF. When kept at 31°C, the infective metacyclic trypanosomes appeared by 14 days PF. By 2–4 days PF, "pear-shaped" epimastigotes were present in leeches (Woo, 1969); from day 3, slender, epimastigotes appeared; transitional forms were observed from day 4 PF, with metacyclic trypanosomes present by day 14. Transmission by bite of infected leeches to *C. picta marginata, Chelydra serpentina*, and *Graptemys geographica* was successful, with blood infections appearing by 4 days.

TRYPANOSOMA SPECIES OF AFRICAN TURTLES

Trypanosoma pontyi Bouet 1909

Diagnosis A slender *Trypanosoma* species with estimated body length 46.8 μm, body width varies 4.5–5.4 μm, and the shape index appears to be 8.7–10.4. The nucleus is tiny, 1.8 μm in diameter. Estimated positions of the kinetoplast and nucleus relative to body length are 69.2% and 61.5%, respectively.

Type Host *Pelusios subniger* Lacèpede (Testudines: Pelomedusiidae) (= *Sternothaerus derbianus*).

Other Hosts None known.

Type Locality West Africa.

Other Localities None known.

Prevalence Unknown.

Morphological Variation Trypomastigotes are slender, with body length estimated at 46.8 μm, width is 4.5–5.4 μm, and the shape index is about 8.7–10.4. The kinetoplast is estimated at 32.4 μm from the anterior end and is 3.6 μm behind the nucleus, 12.6 μm from the posterior end. The nucleus is very small, 1.8 μm in diameter, and is situated 28.8 μm from the anterior end. Positions of the kinetoplast and nucleus relative to body length are 69.2% and 61.5%, respectively. The large undulating membrane terminates in a free flagellum 10.8–12.6 μm in length.

Invertebrate Host Unknown.

Remarks The type description provides minimal information, and the species account is included here only because so few *Trypanosoma* species are known from African turtles.

Trypanosoma leroyi Commes 1919

Diagnosis A large, strongly curved, and slender *Trypanosoma* species, 59.4 × 4.2 μm in dimension, with a well-developed undulating membrane. The large nucleus is situated at 82% and the kinetoplast at 86% relative to body length.

Type Host *Kinixy homeana* Bell (Testudines: Testudinidae).

Other Hosts None known.

Type Locality Senegal.

Other Localities None known.

Prevalence Unknown.

Morphological Variation A large trypanosome with a strongly curved, slender shape. Body length is 50.4 μm, width minus undulating membrane is 4.2 μm, and the free flagellum is 12.6 μm. The large nucleus, no dimensions stated, is located approximately 41 μm behind the anterior margin, 2.1 μm anterior to the kinetoplast. The kinetoplast is situated 7 μm anterior to the posterior body margin and 48 μm from the anterior end. Approximate positions of the kinetoplast and nucleus relative to body length are 86% and 82%, respectively. There are four or five waves in the undulating membrane.

Invertebrate Host Unknown.

Remarks Minimal information was provided in the type description, and the trypanosome has not been reported since.

TRYPANOSOMA SPECIES OF ASIAN TURTLES

Trypanosoma gangetica Sinha 1978

Diagnosis A monomorphic *Trypanosoma* species, elongate and slender, with pointed ends, 33.7–36.7 × 1.8–4.8 μm, and prominent undulating membrane. The cytoplasm contains seven to ten vacuoles. Kinetoplast and nucleus positions relative to body length average 87.3% and 39.5%, respectively. The average body shape index is 13.9.

Type Host *Trionyx gangeticus* Cuvier (Testudines: Trionychidae).

Other Hosts None known.

Type Locality Ganges River, near Bongaoa, 24 Parganas, West Bengal, India.

Other Localities None reported.

Prevalence Five of 15 (33%) *Trionyx gangeticus* were infected by *Trypanosoma gangetica*.

Morphological Variation Trypomastigotes are monomorphic, elongate and slender, averaging 34.7 μm in length, exclusive of the free flagellum, by 2.5 μm (33.7–36.7 × 1.8–4.8) with sharply pointed ends. The kinetoplast is located 27.3 μm (24.9–29.3) from the anterior end, 4.4 μm (4.1–4.6) from the posterior, and 13.6 μm (12.1–14.5) behind the nucleus. The oval nucleus is 13.7 μm (12.8–14.8) from the anterior end and averages 2.3 × 1.5 μm (2.2–2.4 × 1.3–1.8) in size. The nuclear index averages 1.53. Positions of the kinetoplast and nucleus relative to body length average 87.3% and 39.5%, respectively. The body shape index averages 13.9. The undulating membrane forms six to eight waves, ending in a free flagellum 20.9 μm (19.2–21.9). Seven to ten vacuoles are present in the cytoplasm.

Invertebrate Host Unknown.

Trypanosoma balithaensis Ray 1987

Diagnosis A monomorphic *Trypanosoma* species characterized by an elongate and slender form and pointed ends. Body length is 30–35 μm, and width is 1.5–2.5 μm. Positions of the kinetoplast and nucleus relative to body length are 71.7% and 63.1%, respectively. The average body shape index is 16.3. Trypomastigotes ingested by the leech *Helobdella nociva* transform into amastigotes, spheromastigotes, epimastigotes of two forms, and ultimately metacyclic trypanosomes.

Type Host *Lissemys p. punctata* (Bonnaterre) (Testudines: Trionychidae).

Other Hosts None known.

Type Locality Balitha Village, Bankura District, West Bengal, India.

Other Localities None reported.

Prevalence Unknown.

Morphological Variation Trypomastigotes are monomorphic, elongate, and slender, with sharply pointed ends. Body length averages 32.5 μm, without the free flagellum, by 2.0 μm in width (30–35 × 1.5–2.5). The dot-like kinetoplast is located within a halo, 23.3 μm from the anterior end, 9.2 μm from the posterior, and 2.8–3.0 μm behind the nucleus. The nucleus is rounded, 2.0 μm in diameter, and 20.5 μm from the anterior end. The nuclear index averages 1.00. Positions of the kinetoplast and nucleus relative to body length average 71.7% and 63.1%, respectively. The body shape index averages 16.3. The undulating membrane forms five or six waves, ending in a free flagellum that is 18.5 μm (17.5–20.0).

Invertebrate Host The leech *Helobdella nociva*, common on the turtle hosts, is the apparent vector of *T. balithaensis*. By day 2 postingestion, the trypomastigotes become amastigote stages 7.5 μm in diameter. Amastigotes divide twice, producing four small pyriform bodies that develop into spheromastigotes 30 × 8.5 μm with a free flagellum 6.2 μm in length. At 4 days PI, two forms of epimastigotes are present: short stumpy forms 15.5 × 3.5 μm, with a free flagellum of 3.5 μm, and long, slender forms 25.5 × 2.0 with free flagella of 6.2 μm. Some of the latter forms begin division at this time. By 6 days PI, the long forms become attenuated with a pointed posterior and are metacyclic trypanosomes 30 × 2.5 μm, with free flagella 16.5 μm in length. At 9–10 days, the crop and intestinal ceca of the leech host is filled with long, slender epimastigotes and metacyclic trypanosomes.

TRYPANOSOMA SPECIES OF AUSTRALIAN TURTLES

Trypanosoma chelodina Johnston 1907

Diagnosis A monomorphic, elongate, and slender *Trypanosoma* species with pointed ends, 34–43 × 4–8.5 μm, and a prominent undulating membrane. Kinetoplast and nucleus positions relative to body length are 55–87%

and 32–50%, respectively. The body shape index varies between 5.1 and 8.5.

Type Host *Chelodina longicollis* (Shaw) (Testudines: Chelidae).

Other Hosts *Emydura macquarii* (T. H. Johnston and Cleland, 1912), *Emydura latisternum, Emydura krefftii, Elseya dentata* (Mackerras, 1961a).

Type Locality Morgan, South Australia.

Other Localities In Australia, Sydney, New South Wales, and Brisbane, Mt. Nebo, and Eidsvold, Queensland (Mackerras, 1961a), and Townville, Queensland (Breinl, 1913).

Prevalence Two of ten *Chelodina longicollis* from Townville, examined by Breinl (1913), were infected by *T. chelodina*.

Morphological Variation In the type host, the monomorphic trypomastigotes are elongate and slender, 39.5–43 × 5–8.5 μm (T. H. Johnston and Cleland, 1912). The kinetoplast is located 28.5–33.5 μm from the anterior end, 5–8 μm from the posterior, and 10–12 μm behind the nucleus. The nucleus is 18.5–21.5 μm behind the anterior end. Estimated positions of the kinetoplast and nucleus relative to body length are 72.2–77.9% and 46.8–50.0%, respectively. Both ends of the body are pointed, and the undulating membrane is prominent and terminates in a short free flagellum, 3–6 μm in length. The body shape index is estimated to be 5.1–7.9. Nuclear dimensions were not stated by either T. H. Johnston and Cleland (1912) or Mackerras (1961a). In *Emydura krefftii* (Mackerras, 1961a), trypomastigotes are 34–38 × 4–5.8 μm, estimated shape index 6.6–8.5. The kinetoplast is 19–33 μm from the anterior end, 5–9 μm from the posterior, and 8–16 μm behind the nucleus. The nucleus is 11–17 μm from the anterior end. Relative to body length, the kinetoplast is located at 55.9–86.8%, and the nucleus at 32.4–44.7%. The free flagellum is longer than in the type host, 15–19 μm.

Invertebrate Host Unknown.

Remarks Although there are discrepant values for size, kinetoplast and nucleus position, and length of the free flagellum, Mackerras (1961a) appeared to be satisfied that *T. chelodina* from *C. longicollis* and the trypanosomes from *E. krefftii* were conspecific, and this seems to be a valid conclusion, given the variability that can occur on fixed slides.

TRYPANOSOMA SPECIES OF CROCODILIANS

Trypanosoma cecili Lainson 1977

Diagnosis A large, monomorphic *Trypanosoma* species, elongate and slender, 62–84 × 6–9 μm, with an oval nucleus 6–9 × 4–5 μm that lies transversely across midbody. Positions of the kinetoplast and nucleus relative to body length are 62.6% and 55.8%, respectively. The shape index is 8.9, and the nuclear index is 1.5.

Type Host *Caiman c. crocodilus* (Linnaeus) (Crocodylia: Alligatoridae).

Other Hosts None known.

Type Locality Barcarena, Pará State, Brazil.

Other Localities None reported.

Prevalence *Trypanosoma cecili* infections were found in 3 of 60 (5.0%) *C. crocodilus* at the type locality (Lainson, 1977).

Morphological Variation Trypomastigotes are monomorphic, elongate, and slender, 71.3 × 8.0 μm (62–84 × 6–9, N = 4). The kinetoplast is located 44.6 μm (37–55) from the anterior end, 26.8 μm (24–34) from the posterior, and 4.8 μm (3–6) behind the nucleus. The oval nucleus is 7.0 × 4.8 μm (6–9 × 4–5) and is situated 39.8 μm (34–49) from the anterior end. Positions of the kinetoplast and nucleus relative to body length average 62.6% and 55.8%, respectively. The shape index is 8.9, and the nuclear index is 1.5. The undulating membrane is well-developed and terminates in a free flagellum that averages 17.0 μm (15–20) in length. The cytoplasm stains heavily, both ends of the body are finely pointed, and longitudinal striations or myonemes can be discerned in some trypomastigotes. The kinetoplast is associated with a large, irregularly shaped vacuole, and the long axis of the nucleus occupies a transverse position across the midbody.

Invertebrate Host Unknown.

Remarks Only two *Trypanosoma* species, both large in size, have been described from crocodilians, *T. grayi* of African *Crocodylus niloticus* and *T. cecili* of Brazilian caimans. As Lainson (1977) mentioned, a much smaller trypanosome, only 35 μm in body length with a posterior kinetoplast, was reported from *Crocodylus cataphractus* in the Congo by Dutton et al. (1907) but has not been seen since.

Trypanosoma grayi Novy, 1906

Diagnosis A large *Trypanosoma* species with finely acuminate anterior and posterior ends, 58–72 µm in body length, and 6–8 µm in maximum width. The nucleus is oval and large, nearly the breadth of the trypanosome, and situated 33.5–43.5 µm from the anterior end and 18.5–29.5 µm from the posterior margin. The kinetoplast is small, surrounded by a vacuole, and located 2.5–6.5 µm immediately posterior to the nucleus, 16–25 µm from the posterior end of the body, and 36.0–50.0 µm from the anterior end. The location of the nucleus and kinetoplast relative to the anterior margin is 62.2% and 67.9%, respectively. The free flagellum is 10–23 µm. Longitudinal pellicular striations are sometimes visible.

Type Host *Crocodylus niloticus* Laurenti (Crocodilia: Crocodylidae).

Type Locality Restricted here to the vicinity of Entebbe, Uganda.

Other Hosts None known.

Other Localities Nigeria (Lloyd et al., 1924); Upper Volta (Challier, 1973.).

Prevalence Not precisely stated.

Morphological Variation Lainson (1977) tabulated the measurements of *T. grayi* by Hoare (1931b) and calculated from these data other characters useful for comparison with *T. cecili*. Body length, exclusive of the flagellum, averaged 61.6 µm (58–72), and maximum width was 7.4 µm (6–8). The nucleus to the anterior end measures 38.3 µm (33.5–43.5) and to the posterior end is 23.3 µm (18.5–29.5), with no data shown for nucleus dimensions. The kinetoplast is 3.5 µm (2.5–6.5) posterior to the nucleus and 19.5 µm (16–25) from the posterior end. The kinetoplast to anterior end is estimated here at 41.8 µm (37–55). The free flagellum is 14.3 µm (10–23). The position of the nucleus and kinetoplast relative to the anterior end is estimated here to be 62.2% and 67.9%, respectively.

Invertebrate Host On ingestion of infected crocodile blood by the tsetse fly *Glossina palpalis*, the trypanosomes multiply in the midgut as epimastigotes and metatrypanosomes. Developing trypanosomes are contained within the peritrophic membrane while in the midgut, but by 5 days postfeeding are migrating into the hindgut and escaping from within the peritrophic membrane, attaching to the hindgut (Hoare, 1931a). After migrating into the hind gut, they accumulate as small, infective metatrypanosomes (metacyclic trypanosomes), either free in the lumen or attached to the cuticular intima (Molyneux, 1977) throughout the length of the hindgut. By 6–8 days, all have left the intraperitrophic space of the hindgut, migrating forward into the extraperitrophic space of the midgut, continuing to reproduce. Between 13 and 34 days, reverse migration occurs into the ileum, and the midgut is again free of flagellates. Evacuation of feces depletes flagellate numbers, and Hoare (1931a) thought it possible for flies to lose their infections completely. In addition to *G. palpalis, Glossina tachinoides* in Upper Volta is a host to *T. grayi* (Challier, 1973; Molyneux, 1977). Infection from this posterior station occurs either by discharge in the feces onto mucous membranes of the crocodile mouth or when flies are crushed by the crocodile jaws. New infections can appear in experimental juvenile crocodiles as early as 4 days postingestion.

Remarks Hoare, in a series of articles (1928, 1929, 1931a, 1931b, 1932) described the morphology and life history of *T. grayi* from his research at Entebbe, Uganda. Because data from other areas is fragmentary, and in some cases confusing, the type locality for *T. grayi* is restricted here to the vicinity of Entebbe.

Sauroleishmania Parasites of Reptiles

Although acceptance of the genus *Sauroleishmania* is not universal among students of the leishmanial kinetoplastid flagellates, there are distinct differences between the *Leishmania* parasites of mammals and the leishmania of reptiles in nucleic acids, isoenzymes, antigens, and the distance between subpellicular microtubules that are adequate to justify distinction of the reptilian species as a separate genus. Other biological differences, perhaps of less importance, include vertebrate and invertebrate host types, development site within the vectors and mode of transmission, the forms occurring in the vertebrate host, and types of host cell parasitized. The genus *Sauroleishmania* Ranque, 1973 was recognized by Killick-Kendrick et al. (1986) for the leishmanial parasites of reptiles. The genus was defined thus:

> Trypanosomatids of the blood or intestinal tract of reptiles occurring as amastigotes and/or promastigotes; amastigotes in the blood parasitize macrophages and precursors, thrombocytes or erythrocytes; in the Old World, the invertebrate hosts of the blood-inhabiting forms are species of *Sergentomyia* (Diptera, Psychodidae) within the

gut of which the parasites have a hypopylarian or peripylarian development as a promastigote; invertebrate hosts of intestinal forms unknown; transmission assumed to be by bite and/or ingestion of vector.

The pattern of development in the saurian host of the species described here from *Hemidactylus platycephalus* differs so greatly from that of mammalian *Leishmania* species that the separate generic classification is further justified.

Leishmanial flagellates parasitize only lizards and a few snakes among the Reptilia. Blood cells of lizards containing amastigotes were first found by Chatton and Blanc (1914a) in *Tarentola mauritanica* from southern Tunisia. Several isolations of leptomonad flagellates in cultures of blood or organs were made during the next decade from the same lizard species, *T. mauritanica*, in southern Europe or northern Africa by Ed. Sergent et al. (1915), Chatton and Blanc (1918b), Nicolle et al. (1920), A. Pittaluga and Buen (1917), and Laveran and Franchini (1921). During the same period, leptomonad flagellates were found in the cloacas of South African *Chamaeleo pumilis* (Bayon, 1915) and *Chamaeleo vulgaris* in Egypt by Wenyon (1921), who designated them *Leishmania chamaeleonis* and recognized the gecko parasite as *Leishmania tarentolae*. Franchini (1921) found rounded or leishmaniform, oval, and elongate *Herpetomonas* in the feces of a *T. mauritanica* from Sardinia. A *Herpetomonas* species was found in *C. vulgaris* in Tripoli in feces and in cultures of heart blood (Franchini, 1921). Apparent free or intracellular amastigotes were found on smears of liver or blood from 2 of 40 skinks, *Chalcides ocellatus*, in Sicily by Franchini (1921). In the Western Hemisphere, Leger (1918) described *Leptomonas henrici* flagellates found in the blood and large intestine of an *Anolis* species from Martinique, certainly *Anolis marmoratus*, the only anole native to Martinique (Telford, 1995b). Blood obtained from cardiac puncture contained leptomonads and a few rounded forms without a flagellum, resembling amastigotes, in 2 of 30 lizards. Leptomonads were seen in the cloaca of many of the other anoles. These were considered by Wenyon (1926) to be *Leishmania henrici*, possibly obtained from prey insects. Strong (1924) described a flagellate, *Leptomonas davidi*, from *Cnemidophorus lemniscatus*, possibly acquired from ingested insects. Jimenez Ozete (1981) reported a flagellate similar to *Leishmania* in *Anolis lucius* of Cuba. It is doubtful that these three reports adequately demonstrate the presence of *Sauroleishmania* in lizards of the Western Hemisphere, in view of the very large numbers of lizards examined for hemoparasites by J. J. Schall in the Caribbean and western United States, by S. C. Ayala in California and Colombia, and by me in Panama, Costa Rica, Venezuela, Mexico, the Caribbean, and both the southeastern and southwestern United States.

Ovezmukhammedov (1991) recognized ten valid species in *Sauroleishmania*, considered to be a subgenus of *Leishmania*, and included three undescribed forms of possible validity, *Sauroleishmania* spp. 1 and 2 (Telford 1979b), from *Teratoscincus scincus* and *Agama agilis*, respectively, of Pakistan (**Plate 66B**), and *Sauroleishmania* sp. Ovezmukhammedov (1991) from hosts of the families Gekkonidae, Agamidae, Scincidae, Lacertidae, and Varanidae in southern Asia, some of which are known hosts of described *Sauroleishmania* species. Killick-Kendrick et al. (1986) included the three species that occupy the large intestine or cloaca of the host, *S. chamaeleonis, S. davidi*, and *S. henrici,* as *Sauroleishmania* species. Until evidence is obtained from biochemical comparison of these intestinal inhabitants with those cultured from blood or organs and with species of *Herpetomonas* and *Leptomonas*, their exclusion from *Sauroleishmania*, species of which apparently occur only in the bloodstream, is justified in the opinions of Ovezmukhammedov (1991) and Telford (1995b). In addition to *henrici, chamaeleonis*, and *davidi*, Ovezmukhammedov and Saf'janova (1989) earlier excluded the species *zmeevi, sofieffi, phrynocephali*, and *helioscopi* from the subgenus *Sauroleishmania* because their descriptions did not agree with the life cycle of reptilian leishmanial species. Although Paperna et al. (2001) found no ultrastructural differences in amastigotes between the species *Leishmania zuckermani*, described from a South African gecko, and those from mammalian species, their conclusion that "species of *Leishmania* are indistinguishable by ultrastructural details" is not accurate. D. H. Lewis (1975) found, from ultrastructural study of promastigotes from cultures of three *Sauroleishmania* species, substantial differences from the mammalian parasite *Leishmania mexicana* in number of subpellicular microtubules and the distance between them: *L. mexicana* had an average of 118 subpellicular microtubules with a mean separation of 26.5 nm between tubules, while the distance between microtubules of the saurian flagellates was 45.6 nm. *Sauroleishmania hoogstraali* (118) and *S. agamae* (117) had similar or identical numbers of microtubules, 118 and 117, respectively, but in two isolates of *S. adleri* there were considerably fewer microtubules, 78.2 and 81.9. Garnham (1971) cited Avakjan, who found that subpellicular microtubules were 58–67 nm apart in promastigotes from reptiles, compared to 35–42 nm in mammalian parasites, similar to the results of Lewis. Lewis concluded that there was no other difference in basic ultrastructure from the organization of other trypanosomatids or among the *Sauroleishmania* species studied.

The descriptions of most *Sauroleishmania* species are inadequate. Some, described as new because they were found in a new host or geographic area, were synonymized by Ovezmukhammedov (1991. Most species are known only from the cultural form, promastigotes, which are very similar in morphology and size. The use in taxonomy of promastigote size from cultures was criticized by Adler (1929), who thought their behavior would be a better criterion, but promastigote length and width were among the characters used by Heisch (1958) to distinguish *S. adleri* from *Leishmania donovani*. Although disparaging agglutination as useful in differentiating species, Adler and Theodor (1929) distinguished *S. ceramodactyli* from *S. tarentolae* by serological comparison. These species are also distinguished by site of development within experimental sand fly vectors, *S. ceramodactyli* occupying the posterior station, in the hindgut, and *S. tarentolae* the anterior, from the thoracic midgut to the head. McMillan (1965) characterized *S. hoogstraali* by flagellar movements, the effects of serological differences on flagellar motion and development, and growth in media containing immune sera at different dilutions. Taxonomic differences among species of *Sauroleishmania* have not been well defined, but these should be amenable now to genome analysis and comparison of isoenzymes. Morphological differences are nearly useless in comparison to their value among the other kinetoplastid flagellates found in reptiles, the *Trypanosoma* species.

In the vertebrate host, the leishmanial trypanosomatids occur either as an intracellular rounded or oval stage, the amastigote, that lacks a free flagellum but contains a usually prominent kinetoplast, and a slender elongated extracellular stage, the promastigote, in which a kinetoplast situated anteriorly to the nucleus gives rise to a free flagellum. Promastigotes were the "leptomonad" form of early investigators (Garnham, 1971) and are the characteristic stage found in the invertebrate host and in cultures of heart blood or blood from tissues of vertebrates or gut contents from invertebrates. Most leishmanial infections from reptiles have been detected by isolation of promastigotes in culture. Extracellular *Sauroleishmania* parasites in the bloodstream of reptiles must be promastigotes, but evidence of this is scanty. Although Leger (1918) apparently found promastigotes in blood from cardiac puncture, verification is needed that his isolates were of leishmanial parasites. Reports by Russian scientists represent the best evidence that promastigotes of *Sauroleishmania* species occur in the blood of the saurian host. Promastigotes were reported in tissue smears from *Eremias lineolata* in Tajikistan (Zmeyev, 1936), and promastigotes in blood smears from three *Gymnodactylus caspius* were found by Popov (1941) in Azerbaijan. Andrushko and Markov (1955a, 1955b) reported promastigotes from blood smears of *Phrynocephalus interscapularis, Eremias grammica*, and *E. intermedia* in Turkmenia. Promastigotes were found on liver impression smears in four leishmanial infections of *Agama sanguinolenta* and *Phrynocephalus mystaceus* in Tashkent by Belova (1966).

Leishmania occur as amastigotes within monocytes, histiocytes, and macrophages, either in visceral foci or in the skin of mammalian hosts, but there is "no evidence that *Sauroleishmania* species accumulate or reproduce in visceral foci. The culture of promastigotes from various organs is not proof of a focused infection, for blood cells occur in all organs" (Telford, 1995b). The reports of amastigotes in circulating blood cells suggest that *Sauroleishmania* parasitizes only blood cells as amastigotes.

Nine reports describe amastigotes in natural infections of described *Sauroleishmania* species present within blood cells. In *Tarentola mauritanica*, Chatton and Blanc (1914a) found small, elongate leishmaniform bodies of *S. tarentolae* in the erythrocytes, and Rioux et al. (1969) found amastigotes of *S. tarentolae* in groups of three to ten parasites in the monocytes of *T. mauritanica*. In Sudan, Elwasila (1988) found leukocytic amastigotes of *S. tarentolae* in 3 of 16 *Tarentola annularis*. In India, amastigotes of *Sauroleishmania hemidactyli*, earlier described from promastigotes as a *Herpetomonas* by Mackie, Das Gupta, and Swaminath (1923), were found in probable late erythroblasts or early proerythrocytes in peripheral blood of a gecko, *Hemidactylus gleadovii*, in India by Shortt and Swaminath (1928). David (1929a) found a monocyte containing 16 amastigotes of *Sauroleishmania agamae* from *Agama stellio* of Palestine, which she described from blood culture (David, 1929b). Amastigotes, apparently of *S. agamae*, were present on blood films from 16 of 200 (8%) *A. stellio* examined from Lebanon (Edeson and Himo, 1973), without identifying the blood cell type. Although these authors did not identify the blood cells that were parasitized, a cell containing amastigotes was illustrated and identified as a monocyte by Wilson and Southgate (1979) from a slide Edeson provided. Telford (1979b) reported amastigote infections in thrombocytes of two of five *Agama agilis* and three of nine *Teratoscincus scincus* in Pakistan. Ovezmukhammedov (1991) illustrated amastigotes of *Sauroleishmania gymnodactyli* in various types of cells.

Following the entrance into lizards and snakes by promastigotes via ingestion, defecation, or bite of the vector, amastigotes of *Sauroleishmania* species apparently first appear, perhaps briefly, in circulating nonerythroid cells of various types, after which erythroid cells, especially the precursors to erythrocytes, erythroblasts, and proerythrocytes, contain amastigotes that then disappear during chronic infection. Either or both types of cells may be detected containing amastigotes, depending on the point in the cycle at which the blood sample is taken: white cells infected with *S. agamae* (David, 1929a; Wilson and Southgate, 1979) and *S. tarentolae* (Rioux et al., 1969; Elwasila, 1988); thrombocytes by *Sauroleishmania* spp. a and b (Telford, 1979b) (**Plate 66B**); Ovezmukhammedov, 1991); and red blood cells infected with *S. tarentolae* (Chatton and

Blanc, 1914a), *S. hemidactyli* (Shortt & Swaminath, 1928), and *S. zuckermani* (Paperna et al., 2001). Only in the *Sauroleishmania* species from *Hemidactylus platycephalus* of Tanzania (Telford, 1995b, and below) has an initial infection been followed from onset of patency by amastigotes in nonerythroid cells followed by their appearance in erythrocytes and disappearance from leukocytes for the remainder of patent infection. The evident resemblance of erythroblasts or proerythrocytes infected by *S. tarentolae* (Chatton and Blanc, 1914a) to those of the Tanzanian gecko is striking. Amastigotes occupied vacuoles in groups of three to ten at one pole of the cell. *Sauroleishmania tarentolae* and the Tanzanian *Sauroleishmania* species are evidently the only species for which both white and red cell infections have been reported. The isolations from blood obtained from the heart or other organs may have been from lizards with chronic infections, consequently with parasitemias so low that detection of infections microscopically was unlikely. Duration of infection in lizards is apparently limited to the report by Parrot (1937), who found that a gecko positive for *S. tarentolae* by blood culture could infect *Phlebotomus minutus* 9 months later; no sand flies were infected when fed after 21 months.

There has been no successful experimental transmission of a *Sauroleishmania* species to a susceptible lizard or snake by the bite or ingestion of an infected sand fly. Both modes of transmission are apparently possible, depending on the parasite species. Multiplication of the promastigotes begins in the blood meal ingested by the sand fly (Killick-Kendrick, 1979). Species that develop in the hindgut of the sand fly occupy the hypopylarian or posterior position and presumably must be transmitted by ingestion of the insect host (Killick-Kendrick, 1979), while peripylarian species, positioned in the anterior station, would be transmitted by bite of the vector. Promastigotes of peripylarian species multiply primarily in the abdominal midgut and then move forward to the thoracic midgut, esophagus, and pharynx of the sand fly, reaching the head.

Sauroleishmania ceramodactyli (Adler and Theodor, 1929) and *S. agamae* (David (1929b), both develop in the abdominal midgut and hindgut of *Phlebotomus papatasi*, but when this sand fly fed on *S. tarentolae*, multiplication occurred in the stomach, and promastigotes then ascended into the thoracic midgut, pharynx, and proboscis of the flies (Adler and Theodor, 1929). Although it is a known vector of mammalian *Leishmania* species, *P. papatasi* fed readily on geckoes and became infected easily with *S. ceramodactyli*, which suggested to Adler and Theodor (1929) that this sand fly species might be the natural vector, transmitting *S. ceramodactyli* when ingested by the host geckoes. There were conflicting and somewhat confusing results when *Sergentomyia minuta* (actually *Se. antennata*) was fed on *S. tarentolae* by Parrot (1934a, b, 1935). Promastigotes multiplied in the stomach in association with the blood meal within 48 hours postingestion, but were then expelled with the fecal material, and infection did not persist after digestion was completed. Transmission, therefore, had to occur when a newly engorged, infected sand fly was eaten by a gecko, rather than infection by bite of the fly. In Malta, however, Adler and Theodor (1935) fed three *Sergentomyia parroti* (= *S. minuta*) on *Tarentola mauritanica* infected by both *S. tarentolae* and *Trypanosoma platydactyli*, and infections by both species were present in the sand flies, dissected from 4 to 12 days later. Both parasites occupied the stomach and anterior part of the cardia of the infected sand flies. The differing results of Adler and Parrot with *S. tarentolae* may have been caused by factors such as differing sources of promastigotes (from infected hosts or from culture), the use of unnatural vectors, a confusion of *S. tarentolae* and *Trypanosoma platydactyli* in the flies, different *Sauroleishmania* species in *Tarentola mauritanica* from different areas, and examination of insufficient numbers of infected flies (Killick-Kendrick (1979). Similar confusion has also attended studies on *Sauroleishmania gymnodactyli* (reviewed by Telford, 1995b), which commonly infects *Sergentomyia arpaklensis* (Alexeieff, Saf'janova, and Karapetian, 1975). In *Sergentomyia arpaklensis, S. gymnodactyli* appears to develop in the midgut and hindgut only (Saf'janova and Alexeieff, 1967). "Other mentions of *Sergentomyia* species as natural invertebrate hosts of *Sauroleishmania* species, while both logical and reasonable, do not establish this as fact" (Telford review, 1995b).

Studies on cultivation, biochemistry, physiology, molecular biology, antigenicity and immunity, host distribution and specificity, and prevalence of *Sauroleishmania* species were reviewed in detail by Telford (1995b) and do not require discussion here. Only accounts of the three species for which some descriptive data are available for the amastigote stage within their host cells are included below. Other species of *Sauroleishmania*, known only as promastigotes from blood cultures or infected sand flies, are compared in the relevant literature cited by Ovezmukhammedov (1991) and Telford (1995b).

Species Accounts

Sauroleishmania tarentolae (Wenyon) 1921

Diagnosis A *Sauroleishmania* species parasitic in geckoes in Mediterranean Europe and northern Africa in which amastigotes occupy both leukocytic and erythroid cells. In erythrocytes, small groups of three to ten amastigotes occupy parasitophorous vacuoles at one end of the host cell.

Type Host *Tarentola mauritanica* Linnaeus (Sauria: Gekkonidae).

Type Locality Metlaoui, Gafsa Region, southern Tunisia.

Other Hosts *Cyrtodactylus kotschyi* (Pozio et al., 1983), *Tarentola annularis* (Elwasila, 1988).

Other Localities Banyuls and Gard, France (Rioux et al., 1979; Killick-Kendrick, 1979); Malta (Adler and Theodor, 1935); Biskra and Tatouie, Tunisia (Chatton and Blanc, 1918); Algeria (Ed. Sergent et al., 1915; Parrot and Foley, 1939; Italy (Pozio et al., 1983; Sudan (Elwasila, 1988); Spain.

Prevalence In *T. mauritanica*, 2 of 31 (6%) showed amastigotes in blood cells (Chatton and Blanc, 1914a); 2 of 80 (3%) at Banyuls, southern France, were similarly infected (Rioux et al., 1979). Prevalence data determined by xenodiagnosis or blood culture have yielded much greater prevalence: 20% positive at Banyuls (Rioux et al., 1979), 11% on Malta (Adler and Theodor, 1935), and 36% in Tunis (Chatton and Blanc, 1918, vide Adler and Theodor, 1935), but the presence of *Trypanosoma platydactyli* in Tunisia suggests the possibility of mixed infections, both there and in Algeria, where promastigotes were obtained in 15.7% and in 6 of 11 blood cultures by Ed. Sergent et al. (1915) and Parrot and Foley (1934), respectively.

Morphological Variation Adequate morphological data are not available.

Invertebrate Host Adler and Theodor (1929) fed *Phlebotomus papatasi* on infected geckoes and found that *S. tarentolae* multiplied in the stomach and then moved forward into the thoracic midgut, pharynx, and proboscis of infected sand flies. Parrot (1934a, 1934b) found multiplication in the stomach of *Sergentomyia antennata* (previously identified as *Phlebotomus minuta*, then *Sergentomyia minuta*, Killick-Kendrick, 1979). Although a slight tendency to invade anteriorly in the gut was observed, promastigotes were expelled with fecal material, which led Parrot (1934b, 1935) to believe that transmission occurred when geckoes ingested newly engorged sand flies. *Sergentomyia parroti* fed on *T. mauritanica* in Malta by Adler and Theodor (1935) were infected by both *S. tarentolae* and *Trypanosoma platydactyli* in the stomach and anterior cardia 4, 10, and 12 days postfeeding. Killick-Kendrick (1979) found natural infections of promastigotes in *Sergentomyia minuta* at Banyuls and Gard, France. All sand flies had promastigote infections in malpighian tubules. In one fly, the midgut and pharynx were infected heavily, and in two sand flies the midgut was negative but the pylorus and ileum were infected. Because *T. platydactyli* is present in the same localities and can produce promastigotes in culture (Wallbanks et al., 1985), the results obtained by Killick-Kendrick are unclear.

Remarks Although *Sauroleishmania tarentolae* has been studied by many workers, it is remarkable that no one has yet transmitted the infection from cultured promastigotes or experimentally infected sand flies and followed the course of infection from patency in infected geckoes.

Sauroleishmania zuckermani (Paperna, Boulard, Hering-Hagenbeck, and Landau) 2001

Diagnosis A *Sauroleishmania* species that occupies only erythrocytes as individual amastigotes or loose-to-compact aggregates that may contain up to 50 amastigotes within the parasitophorous vacuole.

Type Host *Pachydactylus turneri* Gray (Sauria: Gekkonidae).

Other Hosts None known.

Type Locality Gatueng Province, Republic of South Africa.

Other Localities None known.

Prevalence Unknown.

Morphological Variation Amastigotes are oblong under light microscopy, 2 × 1 µm with an apical vacuole. Aggregates of amastigotes within erythrocytes vary from loose assemblages of 4–8 amastigotes to compact groups of 25–50 within a single parasitophorous vacuole, commonly rounded with a central "hollow" or blank space. Compact amastigote aggregates can reach 3–4 µm, and those with looser arrangement 4–5 µm. Erythrocytes can contain several aggregates as well as single amastigotes. Ultrathin sections of erythrocytes containing three to seven amastigotes within one or two parasitophorous vacuoles were examined by transmission electron microscopy (TEM). Dimensions of amastigotes were 2.0–2.2 × 1.0–1.5 µm. They were bound in a "double periplast and encased beneath by a layer of subpellicular microtubules" (Paperna et al., 2001). The nucleus was central, within a cytoplasm "densely loaded with ribosomes." The nuclear chromatin was concentrated peripherally, and a "single electron-dense nucleolus was in the center." The flagellum occupied a flagellar pocket outlined by a "dense thick sheath enclosing peripheral and central sets of microtubules." The kinetoplast was perpendicular to the flagellum, "with the fibrillar DNA band and the cristae." There was a vaguely traceable Golgi complex, and conspicuous mitochondria alongside and behind the nucleus (see Figures 8 and 9 in Paperna et al., 2001).

Life Cycle Amastigotes evidently divide within parasitophorous vacuoles in the erythrocytes to produce their characteristic aggregations.

Invertebrate Host Unknown.

Remarks Paperna et al. (2001) found no evidence of a leukocytic cycle preceding the appearance of amastigotes within erythrocytes. A few single amastigotes within monocytes of the spleen were attributed to phagocytosis of the amastigotes.

Sauroleishmania platycephala sp. nov. (**Plate 68**)

Diagnosis A *Sauroleishmania* species with a diminishing leukocytic cycle that precedes divisional cycles within erythroid cells in which amastigotes initially parasitize probable erythroid stem cells, erythroblasts, and proerythrocytes, finally erythrocytes. Aggregates of amastigotes within erythroid cells may contain 16–24 parasites.

Type Host *Hemidactylus platycephalus* Peters (Sauria: Gekkonidae).

Other Hosts *Cnemaspis barbouri.*

Type Locality Kimboza Forest, 1 km north of the Ruvu River below Kibungo Village, south side of the Uluguru Mountains, Morogoro Region, Tanzania.

Other Localities Morogoro, Morogoro Region, and Magrotto Mountain, Eastern Usambara Mountains, Tanga Region, Tanzania.

Prevalence Overall, 16 of 123 (13.0%) *H. platycephalus*: 13 of 41 (31.7%) from Kimboza Forest, 1 of 41 (2.4%) from Morogoro, and 2 of 4 from Magrotto Mountain. Two of 23 (8.7%) *C. barbouri* from Kimboza Forest were infected by *S. platycephala*.

Morphological Variation In all types of host cells, amastigotes varied in size from 2 × 1 to 5 × 2 µm. The larger, rounded aggregates of 16–24 amastigotes in immature erythroid cells, notably erythroblasts and basophilic proerythrocytes, usually equaled or exceeded the host cell nucleus in size. The smaller aggregates were in polychromatophilic proerythrocytes or erythrocytes, contained six to eight amastigotes, and were one-third to one-half the host nucleus size, without a distinct parasitophorous vacuole. Amastigotes within leukocytes did not form distinct aggregates within a vacuole but appeared to be scattered within the cytoplasm, sometimes encircling the host cell nucleus.

Developmental Cycle within Blood Cells In an *H. platycephalus* followed by blood smears at 2- to 10-day, usually 5-day, intervals for 150 days, the entire course of initial infection in blood cells was observed (Telford, 1995b). Amastigotes appeared in leukocytes and erythroblasts on day 36 following capture of a gecko thought to be negative for *Sauroleishmania* parasites. Infected white cells included monocytes, lymphocytes, monocytoid and lymphocytoid azurophils, and stem cells, but rarely thrombocytes. Leukocytes comprised 70% of infected cells, stem cells 20%, and erythroblasts 10% on day 1 of patency. On day 5, stem cell infections disappeared, and infected leukocytes and erythroblasts increased their numbers to 76% and 24%, respectively. By day 12, parasitized leukocytes comprised 7.8%, stem cells 3.9%, erythroblasts 70.6%, and proerythrocytes 17.6%. Only 2% of parasitized cells were leukocytic on day 17, 68% were erythroblasts, and 30% proerythrocytes. From days 22–37 of patency, no parasitized leukocytes were found. Erythroblasts comprised 52% of host cells on day 22 and proerythrocytes 48%. On day 32, 16% of host cells were erythrocytes, 44% were proerythrocytes, and 40% were erythroblasts. Infected erythrocytes disappeared by the last day of patency, day 37, with proerythrocytes 70% and erythroblasts 30% of infected cells. From day 41 until death on day 150, only a single infected cell was found, on day 96.

Telford (1995b) described an apparent series of developmental stages from three infected geckoes:

> The average sample sizes (number of infected cells) for each slide examined in these three infections were 493, 1315, and 1481 parasitized cells. The six stages of infection were: (1) One or several amastigotes within the host cell cytoplasm, oriented in no particular relationship to each other, with no evidence of a parasitophorous vacuole (Figs. 12, 13). (2) Associations of 2–4 amastigotes, lying closely together, often apparently touching (Figs. 14, 15). Some of the pairs may have resulted from fission rather than a coming together, for some had two nuclei associated with a single wide kinetoplast.

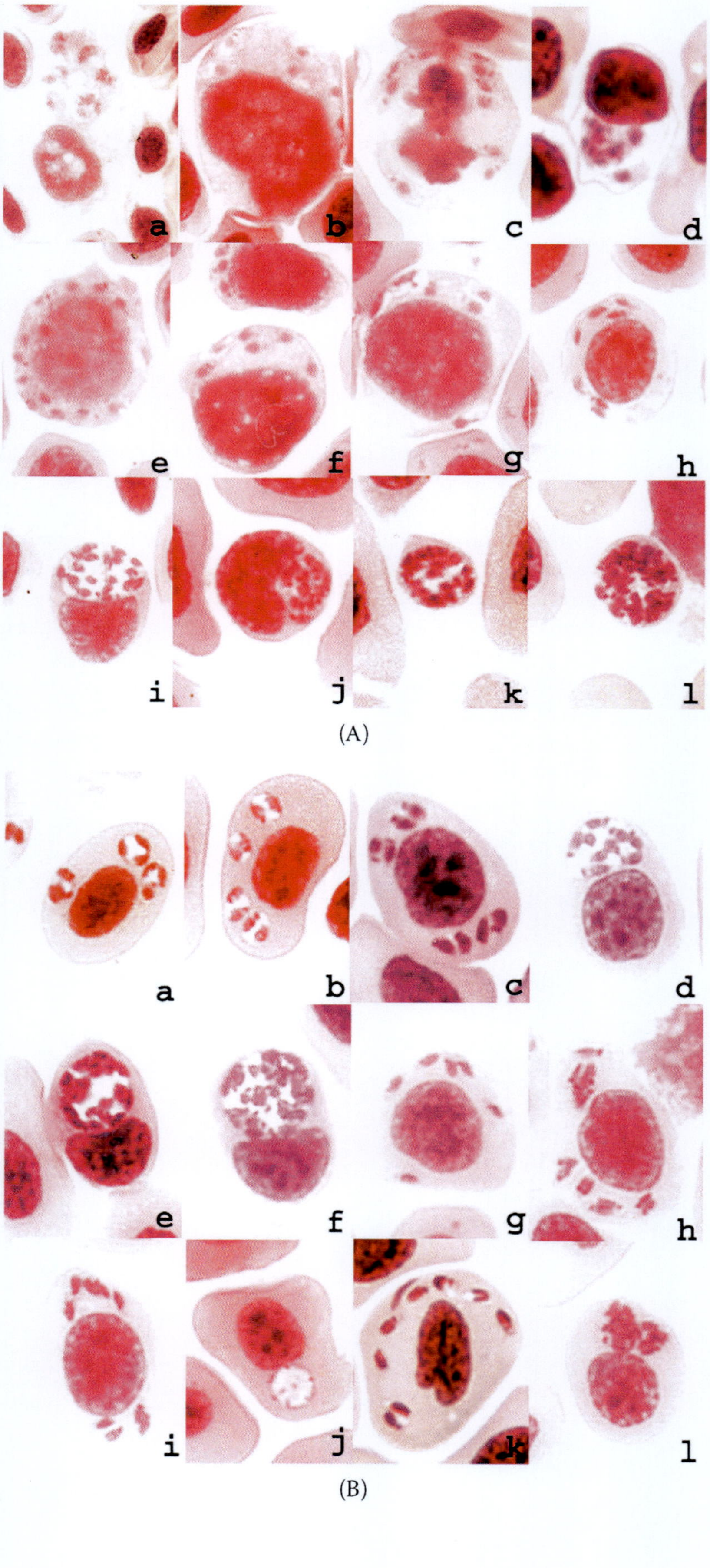

Plate 68 *Sauroleishmania platycephala* sp. nov. from *Hemidactylus platycephalus*, Tanzania. (**A**) Amastigotes in nonerythroid and early erythroid blood cells, **a–j**. Apparently enucleated host cells, **k**, **l**. (**B**) Amastigotes in an apparent developmental sequence within proerythrocytes and erythrocytes in circulating blood of the host.

(3) Development of a parasitophorous vacuole around the associated amastigotes, which numbered from 2–24. These were the most common stage found, accounting for 49–95% of the total amastigote sample from each slide. The vacuole formed from an unstained (white) space between associated amastigotes (16–19), eventually forming a sphere around them. After the vacuole formed, multiplication apparently occurred. There were two or three parasitophorous vacuoles containing amastigotes in some cells (Fig. 20), but most showed single vacuoles (Figs. 21–24). (4) Amastigotes became aligned along the inner periphery of the vacuole, with the normally deeply stained nuclei becoming paler, assuming a pinkish hue (Figs. 25, 26). At this point, amastigotes appeared to flatten themselves against the side of the vacuole, leaving an empty central space. (5) Nuclei became invisible, with only the dark, prominent kinetoplasts marking the location of amastigotes (Figs. 27, 28). Vacuoles appeared as white spheres containing 2–8 or more dark dots. (6) The vacuole disappeared in the final, very uncommon stage, leaving a mass of amastigotes somewhat jumbled together (Figs. 29–31). This stage appeared to precede the rupture of the host cell and release of free amastigotes.

Invertebrate Host Promastigotes developed in a sand fly from a laboratory colony of *Lutzomyia vexator* of Florida origin (Telford, 1995b). The natural vector is almost certainly a species of *Sergentomyia*.

Effects on Host No pathology by *Sauroleishmania* infection was detected among the several lizards followed for varying periods of time. In the gecko examined over 150 days, leukocytemia was less than 0.01% of total blood cells on day 1 of patency, 0.9% on day 6, 5.1% on day 12, 7.9% on day 17, 1.2% on day 22, 2.2% on day 32, and 0.2% on the last day of patency, day 37.

Remarks It is noteworthy that the only other saurian host found infected by sauroleishmanial parasites among more than 1000 Tanzanian lizards examined from 1981 to 1985 was the gecko *Cnemaspis barbouri*, collected from the same pile of boulders in Kimboza Forest, a tract of relict lowland tropical moist forest, from which many of the infected *Hemidactylus platycephalus* were obtained. *Sauroleishmania platycephala* has a considerable geographic range in Tanzania, demonstrated by the presence of two infections among four *H. platycephali* collected on Magrotto Mountain in the Eastern Usambaras of Tanga Region. Hapantotype blood films are deposited in the U.S. National Parasite Collection (USNPC), Beltsville, Maryland, nos. 100339 and 100341.

5
PIROPLASMORIDA

Two genera of apparent haemosporidian nature, *Sauroplasma* Du Toit (1937) and *Serpentoplasma* Pienaar 1962, are presently classified within the Haemohormidiidae Levine 1984. Both are small in size, in early stages often similar in appearance to anaplasmoid bodies, less than 2.5 μm in diameter, although some stages may exceed 4 μm. They commonly are seen as granules of chromatin associated with a vacuole or resemble a ring. When well stained, both chromatin and cytoplasm are visible. Reproduction by *Sauroplasma* takes place by both binary fission and budding. According to Pienaar (1962), in the binary fission mode, "the spherical parasite elongates, and the nuclear material concentrates at the two opposite extremities … a constriction appears in the middle of the elongated parasite. This constriction continues until two separate and approximately equal daughter cells are formed." In the budding process a bud develops at the edge of the rounded parasite, appearing initially to be "solid chromatin." Cytoplasm is extruded to form a "lumen," and the bud grows to nearly the size of the original body, at which point it segments into two halves. Pienaar also found evidence of a third mode of reproduction, "*x-wise* or *Nuttallia-type* of division."

Sauroplasma (**Plate 69**)

Sauroplasma infections have been found in lizards from Asia, Africa, Europe, and North and South America, and in lizards from the families Lacertidae, Cordylidae, Scincidae, Agamidae, Chamaeleonidae, Gekkonidae, Teiidae, and Polychrotidae. Their resemblance to vacuoles in the erythrocyte cytoplasm has certainly caused infections to be overlooked during examination of blood slides. In one instance, the report of a *Sauroplasma* species in the Madagascan gecko *Uroplatus fimbriatus* by Uilenberg and Blanc (1966), reexamination of the type material by Frank (1974) convinced him that vacuoles and perhaps other artifacts were mistaken as parasites. Brygoo (1963a), however, illustrated typical *Sauroplasma* in the blood of *Zonosaurus madagascariensis* and *Chamaeleo verrucosus*, demonstrating the presence of the parasite on Madagascar. Any lingering doubt over the nature of *Sauroplasma* as a parasitic organism should have been resolved by the publication of four electron micrographs of a *Sauroplasma* species in erythrocytes of juvenile Cuban iguanas, *Cyclura nubila* (Alberts et al., 1998). Although further and more detailed study of this material has not been done, the parasites shown have some similarity to piroplasms, and their place as a piroplasmorid appears justified. Svahn (1976) described *Sauroplasma boreale* from Swedish and Danish lizards, and placed it in the Theileridae because she observed no reproduction of *S. boreale* in erythrocytes and considered the reproductive stages described by Du Toit (1937) and Pienaar (1962) as double infections of the host cells. The issue is not settled. It appears that *Sauroplasma* is a piroplasm but with little affinity to the mammalian piroplasms *Babesia* and *Theileria*. Parasitemias by *Sauroplasma* can be very high, up to 43–56% of erythrocytes (Du Toit, 1937), usually with one, rarely two parasites in a host cell. Pienaar (1962) found only 5–10% parasitemias in infected *Cordylus vittifer*. There is no evidence that infection by *Sauroplasma* is pathogenic to the lizard host.

The species accounts next are minimal as insufficient information is available from the sparse descriptions of the three named species.

Species Accounts

Sauroplasma Du Toit 1937

Diagnosis "Small, round or irregularly shaped, unpigmented parasites of the red blood corpuscles of lizards.

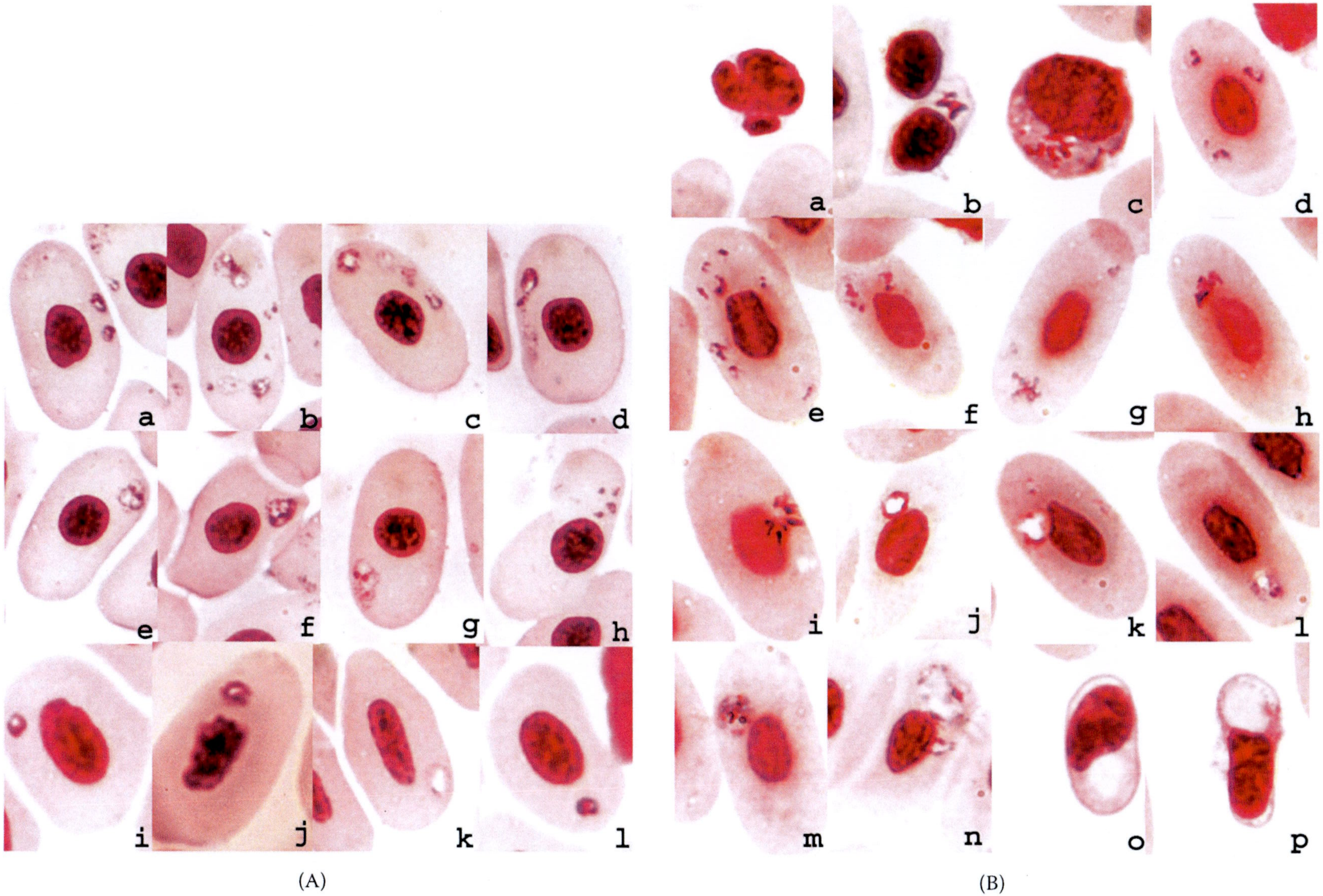

Plate 69 Piroplasmorid parasites of reptilian blood cells. (**A**) *Sauroplasma* spp. of lizard erythrocytes: **a–h** from *Corytophanes cristatus*, Costa Rica; **i–l** from *Takydromus tachydromoides*, Honshu, Japan. (**B**) *Serpentoplasma* sp. from *Trimorphodon biscutatus*, Mexico. Host cells are thrombocytes, **a**, **b**, **o**, and **p**; a probable monocyte, **c**; and erythrocytes, **d–n**. Parasites in **l–p** appear to be gametocytes.

The typical form is that of a ring or signet-ring with a large central vacuole. Multiplication takes place by binary fission or by a process of budding" (Du Toit, 1937).

Sauroplasma thomasi Du Toit, 1937

Type Host *Cordylus giganteus* (Smith) (Sauria: Cordylidae).

Other Hosts *Cordylus jonesi* (Pienaar, 1962).

Type Locality Near Wesselsbron and Odendaalarust, Orange Free State, South Africa.

Other Localities Save River valley, Mozambique (Pienaar, 1962).

Prevalence Unknown. Du Toit (1937) obtained two *C. giganteum,* and both were infected. Pienaar (1962) reported single infected specimens of *C. giganteus* and *C. jonesi.*

Sauroplasma zonurum Pienaar 1962

Type Host *Cordylus vittifer* (Reichenow) (Sauria: Cordylidae).

Other Hosts None known.

Type Locality Potchefstroom District, southwest Transvaal, South Africa.

Other Localities None reported.

Prevalence Not stated by Pienaar (1962); 2 of 56 (3.6%) *C. vittifer* from the vicinity of the type locality were infected (Telford).

Sauroplasma boreale Svahn 1976

Type Host *Lacerta agilis* Linnaeus (Sauria: Lacertidae).

Other Hosts None known.

Type Locality North shore of Lake Snogeholmssjön, Scania, Sweden.

Other Localities Jydelejet, Mon Island, Denmark (Svahn, 1976).

Prevalence At the type locality 32 of 98 (67%) and at Jydelejet 2 of 4 *L. agilis* were infected (Svahn, 1976).

Distribution and Prevalence of Sauroplasma Species among Several Lizard Faunas

Sauroplasma sp. in Panamanian lizards (**Plate 69**)

Among 2017 lizards of 36 species examined, only 3 (0.1%) were infected: 2 of 100 (2%) *Corytophanes cristatus* and 1 of 248 (0.4%) *Anolis lionotus* (Telford, unpublished).

Sauroplasma sp. in Japanese Lizards (**Plate 69**)

Among 2138 Japanese lizards of ten species examined, only two species were infected by *Sauroplasma.* Overall prevalence was 68 of 1943 (3.5%) *Takydromus tachydromoides*, varying annually at the primary study site at Hanno, Saitama Prefecture, Honshu (Telford, 1997b) from 2.7% in 1965, to 0.2% in 1966, and to 1.1% in 1967. Prevalence was age specific, highest among juvenile lizards, in which 96% of infections were found. Infections were detected from March through June only, with none from August to November. There was little difference among microhabitats. The seasonal prevalence pattern coincided with the presence of nymphal ticks, *Ixodes nipponensis*, on lizards. A single infection was found in 1 of 13 *Gekko japonicus* from a building in Tokyo.

Sauroplasma sp. in Southeast Asian Lizards

Infections were found in 14 of 875 lizards (1.6%) of 33 species examined. Eight species were infected: *Gekko gecko*, 1 of 38 (2.6%); *Hemidactylus* sp., 2 of 34 (5.9%); *Calotes mystaceus*, 3 of 135 (2.2%); *Calotes versicolor*, 2 of 163 (1.2%); 1 of 2 *Calotes cristatellus*; 1 of 7 *Physignathus concinnus*; 1 of 161 *Mabuya multifasciata* (0.6%); and 3 of 7 *Sphenomorphus sabanus* (Telford, unpublished).

Sauroplasma sp. in Tanzanian Lizards

Among 1030 lizards of 44 species examined, infections were found in 8 (0.8%). Six species were infected: *Chamaeleo deremensis*, 1 of 23 (4.3%); *Chamaeleo dilepis*, 1 of 50 (2%); *Chamaeleo tempeli*, 2 of 30 (6.7%); 2 of 3 *Rhampholeon brevicaudatus*; 1 of 6 *Gerrhosaurus major*; and 1 of 167 (0.5%) *Mabuya striata* (Telford, unpublished).

Sauroplasma sp. in Venezuelan Lizards

Infections were found in 8 of 828 (0.9%) lizards of 20 species examined. Two host species only were positive, 1 of 22 (4.5%) *Thecadactylus rapicaudus* and 7 of 15 (46.7%) *Euspondylus brevifrontalis* (Telford, unpublished).

Sauroplasma sp. in Florida Lizards

In 22 species of Florida lizards, 10 of 2755 (0.4%) were infected. All were *Anolis carolinensis*, with a prevalence of 10 in 1546 (0.6%) examined (Telford, unpublished).

Remarks Overall, among the lizard faunas surveyed, *Sauroplasma* sp. infections have to be considered rare. Local populations with relatively small samples, that is, *Lacerta agilis* in Sweden (32 of 48; Svahn, 1976), *Sphenomorphus sabanus* in Borneo (3 of 7), and *Euspondylus brevifrontalis* (7 of 15) from a single Andean population, may show unusually high prevalence. Generally, it appears that prevalence of less than 0.1% to 6% is typical wherever these parasites occur. A comparable survey of 855 Pakistani lizards of 38 species found no *Sauroplasma* sp. infections (Telford, unpublished).

Serpentoplasma Pienaar 1962

Diagnosis An apparent piroplasmorid parasite of snakes in which infection begins in thrombocytes, where dense chromatin masses produce nuclei that then infect erythroid cells. Division in both types of cells is by budding, binary fission, or perhaps by merogony, sometimes resulting in *Nuttallia*-like tetrads of nuclei. The parasites invade erythroid cells, where they may appear as vacuoles with a dot of chromatin at the margin, or as irregularly shaped nuclei within the cytoplasm.

Serpentoplasma najae was described from a single *Naja nigricollis* collected in the Western Transvaal by Pienaar (1962). The description provides little distinction from *Sauroplasma*, and perhaps the cobra parasite could belong to that genus. In Pienaar's words:

> The intracorpuscular forms occurred as tiny spherical or oval bodies (less than 1 μm in diameter), with typical signet-ring configuration (clear, central vacuole and prominent peripheral chromatin dot). … Anaplasmoid forms, consisting merely of a chromatin granule were not infrequently encountered. … Multiplication is most commonly effected by the budding process … but binary fission and a *Nuttalia-type* of division were also in evidence. … Apart from mild anaemic changes and polychromasia, no other pathological features could be associated with the relatively heavy *Serpentoplasma* infection.

There have been no further reports of a *Serpentoplasma* infection in snakes since Pienaar's 1962 description, except by Telford (1984b) in *Trimorphodon biscutatus* (discussed in a separate section). Infections reported by Fletch and Karstad (1968) in *Thamnophis sirtalis* collected on the north shore of Lake Erie in Ontario may represent *Serpentoplasma,* but the inclusions are not very similar to those described here.

During the last 45 years, I have examined a minimum of 827 snakes of 132 species from the six geographical areas where my major surveys were conducted: Mexico (149 snakes, 18 species); Panama (219 snakes, 40 species); Venezuela (48 snakes, 15 species); Florida (411 snakes, 34 species); Tanzania (32 snakes, 25 species); Pakistan (37 snakes, 11 species); and Japan (183 snakes, 16 species). Perhaps 200 or more snakes from other areas, such as the southeastern and southwestern United States, East and West Africa, Burma, Thailand, and Indonesia, have also been examined, only 1 of which had a *Serpentoplasma* infection.

A total of ten infections in six host species from the listed areas have been found, indicating an overall prevalence of approximately 1%. All showed the basic structure described by Pienaar for *Serpentoplasma*, but in each infection additional stages were seen, thus indicating the need for an expansion of the generic definition. No species are named next, but the parasites found in each of the six host species are briefly described, along with locality data and prevalence.

Serpentoplasma sp. of *Trimorphodon biscutatus* (Colubridae), Mexico (**Plate 69**)

In 13 *T. biscutatus*, one (7.7%), collected 3.6 km east of Manzanillo, Colima State, Mexico, was parasitized by a *Serpentoplasma* sp. quite different in appearance from the species found in *Boa constrictor* from Colima. Thrombocytes were less commonly infected, host erythrocytes often had multiple infections of two to nine small parasites, and only a single parasite contained verified pigment. Although no parasites were observed when captured in July 1963, 4.5% of the erythrocytes were infected by 18 January 1964, after which parasitemia declined rapidly. Parasites of all stages were still present in March 1964. Thrombocytic parasites appeared as one to three dense chromatin masses and multiplied by fission or budding, even forming as a *Nuttallia*-type tetrad of nuclei. All three modes of division were seen in erythrocytes. The smaller parasites seldom showed prominent vacuoles, often appearing to lack them even as their size reached 3–5 μm in diameter. Occasional parasites resembled meronts with up to eight nuclei. Stages resembling gametocytes appeared to form from the parasites with prominent vacuoles. The largest observed was roughly ovoid in shape, 6 × 4.5 μm in size, and contained a prominent, elongate pigment mass, confirmed under polarized light. A single infected erythroblast containing four dividing nuclei was found. Some thrombocytes contained prominent oval or round white masses in their cytoplasm that resembled gametocytes.

Serpentoplasma sp. of *Boa constrictor* (Boidae), Mexico (**Plate 70**)

One of 57 *B. constrictor* (1.8%), collected in the vicinity of Colima, Colima State, Mexico, was found infected by a *Serpentoplasma*. Thrombocytic parasites appeared singly or in pairs as dense chromatin masses. Two such masses, 2.5 × 2.5 µm and 3.5 × 3 µm in size, contained about six and eight nuclei, respectively. Erythrocytic parasites formed as vacuoles with chromatin or with dark granules along one or two sides and usually appeared as one inclusion per host cell. Uninucleate parasites were 1 × 1–1.5 × 1.3 µm, increasing in size as nuclei budded off or simply divided by fission, sometimes appearing roughly fan shaped with cytoplasm at their base. Small pigmented parasites were common, verified under polarized light. Two apparent

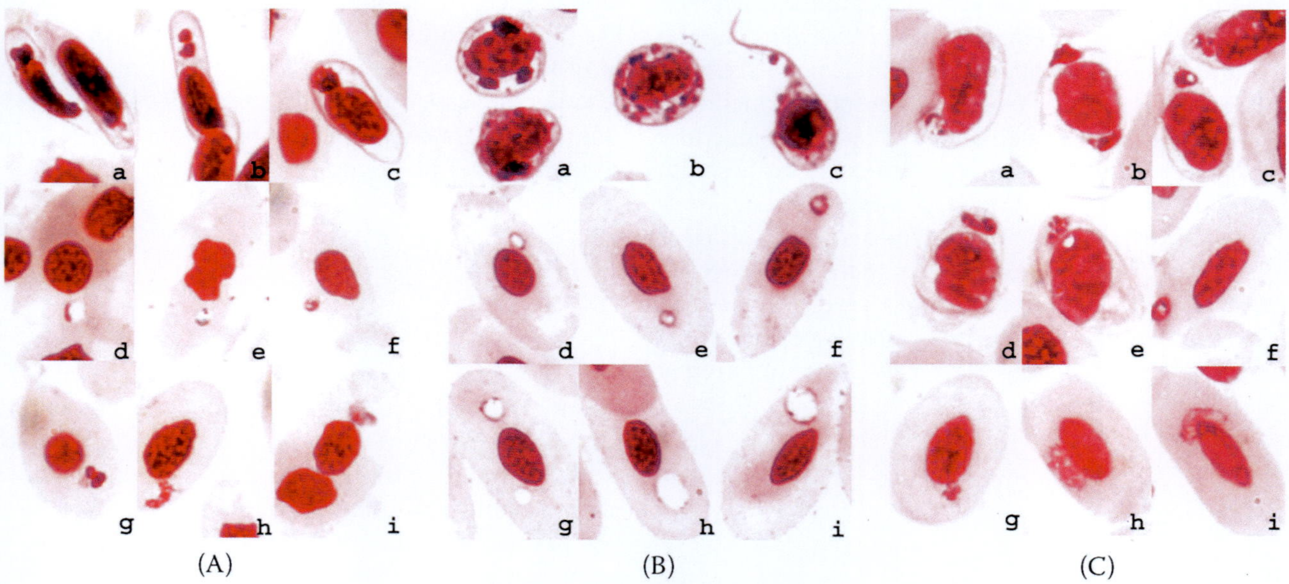

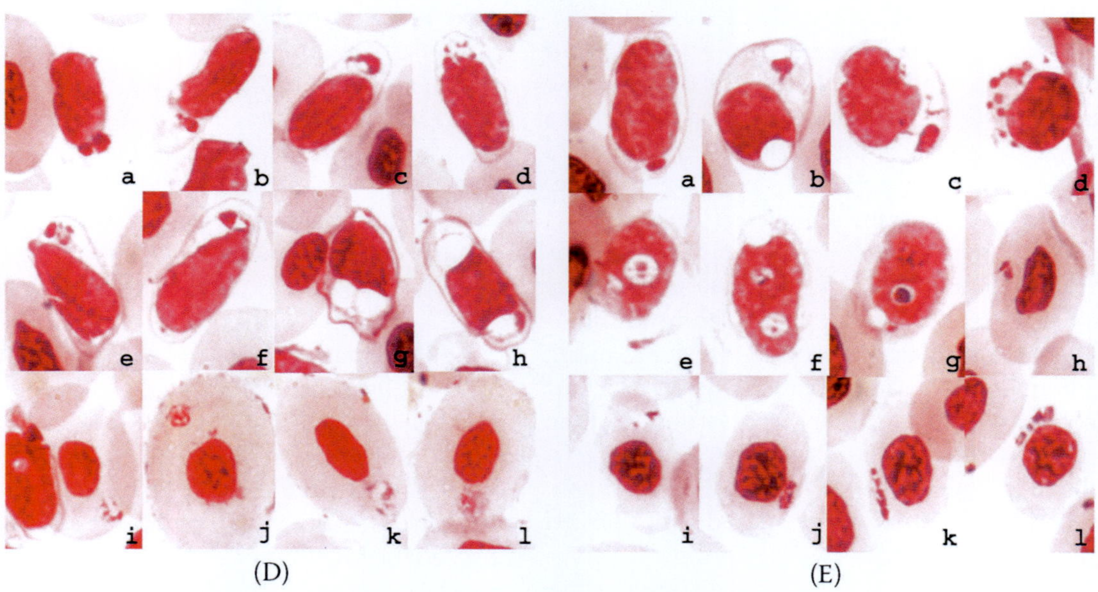

Plate 70 *Serpentoplasma* spp. from snakes. (**A**) *Serpentoplasma* sp. of *Boa constrictor,* Mexico. Host cells are thrombocytes, **a–c**; and erythrocytes, **d–i**. Pigment in **e**, **f**, and **i** verified under polarized light. (**B**) *Serpentoplasma* sp. of *Python regius,* Ghana. Host cells are thrombocytes, **a–c**; and erythrocytes, **d–i**. Parasites in **h** and **i** appear to be gametocytes. (**C**) *Serpentoplasma* sp. of *Opheodrys semicarinatum,* Ryukyu Islands, Japan. Host cells are thrombocytes, **a–e**; and erythrocytes, **f–i**. Parasite in **i** appears to be a gametocyte. (**D**) *Serpentoplasma* sp. of *Nerodia c. clarkii,* Florida. Host cells are thrombocytes, **a–h**; and erythrocytes, **i–l**. (**E**) *Serpentoplasma* sp. of *Storeria occipitomaculata,* Florida. Host cells are thrombocytes, **a–g**; and erythrocytes, **h–l**.

gametocytes, 4 × 2 µm and 3.5 × 3 µm, were found, the latter containing a large, verified pigment granule.

Serpentoplasma sp. of *Python regius* (Boidae), Ghana (**Plate 70**)

One of 18 *P. regius* (5.6%) was infected by *Serpentoplasma*. The parasites were most often found in thrombocytes, appearing as large, dense chromatin masses, usually in an elongated ovoid shape, 3–4 × 2–3 µm in size, and numbered 1–4 per cell. These divided to form up to 12 daughter masses, apparently nuclei, 1.5–2 µm in diameter. Erythrocytes were less commonly infected and contained round-to-oval inclusions from small, anaplasmoid-like bodies, vacuoles with a dot or two of apparent chromatin on their margin, larger forms with a central vacuole ringed by chromatin, and still larger, 3- to 4-µm spherical bodies with or without chromatin dots. The largest stages were 4–5 × 3–4 µm, which resembled gametocytes and often showed one to five dark pigment-like granules that did not refract under polarized light. Multiply infected erythrocytes with one to three parasites were unusual, with most host cells containing a single inclusion. Host cells were apparently not affected by the parasites.

Serpentoplasma sp. of *Opheodrys semicarinatum* (Colubridae), Japan (**Plate 70**)

Four of five *O. semicarinatum* collected on Amami Oshima, Ryukyu Islands, Japan, were infected by *Serpentoplasma* in both thrombocytes and erythrocytes. Thrombocytic parasites usually formed as one or two chromatin masses of irregular shape, and some appeared to be budding off a second nucleus, separated by a vacuole or light space. One parasite contained five nuclei. Dimensions ranged from about 1 µm to 4–4.5 × 1.5–2 µm. Single-parasite infections were most common in both types of cells, but double infections were not rare. In erythrocytes, oval parasites commonly showed a central vacuole surrounded by chromatin, and both budding into two parasites and division into four nuclei in a *Nuttallia*-like arrangement were observed. One parasite, 6 × 3 µm, contained seven nuclei. A single apparent gametocyte was found, roughly triangular in shape, 6 × 3.5 µm. Pigment, verified under polarized light, formed a single granule in an ovoid parasite 3 × 2.5 µm.

Serpentoplasma sp. of *Nerodia c. clarkii* (Colubridae), Florida (**Plate 70**)

Two of six *N. clarkii* collected at Cedar Key, Levy County, Florida, were found with *Serpentoplasma* infections. In both infections, thrombocytes were more heavily parasitized than erythrocytes, often with two to four chromatin masses present, in contrast to single parasites within red blood cells. The chromatin masses were 1 × 1 µm to 4 × 2.5 µm. Some showed the budding mode of division, while in others a merogony appeared responsible for producing seven or eight nuclei. These were larger in size than most thrombocytic parasites, with the 2 seven-nucleate forms 4 × 2.5–3 µm. A single, apparently segmenting meront, had eight nuclei. One or two ovoid or round, unstained bodies without prominent chromatin were commonly seen in thrombocytes, usually polar in position. Erythrocytic parasites were usually 3 × 2.5–3 µm, with two or more chromatin masses and up to three dark granules. A single parasite, 4 × 3 µm, resembled a gametocyte with a centrolateral nucleus and granules that were verified as pigment under polarized light.

Serpentoplasma sp. of *Storeria occipitomaculata* (Colubridae), Florida (**Plate 70**)

One of two *S. occipitomaculata* from Gainesville, Alachua County, Florida, had an infection by a *Serpentoplasma*. Thrombocytes were more commonly infected than were erythrocytes. The earliest stages were round-to-oval dots of chromatin, less than 1.0 in diameter to 2 × 1.5 µm. As size of the chromatin mass increased, one to three or more prominent oval or round vacuoles appeared in the cytoplasm of the thrombocytes adjacent to the chromatin bodies, some of which began division. As many as ten nuclei were seen, some associated with a dark, smaller granule. In a possibly later stage of development, round vacuoles appeared within thrombocyte nuclei in which a rounded, dense mass of chromatin divided, perhaps by fission, into two elongate, narrow progeny with a central nucleus in each, resembling sporozoites. Some thrombocytic nuclear vacuoles appeared to contain more than two such "sporozoites." Within erythrocytes the smallest parasites were round, less than 1.0 µm in diameter. As these grew, some became elongated with a terminal chromatin body separated by a prominent vacuole from a smaller chromatin area or a dark dot at the other end. This form could reach a size of 3.5 × 1–1.5 µm. Other small parasites were irregular in shape, 2 × 1.5 µm, without a vacuole, and appeared often to be dividing. Central-to-lateral dark dots did not refract under polarized light. At a size of 3.5 × 1.5 µm, elongate bodies contained four distinct chromatin masses. Two erythrocytes were observed with elongate parasites 6 × 1–2 µm that each contained five apparent nuclei arranged linearly. No stages resembling possible gametocytes were present.

Remarks Generic identity of the infections in the six host species described seems certain because of the parasite characteristics shared among them. The absence of some stages probably relates to stage of infection at the time the (usually) single sample of blood was obtained. In the original description of the genus from a South African cobra, there was no mention of parasitized thrombocytes by Pienaar (1962), which again could have been due to the stage of infection that he sampled. Otherwise, the common presence of reproduction by binary fission, budding, and possible merogony that resulted in *Nuttallia*-like tetrads of nuclei is a strong point of similarity. The presence in most infections of occasional dark dots within parasites that give a positive reaction for pigment when examined under polarized light suggests that pigment may be variably present, but this is not included in the revision of the generic definition.

Given the rarity of *Serpentoplasma* infections, it will be difficult to obtain adequate material for ultrastructural and genomic studies, which must be done to determine the systematic position of these parasites. It is futile to speculate on the mode of transmission. The distribution of *Serpentoplasma* is cosmopolitan, as indicated by infections in elapid, boid, and colubrid snakes from South and West Africa, Japan, Mexico, and Florida. At least two areas (Colima State, Mexico, and north-central Florida) may have two species present in each area: *Boa constrictor* and *Trimorphodon biscutatus* in Mexico and *Storeria occipitomaculata* and *Nerodia clarkii* in Florida. The host ecology suggests a vector group with a broad distribution: *P. regius, B. constrictor*, and *O. semicarinatum* are semiarboreal/terrestrial species, *T. biscutatus* is terrestrial, *S. occipitomaculata* is secretive, and *N. clarkii* occupies salt marsh and mangroves along the Florida coast. It is appropriate to comment that an infected snake might be encountered anywhere, at least in tropical, subtropical, and warm temperate areas; the infected *S. occipitomaculata* was found under a board about 10 feet away from the Wildlife Parasitology Laboratory on the University of Florida campus.

Parasites of Uncertain Relationship

A number of intraerythrocytic inclusions have been described during the last century from lizards, snakes, and turtles, and these cannot be classified on the basis of present knowledge. The little information on them was reviewed by Telford (1984b) and is not repeated here except for the few for which additional study has been reported. Three taxa may be piroplasmid apicomplexans and are similar to *Sauroplasma* and *Serpentoplasma*: *Nuttallia guglielmi* Carpano (1939) from the tortoise *Testudo campanulata, Cingula boodontis* of the snake *Boaedon lineatus* Awerinzew (1914), and *Aegyptianella carpani* Battelli (1947) from *Naja nigricollis*. This last organism does not resemble the rickettsial genus but rather may be related to *Serpentoplasma*, although the presence up to 30 or more inclusions in individual erythrocytes is certainly different. Slides from Galapagos tortoises at the San Diego Zoo sent to me contained small erythrocytic inclusions, some of which formed tiny tetrads of chromatin about 3 μm in diameter, resembling *Nuttallia* reported from various mammals and similar in appearance to *N. guglielmi* described by Carpano. Because they were present in hosts long captive in a zoological park collection, there is no certainty that the infections were acquired in nature.

6

PROKARYOTIC PARASITES OF REPTILIAN BLOOD CELLS

Rickettsiales: *Aegyptianella* (**Plate 71**)

Desser (1987) demonstrated by ultrastructure that inclusions in erythrocytes of Canadian anurans were rickettsiae and designated them *Aegyptianella ranarum*. Under light microscopy, the roughly spherical inclusions were 3–11 µm in diameter, densely stained when small, and bordered by a darker stained margin. In the smaller inclusions, "several closely packed rickettsiae arranged in parallel" were visible, while "lighter staining, filamentous material" was visible in the larger inclusions. Electron microscopy confirmed the rickettsial identity. Small inclusions were contained within a "membrane-bound vacuole in the cytoplasm of the erythrocyte." A "moderately dense, irregularly arranged granular matrix" was contained within them, associated with "large spherical bodies of variable density and smaller patches of electron dense material." Within the granular matrix, rod-shaped microorganisms were embedded. In larger inclusions, there were "90–120 closely spaced organisms, usually arranged in parallel, within a lucent matrix consisting of scattered granular material." Their rickettsial identity was evident at higher magnifications, and their dimensions were "1–1.7 µm in length by 200–300 nm in diameter" (Desser, 1987).

Ultrastructural study of inclusions in erythrocytes from Corsican anurans, identified as *Cytamoeba bactifera* Labbé 1894, proved their identity as rickettsia also, in the same genus, *Aegyptianella* (Desser and Barta, 1989). These inclusions contained somewhat larger organisms in comparison with *A. ranarum*, measuring under light microscopy 2.3–4.9 µm by about 0.5 µm, "about four times the size of *A. ranarum*" and "about one-tenth the number of organisms per inclusion," that is, about 12 rickettsiae (Desser and Barta, 1989). Similar inclusions, probably all species of *Aegyptianella*, have been reported from anurans elsewhere, in New York (Stebbins, 1904; Hegner, 1921), West Africa (Dutton et al., 1907), Madagascar (Brygoo, 1963a), Brazil (De Sousa and Freire, 1975), Panama (Telford, 1984b), and Japan (Telford, 1997b).

The first report of an apparent rickettsia in reptilian erythrocytes appears to be that of *Tunetella emydis* from the turtle *Mauremys leprosa* by Brumpt and Lavier (1935), since considered to be an *Aegyptianella* by Carpano (1939) and Gothe and Kreier (1977). *Aegyptianella carpani* of *Naja nigricollis* (Battelli, 1947), however, appears from illustrations more likely to be a piroplasm, similar to *Sauroplasma* and *Serpentoplasma* (if they are indeed piroplasms). Other inclusions reported in the literature as *Tunetella*, *Bertariella*, *Grahamella*, and *Sauromella* could also be rickettsial.

Although the ultrastructure has not been studied, *Aegyptianella* (**Plate 71A**) apparently infects the scincid lizard *Eumeces inexpectatus* in Florida (reported as *Cytamoeba* by Telford, 1984b), the agamid *Laudakia caucasica* in Pakistan, and the lacertid *Takydromus tachydromoides* in Japan (Telford, 1984b, 1997b). In a community with infected aquatic anurans, overall prevalence of an *Aegyptianella* species in *T. tachydromoides* was 5.7%. Seasonal and annual prevalence varied from 1.7% to 8.9%, but in one sampling period, July–August, reached 22.6%. Prevalence was twice as high in male lizards compared to females and far higher among mature lizards than in the immature cohort of the population: 73.5% of total infections (68) in mature male lizards, 25.0% in mature females, and 1.5% in immature males, with no immature females infected (Telford, 1997b). There was no clear pattern of seasonal prevalence: In males, peaks occurred at 20.0% in March–April 1965, 33.3% in May–June 1966, and 55.6% in July–August

Plate 71 Rickettsial, chlamydial, and viral parasites in blood cells of lizards. (**A**) *Aegyptianella* spp. in erythrocytes of *Eumeces inexpectatus,* Florida, **a–d**; *Laudakia* (*Agama*) *caucasica*, Pakistan, **e–h**; and *Takydromus tachydromoides*, Honshu, Japan, **i-l**. (**B**) Pox virus and *Chlamydia* infections in monocytes of *Chamaeleo dilepis*, Tanzania. Monocytes with both infections, **a–h**.

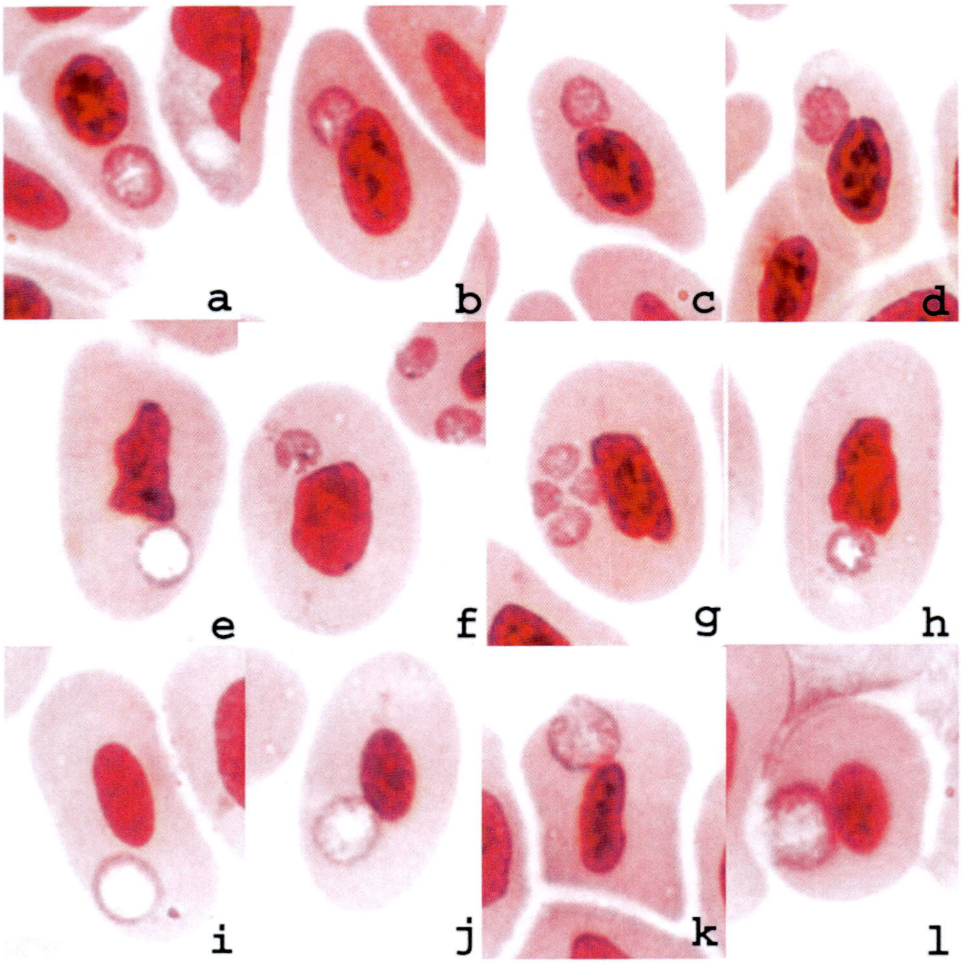

(A)

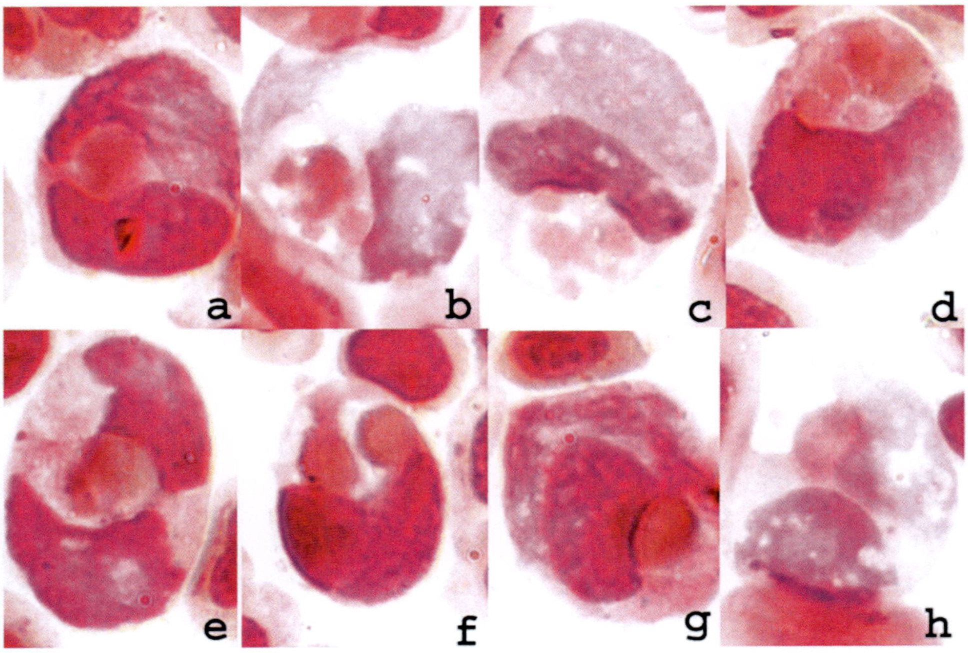

(B)

1967. Prevalence among females was at a maximum in September–October 1966, 23.1%, and diminished slightly to 15.8% in March–April 1967, following hibernation. Depending on the year, prevalence could increase or decrease over the hibernating period. The tick *Ixodes nipponensis* was suspected to be the vector of the *Aegyptianella* sp. among the lizards because of similarity in proportionate distributions of infections by host age and tick infestation. Desser (1987) suggested that leeches were responsible for transmission of *A. ranarum* among Canadian anurans, but tick transmission appeared to be more likely in the Japanese lizard population.

Chlamydial Infection of Reptilian Leukocytes (**Plate 71**)

Chlamydial infection of vertebrates is widely distributed, especially among birds, in which *Chlamydia psittaci*, causative agent of psittacosis, is capable of infecting humans and anurans as well (Moulder, 1984; Newcomer et al., 1982). In reptiles, Jacobson et al. (1989) reported infections in captive-born puff adders (*Bitis arietans*). No infections within circulating blood cells were known until an infection in 1 of 50 (2%) *Chamaeleo dilepis*, collected in Tanzania, was reported by Jacobson and Telford (1990). Monocytes of the chameleon contained poxvirus at time of capture, but 46 days later chlamydiae were present also in the monocytes (**Plate 71B**), either alone or in mixed infection with poxvirus in equal proportions of 1% for single chlamydial infection or mixed infections. On day 55 of captivity, the host became moribund, showing a monocytemia of 36%, with 11% of monocytes infected by both pathogens and 1% with chlamydial infection alone. Macrophages in liver and spleen were infected by both agents as well.

Under transmission electron microscopy, monocytes infected with chlamydiae showed pleomorphic bodies representing three developmental stages: "large oval, 800 to 900 nm reticulate bodies"; "contracted 440 to 680 nm intermediate bodies containing an electron-dense center"; and "small, round, dense 400–440 nm elementary bodies." These were "compatible with the three stages of chlamydiae (Moulder, 1984)" (Jacobson and Telford, 1990). The membrane surrounding chlamydiae was usually ruptured, releasing the bodies into the cytoplasm.

7

VIRAL INFECTIONS OF CIRCULATING BLOOD CELLS

Poxvirus Infection of Leukocytes (**Plate 71**)

Although poxvirus infections are known to occur in captive reptiles, there have been no reports of the virus infecting them in nature (Jacobson and Telford, 1990). In Tanzania, 3 of 50 (6%) *Chamaeleo dilepis* were found with inclusions containing poxvirus, primarily in monocytes (**Plate 71B**), although one *C. dilepis* also had a few thrombocytes and lymphocytes that showed infections. Under light microscopy, the "inclusions stained lightly eosinophilic, had a lacy appearance, and often contained a blue-green staining body" (Jacobson and Telford, 1990). At capture, in the lizard with the heaviest infection, monocytes comprised about 13% of circulating leukocytes, and inclusions were present in about 96% of the monocytes. On day 46, monocytes also contained chlamydiae. At death on day 55 post-capture, the monocytemia had risen to 36%, with 88% of monocytes infected with poxvirus alone, 11% with mixed infection with chlamydiae, and 1% singly infected by the latter parasite. Both types of parasites were present in macrophages of spleen and liver.

When examined by transmission electron microscopy, the membrane-bound inclusions were "composed of a uniform population of 340 to 390 nm by 210 to 235 nm oval-shaped virions, containing dumbbell shaped nucleoids and lateral bodies ... typical of poxvirus (Dale and Pogo, 1981)" (Jacobson and Telford, 1990). The blue-green staining bodies observed under light microscopy proved under electron microscopy to be uniformly radiodense, contained no virions, and were surrounded by vacuoles containing poxvirus.

Brygoo (1963a) reported similar inclusions in *Chamaeleo lateralis* of Madagascar, considered them to be viral, but did not identify the inclusions by ultrastructural study. Both oral and intraperitoneal routes of infection were successful, but other *Chamaeleo* species could not be infected. Prevalence was 16 of 248 (6.4%) in *C. lateralis*.

Iridovirus Infections of Erythrocytes

Lizard Erythrocytic Virus (**Plate 72**)

Inclusions in the erythrocytes of *Tarentola mauritanica*, consisting of reddish dots about 1 μm in diameter that increased to 3–4 μm, circular in shape, and accompanied by globular albuminoid bodies, were designated *Pirhemocyton tarentolae* by Chatton and Blanc (1914b). Another species, *P. lacertae*, was recognized by Brumpt and Lavier (1935) in *Lacerta viridis*. This infection was similar to *P. tarentolae,* but the albuminoid bodies were not present. *Pirhemocyton* species designations continued thereafter until 1966, with perhaps the most egregious example that of Rousselot (1953), who assigned seven specific names to infections in West African lizards, apparently on the basis of "different host, therefore different parasite." *Pirhemocyton* was generally considered to be a protozoan parasite, perhaps a piroplasm, but Blanc and Ascione (1958) suggested that it might be a type of virus. Stehbens and Johnston (1966) demonstrated that the ultrastructure of a *Pirhemocyton* in the Australia gecko *Gehyra variegata* was that of a DNA-type icosahedral virus, similar morphologically to *Sericesthis* and *Tipula* iridescent viruses. Even after this clear resolution of its identity, some workers continued to treat *Pirhemocyton* as a protozoan, notably Arcay de Peraza and de La Roca (1971), whose cytochemical analysis of "*Pirhemocyton iguanae*" found both RNA and DNA present as well as respiratory activity.

Plate 72 Lizard erythrocytic virus (LEV) infections of erythrocytes. (**A**) LEV infections of *Chamaeleo dilepis*, **a–f**; and *Chamaeleo fischeri*, **g–l**, both from Tanzania. (**B**) LEV infections of *Calotes versicolor*, Burma, **a–f**; and *Takydromus tachydromoides*, Honshu, Japan, **g–l**.

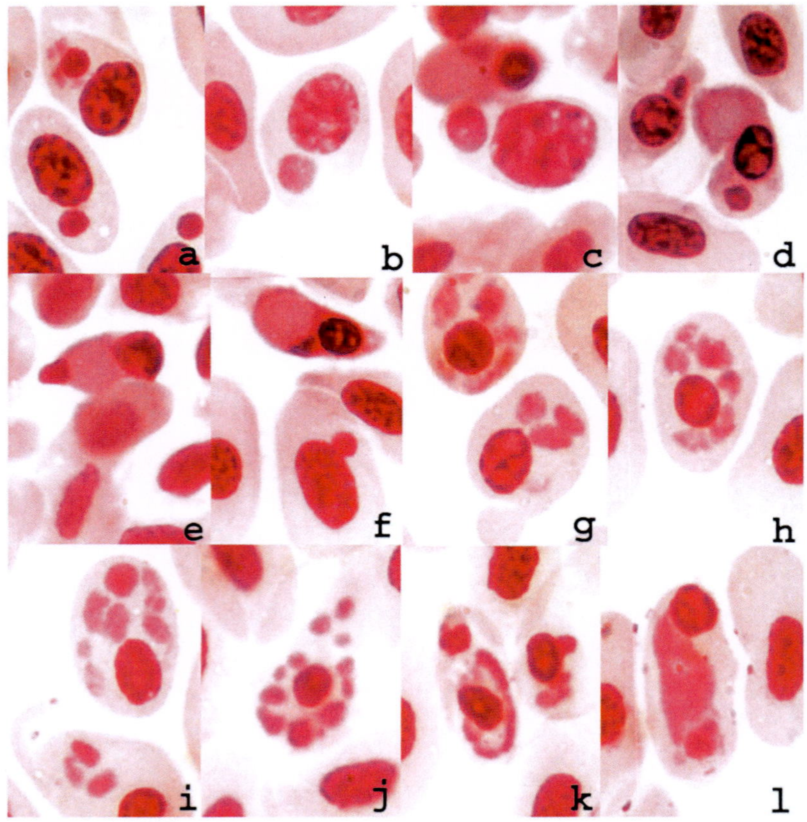

(A)

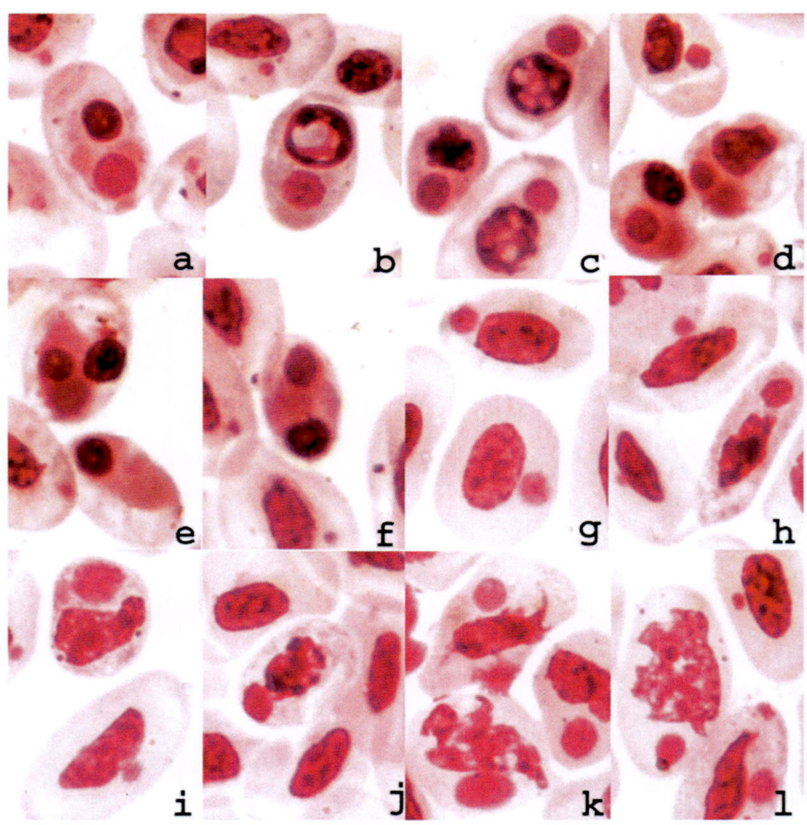

(B)

The ultrastructural study by Stehbens and Johnston (1966) proved unequivocally the viral structure of *Pirhemocyton*: Within the erythrocyte cytoplasm, polygonal bodies, mostly hexagonal in shape and 200 to 240 nm in diameter and most not exceeding 220 nm, were present. Two unit membranes, about 70 Å in width, were separated by a space 30- to 50-Å wide and formed a pellicle for particles.

> Many particles were filled with granular and fibrillar material, occasionally with a whorled arrangement. A denser central core (0.12–0.16 μ in diameter) was at times observed. Its density varied and irregular arrangements as well as target and cartwheel forms were seen. Within a few incompletely filled shells there were circular profiles 150–200 Å in diameter and branching strands of filamentous material. (Stehbens and Johnston, 1966)

> The *Pirhemocyton* body corresponded to an area of lighter density … than the cytoplasm of the affected erythrocyte. … These areas measured up to 3.5 μ in diameter and closely resembled the cytoplasmic assembly pool or factory of virus production. … They contained numerous incompletely formed polygonal bodies … entirely restricted to these sites. The matrix of the light density zones consisted of granular amorphous material in which were embedded fibrils less than 30 Å thick and a few circular or oval membranous particles.

Complicated polygonal particles were also present in the assembly pools, often near the periphery. Under light microscopy, more densely staining regions may be seen in *Pirhemocyton* inclusions, which corresponded to "dense amorphous filamentous aggregates … structurally similar to the core of the polygonal particles." The "albuminoid" bodies reported to be present in some *Pirhemocyton* species proved to be vacuoles of two types, one type containing "homogeneous material of moderate electron density" and the other apparently empty "except for small quantities of granular material of very light density, its limiting membrane at times being indefinite." There was no ultrastructural evidence of "their association with the *Pirhemocyton* particles."

The viral identity of *Pirhemocyton* was confirmed by Telford and Jacobson (1993) in the ultrastructure of infections from Tanzanian chameleons that presented very different forms under light microscopy (**Plate 72A**). Infections in *Chamaeleo dilepis* showed a very typical *Pirhemocyton* appearance: usually one, occasionally two, small, red-staining (acidophilic) inclusions varying among infections in size: 0.5 to 1.5 μm in infections with high viremias of 86–89%, to 2 to 3 μm in viremias of 6–13%. Albuminoid vacuoles, some 4 to 6 μm in diameter, were present in 11–41% of infected erythrocytes when the acidophilic bodies were larger in size, and were absent when the inclusions were 2.0 μm or smaller in diameter. Identity of the red-staining inclusions as assembly pools for viral particles in various stages was confirmed, with ultrastructure corresponding to that found by Stehbens and Johnston (1966), although the icosahedral particles were smaller, 140 to 180 nm. Only about 5% of the infected erythrocytes contained two inclusions. Prevalence in *C. dilepis* was 9 of 50 (18%).

In one of three *Bradypodion fischeri* collected in the Western Usambara Mountains, a very different parasitemia was present. In the positive chameleon when first examined, 17% of erythrocytes contained multiple, often large, irregularly shaped, red-staining inclusions. No albuminoid bodies were present in erythrocytes. From 2 to 30 inclusions were found in infected cells, with the size of inclusions 2 to 7 μm in diameter. Some inclusions were larger than the host cell nucleus, with inclusions larger when only two or three were present. Viremia declined on slides taken at weekly intervals, reaching less than 0.01% 25 days after capture, and thereafter the lizard was apparently negative. Under electron microscopy, multiple aggregations of viral particles were found, corresponding to the red-stained areas visible under light microscopy. The particles were tightly packed and ranged in size from 156 to 200 nm, larger than those in *C. dilepis*, with average sizes of 179 nm in *B. fischeri* and 159 nm in *C. dilepis*. The particles in both species were smaller than those in the Australian gecko examined by Stehbens and Johnston (1966). The appearance of the viral inclusions in *C. dilepis* under light microscopy was very similar to the "*Pirhemocyton chamaeleonis*" found by Brygoo (1963a) in Madagascan chameleons, particularly with respect to narrow, spindle-shaped infected erythrocytes present at peak of infection, which disappeared as infection declined.

Lizard erythrocytic virus (LEV) infections have been reported many times from lizards of the families Lacertidae, Gekkonidae, Agamidae, Chamaeleonidae, Cordylidae, Iguanidae, Scincidae, Teiidae, Varanidae, and Phrynosomatidae (M. R. L. Johnston, 1975; Telford, 1984b; Paperna and Alves de Matos, 1993). It appears to be cosmopolitan in its distribution.

Prevalence of LEV can be very high in lizard populations: 18 of 22 (82%) *Takydromus sexlineatus* from Thailand were infected. In the population study of *Takydromus tachydromoides* in Japan over a 3-year period (Telford, 1997b), annual prevalence varied from 2.2% to 6.1%, with maximum prevalence within a year occurring during July–August at 9.2–16.1%, in comparison to March–April at 1.5–4.3%, May–June at 0.8–5.6%, and September–October at 0.8–4.6%. There were no sexual differences in prevalence, but LEV ranged from 2.4% to 12.9% among adults, in comparison to 0–1.6% in immature lizards. There was a very low correlation between exposure to tick infestation and presence of LEV infection, .02, and infections were present in other localities with near absence of *Ixodes nipponensis* at similar levels to the main study area. The peak seasonal prevalence, July–August, occurred when prey arthropod populations were at maximum, and the high tick prevalence at that time was comprised predominantly by larvae, suggesting that ingestion of LEV-infected prey insects might be the source of infections in lizards.

Infection by intraperitoneal inoculation was easily accomplished in the Japanese lizards, with peak viremias varying 41.8–94.9% of erythrocytes. Mortality did occur in some experimental lizards, but others with massive infections recovered. Brygoo (1963a) infected Madagascan chameleons with LEV by inoculation of infected blood or by a broth of mosquitoes fed on infected lizards 6 hours earlier. Chameleons infected died in 16 days. Naturally acquired infections were thought to have lower mortality because viremias were lower, although experimentally infected *Lacerta monticola* and *L. schreiberi* could survive LEV infections restricted to erythrocytes, even with 98% of erythrocytes infected. If the LEV infection became systemic, spreading to leukocytes, mortality occurred (Alves de Matos et al., 2002).

Snake Erythrocytic Virus (**Plate 73**)

Red-staining erythrocytic inclusions associated with crystalloid bodies that are rectangular, square, hexagonal, or rod-like have been reported as *Toddia* from anurans and snakes since França (1911) described them from a West African anuran. Since demonstration of its viral nature (Smith et al., 1994), it has been known as snake erythrocyte virus (SEV). Until Marquardt and Yeager (1967) found *Toddia* infections in 4 of 163 (2.5%) *Agkistrodon piscivorous leucostoma* in Louisiana, the only records of *Toddia* were in frogs from Africa and Venezuela (M. R. L. Johnston, 1975), although Mathis and Leger (1911) apparently found it in a Vietnamese toad. De Sousa et al. (1973) reported 15 infections in four viperid and one colubrid species among 683 Brazilian snakes. In four *Bothrops* species (*B. moojeni, B. pradoi, B. lateralis, B. jararaca*), the crystalloid inclusions were mostly hexagonal in shape, in *Chironius flavolineatus* they were quadrangular, and in *B. alternatus* they were quadrangular or rectangular. Another species of *Toddia, T. lacertiliarum,* was described from the lizard *Iguana iguana* in Venezuela by Peraza et al. (1969). Peraza and de la Roca (1971) found that the crystalloid body in *T. lacertiliarum* was insoluble in methanol and stained pale pink in pyronine, in comparison with the globules of *Pirhemocyton* that dissolved in methanol. *Toddia* infections in Asiatic anurans were confirmed by Miyata and Miyagi (1977) in an Okinawan frog, *Rana holsti.*

Booker and Yongue (1982) reported *Toddia* inclusions containing square-shaped crystalloid bodies in the erythrocytes of *Nerodia s. sipedon* from Virginia and Michigan. Smith et al. (1994a) described *Toddia* infections in the same host species from Ontario by light and electron microscopy. The crystalloid bodies were smaller, about 2 μm, than in reports from other hosts and were square and bluish in color, membrane bound, with an interior "which had an unstructured consistency similar to the cytoplasm of the host erythrocyte." In addition to the crystalloids, two other types of inclusions were visible in erythrocytes: irregular translucent structures that occupied portions of the host cell cytoplasm and were often surrounded by a "diffuse, darker staining region," and a smaller, acidophilic body often adjacent to the translucent inclusions. Early stages of infection in an erythrocyte were apparently characterized by presence of these two different inclusions, while the crystalloid bodies typically appeared as the infection progressed, "associated with flocculent masses of cytoplasmic material." Infected cells became rounded, and their nuclei were pycnotic. Under electron microscopy, mature icosahedral viral particles (195–200 nm) were present in erythrocyte cytoplasm. These had an electron-dense nucleoid surrounded by "a less electron-dense hexagonal capsid." The particles were surrounded by a trilaminar membrane, and pairs or clusters of particles were often enclosed "by a common trilaminar envelope. … Virus particles appeared to form singly from a region on the periphery of a membrane-bound inclusion in the cell cytoplasm." A material "only slightly more granular" than the erythrocyte cytoplasm filled the interior of the "apparent viral assembly site." Infected cells showed concentric membranes near their periphery, contained "irregular membrane-bounded masses of amorphous material similar in consistency to the cell cytoplasm," and, rarely, showed extensive cytoplasmic destruction accompanied by nuclear degeneration. Prevalence of SEV in *Nerodia sipedon* was high in Ontario: 15 of 26 (58%) were infected, with viremias ranging from 0.1% to 28%.

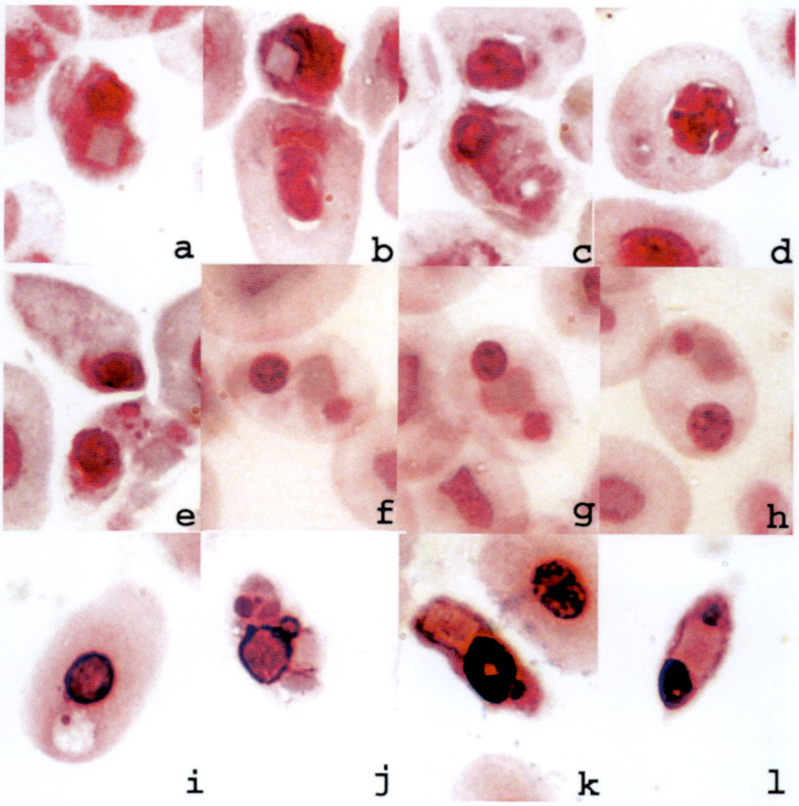

(A)

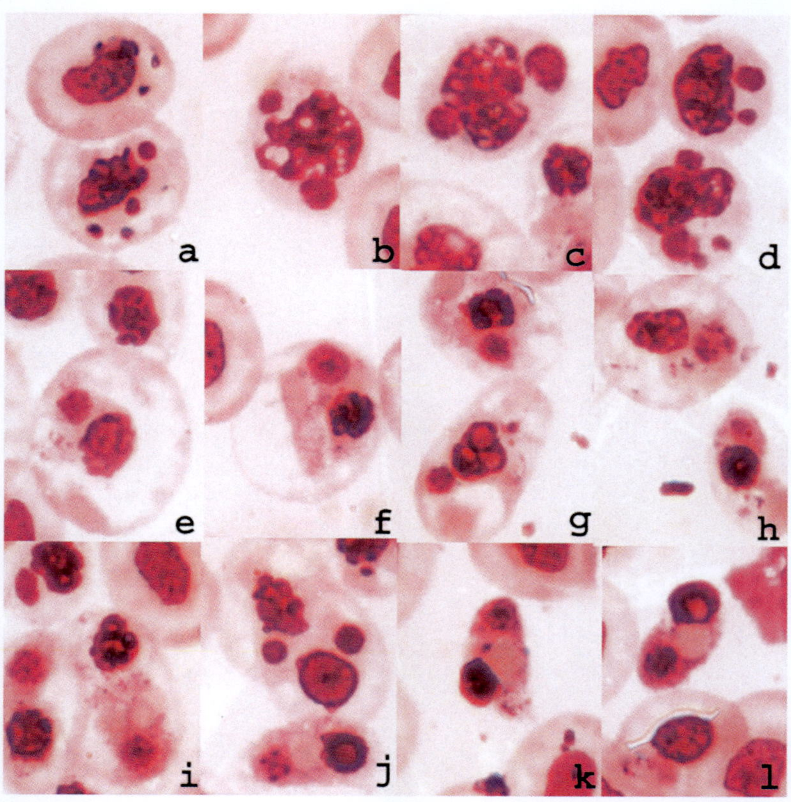

(B)

Plate 73 Snake erythrocytic virus (SEV) infections of erythrocytes. (**A**) SEV infections of *Nerodia sipedon,* Canada, **a–e**; *Oxybelis aeneus,* Honduras, **f–h**; and *Nerodia rhombifera,* Mexico, **i–l**. (**B**) SEV infection in *Thamnophis sauritus sackenii,* Florida.

LITERATURE CITED

Abdel-Ghaffar, F. A., A. R. Bashtar, and M. Shazly. 1991. Life cycle of *Hepatozoon seurati* com. Nov. 1. Gamogony and sporogony inside the vector *Culex pipiens molestus*. *J. Egyptian-German Soc. Zool.* 3: 211–226.

Adler, S. 1924. A note on *Plasmodium agamae* (Wenyon, 1908). *Ann. Trop. Med. Parasitol.* 18: 131–133.

Adler, S. and O. Theodor. 1925. A sporozoon of *Phlebotomus papatasi*. *Ann. Trop. Med. Parasitol.* 19: 309–313.

Adler, S. and O. Theodor. 1929. Observations on *Leishmania ceramodactyli* n. sp. *Trans. Roy. Soc. Trop. Med. Hyg.* 22: 343–356.

Adler, S. and O. Theodor. 1931. Investigations on Mediterranean kala azar. IV. Infection of sandflies with *Leishmania infantum*. Natural parasites of wild sandflies in Catania. Observations on *Trypanosoma ptyodactyli* Catouillard. *Proc. R. Soc. Lond., B*, 108: 481-493.

Adler, S. and O. Theodor. 1935. Investigations on Mediterranean Kala Azar. X.—A note on *Trypanosoma platydactyli* and *Leishmania tarentolae*. *Proc. R. Soc. Lond., B*, 116: 543-544.

Aikawa, M. 1971. Parasitological review. *Plasmodium*: the fine structure of malarial parasites. *Exp. Parasitol.* 30: 284–320.

Aikawa, M., C. G. Huff, and H. Sprinz. 1969. Comparative fine structure study of the gametocytes of avian, reptilian, and mammalian malarial parasites. *J. Ultrastruct. Res.* 26: 316–331.

Aikawa, M. and H. B. Jordan. 1968. Fine structure of a reptilian malarial parasite. *J. Parasitol.* 54: 1023–1033.

Alberts, A. C., M. L. Oliva, M. B. Worley, S. R. Telford, Jr., P. J. Morris, and D. L. Janssen. 1998. The need for prerelease screening in animal translocations: a case study of the Cuban iguana (*Cyclura nubila*). *Anim. Conserv.* 1: 165–172.

Alexeieff, A. N., V. M. Saf'janova, and A. B. Karapetian. 1975. On the viability of sandflies (Diptera, Psychodidae, Phlebotominae) after infection with promastigotes of different species of *Leishmania*. *Parazitologiya* 9: 271–277.

Allison, B. and S. S. Desser. 1981. Developmental stages of *Hepatozoon lygosomarum* (Dore, 1919) comb. n. (Protozoa: Haemogregarinidae), a parasite of a New Zealand skink, *Leiolopisma nigriplantare*. *J. Parasitol.* 67: 852–858.

Alvarez Calvo, J. A. 1975. Nueva especies de Hemococcidides en lacertidas españoles. *Cuad. Cienc. Biol.* 2–4: 207–222.

Alves de Matos, A. P., I. Paperna, and E. Crespo. 2002. Experimental infection of lacertids with lizard erythrocytic viruses. *Intervirology* 45: 150–159.

Andrushko, A. M. and G. S. Markov. 1955a. The new finding of *Leishmania* in reptiles of Central Asia. *Westn. Leningr. Univ.* 1: 51–59.

Andrushko, A. M. and G. S. Markov. 1955b. Incidence of infection with blood parasites in reptiles of various biotopes in the Kara-Kum desert. *Dokl. Akad. Nauk SSR* 104: 674-677.

Aragão, H. de B. and A. Neiva. 1909. A contribution to the study of the intraglobular parasites of lizards. Memorias Inst. Oswaldo Cruz 1: 44–50.

Arantes, J. B. and F. da Fonseca. 1931a. Pesquisas sobre trypanosomas. I. *Trypanosoma butantanense* sp. n. parasita da serpente *Ophis merremi* Wagler, 1824. *Mem. Inst. Butantan* 6: 215–222.

Arantes, J. B. and F. da Fonseca. 1931b. Pesquisas sobre trypanosomas. III. *Trypanosoma merremi* sp. n. parasita da serpente Ophis merremi Wagler, 1824. *Mem. Inst. Butantan* 6: 225–229.

Ashford, R., M. A. Bray, and W. A. Foster. 1973. Observations on *Trypanosoma boueti* in the skink *Mabuya striata* (Reptilia) and the sandfly *Sergentomyia bedfordi* in Ethiopia. *J. Zool.* 171: 285–292.

Austin, C. C. and S. L. Perkins. 2006. Parasites in a biodiversity hotspot: a survey of hematozoa and a molecular phylogenetic analysis of *Plasmodium* in New Guinea skinks. *J. Parasitol.* 92: 770–777.

Awerinzew, S. V. 1914. Beitrage zur Morphologie und Entwicklungs geschichte der Protozoen von Deutsch-Ost-Afrika. *Z. Mikrobiol.* 1:1–9.

Ayala, S. C. 1970a. Haemogregarine from sandfly infecting both lizards and snakes. *J. Parasitol.* 56: 387–388.

Ayala, S. C. 1970b. *Plasmodium mexicanum* in California: natural history and development in Phlebotomine sandflies (Diptera: Psychodidae). *J. Parasitol.* 56, Sect. II: 13 (abstr.).

Ayala, S. C. 1970c. Two new trypanosomes from California toads and lizards. *J. Protozool.* 17:370–373.

Ayala, S. C. 1970d. Protozoan Parasites Associated with California Sandflies. Ph.D. Dissertation, Univ. California, Berkeley, 99 pp.

Ayala, S. C. 1971a. Sporogony and experimental transmission of *Plasmodium mexicanum*. *J. Parasitol.* 57: 598–602.

Ayala, S. C. 1971b. Trypanosomes in wild California sandflies, and extrinsic stages of *Trypanosoma bufophlebotomi*. *J. Protozool.* 18: 433–436.

Ayala, S. C. 1973. The Phlebotomine sandfly-protozoan parasite community of central California grasslands. *Am. Midl. Nat.* 89: 266–280.

Ayala, S. C. 1975. Malaria and hemogregarines from lizards of the western Caribbean islands of San Andres and Providencia. *Rev. Inst. Med. Trop. São Paulo* 17: 218–224.

Ayala, S. C. 1977. Plasmodia of reptiles. In: J. P. Kreier (Ed.), *Parasitic Protozoa*, Academic Press, New York, Vol. 3, pp. 267–309.

Ayala, S. C. 1978. Checklist, host index, and annotated bibliography of *Plasmodium* from reptiles. *J. Protozool.* 25: 87–100.

Ayala, S. C., A. D'Alessandro, R. Mackenzie, and D. Angel. 1973. Hemoparasite infections in 830 wild animals from the eastern llanos of Colombia. *J. Parasitol.* 59: 52–59.

Ayala, S. C., C. Atkinson, and N. Vakalis. 1983. Two new trypanosomes from North American snakes. *J. Parasitol.* 69: 391–396.

Ayala, S. C. and P. E. Hertz. 1981. Malaria infection in *Anolis* lizards on Martinique, Lesser Antilles. *Rev. Inst. Med. Trop. São Paulo* 23: 12–17.

Ayala, S. C. and D. Lee. 1970. Saurian malaria: development of sporozoites in two species of phlebotomine sandflies. *Science* 167: 891–892.

Ayala, S. C. and J. G. McKay. 1971. *Trypanosoma gerrhonoti* n. sp. and extrinsic development of lizard trypanosomes in California sandflies. *J. Protozool.* 18: 430–433.

Ayala, S. C., E. Moreno, and R. Balaños. 1978. *Plasmodium pessoai* sp. n. from two Costa Rican snakes. *J. Parasitol.* 64: 330–335.

Ayala, S. C. and J. L. Spain. 1976. A population of *Plasmodium colombiense*, sp. n. in the iguanid lizard *Anolis auratus*. *J. Parasitol.* 62: 177–189.

Baker, J. R. 1961. Attempts to find the vector of the Plasmodiidae of the lizard *Agama agama agama* in Liberia. *Ann. Rept. Res. Activ. Liberian Inst. Am. Found. Trop. Med.* 28–40.

Ball, G. H. 1967a. Some blood sporozoa from East African reptiles. *J. Protozool.* 14: 198–210.

Ball, G. H. 1967b. Blood sporozoans from East African Amphibia. *J. Protozool.* 14: 521–527.

Ball, G. H. and J. Chao. 1973. The complete development of the sporogonous stages of *Hepatozoon rarefaciens* cultured in a *Culex pipiens* cell line. *J. Parasitol.* 59: 513–515.

Ball, G. H., J. Chao, and S. R. Telford, Jr. 1967. The life history of *Hepatozoon rarefaciens* (Sambon and Seligmann, 1907) from *Drymarchon corais* (Colubridae), and its experimental transfer to *Constrictor constrictor* (Boidae). *J. Parasitol.* 53: 897–909.

Ball, G. H., J. Chao, and S. R. Telford, Jr. 1969. *Hepatozoon fusifex* sp. n., a hemogregarine from *Boa constrictor* producing marked morphological changes in infected erythrocytes. *J. Parasitol.* 55: 800–813.

Ball, G. H. and S. N. Oda. 1971. Sexual stages in the life history of the hemogregarine *Hepatozoon rarefaciens* (Sambon and Seligmann, 1907). *J. Protozool.* 18: 697–700.

Ball, G. H. and G. Pringle. 1965. *Plasmodium fischeri* n. sp. from *Chamaeleo fischeri*. *J. Protozool.* 12: 479–482.

Ballinger, R. E. 1973. Comparative demography of two viviparous iguanid lizards (*Sceloporus poinsetti* and *Sceloporus jarrovi*). *Ecology* 54: 269–283.

Barta, J. R. and S. S. Desser. 1989. Development of *Babesioma stableri* (Dactylosomidae; Adeleina; Apicomplexa) in its leech vector (*Batrachobdella picta*) and the relationship of the dactylosomatids to the piroplasms of higher vertebrates. *J. Protozool.* 36: 241–253.

Bashtar, A. R. and F. A. Abdel-Ghaffar. 1987. Light microscope study of the life cycle of *Haemogregarina najae* infecting the snake *Naja nigricollis* (Elapidae, Proteroglypha, Squamata) from Egypt. *Proc. Zool. Soc. A. R. Egypt* 14: 33–39.

Bashtar, A. R., F. A. Abdel-Ghaffar, and M. A. Shazly. 1987. Developmental stages of *Hepatozoon gracilis* (Wenyon, 1909) comb. nov., a parasite of the Egyptian skink, *Mabuya quinquetaeniata*. *Parasitol. Res.* 73: 507–514.

Bashtar, A. R., F. A. Abdel-Ghaffar, and M. A. Shazly. 1991. Life cycle of *Hepatozoon mehlhorni* sp. nov. in the viper *Echis carinatus* and the mosquito *Culex pipiens*. *Parasitol. Res.* 77: 402–410.

Bashtar, A. R., R. Boulos, and H. Mehlhorn. 1984a. *Hepatozoon aegypti* nov. sp.: 1. Life cycle. *Z. Parasitenkd.* 70: 29–41.

Bashtar, A. R., F. A. Ghaffar, and H. Mehlhorn. 1984b. *Hepatozoon aegypti* nov. sp.: 2. Electron microscopic studies on the erythrocytic stages and schizogony inside the snake, *Spalaerosophis diadema*. *Z. Parasitenkd.* 70: 43–52.

Bashtar, A. R., F. A. Ghaffar, and H. Mehlhorn. 1984c. *Hepatozoon aegypti* nov. sp.: 3. Electron microscopic studies on the gamogony and sporogony inside the vector *Culex pipiens molestus*. *Z. Parasitenkd.* 70: 53–65.

Battelli, C. 1947. Su di un piroplasma della *Naia nigricollis* (*Aegyptianella carpani* n. sp.). *Riv. Parassitol.* 8: 205–212.

Bayon, H. 1915. Herpetomonidae found in *Scatophaga hottentota* and *Chamaeleon pumilus*. *Trans. R. Soc. South Afr.* 5: 61–63.

Belova, E. M. 1966. Detection of leptomonads in some species of lizards in the Turkmen S. S. R. *Med. Parazitol. (Mosk)* 35:281–282.

Beyer, T. V. and N. V. Sidorenko. 1984. *Karyolysus* sp. (Haemogregarinidae, Adeleida, Apicomplexa) host-parasite relationships of persisting stages. *J. Protozool.* 31: 513–517.

Blanc, G. and L. Ascione. 1958. Sur un parasite endoglobulaire du lézard *Eremias guttulatus olivieri* Augoin den la region de Marrakech (Moroc). *Bull. Soc. Pathol. Exot.* 51: 508–511.

Bonorris, J. S. and G. H. Ball. 1955. *Schellackia occidentalis* n. sp., a blood-inhabiting coccidian found in lizards in Southern California. *J. Protozool.* 2: 31–34.

Booden, T., J. Chao, and G. H. Ball. 1970. Transfer of *Hepatozoon* sp. from *Boa constrictor* to a lizard, *Anolis carolinensis*, by mosquito vectors. *J. Parasitol.* 56: 832–833.

Booker, K. A. and W. H. Yongue, Jr. 1982. *Cytotoddia* (=*Toddia*) infection of Serpentes and its incidence in two geographic areas. *Va. J. Sci.* 33: 11–21.

Borner, C. 1901. Untersuchungen über Hämosporidien. I. Ein Beitrag zur Kenntniss des Genus *Haemogregarina* Danilewsky. *Z. Wissenschaft. Zool.*, 69: 398–416.

Bouet, G. 1909a. Sur deux hémocytozoaires pigmentés des reptiles. *C. R. Soc. Biol. Paris* 66: 43–45.

Bouet, G. 1909b. Hemogrégarines de l'Afrique occidentale francaise. *C. R. Soc. Biol. Paris* 66: 741–743.

Boulard, Y., I. Landau, D. Baccam, G. Petit, and R. Lainson. 1987. Observations ultrastructurales sur les formes sanguines de Garniidés (*Garnia gonatodi, G. uranoscodoni* et *Fallisia effusa*) parasites de lézards sud-americains. *Eur. J. Protistol.* 22: 66–75.

Boulard, Y., G. Petit, I. Landau, A. F. Gomes, and L. Touratier. 1983. Sporogonie de *Plasmodium agamae* chez *Culicoides nubeculosus*: II—Observations ultrastructurales. *Protistologica* 19: 543–551.

Boulard, Y., I. Paperna, G. Petit, and I. Landau. 2001. Ultrastructure of developmental stages of *Hemolivia stellata* (Apicomplexa: Haemogregarinidae) in the cane toad *Bufo marinus* and in its vector tick *Amblyomma rotondatum*. *Parasitol. Res.* 87: 598–604.

Bradbury, P. C. and B. B. Galucci. 1971. The fine structure of differentiating merozoites of *Haemoproteus columbae* Kruse. *J. Protozool.* 18: 679–686.

Bray, R. S. 1959. On the parasitic protozoa of Liberia. II. The malaria parasites of agamid lizards. *J. Protozool.* 6: 13–18.

Breinl, A. 1913. Parasitic protozoa encountered in the blood of Australian native animals. *Rep. Aust. Inst. Trop. Med. 1911*: 30–38.

Bristovetzky, M. and I. Paperna. 1990. Life cycle and transmission of *Schellackia cf. agamae*, a parasite of the starred lizard *Agama stellio*. *Int. J. Parasitol.* 20: 883–892.

Bromwich, C. R. and J. J. Schall. 1986. Infection dynamics of *Plasmodium mexicanum*, a malarial parasite of lizards. *Ecology* 67: 1227–1235.

Brumpt, E. 1914. Le xenodiagnostic. Application au diagnostic de quelques infections parasitaires et en particulier à la Trypanosomiase de Chagas. *Bull. Soc. Pathol. Exot.* 7: 706-710.

Brumpt, E. 1938. Formes évolutives d'*Haemogregarina mauritanica* chez la tique *Hyalomma syriacum*. *Ann. Parasitol. Hum. Comp.* 16: 350–361.

Brumpt, E. and G. Lavier. 1935. Sur un piroplasmidé nouveau parasite de tortue, *Tunetella emydis* n. g., n. sp. *Ann. Parasitol. Hum. Comp.* 13: 544–550.

Brygoo, E. R. 1962. Un nouveau *Plasmodium* de caméléon *Haemamoeba robinsoni* n. sp. *Arch. Inst. Pasteur Madagascar* 30: 161–169.

Brygoo, E. R. 1963a. Contribution à la connaissance de la Parasitologie des Caméléons malgaches. Pt. 2. *Ann. Parasitol. Hum. Comp.* 38: 525-739.

Brygoo, E. R. 1963b. Hematozoaires de reptiles Malgaches. I. *Trypanosoma therezieni* n. sp. parasite des caméléons de Madagascar. *Arch. Inst. Pasteur Madagascar* 31: 133–141.

Brygoo, E. R. 1965a. Description de *Trypanosoma brazili* E. Brumpt 1914. *Arch. Inst. Pasteur Madagascar* 34: 41–46.

Brygoo, E. R. 1965b. Hematozoaires de reptiles Malgaches. III. Deux Trypanosomes nouveaux: *Trypanosoma haranti* n. sp. d'Ophidien et *Trypanosoma domerguei* n. sp. parasite d'Iguane. *Arch. Inst. Pasteur Madagascar* 34: 47–54.

Brygoo, E. R. 1966a. Hematozoaires de reptiles Malgaches. V. Notes sur les hemoparasites des Gerrhosauridés de Madagascar avec description d'un trypanosome nouveau: *Trypanosoma betschi* n. sp., parasite de Zonosaurus. *Arch. Inst. Pasteur Madagascar* 35: 165–170.

Brygoo, E. R. 1966b. Hematozoaires de reptiles Malgaches. VI. *Trypanosoma petteri* n. sp. parasite de *Phelsuma*. Liste des Trypanosomes de Reptiles. *Arch. Inst. Pasteur Madagascar* 35:171–184.

Campbell, H. W. 1973. Ecological observations on *Anolis lionotus* and *Anolis poecilopus* (Reptilia, Sauria) in Panama. *Am. Mus. Novit.* 2516, 29 pp.

Carini, A. 1909. Sur une Hemogregarine du Caiman latirostris Daud. *Bull. Soc. Pathol. Exot.* 2: 471–472.

Carini, A. 1941. Sobre um plasmodium endoglobular de um largarto. *Arq. Biol. São Paulo* 25: 46–47.

Carini, A. 1942. Sobre uma haemogregarina e um plasmodium da *Iguana iguana*. *Arq. Biol. São Paulo* 26: 6–7.

Carini, A. 1945. Considerações sobre o *Plasmodium rhadinurum* (Thompson and Huff 1944) da *Iguana*. *Arq. Biol. São Paulo* 29: 147–149.

Carini, A. and M. Rudolph. 1912. Sur quelques hématozoaires de lézards au Brésil. *Bull. Soc. Pathol. Exot.* 5: 592–595.

Carpano, M. 1939. Sui piroplasmidi dei cheloni e su una nuova species rinvenuta nelle tartarughe *Nuttallia guglielmi*. *Riv. Parassitol.* 3: 267–276.

Castellani, A. and A. Willey. 1904. Observations on the haematozoa of vertebrates in Ceylon. *Spolia Zeylanica* 2: 78–92.

Catouillard, G. 1909. Sur un trypanosome du gecko commun de Tunisie (*Platydactylus muralis*). *C. R. Soc. Biol.* 67: 804–805.

Challier, A. 1973. Ecologie de *Glossina palpalis gambiense* Vanderplank 1949 (Diptera Muscidae) en Savane d'Afrique Occidentale. Mem. ORSTom No. 64, thesis, Paris 1973.

Chao, J. and G. H. Ball. 1969. Transfer of *Hepatozoon rarefaciens* (Sambon and Seligmann, 1907) from the indigo snake to a gopher snake by a mosquito vector. *J. Parasitol.* 55: 681–682.

Chatton, E. and G. Blanc. 1914a. Existence de corps leishmaniformes dans les hématoblastes d'un Gecko barbaresque, *Tarentola mauritanica* L. Gunth. *C. R. Soc. Biol.* 77: 430–433.

Chatton, E. and G. Blanc. 1914b. Sur un hématozoaire nouveau, *Pirhemocyton tarentolae*, du Gecko, *Tarentola mauretanica*, et sur les altéracions globulaires qu'il determine. *C. R. Soc. Biol.* 77: 496–498.

Chatton, E. and G. Blanc. 1918a. Culture du trypanosome due gecko chez la punaise des lits. *Bull. Soc. Pathol. Exot.* 11: 387–391.

Chatton, E. and G. Blanc. 1918b. Le *Leptomonas* de la tarente dans un region indemne de bouton d'orient. Observations et experiences. *Bull. Soc. Pathol. Exot.* 11: 595–609.

Chatton, E. P. L. and E. Roubaud, 1913. Sporogonie d'une hémogrégarine chez une tsetsé (*Glossina palpalis* R. Desv.). *Bull. Soc. Pathol. Exot.* 6: 226–233.

Chiriac, E. and I. Steopoe. 1977. *Lankesterella baznosanui* nov. sp.—parasite endoglobulaire de *Lacerta vivipara*. *Rev. Roum. Biol. Ser. Zool.* 22: 139.

Christensen, H. A. and S. R. Telford, Jr. 1972. *Trypanosoma thecadactyli* sp. n. from forest geckoes in Panama, and its development in the sandfly *Lutzomyia trinidadensis* (Newstead) (Diptera, Psychodidae). *J. Protozool.* 19: 403–406.

Clark, A. M., B. W. Bowen, and L. C. Branch. 1999. Effects of natural habitat fragmentation on an endemic scrub lizard (*Sceloporus woodi*): an historical perspective based on a mitochondrial DNA gene genealogy. *Mol. Ecol.* 8: 1093–1104.

Commes, C. 1919. Hémogrégarine et trypanosome d'un chélonien (*Cinixys homeana*). *Bull. Soc. Pathol. Exot.* 12: 14–16.

Daly, J. J., M. Mayhue, H. H. Menna, and C. H. Calhoun. 1980. Virus-like particles associated with *Pirhemocyton* inclusion bodies in the erythrocytes of a water snake, *Nerodia erythrogaster flavigaster*. *J. Parasitol.* 66: 82–87.

Danilewskii, V. I. 1885. De Hämatozoën der Kaltblüter. *Arch. Mikr. Anat.* 24: 588–598.

Danilewskii, V. I. 1886. Recherches sur la parasitologie du sang. Les Hematozoaires des lezards (avec 2 planches). *Arch. slaves Biol.* 1: 364–396.

da Silva Cordeiro, N. S. 1977. Verificação do parasitismo do *Polychrus acutirostris* Spix, 1821 (Sauria: Iguanidae), novo hospedeiro natural do *Plasmodium (Carinamoeba) minasense* Carini and Rudolph, 1912. *Mem. Inst. Butantan* 40/41: 299–304.

David, A. 1929a. Recherche experimentale sur un hématozoaire du genre *Leishmania* (*L. agamae* A. David). These Doctorat en Sciences, Universite de Paris. Presses Universitaires de France, Paris, 1929. p. 54.

David, A. 1929b. Note préliminaire sur un *Leishmania* trouvé chez le lézard gris de la région de Tibériade (Basse-Galilée, Palestine). *Ann. Parasitol. Hum. Comp.* 7: 190–192.

De Biasi, P., S. B. Pessôa, and H. E. Belluomini. 1972. Novas observacoes sobre transmissão congênita de hemoatozoarios de serpentes peçonhentas viviparas. *Mem. Inst. Butantan* 36: 245–249.

De Biasi, P., S. B. Pessôa, G. Puorto, and W. Fernandes. 1975. Nota sobre formas evolutivas de *Trypanosoma* de serpentes em meio de cultura. *Mem. Inst. Butantan* 39: 85–101.

DeGiusti, D. L. 1965. *Haemoproteus metchnikovi* of chelonians, its tissue phase, peripheral parasitemia, host specificity and geographical distribution. In: *Progress in Parasitology 176*. Excerp. Med. Found., Int. Ser.: 91.

DeGiusti, D. L. and D. Dobrzechowski. 1974. The biology of the chelonian haemoproteid *Haemoproteus metchnikovi* in turtle hosts and in the intermediate host, *Chrysops callidus*. *Proc. 3rd Int. Congr. Parasitol.*, 1: 80–81 (abstr.).

DeGiusti, D. L., C. R. Sterling, and D. Dobrzechowsky. 1973. Transmission of the chelonian haemoproteid *Haemoproteus metchnikovi* by a tabanid fly *Chrysops callidus*. *Nature* 242(5392): 50–51.

De Mello, F. 1916. Quelques considérations sur les affinités zoologiques du genre "*Haemocystidium*" avec description d'une espèce nouvelle. *An. Fac. Med. Porto*, 3: 33–39.

De Mello, F. 1934. Contribution à l'étude du cycle évolutif des *Haemocystidium*. *Arq. Esc. Med. Cirurg. Nova Goa A*, 9: 1785–1799.

De Mello, I. F. 1932. Contribution a l'étude des hémogrégarinides des tortues indiennes. I. *Haemogregarina malabarica* n. sp. et son cycle évolutif chez *Emyda granosa*, subspecies *vittata* Peters. *Arq. Escola Med. Cirurg. Nova Goa A* 8: 1411–1425.

De Mello, I. F. and L. Lobo. 1938. On the morphology and identication of a trypanosome found in the blood of *Hemidactylus brooki* Gray. *Proc. Ind. Acad. Sci.* 7 B: 104–108.

De Sousa, M. A., P. de Biasi, and S. B. Pessôa. 1973. Protistas "incerta sedis" de ofidios do Brasil: *Toddia* Franca, 1912 e *Pirhemocyton* Chatton and Blanc, 1914—Estudo comparativo. *Mem. Inst. Oswaldo Cruz* 71: 443–468.

De Sousa, M. A. and E. G. Freire. 1975. Considerações sobre a natureza desenvolvimento e transmissão de *Cytamoeba bacterifera* Labbé, 1894 de *Leptodactylus ocellatus*. *Mem. Inst. Oswaldo Cruz* 73: 109–119.

Desser, S. S. 1973. A description of intraerythrocytic schizonts and gametocytes of a haemogregarine of the snapping turtle *Chelydra serpentina*. *Can. J. Zool.* 51: 431–432.

Desser, S. S. 1978. Morphological, cytochemical, and biochemical observations on the blood of the tuatara, *Sphenodon punctatus*. *New Zealand J. Zool.* 5: 503–508.

Desser, S. S. 1987. *Aegyptianella ranarum* sp. n. (Rickettsiales, Anaplasmataceae): ultrastructure and prevalence in frogs from Ontario. *J. Wildlife Dis.* 23: 52–59.

Desser, S. S. 1993. The Haemogregarinidae and Lankesterellidae. In: J. P. Kreier (Ed.), *Parasitic Protozoa*, 2nd ed., Academic Press, San Diego, CA, Vol. 4, pp. 247–272.

Desser, S. S. and J. R. Barta. 1989. The morphological features of *Aegyptianella bacterifera*: an intraerythrocytic rickettsia of frogs from Corsica. *J. Wildlife Dis.* 25: 313–318.

Desser, S., A. M. Fallis, and P. C. C. Garnham. 1968. Relapses in ducks chronically infected with *Leucocytozoon simondi* and *Parahaemoproteus nettionis*. *Can. J. Zool.* 46: 281–285.

Desser, S. S., H. Hong, and D. S. Martin. 1995. The life history, ultrastructure, and experimental transmission of *Hepatozoon catesbianae* n. comb., an apicomplexan parasite of the bullfrog, *Rana catesbeiana* and the mosquito, *Culex territans* in Algonquin Park, Ontario. *J. Parasitol.* 81: 212–222.

Desser, S. S., S. B. McIver, and A. Ryckman. 1973. *Culex territans* as a potential vector of *Trypanosoma rotatorium*. 1. Development of the flagellate in the mosquito. *J. Parasitol.* 59: 353–358.

Desser, S. S. and D. Yekutiel. 1986/87. Blood parasites of amphibians and reptiles in Israel. *Israel J. Zool.* 34: 77–90.

Diniz-Jose, A. P., O. Silva-Edilene, R. Lainson, and W. de Souza. 2000. The fine structure of *Garnia gonadati* and its association with the host cell. *Parasitol. Res.* 86: 971–977.

Dobell, C. C. 1910a. On some parasitic protozoa from Ceylon. *Spolia Zeylanica Colombo* 7: 65–87.

Dobell, C. C. 1910b. Contributions to the life-history of *Haemocystidium simondi* Castellani et Willey. *Festschrift 60 Geburst. R. Hertwigs* 2: 124–132.

Dollahon, N. R., A. M. Bauer, and A. D. Russell. 1996. Haemoparasite protozoans (Plasmodiidae and Haemogregarinidae) from a New Caledonian gekkonid lizard, *Rhacodactylus leachianus*. *Can. J. Zool.* 74: 2104–2108.

Doré, A. B. 1919. The occurrence of *Haemogregarina* in New Zealand lizards. *N. Z. J. Sci. Technol.* 2: 163–164.

Dunlap, K. D. and Schall, J. J. 1995. Hormonal alterations and reproductive inhibition in male fence lizards (*Sceloporus occidentalis*) infected with the malarial parasite *Plasmodium mexicanum*. *Physiol. Zool.* 68: 608–621.

Du Toit, P. J. 1937. A new piroplasm (*Sauroplasma thomasi* n. g., n. sp.) of a lizard (*Zonurus giganteus*, Smith). *Onderstepoort J. Vet. Sci. Anim. Ind.* 9: 289–299.

Dutton, J., J. T. Todd, and E. N. Tobey. 1907. Concerning certain parasitic protozoa observed in Africa. *Ann. Trop. Med. Parasitol.* 1: 287–348.

Ebraheem, M. H., N. A. Rashdan, H. M. Fayed, and F. H. Galal. 2006. Laboratory studies on the possibility of *Culex (Culex) pipiens* L. to harbour and transmit *Hepatozoon matruhensis* to the Egyptian snake *Psammophis schokari*. *J. Egypt. Soc. Parasitol.* 36: 241–250.

Edeson, J. F. B. and J. Himo. 1973. *Leishmania* sp. in the blood of a lizard (*Agama stellio*) from Lebanon. *Trans. R. Soc. Trop. Med. Hyg.* 67: 27.

Eisen, R. J. and D. F. DeNardo. 2000. Life history of a malaria parasite (*Plasmodium mexicanum*) in its host, the western fence lizard (*Sceloporus occidentalis*): host testosterone as a source of seasonal and among-host variation? *J. Parasitol.* 86: 1041–1045.

Elwasila, M. 1988. *Leishmania tarentolae* Wenyon, 1921 from the gecko *Tarentola annularis* in the Sudan. *Parasitol. Res.* 74: 591–592.

Fantham, H. B. and A. Porter. 1950. The endoparasites of certain South African snakes, together with some remarks on their structure and effects on their hosts. *Proc. Zool. Soc. Lond.* 120: 599–647.

Fialho, R. F. and J. J. Schall. 1995. Thermal ecology of a malarial parasite and its insect vector: consequences for the parasite's transmission success. *J. Anim. Ecol.* 64(5): 553–562.

Finkelman, S. and I. Paperna. 1998. *Schellackia calotesi* n. sp. from agamid lizards of the genus *Calotes* in Thailand. *Parasite* 5: 23–26.

Foley, H. and A. Catanei. 1925. L'hémogrégarine de la vipère à cornes (*Haemogregarina seurati*, Laveran et Pettit, 1911): ses formes de multiplication. *Bull. Soc. Pathol. Exot.* 18: 393–397.

França, C. 1911. Notes sur les hématozoaires de la Guinée Portugaise. *Arq. R. Inst. Bact. Camara Pestana* 3: 229–238.

Franchini, G. 1921. Sur les flagellés intestinaux du type *Herpetomonas* du *Chamaeleon vulgaris* et leur culture, et sur les flagellés du type *Herpetomonas* du *Chalcides (Gongylus) ocellatus* et *Tarentola mauritanica*. *Bull. Soc. Pathol. Exot.* 14: 641–645.

Frank, W. 1974. *Sauroplasma* and other so-called parasites of reptile blood corpuscles. *Proc. Int. Congr. Parasitol. III*, 3: 1666.

Gabaldon, A., G. Ulloa, and N. Zerpa. 1985. *Fallisia (Plasmodioides) neotropicalis* subgen. nov., sp. nov. from Venezuela. *Parasitology* 90: 217–225.

Garnham, P. C. C. 1950. Blood parasites of East African vertebrates with a brief description of exo-erythrocytic schizogony in *Plasmodium pitmani*. *Parasitology* 40: 328–337.

Garnham, P. C. C. 1954. A haemogregarine in *Argas brumpti. Riv. Parassitol.* 15: 425–435.

Garnham, P. C. C. 1965. *Plasmodium wenyoni* sp. nov., a malaria parasite of a Brazilian snake. *Trans. R. Soc. Trop. Med. Hyg.* 59: 277–279.

Garnham, P. C. C. 1966. *Malaria Parasites and Other Haemosporidia*, Blackwell Science, Oxford, U.K. 1116 pp.

Garnham, P. C. C. 1971. The genus *Leishmania. Bull. WHO* 44: 477–489.

Garnham, P. C. C. and A. J. Duggan. 1986. *Catalogue of the Garnham Collection of Malaria Parasites and Other Haemosporidia,* Press Syndicate, University of Cambridge, Cambridge, U.K. 191 pp.

Garnham, P. C. C. and B. O. L. Duke. 1953. Certain parasitic protozoa from the Gambia. *Trans. R. Soc. Trop. Med. Hyg.* 47: 7–8.

Garnham, P. C. C. and S. R. Telford, Jr. 1984. A new malaria parasite *Plasmodium (Sauramoeba) heischi* in skinks (*Mabuya striata*) from Nairobi, with a brief discussion of the distribution of malaria parasites in the family Scincidae. *J. Protozool.* 31: 518–521.

Goodwin, M. H. 1951. Observations on the natural occurrence of *Plasmodium floridense*, a saurian malaria parasite, in *Sceloporus undulatus undulatus. J. Natl. Malar. Soc.* 10: 57–67.

Goodwin, M. H. and T. K. Stapleton. 1952. The course of natural and induced infections of *Plasmodium floridense* Thompson and Huff in *Sceloporus undulatus undulatus* (Latreille). *Am. J. Trop. Med. Hyg.* 1: 773–783.

Gorgas Memorial Laboratory. 1964. Blood parasites of reptiles from Almirante. In: 35th Annual Report (1963), U.S. Government Printing Office, Washington, DC, pp. 14–15.

Gothe, R. and J. D. Kreier. 1977. *Aegyptianella, Eperythrozoon* and *Haemobartonella.* In: J. D. Kreier (Ed.), *Parasitic Protozoa*, Academic Press, New York, Vol. 4, pp. 251–294.

Greiner, E. C. and P. M. Daggett. 1973. A saurian *Plasmodium* in a Wyoming population of *Sceloporus undulatus. J. Herpetol.* 7: 303–304.

Grewal, M. S. 1955. On a new trypanosome from the blood of an African gecko, *Hemidactylus brookii angulatus* Gray, 1845. *Panjab Univ. Res. Bull.* 106: 269–281.

Guerrero, S. and S. C. Ayala. 1977. Hemoparasitos de algunos reptiles y anfibios de la selva amazonica del Peru. *Rev. Inst. Med. Trop. São Paulo* 19: 283–288.

Guerrero, S. and J. Pickering. 1984. Malaria survey in *Anolis* lizards of Puerto Rico. *J. Parasitol.* 70: 162–163.

Guerrero, S., C. Rodriguez, and S. C. Ayala. 1977. Hemoparásitos en lagartijas en la isla de Barro Colorado, Panamà. *Biotropica* 9: 118–123.

Hahn, C. W. 1909. The stages of *Haemogregarina stepanowi* Danilewsky found in the blood of turtles, with special reference to changes in the nucleus. *Arch. Protistenkunde* 17: 307–376.

Hardy, G. S. 1977. The New Zealand Scincidae (Reptilia: Lacertilia) a taxonomic and zoogeographic study. *N. Z. J. Zool.*, 4: 221–325.

Hartmann, M. and C. Chagas. 1910. Vorläufige Mitteilung über Untersuchungen au Schalangenhämogregarinen nebst Bemerkungen zu der vorstehenden Arbeit von E. Reichenow uber *Haemogregarina stepanowi. Arch. Protistenkunde* 20: 351–360.

Hegner, R. W. 1921. *Cytamoeba bactifera* in the red blood cells of the frog. *J. Parasitol.* 7: 157–161.

Heisch, R. B. 1958. On *Leishmania adleri* sp. nov. from lacertid lizards (*Latastia* sp.) in Kenya. *Trans. R. Soc. Trop. Med. Hyg.*, 52: 68-71.

Herban, N. L. and G. R. Coatney. 1969. *Plasmodium basilisci* from El Salvador-Honduras. *J. Parasitol.* 55: 225–226.

Hewitt, R. 1940. *Haemoproteus metchnikovi* (Simond, 1901) from the yellow-bellied terrapin (*Pseudemys elegans*). *J. Parasitol.* 26: 287–295.

Hoare, C. A. 1928. Studies on *Trypanosoma grayi* I. The effect of goat's blood. *Trans. R. Soc. Trop. Med. Hyg.* 22: 131–136.

Hoare, C. A. 1929. Studies on *Trypanosoma grayi.* 2. Experimental transmission to the crocodile. *Trans. R. Soc. Trop. Med. Hyg.* 23: 39–56.

Hoare, C. A. 1931a. The peritrophic membrane of *Glossina* and its bearing upon the life-cycle of *Trypanosoma grayi. Trans. R. Soc. Trop. Med. Hyg.* 25: 57–64.

Hoare, C. A. 1931b. Studies on *Trypanosoma grayi.* III. Life cycle in the tsetse-fly and in the crocodile. *Parasitology* 23: 449–484.

Hoare, C. A. 1932. On protozoal blood parasites collected in Uganda. With an account of the life cycle of the crocodile haemogregarine. *Parasitology* 24: 210–224.

Hoare, C. A. 1972. *The Trypanosomes of Mammals*, a Zoological Monograph, Blackwell Science, Oxford, U.K. 749 pp.

Hoge, A. R. 1965. Preliminary account on neotropical Crotalinae (Serpentes-Viperidae). *Mem. Inst. Butantan* 32: 109–184.

Huff, C. G. 1941. Saurian malaria. *J. Parasitol.* 27 (Suppl.): 29.

Huff, C. G. and D. F. Marchbank. 1953. Saurian malaria in Panama. *Nav. Med. Res. Inst. Rep.* 11: 509–516.

Huygelen, C. and J. Mortelmans. 1958. Au sujet d'un *Haemoproteus* chez la vipère *Bitis arietans. Bull. Soc. Pathol. Exot.* 51: 506–507.

Jacobson, E. R., J. M. Gaskin, and J. Mansell. 1989. Chlamydial infection in puff adders (*Bitis arietans*). *J. Zoo Wildlife Med.* 20: 364–369.

Jacobson, E. R. and S. R. Telford. 1990. Chlamydial and poxvirus infections of circulating monocytes of a flap-necked chamaeleon (*Chamaeleo dilepis*). *J. Wildlife Dis.* 26: 572–577.

Jefferson, N. H. 1965. The biology of *Trypanosoma chrysemydis* Roudabush and Coatney. Ph.D. dissertation, University of Minnesota. 110 pp.

Jimenez Ozete, H. 1981. Informe de *Leishmania* (Zoomastigophorea: Kinetoplastida) en *Anolis lucius* en Cuba. *Rev. Cubana Med. Trop.* 33: 106–109.

Johnston, M. R. L. 1975. Distribution of *Pirhemocyton* Chatton and Blanc and other, possibly related infections of poikilotherms. *J. Protozool.* 22: 529–535.

Johnston, T. H. 1907. A trypanosome found in the river Murray turtle (*Chelodina longicollis*). Austral. Med. Gaz. 26: 26.

Johnston, T. H. and J. B. Cleland. 1909. On a new melanin-producing haematozoon from an Australian tortoise. *J. Proc. R. Soc. N.S.W.* 43: 97–103.

Johnston, T. H. and J. B. Cleland. 1912. The haematozoa of Australian reptilia. No. 2. *Proc. Linn. Soc. N.S.W.* 36: 479–491.

Jordan, H. B. 1964. Lizard malaria in Georgia. *J. Protozool.* 11: 562–566.

Jordan, H. B. 1970a. The occurrence and development of *Plasmodium mexicanum* in the western fence lizard *Sceloporus occidentalis*. *J. Protozool.* 17: 86–89.

Jordan, H. B. 1970b. Asexual reproduction of *Plasmodium floridense* in *Anolis carolinensis*. *J. Parasitol.* 56, Sect. II, Pt. 1: 176 (abstr.).

Jordan, H. B. 1975. The effect of host constitution on the development of *Plasmodium floridense*. *J. Protozool.* 22: 241–244.

Jordan, H. B. 1986. Strain differences in *Plasmodium floridense*. *Soc. Protozool. 39th Ann. Mtg.*, Kingston, RI, abstr. 31, p. 13.

Jordan, H. B. and M. Friend. 1971. The occurrence of *Schellackia* and *Plasmodium* in two Georgia lizards. *J. Protozool.* 18: 485–487.

Joyeux, C. E. 1913. Note sur quelques protozaires sanguicoles et intestinaux observés en Guinée francaise. *Bull. Soc. Pathol. Exot.* 6: 612–615.

Khan, R. A., D. J. Forrester, T. M. Goodwin, and C.A. Ross. 1980. A haemogregarine from the American alligator (*Alligator mississippiensis*). *J. Parasitol.* 66: 324–328.

Killick-Kendrick, R. 1979. Biology of *Leishmania* in phlebotomine sandflies. In: W. H. R. Lumsden and D. A. Evans (Eds.), *Biology of the Kinetoplastida*, Academic Press, London, Vol. 2, pp. 396–460.

Killick-Kendrick, R., R. Lainson, J.-A. Rioux, and V. M. Saf'janova. 1986. The taxonomy of *Leishmania*-like parasites of reptiles. In: *Leishmania. Taxonomie et Phylogénèse. Applications Eco-epidéemiologiques*, Colloq. int. CNRS/INSERM, 1984, IMEEE, Montpellier, pp. 143–148.

Killick-Kendrick, R. and MacW. Warren. 1968. Primary exoerythrocytic schizonts of a mammalian *Plasmodium* as a source of gametocytes. *Nature* 220 (5163): 191–192.

Kimsey, R. B. 1992. Host association and the capacity of sand flies as vectors of lizard malaria in Panama. *Int. J. Parasitol.* 22: 657–664.

Klein, T. A. 1985. Development and transmission of saurian *Plasmodium* and *Schellackia* in bloodfeeding arthropods. Ph.D. dissertation, University of Florida, unpublished.

Klein, T. A., D. G. Young, E. C. Greiner, S. R. Telford, Jr., and J. F. Butler. 1988a. Development and experimental transmission of *Schellackia golvani* and *Schellackia occidentalis* by ingestion of infected blood-feeding arthropods. *Int. J. Parasitol.* 18: 259–267.

Klein, T. A., D. C. Akin, D. G. Young, S. R. Telford, Jr., and J. F. Butler. 1988b. Sporogony, development and ultrastructure of extrinsic stages of *Plasmodium mexicanum*. *Int. J. Parasitol.* 18: 463–476.

Klein, T. A., D. C. Akin, D. G. Young, and S. R. Telford, Jr. 1988c. Sporogony, development, and ultrastructure of *Plasmodium floridense* in *Culex erraticus*. *Int. J. Parasitol.* 18: 711–719.

Klein, T. A., D. G. Young, and S. R. Telford, Jr. 1987a. Vector incrimination and experimental transmission of *Plasmodium floridense* by bite of infected *Culex (Melanoconion) erraticus* (Diptera: Culicidae). *J. Am. Mosq. Control Assoc.* 3: 165–175.

Klein, T. A., D. G. Young, S. R. Telford, Jr., and R. Kimsey. 1987b. Experimental transmission of *Plasmodium mexicanum* by bites of infected *Lutzomyia vexator* (Diptera: Psychodidae). *J. Am. Mosq. Control Assoc.* 3: 154–164.

Koidzumi, M. 1910. On the development of *Haemogregarina* sp. parasitic in common Japanese tortoise, *Clemmys japonica*. *Arch. Protistenkunde* 18: 260–274.

Krasil'nikov, E. N. 1965. [Blood parasites of turtles of southeast Georgia.] *Zool. Z.* 44: 1454–1460.

Krotoski, W. A., W. E. Collins, J. R. Broderson, M. Warren, and D. M. Krotoski. 1981. The 48-hour exoerythrocytic stage of *Plasmodium cynomolgi bastianelli*. *Am. J. Trop. Med. Hyg.* 30: 31–37.

Krotoski, W. A., R. S. Bray, P. C. Garnham, et al. 1982a. Observations on early and late post-sporozoite tissue stages in primate malaria I. Discovery of a new latent form of *Plasmodium cynomolgi* (the hymnozoite) and failure to detect hepatic forms within the first 24 hours after infection. *Am. J. Trop. Med. Hyg.* 31: 24–35.

Krotoski, W. A., W. E. Collins, R. S. Bray, et al. 1982b. Demonstration of hypnozoites in sporozoite-transmitted *Plasmodium vivax* infection. *Am. J. Trop. Med. Hyg.* 31: 1291–1293.

Labbé, A. 1894. Recherches zoologique et biologiques sur les parasites endoglobulaires du sang des Vertébrés. Arch. Zool. Expér. Gen. 2: 55–258.

Labbé, A. 1899. Sporozoa. In: *Das Tierreich*, Berlin, 5. 180 pp.

Lacépède, B. G. E. L. 1789. Histoire naturelle des quadrupeds oviparas et des serpens. Tome second. Histoire naturelle des serpens. Paris, Imprimerie de Roi. 527 pp.

Lainson, R. 1977. *Trypanosoma cecili* n. sp., a parasite of the South American cayman, *Caiman crocodilus crocodilus* (Linnaeus, 1758)(Crocodilia: Alligatoridae). *Protozool.* 3: 87–93.

Lainson, R. 1995. *Progarnia archosauriae* nov. gen., nov. sp. (Haemospsorina: Garniidae), a blood parasite of *Caiman crocodilus crocodilus* (Archosauria: Crocodilia) and comments on the evolution of reptilian and avian haemosporines. *Parasitology* 110: 513–519.

Lainson, R., M. C. de Souza, and C. M. Franco. 2003. Hematozoan parasites of the lizard *Ameiva ameiva* (Teiidae) from Amazonian Brazil: a preliminary note. *Mem. Inst. Oswaldo Cruz*, 98: 1067–1070.

Lainson, R., I. Landau, and I. Paperna. 2001. *Plasmodium kentropyxi* n. sp. (Apicomplexa: Haemosporina: Plasmodiidae) and a *Plasmodium tropiduri*-like parasite in the lizard *Kentropyx calcarata* (Lacertilia: Teiidae) in north Brazil. *Parasite* 8: 107–113.

Lainson, R., I. Landau, and J. J. Shaw. 1971. On a new family of non-pigmented parasites in the blood of reptiles: Garniidae fam. nov. (Coccidiida: Haemosporidiidae). Some species of the new genus *Garnia. Int. J. Parasitol.* 1: 241–250.

Lainson, R., I. Landau, and J. J. Shaw. 1974a. Further parasites of the family Garniidae (Coccidiida: Haemosporidiidea) in Brazilian lizards. *Fallisia effusa* gen. nov., sp. nov. and *Fallisia modesta* gen. nov., sp. nov. *Parasitology* 68: 117–125.

Lainson, R., I. Landau, and J. J. Shaw. 1974b. Observations on non-pigmented haemosporidia of Brazilian lizards, including a new species of *Saurocytozoon* in *Mabuya mabouya* (Scincidae). *Parasitology* 69: 215–233.

Lainson, R. and Naiff, R. D. 1998. *Haemoproteus* (Apicomplexa: Haemoproteidae) of tortoises and turtles. *Proc. R. Soc. Lond. B. Biol. Sci.* 265 (1400): 941–949.

Lainson, R. and Naiff, R. D. 1999. *Garnia karyolytica* n. sp. (Apicomplexa: Haemosporina: Garniidae), a blood parasite of the Brazilian lizard *Thecodactylus rapicaudus* (Squamata: Gekkonidae). *Parasite* 6: 209–215.

Lainson, R., I. Paperna, and R. D. Naiff. 2003. Development of *Hepatozoon caimani* (Carini, 1909) Pessoa, de Biasi and de Souza, 1972 in the caiman *Caiman c. crocodilus*, the frog *Rana catesbeiana* and the mosquito *Culex fatigans. Mem. Inst. Oswaldo Cruz* 98: 103–113.

Lainson, R. and J. J. Shaw. 1969a. A new haemosporidian of lizards, *Saurocytozoon tupinambi* gen. nov., sp. nov. in *Tupinambis nigropunctatus* (Teiidae). *Parasitology* 59: 159–162.

Lainson, R. and J. J. Shaw. 1969b. New host records for *Plasmodium diploglossi, P. tropiduri* Aragão and Neiva 1909, and *P. cnemidophori* Carini, 1941. *Parasitology* 59: 163–170.

Lainson, R., J. J. Shaw, and I. Landau. 1975. Some blood parasites of the Brazilian lizards *Plica umbra* and *Uranoscodon superciliosa* (Iguanidae). *Parasitology* 70: 119–141.

Lainson, R., J. J. Shaw, and R. D. Ward. 1976. *Schellackia landaue* sp. nov. (Eimeriorina: Lankesterellidae) in the Brazilian lizard *Polychrus marmoratus* (Iguanidae): experimental transmission by *Culex pipiens fatigans. Parasitology* 72: 225–243.

Laird, M. 1950. *Haemogregarina tuatarae* sp. n. from the New Zealand Rhynchocephalian *Sphenodon punctatus* (Gray). *Proc. Zool. Soc. Lond.* 120: 529–533.

Laird, M. 1951. *Plasmodium lygosomae* n. sp., a parasite of a New Zealand skink, *Lygosoma moco* (Gray, 1839). *J. Parasitol.* 37: 183–189.

Laird, M. 1960. Malayan protozoa. 3. Saurian malaria parasites. *J. Protozool.* 7: 245–250.

Landau, I. 1973. Diversité des mécanismes assurant la pérennité de l'infection chez les sporozoaires coccidiomorphes. *Mem. Mus. Nat. Hist. Natur. A Zool.* 77: 1–62.

Landau, I., A. G. Chabaud, J. C. Michel, and E. R. Brygoo. 1970. Mise en évidence d'un double mode de transmission chez un *Hepatozoon* de Reptiles malgaches. *C. R. Acad. Sci.* 270: 2308–2310.

Landau, I., R. Lainson, Y. Boulard, Y. Michel, and J. J. Shaw. 1973. Developpement chez *Culex pipiens* de *Saurocytozoon tupinambi* (Sporozaire, Leucocytozoidae), parasite de lézards brésiliens. *C. R. Hebd. Acad. Sci.* 276: 2449–2452.

Landau, I., R. Lainson, Y. Boulard, and J. J. Shaw. 1974. Transmission au laboratoire et description de l'Hémogrégarine *Lainsonia legeri* n. sp. (Lankesterellidae) parasite de lézards brésiliens. *Ann. Parasitol. Hum. Comp.* 49: 253–263.

Landau, I., J. C. Michel, A. G. Chabaud, and E. R. Brygoo. 1972. Cycle biologique d'*Hepatozoon domergui*; discussion sur les caractères fondamentaux d'un cycle de Coccidie. *Z. Parasitenkd.* 38: 250–270.

Landau, I. and I. Paperna. 1997. The assignment of *Hepatozoon mauritanicum*, a tick-transmitted parasite of tortoises, to the genus *Hemolivia. Parasite* 4: 365–367.

Langmann, G. 1899. On haemosporidia in American reptiles and batrachians. *N.Y. Med. J.* 69:1–6.

Laveran, C. L. A. 1902. Sur quelques hémogrégarines des ophidiens. *C. R. Hebd. Seances Acad. Sci.* 135: 1036–1040.

Laveran, C. L. A. 1905. Sur hémamibe nouvelle de *Testudo pardalis. C. R. Soc. Seances Biol.* 59: 176–178.

Laveran, C. L. A. 1909. Au sujet des hémogrégarines de *Tupinambis teguixin* L. *C. R. Soc. Biol.* 67: 9–10.

Laveran, C. L. A. and G. Franchini. 1921. Des hématozoaires du gecko et spécialement de *Herpetomonas tarentolae.* Procédé simple de culture des *Herpetomonas. Bull. Soc. Pathol. Exot.* 14: 323–326.

Laveran, C. L. A. and A. Pettit. 1909. Contribution à l'étude des Hémogrégarines de *Lacerta ocellata* Daud. *Bull. Soc. Pathol. Exot.* 2: 295–298.

Laveran, C. L. A. and A. Pettit. 1911. Sur une hémogrégarine de la vipère à cornes. *C. R. Soc. Biol.* 70: 95–96.

Laveran, C. L. A. and A. T. Salimbeni. 1909. Sur une hémogrégarine de *Tupinambis teguixin*. *C. R. Acad. Sci.* 148: 132–134.

Lavier, G. 1943. L'evolution de la morphologie dans le genre *Trypanosoma*. *Ann. Parasitol. Hum. Comp.* 19: 168–202.

Lee, J. J., G. F. Leedale, and P. Bradbury (Eds.). 2000. *Illustrated Guide to the Protozoa*, 2nd ed., Society of Protozoologists, Blackwell, Oxford, U.K. 1475 pp.

Leger, M. 1918. Infection sanguine par *Leptomonas* chez un saurien. *C. R. Soc. Biol. Paris* 81: 772–774.

Leger, M. 1919. Hémogrégarine et *Plasmodium* du *Tupinambis nigropunctatus*. *Bull. Soc. Pathol. Exot.* 12: 217–220.

Leger, M. and A. Leger. 1914. Hématozoaires des reptiles du Haut-Sénégal-Niger. *Bull. Soc. Pathol. Exot.* 7: 488–493.

Leger, M. and P. Mouzels. 1917. *Plasmodium* d'*Iguana nudicollis*. *Bull. Soc. Pathol. Exot.* 10: 95–98.

Levine, N. D. 1980. Some corrections of coddidian (Apicomplexa: Protozoa) nomenclature. *J. Parasitol.* 66: 830–834.

Levine, N. D. 1984. Nomenclatural corrections and new taxa in the apicomplexan protozoa. *Trans. Am. Microsc. Soc.* 103: 195–204.

Lewis, D. H. 1975. Ultrastructural study of promastigotes of *Leishmania* from reptiles. *J. Protozool.* 22: 344–352.

Lewis, J. E. and E. D. Wagner. 1964. *Hepatozoon sauromali* sp. n., a hemogregarine from the chuckwalla (*Sauromalus* spp.) with notes on the life history. *J. Parasitol.* 50: 11–14.

Lloyd, L. L. and W. B. Johnson. 1924. The trypanosome infections of tsetse flies in northern Nigeria and a new method of estimation. *Bull. Entomol. Res.* 14: 265–288.

Lloyd, L. L., W. B. Johnson, W. A. Young, and H. Morrison. 1924. Second report of the tsetse fly investigation in the northern province of Nigeria. *Bull. Entomol. Res.* 15: 1–27.

Lowichik, A., N. L. Lanners, R. C. Lowrie, and N. E. Meiners. 1993. Gametogenesis and sporogony of *Hepatozoon mocassini* (Apicomplexa: Adeleina: Hepatozoidae) in an experimental mosquito host. *J. Eukaryot. Microbiol.* 40: 287–297.

Lowichik, A., L. Nystuen, and M. Palma. 1988. Lizard haemoparasites from "Los Tuxtlas" Tropical Biology Station, Veracruz, Mexico. *J. Parasitol.* 74: 885–888.

Lowichik, A. and R. G. Yaeger. 1987. Ecological aspects of snake hemogregarine infections from two habitats in southeastern Louisiana. *J. Parasitol.* 73: 1109–1115.

Macfie, J. W. S. 1914. Notes on some blood parasites collected in Nigeria. *Ann. Trop. Med. Parasitol.* 8: 439–463.

Macfie, J. W. S. 1916. The results of dissection of tsetse flies at Accra. *Report Accra Laboratory for 1915*: 49.

Macfie, J. W. S. 1919. Two parasites of *Naja nigricollis*. *Ann. Trop. Med. Parasitol.* 13: 23–30.

Mackerras, M. J. 1961a. The haematozoa of Australian reptiles. *Aust. J. Zool.* 9: 61–122.

Mackerras, M. J. 1961b. The haematozoa of Australian frogs and fish. *Aust. J. Zool.* 9: 123–139.

Mackerras, M. J. 1962. The life history of a *Hepatozoon* (Sporozoa: Adeleidea) of varanid lizards in Australia. *Aust. J. Zool.* 10: 35–44.

Mackie, F. P., B. M. Das Gupta, and C. S. Swaminath. 1923. Progress report on kala-azar. *Ind. J. Med. Res.* 11: 591–599.

Mahrt, J. L. 1979. Hematozoa of lizards from southeastern Arizona and Isla San Pedro Nolasco, Gulf of California, Mexico. *J. Parasitol.* 65: 972–975.

Mahrt, J. L. 1987. Lizard malaria in Arizona: island biogeography of *Plasmodium chiricahuae* and *Sceloporus jarrovi*. *Southwest. Nat.* 32: 347–350.

Mahrt, J. L. 1989. Prevalence of malaria in populations of *Sceloporus jarrovi* (Reptilia: Iguanidae) in southeastern Arizona. *Southwest. Nat.* 34: 436–438.

Malagon, F. and M. Salmeron. 1988. *Plasmodium (Sauramoeba) pelaezi* n. sp., a malaria parasite of the Mexican iguanid lizard *Urosaurus bicarinatus bicarinatus* (Dumeril, 1856) (Sauria: Iguanidae). *Syst. Parasitol.* 12: 141–148.

Manawadu, B. R. 1972. A new saurian malaria parasite *Plasmodium clelandi* sp. n. from Ceylon. *J. Protozool.* 19: 587–589.

Marinkelle, C. J. and S. K. Al-Mahdawi. 1981. *Trypanosoma turcici* sp. n. (Trypanosomatidae) from the Mediterranean gecko *Hemidactylus turcicus* from Iraq. *Bull. Nat. Hist. Res. Cent. Univ. Baghdad* 7: 73–79.

Maroli, M., M. Grammicia, L. Gradoni, P. D. Ready, D. F. Small, and C. Aquino. 1983. Natural infections of phlebotomine sandflies with Trypanosomatidae in central and south Italy. *Trans. R. Soc. Trop. Med. Hyg.* 82: 227–228.

Maroli, M., E. Pozio, L. Gradoni, and M. Gramiccia. 1987. Sandfly species (Diptera, Psychodidae) colonizing new formed rock crevices in Apulia (Italy). *Ann. Parasitol. Hum. Comp.* 62: 185–187.

Marquardt, W. C. 1966. Haemogregarines and *Haemoproteus* in some reptiles in southern Illinois. *J. Parasitol.* 52: 823–824.

Marquardt, W. C. and R. G. Yaeger. 1967. The structure and taxonomic status of *Toddia* from the cottonmouth snake *Agkistrodon piscivorus leucostoma*. *J. Protozool.* 14: 726–731.

Martin, G. 1907. Sur un trypanosome de Saurien (*Trypan. boueti* n. sp.). *C. R. Soc. Biol.* 62: 594–596.

Mathis, C. and M. Leger. 1909. Sur un nouveau trypanosome des Serpents du Tonkin. *C. R. Soc. Biol.* 67: 572–574.

Mathis, C. and M. Leger. 1911. Recherches de parasitologie et de pathologie humaines et animales au Tonkin. Paris. 451 pp.

Maximow, A. A. and W. Bloom. 1953. *A Textbook of Histology.* 6th ed., reprinted 1953. W. B. Saunders Company, Philadelphia. 616 pp.

McKay, J. G. 1936. Unpublished M.S. thesis, Department of Zoology, University of California, Berkeley.

McMillan, B. 1965. Leishmaniasis in the Sudan Republic. 22. *Leishmania hoogstraali* sp. n. in the gecko. *J. Parasitol.* 51: 336–339.

Michel, J.-C. 1973. *Hepatozoon mauritanicum* (Et. et Ed. Sergent, 1904) n. comb., parasite de *Testudo graeca*: redescription de la sporogonie chez *Hyalomma aegyptium* et de la schizogonie tissulaire d'après le matériel d'E. Brumpt. *Ann. Parasitol. Hum. Comp.* 48: 11–21.

Miller, W.W. 1908. *Hepatozoon perniciosum* (n.g., n.sp.); a haemogregarine pathogenic for white rats, with a description of the sexual cycle in the intermediate host, a mite (*Laelaps echidninus*). *Hyg. Lab. Bull.*, 46: 1–51.

Minter-Goedbloed, E., C. J. Leake, D. M. Minter, J. McNamara, C. Kimber, P. Bastien, D. A. Evans, and D. LeRay. 1993. *Trypanosoma varani* and *T. grayi*-like trypanosomes: development in vitro and in insect hosts. *Parasitol. Res.* 79: 329–333.

Misra, K. K. 1976. *Haemogregarina gangetica*, a new name for *Haemogregarina simondi* of a river turtle *Trionyx gangeticus*. *Acta Protozool.* 15: 21–22.

Misra, K. K., N. C. Nandi, S. Raut, and A. Choudhury. 1974. *Haemogregarina simondi* sp.n., a new haemogregarine from a river turtle, *Trionyx gangeticus* Cuvier. *Acta Protozool.* 12: 345–350.

Mitchell, O. G. 1965. *Hepatozoon sauromali* from the chuckwalla, *Sauromalus obesus*, in California. *J. Parasitol.* 51: 36.

Miyata, A. 1977. *Trypanosoma ryukyuense* n. sp. (Protozoa: Trypanosomatidae) detected from *Eublepharis kuroiwae kuroiwae* (Namie, 1912) (Reptilia: Gekkonidae) in Okinawa Island. *Trop. Med.* 19: 157–167.

Miyata, A. and I. Miyagi. 1977. *Toddia* detected from *Rana (Babina) holsti* Boulenger, 1892, in Okinawa Island. *Trop. Med.* 19: 123–128.

Miyata, A., I. Miyagi, and T. Toma. 1987. Detection of haemoprotozoa from the cold-blooded animals collected in the Ryukyu Islands. *Biol. Mag. Okinawa* 25: 21–34.

Mohiuddin, A., R. A. Pal, and A. A. Warsi. 1967. *Haemogregarina echisi* n. sp. from the saw-scaled viper *Echis carinatus* of the Sind Region of West Pakistan. *J. Protozool.* 14: 255–259.

Molyneux, D. H. 1977a. The attachment of *Trypanosoma grayi* in the hindgut of *Glossina*. *Protozool.* 3: 83–86.

Molyneux, D. H. 1977b. Vector relationships in the Trypanosomatidae. *Advances in Parasitol.* 15: 1–82.

Moore, J. and R. E. Sinden. 1973. The fine structure of *Plasmodium mexicanum*. *Trans. R. Soc. Trop. Med. Hyg.* 67: 13–14 (abstr.).

Moulder, J. W. 1984. Chlamydiaceae. In: N. R. Krieg and J. G. Holt (Eds.), *Bergey's Manual of Systematic Bacteriology*, William and Wilkins, Baltimore, MD, pp. 729–739.

Nadler, S. A. and J. H. Miller. 1984. A redescription of *Hepatozoon moccasini* (Laveran, 1902) n. comb. from *Agkistrodon piscivorus leucostoma* Troost, 1836. *J. Protozool.* 31: 321–324.

Nadler, S. A. and J. H. Miller. 1985. Fine structure of *Hepatozoon moccasini* (Apicomplexa, Eucoccidiorida) gamonts and modifications of the infected erythrocyte plasmalemma. *J. Protozool.* 32: 275–279.

Newcomer, C. E., M. R. Anver, J. L. Simons, W. Wilcke, and G. W. Nace. 1982. Spontaneous and experimental infections of *Xenopus laevis* with *Chlamydia psittaci*. *Lab. Anim. Sci.* 32: 680–686.

Nicolle, C., G. R. Blanc, and M. Langeron. 1920. Recherches expérimentales sur le rôle du gecko (*Tarentola mauritanica*) dans l'étiologie du bouton d'orient. Mission de Tamerza (octobre 1919). Note préliminaire. *Bull. Soc. Pathol. Exot.* 13: 508–51.

Noller, W. 1920. Kleine Beobachtungen an parasitischen Protozoen. (Zugleich vorläufige Mitteilung über die Befruchtung und Sporogonie von *Lankesterella minima* Chaussat. *Arch. Protistenkunde* 41: 169–189.

Novy, F. G. 1906. The trypanosomes of tsetse flies. *J. Infect. Dis.* 3: 394–411.

Oda, S. N., Chao, J. and G. H. Ball. 1971. Additional instances of transfer of reptile hemogregarines to foreign hosts. *J. Parasitol.* 57: 1377–1378.

Omran, L. A. A., A. M. Mandour, and M. K. El-Naffar. 1981. *Schellackia agami*, a probably new species from the reptile *Agama stellio* captured in Assiut locality. *J. Egypt. Soc. Parasitol.* 11: 429–431.

Osimani, J. J. 1942. *Haemogregarina triatomae* n. sp. from a South American lizard *Tupinambis teguixin* transmitted by the Reduviid *Triatoma rubrovaria*. *J. Parasitol.* 28: 147–154.

Ostrovska, K. and I. Paperna. 1987a. Fine structure of gamont stages of *Schellackia cf. agamae* (Lankesterellidae, Eucoccidia) from the starred lizard *Agama stellio*. *Parasitol. Res* 73: 492–499.

Ostrovska, K. and I. Paperna. 1987b. Ultrastructural studies on the merogony of *Schellackia cf. agamae* (Lankesterellidae, Apicomplexa) from the starred lizard *Agama stellio*. *Ann. Parasitol. Hum. Comp.* 62: 380–386.

Ovezmukhammedov, A. 1987. [*Protistofauna of Reptiles*], A. Berdiev (Ed.), Ashkabad Turkmen S.S.R. 373 pp.

Ovezmukhammedov, A. 1991. [*Leishmania* of Reptiles], Izdatelstvo, Yilim, Ashkabad. Turkmen S.S.R. 355 pp.

Ovezmukhammedov, A. and V. M. Saf'janova. 1989. [Certain problems of taxonomy of *Leishmania* from reptiles.] *Parazitologiia* 23: 334–343.

Paperna, I. 1989. Developmental cycle of chelonian haemogregarines in leeches with extra-intestinal multiple sporozoite oocysts and a note on the blood stages in the chelonian hosts. *Dis. Aquat. Org.* 7: 149–153.

Paperna, I. 1993. Electron microscopic study of *Schellackia cf. agamae* sporozoite infection in mosquitoes. *Int. J. Parasitol.* 23: 187–190.

Paperna, I. and A. P. Alves de Matos. 1993. Erythrocytic viral infections of lizards and frogs: new hosts, geographical locations and description of the infection process. *Ann. Parasitol. Hum. Comp.* 68: 11–23.

Paperna, I. and Y. Boulard. 1990. Fine structure of a macrogametocyte of *Fallisia copemani* Paperna and Landau 1990 (Haemosporidia: Garniidae). *Ann. Parasitol. Hum. Comp.* 65: 214–217.

Paperna, I. and Y. Boulard. 2000. Fine structure of gametocytes in five species of *Haemoproteus* (Haemosporidia) from geckos and agamid lizards. *Zoosystema* 22: 443–457.

Paperna, I., Y. Boulard, S. H. Hering-Hagenbeck, and I. Landau. 2001. Description and ultrastructure of *Leishmania zuckermani* n. sp. amastigotes detected within the erythrocytes of the South African gecko *Pachydactylus turneri* Gray, 1864. *Parasite* 8: 349–35.

Paperna, I. and S. Finkelman. 1996. *Schellackia ptyodactyli* sp. n. of the fan-footed gecko *Ptyodactylus hasselquistii* from the rift escarpment of the lower Jordan Valley. *Folia Parasitol.* 43: 161–172.

Paperna, I. and S. Finkelman. 1999. Exoerythrocytic stage of *Haemoproteus ptyodactyli* in the spleen of its host, the gecko *Ptyodactylus hasselquistii*. *Folia Parasitol.* 46: 154.

Paperna, I., T. Kremer-Mecabell, and S. Finkelman. 2002. *Hepatozoon kisrae* n. sp. infecting the lizard *Agama stellio* is transmitted by the tick *Hyalomma cf. aegyptium*. *Parasite* 9: 17–27.

Paperna, I. and R. Lainson. 2002. Fine structure of erythrocytic stages of a *Plasmodium tropiduri*-like malaria parasite found in the lizard *Kentropyx calcarata* (Teiidae) from north Brazil. *Folia Parasitol.* 49: 2–8.

Paperna, I. and R. Lainson. 2003. Ultrastructural studies on the sporogony of *Hepatozoon* spp. in *Culex quinquefasciatus* Say, 1823 fed on infected *Caiman crocodilus* and *Boa constrictor* from northern Brazil. *Parasitology* 127: 147–154.

Paperna, I. and R. Lainson. 2004. *Hepatozoon cf. tarzii* (Sambon and Seligman, 1907) infection in the snake *Boa constrictor constrictor* from north Brazil: transmission to the mosquito *Culex quinquefasciatus* and the lizard *Tropidurus torquatus*. *Parasite* 11: 175–181.

Paperna, I. and I. Landau. 1990a. *Fallisia copemani* n. sp. (Haemosporidia: Garniidae) from the Australian skink *Carlia rhomboidalis*. *Ann. Parasitol. Hum. Comp.* 65: 16–21.

Paperna, I. and I. Landau. 1990b. *Billbraya australis* n. gen. n. sp. from the south Australian gecko *Phyllodactylus marmoratus*. *Ann. Parasitol. Hum. Comp.* 65: 170–176.

Paperna, I. and I. Landau. 1991. *Haemoproteus* (Haemosporidia) of lizards. *Bull. Mus. Natl. Hist. Nat. Paris, 4* 13: 309–349.

Parrot, L. 1927. Sur un parasite intraglobulaire pigmenté de *Tarentola mauritanica* (L). *Arch. Inst. Pasteur Alger.* 5: 1–8.

Parrot, L. 1934a. L'évolution de Leishmania tarentolae Wenyon chez Phlebotomus minutus Rond. *Bull. Soc. Pathol. Exot.* 27: 839–843.

Parrot, L. 1934b. Evolution d'un hématozoaire de gecko (*Leishmania tarentolae*) chez un moucheron piqueur du groupe des phlébotomes (*Phlebotomus minutus*). *C. R. Acad. Sci. Paris* 199: 1073–1074.

Parrot, L. 1935. Nouvelles recherches sur l'évolution de *Leishmania tarentolae* chez *Phlebotomus minutus* Rondani. *Bull. Soc. Pathol. Exot.* 28: 958–960.

Parrot, L. 1937. Sur la durée de l'infection du Gecko *Tarentola mauritanica* par *Leishmania tarentolae* Wenyon. *C. R. Soc. Biol.* 124: 337–338.

Parrot, L. and H. Foley. 1939. Sur la fréquence de la leishmaniose du gecko dans le Sud Oranais. *Arch. Inst. Pasteur Alger.* 17: 231–232.

Paterson, W. B. and S. S. Desser. 1976. Observations of *Haemogregarina balli* sp. n. from the common snapping turtle, *Chelydra serpentina*. *J. Protozool.* 23: 294–301.

Paterson, W. B., S. S. Desser, and J. R. Barta. 1988. Ultrastructural features of the apical complex, pellicle, and membranes investing the gamonts of *Haemogregarina magna* (Apicomplexa: Adeleina). *J. Protozool.* 35: 73–80.

Peláez, D. 1967. Estudios sobre hematozoarios. XIII. Un nuevo *Plasmodium* de *Ameiva* en México. *Ciencia* 25: 121–129.

Peláez, D., A. Barrera, F. De la Jara, and R. Perez-Reyes. 1951. Estudios sobre hematozoarios. II. Interés de las investigaciones sobre el paludismo en los animales. *Rev. Palud. Med. Trop.* 3: 59–76.

Peláez, D. and R. Perez-Reyes. 1952. Estudios sobre hematozoarios. III. Las especies americanas del género *Plasmodium* en reptiles. *Rev. Palud. Med. Trop.* 3/4: 137–160.

Peláez, D. and R. Perez-Reyes. 1959. Estudios sobre hematozoarios. IX. Una nueva especies de *Plasmodium* de reptiles mexicanos, parasita de *Basiliscus vittatus*. *Rev. Latinoam. Microbiol.* 2: 41–50.

Peláez, D., R. Perez-Reyes and A. Barrera. 1948. Estudios sobre hematozoarios. I. *Plasmodium mexicanum* Thompson and Huff, 1944 en sus huéspedes naturales. *Anal. Esc. Nac. Cienc. Biol.* 5: 197–215.

Peláez, D. and F. Streber. 1955. Estudios sobre hematozoarios. V. *Trypanosoma serveti nov. sp.* parásito de un *Sceloporus* de México. *Anal. Esc. Nac. Cienc. Biol.* 8: 147-152.

Peraza, L. A. de, P. Nasir, and M. T. Diaz. 1969. The "paranuclear corpuscles" in poikilothermal vertebrates: Description of a new species of *Toddia* from *Iguana iguana* in Venezuela. *Acta Biol. Venezuélica.* 7: 191–199.

Peraza, L. A. de and C. D.-M. de La Roca. 1971. The "paranuclear corpuscles" in poikilothermic vertebrates: description of a new species of *Pirhemocyton* in *Iguana iguana* of Venezuela, with remarks on the nature of these organisms, and their relation to allied parasites. *Mem. Inst. Oswaldo Cruz,* 69: 57–67.

Perkins, S. L. 2000a. Phylogeography of Caribbean lizard malaria: tracing the history of vector-borne parasites. *J. Evol. Biol.* 14: 22–33.

Perkins, S. L. 2000b. Species concepts and malaria parasites: detecting a cryptic species of Plasmodium. *Proc. R. Soc. London B,* 267: 2345–2350.

Perkins, S. L. and J. J. Schall. 2002. A molecular phylogeny of malarial parasites recovered from cytochrome b gene sequences. *J. Parasitol.* 88: 972–978.

Pessôa, S. B. 1967. Notas sôbre hemogregarinas de serpentes Brasileiras. II. Hemogregarinas de algumas especies de serpentes das familias Colubridae e Boidae. *Rev. Brasil Biol.* 27: 49–56.

Pessôa, S. B. 1968. Notas sôbre hemogregarinas de serpents Brasileiras. VI. *Rev. Brasil. Biol.* 28: 71–76.

Pessôa, S. B., H. E. Belluomini, P. de Biasi, and D. M. de Sousa. 1971. Notas sôbre hemogregarinas de serpents Brasileiras. XV. Esporogonia da hemogregarina da Bothrops moojeni Hoge, 1965 no *Culex (Culex) dolosus,* (L. Arribálzaga, 1891). *Arquiv. Inst. Biol. São Paulo* 38: 253–258.

Pessôa, S. B. and J. Cavalheiro. 1969a. Notas sôbre Hemogregarinas de serpentes Brasileiras. VII. *Rev. Bras. Biol.* 29: 351–354.

Pessôa, S. B. and J. Cavalheiro. 1969b. Notas sôbre Hemogregarinas de serpentes Brasileiras. VIII. Sôbre a evolução da *"Haemogregarina miliaris"* na sanguessuga *"Haementeria lutzi."* *Rev. Bras. Biol.* 29: 451–458.

Pessôa, S. B. and J. Cavalheiro. 1969c. Notas sôbre Hemogregarinas de serpentes Brasileiras. IX. Sôbre a haemogregarina da *Helicops carinicauda* (Wied.). *Rev. Goiana Med.* 15: 161–168.

Pessôa, S. B., J. Cavalheiro, and D. M. de Souza. 1970. Notas sobre Hemogregarinas de serpentes Brasileiras. XIII. Evolução esporogonica da hemogregarina da Thamnodynastes strigatus (Colubridae). *Arq. Inst. Biol. São Paulo* 37: 213–217.

Pessôa, S. B. and P. de Biasi. 1973. Plasmódio de uma lagartixa, *Urostrophus vautieri* D. and B. (Sauria, Iguanidae). *Mem. Inst. Butantan* 37: 309–316.

Pessôa, S. B., P. de Biasi, and D. de Souza. 1972. Esporulação do *Hepatozoon caimani* (Carini, 1909), parasita do jacaré-e-papo-amarelo: *Caiman latirostris* Daud, no *Culex dolosus* (L. Arribálzaga). *Mem. Inst. Oswaldo Cruz* 70: 379–383.

Pessôa, S. B., P. de Biasi, and G. Puorto. 1974a. Nota sôbre a frequencia de hemoparasitas em serpentes do Brasil. *Mem. Inst. Butantan* 38: 69–118.

Pessôa, S. B., P. de Biasi, and G. Puorto. 1974b. Transferencia do *Hepatozoon tupinambis* parasita do lagarto *Tupinambis teguixin,* para a serpente cascavel (*Crotalus durissus terrificus*) por intermédio do mosquito *Culex fatigans. Mem. Inst. Oswaldo Cruz* 72: 295–299.

Pessôa, S. B., P. de Biasi, and L. Sacchetta. 1974c. Evolução do *Hepatozoon* sp. parasita do *Leptophis ahaetulla* (Lineu) (Serpentes, Colubridae) no *Culex fatigans. Mem. Inst. Butantan* 38: 119–122.

Pessôa, S. B., P. de Biasi, and L. Sacchetta. 1974d. Nota sôbre o *Hepatozoon tupinambis* (Laveran e Salibeni, 1909) (Protozoa, Apicomplexa), parasita do teju (*Tupinambis teguixin* Lineu, 1758) (Sauria, Teiidae). *Mem. Inst. Butantan* 38: 123–130.

Pessôa, S. B. and G. C. Fleury. 1968. *Plasmodium tomodoni* sp. n. parasita da serpente *Tomodon dorsatus* D. and B. *Rev. Bras. Biol.* 28: 525–530.

Pessôa, S. B. and G. C. Fleury. 1969. Duas novas espécies de Tripanosomas parasites de serpents do Brasil. *Rev. Bras. Biol.* 29:81–86.

Pessôa, S. B. and J. A. S. Lopes. 1963. Nota sôbre alguns hematozoarios de lagartixas de Jacobina (Bahia, Brasil). *Rev. Inst. Med. Trop. São Paulo* 5: 133–139.

Pessôa, S. B., L. Sacchetta, and J. Cavalheiro. 1971. Notas sôbre Hemogregarinas de serpentes Brasileiras. XV. Sôbre uma nova especie do genero *Haemogregarina* (S.S.) parasita da *Thamnodynastes pallidus nattereri* (Thumberg) e sua evolução em mosquitos. *Rev. Latin Am. Microbiol.* 13: 29–32.

Petit, G., I. Landau, D. Baccam, and R. Lainson. 1990. Description et cycle biologique d'*Hemolivia stellata* n.g., n.sp., hémogrégarine de crapauds brésiliens. *Ann. Parasitol. Hum. Comp.* 65: 3–15.

Petit, G., I. Landau, Y. Boulard, A. Gomes, and L. Touratier. 1983. Sporogonie de *Plasmodium agamae* chez *Culicoides nubeculosus* au laboratoire: I—Expérimentation et description du cycle. *Protistologica* 19: 537–541.

Pienaar, U. de V. 1962. *Haematology of Some South African Reptiles,* Witwatersrand University Press, Johannesburg. 298 pp.

Pittaluga, A. and S. Buen. 1917. Nota sobre los dipteros del género Phlebotomus en España. *Bol. Inst. Nac. Hig. Alfonso XIII.* 13: 137.

Pittaluga, G. 1912. Ein neuer Blutparasit der afrikanischer Schildkröte, *Clemmys africana*, "*Haemoproteus cajali*" n. sp. *Zentralbl. Bakteriol. Parasitkunde I*, 63: 241–242.

Poinar, G., Jr. and S. R. Telford, Jr. 2005. *Paleohaemoproteus burmacis* gen.n., sp. n. (Haemosporidia: Plasmodiidae) from an early Cretaceous biting midge (Diptera: Ceratopogonidae). *Parasitology* 131: 79–84.

Popov, P. P. 1941. [Cutaneous leishmaniasis in Azerbardzhan.] In: [*Problemy Koznogo Lejsmanioza, Ashkabad*], pp. 107–112.

Pozio, E., M. Grammica, L. Gradoni, and M. Maroli. 1983. Hemoflagellates in *Cyrtodactylus kotschyi* (Steindachner, 1870) (Reptilia, Gekkonidae) in Italy. *Acta Trop.* 40: 399–400.

Pringle, G. 1960. Two new malaria parasites from East African vertebrates. *Trans. R. Soc. Trop. Med. Hyg.* 54: 411–414.

Ramanandan Shanavas, K. and P. Ramachandran. 1990. Life history of *Hepatozoon octosporei* sp. n., a new Haemogregarine from the Skink, *Mabuya carinata* (Schneider), with Notes on the in vitro Excystment of its Oocysts. *Arch. Protistenkunde* 138: 127–137.

Rand, A. S., S. Guerrero, and R. M. Andrews. 1984. The ecological effects of malaria on populations of the lizard *Anolis limifrons* on Barro Colorado Island, Panama. In: *Advances in Herpetology and Evolutionary Biology*, A. G. J. Rhodin and K. Miyata (Eds.), pp. 455–471. Harvard University Press, Cambridge, MA.

Rashdan, N. A. and S. E. El-Sebaii. 2006. *Culex neavei* Theobald, as a possible transmitter of *Hepatozoon matruhensis* to the Egyptian snake *Psammophis schokari*. *J. Egypt. Soc. Parasitol.* 36: 1–6.

Ray, H. N. and A. C. Sarkar. 1969. A new haemosporidian from the garden lizard (*Calotes versicolor*). *Prog. Protozool.* 3: 350–351.

Ray, R. 1987. *Trypanosoma balathaensis* sp. nov. from a pond water turtle, *Lissemys punctata* (Bonnaterre) and its development in the leech vector *Helobdella nociva* Harding. *Acta Protozool.* 26: 63–67.

Ray, R. and A. Bhattacharjee. 1984. *Haemogregarina choudhuryi* sp. n. (Apicomplexa: Haemogregarinidae) in common pond water turtle, *Lissemys punctata punctata* (Bonnaterre) from West Bengal. *Acta Protozool.* 23: 67–74.

Reichenow, E. 1910. *Haemogregarina stepanowi*. Die Entwicklungsgeschichte einer Hämogregarine. *Arch. Protistenkunde* 20: 251–350.

Reichenow, E. 1913. *Karyolysus lacertae*, ein wirtwechselndes Coccidium der Eidechse *Lacerta muralis* und der Milbe *Liponyssys saurarum*. *Arb. K. Gesundhtsamte Berlin* 45: 317–363.

Reichenow, E. 1920. Der Entwicklungsgang der Häamococcidien *Karyolysus* und *Schellackia* nov. gen. Sitzungsb. *Natur. Fr. Berlin* 10: 440–442.

Reichenow, E. 1921. Die Hämococcidien der Eidechsen. Vorbemerkungen und I. Theil. Die Entwicklungsgeschichte von *Karyolysus*. *Arch. Protistenkunde* 42: 179–291.

Ressel, S. and J. J. Schall. 1987. Parasites and showy males: malarial infection and color variation in fence lizards. *Oecologia* 78: 158–164.

Riding, D. 1930. *Haemoproteus* of *Tarentola annularis*. *Trans. R. Soc. Trop. Med. Hyg.* 23: 635–636.

Rioux, J.-A., R. Killick-Kendrick, and P. C. C. Garnham. 1979. *Leishmania tarentolae* and other blood parasites of geckoes at Banyuls in the south of France. *Trans. R. Soc. Trop. Med. Hyg.* 73: 319.

Rioux, J.-A., L. P. Knoepfler, and A. Martini. 1969. Presence en France de *Leishmania tarentolae* Wenyon, 1921 parasite du gecko *Tarentola mauritanica* (L., 1758). *Ann. Parasitol. Hum. Comp.* 44: 115–116.

Robertson, M. 1908. A preliminary note on Haematozoa from some Ceylon reptiles. *Spolia Zeylanica* 5: 178–185.

Robertson, M. 1909. Studies on Ceylon haematozoa. No. 1. The life cycle of *Trypanosoma vittatae*. *Q. J. Microsc. Soc.* 53: 665–695.

Robin, L. A. 1936a. Cycle évolutif d'un *Hepatozoon* de *Gecko verticillatus*. *Ann. Inst. Pasteur Paris* 56: 376–394.

Robin, L. A. 1936b. Évolution d'une hémogrégarine de saurien chez des Moustiques. *C. R. Acad. Sci.* 202: 512–515.

Rocha e Silva, E. O. 1975. Ciclo evolutivo do *Hepatozoon triatomae* (Sporozoa, Haemogregarinidae) parasita de triatomineos. *Rev. Saude Publ. São Paulo* 9: 383–391.

Rocha e Silva, E. O. and D. de C. Rodrigues. 1974. Encontro do *Plasmodium (S.) tropiduri* no estado de São Paulo, Brasil. *Rev. Saude Publ. São Paulo* 8: 163–170.

Rogier, E. 1977. Description et cycle biologique de *Schellackia agamae* (Laveran et Petit, 1909) Lankesterellidae parasite d'agames de République Centre Africaine. *Protistologica* 13: 9–13.

Rogier, E. and I. Landau. 1975. Description de *Schellackia golvani* n. sp. (Lankesterellidae), parasite de Lézards de Guadeloupe. *Bull. Mus. Nat. Hist. Natur. 3*, 284: 91–97.

Roudabush, R. L. and G. R. Coatney. 1937. On some blood protozoa of reptiles and amphibians. *Trans. Am. Microsc. Soc.* 56: 291-297.

Rousselot, R. 1953. *Notes de Parasitologie Tropical. I. Parasites du Sang des Animaux*, Vigot Freres (Ed.), Paris, pp. 22, 25, 34, 35.

Saf'janova, V. M. and A. N. Alexeieff. 1967. [Susceptibility of *Phlebotomus papatasi* Sc. And *Sergentomyia arpaklensis* Perf. to leptomonads of different serological groups experimentally.] *Med. Parazitol.* 36: 580–586.

Sambon, L. W. 1909a. The haemogregarines of snakes. *J. Trop. Med. Hyg.* 12: 22–24.

Sambon, L. W. 1909b. The haemogregarines of snakes. *J. Trop. Med. Hyg.* 12: 38–41.

Sambon, L. W. 1909c. The haemogregarines of snakes. *J. Trop. Med. Hyg.* 12: 48–55.

Sambon, L. W. 1909d. The haemogregarines of snakes. *J. Trop. Med. Hyg.* 12: 70–75.

Sambon, L. W. and C. G. Seligman. 1907. Descriptions of five new species of haemogregarines from snakes. *Proc. Zool. Soc. Lond.* 19: 283–284.

Sambon, L. W. and C. G. Seligman. 1908. The haemogregarines of snakes. *J. Trop. Med. Hyg.* 11: 355–358.

Schall, J. J. 1983. Lizard malaria: parasite-host ecology. In: *Lizard Ecology*. R. B. Huey, E. R. Pianka, and T. W. Schoener (Eds.), pp. 84–100. Harvard University Press, Cambridge, MA.

Schall, J. J. 1989. The sex ratio of *Plasmodium* gametocytes. *Parasitology* 98: 343–350.

Schall, J. J. 1990a. The ecology of lizard malaria. *Parasitology Today* 6: 264–269.

Schall, J. J. 1990b. Virulence of lizard malaria: the evolutionary ecology of an ancient parasite-host association. *Parasitology* 100 (Suppl.): 35–52.

Schall, J. J. 1992. Parasite-mediated competition in *Anolis* lizards. *Oecologia* 92(1): 58–64.

Schall, J. J., A. F. Bennett, and R. W. Putnam. 1982. Lizards infected with malaria: physiological and behavioural consequences. *Science* 217: 1057–1059.

Schall, J. J. and Bromwich, C. R. 1994. Interspecific interactions tested: two species of malarial parasite in a West African lizard. *Oecologia* 97: 326–332.

Schall, J. J. and M. D. Dearing. 1987. Malarial parasitism and male competition for mates in the western fence lizard, *Sceloporus occidentalis*. *Oecologia* 73: 389–392.

Schall, J. J. and P. R. Houle. 1992. Malarial parasitism and home range and social status of male western fence lizards, *Sceloporus occidentalis*. *J. Herpetol.* 26: 74–76.

Schall, J. J. and Marghoob, A. B. 1995. Prevalence of a malarial parasite over time and space: *Plasmodium mexicanum* in its vertebrate host, the western fence lizard *Sceloporus occidentalis*. *J. Anim. Ecol.* 64: 177–185.

Schall, J. J., A. R. Pearson, and S. L. Perkins. 2000. Prevalence of malaria parasites (*Plasmodium floridense* and *Plasmodium azurophilum*) infecting a Puerto Rican lizard (*Anolis gundlachi*): a nine-year study. *J. Parasitol.* 86: 511–515.

Schall, J. J. and G. A. Sarni. 1987. Malarial parasitism and the behavioural time budget of the lizard, *Sceloporus occidentalis*. *Copeia* 1987: 84–95.

Schall, J. J., and T. C. Smith. 2006. Description of a malarial parasite (*Plasmodium mexicanum*) in ectoparasites (mites and ticks), and possible significance for transmission. *J. Parasitol.* 92: 413–415.

Schall, J. J., and S. P. Vogt. 1993. Distribution of malaria in *Anolis* lizards of the Luquillo forest, Puerto Rico: implications for host community ecology. *Biotropica* 25: 229–235.

Schwetz, J. 1931. Sur quelques hématozoaires des lézards de Stanleyville et du Lac Albert. *Ann. Parasitol. Hum. Comp.* 9: 193–201.

Scorza, J. V. 1970a. Lizard Malaria. Unpublished Ph.D. dissertation, University of London.

Scorza, J. V. 1970b. Unpigmented gametocytes of *Plasmodium (Sauramoeba) tropiduri* in thrombocytes of *Tropidurus torquatus* (Iguanidae). *J. Parasitol.* 56, Sect. 2, pt. 2: 470 (abstr.).

Scorza, J. V. 1971a. Electron microscope study of the blood stages of *Plasmodium tropiduri*, a lizard malaria parasite. *Parasitology* 63: 1–20.

Scorza, J. V. 1971b. Anaemia in lizard malaria infections. *Parasssitologia* 13: 391–405.

Scorza, J. V. 1971c. Asexual and sexual stages of a malaria parasite in the thrombocytes of *Tropidurus torquatus* (Iguanidae) infected with *Plasmodium tropiduri*. *J. Protozool.* 18: 403–410.

Scorza, J. V. 1972. Morphological and cytochemical observations on the phagotrophy of *Plasmodium (Sauramoeba) tropiduri* and two other species of malaria parasite. *Rev. Inst. Med. Trop. São Paulo* 14: 97–105.

Scorza, J. V. and C. Dagert. 1956. *Plasmodium pifanoi* nov. sp. parásito de *Ameiva ameiva ameiva*. *Nov. Cient. Mus. Hist. Nat. La Salle (Caracas), Ser. Zool.* 20: 3–7.

Scorza, J. V. and C. Dagert. 1957. Resultados zoológicos de la expedición de la Universidad Central de Venezuela a la región del Auyantepui en la Guyana Venezolana, Abril de 1956. I. Sobre un nuevo *Plasmodium* en *Anolis* sp. del estado Bolivar. *Acta Biol. Venez.* 2: 109–114.

Scorza, J. V., C. C. Monteiro, and C. de Scorza. 1971. Cytolysomes in the erythrocytes of lizards infected with *Plasmodium* spp. *Parasitol.* 62: 63–69.

Scorza, J. V., C. de Scorza, and M. C. C. Monteiro. 1972. Cytochemical observations of three acid hydrolases in blood stages of malaria parasites. *J. Trop. Med. Parasitol.* 66: 167–172.

Sergent, Ed. and Et. Sergent. 1904. Sur une hemogrégarine, parasite de *Testudo mauritanica*. *C. R. Soc. Biol.* 56: 130–131.

Sergent, Ed., Et. Sergent, G. Lemaire, and G. Senevet. 1915. Hypothèse sur le phlébotome "transmetteur" et la tarente "reservoir de virus" du bouton d'orient. *Ann. Inst. Pasteur Paris* 29: 309–322.

Sergent, Et. 1937. Transmission de *Plasmodium relictum* selon des modes non habituels. *Arch. Inst. Pasteur Alger.* 15: 11–17.

Sergent, Et. and Ed. Sergent. 1912. Paludisme des oiseau (*Plasmodium relictum*). L'infection peut se faire par simple frottis du thorax du moustique sur le peau. *C. R. Seances Soc. Biol. Ses. Fil.* 72: 36.

Shortt, H. E. 1922. Review of the position of the genus *Haemocystidium* (Castellani and Willey, 1904), with a description of two new species. *Ind. J. Med. Res.* 9: 814–826.

Shortt, H. E. 1962. Exo-erythrocytic schizont of *Haemoproteus*, parasite of the gecko *Hemidactylus* sp. in the liver. *Trans. R. Soc. Trop. Med. Hyg.* 56: 3.

Shortt, H. E. and C. S. Swaminath. 1928. Preliminary note on three species of Trypanosomidae. *Ind. J. Med. Res.* 16: 241–244.

Shortt, H. E. and C. S. Swaminath. 1931. Life-history and morphology of *Trypanosoma phlebotomi* (Mackie, 1914). *Ind. J. Med. Res.* 19: 541–564.

Siddall, M. E. 1995. Phylogeny of adeleid blood parasites with a partial systematic revision of the haemogregarine complex. *J. Eukaryot. Microbiol.* 42: 116–125.

Siddall, M. E. and S. S. Desser. 1990. Gametogenesis and sporogonic development of *Haemogregarina balli* (Apicomplexa: Adeleina: Haemogregarinidae) in the leech *Placobdella ornata. J. Protozool.* 37: 511–520.

Siddall, M. E. and S. S. Desser. 1991. Merogonic development of *Haemogregarina balli* (Apicomplexa: Haemogregarinidae) in the leech *Placobdella ornata* (Glossiphonidae), its transmission to a chelonian intermediate host and phylogenetic implications. *J. Parasitol.* 77: 426–436.

Siddall, M. E. and S. S. Desser. 1992. Prevalence and intensity of *Haemogregarina balli* (Apicomplexa: Adeleina: Haemogregarinidae) in three turtle species from Ontario, with observations on intraerythrocytic development. *Can. J. Zool.* 70: 123–128.

Siddall, M. E. and S. S. Desser. 2001. Transmission of *Haemogregarina balli* from painted turtles to snapping turtles through the leech, *Placobdella ornata. J. Parasitol.* 87: 1217–1218.

Silva, E. O., J. P. Diniz, S. Alberio, R. Lainson, W. de Sousa, and R. A. DaMatta. 2004. Blood monocyte alteration caused by a hematozoan infection in the lizard *Ameiva ameiva* (Reptilia: Teiidae). *Parasitol. Res.* 90: 448–456.

Silva, E. O., J. A. P. Diniz, R. Lainson, R. A. DaMatta, and W. de Souza. 2005. Ultrastructural aspects of *Fallisia effusa* (Haemosporina: Garniidae) in thrombocytes of the lizard *Neusticurus bicarinatus* (Reptilia: Teiidae). *Protistology* 156: 35–43.

Simond, P. L. 1901. Contribution a l'étude des hématozoaires endoglobulaires des reptiles. *Ann. Inst. Pasteur* 15: 319–351.

Sinha, C. K. 1978. *Trypanosoma gangetica* new species from a fresh water turtle *Trionyx gangeticus. Acta Protozool.* 17: 9–12.

Široký, P., M. Kamler, F. L. Frye, P. Fictum, and D. Modry. 2007. Endogenous development of *Hemolivia mauritanica* (Apicomplexa: Adeleina: Haemogregarinidae) in marginated tortoise *Testudo marginata* (Reptilia: Testudinidae): evidence from experimental infections. *Folia Parasitol.* 54: 13–18.

Široký, P., M. Kamler, and D. Modrý. 2004. Long-term occurrence of *Hemolivia c.f. mauritanica* (Apicomplexa: Adeleina: Haemogregarinidae) in captive *Testudo marginata* (Reptilia: Testudinidae: evidence for cyclic merogony? *J. Parasitol.* 90: 1391–1393.

Široký, P., M. Kamler and D. Modry. 2005. Prevalence of *Hemolivia mauritanica* (Apicomplexa: Adeleina: Haemogregarinidae) in natural populations of tortoises of the genus *Testudo* in the east Mediterranean region. *Folia Parasitol.* 52: 359–361.

Sloboda, M., M. Kamler, J. Bulantova, J. Votypka, and D. Modry. 2007. A new species of *Hepatozoon* (Apicomplexa: Adeleorina) from *Python regius* (Serpentes: Boidae) and its experimental transmission by a mosquito vector. *J. Parasitol.* 93: 1189–1198.

Smallridge, C. J and C. M. Bull. 1999. Transmission of the blood parasite *Hemolivia mariae* between its lizard and tick hosts. *Parasitol. Res.* 85: 858–863.

Smallridge,C. J. and C. M. Bull. 2001a. Prevalence of infection by the protozoan *Hemolivia mariae* in ticks. *Parasitol. Res.* 87: 169–172.

Smallridge, C. J. and C. M. Bull. 2001b. Infection dynamics of *Hemolivia mariae* in the sleepy lizard *Tiliqua rugosa. Parasitol. Res.* 87: 657–661.

Smallridge, C. and I. Paperna. 1997. The tick-transmitted haemogregarinid of the Australian sleepy lizard *Tiliqua rugosa* belongs to the genus *Hemolivia. Parasite* 4: 359–363.

Smallridge, C. and I. Paperna. 2000a. Ultrastructure studies on post-oocyst development of the lizard hemogregarine *Hemolivia mariae* in the tick *Amblyomma limbatum. Parasitol. Res.* 86: 467–471.

Smallridge, C. and I. Paperna. 2000b. Ultrastructure of *Hemolivia mariae* gamonts in the blood of the lizard *Tiliqua rugosa* and their development to oocyst stage in the tick *Amblyomma limbatum. Parasitol. Res.* 86: 563–569.

Smith, T. G. 1996. The genus *Hepatozoon* (Apicomplexa: Adeleina). *J. Parasitol.* 82: 565–585.

Smith, T. G. and S. S. Desser. 1997a. Phylogenetic analysis of the genus *Hepatozoon* Miller, 1908 (Apicomplexa: Adeleorina). *Syst. Parasitol.* 36: 213–221.

Smith, T. G. and Desser, S. S. 1997b. Ultrastructural features of the gametogenic and sporogonic development of *Hepatozoon sipedon* (Apicomplexa: Adeleorina): The applicability of ultrastructural data in differentiating among *Hepatozoon* species. *Parasite* 4: 359–363.

Smith, T. G. and Desser, S. S. 1998. Ultrastructural features of cystic and merogonic stages of *Hepatozoon sipedon* (Apicomplexa: Adelerorina) in Northern Leopard frogs (*Rana pipiens*) and northern water snakes (*Nerodia sipedon*) from Ontario, Canada. *J. Eukaryot. Microbiol.* 45: 419–425.

Smith, T. G., S. S. Desser, and H. Hong. 1994a. Morphology, ultrastructure and taxonomic status of *Toddia* sp. in northern water snakes *Nerodia sipedon sipedon* from Ontario, Canada. *J. Wildlife Dis.* 30: 169–175.

Smith, T. G., S. S. Desser, and D. S. Martin. 1994b. The development of *Hepatozoon sipedon* (Apicomplexa: Adeleina: Hepatozoidae) in its natural host, the Northern water snake (*Nerodia sipedon sipedon*), the culicine vectors, *Culex pipiens* and *Culex territans*, and an intermediate host, the Northern leopard frog (*Rana pipiens*). *Parasitol. Res.* 80: 559–568.

Southgate, B. A. 1970. *Plasmodium (Sauramoeba) giganteum* in *Agama cyanogaster*: a new host record. *Trans. R. Soc. Trop. Med. Hyg.* 64: 12–13 (abstr.).

Spawls, S., K. Howell, R. Drewes and J. Ashe. 2002. *A Field Guide to the Reptiles of East Africa*, Academic Press, New York. 543 pp.

Staats, C. M. and Schall, J. J. 1996a. Malarial parasites (*Plasmodium*) of *Anolis* lizards: biogeography in the Lesser Antilles. *Biotropica* 28: 388–393.

Staats, C. M. and Schall, J. J. 1996b. Distribution and abundance of two malarial parasites of the endemic *Anolis* lizard of Saba Island, Netherlands Antilles. *J. Parasitol.* 82: 409–413.

Stebbins, J. H., Jr. 1904. Upon the occurrence of haemosporodia in the blood of *Rana catesbeiana*, with an account of their probable life history. *Trans. Am. Microsc. Soc.* 25: 55–62.

Stehbens, W. E. and M. R. L. Johnston. 1966. The viral nature of *Pirhemocyton tarentolae*. *J. Ultrastruct. Res.* 15: 543–554.

Stehbens, W. E. and M. R. L. Johnston. 1967. The ultrastructure of a haemogregarine parasitic in *Gehyra variegata* (Dumeril and Bibron, 1836). *Parasitology* 57: 251–261.

Sterling, C. R. 1972. Ultrastructure study of gametocytes and gametogenesis of *Haemoproteus metchnikovi*. *J. Protozool.* 19: 69-76.

Sterling, C. R. and D. L. DeGiusti. 1972. Ultrastructural aspects of schizogony, mature schizonts, and merozoites of *Haemoproteus metchnikovi*. *J. Parasitol.* 58: 641–652.

Sterling, C. R. and D. L. DeGiusti. 1974. Fine structure of differentiating oocysts and mature sporozoites of *Haemoproteus metchnikovi* in its intermediate host *Chrysops callidus*. *J. Protozool.* 21: 276–283.

Strong, R. P. 1924. Investigations upon flagellate infections. *Am. J. Trop. Med.* 4: 345–385.

Svahn, K. 1974. Incidence of blood parasites of the genus *Karyolysus* (Coccidia) in Scandinavian lizards. *Oikos* 25: 43–53.

Svahn, K. 1975a. *Pirhemocyton* a blood parasite in *Lacerta agilis*. *Norveg. J. Zool.* 23: 199–220.

Svahn, K. 1975b. Blood parasites of the genus *Karyolysus* (Coccidia, Adeleidae) in Scandinavian lizards. Description and life cycle. *Norweg. J. Zool.* 23: 277–295.

Svahn, K. 1976. A new piroplasm *Sauroplasma boreale* sp. n. (Haemosporidia, Theileriidae) from the sand lizard *Lacerta agilis* L. Norwegian *J. Zool.* 24:1–6.

Talice, R. V. 1929. Parasitismo de *Triatoma rubrovaria* per un sporozoire. *Ann. Parasitol. Hum. Comp.* 7: 257–261.

Telford, S. R., Jr. 1966. Morphological observations on haemosporidian parasites of some southern California and Mexican lizards. *Jpn. J. Exp. Med.* 36: 237–250.

Telford, S. R., Jr. 1969. A new saurian malarial parasite *Plasmodium balli* from Panama. *J. Protozool.* 16: 431–437.

Telford, S. R., Jr. 1970a. Exoerythrocytic gametocytes of saurian malaria. *Q. J. Florida Acad. Sci.* 33: 77–79.

Telford, S. R., Jr. 1970b. *Plasmodium chiricahuae* sp. nov. from Arizona lizards. *J. Protozool.* 17: 400–405.

Telford, S. R., Jr. 1970c. Saurian malaria parasites in eastern Panama. *J. Protozool.* 17: 566–574.

Telford, S. R., Jr. 1970d. A comparative study of endoparasitism among some southern California lizard populations. *Am. Midl. Nat.* 83: 516–554.

Telford, S. R., Jr. 1971. A malaria parasite, *Plasmodium aurulentum* sp. nov. from the neotropical forest gecko *Thecadactylus rapicaudus*. *J. Protozool.* 18: 308–311.

Telford, S. R., Jr. 1972a. The course of infection of Japanese saurian malaria (*Plasmodium sasai* Telford and Ball) in natural and experimental hosts. *Jpn. J. Exp. Med.* 42: 1–21.

Telford, S. R., Jr. 1972b. Malarial parasites of the "Jesu Cristo" lizard *Basiliscus basiliscus* (Iguanidae) in Panama. *J. Protozool.* 19: 77–81.

Telford, S. R., Jr. 1973a. Malaria parasites of the "borriguerro" lizard, *Ameiva ameiva* (Sauria: Teiidae) in Panama. *J. Protozool.* 20: 203–207.

Telford, S. R., Jr. 1973b. Saurian malarial parasites from Guyana: their effect upon the validity of the family Garniidae and the genus *Garnia*, with descriptions of two new species. *Int. J. Parasitol.* 3: 829–842.

Telford, S. R., Jr. 1974. The malarial parasites of *Anolis* species (Sauria: Iguanidae) in Panama. *Int. J. Parasitol.* 4: 91–102.

Telford, S. R., Jr. 1975. Saurian malaria in the Caribbean: *Plasmodium azurophilum* sp. nov., a malaria parasite with schizogony and gametogony in both red and white cells. *Int. J. Parasitol.* 5: 383–394.

Telford, S. R., Jr. 1977. The distribution, incidence and general ecology of saurian malaria in Middle America. *Int. J. Parasitol.* 7: 299–314.

Telford, S. R., Jr. 1978a. Intralymphocytic schizonts associated with an initial infection of *Saurocytozoon tupinambi* in *Tupinambis teguixin*. *Int. J. Parasitol.* 8: 133–138.

Telford, S. R., Jr. 1978b. Saurian malaria in Texas. *J. Parasitol.* 64: 553–554.

Telford, S. R., Jr. 1978c. A haemoparasite survey of Florida lizards. *J. Parasitol.* 64: 1126–1127.

Telford, S. R., Jr. 1978d. The saurian malarias of Venezuela: haemosporidian parasites of gekkonid lizards. *Int. J. Parasitol.* 8: 341–353.

Telford, S. R., Jr. 1979a. A malarial parasite of Australian skinks, *Plasmodium mackerrasae* sp. n. *J. Parasitol.* 65: 409–413.

Telford, S. R., Jr. 1979b. Evolutionary implications of *Leishmania* amastigotes in circulating blood cells of lizards. *Parasitology*, 79: 317–324.

Telford, S. R., Jr. 1979c. Two new trypanosomes from Neotropical gekkonid lizards. *J. Parasitol.* 65: 898–901.

Telford, S. R., Jr. 1979d. A taxonomic reconsideration of some *Plasmodium* species from iguanid lizards. *Ann. Parasitol. Hum. Comp.* 54: 129–144.

Telford, S. R., Jr. 1979e. A taxonomic revision of small Neotropical saurian malarias allied to *Plasmodium minasense. Ann. Parasitol. Hum. Comp.* 54: 409–422.

Telford, S. R., Jr. 1980. The saurian malarias of Venezuela: *Plasmodium* species from iguanid and teiid hosts. *Int. J. Parasitol.* 10: 365–374.

Telford, S. R., Jr. 1982a. Interpopulation variation of a saurian malaria, *Plasmodium sasai* Telford and Ball, 1969, in three host species distributed within a range of 24° north latitude. *Int. J. Parasitol.* 12: 17–22.

Telford, S. R., Jr. 1982b. New saurian trypanosomes from Southeast Asian skinks and lacertids. *J. Parasitol.* 68: 319–324.

Telford, S. R., Jr. 1982c. *Plasmodium lionatum* sp. n., a parasite of the flying gecko, *Ptychozoon lionatum*, in Thailand. *J. Parasitol.* 68: 1154–1157.

Telford, S. R., Jr. 1982d. Re-discovery of *Haemoproteus kopki* (De Mello, 1916) in two geckoes from the Afghan border of Pakistan. *Parasitology* 84: 527–536.

Telford, S. R., Jr. 1983a. *Saurocytozoon* parasites (Haemosporidia: Plasmodiidae) from southeast Asian skinks. *J. Parasitol.* 69: 1141–1145.

Telford, S. R., Jr. 1983b. Studies on African saurian malarias: redescription of *Plasmodium mabuiae* Wenyon, 1909 from Tanzanian skinks. *J. Parasitol.* 69: 1146–1149.

Telford, S. R., Jr. 1983c. *Plasmodium saurocaudatum* sp.n., a parasite of *Mabuya multifasciata* in southeast Asia. *J. Parasitol.* 69: 1150–1155.

Telford, S. R., Jr. 1984a. Studies on African saurian malarias: three *Plasmodium* species from gekkonid hosts. *J. Parasitol.* 70: 343–354.

Telford, S. R., Jr. 1984b. Haemoparasites of reptiles. In: G. L. Huff, F. Frye, and E. R. Jacobson (Eds.), *Diseases of Amphibians and Reptiles*, Plenum Press, New York, pp. 385–517.

Telford, S. R., Jr. 1986a. Studies on African saurian malarias: *Plasmodium holaspi* n. sp. from the flying lacertid *Holaspis guentheri*. *J. Parasitol.* 72: 271–275.

Telford, S. R., Jr. 1986b. *Fallisia* parasites (Haemosporidia: Plasmodiidae) from the flying lizard, *Draco maculatus* (Agamidae) in Thailand. *J. Parasitol.* 72: 766–769.

Telford, S. R., Jr. 1987. Studies on African saurian malarias: *Plasmodium* parasites of cordylid lizards. *Ann. Parasitol. Hum. Comp.* 62: 405–417.

Telford, S. R., Jr. 1988a. A contribution to the systematics of the reptilian malaria parasites, Family Plasmodiidae (Apicomplexa: Haemospororina). *Bull. Florida State Mus. Biol. Sci.* 34: 65–96.

Telford, S. R., Jr. 1988b. Studies on African saurian malarias: five *Plasmodium* species from the Uzungwe Mountains of Tanzania. *Int. J. Parasitol.* 18: 197–219.

Telford, S. R., Jr. 1989. Discovery of the pre-erythrocytic stages of a saurian malaria parasite, hypnozoites, and a possible mechanism for the maintenance of chronic infections throughout the life of the host. *Int. J. Parasitol.* 19: 597–616.

Telford, S. R., Jr. 1993. A species of *Schellackia* (Apicomplexa: Lankesterellidae) parasitising east and southeast Asian lizards. *Syst. Parasitol.* 25: 109–117.

Telford, S. R., Jr. 1994. Plasmodia of reptiles. In: J. P. Kreier (Ed.), *Parasitic Protozoa*, 2nd ed., Academic Press, New York, Vol. 7, pp 1–71.

Telford, S. R., Jr. 1995a. A review of trypanosomes of gekkonid lizards, including the description of five new species. *Syst. Parasitol.* 31: 37–52.

Telford, S. R., Jr. 1995b. The kinetoplastid hemoflagellates of reptiles. In: J. P. Kreier (Ed.), *Parasitic Protozoa*, Vol. 10, pp. 161–223.

Telford, S. R., Jr. 1995c. Two new trypanosome species from East African cordylid lizards. *Syst. Parasitol.* 31: 61–65.

Telford, S.R., Jr. 1995d. *Plasmodium* spp. (Apicomplexa: Plasmodiidae) of the flying lizard *Draco volans* (Agamidae). *J. Parasitol.* 31: 53–60.

Telford, S. R., Jr. 1996a. Two new species of *Haemocystidium* Castellani and Willey (Apicomplexa: Plasmodiidae) from Pakistani lizards, and the support their meronts provide for the validity of the genus. *Syst. Parasitol.* 34: 197–214.

Telford, S. R., Jr. 1996b. Epizootiology of the Japanese saurian malaria parasite, *Plasmodium sasai. J. Parasitol.* 82: 226–232.

Telford, S. R., Jr. 1996c. A review of the trypanosomes from lizards of the family Iguanidae (*sensu lato*), including the descriptions of five new species, and an evaluation of the effect of host difference upon taxonomic characters of saurian trypanosomes. *Syst. Parasitol.* 34: 215–237.

Telford, S. R., Jr. 1997a. *The Ecology of a Symbiotic Community.* Vol. 1. *Population Biology of the Japanese Lizard* Takydromus tachydromoides *(Schlegel) (Lacertidae),* Krieger, Malabar, FL. 125 pp.

Telford, S. R., Jr. 1997b. *The Ecology of a Symbiotic Community.* Vol. 2. *The Component Symbiote Community of the Japanese Lizard* Takydromus tachydromoides *(Schlegel) (Lacertidae),* Krieger, Malabar, FL. 153 pp.

Telford, S. R., Jr. 1998a. Additional Fallisia spp. (Apicomplexa: Plasmodiidae) of Neotropical lizards. *Syst. Parasitol.. Parasitology* 40: 185–193.

Telford, S. R., Jr. 1998b. The development and persistence of phanerozoites in experimental infections of *Plasmodium sasai. Int. J. Parasitol.* 28: 475–484.

Telford, S. R., Jr. 2005. A *Haemocystidium* species from the East African gecko *Lygodactylus capensis grotei. J. Parasitol.* 91: 135–138.

Telford, S. R., Jr. 2007. Redescription of *Haemoproteus mesnili* (Apicomplexa: Plasmodiidae) and its meronts, with description of a second haemosporidian parasite of African cobras. *J. Parasitol.* 93: 673–679.

Telford, S. R., Jr. and G. H. Ball. 1969. *Plasmodium sasai* n. sp. from the Japanese lizard *Takydromus tachydromoides. J. Protozool.* 16: 312–317.

Telford, S. R., Jr., J. F. Butler, and R. S. Telford. 2002. *Hepatozoon* species (Apicomplexa: Hepatozoidae) of the corn snake, *Elaphe guttata* (Serpentes: Colubridae) and the pigmy rattlesnake, *Sistrurus miliarius barbouri* (Serpentes: Viperidae) in south Florida. *J. Parasitol.* 88: 778–782.

Telford, S. R., Jr., J. F. Butler, and P. E. Moler. 2005a. Two additional *Hepatozoon* species (Apicomplexa: Hepatozoidae) from the southern black racer, *Coluber constrictor priapus* (Serpentes: Colubridae) in northern Florida. *J. Parasitol.* 91: 139–143.

Telford, S. R., Jr., J. F. Butler, and R. S. Telford. 2005b. *Hepatozoon polytopis* n. sp. parasitic in two genera and species of colubrid snakes in southern Florida. *J. Parasitol.* 91: 144–147.

Telford, S. R., Jr., J. A. Ernst, A. M. Clark, and J. F. Butler. 2004. *Hepatozoon sauritus*: a polytopic hemogregarine of three genera and four species of snakes in northern Florida, with specific identity verified from genome analysis. *J. Parasitol.* 90: 352–358.

Telford, S. R., Jr. and E. R. Jacobson. 1993. Lizard erythrocytic virus in East African chamaeleons. *J. Wildlife Dis.* 29: 57–63.

Telford, S. R., Jr., R. N. Johnson, and D. G. Young. 1989. Additional *Plasmodium* species from *Anolis* lizards of Hispaniola and Panama. *Int. J. Parasitol.*19: 275–284.

Telford, S. R., Jr. and I. Landau. 1987. *Plasmodium brygooi* sp. n. from *Chamaeleo brevicornis* of Madagascar, with a redescription of *Plasmodium robinsoni* (Brygoo) 1962 from its type host and an additional host, *Chamaeleo parsoni crucifer. Ann. Parasitol. Hum. Comp.* 62: 395–404.

Telford, S. R., Jr., P. E. Moler, and J. F. Butler. 2008. *Hepatozoon* species of the timber rattlesnake in northern Florida: evidence of salivary gland oocysts and a natural cross-familial transmission of an *Hepatozoon* species. *J. Parasitol.* in press.

Telford, S. R., Jr., T. M. Norton, P. E. Moler, and J. B. Jensen. 2009. A *Haemogregarina* species of the alligator snapping turtle, *Macrochelys temminckii* (Testudines: Chelydridae) in Georgia and Florida that produces macromeronts in circulating erythrocytes. *J. Parasitol.* submitted.

Telford, S. R., Jr. and J. Stein. 2000. Two malaria parasites (Apicomplexa: Plasmodiidae) of the Australia skink *Egernia stokesii*. *J. Parasitol.* 86: 395–406.

Telford, S. R., Jr. and S. R. Telford III. 2003. Rediscovery and redescription of *Plasmodium pifanoi* and description of two additional *Plasmodium* parasites of Venezuelan lizards. *J. Parasitol.* 89: 362–368.

Telford, S. R., Jr., S. R. Telford III, and J. F. Butler. 2002. The status of *Haemogregarina mansoni* Sambon and Seligmann from *Zamenis flagelliformis* Laurenti. *J. Parasitol.* 88: 783–785.

Telford, S. R., Jr. and J. F. X. Wellehan, Jr. 2005. Two *Plasmodium* species of the crocodile skink *Tribolonotus gracilis* from Irian Jaya, Indonesia. *J. Parasitol.* 91: 148–151.

Telford, S. R., Jr., E. J. Wozniak, and J. F. Butler. 2001. Haemogregarine specificity in two communities of Florida snakes, with descriptions of six new species of *Hepatozoon* (Apicomplexa: Hepatozoidae) and a possible species of *Haemogregarina* (Apicomplexa: Haemogregarinidae). *J. Parasitol.* 87: 890–905.

Theiler, M. 1930. Special protozoological studies of the blood: protozoological studies of smaller animals in Liberia. In: R. P. Strong (Ed.), *The African Republic of Liberia and the Belgian Congo, Harvard African Expedition*, Harvard University Press, Cambridge, MA, pp. 490–498.

Thiroux, A. 1910. Une hémogrégarine de *Crocodilus niloticus. C. R. Soc. Biol.* 69: 577–578.

Thiroux, A. 1913. Les formes de reproduction par schizogonie et sporogonie d'*Haemogregarina pettiti* (Thiroux, 1910), chez *Crocodilus niloticus. Bull. Soc. Pathol. Exot.* 6: 327–330.

Thompson, P. E. 1944. Changes associated with acquired immunity during initial infections in saurian malaria. *J. Infect. Dis.* 74: 138–150.

Thompson, P. E. and T. A. Hart. 1946. *Plasmodium lacertiliae* n. sp.and other saurian blood parasites from the New Guinea area. *J. Parasitol.* 32: 79–82.

Thompson, P. E. and C. G. Huff. 1944a. A saurian malarial parasite, *Plasmodium mexicanum*, n. sp., with both *elongatum-* and *gallinaceum*-types of exoerythrocytic stages. *J. Infect. Dis.* 74: 48–67.

Thompson, P. E. and C. G. Huff. 1944b. Saurian malarial parasites of the United States and Mexico. *J. Infect. Dis.* 74: 68–79.

Travassos Santos Dias, J.A. 1953. [*Haemoproteus balazuci* sp. n. from tortoise, Mocambique.] *Bol. Soc. Estud.* 23 (82): 41–73.

Uilenberg, G. and Ch.-P. Blanc. 1966. Note sur un Hématozoaire d'un Reptile malgache, *Uroplatus fimbriatus* (Gekkonidae): *Sauroplasma* sp. (Protozoa Incertae sedis). *Ann. Parasitol. Hum. Comp.* 41: 209–212.

van den Berghe, L., M. Chardome, and E. Peel. 1964. *Trypanosoma mochli*, a trypanosome from an African lizard. *Parasitology* 54:451–452.

Vardo, A. M., K. D. Kaufhold, and J. J. Schall. 2007. Experimental test for premunition in a lizard malaria parasite (*Plasmodium mexicanum*). *J. Parasitol.* 93: 280–282.

Vivier, E., A. Petitprez, and I. Landau. 1972. Observations ultrastructurales sur la sporoblastogenese de l'hemogregarine, *Hepatozoon domerguei*, Coccidie, Adeleidea. *Protistologica* 8: 315–333.

Wallbanks, K. R., R. Maazoun, E. U. Canning, and J. A. Rioux. 1985. The identity of *Leishmania tarentolae* Wenyon 1921. *Parasitology* 90: 67–78.

Walliker, D. 1965. *Trypanosoma superciliosae* sp. nov. from the lizard *Uranoscodon superciliosa* L. *Parasitology* 55: 601–606.

Walliker, D. 1966. Malaria parasites of some Brazilian lizards. *Parasitology* 56: 39–44.

Wang, C. C. and S. H. Hopkins. 1965. *Haemogregarina* and *Haemoproteus* (Protozoa, Sporozoa) in blood of Texas freshwater turtles. *J. Parasitol.* 51: 682–683.

Wenyon, C. M. 1909a (dated 1908). Report of travelling pathologist and protozoologist. In: A. Balfour (Ed.), *Third Report, Wellcome Research Laboratory at the Gordon Memorial College, Khartoum*, Bailliere, Tindall and Cox, London, pp. 146–150.

Wenyon, C. M. 1909b. A trypanosome and a haemogregarine of a Tropical American snake. *Parasitology* 1: 314–317.

Wenyon, C. M. 1915. The pigmented parasites of cold-blooded animals, with some notes on a plasmodium of the Trinidad iguana. *J. Trop. Med. Hyg.* 18: 133–140.

Wenyon, C. M. 1921. Observations on the intestinal protozoa of three Egyptian lizards, with a note on a cell-invading fungus. *Parasitology* 12: 350–365.

Wenyon, C. M. 1926. *Protozoology, a Manual for Medical Men, Veterinarians and Zoologists*, Bailliere, Tyndall and Cox, London, Vols. 1 and 2.

Williams, E. E. 1983. Ecomorphs, faunas, island size, and diverse end points in island radiation of *Anolis*. In: R. B. Huey, E. R. Pianka, and T. W. Schoener (Eds.), *Lizard Ecology, Studies of a Model Organism*, Harvard University Press, Cambridge, MA, pp. 326–370.

Wilson, V. C. L. C. and B. A. Southgate. 1978. Lizard *Leishmania*. In: W. H. R. Lumsden and D. A. Evans, (Eds.), *Biology of the Kinetoplastida,* Academic Press, London, Vol. 2, pp. 241–267.

Woo, P. T. K. 1969. The life cycle of *Trypanosoma chrysemydis*. *Canadian J. Zool.* 47: 1139–1151.

Wood, S. F. 1935. Variations in the cytology of the blood of geckos (*Tarentola mauritanica*) infected with *Haemogregarina platydactyli*, *Trypanosoma platydactyli*, and *Pirhemocyton tarentolae*. *Univ. California. Publ. Zool.* 41: 9–22.

Wood, S. F. and F. D. Wood. 1936. Occurrence of haematozoa in some California cold-blooded vertebrates. *J. Parasitol.* 22: 518–520.

Wozniak, E. J., S. R. Telford, Jr., and G. L. McLaughlin. 1996. Molecular differentiation of two New World saurian *Plasmodium* species in different host species using the polymerase chain reaction. *J. Parasitol.* 82: 372–375.

Young, M. D. 1941. The oral transmission of *Plasmodium relictum* in the pigeon. *Public Health Rep.* 56: 1439–1440.

Zmeyev, G. Ja. 1936. [Haemoparasite fauna of wild vertebrates in some southern areas of Tadzhikistan.] *Trans. Tadzhik Acad. Sci. USSR* 6: 249–266.

APPENDIX

The Identification of Reptilian Hemoparasites

Although this work is not intended to be a manual for the identification of hemoparasites, this section was prepared for the benefit of those nonspecialists who may encounter infections within blood cells of reptiles. The increasing exposure of veterinarians to exotic reptiles acquired by hobbyists through the burgeoning pet trade will certainly provide the veterinarians opportunity to expand their knowledge of parasitology from the usual information relating to the species of domestic and economic importance that form the core of training in veterinary parasitology. Usually, hemoparasite presence will have no causative relationship to the observed disease, but there can be exceptions. Virtually all of the hemoparasites could better be termed *symbiotes* as the result of a long coevolutionary history with their hosts, in which the problems of deleterious effects on the host (i.e., disease) were solved long ago by the survival and perpetuation of resistant individual hosts and less-virulent strains of parasites. The physiological stress of other infective agents, however, would be expected to produce some disease effects in natural populations, as would overwhelming infections that can develop in immature hosts, as reported for *Plasmodium mexicanum* in its California lizard hosts. Certainly, some of the disease cases evident to the clinician have resulted from the stress of the captive environment on the reptile and from the exposure in captivity to nonsympatric hosts and their symbiotes.

Methods and Procedures

Blood samples can be collected with less-discernible effect on the host by venipuncture, especially from the caudal vein. However, in field studies of lizards where some form of individual marking is needed, it is simpler to clip off the terminal segment or two of toes, in a pattern that will identify the individual. After severing the toe portion, pause briefly and then squeeze the toe area gently. In most lizards, an adequate amount of blood will appear at the tip of the toe that was clipped. Some lizards bleed more readily than others (i.e., skinks and geckoes often seem to vasoconstrict the clipped toe before the blood drop appears). I have never seen infection of a clipped toe among the many thousands of lizards I have bled in the field and laboratory by this procedure. In addition to venipuncture, it is sometimes necessary to obtain blood directly from the heart of snakes, in particular when holding a venomous species behind the head with one hand, its body secured between one's legs, and the other hand holding the syringe. This technique is not recommended if PVC tubes of appropriate sizes are available. The snake is prodded appropriately to enter the tube, its body grasped firmly at the tube entrance (with the head and neck safely within the tube), and blood is withdrawn by puncture of the caudal vein. This procedure is best done with assistance. Cardiac puncture of small snakes, lizards, and turtles is not recommended due to the possibility of lacerating the heart tissue. I usually use a blood lancet or needle to stick into the middle of the rostral plate of snakes to obtain the necessary amount of blood. This procedure also works well with legless lizards. Again, however, some snakes bleed more readily from the rostral than do others: Natricine snakes are easier bled than rat snakes, but the racers bleed well. The method to be employed with crocodilians will vary with the size of the animal to be bled; perhaps puncture of the caudal vein is safest for the investigator if there is assistance. In survey work, it is often necessary to obtain blood from specimens killed by vehicles or other means. If freshly killed, the heart can be exposed and gashed, releasing the blood for collection. If some hours have elapsed since death of the animal and before bile stains have appeared on the abdomen of snakes and lizards, it may be possible to find satisfactory blood within the liver if the heart appears to have none. Temperature can affect both the condition of blood cells and the contained intracellular parasites. Diagnosis of parasite presence is possible sometimes from frozen hosts, but identification is seldom possible.

Once a sample is available, thin blood smears only should be prepared as quickly as possible after drawing the blood. Immediately after the blood is spread along the slide, the slide should be waved vigorously in the air to dry the cells or else intracellular parasites may become extracellular through lysis of the cells or their own motility. If samples are obtained by syringe, citrate-washed syringes and needles produce better stain reactions on slides than does the use of heparin-washed syringes. In recent years, it has become useful to obtain a filter paper sample of the blood for possible DNA extraction should something of interest be present in the blood. And, when a possible viral or bacterial parasite is apparent, it will be necessary to put a drop of blood in an appropriate fixative for electron microscopy. Staining will be much better if slides are fixed and stained within the same day of collection, but fixed slides sometimes yield good stains even some months after fixation. Fixation on the day of preparation is highly desirable. Absolute methanol is the fixative of choice, and in areas of high humidity should be monitored for its effect on blood cells as these will show vacuoles if the alcohol has absorbed too much water vapor before use (so keep the bottle tightly closed). Fixation for 1–2 minutes is adequate, followed by drying in a standing-on-end position. The blood stain of choice is Giemsa, and the quality of the Giemsa used will affect the staining reaction. Harleco Giemsa® (619/71, EM Science) is probably the best stain available, but the older preparation sold by Fisher is also useful. The stain should be diluted in a ratio of 1 part stain to 9 parts distilled water at pH 7.0, and the staining time is longer than for mammal and bird blood, at least 55 minutes, followed by washing with distilled water, with no great damage done by tap water. The rapid-acting solutions combining fixative and stain used commonly today can provide a diagnosis, but for serious study a Giemsa-stained slide is preferable. Trypanosome presence may be easier to detect by placing a drop of blood in citrate, applying a coverslip, and examining at ×100 or ×400 for detection of motion by the flagellate. This also works well for microfilariae.

Stained trypanosomes from cultured blood have little use for species determination as the cultural forms are all rather similar, but cultured samples will reveal prevalence much more accurately than slides of peripheral blood. With the larger intracellular parasites, examination of a dried unstained blood smear with reduced light can provide a quick determination of parasite presence (they are usually brighter than the host cell cytoplasm) but is of little help in identification except as "hemogregarine." With some parasites, such as the hemogregarines, knowledge of the stages that occur in the vector is essential for proper generic identification, research that is both time consuming and labor intensive, and is probably unnecessary for clinical needs.

The pictorial key included here is designed only to steer the clinician toward the correct taxonomic groups to which hemoparasites belong. Once this is apparent, then the text accounts of the various genera can be consulted for information on the geographic and host range, morphological variation, life cycles and pathology where known, and additional pictures of species. Variability in morphology observable under light microscopy greatly diminishes the value of taxonomic keys for identification, and none have been prepared for species identification here. The pictures should be adequate, along with the few structural features highlighted within the figures (**Plates 74–83**).

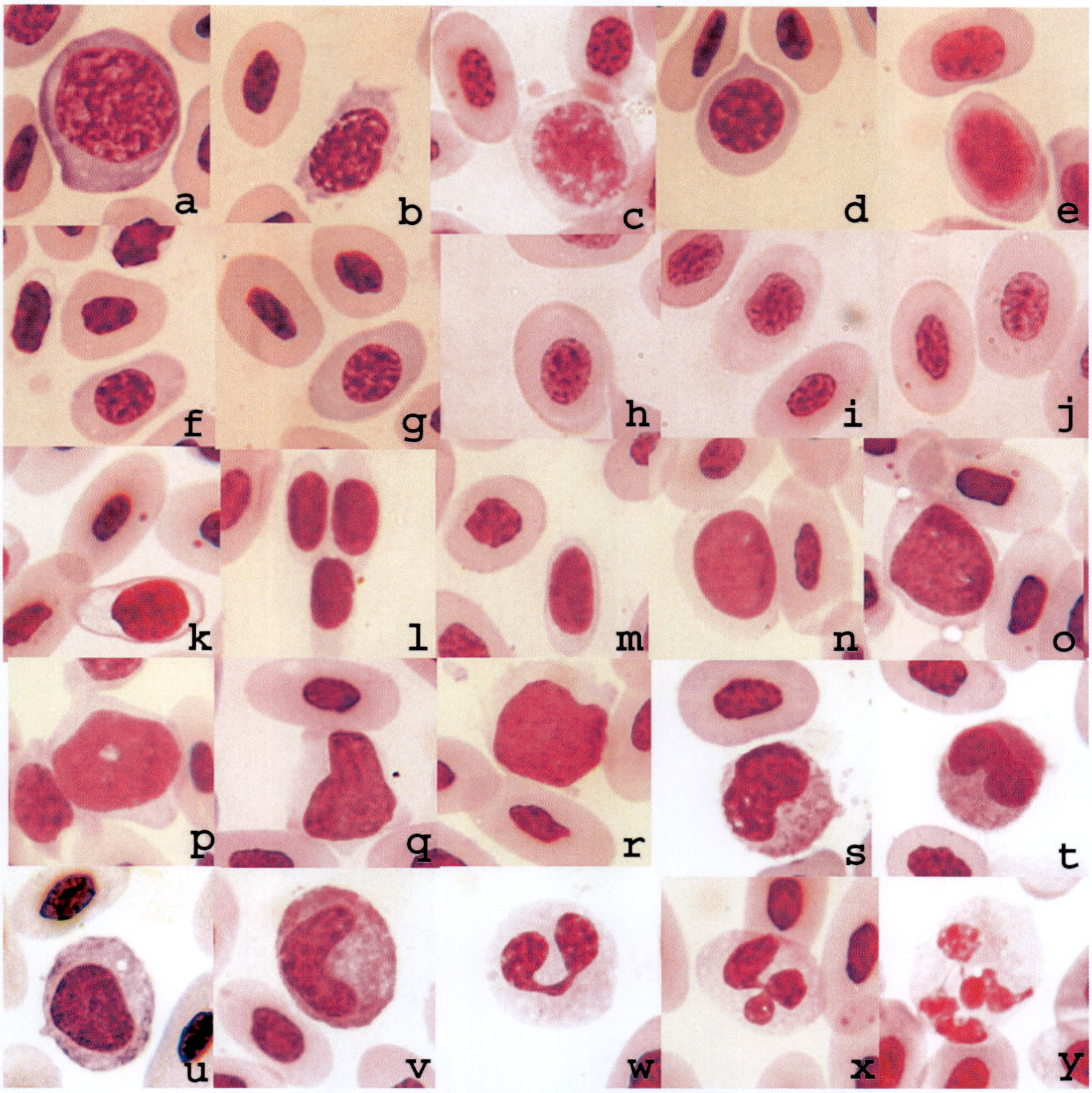

Plate 74 Saurian blood cell types most often host to hemoparasites. Probable stem cells, **a–c**; erythroblasts, **d**, **e**; basophilic proerythrocytes, **e–g**, **m**; polychromatophilic proerythrocytes, **c**, **h–j**, **u**; erythrocytes, **b**, **g**, **i-k**, **n**, **o**, **q–s**, **u**, **x**; thrombocytes, **f**, **k–m**; monocytes, **n–p**; large lymphocytes, **q**, **r**; azurophil granulocytes, **s–v**; polymorphonuclear amphophilic leukocytes (Pienaar, 1962; Telford, 1975) or, commonly, "neutrophils," **w–y**. Lizard species: Polychrotidae, *Anolis cybotes* (**l**, **m**, **s**, **t**, **w**); Teiidae, *Tupinambis teguixin* (**n**, **q**, **r**, **u**, **v**), *Kentropyx calcarata* (**a**, **b**, **d**, **f**, **g**); Lacertidae, *Takydromus tachydromoides* (**c**, **h–j**); Gekkonidae, *Hemidactylus platycephalus* (**e**, **y**); Chamaeleonidae, *Chamaeleo dilepis* (**k**, **o**, **p**, **x**).

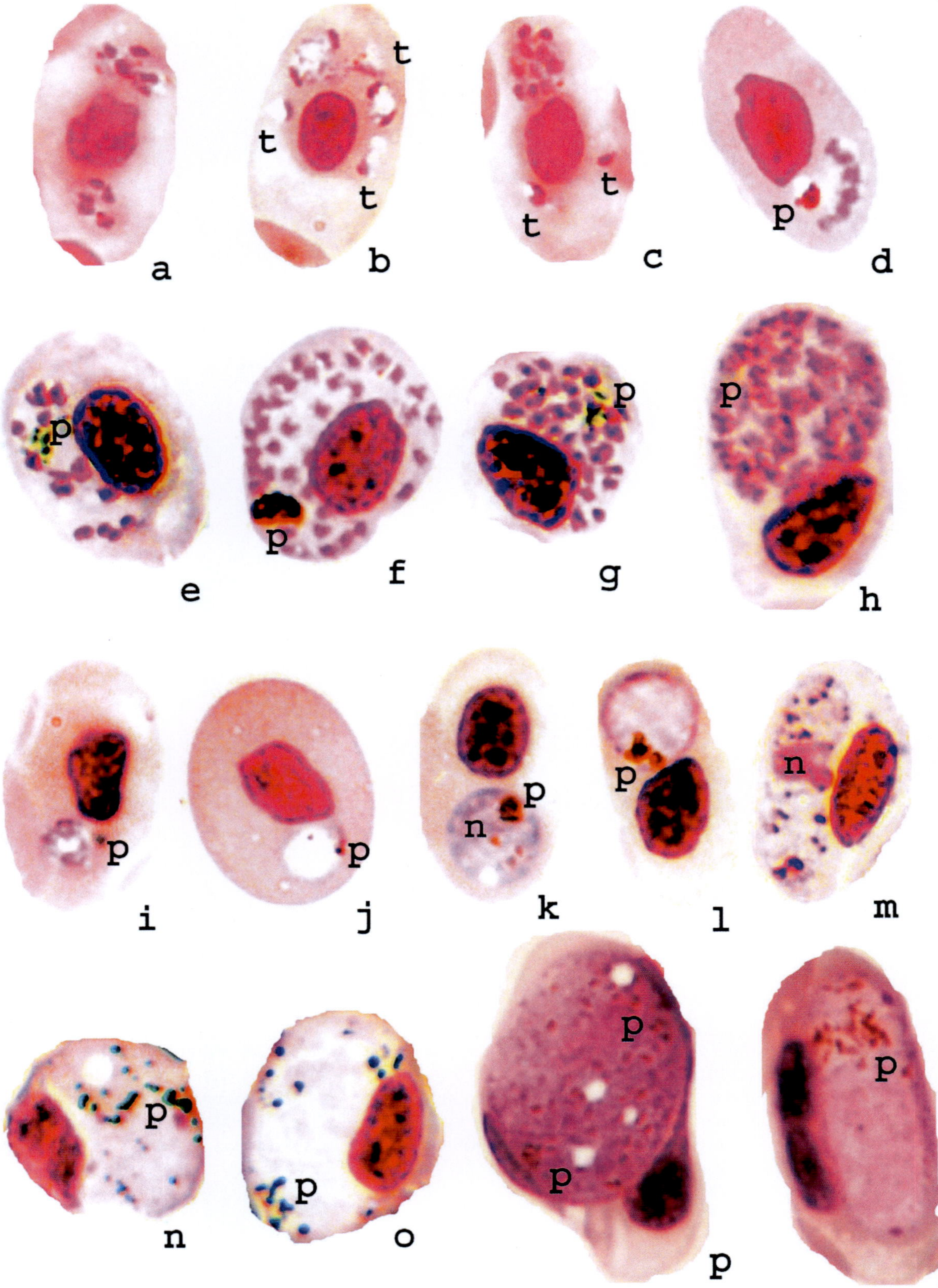

Plate 75 Life history stages of several *Plasmodium* species in lizards and snakes. Asexual stages **a–h**: **a–c**, trophozoites (t) and young meronts or schizonts; **c–h**, nearly mature or mature meronts; **g**, fully mature meront or segmenter. Gametocytes, the sexual stages, of various sizes and shapes, **i–q**: macrogametocytes (female) **i**, **k**, **m**, **p**; microgametocytes (male) **j**, **l**, **n**, **o**, **q**. Pigment (p) or hemozoin is present in older meronts and mature gametocytes, and host cells are mature or immature erythrocytes.

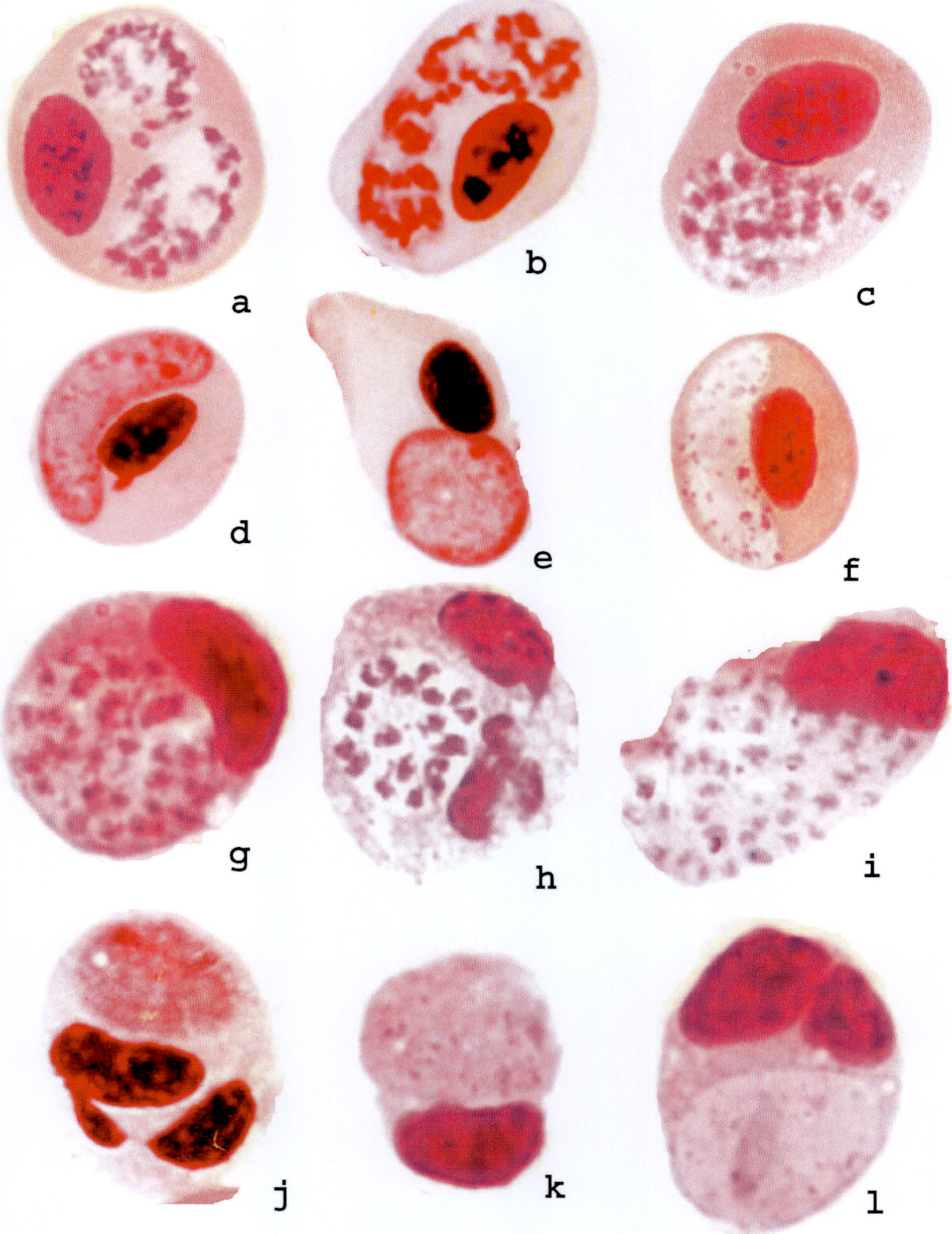

Plate 76 Meronts and gametocytes of some unpigmented *Plasmodium* species of the subgenus *Garnia* in erythrocytes (**a–f**) and leukocytes (**g–l**) of neotropical lizards. Figures **a–c** and **g–i** are meronts; **c** and **i** are fully mature and segmenting with cytoplasm formed around each nucleus. Macrogametocytes are **d**, **e**, **j**; **f**, **k**, and **l** are microgametocytes. Leukocytic host cells are "neutrophils" (**h**, **j**), an azurophil granulocyte (**l**), a lymphocyte (**k**), and two possible monocytes (**g**, **i**).

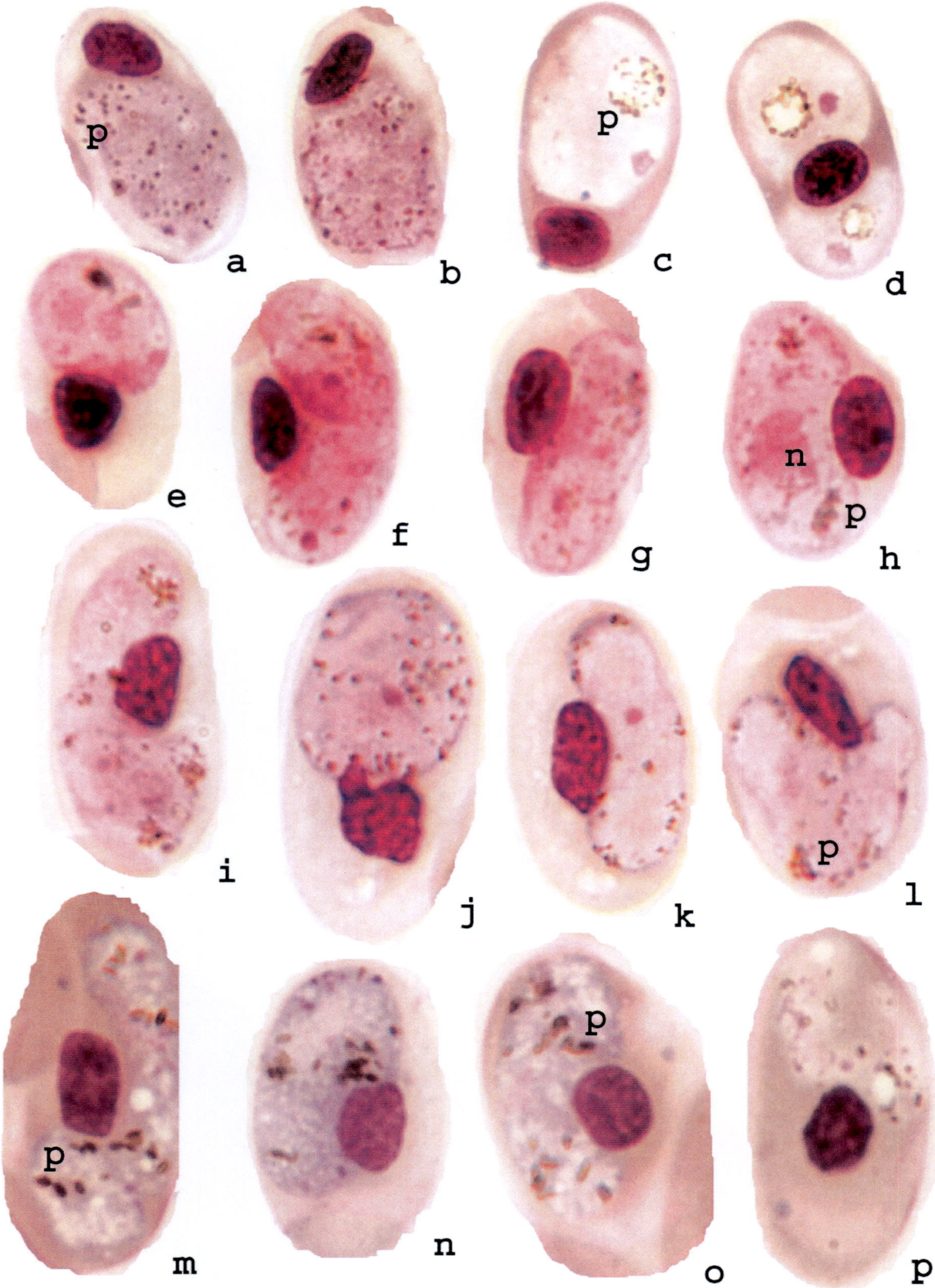

Plate 77 Gametocytes of some *Haemocystidium* species of Old World lizards (**a–h**) and *Haemoproteus* species in African cobras (**i–l**) and freshwater turtles (**m–p**) or tortoises. All show pigment granules (p); nuclei (n) may be distinct in macrogametocytes; multinucleate stages are absent from *Haemoproteus* and only rarely seen in *Haemocystidium* infections. Figures **a**, **b**, **e**, **f**, **h**, **i**, **m–o** are macrogametocytes; **c**, **d**, **j–l**, **p** are microgametocytes.

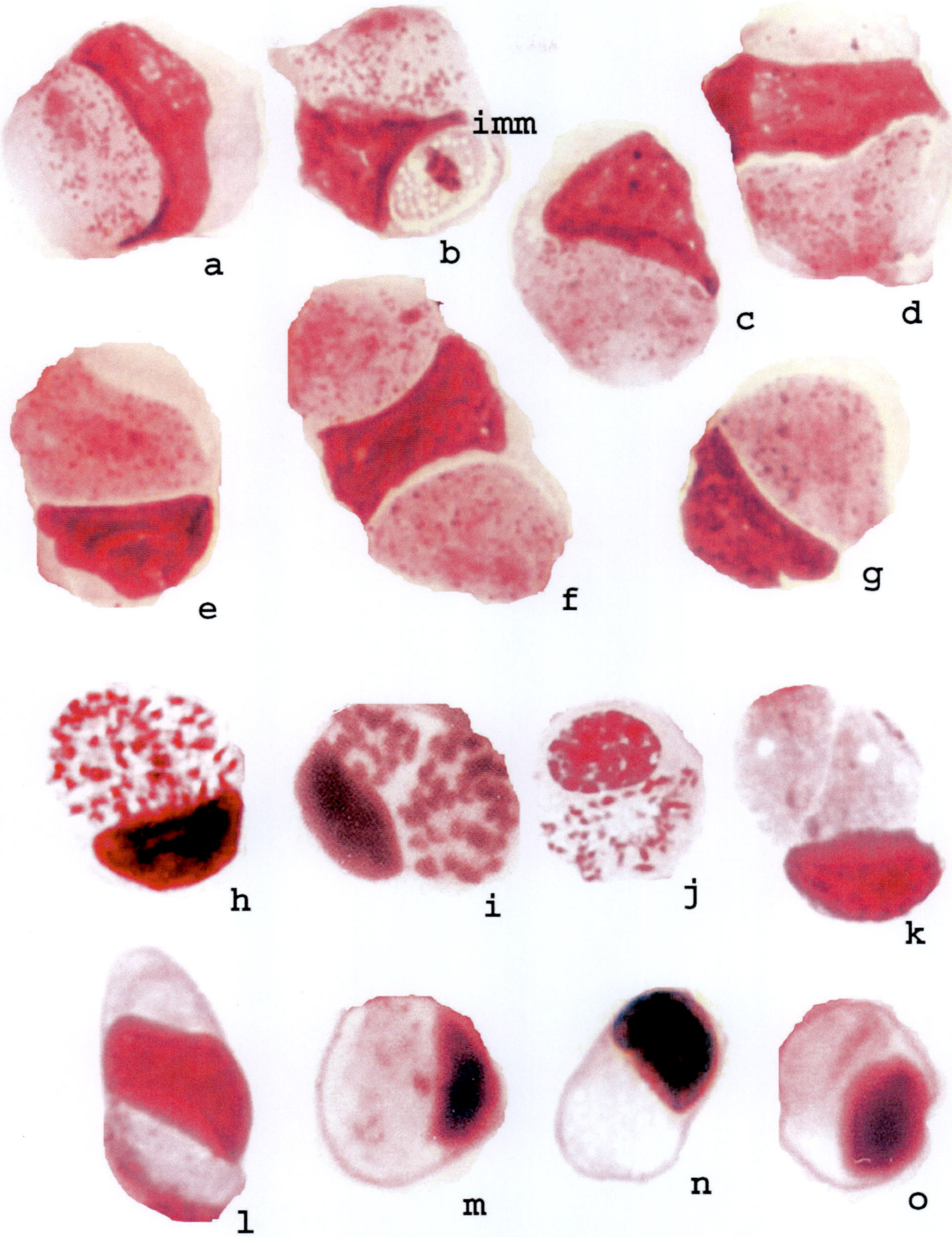

Plate 78 Unpigmented plasmodiid parasites of leukocytes and thrombocytes of lizards. *Saurocytozoon* gametocytes, **a–g**, occur in lymphocytes and, uncommonly, immature erythrocytes of South American teiid lizards and in lymphocytes of skinks in Brazil and southeast Asia. Rare meronts may be seen in circulating blood cells only early, if at all, in infections. Figures **a**, **b**, **d–f** (bottom) are macrogametocytes; **e**, **f** (top), **g** are microgametocytes. Asexual and sexual stages of *Fallisia* species, **h–o**, parasitize thrombocytes, lymphocytes, and in one species, erythroblasts, in neotropical, southeast Asian, and Japanese lizards. Meronts in figures are in a lymphocyte (**h**), thrombocyte (**i**), and erythroblast (**j**), macrogametocytes **k**, **l**, **o**, and microgametocytes **m**, **n** are in thrombocytes.

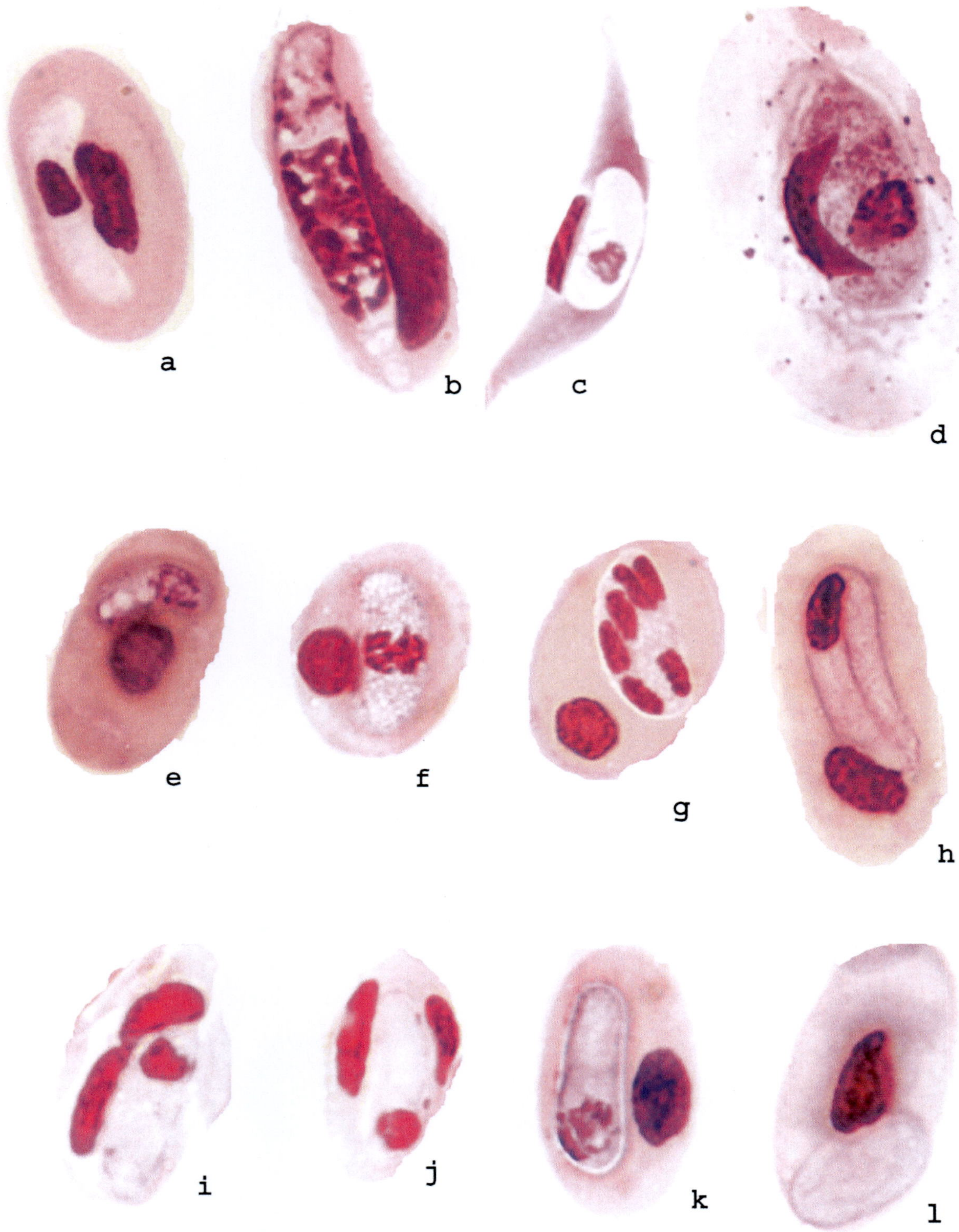

Plate 79 Hemogregarines of four genera, all in erythrocytic host cells. Gamonts of *Hepatozoon* species (**a–d**) usually cause little distortion of host cells (**a**), but some species significantly alter erythrocyte shape (**b–d**). *Hepatozoon* parasitizes all reptilian groups worldwide; only young and mature gamonts are seen in peripheral blood. *Haemogregarina* species of freshwater turtles and alligators often show the four stages depicted in **e–h**: trophozoite (**e**), premeront (**f**), meront (**g**), and gamont (**h**). Host cells are erythrocytes and are seldom significantly distorted. *Karyolysus* species (**i**, **j**) occur in European, Asian, and Australian lizards and commonly cause severe distortion of erythrocyte nuclei. Only young and mature gamonts occur in circulating blood. *Hemolivia* (**k**, **l**) is known only from neotropical toads, European/North African tortoises, and Australian skinks. Gamont nuclei are usually apical in position; the parasite often occupies an opaque sheath.

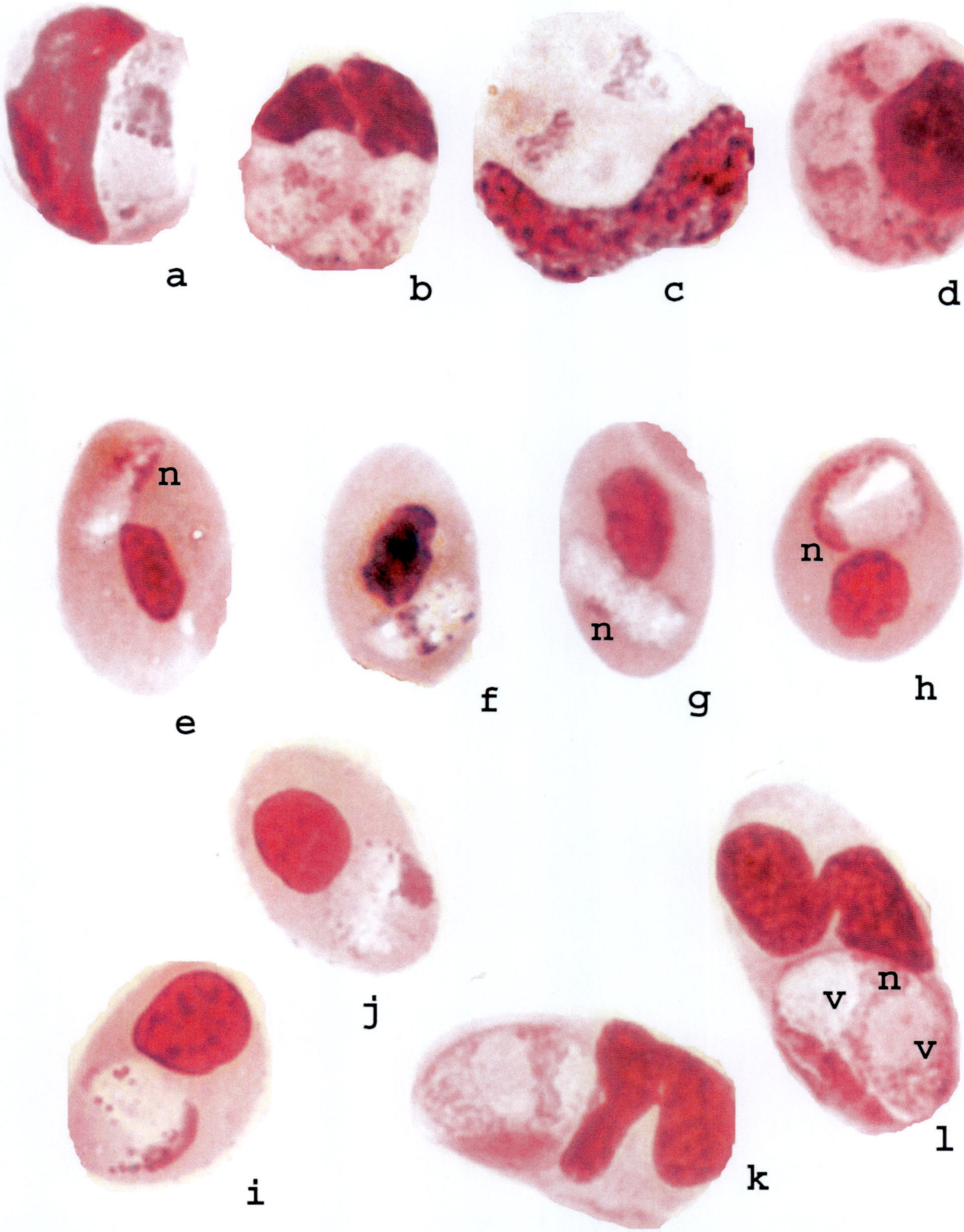

Plate 80 Hemococcidia of two genera are common leukocytic and erythrocytic parasites of lizards worldwide. Only sporozoites are seen in circulating erythrocytes, monocytes, or lymphocytes, uncommonly in thrombocytes. The sporozoite nucleus (n) is often barely visible and is commonly preceded or followed by a bluish-to-white crystalloid body (v) that resembles a vacuole. Sporozoite shape may be narrow and pointed at one end or rounded. *Schellackia* species (**a–h**) are cosmopolitan; the two *Lainsonia* species are neotropical, erythrocytic in green iguanas (**i**, **j**) or leukocytic in tegu lizards (**k**, **l**).

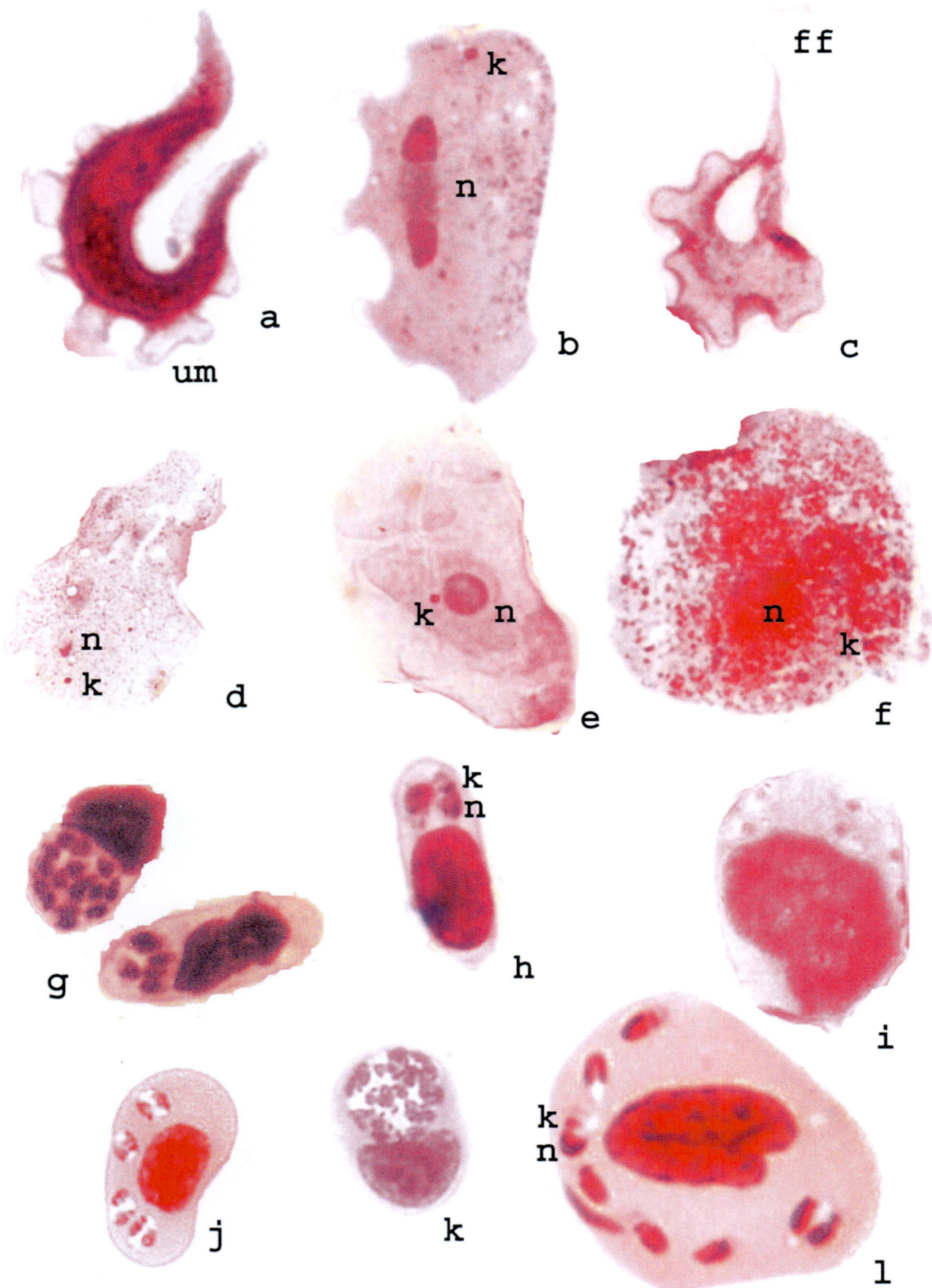

Plate 81 Kinetoplastid flagellates are characterized by the presence of a dot-like kinetoplast (k) near or behind the nucleus (n). The extracellular *Trypanosoma* species vary from very slender (**a**, **c**) to leaf-like (**b**, **d**, **e**) or broad, rounded shapes (***f***). An undulating membrane (um) arises from the kinetoplast and extends anteriorly along the margin or surface of the parasite to form, in some species, a free flagellum (ff). Trypanosomes occur in all reptilian groups except the tuatara, worldwide. *Sauroleishmania* amastigotes (**g–l**) parasitize thrombocytes (**g**, **h**), leukocytes (**i**), and erythroid cells (**j–l**) of Old World lizards and snakes. There are no proven records from the Western Hemisphere. Free-living forms, promastigotes, have been isolated from blood or tissue cultures without reports of intracellular amastigotes. These are ovoid to elongate with a single, prominent flagellum arising from the kinetoplast and extending anteriorly and a prominent nucleus.

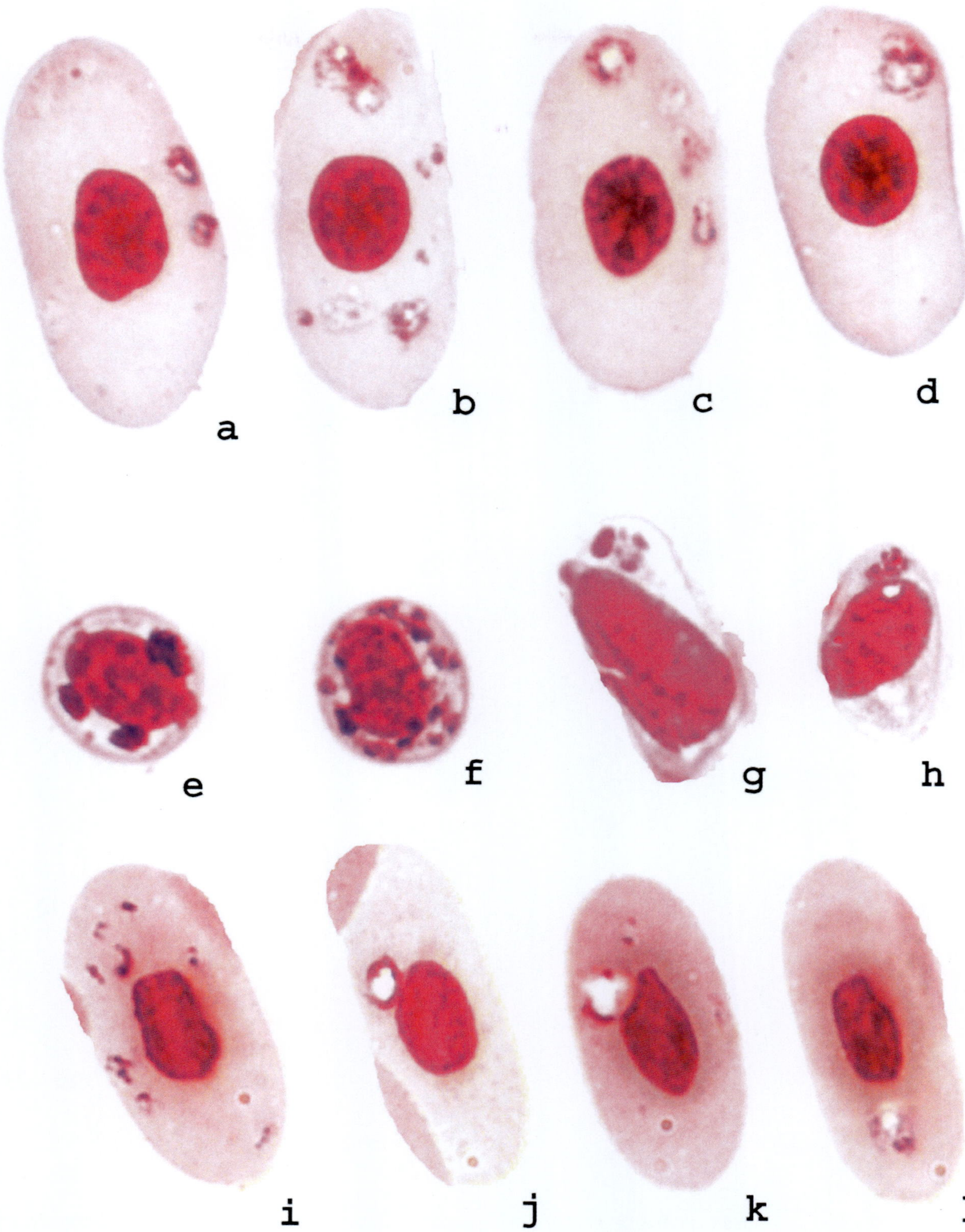

Plate 82 Piroplasmorid parasites of two genera, *Sauroplasma* (**a–d**) in lizards and *Serpentoplasma* (**e–l**) in snakes, occur worldwide but are very poorly known. They form as small inclusions containing chromatin granules associated often with a vacuole and appear to multiply by binary fission or budding, sometimes forming an X. *Serpentoplasma* infection of snakes apparently begins in thrombocytes (**e–h**), where dense chromatin masses divide by binary division or budding, and possibly by a type of merogony. Infection of erythrocytes (**i–l**) results in small chromatin bodies that later appear along the margins of vacuoles. Larger forms bear some resemblance to meronts and gametocytes of plasmodiids but remain very small.

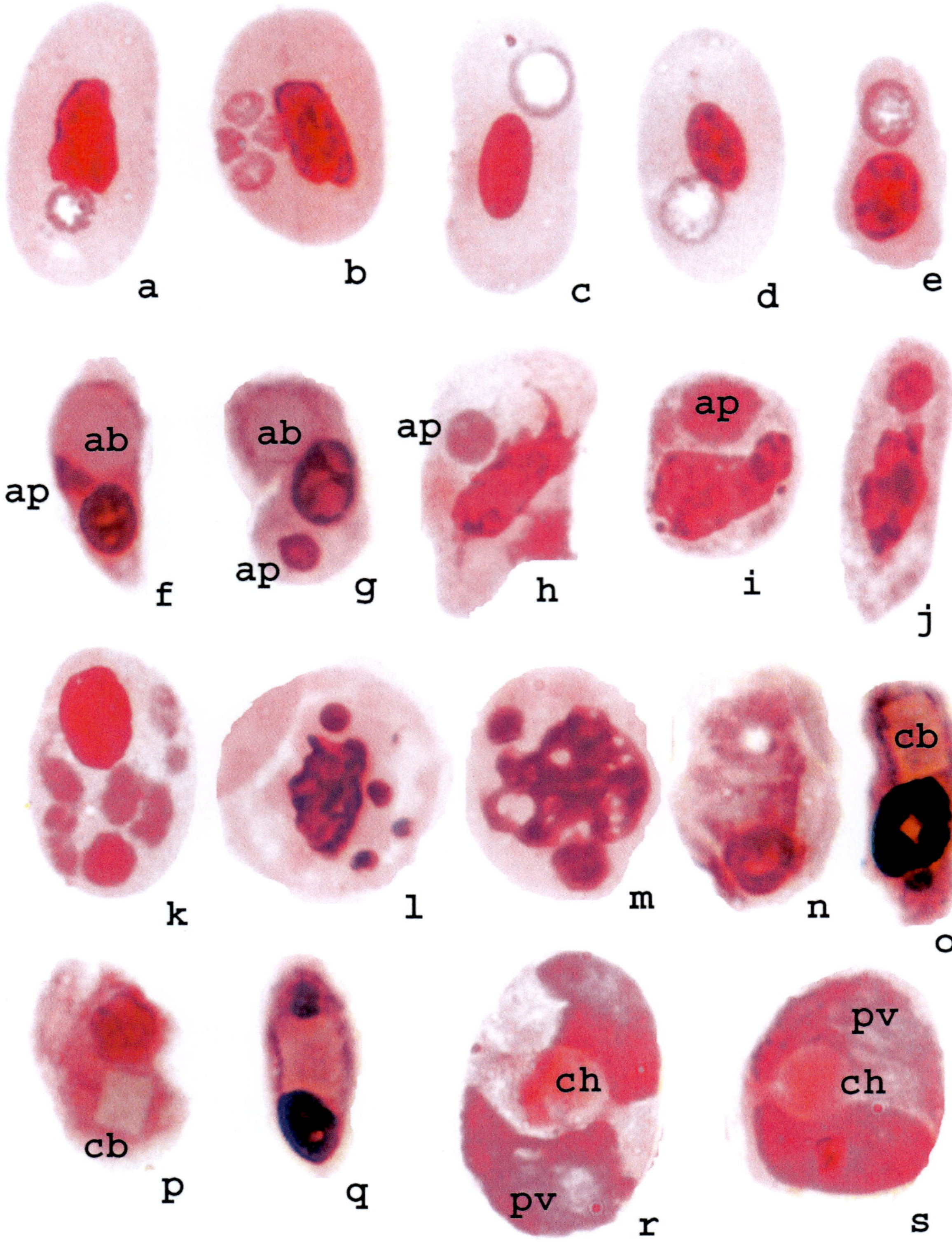

Plate 83 Bacterial and viral inclusions in erythrocytes of lizards and snakes. *Aegyptianella* (**a–e**) is known from some lizards in Asia and North America, as well as frogs in several areas, and may occur in turtles as well. The nearly spherical inclusions in erythrocyte cytoplasm stain densely when small, but as size increases, the center lightens, and the entire inclusion may become nearly colorless with a dark-stained margin. Two types of iridoviruses parasitize red blood cells of reptiles, lizard erythrocytic virus (LEV, or *Pirhemocyton*) (**f–k**) in lizards and snake erythrocyte virus (SEV, or *Toddia*) in snakes (**l–q**). Both appear most commonly as rounded or sometimes variably shaped dark red-stained areas (ap) in the cytoplasm, which represent the viral "assembly pools." LEV is usually accompanied by a circular "albuminoid body" (ab), often resembling a vacuole, and SEV by a square-to-rectangular "crystalloid body" (cb). Single or mixed infections of poxvirus (pv) and chlamydia (ch) have been seen in monocytes of East African chameleons (**r**, **s**).

INDEX

AMPHIBIANS

CROCODILIANS

LIZARDS

SNAKES

TUATARA

TURTLES